Plant Physiological Ecology

D0087111

Springer
New York
Berlin
Heidelberg
Barcelona
Hong Kong
London
Milan
Paris
Singapore
Tokyo

Plant Physiological Ecology

**Hans Lambers   F. Stuart Chapin III
Thijs L. Pons**

With 356 Illustrations

Springer

Hans Lambers, Ph.D.
Department of Plant Ecology and
  Evolutionary Biology
Utrecht University
3584 CA Utrecht, The Netherlands
and
Plant Sciences
Faculty of Agriculture
University of Western Australia
Nedlands, WA 6907, Australia

F. Stuart Chapin III, Ph.D.
Institute of Arctic Biology
University of Alaska
Fairbanks, AK 99775, USA

Thijs L. Pons, Ph.D.
Department of Plant Ecology and
  Evolutionary Biology
Utrecht University
3584 CA Utrecht, The Netherlands

Cover art from Lourens Poorter, Utrecht University.

Library of Congress Cataloging-in-Publication Data
Lambers, H.
    Plant physiological ecology/Hans Lambers, F. Stuart Chapin, Thijs L. Pons.
      p.   cm.
    Includes bibliographical references and index.
    ISBN 0-387-98326-0 (alk. paper)
    1. Plant ecophysiology.   I. Chapin, F. Stuart (Francis Stuart),
III.   II. Pons, Thijs Leendert, 1948– .   III. Title.
  QK717.L35   1998
  571.2—dc21   97-33273

Printed on acid-free paper.

© 1998 Springer-Verlag New York Inc.
All rights reserved. This work may not be translated or copied in whole or in part without
the written permission of the publisher (Springer-Verlag New York, Inc., 175 Fifth Avenue, New York, NY 10010, USA), except for brief excerpts in connection with reviews or
scholarly analysis. Use in connection with any form of information storage and retrieval,
electronic adaptation, computer software, or by similar or dissimilar methodology now
known or hereafter developed is forbidden.
The use of general descriptive names, trade names, trademarks, etc., in this publication,
even if the former are not especially identified, is not to be taken as a sign that such names,
as understood by the Trade Marks and Merchandise Marks Act, may accordingly be used
freely by anyone.

Production coordinated by Chernow Editorial Services, Inc. and managed by
Bill Imbornoni; manufacturing supervised by Joe Quatela.

Typeset by Best-set Typesetter Ltd., Hong Kong.
Printed and bound by Maple-Vail Book Manufacturing Group, York, PA.
Printed in the United States of America.

9 8 7 6 5 4 3 2 (Corrected second printing, 2000)

ISBN 0-387-98326-0 Springer-Verlag New York Berlin Heidelberg SPIN 10753354

QK
717
.L35
1998

030200—4495 NF 1

Foreword

The individual is engaged in a struggle for existence (Darwin). That struggle may be of two kinds: The acquisition of the resources needed for establishment and growth from a sometimes hostile and meager environment and the struggle with competing neighbors of the same or different species. In some ways, we can define *physiology* and *ecology* in terms of these two kinds of struggles. Plant ecology, or plant sociology, is centered on the relationships and interactions of species within communities and the way in which populations of a species are adapted to a characteristic range of environments. Plant physiology is mostly concerned with the individual and its struggle with its environment. At the outset of this book, the authors give their definition of *ecophysiology*, arriving at the conclusion that it is a point of view about physiology. A point of view that is informed, perhaps, by knowledge of the real world outside the laboratory window. A world in which, shall we say, the light intensity is much greater than the 200 to 500 μmol photons $m^{-2}s^{-1}$ used in too many environment chambers, and one in which a constant 20°C day and night is a great rarity. The standard conditions used in the laboratory are usually regarded as treatments. Of course, there is nothing wrong with this in principle; one always needs a baseline when making comparisons. The idea, however, that the laboratory control is the norm is false and can lead to misunderstanding and poor predictions of behavior.

The environment from which many plants must acquire resources is undergoing change and degradation, largely as a result of human activities and the relentless increase in population. This has thrown the spotlight onto the way in which these changes may feed back on human well-being. Politicians and the general public ask searching questions of biologists, agriculturalists, and foresters concerning the future of our food supplies, building materials, and recreational amenities. The questions take on the general form, "Can you predict how 'X' will change when environmental variables 'Y' and 'Z' change?" The recent experience of experimentation, done at high public expense, on CO_2 enrichment and global warming, is a sobering reminder that not enough is known about the underlying physiology and biochemistry of plant growth and metabolism to make the confident predictions that the customers want to hear. Even at the level of individual plants, there seems to be no clear prediction, beyond that the response depends on species and other ill-defined circumstances. On the broader scale, predictions about the response of plant communities are even harder to make. In the public mind, at least, this is a failure. The only way forward is to increase our understanding of plant metabolism, of the mechanisms of resource capture, and the way in which the captured resources are allocated to growth or storage in the plant. To this extent, I can see no distinction between plant physiology and ecophysiology. There are large numbers of missing pieces of information about plant physiology—period. The approach of the new millennium, then, is a good time to recognize the need to study plant physiology anew, bringing to bear the impressive new tools made available by gene cloning and recombinant DNA technology. This book is to be welcomed if it will encourage ecologists to come to grips with the processes which determine the behavior of "X" and encourage biochemistry and physiology students to take a more realistic view of the environmental variables "Y" and "Z."

The book starts, appropriately, with the capture of carbon from the atmosphere. Photosynthesis is obviously the basis of life on earth, and some of the most brilliant plant scientists have made it their life's work. As a result, we know more about the molecular biophysics and biochemistry of photosynthesis than we do about any other plant process. The influence of virtually every environmental variable on the physiology of photosynthesis and its regulation has been studied. Photosynthesis, however, occurs in an environment over which the individual plant has little control. In broad terms, a plant must cope with the range of temperature, rainfall, light intensity, and CO_2 concentration to which its habitat is subjected. It cannot change these things. It must rely on its flexible physiological response to mitigate the effects of the environment. At a later stage in the book, the focus shifts below ground, where the plant has rather more control over its options for capturing resources. It may alter the environment around its roots in order to improve the nutrient supply. It may benefit from microbial assistance in mobilizing resources or enter into more formal contracts with soil fungi and nodule-forming bacteria to acquire nutrient resources that would otherwise be unavailable or beyond its reach. Toward its close, the book turns to such interactions between plants and microbes and to the chemical strategies that have evolved in plants that assist them in their struggles with one another and against browsing and grazing animals. The authors end, then, on a firmly ecological note, and introduce phenomena that most laboratory physiologists have never attempted to explore. These intriguing matters remind us, as if reminders were needed, of "how little we know, how much to discover" (Springer and Leigh).

DAVID T. CLARKSON
IACR-Long Ashton Research Station
University of Bristol
April 1997

Acknowledgments

Numerous people have contributed to the text and illustrations in this book by commenting on sections and chapters, providing photographic material, drawing numerous figures, and so on. Apart from many undergraduate students and some anonymous reviewers, we wish to thank the following colleagues: Rien Aerts, Owen Atkin, Fraser Bergersen, Henny Blom, Ad Borstlap, Tjeerd Bouma, Neil Bridson, Lieve Bultynck, Marion Cambridge, Zoe Cardon, Melissa Chapin, Christa Critchley, David Clarkson, Brent Clothier, Joseph Craine, David Day, Margriet Dekker, Manny Delhaize, David de Pury, Inge Dörr, Neil Emery, John Evans, Tatsuhiro Ezawa, Alastair Fitter, Eric Garnier, James Graham, Brian Gunning, Ellis Hoffland, Arian Jacobs-Brouwer, Tibor Kalapos, Gamini Keerthisinghe, Ronald Kempers, Marga Knoester, Marjolein Kortbeek-Smithuis, John Kuo, Catarina Mata, Margareth McCully, John Milburn, Frank Millenaar, Harvey Millar, Frank Minchin, Rana Munns, Oscar Nagel, John Passioura, Kristel Perreijn, Carol Peterson, Larry Peterson, Corné Pieterse, Hendrik Poorter, Lourens Poorter, Pieter Poot, Hidde Prins, Peter Ryan, Peter Ryser, Ingeborg Scheurwater, Heather Sherwin, Sally Smith, Ichiro Terashima, Aart van Bel, Adrie van der Werf, Liz van Volkenburgh, Rens Voesenek, Susanne von Caemmerer, Michelle Watt, Maria Weiper, Mark Westoby, Lynne Whitehead, and Piet Wolswinkel.

HANS LAMBERS
F. STUART CHAPIN III
THIJS L. PONS

Units and Conversions

Concentration, amount
Concentration equals content per unit volume, mass or area
Molarity = $mol \cdot l^{-1}$ solution
Molality = molarity $\times$ activity
$1\,mol \cdot m^{-3} = 1\,mmol \cdot kg^{-1} = 1\,mM$
$1\,\mu mol\ CO_2 \cdot m^{-3} = 0.00243$ Pa = 2.43 kPa (at 293 K and 101.3 kPa atmospheric pressure
$1\,dalton = 1.6605 \cdot 10^{-24}\,g$

Energy
$1\,J = 1\,N \cdot m = 1\,kg \cdot m^2 \cdot s^{-2} = 1\ W \cdot s$
$1\,W \cdot h = 3.6\ kW \cdot s = 3.6\ kJ$
$1\,MJ = 0.278\ kW \cdot h$
$1\,cal = 4.1868\ J$
$1\,kcal = 1.163\ W \cdot h$

Electric conductance
$1\,S = 1\,ohm^{-1}$

Gas exchange, conductance
$1\,mol \cdot m^{-2} \cdot s^{-1} = 0.0224 \cdot (T/273) \cdot (101.3/P)\ m \cdot s^{-1}$ (T = absolute temperature; P = atmospheric pressure in kPa)
$1\,mol \cdot m^{-2} \cdot s^{-1} = 0.024\,m \cdot s^{-1}$ (at 293 K and 101.3 kPa)
$1\,mm \cdot s^{-1} = 41.7\,mmol \cdot m^{-2} \cdot s^{-1}$ (at 293 K and 101.3 kPa)

Pressure
$1\,MPa = 10^6\ Pa = 10^6\ N \cdot m^{-2} = 10^6\ J \cdot m^{-3}$ (= 10 bar, a discarded metric unit)

Radiation (McCree 1991)
$1\,W \cdot m^{-2} = 1\ J \cdot m^{-2} \cdot s^{-1}$
$1\,mol\ photons = 1 \cdot 8 \cdot 10^5\ J$ (at λ 650 nm) to $2.7 \cdot 10^5\ J$ (at λ 450 nm)
$1\,W \cdot m^{-2}$ (PAR) = 4.6 μmol (photons) $m^{-2} s^{-1}$ (daylight, sunny)

$1\,\text{W}\cdot\text{m}^{-2}$ (PAR) = 4.2 µmol (photons) $\text{m}^{-2}\text{s}^{-1}$ (daylight, diffuse)
$1\,\text{W}\cdot\text{m}^{-2}$ (PAR) = 4.6/4.6/5.0 µmol (photons) $\text{m}^{-2}\text{s}^{-1}$ (metal halide lamp/ white fluorescent tube/incandescent lamp)

Water potential
Ψ_π (MPa) = $-RTc_s$ (c_s = concentration in $\text{mol}\,\text{m}^{-3}$)
Ψ_{air} (MPa) = $-RT/V_w^0 \cdot \ln RH$
V_w^0 = partial molar volume of water
RH (relative humidity of the air, %) = $p/p_0 \cdot 100$
p = water vapor pressure
p_0 = saturated water vapor pressure
$1\,\text{M}$ = $10^3\,\text{mol}\cdot\text{m}^{-3}$ ~ $-2.4\,\text{MPa}$ (at 20°C)

Reference

McCree, K.J. (1981) Photosynthetically active radiation. In: Encyclopedia of plant physiology, N.S., Vol 12A, O.L. Lange, P.S. Nobel, C.B. Osmond, & H. Ziegler (eds). Springer-Verlag, Berlin, pp. 41–55.

Abbreviations

a	radius of a root (a_r) or root plus root hairs (a_e)
A	rate of CO_2 assimilation; also total root surface
A_f	foliage area
A_{max}	light-saturated rate of net CO_2 assimilation at ambient p_a
A_s	sapwood area
ABA	abscisic acid
ADP	adenosine diphosphate
AM	arbuscular mycorrhiza
AMP	adenosine monophosphate
APAR	absorbed photosynthetically active radiation
ATP	adenosine triphosphate
b	individual plant biomass; buffer power of the soil
B	stand biomass
c_s	concentration of the solute
C	nutrient concentration in solution; also convective heat transfer
C_3	photosynthetic pathway utilizing 3-carbon intermediate
C_4	photosynthetic pathway utilizing 4-carbon intermediate
C_{li}	initial nutrient concentration
C_{min}	solution concentration at which uptake is zero
C:N	carbon:nitrogen ratio
CAM	Crassulacean Acid Metabolism
CC	carbon concentration
CE	carbohydrate equivalent
chl	chlorophyll
CPF	carbon dioxide production value
d	plant density; also, leaf dimension
D	diffusivity of soil water
D_e	diffusion coefficient of ion in soil
DHAP	dihydroxyacetone phosphate
DM	dry mass
DNA	deoxyribonucleic acid
e	water vapor pressure in the leaf (e_i; or e_l in Sect. 2.5 of the chapter on the plant's energy balance) or atmosphere (e_a); also emissivity of a surface
E	leaf transpiration rate
f	tortuosity

F	rate of nutrient supply to the root surface; also fluorescence, in dark-incubated leaves: minimal (F_0) and maximal in a pulse of saturating light (F_m), in leaves in light symbols are F_0' and F_m' (Box 2)
FAD(H$_2$)	flavine adenine dinonucleotide (reduced form)
FM	fresh mass
FR	far-red
g	diffusive conductance for CO_2 (g_c) and water vapor (g_w); boundary layer conductance (g_a); internal conductance (g_i); stomatal conductance (g_s); boundary layer conductance for heat transport (g_{ah})
GA	gibberellic acid
GE	glucose equivalent
GOGAT	glutamine 2-oxoglutarate aminotransferase
HCH	hydroxycyclohexenone
HIR	high-irradiance response
I	irradiance above (I_o) or beneath (I) the canopy; irradiance absorbed (Box 2); also nutrient inflow
I_{max}	maximum rate of nutrient inflow
IAA	indoleacetic acid
IR$_s$	short-wave infrared radiation
J	rate of photosynthetic electron flow
J_{max}	rate of photosynthetic electron flow at saturating irradiance and p_a
J_v	water flow
k	rate of root elongation; extinction coefficient for light
K	carrying capacity (e.g., K species)
k_{cat}	catalytic constant of an enzyme
K_i	inhibitor concentration giving half-maximum inhibition
K_m	substrate concentration at half V_{max} (or I_{max})
l	leaf area index
L	rooting density; also, conductance in Sect 5.1 of the chapter on plant water relations; also, length of xylem element in Sect 5.3.2 of the chapter on plant water relations;
$L_{(I)}$	relative limitation of effective PSII quantum yield at a particular irradiance
LAI	leaf area index
LAR	leaf area ratio
LFR	low-fluence response
LHC	light-harvesting complex
LMA	leaf mass per unit area
LMR	leaf mass ratio
LR	long-wave infrared radiation that is incident (LR_{in}), reflected (LR_r), emitted (LR_{em}), absorbed (LR_{abs}), or net incoming (LR_{net}); also leaf respiration on an area (LR_a) and mass (LR_m) basis
mRNA	messenger ribonucleic acid
M	energy dissipated by metabolic processes
ME	malic enzyme
MRT	mean residence time
N_w	mol fraction, that is, the number of moles of water divided by the total number of moles
NAD(P)	nicotinamide adenine dinucleotide(phosphate) (in its oxidized form)
NAD(P)H	nicotinamide adenine dinucleotide(phosphate) (in its reduced form)
NAR	net assimilation rate
NDVI	normalized difference vegetation index
NEP	net ecosystem production
NIR	near-infrared reflectance; net rate of ion uptake

NMR	nuclear magnetic resonance
NP	nitrogen productivity
NPP	net primary production
NPQ	fluorescence quenching due to nonphotochemical processes
NUE	nitrogen-use efficiency
p	vapor pressure (Box 3)
p_o	vapor pressure of air above pure water
p_a	atmospheric partial pressure of CO_2
p_i	intercellular partial pressure of CO_2
P	atmospheric pressure
P_{fr}	far-red-absorbing configuration of phytochrome
P_i	inorganic phosphate
P_r	red-absorbing configuration of phytochrome
PAR	photosynthetically active radiation
PC	phytochelatin
PEP	phospho*enol*pyruvate
PEPCK	phospho*enol*pyruvate-carboxykinase
pH	hydrogen ion activity; negative logarithm of the H^+ concentration
PFD	photon flux density
PGA	phosphoglycerate
PNC	plant nitrogen concentration
PNUE	photosynthetic nitrogen-use efficiency
PQ	photosynthetic quotient
PS	photosystem
PV'	amount of product produced per gram of substrate
q_N	fluorescence quenching due to nonphotochemical processes
qP	photochemical quenching
Q	ubiquinone; in reduced state (Q_r = ubiquinol) or total quantity (Q_t)
Q_{10}	temperature coefficient
Q_A	electron acceptor in photosynthetic electron transport chain (Box 2)
r	diffusive resistance for CO_2 (r_c) and water vapor (r_w); stomatal resistance for water vapor (r_s); internal resistance (r_i); also radial distance from the root axis; also, in Sect 5.2 of respiration, it is used as respiration; also, growth rate (in volume) in the Lockhart equation; also, proportional root elongation; intrinsic rate of population increase (e.g., r species)
r_i	spacing between roots
r_o	root diameter
R	red; also radius of a xylem element; also, gas constant
R_a	molar abundance ratio of $^{13}C/^{12}C$ in the atmosphere
R_d	dark respiration
R_{day}	dark respiration during photosynthesis (Box 1)
R_e	ecosystem respiration
R_p	plant respiration; also, molar abundance ratio of $^{13}C/^{12}C$ in plants
R_h	heterotrophic respiration
R*	minimal resource level utilized by a species
RGR	relative growth rate
RH	relative humidity of the air
RMR	root mass ratio
RNA	ribonucleic acid
RQ	respiratory quotient
RR	rate of root respiration
RuBP	ribulose-1,5-bisphosphate
Rubisco	ribulose-1,5-bisphosphate carboxylase/oxygenase
RWC	relative water content

S	nutrient uptake by roots (Sect. 2.3 of the chapter on mineral nutrition)
$S_{c/o}$	specificity of carboxylation by Rubisco
SHAM	salicylichydroxamic acid
SLA	specific leaf area
SMR	stem mass ratio
SR	short-wave solar radiation that is incident (SR_{in}), reflected (SR_r), transmitted (SR_{tr}), absorbed (SR_{abs}), used in photosynthesis (SR_A), emitted in fluorescence (SR_{FL}), or net incoming (SR_{net}); also rate of stem respiration
SRL	specific root length
t^*	time constant
t RNA	transfer ribonucleic acid
T	temperature
TCA	tricarboxylic acid
TR	total radiation that is absorbed (TR_{abs}) or net incoming (TR_{net})
u	wind speed
UV	ultraviolet
V	volume
V_c	rate of carboxylation (Box 1)
V_o	rate of oxygenation
V_{cmax}	maximum rate of carboxylation
V_w^o	molar volume of water
VAM	vesicular arbuscular mycorrhiza
VIS	visible reflectance
VLFR	very low fluence response
V_{max}	substrate-saturated enzyme activity
VPD	vapor pressure difference (between leaf and air)
w	mole fraction of water vapor in the leaf (w_i) or atmosphere (w_a)
WUE	water-use efficiency
Y	yield threshold (in the Lockhart equation)
γ	surface tension
Γ	CO_2-compensation point
Γ^*	CO_2-compensation point in the absence of dark respiration
δ	boundary layer thickness; also isotopic content
Δ	isotopic discrimination
ΔT	temperature difference
ε	elastic modulus; also, emissivity
η	viscosity constant
θ	curvature of the irradiance response curve; also, volumetric moisture content (mean value, θ', or at the root surface, θ_a)
λ	latent heat of evaporation
μ_w	chemical potential of water
μ_{wo}	chemical potential of pure water under standard conditions
σ	Stefan-Boltzman constant
φ	quantum yield (of photosynthesis); also, yield coefficient (in the Lockhart equation); also, leakage of CO_2 from the bundle sheath to the mesophyll
Ψ	water potential
Ψ_{air}	water potential of the air
Ψ_m	matric potential
Ψ_p	pressure potential; hydrostatic pressure
Ψ_π	osmotic potential

Contents

Contents

1
Assumptions and Approaches

Introduction—History, Assumptions, and Approaches

1. What Is Ecophysiology?

Plant ecophysiology is an experimental science that seeks to describe the **physiological mechanisms** that underlie ecological observations. In other words, ecophysiologists, or physiological ecologists, address ecological questions about the controls over the growth, reproduction, survival, abundance, and geographical distribution of plants as these processes are affected by the interactions between plants with their physical, chemical, and biotic environment. These ecophysiological patterns and mechanisms can help us to understand the functional significance of specific plant traits and their evolutionary heritage.

Although the questions addressed by ecophysiologists are derived from a higher level of integration (ecology, agronomy), the ecophysiological explanations often require mechanistic understanding at a lower level of integration (physiology, biochemistry, biophysics, molecular biology). It is quintessential, therefore for an ecophysiologist to have an appreciation for both ecological questions and biophysical, biochemical, and molecular methods and processes. The questions for an ecophysiologist are derived from "ecology" in its widest sense, including, questions that originate from agriculture, horticulture, forestry, and environmental sciences. Such questions and problems often need an ecophysiological approach.

Although ecophysiology is clearly different from other aspects of plant physiology with respect to the ecological problems it addresses, there is a great similarity in methodology. Processes such as photosynthesis, assimilate transport, or specific aspects of a plant's hormone metabolism can be studied for their own merit as well as from an ecophysiological perspective. Ecophysiological research can be carried out for its own sake (i.e., to further our understanding of ecophysiology); however, there are also a large number of applied aspects, often pertaining to agriculture, environmental issues, or nature conservation that benefit from an ecophysiological perspective. A modern ecophysiologist, therefore requires a good understanding of both the molecular aspects of plant processes and the functioning of the intact plant in its environmental context.

2. The Roots of Ecophysiology

As pointed out previously plant ecophysiology aims to provide causal mechanistic explanations for ecological questions that relate to survival, distribution, abundance, and interactions of plants with other organisms. Why does a particular species live where it does? How does it manage to grow there successfully and why is it absent from other environments? These questions were initially asked by geographers who described the global distributions of plants (Schimper 1898, Walter 1974). They observed consistent patterns of morphology associated with different environments and concluded that these differences in morphology must be important in explaining plant distributions. Geographers who know climate, could therefore predict the predominant life forms of plants (Holdridge 1947). For example, many desert plants have small thick leaves that minimize the heat load and danger of overheating in hot environments, whereas shade plants often have large thin leaves that maximize light interception. These observations of morphology provided the impetus to investigate the physiological traits of plants from contrasting physical environments (Blackman 1919, Ellenberg 1953, Pearsall 1938).

Although ecophysiologists initially emphasized physiological responses to the abiotic environment [e.g., to calcareous vs. acidic substrates (Clarkson 1967) or dry vs. flooded soils (Crawford 1978)], physiological interactions with other plants, animals, and microorganisms benefit from an understanding of ecophysiology. As such, ecophysiology is an essential element of every ecologist's training.

A second impetus for the development of ecophysiology came from agriculture and physiology. Even today, agricultural production in industrialized nations is limited to 25% of its potential by drought, infertile soils, and other environmental stresses (Boyer 1982, 1985). A major objective of agricultural research has always been to develop crops that are less sensitive to environmental stress so that they can withstand periods of unfavorable weather or be grown in less favorable habitats. For this reason agronomists and physiologists have studied the mechanisms by which plants respond to or resist environmental stresses. Because some plants grow naturally in extremely infertile, dry, or salty environments, physiologists were curious to know the mechanisms by which this is accomplished.

Plant ecophysiology is the study of physiological responses to environment. The field developed rapidly as a relatively unexplored interface between ecology and physiology. Ecology provided the questions, and physiology provided the tools to determine mechanism. Techniques that measured the microenvironment of plants, their water relations, and their patterns of carbon exchange became typical tools of the trade in plant ecophysiology. With time, these studies have explored the mechanisms of physiological adaptation at ever finer levels of detail from the level of the whole plant to its biochemical and molecular bases. For example, plant growth was initially described in terms of changes in plant mass. Development of portable equipment for measuring leaf gas exchange enabled ecologists to measure rates of carbon gain and

loss by individual leaves. Growth analyses documented carbon and nutrient allocation to roots and leaves, as well as rates of production and death of individual tissues. These processes together provide a more thorough explanation for differences in plant growth in different environments (Lambers & Poorter 1992, Mooney 1972). Studies of plant water relations and mineral nutrition provide additional insight into controls over rates of carbon exchange and tissue turnover. We have more recently learned many details about the biochemical basis of photosynthesis and respiration in different environments, and, finally, about the molecular basis for differences in key photosynthetic and respiratory proteins. This mainstream of ecophysiology has been highly successful in explaining why plants are able to grow where they do.

3. Physiological Ecology and the Distribution of Organisms

Although there are 270,000 species of land plants (Hammond 1995), a series of filters eliminates most of these species from any given site and restricts the actual vegetation to a relatively small number of species (Fig. 1). Many species are absent from a given plant community for historical reasons. They may have evolved in a different region and never dispersed to the study site. For example, the tropical alpine of South America has few species in common with the tropical alpine of Africa, despite similar environmental conditions, whereas eastern Russia and Alaska have very similar species composition because of extensive migration of species across a land bridge that connected these regions when Pleistocene glaciations lowered sea level 20,000 to 100,000 years ago.

Of those species that arrive at a site, many lack the appropriate physiological traits to survive the physical environment. For example, whalers and other ships brought supplies, ballast, and the same types of weed seeds to Svalbard, north of Norway, and to Barrow, in northern Alaska, as those that have invaded temperate islands. There are currently no exotic weed species in common, however, between these arctic sites and temperate shipping ports (Billings 1973). The physical environment has clearly filtered out many species that may have arrived but lacked the physiological traits to grow, survive, and reproduce in the Arctic.

Biotic interactions exert an additional filter that eliminates many species that may have arrived and are capable of surviving the physical environment.

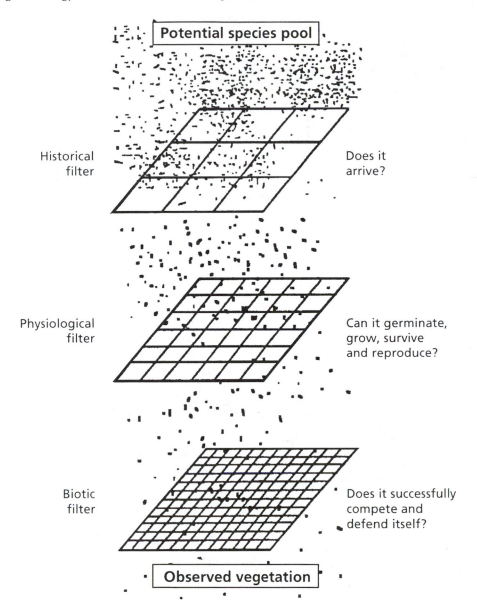

FIGURE 1. Historical, physiological, and biotic filters that determine the species composition of vegetation in a particular site.

During historical time, introduction of the European chestnut virus to North America eliminated *Castanea* (chestnut), a former dominant tree species in the forests of eastern North America. On the other hand, when a plant species is introduced to a new place without the diseases or herbivores that restricted its distribution in its native habitat, this species often becomes an aggressive invader, such as *Opuntia* (prickly pear) in Australia, *Solidago* (golden rod) in Europe, *Cytisus* (Scotch broom) in North America, and *Acacia* (wattle) in South Africa. Because of biotic interactions, the actual distribution of a species (realized niche—as determined by ecological amplitude) is more restricted than the range of conditions where it can grow and reproduce (its fundamental niche as determined by physiological amplitude).

Historical, physiological, and biotic filters are constantly changing and interacting. Human and natural introductions or extinctions of species,

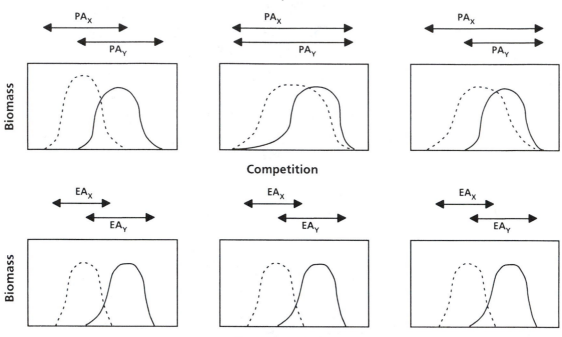

FIGURE 2. Biomass production of two hypothetical species (x and y) as a function of resource supply. In the absence of competition (upper panels), the physiological amplitude of species x and y (PA_x and PA_y, respectively) define the range of conditions over which each species can grow. In the presence of competition (lower panels), plants grow over a smaller range of conditions (their ecological ampli- tude, EA_x and EA_y) that is constrained by competition from other species. A given pattern of species distribution (e.g., that shown in the bottom panels) can result from species that differ in their maximum biomass achieved (left-hand pair of graphs), shape of resource response curve (center pair of graphs), or physiological amplitude (right-hand pair of graphs). Adapted from Walter (1973).

chance dispersal events, and extreme events such as volcanic eruptions or floods change the species pool present at a site. Changes in climate, weathering of soils, and introduction or extinction of species change the physical and biotic environment. Those plant species, which can grow and reproduce under the new conditions or which respond evolu- tionarily so that their physiology provides a better match to this environment, will persist. Because of these interacting filters, the species present at a site are simply those that arrived and survived. There is no reason to assume that the species present at a site perform optimally under those conditions (Vrba & Gould 1986). In fact, controlled-environment stud- ies typically demonstrate that a given species is most common under environmental conditions that are distinctly suboptimal for most physiological processes because biotic interactions prevent most species from occupying those habitats that are most favorable (Fig. 2).

4. Time Scale of Plant Response to Environment

We define **stress** as an environmental abiotic or biotic factor that reduces the rate of some physi- ological process (e.g., growth or photosynthesis) below the maximum rate that the plant could other- wise sustain. Examples of stress include low nitro- gen availability, heavy metals, high salinity, and shading by neighboring plants. The immediate response of the plant to stress is a reduction in performance (Fig. 3). Plants compensate for the detrimental effects of stress through many mecha- nisms that operate over different time scales, depending on the nature of the stress and the physiological processes that are affected. These compensatory responses together enable the plant to maintain a relatively constant rate of physiologi- cal processes, despite the occurrence of stresses that

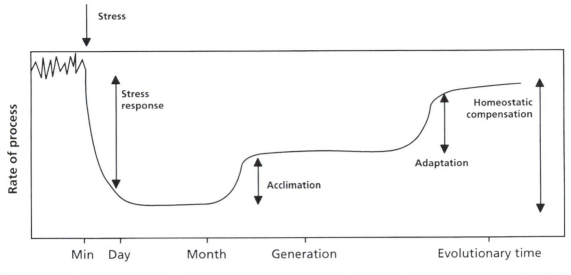

FIGURE 3. Typical time course of plant response to environmental stress. The immediate response to environmental stress is a reduction in physiological activity. Through *acclimation*, individual plants compensate for this stress such that activity returns toward the control level. Over evolutionary time populations *adapt* to environmental stress, resulting in a further increase in activity level toward that of the unstressed unadapted plant. The total increase in activity resulting from acclimation and adaptation is the in situ activity observed in natural populations and represents the total homeostatic compensation in response to environmental stress.

periodically reduce performance. If a plant is going to be successful in a stressful environment, then there must be some degree of stress **resistance**. Mechanisms of stress resistance differ widely between species. They range from **avoidance** of the stress (e.g., in deep-rooting species growing in a low-rainfall area) to stress **tolerance** (e.g., in Mediterranean species that can cope with a low leaf water content).

Physiological processes differ in their sensitivity to stress. The most meaningful processes to consider are probably growth and reproduction, which are processes that integrate the stress effects on individual physiological processes. To understand the mechanism of plant response, however, we must consider the response of individual processes at a finer scale (e.g., the response of photosynthesis or of light-harvesting pigments to a change in light intensity). We recognize at least three distinct time scales of plant response to stress:

1. The **stress response** is the immediate detrimental effect of a stress on a plant process. This generally occurs over a time scale of seconds to days, during which time the net effect on the process is a decline in performance.

2. **Acclimation** is the morphological and physiological adjustment by individual plants to compensate for the decline in performance following exposure to stress. This homeostatic adjustment occurs through changes in the activity or synthesis of new biochemical constituents such as enzymes. These biochemical changes then cause a cascade of effects that are observed at other levels, such as changes in rate or environmental sensitivity of a specific process (e.g., photosynthesis), growth rate of whole plants, and morphology of organs or the entire plant. Acclimation to stress always occurs within the lifetime of an individual, usually within days to weeks. Acclimation can be demonstrated by comparing genetically similar plants that are growing in different environments. We can distinguish between **acclimation** (the homeostatic response to a single well-defined stress; e.g., low temperature) and **acclimatization** (the homeostatic response to complex changes in multiple environmental parameters, such as what occurs when a plant is transplanted from a lowland meadow to an alpine environment).

3. **Adaptation** is the evolutionary response that results from genetic changes in populations leading to morphological and physiological compensation

for the decline in performance caused by stress. The physiological mechanisms of response are often similar to those of acclimation because both require changes in the activity or synthesis of biochemical constituents and cause changes in rates of individual physiological processes, growth rate, and morphology. Adaptation, however, as we define it, differs from acclimation in that it requires genetic changes in populations; therefore, it typically requires many generations to occur. We can study adaptation by comparing genetically distinct plants grown in a common environment.

Not all genetic differences among populations reflect adaptation. Plants may differ genetically because their ancestral species or populations were genetically distinct before they arrived in the habitat we are studying. Evolutionary biologists have often criticized ecophysiologists for promoting the "Panglossian paradigm" (i.e., the idea that just because a species exhibits certain traits in a particular environment, these traits must be beneficial and must have resulted from natural selection in that environment) (Gould & Lewontin 1979).

There are at least two additional processes that can cause particular traits to be associated with a given environment:

1. Through the quirks of history, the ancestral species or population that arrived at the site may have been preadapted (i.e., exhibited traits that allowed continued persistence in these conditions). Natural selection for these traits may have occurred under very different environmental circumstances. For example, the tree species that currently occupy the mixed deciduous forests of Europe and North America were associated with very different species and environments during the Pleistocene, 100,000 years ago. They cooccur now because they migrated to the same place some time in the past (the **historical filter**), can grow and reproduce under current environmental conditions (the **physiological filter**), and outcompeted other potential species in these communities and successfully defended themselves against past and present herbivores and pathogens (the **biotic filter**).

2. Once species arrive in a given geographic region, their distribution is fine-tuned by **ecological sorting**, in which each species tends to occupy those habitats where it most effectively competes with other plants and defends itself against natural enemies (Vrba & Gould 1986).

5. Conceptual and Experimental Approaches

Documentation of the correlation between physiological and morphological traits and environmental conditions is the raw material for many ecophysiological questions. Plants in the high alpine of Africa are strikingly similar in morphology and physiology to those of the alpine of tropical South America and New Guinea, despite very different phylogenetic histories. The similarity of physiology and morphology of shrubs from mediterranean regions of western parts of Spain, Chile, Australia, and the United States suggests that the distinct floras of these regions have undergone convergent evolution in response to similar climatic regimes (Mooney & Dunn 1970). For example, evergreen shrubs are common in each of these regions. These shrubs have small thick leaves, which continue to photosynthesize under conditions of low water availability during the warm, dry summers characteristic of mediterranean climates. The shrubs of all mediterranean regions effectively retain nutrients when leaves are shed, which is a trait that could be important on the infertile soils. They also often resprout after fire, which is common in these regions. Documentation of a correlation of traits with environment, however, can never determine the relative importance of adaptation to these conditions and other factors, such as preadaptation of the ancestral floras and ecological sorting of ancestral species into appropriate habitats. In addition, traits that are measured under field conditions reflect the combined effects of differences in magnitude and types of environmental stresses, genetic differences among populations in stress response, and acclimation of individuals to stress. Thus, documentation of correlations between physiology and environment in the field provides a basis for interesting ecophysiological hypotheses, but these hypotheses can rarely be tested without complementary approaches such as growth experiments or phylogenetic analyses.

Growth experiments allow one to separate the effects of acclimation by individuals and genetic differences among populations. Acclimation can be documented by measuring the physiology of genetically similar plants grown under different environmental conditions. For example, such experiments show that plants grown at low temperatures generally have a lower optimum temperature for photosynthesis than do warm-grown plants (Billings et al. 1971). We can demonstrate genetic differences by growing plants collected from alpine and low-elevation habitats under the same environmental conditions. The alpine plant generally has a lower temperature optimum for photosynthesis than does the low-elevation population. Thus, many alpine plants photosynthesize just as rapidly as do their low-elevation counterparts, due to both acclimation and adaptation. Controlled-environment experiments are an important complement to field observations. On the other hand, field observations and experiments provide a context for interpreting the significance of laboratory experiments.

Both acclimation and adaptation reflect complex changes in many plant traits, which makes it difficult to evaluate the importance of changes in any particular trait. Ecological modeling and molecular modification of specific traits are two approaches that have been used to explore the ecological significance of specific traits. **Ecological models** can range from simple empirical relationships (e.g., the temperature response of photosynthesis) to complex mathematical models that incorporate many indirect effects, such as negative feedbacks of sugar accumulation to photosynthesis, and changes in plant nitrogen status or leaf area. A common feature of these models is the assumption that there are both **costs** and **benefits** associated with a particular trait, such that no trait enables a plant to perform best in all environments. (i.e., there are no "super-plants" or "Darwinian demons" that are superior in all environments). That is presumably why there are so many interesting physiological differences among plants. These models seek to identify the conditions under which a particular trait allows optimal performance or compares performance of two plants that differ in traits. The theme of **trade-offs**, (i.e., the costs and benefits of particular traits) is one that will recur commonly in this book.

A second, more experimental approach to the question of optimality, which has only recently become available, is **molecular modification** of the gene that codes for a trait. In this way we can explore the consequences of a change in photosynthetic capacity, sensitivity to a specific hormone, or response to shade. This molecular approach is a refinement of comparative ecophysiological studies in which plants from different environments that are as similar as possible, except with respect to the trait of interest, are grown in a common environment. For example, comparison of herbaceous species from fertile and infertile soils has shown that a large leaf area allows rapid growth in plants from fertile soils, but that this benefit is diminished under conditions of low nutrient supply (Lambers et al. 1998). Molecular modification of single genes allows evaluation of the physiological and ecological consequences of a trait, while holding constant the rest of the biology of the plant.

6. New Directions in Ecophysiology

Plant ecophysiology has several new and potentially important contributions to make to biology. The exponentially growing human population requires increasing supplies of food and fiber. This comes at a time when the best agricultural lands are already in production or being lost to urban development. It is thus increasingly critical that we identify traits or suites of traits that maximize food and fiber production on both productive and unproductive sites. The development of varieties that grow effectively with inadequate supplies of water and nutrients is particularly important in less-developed countries that often lack the economic and transportation resources to support high-intensity agriculture. Although molecular biology and traditional breeding programs provide the tools to develop new combinations of traits in plants, ecophysiology is perhaps the field that is best suited to determine the costs, benefits, and consequences of changes in these traits, as whole plants interact with complex environments.

Past ecophysiological studies have described important physiological differences among plant species and have demonstrated many of the mechanisms by which plants can live where they occur. These same physiological processes, however, have important effects on the environment in shading the soil, removing nutrients that might otherwise be available to other plants or soil microorganisms and transporting water from the soil to the atmosphere, thus both drying the soil and increasing atmospheric moisture. These plant effects can be large and provide a mechanistic basis for understanding processes at larger scales, such as community, ecosystem, and climatic processes (Chapin 1993). For example, forests that differ only in species composition can differ substantially in productivity and rates of nutrient cycling. Simulation models suggest that species difference in stomatal conductance and rooting depth could significantly affect the climate at regional and continental scales (Shukla et al. 1990). As human activities increasingly alter the species composition of large portions of the globe, it is critical that we understand the ecophysiological basis of community, ecosystem, and global processes.

7. The Structure of the Book

The first chapters in this book deal with the primary processes of carbon metabolism and transport. After introducing some biochemical and physiological aspects of photosynthesis, we will discuss differences in photosynthetic traits among species and link these with the species' natural habitat. Trade-offs are discussed, like that between a high water-use efficiency and a high efficiency of nitrogen use in the photosynthetic machinery. In the following chapter we will analyze carbon use in respiration and explore its significance for the plant's carbon balance in different species and environments. Species differences in the transport of photosynthates from the site of production to various sinks are discussed next (i.e., phloem loading in the minor veins of the leaves and the types of carbohydrates that are transported in the sieve tubes are pivotal plant traits for the geographic distribution of species and for their performance in different environments). The phloem transport system in climbing plants also involves an interesting trade-off between transport capacity and safety of the system. A similar trade-off is encountered in the following chapter, which deals with plant water relations. Plant energy balance and the effects of radiation and temperature are subsequently discussed. Having restricted the discussion of photosynthesis, water use, and energy balance to individual leaves and whole plants so far, we will then scale the processes up to the level of an entire canopy, demonstrating that processes at the level of a canopy are not necessarily the sum of what happens in single leaves, because of the effects of surrounding leaves. The following chapter discusses the plant's mineral nutrition and the numerous ways plants cope with soils in which nutrients have low availability or where metals occur in toxic concentrations (e.g., sodium, aluminum, and heavy metals). These first chapters emphasize those aspects that help us to analyze ecological problems. They also provide a sound basis for later chapters in the book, which deal with a higher level of integration.

A later set of chapters will deal with patterns of growth and allocation, life-history traits, and interactions of individual plants with other organisms: surrounding plants, herbivores and pathogens, animals used as a prey by carnivorous plants, parasitic plants, and symbiotic microorganisms. These chapters build on information provided in the first chapters.

The final chapters deal with ecophysiological traits that affect decomposition of plant material in contrasting environments, and with the role of plants in ecosystem and global processes. Many topics in the first two sets of chapters will be addressed again, now from a different perspective.

Throughout the text, "boxes" are used to elaborate on specific problems, without cluttering up the text. They are meant for the student who wishes to gain a deeper understanding of problems discussed in the various chapters. A glossary has been added to be able to check the meaning of numerous terms used throughout this text or in the vast literature on plant physiological ecology quickly. The numerous references added to each chapter should help the reader to access the relevant literature used for this text. We have followed the approach to provide few references whenever the information is at the level of first-year textbooks in physiology or ecology, but added several when dealing with more recent developments ("the cutting edge").

References and Further Reading

Billings, W.D. (1973) Arctic and alpine vegetation: Similarities, differences, and susceptibility to disturbance. BioScience 23:697–704.

Billings, W.D., Godfrey, P.J., Chabot, B.F., & Bourque, D.P. (1971) Metabolic acclimation to temperature in arctic and alpine ecotypes of *Oxyria digyna*. Arct. Alp. Res. 3:277–289.

Blackman, V.H. (1919) The compound interest law and plant growth. Ann. Bot. 33:353–360.

Boyer, J.S. (1982) Plant productivity and environment. Science 218:443–448.

Boyer, J.S. (1985) Water transport. Annu. Rev. Plant Physiol. 36:473–516.

Chapin III, F.S. (1993) Functional role of growth forms in ecosystem and global processes. In: Scaling Physiological Processes: Leaf to Globe, J.R. Ehleringer & C.B. Field (eds). Academic Press, San Diego, pp. 287–312.

Clarkson, D.T. (1967) Phosphorus supply and growth rate in species of *Agrostis* L. J. Ecol. 55:111–118.

Crawford, R.M.M. (1978) Biochemical and ecological similarities in marsh plants and diving animals. Naturwissenschaften 65:194–201.

Ellenberg, H. (1953) Physiologisches und ökologisches Verhalten derselben Pflanzanarten. Ber. Deut. Botan. Ges. 65:351–361.

Gould, S.J. & Lewontin, R.C. (1979) The spandrels of San Marco and the Panglossian paradigm: A critique of the adaptationist programme. Proc. R. Soc. Lond. B. 205:581–598.

Hammond, P.M. (1995) The current magnitude of biodiversity. In: Global biodiversity assessment, V.H. Heywood (ed). Cambridge University Press, Cambridge, pp. 113–138.

Holdridge, L.R. (1947) Determination of world plant formations from simple climatic data. Science 105:367–368.

Lambers, H. & Poorter, H. (1992) Inherent variation in growth rate between higher plants: A search for physiological causes and ecological consequences. Adv. Ecol. Res. 22:187–261.

Lambers, H., Poorter, H., & Van Vuuren, M.M.I. (eds). (1998) Inherent variation in plant growth. Physiological mechanisms and ecological consequences. Backhuys, Leiden.

Mooney, H.A. (1972) The carbon balance of plants. Annu. Rev. Ecol. Syst. 3:315–346.

Mooney, H.A. & Dunn, E.L. (1970) Convergent evolution of mediterranean-climate sclerophyll shrubs. Evolution 24:292–303.

Pearsall, W.H. (1938) The soil complex in relation to plant communities. J. Ecol. 26:180–193

Schimper, A.F.W. (1898) Pflanzengeographie und Physiologische Grundlage. Verlag von Gustav Fischer, Jena.

Shukla, J., Nobre, C., & Sellers, P. (1990) Amazon deforestation and climate change. Science 247:1322–1325.

Vrba, E.S. & Gould, S.J. (1986) The hierarchical expansion of sorting and selection: Sorting and selection cannot be equated. Paleobiology 12:217–228.

Walter, H. (1974) Die Vegetation der Erde. Gustav Fisher Verlag, Jena.

2
Photosynthesis, Respiration, and Long-Distance Transport

2A. Photosynthesis

1. Introduction

Approximately 40% of a plant's dry mass consists of carbon, fixed in photosynthesis. This process is vital for growth and survival of virtually all plants during the major part of their growth cycle. In fact, life on earth in general, not just that of plants, totally depends on current and/or past photosynthetic activity. Leaves are beautifully specialized organs that enable plants to intercept the light necessary for photosynthesis. The light is captured by a large array of chloroplasts that are in close proximity to air and not too far away from vascular tissue, which supplies water and exports the products of photosynthesis. CO_2 uptake occurs through leaf pores, the stomata, which are able to rapidly change their aperture (see Sect. 5.4 in plant water relations). Once inside the leaf, CO_2 diffuses from the intercellular air spaces to the sites of carboxylation in the chloroplast (C_3 species) or in the cytosol (C_4 and CAM species).

Ideal conditions for photosynthesis include an ample supply of water and nutrients to the plant, as well as optimal temperature and light conditions. Even when the environmental conditions are less favorable, however, such as in a desert or the understory of a forest, photosynthesis—at least of the adapted plants—continues. This chapter addresses how such plants manage to continue to photosynthesize and/or protect their photosynthetic machinery in such adverse environments.

2. General Characteristics of the Photosynthetic Apparatus

2.1 The Light and Dark Reactions of Photosynthesis

The primary processes of photosynthesis occur in the chloroplast. In C_3 plants most of the chloroplasts are located in the mesophyll cells of the leaves (Fig. 1). Three main processes are distinguished:

1. Absorption of photons by pigments, mainly chlorophylls, associated with two photosystems. The pigments are embedded in internal membrane structures (**thylakoids**) and absorb a major part of the energy of the photosynthetically active radiation (PAR; 400 to 700 nm). They transfer the excitation energy to the reaction centers of the photosystems where the second process starts.

2. Electrons derived from the splitting of water with the simultaneous production of oxygen are transported along an electron-transport chain embedded in the thylakoid membrane. NADPH and ATP produced in this process are used in the third process. Because these two reactions depend on light energy, they are called the **light reactions** of photosynthesis.

3. The NADPH and ATP are used in the photosynthetic carbon-reduction cycle (Calvin cycle), in which CO_2 is assimilated, leading to the synthesis of C_3 compounds (triose-phosphates). This process

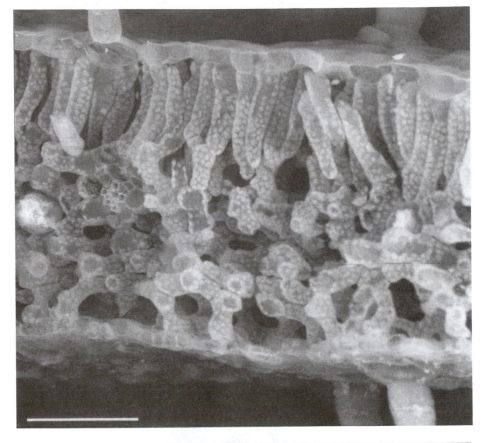

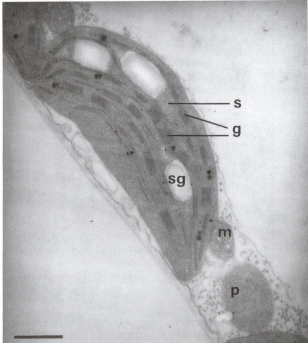

FIGURE 1. *Above*: Scanning electron microscope cross-sectional view of a leaf of *Nicotiana tabacum* (tobacco) that shows palisade tissue beneath the upper (adaxial) epidermis and spongy tissue adjacent to the (lower) abaxial epidermis. *Below*: Transmission electron microscope micrograph of a tobacco chloroplast that shows stacked (grana, g) and unstacked regions of the thylakoids, stroma (s), and starch granules (sg). Note the close proximity of two mitochondria (m, top and bottom) and one peroxisome (p) (scale bar is 1 µm) (courtesy J.R. Evans, Research School of Biological Sciences, Australian National University, Canberra, Australia).

can proceed in the absence of light and is referred to as the **dark reaction** of photosynthesis.

2.1.1 Absorption of Photons

The reaction center of **photosystem I** (PSI) is a chlorophyll dimer with an absorption peak at 700 nm, hence called P_{700}. There are about 110 "ordinary" chlorophyll a (chl a) molecules per P_{700}, a few chlorophyll b (chl b) molecules, as well as about 11 different protein molecules to keep the chlorophyll molecules in the required position in the thylakoid membranes. The number of PSI units can be quantified by determining the amount of P_{700} molecules.

The reaction center of **photosystem II** (PSII) is a chlorophyll molecule with an absorption peak at 680 nm, called P_{680}. There are about 30 times more chl a molecules than chl b; several protein molecules keep the chlorophyll molecules in the required position in the thylakoid membranes. In vitro, P_{680} is too unstable to be used to quantify the amount of PSII. Atrazine, however, binds specifically to one of the complexing protein molecules of PSII; when using ^{14}C-labeled atrazine, this binding can be quantified and used to determine the total amount of PSII.

A large part of the chlorophyll is located in the **light-harvesting complex** (LHC). These chlorophyll molecules act as antennae to trap light and to transfer their excitation energy to the reaction centers of one of the photosystems. The reaction centers are strategically located to transfer electrons along the electron-transport chains. The ratio of chl a to chl b is about 1.12 for LHC. Chl b, therefore, is predominantly associated with the LHC and little with PSI and PSII core complexes.

Leaves appear green in white light because chlorophyll absorbs more efficiently in the blue and red than it does in the green portions of the spectrum. The **absorption spectrum** of intact leaves differs from that of chlorophyll because intact leaves absorb a significant portion of the radiation in regions where chlorophyll absorbs very little in vitro (Fig. 2). This is due to (1) the modification of the absorption spectra of the chlorophyll molecules in vivo, (2) the presence of accessory pigments, such as carotenoids, in the chloroplast, and (3) light scattering within the leaf.

2.1.2 Fate of the Excited Chlorophyll

Each quantum of red light absorbed by a chlorophyll molecule raises an electron from a ground

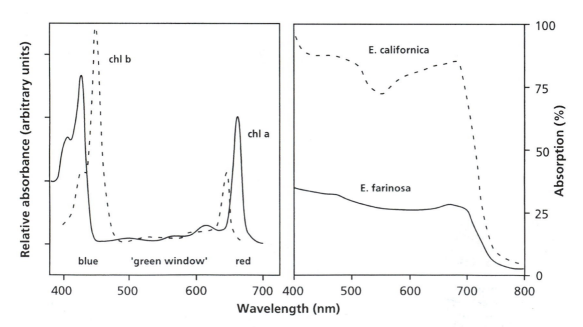

FIGURE 2. The absorption spectrum of chlorophyll a and chlorophyll b (left; Anderson & Beardall 1991; copyright Blackwell Science Ltd) and that of an intact green leaf of *Encelia californica*; for comparison, the absorption spectrum of an intact white (pubescent) leaf of *Encelia farinosa* is also given (right; Ehleringer et al. 1976). Copyright by the American Association for the Advancement of Science.

state to an excited state. Absorption of light of shorter wavelengths (e.g., blue light) excites the chlorophyll to an even higher energy state. In the higher-energy state after absorption of blue light, however, chlorophyll is unstable and rapidly gives up some of its energy to the surroundings as heat so that the elevated electron immediately falls back into the orbit of the electron excited by red light. Thus, chlorophyll reaches the same excitation state upon photon capture whatever the wavelength of the light absorbed. In this excitation state, chlorophyll is stable for 10^{-9} seconds, after which it disposes of its available energy in one of three ways (Krause & Weis 1991):

1. The energy may be transferred to other chlorophyll molecules, so that it can reach the reaction center and be used in **photochemistry**, driving biochemical reactions. This process is highly efficient, accounting for about 90% of the energy transfer under favorable environmental conditions.

2. The excited chlorophyll can return to its ground state by converting its excitation energy into **heat**. No photon is emitted in this case.

3. It can re-emit a photon and thereby return to its ground state; this process is called **fluorescence**. Most fluorescence is emitted by chl a of PSII. The wavelength of fluorescence is slightly longer than that of the absorbed light because a portion of the excitation energy is lost before the fluorescence photon is emitted. Chlorophylls usually fluoresce in the red; it is a deeper red (the wavelength is about 10 nm longer) than the red absorption peak of chlorophyll. Fluorescence can be quantified by measuring emission of this deep-red wavelength during irradiance with blue or red light. Fluorescence is a measure of the efficiency of electron transfer. It increases under conditions of excessive light, inadequate CO_2 supply, or stresses that negatively affect the photochemical reactions.

The process with the highest rate will be the one most likely to deactivate the excited chlorophyll molecule.

2.1.3 Membrane-Bound Photosynthetic Electron Transport and Bioenergetics

The excitation energy captured by the pigments is transferred to the reaction centers of PSI and PSII. PSI and PSII are largely associated with the "**unstacked**" and with the "**stacked**" regions of the thylakoids, respectively (Fig. 1). In PSII an electron derived from the splitting of water is transferred to the first electron acceptor of the photosynthetic electron-transport chain (predominantly located in the "unstacked" regions of the thylakoid membrane). In the process protons are transported across the membrane into the thylakoid lumen, in addition to the protons released during splitting of water. As a result, the lumen becomes acidified and positively charged. The electrochemical gradient across the thylakoid membrane, which represents a **proton-motive force**, is subsequently used to phosphorylate ADP, thus producing **ATP**. This reaction is catalyzed by an **ATPase**, or **coupling factor** (located in the "unstacked" regions of the thylakoids). In **noncyclic electron transport**, NADP is the terminal acceptor of electrons from PSI, which results in formation of NADPH. In **cyclic electron transport**, electrons are transferred from PSI to cytochrome in the thylakoid lumen, thus contributing to proton extrusion and ATP synthesis. NADPH and ATP are subsequently used in the carbon-reduction cycle.

The various components of the light reaction of photosynthesis are therefore located in different regions of the thylakoids: exposed to stroma (unstacked) or exposed to other thylakoids (stacked). More details are given in Figure 3.

2.1.4 Photosynthetic Carbon Reduction

Ribulose-1,5-bisphosphate (RuBP) and CO_2 are the substrates for the principal enzyme of the carbon-reduction or Calvin cycle: ribulose-1, 5-bisphosphate carboxylase/oxygenase (**Rubisco**) (Fig. 4). The first product of carboxylation of RuBP by Rubisco is phosphoglyceric acid (PGA), which is a compound with three carbon atoms; hence, the name, C_3 **photosynthesis**. With the consumption of the ATP and NADPH produced in the light reaction, PGA is reduced to a triose-phosphate (triose-P), some of which is exported to the cytosol in exchange for inorganic phosphate (P_i). In the cytosol, triose-P is used to produce sucrose and other metabolites that are exported via the phloem or used in the leaves. Most of the triose-P remaining in the chloroplast is used to regenerate RuBP through a series of reactions that are part of the **Calvin cycle** in which ATP and NADPH are consumed (Fig. 4). Some of the triose-P remaining in the chloroplast may be used to produce starch, which is stored in the chloroplast. Starch may be hydrolyzed during the night, and the product of this reaction, triose-P, is exported to the cytosol. The photosynthetic carbon-reduction cycle has various control points and factors that function as stabilizing mechanisms under changing environmental conditions.

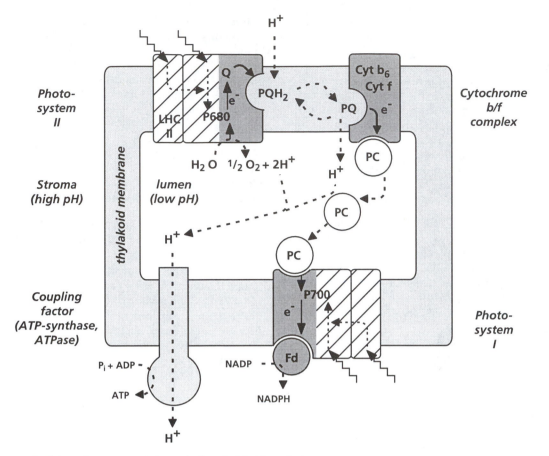

FIGURE 3. Schematic representation of the thylakoid membrane, enclosing the thylakoid lumen, that shows the transfer of excitation energy and of electrons, migration of molecules, and chemical reactions. P_{700}: reaction center of photosystem I; P_{680}: reaction center of photosystem II; LHC: light-harvesting complex; Q: quinone; PC, plastocyanin; Fd: ferredoxin; cyt: cytochrome.

2.1.5 Oxygenation and Photorespiration

Rubisco catalyzes both the **carboxylation** of RuBP and its **oxygenation**. The ratio of the carboxylation and the oxygenation reaction strongly depends on the relative concentrations of CO_2 and O_2, and on leaf temperature. The products of the carboxylation reaction are two C_3 molecules (PGA), whereas the oxygenation reaction produces only one PGA and one C_2 molecule: phosphoglycolate (GLL-P). This C_2 molecule is first dephosphorylated in the chloroplast, producing glycolate (GLL) (Fig. 5). This is exported to the peroxisomes, where it is metabolized to glyoxylate and then **glycine**. Glycine is exported to the mitochondria, where two molecules are converted to produce one serine with the release of one CO_2 and one NH_3. Serine is exported back to the peroxisomes, where a transamination occurs, producing one molecule of hydroxypyruvate and

then glycerate. Glycerate moves back to the chloroplast to be converted into PGA. Out of two phosphoglycolate molecules, therefore, one glycerate is made and one C-atom is lost as CO_2. The entire process, starting with the oxygenation reaction, is called **photorespiration** because it is a respiratory process that depends on light, as opposed to "dark respiration", which largely consists of mitochondrial decarboxylation processes that proceed independent of light. Dark respiration is discussed in various sections in the chapter on plant respiration.

2.2 Supply and Demand of CO_2 in the Photosynthetic Process

The rate of photosynthetic carbon assimilation is determined by both the supply and demand for

CO_2. The **supply of CO_2** to the chloroplast is governed by diffusion in the gas and liquid phases. It can be limited at several points in the pathway from the air surrounding the leaf to the site of carboxylation inside (see Fig. 5 in the plant's energy balance). The **demand for CO_2**, is determined by the rate of processing the CO_2 in the chloroplast, which is governed by the structure and biochemistry of the chloroplast (Sect. 2.1), by environmental factors such as irradiance, and by factors that affect plant demand for carbohydrates (Sect. 4.2). Limitations imposed by either supply or demand can regulate the overall rate of carbon assimilation, as explained shortly.

2.2.1 The CO_2-Response Curve

The response of photosynthetic rate to CO_2 concentration is the principal tool to analyze the demand for CO_2 (Farquhar & Sharkey 1982) (Fig. 6). The graph giving CO_2 assimilation (A) as a function of intercellular CO_2 partial pressure (p_i) is generally referred to as the **A–p_i curve**, or A–c_i when CO_2 concentration is given as mole fraction in air. There is no net CO_2 assimilation until the production of CO_2 in respiration (mainly photorespiration, but also some dark respiration that occurs in the light) is fully compensated by the fixation of CO_2 in photosynthesis. The CO_2 pressure at which this is

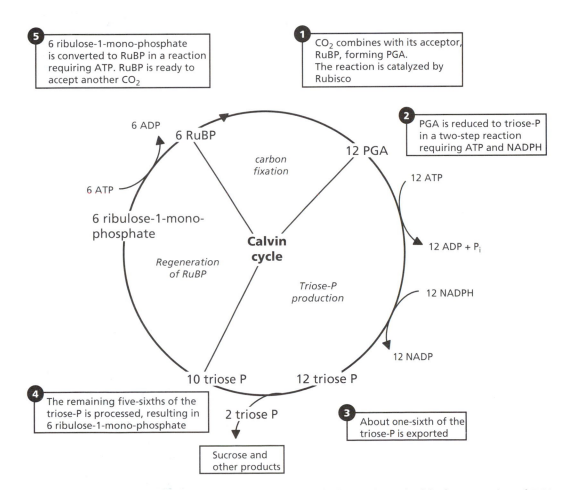

FIGURE 4. Schematic representation of the photosynthetic carbon reduction cycle (Calvin cycle) that shows major steps: carbon fixation, triose-P production, and regeneration of RuBP. 1: CO_2 combines with its substrate, ribulose-1,5-bisphosphate (RuBP), catalyzed by ribulose bisphosphate carboxylase/oxygenase (Rubisco), producing phosphoglyceric acid (PGA). 2: PGA is reduced to triose-phosphate (triose-P), in a two-step reaction; the reaction for which ATP is required is the conversion of PGA to 1,3-bisphosphoglycerate, catalyzed by phosphoglycerate kinase. 3 & 4: Part of the triose-P is exported to the cytosol, in exchange for P_i; the remainder is used to regenerate ribulose-1-monophosphate. 5: ribulose-1-monophosphate is phosphorylated, catalyzed by ribulose-5-phosphate kinase, producing RuBP.

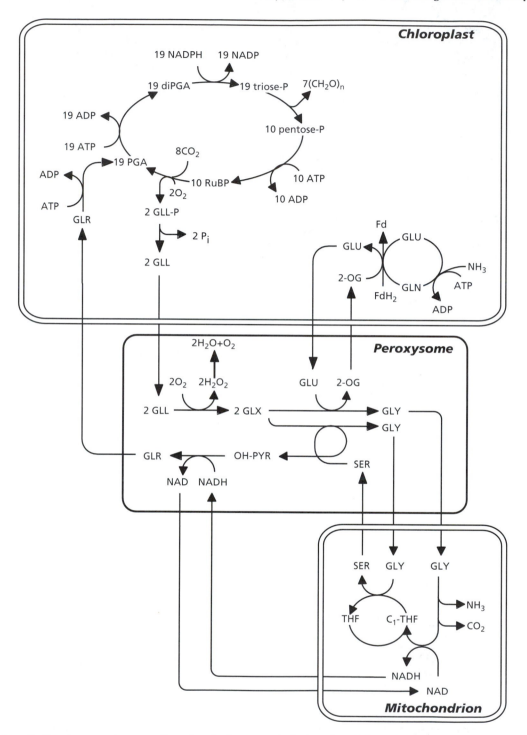

FIGURE 5. Reactions and organelles involved in photorespiration. In C$_3$ plants, at 20°C, 20,000 Pa O$_2$ and 25 Pa CO$_2$ in the chloroplast, 2 out of 10 RuBP molecules are oxygenated, rather than carboxylated. The oxygenation reaction produces phosphoglycolate (GLL-P), which is dephosphorylated to glycolate (GLL). Glycolate is subsequently metabolized in peroxisomes and mitochondria, in which glyoxylate (GLX) and the amino acids glycine (GLY) and serine (SER) play a role. Serine is exported from the mitochondria and converted to hydroxypyruvate (OH-PYR) and then glycerate (GLR) in the peroxisomes, after which it returns to the chloroplast (after Ogren 1984). With permission, from the Annual Review of Plant Physiology, Vol. 35, copyright 1984, by Annual Reviews Inc.

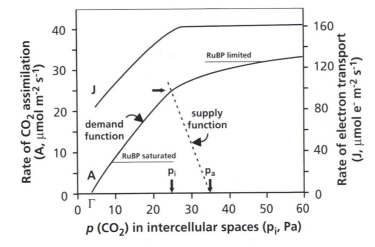

FIGURE 6. The relationship between the rate of CO_2 assimilation, A, and the partial pressure of CO_2 in the intercellular spaces, p_i, for a C_3 leaf: the "demand function." The concentration at which A = 0 is the CO_2-compensation point (Γ). In the "linear" region at low values of p_i, the concentration of CO_2 is limiting, whereas that of ribulose-1,5-bisphosphate (RuBP) is saturating the activity of Rubisco. At higher values of p_i, in the "saturating" region, the concentration of CO_2 is saturating the activity of Rubisco, whereas the regeneration of RuBP limits carboxylation. The rate of photosynthetic electron transport, J, is given by the upper curve in the figure. It can be determined with the fluorescence technique, whereas the rate of CO_2 assimilation is measured with a gas exchange analyzer. The increase of A in the "saturating" region, where electron transport does not change, is due to the suppression of the oxygenation reaction of Rubisco. The rate of diffusion of CO_2 from the atmosphere to the intercellular spaces is given by the "supply function" (the broken line). The slope of this line is the leaf conductance. The intersection of the "supply function" with the "demand function" is the actual rate of net CO_2 assimilation (indicated by a horizontal arrow) at a value of p_i (indicated by a vertical arrow) that occurs in the leaf at the value for p_a in normal air (indicated by the right-hand vertical arrow). Values for p_i for well-watered C_3 plants are around 20 to 25 Pa when measurements are made in normal air and at fairly high irradiance.

reached is the CO_2-compensation point (Γ). In C_3 plants this is largely determined by the kinetic properties of Rubisco, with values for Γ in the range of 4 to 5 Pa.

Two regions of the CO_2-response curve above the compensation point can be distinguished. At low p_i, which is below values normally found in leaves (approximately 25 Pa), photosynthesis increases steeply with increasing partial pressure of CO_2. This is the region where CO_2 limits the rate of functioning of Rubisco, whereas RuBP is present in saturating quantities (**RuBP-saturated** or **CO_2-limited region**). This part of the A–p_i relationship is also referred to as the **initial slope** or the **carboxylation efficiency**. At light saturation and with a fully activated enzyme (see Sect. 3.4.2 in this chapter for details on "activation"), the initial slope governs the carboxylation capacity of the leaf, which in turn depends on the amount of active Rubisco.

In the region at high p_i, the increase in A with increasing p_i levels off. CO_2 no longer restricts the carboxylation reaction, but now the rate at which RuBP becomes available limits the activity of Rubisco (**RuBP-limited region**). This rate, in turn, depends on the activity of the Calvin cycle, which ultimately depends on the rate at which ATP and NADPH are produced in the light reactions. In this region, photosynthetic rates are limited by the rate of electron transport (J in Fig. 6). This may either be due to limitation by light or, at light saturation, by a limited capacity of electron transport. Even at a high p_i, in the region where the rate of electron transport, J, no longer increases with increasing p_i, the rate of net CO_2 assimilation continues to increase slightly because the oxygenation reaction of Rubisco is increasingly suppressed with increasing partial pressure of CO_2 in favor of the carboxylation reaction. At a normal atmospheric partial pressure of CO_2 and O_2 (35 and 21,000 Pa, respectively) and at a temperature of 20°C, the ratio between the carboxylation and oxygenation reaction is about 4:1. The method by which this ratio and various other parameters of the A–p_i curve can be assessed is further explained in Box 1. Plants typically operate at a p_i where **CO_2** and **electron transport co-limit** the rate of CO_2 assimilation (i.e., the point

Box 1
Mathematical Description of the CO_2 Response and Further Modeling of Photosynthesis

Based on known biochemical characteristics of Rubisco and the requirement of $NADPH_2$ and ATP for CO_2 assimilation, Farquhar and Von Caemmerer (1982) have developed a model of photosynthesis of C_3 plants. This model is now widely used in ecophysiological research and is introduced briefly here.

Net CO_2 assimilation (A) is the result of the rate of carboxylation (V_c) minus photorespiration and other respiratory processes. In photorespiration, one CO_2 is produced per two oxygenation reactions (V_o) (Fig. 5 in the chapter on photosynthesis). The rate of dark respiration during photosynthesis may differ from normal dark respiration (night respiration) and is called "day respiration" (R_{day}).

$$A = V_c - 0.5 V_o - R_{day} \qquad (1)$$

CO_2-limited and O_2-limited rates of carboxylation and oxygenation are described with standard Michaelis-Menten kinetics. When both substrates are present, however, they competitively inhibit each other. An effective Michaelis-Menten constant for the carboxylation reaction (K_m) that takes into account competitive inhibition by O_2 is described as:

$$K_m = K_c(1 + O/K_o) \qquad (2)$$

where K_c and K_o are the Michaelis-Menten constants for the carboxylation and oxygenation reaction, respectively, and O is the partial pressure of oxygen.

The rate of carboxylation in the CO_2-limited part of the CO_2-response curve can then be described as:

$$V_c = \frac{V_{cmax}\, p_i}{p_i + K_m} \qquad (3)$$

where V_{cmax} is the rate of CO_2 assimilation at saturating intercellular partial pressure of CO_2 (p_i) (note that the subscript "max" refers to the rate at *saturating* p_i, which is at variance with the way it is used in Sect. 3.2.1 of the chapter on photosynthesis).

The ratio of oxygenation and carboxylation depends on the specificity of Rubisco for CO_2 relative to O_2 ($S_{c/o}$). This specificity varies widely between photosynthetic organisms (Bainbridge et al. 1995), but is remarkably similar among higher plants. Increasing temperature, however, decreases the specificity, because K_o decreases faster with increasing temperature than K_c does (Fig. 35 in the chapter on photosynthesis).

The CO_2-compensation point in the absence of R_{day} (Γ^*) depends on the specificity factor and the O_2 concentration (O):

$$\Gamma^* = 0.5 O/(S_{c/o} \cdot s_c/s_o) \qquad (4)$$

Γ^* increases more strongly with rising temperature than what would be expected from the decrease in $S_{c/o}$, because the solubility in water for CO_2 (s_c) decreases more with increasing temperature than does that for O_2 (s_o).

The CO_2 compensation point (Γ^*) shows little variation between C_3 angiosperms as follows from the similarity of $S_{c/o}$. Γ^* is determined experimentally and used to calculate the rates of carboxylation and oxygenation from photosynthetic rates using:

$$V_o/V_c = 2\Gamma^*/p_i \qquad (5)$$

thus avoiding the need for incorporating the specificity factor and solubilities (Eq. 4).

In the RuBP-limited part of the CO_2-response curve, the rate of electron transport (J) is constant. Increasing the partial pressure of CO_2 increases the rate of carboxylation at the expense of the rate of oxygenation. There is a minimum requirement of four electrons per carboxylation or oxygenation reaction. Hence, the minimum electron transport rate (J) required for particular rates of carboxylation and oxygenation is:

$$J = 4(V_c + V_o) \qquad (6)$$

Using Equations 5 and 6, the rate of carboxylation can then be expressed as:

$$V_c = J/\{4(1 + 2\Gamma^*/p_i)\} \qquad (7)$$

The CO_2-limited and RuBP-saturated rate of photosynthesis {A(c)} can then be calculated using Equations 1, 3, and 5 as:

$$A(c) = \frac{V_{cmax}(p_i - \Gamma^*)}{p_i + K_m} - R_{day} \qquad (8)$$

The RuBP-limited rate of photosynthesis {A(j)} can be calculated using Equations 1, 5, and 7 as:

continued

Box 1. *Continued*

FIGURE 1. The CO_2 response curve of net photosynthesis at 25°C (solid line) calculated, as explained in the text, with values for V_{cmax} and J_{max} of 100 and 140 μmol m^{-2} s^{-1}, respectively. The lower part of the relationship is limited by the carboxylation capacity {A(c)} and the upper part by the electron-transport capacity {A(j)}. The rate of electron transport (J/4) is also shown.

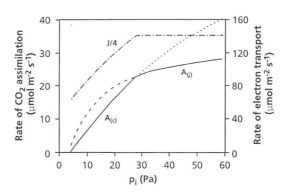

$$A(j) = \frac{J(p_i - \Gamma^*)}{4(p_i + 2\Gamma^*)} - R_{day} \qquad (9)$$

The minimum of the Equations 8 and 9 describes the full CO_2-response curve as shown in Figure 1.

Parameter values for the preceding formulas are normally given for 25°C. Values for other temperatures can be determined from the temperature dependence of the parameter values. Different empirical formulas are used for that purpose. The following examples are derived from Farquhar and Von Caemmerer (1982), Kirschbaum and Farquhar (1984), Brooks and Farquhar (1985), and Von Caemmerer et al. (1994).

$$K_c(25°C) = 40.4\,Pa \qquad K_c = K_{c(25)}\exp^{(59356\,x)}$$

$$K_o(25°C) = 24,800\,Pa \qquad K_o = K_{o(25)}\exp^{(35948\,x)}$$

where $x = 0.000404\,(T - 25)/(T + 273)$

$$\Gamma^*\,(25°C) = 3.7\,Pa$$

$$\Gamma^* = \Gamma^*_{(25)} + 0.188\,(T - 25) + 0.0036\,(T - 25)^2$$

$$V_{cmax} = V_{cmax(25)}\{1 + 0.0505\,(T - 25)$$
$$- 0.248\,10^{-3}(T - 25)^2$$
$$- 8.09\,10^{-5}(T - 25)^3\}$$

$$J_{max} = J_{max(25)}\{1 + 0.0409\,(T - 25)$$
$$- 1.24\,10^{-3}(T - 25)^2$$
$$- 9.42\,10^{-5}(T - 25)^3\}$$

The above values and temperature relations for K_o, K_c, and Γ^* do not vary much between species and growth conditions, but the temperature dependence of V_{cmax}, and particularly J_{max}, may vary considerably. Information on the temperature dependencies of parameter values makes it possible to use the model for a wider temperature range.

Intercellular $p(CO_2)$ varies due to a shift in balance between supply and demand for CO_2. The demand function has already been described; the supply function has been described in Section 2.2.2 of the chapter on photosynthesis. The dynamics of stomatal movement are not very well understood, but empirical descriptions are available for modeling work (e.g., Farquhar & Wong 1984). Electron-transport rate varies due to variation in irradiance as described in Section 3.2.1 of the chapter on photosynthesis, where the formula describing net CO_2 assimilation as a function of irradiance can be used to calculate J by replacing J and J_{max} for A and A_{max}, respectively, and omitting R_{day}. A combination of these mathematical formulations makes it possible to model C_3 photosynthesis over a wide range of environmental conditions. For examples, see Evans and Farquhar (1991) and Harley et al. (1992).

References and Further Reading

Bainbridge, G., Madgwick, P., Parmar, S., Mitchell, R., Paul, M., Pitts, J., Keys, A.J., & Parry, M.A.J. (1995) Engeneering rubisco to change its catalytic properties. J. Exp. Bot. 46:1269–1276.

continued

Box 1. *Continued*

Brooks, A. & Farquhar, G.D. (1985) Effect of temperature on the CO_2/O_2 specificity of ribulose-1,5-
bisphosphate carboxylase/oxygenase and the rate
of respiration in the light. Planta 165:397–406.

Evans, J.R. & Farquhar, G.D. (1991) Modeling canopy
photosynthesis from biochemistry of the C_3
chloroplast. In: Modeling crop photosynthesis-from
biochemistry to canopy, Special Publication No. 19,
K.J. Boote, & R.S. Loomis (eds). Crop Science Society
of America, Madison, pp. 1–15.

Farquhar, G.D. & Von Caemmerer (1982) Modelling of
photosynthetic response to environmental conditions. In: Encyclopedia of plant physiology, N.S.,
Vol 12B, O.L. Lange, P.S. Nobel, C.B. Osmond, & H.
Ziegler (eds). Springer-Verlag, Berlin, pp. 549–587.

Farquhar, G.D. & Wong, S.C. (1984) An empirical
model of stomatal conductance. Aust. J. Plant
Physiol. 11:191–210.

Harley, P.C., Thomas, R.B., Reynolds, J.F., & Strain,
B.R. (1992) Modelling photosynthesis of cotton
grown in elevated CO_2. Plant Cell Environ. 15:271–
282.

Kirschbaum, M.U.F. & Farquhar, G.D. (1984) Temperature dependence of whole-leaf photosynthesis
in *Eucalyptus pauciflora* Sieb. ex Spreng. Aust. J. Plant
Physiol. 11:519–538.

Von Caemmerer, S., Evans, J.R., Hudson, G.S., &
Andrews, T.J. (1994) The kinetics of ribulose-
1,5-bisphosphate carboxylase/oxygenase *in vivo*
inferred from measurements of photosynthesis in
leaves of transgenic tobacco. Planta 195:88–97.

where the RuBP—saturated and the RuBP—limited
part of the CO_2-response curve intersect). This
allows effective utilization of all components of
the light and dark reactions.

2.2.2 Supply of CO_2—Stomatal and Boundary Layer Conductances

As mentioned earlier, the rate of supply of CO_2 to
the chloroplast can be described as a diffusion process ("**supply function**" in Fig. 6). To analyze diffusion limitations it is convenient to use the term
resistance because resistances can be summed to
arrive at the total resistance for the pathway. When
considering fluxes, however, it is more convenient
to use **conductance**, which is the reciprocal of resistance, because the flux varies in proportion to the
conductance.

In a steady state, the rate of CO_2 assimilation (A)
equals the rate of CO_2 diffusion into the leaf. The
rate of CO_2 diffusion can be described by **Fick's
law**. Hence:

$$A = g_c \left(p_a - p_i\right)/P = g_c \left(c_a - c_i\right) = \left(c_a - c_i\right)/r_c \tag{1}$$

where, g_c is the leaf conductance for CO_2 transport;
p_a and p_i are the partial pressure of CO_2 in the
atmosphere and the intercellular spaces, respectively; P is the atmospheric pressure; c_a and c_i are
the corresponding mole or volume fractions in air;
r_c is the inverse of g_c (i.e., the leaf resistance for CO_2
transport).

The **leaf conductance** for CO_2 transport, g_c, can
be derived from measurements on leaf transpiration, which can also be described by Fick's law in a
similar way:

$$E = g_w \left(e_i - e_a\right)/P = g_w \left(w_i - w_a\right)$$
$$= \left(e_i - e_a\right)/P \cdot r_w \tag{2}$$

where g_w is the leaf conductance for water vapor
transport; e_i and e_a are the water vapor pressures in
the leaf and in the atmosphere, respectively; w_i and
w_a are the corresponding mole fractions of water
vapor; and E is the rate of leaf transpiration. E, P,
and e_a can be measured directly. The water vapor
pressures in the leaf can be calculated from measurements of the leaf's temperature, assuming a
saturated water vapor pressure inside the leaf.
Under most conditions this is a valid assumption.
The leaf conductance for water vapor transport
can therefore be determined.

The total **leaf resistance** for water vapor transfer,
r_w, is largely composed of two components that are
in series: the **boundary layer resistance**, r_a, and the
stomatal resistance, r_s (see Fig. 6, in the plant's
energy balance). The boundary layer is the layer of
air adjacent to the leaf that is modified by the leaf.
Its limit is commonly defined as the point at which
the properties of the air are 99% of the values in
ambient air. The boundary layer resistance can be
determined by measuring the rate of evaporation
from a water-saturated piece of filter paper of exactly the same shape and size as that of the leaf.
Conditions that affect the boundary layer, such as
wind speed, should be identical to those during
measurements of the leaf resistance. The stomatal
resistance for water vapor transfer (r_s) can now be
calculated because r_w and r_a are known:

$$r_w = r_a + r_s \tag{3}$$

The resistance for CO_2 transport (r_c) can be calculated from r_w, taking into account that the diffusion

coefficients of the two molecules differ. The ratio H_2O diffusion/CO_2 diffusion in air is approximately 1.6 because water is smaller and diffuses more rapidly than CO_2. This value pertains only to the movement of CO_2 inside the leaf and through the stomata. For the boundary layer above the leaf, where both turbulence and diffusion influence flux, the ratio is approximately 1.37.

$$r_c = (r_a \cdot 1.37) + (r_s \cdot 1.6) = 1/g_c \qquad (4)$$

The intercellular CO_2 partial pressure, p_i, as used in Figure 6, can now be calculated from Equation 1. If calculated according to this, p_i is the partial pressure of CO_2 at the point where evaporation occurs inside the leaf (i.e., largely the mesophyll cell walls that border the substomatal cavity).

The "supply function" (Eq. 1) tends to intersect the "demand function" in the region where carboxylation and electron transport are colimiting (Fig. 6). In C_3 plants, p_i is generally maintained around 25 Pa, but it may increase to higher values at a low irradiance and higher humidity of the air, and decrease to lower values at high irradiance, low water availability, and low air humidity. For C_4 plants, the CO_2 partial pressure in the intercellular spaces is around 10 Pa (Morison 1987).

Under most conditions, the stomatal conductance is considerably less than the boundary layer con-

ductance (g_a is up to $10 \, mol \, m^{-2} s^{-1}$, at wind speeds of up to $5 \, m \, s^{-1}$; g_s has values of up to $1 \, mol \, m^{-2} s^{-1}$ at high stomatal density and widely open stomata), so that stomatal conductance strongly influences CO_2 diffusion into the leaf. For large leaves in still humid air, however, where the boundary layer is thick, the situation may be different, with g_s being closer to g_a.

2.2.3 The Internal Conductance

For the transport of CO_2 from the substomatal cavity to the chloroplast, an **internal conductance**, g_i (or resistance, r_i) also plays a role. Hence, we can describe the net rate of CO_2 assimilation by:

$$A = (p_a - p_c)/(r_a + r_s + r_i) \qquad (5)$$

where p_c is the partial pressure of CO_2 in the chloroplasts. The internal resistance may be small compared with the stomatal resistance, but it cannot invariably be ignored.

As a consequence of the internal resistance, the CO_2 partial pressure in the chloroplast, p_c, is less than p_i. The evidence for this statement largely comes from observations on the carbon isotope ratio of the products of photosynthesis (see Box 2 for further information on this technique). The data

---

Box 2
Discrimination of Carbon Isotopes in Plants

The carbon dioxide in the earth's atmosphere is composed of different carbon isotopes. The majority is $^{12}CO_2$; only approximately 1% of the total amount of CO_2 in the atmosphere is $^{13}CO_2$; an even smaller fraction is the radioactive species $^{14}CO_2$ (which will not be dealt with in the present context). Modern ecophysiological research makes abundant use of the fact that the isotope composition of plant biomass differs from that of the atmosphere. It is of special interest that isotope composition differs between plants which differ in photosynthetic pathway and between plants with a different water-use efficiency. How can we account for that?

The molar abundance ratio of the two carbon isotopes, R, is the ratio between ^{13}C and ^{12}C. The constants k^{12} and k^{13} refer to the rate of processes and reactions in which ^{12}C and ^{13}C participate, respectively. The "isotope effect" is described as:

$$R_{source}/R_{product} = k^{12}/k^{13} \qquad (1)$$

For plants, the isotope effect is to a small extent due to the slower diffusion in air of $^{13}CO_2$, when compared with that of the lighter isotope $^{12}CO_2$ (1.0044 times slower; during diffusion in water, there is little fractionation) (Table 1). The isotope effect is largely due to the biochemical properties of Rubisco, which reacts more readily with $^{12}CO_2$ than it does with $^{13}CO_2$. As a result, Rubisco *discriminates* against the heavy isotope. For Rubisco from spinach, the discrimination is 30.3‰, whereas smaller values are found for this enzyme from bacteria (Guy et al. 1993).

On the path from intercellular spaces to Rubisco a number of additional steps take place, where some isotope fractionation can occur.

continued

Box 2. *Continued*

TABLE 1. **The magnitude of the fractionation during CO_2 uptake.**

Process or enzyme	Fractionation (‰)
Diffusion in air	4.4
Diffusion through the boundary layer	2.9
Dissolution of CO_2	1.1
Diffusion of aqueous CO_2	0.7
CO_2 and HCO_3^- in equilibrium	−8.5 at 30°C
	−9.0 at 25°C
$CO_2 - HCO_3^-$ catalyzed by carbonic anhydrase	1.1 at 25°C
$HCO_3^- - CO_2$ in water, catalyzed by carbonic anhydrase	10.1 at 25°C
PEP carboxylase	2.2
Combined process	−5.2 at 30°C
	−5.7 at 25°C
Rubisco	30 at 25°C

Source: Henderson et al. 1992.

Taken together, the isotope effect in C_3 plants is approximated by the following empirical equation:

$$R_a/R_p = 1.0044\left[(p_a - p_i)/p_a\right] + 1.027\,p_i/p_a \quad (2)$$

where R_a and R_p are the molar abundance ratios of the atmospheric CO_2 and of the C fixed by the plant, respectively. The symbols p_a and p_i are the atmospheric and the intercellular partial pressure of CO_2, respectively. The value 1.027 is an empirical value, incorporating the major discrimination by Rubisco and the minor discrimination by PEP carboxylase (Table 1), as well as accounting for the drop in partial pressure of CO_2 between the intercellular spaces and the chloroplasts.

Since values for R_a/R_p appear rather "clumsy," it is more common to express the data as fractionation values, Δ ("capital delta"). This Δ is defined as $(R_a/R_p - 1)$, or:

$$\Delta = \left[(1.0044\,p_a - 1.0044\,p_i + 1.027\,p_i)/p_a\right] - 1$$
$$= \left[(1.0044\,p_a + 0.0226\,p_i)/p_a\right] - 1$$
$$= (4.4 + 22.6\,p_i/p_a) \times 10^{-3} \quad (3)$$

The isotope composition is described as $\delta^{13}C$ ("lower case delta"):

$$\delta^{13}C(‰) = (R_{sample}/R_{standard} - 1) \times 1000 \quad (4)$$

Values for $\Delta^{13}C$ and $\delta^{13}C$ are related as:

$$\Delta = \delta_{source} - \delta_{plant} \quad (5)$$

where $\delta_{source} = -8‰$ if the source is air (δ_{air}). The standard is the carbon dioxide generated from a cretaceous limestone from South Carolina (USA) consisting mostly of the fossil carbonate skeleton of *Belemnitella americana* (referred to as PDB-belemnite). By definition, it has a $\delta^{13}C$ equal to zero. The $\delta^{13}C$ values of plants are negative, as the processes (diffusion and carboxylation) discriminate against $^{13}CO_2$. The δ-values for C_3 plants are approximately −27‰, indicating that Rubisco is the predominant factor accounting for the observed values and that diffusion is less important. Therefore the physical barrier between the stomatal cavity and the chloroplasts cannot be a major one (the drop in the partial pressure of CO_2 is less than 30%).

For C_4 plants, the following empirical equation has been derived:

$$\Delta = 4.4 + \left[-5.7 + (30 - 1.8)\phi - 4.4\right]p_i/p_a \quad (6)$$

where ϕ refers to the leakage of CO_2 from the bundle sheath to the mesophyll.

Where do these complicated equations lead us? There are many applications. For example, within C_3 plants the carbon isotope composition will give us a very good indication of p_i over a longer time interval than can readily be obtained from gas exchange measurements. The value of p_i in itself is a reflection of the stomatal conductance, relative to the photosynthetic activity. As such, carbon isotope composition provides information on a plant's water use efficiency (Sect. 5.2 of the chapter on plant water relations). How do we arrive there?

From our equations we can deduce that, for a given p_a, there will be relatively less $^{13}CO_2$ fixed by Rubisco at higher, compared with lower, CO_2 concentrations. This is largely due to the discrimination by Rubisco, which has a greater affinity for $^{12}CO_2$ than it does for $^{13}CO_2$. Plant material therefore contains relatively less ^{13}C than does the atmosphere. The extent of the fractionation of carbon isotopes depends on the partial pressures of CO_2 at the site of Rubisco, relative to that in the atmosphere (p_i/p_a). If p_i/p_a is high, much of the $^{13}CO_2$ discriminated against by Rubisco diffuses back to the atmosphere, and the fractionation is large. Whenever p_i/p_a is small, fractionation by Rubisco is less because stomata restrict the outward diffusion of $^{13}CO_2$. Figure 1 illustrates that the discrimination in C_3

continued

Box 2. *Continued*

FIGURE 1. The relationship between the ratio of the internal and the atmospheric CO_2 concentration, at a constant p_a of 34 Pa. Data for both C_3 and C_4 species are presented; the lines are drawn on the basis of a number of assumptions, relating to the extent of leakage of CO_2 from the bundle sheath back to the mesophyll (Evans et al. 1986). Copyright CSIRO, Australia.

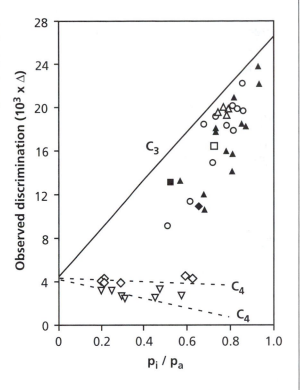

plants is positively correlated with the ratio between the internal and the atmospheric carbon dioxide concentration; this ratio reflects the concentration of CO_2 in the chloroplast if p_a is kept constant.

Carbon isotope discrimination values also differ between species with different photosynthetic pathways: C_3, C_4, or CAM (see the pertinent sections of the chapter on photosynthesis). In C_4 plants most of the $^{13}CO_2$ that is discriminated against by Rubisco does not diffuse back to the atmosphere. This is prevented first by the diffusion barrier between the vascular bundle sheath and the mesophyll cells. Second, the mesophyll cells contain large quantities of carbonic anhydrase and of PEP-carboxylase, which discriminate less against $^{13}CO_2$ (Table 1), and scavenge most of the CO_2 that might escape from the bundle sheath. There is also little discrimination in CAM plants, where the heavy isotopes discriminated against cannot readily diffuse out of the leaves because the stomates are closed for most of the day.

Aquatic C_3 plants also show relatively little discrimination, but this is due to unstirred layers surrounding the leaf, rather than to a different photosynthetic pathway (cf. Sect. 11.6 of the chapter on photosynthesis). These unstirred boundary layers cause diffusion to be a major limitation for their photosynthesis. As a result, fractionation in these plants tends toward the value found for the diffusion process, and not to values determined by Rubisco, such as in terrestrial plants.

Environmental factors, such as temperature and water stress, which affect p_i, also affect the relative concentration of ^{13}C in the plant (cf. pertinent sections of the chapters on photosynthesis, plant water relations, and parasitic associations).

Considering the difference in carbon isotope "signature" of plants with different photosynthetic pathways, the actual values of root biomass of plants in a mixed population has also been used to assess competitive ability between C_3 and C_4 plants (See Sec. 5.4 of the chapter on interactions among plants). Without the information on carbon isotope composition it would be very hard, if at all possible, to collect reliable information on the relative amount of root biomass of different species growing together. Other ecological questions can also be addressed using the carbon isotope "signature", for example, questions relating to food chains: Which species have been consumed? How much carbon is deposited in the soil?

continued

Box 2. *Continued*

References and Further Reading

Brugnoli, E., Hubick, K.T., Von Caemmerer, S., Wong, S.C., & Farquhar, G.D. (1988) Correlation between the carbon isotope discrimination in leaf starch and sugars of C_3 plants and the ratio of intercellular and atmospheric partial pressure of carbon dioxide. Plant Physiol. 88:1418–1424.

Ehleringer, J. (1990) Correlations between carbon isotope discrimination and leaf conductance to water vapor in common beans. Plant Physiol. 93:1422–1425.

Evans, J.R., Sharkey, T.D., Berry, J.A., & Farquhar, G.D. (1986) Carbon isotope discrimination measured with gas exchange to investigate CO_2 diffusion in leaves of higher plants. Aust. J. Plant Physiol. 13:281–292.

Farquhar, G.D., O'Leary, M.H., & Berry, J.A. (1982) On the relationship between carbon isotope discrimination and the intercellular carbon dioxide concentration in leaves. Aust. J. Plant Physiol. 9:131–137.

Francey, R.J., Gifford, R.M., Sharkey. T.D., & Weir. B. (1985) Physiological influences on carbon isotope discrimination in Huon pine (*Lagarostrobos franklinii*). Oecologia 66:211–218.

Guy, R.D., Fogel, M.L., & Berry, J.A. (1993) Photosynthetic fractionation of the stable isotopes of oxygen and carbon. Plant Physiol. 101:37–47.

Henderson, S.A., Von Caemmerer, S., & Farquhar, G.D. (1992) Short-term measurements of carbon isotope discrimination in several C_4 species. Aust. J. Plant Physiol. 19:263–285.

O'Leary, M.H., Madhavan, S., & Paneth, P. (1992) Physical and chemical basis of carbon isotope fractionation in plants. Plant Cell Environ. 15:1099–1104.

---

indicate that for many species p_c is about 30% lower than p_i, when plants are rapidly photosynthesizing. p_c may approximate the partial pressure of CO_2 in the intercellular spaces only when the photosynthetic rates are relatively low (Evans & Von Caemmerer 1996).

The internal conductance varies widely between species and roughly correlates with the photosynthetic capacity of the leaf (Fig. 7). It is interesting that the relationship between internal conductance and photosynthesis is rather similar for scleromorphic and mesophytic leaves. This suggests that if the intercellular air space conductance is smaller in sclerophyllous leaves, as is expected in such thick leaves with stomata at the abaxial side only, then it is offset by greater conductance in the rest of the diffusion path to the site of Rubisco. On the other hand, the low rates of CO_2 assimilation in sclerophyllous leaves might also account for the small decline in partial pressure of CO_2.

The internal conductance is a complicated trait that involves diffusion of CO_2 in the gas phase, dissolving of CO_2 in the liquid phase, and conversion of CO_2 into HCO_3^- catalyzed by carbonic anhydrase and diffusion in the liquid phase. For amphistomatous leaves, which have stomata on both sides, the internal conductance is probably proportional to the surface of chloroplast exposed to intercellular air spaces per unit of leaf area. The ratio of chloroplast surface to leaf area is around 20; however, this ratio varies by an order of magnitude between species, as illustrated in

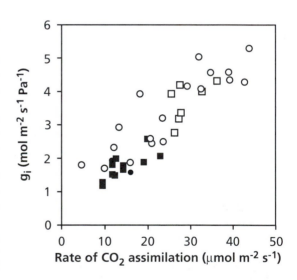

FIGURE 7. The relationship between the internal conductance, g_i, and the rate of CO_2 assimilation as measured at 1 mmol quanta $m^{-2} s^{-1}$ and 35 Pa CO_2 at 25°C. Filled and open symbols refer to scleromorphic and mesophytic leaves, respectively. The units of conductance as used in this graph differ from those used elsewhere in this text. The reason is that when CO_2 is dissolving to reach the sites of carboxylation, the amount depends on the partial pressure of CO_2, and conductance has the units used in this graph. For air space conductance the units could be the same as used elsewhere: moles per square meter per second, if CO_2 is given as a mole fraction. The present range of internal conductances and rates of photosynthesis are typical when plants are grown under optimum conditions and measurements are made under the conditions described in these legends (Evans & Von Caemmerer 1996). Copyright American Society of Plant Physiology.

TABLE 1. The area of the chloroplast in palisade (P) and spongy (S) mesophyll (Area$_{chlor}$) expressed per unit leaf area (Area$_{leaf}$) for species from the mountain range of the East Pamirs, Tadjikistan (3500–4500 m).

			(Area$_{chlor}$)/Area$_{leaf}$		
	P	S	P + S	Lowest (P + S)	Highest (P + S)
Perennial dicotyledonous herbs (54)	12	9	18	3	41
Cushion plants (4)	20	11	26	12	40
Dwarf semishrubs (12)	16	6	21	5	48
Subshrubs (8)	9	7	15	7	24

Source: Pyankov & Kondratchuk 1995, 1998.
*The number of investigated species is given in brackets. The sum P + S differs from P + S because data pertain to both dorsiventral (P + S) and isopalisade (P) species.

Table 1. The variation is not associated with plant functional type and it remains to be established if it correlates with the internal conductance for CO_2 transport.

3. Response of Photosynthesis to Light

The level of irradiance is an important ecological factor on which all photoautotrophic plants depend. Only the photosynthetically active part of the spectrum (PAR; 400 to 700 nm) directly drives photosynthesis. Other effects of radiation pertain to the photoperiod, which triggers flowering and other developmental phenomena in many species, the direction of the light, phototropism, and the spectral quality, which is characterized by the red/far-red ratio, which is of major importance for many aspects of morphogenesis. These effects are discussed in different sections of the chapter on growth and allocation. Effects of infrared radiation are discussed in the chapter on the plant's energy balance; its significance through temperature effects on photosynthesis are discussed in Section 7 of this chapter. Effects of ultraviolet radiation are treated briefly in Section 2.2 of the chapter on effects of radiation and temperature.

Low light intensities pose stresses on plants because irradiance limits photosynthesis and thus net carbon gain and plant growth. Responses of the photosynthetic apparatus to shade can be at two levels: the structural level or the level of the biochemistry. Leaf anatomy, and biochemistry of the photosynthetic apparatus are treated in Section 3.2.2; aspects of morphology at the whole plant level are discussed in Section 5.1 of the chapter on growth and allocation.

High light intensities may also be a stress for plants, particularly if other factors are not optimal. Damage to the photosynthetic apparatus may be the result. The kind of damage to the photosynthetic apparatus that may occur and the mechanisms of plants to cope with excess irradiance will be treated in Section 3.3.

To analyze the response of photosynthesis to irradiance, we will distinguish between the dynamic response of photosynthesis to light (or any other environmental factor) and the steady-state response. A steady-state response is achieved after exposure of a leaf to constant irradiance for some time until a constant response is reached. Dynamic responses are the result of perturbations of steady-state conditions due to sudden changes in light conditions resulting in changes in photosynthetic rates.

Certain genotypes have characteristics that are adaptive in a shady environment (shade-adapted plants). In addition, all plants have the capability to acclimate to a shade environment, to a greater or lesser extent, and form a shade-plant phenotype (shade form). The term **shade plant** may therefore refer to an "adapted" genotype or an "acclimated" phenotype. The term **sun plant** similarly refers normally to a plant grown in high-light conditions, but it is also used to indicate a shade-avoiding species or ecotype. The terms **sun leaf** and **shade leaf** are used more consistently. They refer to leaves that have developed at high and low irradiance, respectively.

3.1 Characterization of the Light Climate Under a Leaf Canopy

The average **irradiance** decreases exponentially through the plant canopy, with the extent of light attenuation depending on both the amount and the arrangement of leaves (Monsi & Saeki 1953):

$$I = I_o e^{-kL} \qquad (6)$$

where I is the irradiance beneath the canopy; I_o is the irradiance at the top of the canopy; k is the extinction coefficient; and L is the leaf area index (total leaf area per unit ground area). The extinction coefficient is low for vertically inclined leaves (e.g., 0.3 to 0.5 for grasses), but high for a horizontal leaf arrangement (about 1.0 for randomly distributed horizontal leaves); a clumped leaf arrangement and deviating leaf angles result in intermediate values for k. A low extinction coefficient allows more effective light transfer through canopies dominated by these plants. Leaves are more vertically inclined in **high-light** habitats than they are in cloudy or shaded environments. This minimizes the probability of photoinhibition and increases light penetration to lower leaves in high-light environments, thereby maximizing whole-canopy photosynthesis (Terashima & Hikosaka 1995). Leaf area index ranges from less than 1 in sparsely vegetated communities like deserts or tundra to 5 to 7 for crops and 5 to 10 for forests (Schulze et al. 1994).

The **spectral composition** of shade light differs from that above a canopy because of the selective absorption of photosynthetically active radiation (PAR) by leaves. Transmittance of photosynthetically active radiation (400 to 700 nm) is typically less than 10%, whereas transmittance of far-red (FR, 730 nm) light is substantial (Terashima & Hikosaka 1995). As a result, the ratio of red (R, 660 nm) to far-red (the R/FR ratio) is lower in the shade. This affects the photoequilibrium of phytochrome, which is a pigment that allows a plant to perceive shading by other plants (see Box 7 in the chapter on growth and allocation).

Another characteristic of the light climate under a leaf canopy is that direct sunlight may arrive as "packages" of high intensity: "**sunflecks.**" There are short spells, therefore, of high irradiance against a background of a low irradiance. Such sunflecks are due to the flutter of leaves, movement of branches, and the changing angle of the sun. Their duration ranges from less than a second to minutes. Sunflecks typically have lower irradiance than direct sunlight due to penumbral effects, but large sunflecks (i.e., those greater than an angular size of 0.5 degrees) can approach irradiances of direct sunlight (Chazdon & Pearcy 1991).

3.2 Physiological, Biochemical, and Anatomical Differences Between Sun and Shade Leaves

Shade leaves exhibit a number of traits that makes them quite distinct from leaves that are exposed to full sunlight. We will first discuss these traits and then some problems that may arise in leaves from exposure to high irradiance. In the last section, we discuss signals and transduction pathways that allow the formation of sun versus shade leaves.

3.2.1 The Light-Response Curve of Sun and Shade Leaves

The steady-state rate of CO_2 assimilation increases asymptotically with increasing irradiance. Below the **light-compensation point** (A = 0), there is insufficient light to compensate for respiratory carbon dioxide release in photorespiration and dark respiration (Fig. 8). At low light intensities, A increases linearly with irradiance, with the light-driven electron transport limiting photosynthesis. The initial slope of the light-response curve based on *absorbed* light (**quantum yield**) describes the efficiency with which light is converted into fixed carbon (typically about 0.06 moles CO_2 fixed per mole of quanta under favorable conditions and a normal atmospheric CO_2 concentration). When the light-response curve is based on *incident* light, the leaf's absorptance also determines the quantum yield; this initial slope is called the **apparent quan-**

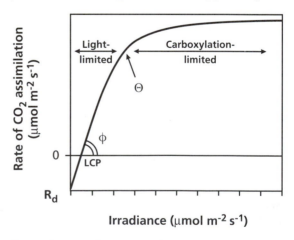

FIGURE 8. Typical response of photosynthesis to irradiance, drawn according to Equation 7 in the text. The intercept with the x-axis is the light-compensation point (LCP), the initial slope of the line gives the quantum yield (φ) and the intercept with the y-axis is the rate of dark respiration (R_d). The curvature of the line is described by θ. At low irradiance, the rate of CO_2 assimilation is light-limited; at higher irradiance A is carboxylation limited. A_{max} is the light-saturated rate of CO_2 assimilation at ambient p_a.

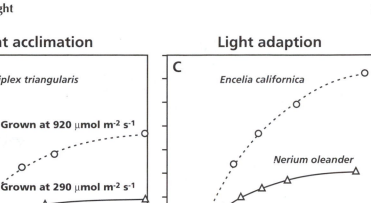

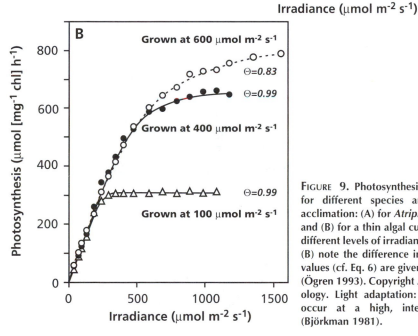

FIGURE 9. Photosynthesis as a function of irradiance for different species and growing conditions. Light acclimation: (A) for *Atriplex triangularis* (Björkman 1981) and (B) for a thin algal culture (*Coccomyxa* sp.) grown at different levels of irradiance 100, 400, or 600 μmol m^{-2}s^{-1} (B) note the difference in "curvature," for which the Θ values (cf. Eq. 6) are given in B, between the three curves (Ögren 1993). Copyright American Society of Plant Physiology. Light adaptation: (C) for species that naturally occur at a high, intermediate, or low irradiance (Björkman 1981).

tum yield. At high irradiance, photosynthesis becomes light-saturated and is limited by the carboxylation rate, which is governed by some combination of CO_2 diffusion into the leaf and carboxylation capacity. The shape of the light-response curve can be satisfactorily described by a nonrectangular hyperbola (Fig. 8):

$$A = \frac{\phi.I + A_{max} - \sqrt{\{(\phi.I + A_{max})^2 - 4\Theta.\phi.I.A_{max}\}}}{2\Theta} - R_d \qquad (7)$$

where A_{max} is the light-saturated rate of gross CO_2 assimilation (net rate of CO_2 assimilation plus dark respiration) at infinitely high irradiance, ϕ is the (apparent) quantum yield (on the basis of either incident or absorbed photons), Θ is the curvature factor, which can vary between 0 and 1, and R_d is the dark respiration. The equation can also be used to describe the light dependence of electron transport, where A is then replaced by J and A_{max} by J_{max} (Box 1). This mathematical description is useful because it contains variables with a clear physiological meaning that can be derived from

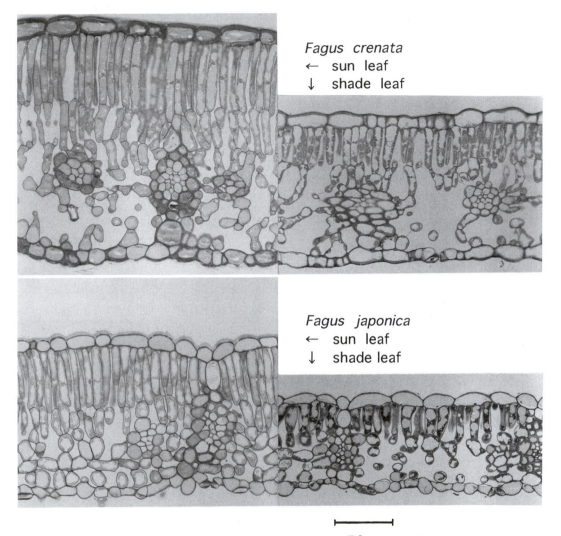

Fagus crenata
← sun leaf
↓ shade leaf

Fagus japonica
← sun leaf
↓ shade leaf

50 μm

FIGURE 10. Light microscopic transverse sections of sun and shade leaves of two tree species: *Fagus japonica* and *Fagus crenata*. Note that the sun leaves of *F. crenata* have two cell layers for the palisade tissue while those of *F. japonica* have only one layer. Shade leaves of both species have only one cell layer. In both species, the number of cell layers is determined in the winter buds. Thus, *F. crenata* cannot develop typical sun leaves on abrupt changes in light environment. Scale bar = 50 μm (courtesy I. Terashima, Institute of Biological Sciences, University of Tsukuba, Tsukuba, Japan and A. Uemura, A. Ishida, & Y. Matsumoto, Forestry and Forest Products Research Institute, Kukizaki, Japan).

light-response curves and used to model photosynthesis.

Sun leaves differ from shade leaves primarily in their higher light-saturated rates of photosynthesis (A_{max}) (Fig. 9). The rate of dark respiration typically covaries with A_{max}. The initial slope of the light-response curves of light-acclimated and shade-acclimated plants (the **quantum yield**) is the same, except when shade-adapted plants become inhib-ited or damaged at high irradiance (photoinhibition or photodestruction), which reduces the quantum yield. The apparent quantum yield (i.e., based on incident photon irradiance) may also vary due to differences in absorptance due to differences in chlorophyll concentration per unit leaf area. This is typically not important in the case of acclimation to light (Sect. 3.2.3), but it cannot be ignored when such factors as nutrient availability and senescence play a role. The transition from the light-limited part to the light-saturated plateau is abrupt in shade leaves, but it is more gradual in sun leaves (higher A_{max} and lower Θ in sun leaves). Although shade leaves typically have a low A_{max}, they have lower light-compensation points and higher rates of photosynthesis at low light because of their lower respiration rates per unit leaf area (Fig. 9).

Just as in acclimation, most plants that have evolved under conditions of high light have higher light-saturated rates of photosynthesis (A_{max}), higher light-compensation points, and lower rates of photosynthesis at low light than do shade-adapted plants when grown under the same conditions (Fig. 9).

3.2.2 Anatomy and Ultrastructure of Sun and Shade Leaves

One mechanism by which sun-grown plants, or sun leaves on a plant, achieve a high A_{max} (Fig. 9) is by producing **thicker** leaves (Fig. 10). The variation in thickness is largely due to the formation of taller palisade cells and/or an increase in the number of layers of palisade cells in sun leaves (Hanson 1917). Plants that naturally occur in high-light environments (e.g., *Eucalyptus* species) may have palisade parenchyma on both sides of the leaf. Such leaves

are naturally positioned (almost) vertically, so that both sides of the leaf receive a high irradiance. Anatomy constrains the potential of leaves to acclimate, so full acclimation to a new light environment typically requires the production of new leaves.

The spongy mesophyll in dorsiventral leaves of dicotyledons increases the **path length** of light in leaves by reflection at the gas–liquid interfaces of these irregularly oriented cells. The relatively large proportion of spongy mesophyll in shade leaves therefore enhances leaf absorptance because of the greater internal **light scattering** (Vogelmann et al. 1996). When air spaces of shade leaves of *Hydrophyllum canadense* or *Asarum canadense* are infiltrated with mineral oil to eliminate this phenomenon, light absorptance at 550 and 750 nm is reduced by 25 and 30%, respectively (Fig. 11). In sun leaves, which have relatively less spongy mesophyll, the effect of infiltration with oil is much smaller. The optical pathlength in leaves ranges from 0.9 to 2.7 times that of an equivalent amount of pigment in water, greatly increasing the effectiveness of light absorption in thin leaves of shade plants (Terashima & Hikosaka 1995).

Leaves of obligate shade plants such as those found in the understory of a tropical rain forest, may have specialized anatomical structures that enhance light absorption even further. Epidermal

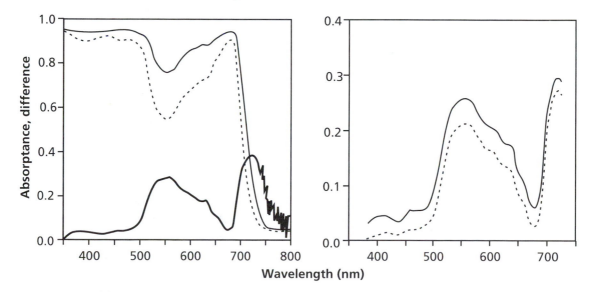

FIGURE 11. A. Light absorptance in a shade leaf of *Hydrophyllum canadense*. The thin solid line gives the absorptance of a control leaf. The broken line shows a leaf infiltrated with mineral oil, which reduces light scattering. The difference between the two lines is given as the thick solid line. B. The difference in absorptance between an oil-infiltrated leaf and a control leaf of *Acer saccharum*. The solid line gives the difference for a shade leaf, the broken line for a sun leaf (DeLucia et al. 1996). Copyright Blackwell Science Ltd.

and hydrenchyma cells may act as **lenses** that concentrate light in a thin layer of mesophyll.

There are fewer chloroplasts per unit area in shade leaves as compared with sun leaves due to the reduced thickness of mesophyll. The **ultrastructure** of the chloroplasts of sun and shade leaves shows distinct differences (Fig. 12). Shade chloroplasts have a smaller volume of stroma, where the Calvin-cycle enzymes are located, but larger grana, which contain the major part of the chlorophyll. Such differences are found between plants grown under different light conditions, or between sun and shade leaves on a single plant, as well as when comparing chloroplasts from the upper and lower side of one, relatively thick, leaf (e.g. Schefflera arboricala, dwarf umbrella plant; Fig. 12). The adaxial (upper) regions have a chloroplast ultrastructure like sun leaves, whereas shade acclimation is found in the abaxial (lower) regions cells of the leaf (Box 3).

3.2.3 Biochemical Differences Between Shade and Sun Leaves

Shade leaves **minimize light limitation** through increases in capacity for light capture on the

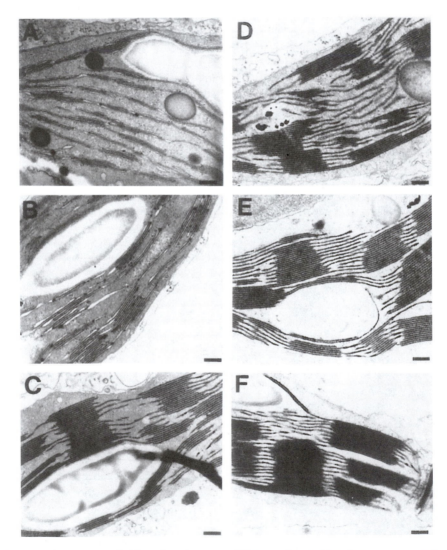

FIGURE 12. Electron micrographs of chloroplasts in sun (A–C) and shade (D–F) leaves of *Schefflera arboricola* (dwarf umbrella plant). Chloroplasts found in upper palisade parenchyma tissue (A, D), lower palisade parenchyma tissue (B, E), and spongy mesophyll tissue (C, F). Note the difference in stacking between sun and shade leaves and between the upper and lower layer inside the leaf. Scale bar = 0.2 μm (courtesy A.M. Syme and C. Critchley, Department of Botany, The University of Queensland, Australia).

Box 3
Carbon-Fixation and Light-Absorption Profiles Inside Leaves

We are already familiar with differences in biochemistry and physiology *between* sun and shade leaves (Sect. 4.2 of the chapter on photosynthesis). If we consider the gradient in the level of irradiance inside a leaf, however, then should we not expect similar differences *within* leaves? Indeed, palisade mesophyll cells at the adaxial (upper) side of the leaf do tend to have characteristics associated with acclimation to high irradiance: a high Rubisco/chlorophyll and chl a/chl b ratio, high levels of xanthophyll-cycle carotenoids, and less stacking of the thylakoids (Robinson & Osmond 1994, Terashima & Hikosaka 1995, Terashima & Inoue 1985). Conversely, the spongy mesophyll cells at the abaxial (lower) side of the leaf have chloroplasts with a lower Rubisco/chlorophyll and chl a/chl b ratio, characteristic for acclimation to low irradiance. What are the consequences of such profiles within the leaf for the exact location of carbon fixation in the leaf?

When trying to answer this question we encounter some problems. First, we need to know the light profile (space irradiance) within a leaf. This can be done with a fiberoptic microprobe, which is moved through the leaf at different angles, taking light readings at different wavelengths (Vogelmann 1993). If we were to deal with a homogeneous solution, then the profile of light absorption would be the same as the space irradiance profile. Chlorophyll, however, is not homogeneously distributed in a cell; rather, it is concentrated in the chloroplasts. In addition, inside the leaf, absorption and scattering properties vary (e.g., going from the palisade to the spongy mesophyll) (cf. Sect. 4.2.4 of the chapter on photosynthesis). How can we obtain information on light absorption?

To calculate light absorption, we need information on the concentration of light-absorbing pigments in the different layers inside the leaf. Knowing the pigment profile, we can use the Lambert-Beer equation to calculate light absorption. The gradient in light absorption can then be compared with the profile of photosynthetic CO_2 assimilation, which has been measured using $^{14}CO_2$ labeling of a leaf, followed by careful sectioning of the leaf (Nishio et al. 1993). How closely does it match the pigment profile?

To calculate this, we first use the information from Section 4.2.2 of the chapter on photosynthesis that sun leaves absorb approximately 85% of all incident radiation to calculate the irradiance at each depth in the leaf:

$$I_0 = {}^{14}C_{tot}/0.85 \qquad (1)$$

where I_0 is the radiation incident on the leaf and $^{14}C_{tot}$ is the total amount of ^{14}C fixed in the profile. At a later stage we will use the Lambert-Beer law to calculate the light absorption profile. As a first approximation, we now relate the ^{14}C absorbance of the leaf exposed to $^{14}CO_2$ from the ^{14}C fixation profile. The light remaining at any depth, I, is given by I_0 minus the ^{14}C in the layers that have been passed through ($\Sigma^{14}C$), or:

Absorbance

$$= \log\left[\left({}^{14}C_{tot}/0.85\right)/\left({}^{14}C_{tot}/0.85 - \Sigma^{14}C\right)\right] \qquad (2)$$

This calculated absorbance increases with increasing depth in the leaf. The apparent extinction coefficient of each layer can be calculated as the ratio of this profile of absorbance and that of the chlorophyll concentration (Fig. 1). This coefficient rises steadily with increasing depth in the leaf. One of the reasons for the increase in the apparent extinction coefficient is the structure of the palisade and spongy parenchyma. Due to the orientation of chloroplasts in the palisade parenchyma, light transmission is quite high, reducing the apparent extinction coefficient. On the other hand, much of the light arriving in the spongy mesophyll is scattered, thereby increasing the apparent extinction coefficient (cf. Sect. 4.2.4 of the chapter on photosynthesis).

We also know the profile of the Rubisco content per unit chlorophyll and can use this as an estimate for the light-saturated rate of photosynthesis. We can then calculate the light-response curve for each layer, using virtually the same equation as introduced in Section 3.2.1 of the chapter on photosynthesis (the only difference being that R_{day} is left out):

$$A = \frac{\phi.I + A_{max} - \left\{\left(\phi.I + A_{max}\right)^2 - 4\theta.\phi.I.A_{max}\right)^{0.5}\right\}}{(2\theta)} \qquad (3)$$

continued

Box 3. *Continued*

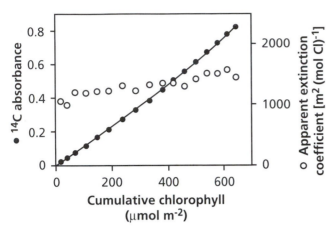

FIGURE 1. The profile of photosynthetic $^{14}CO_2$ assimilation (absorbance of ^{14}C) calculated as a function of cumulative chlorophyll from the adaxial (upper) surface for a sun leaf of *Spinacia oleracea* (spinach). The apparent extinction coefficient was calculated for each layer from the data on absorbance and chlorophyll concentration. It averages at $1500\,m^2$ $(mol\,chl)^{-1}$ (Evans 1995, Nishio et al. 1993). Copyright CSIRO, Australia and American Society of Plant Physiologists

where A is the actual rate of photosynthesis, ϕ is the maximum quantum yield, I is the absorbed irradiance, A_{max} is the light-saturated rate of photosynthesis at normal p_i, and θ describes the curvature. The calculated light-response curves of the adaxial (upper) layers are like those of sun leaves, whereas those for the abaxial (lower) layers are like the ones of shade leaves.

Taking an apparent extinction coefficient of $1500\,m^2$ $(mol\,chl)^{-1}$ (Fig. 1), we can calculate that the lower layer absorbs a quarter of the light of that of the upper layer and yet has half the photosynthetic capacity. As a result, the calculated quantum yield of photosynthesis is greatest for the lower layers. Confirmation of this conclusion comes from fluorescence measurements, used as a probe for photochemical efficiency: Quantum yields are lower for the adaxial side of the leaf, especially at high irradiance (cf. Sect. 2.1.1 of the chapter on photosynthesis and Box 4 for the use

of the fluorescence technique). The quantum yield can be calculated for each layer at any particular irradiance. Combining these values of quantum yield with the light-absorption profile gives us the CO_2 fixation at each layer (Fig. 2). The CO_2 fixation profile peaks about $200\,\mu m$ from the adaxial surface, whereas light absorption, calculated from the light and chlorophyll profiles, peaks at $150\,\mu m$. When light absorption is multiplied by the quantum yield for each layer to calculate the CO_2 fixation profile, however, a very close match with the experimental data of the ^{14}C profile is found (Fig. 2).

The good agreement between calculated and experimental values suggests that the simple assumption of a constant apparent extinction coefficient approximates the optical environment inside the leaf quite well. It also stresses that the CO_2 fixation profile should not be related to the space irradiance inside the leaf; rather, it should be related to the absorbed irradiance.

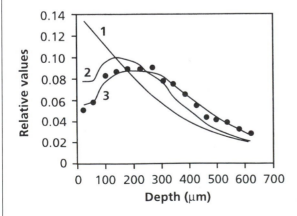

FIGURE 2. The profile of ^{14}C fixation through a sun leaf of *Spinacia oleracea* (spinach) measured at $2\,mol$ quanta $m^{-2}\,s^{-1}$ (dots). Line 1 gives the relative profile of the space irradiance, calculated using Lambert-Beer's law, combined with light scattering. Line 2 gives the relative absorbed irradiance, calculated from line 1 and the pigment profile. Line 3 gives the predicted CO_2 fixation profile, calculated from line 2 and the calculated quantum yield for each layer (Evans 1995, Nishio et al. 1993). Copyright CSIRO, Australia and American Society of Plant Physiologists.

continued

Box 3. *Continued*

References and Further Reading

Evans, J.R. (1995) Carbon fixation profiles do reflect light absorption profiles in leaves. Aust. J. Plant Physiol. 22:865–873.

Nishio, J.N., Sun, J., & Vogelmann, T.C. (1993) Carbon fixation gradients across spinach leaves do not follow internal light gradients. Plant Cell 5:953–961.

Robinson, S.A. & Osmond, C.B. (1994) Internal gradients of chlorophyll and carotenoid pigments in relation to photoprotection in thick leaves of plants with Crassulacean Acid Metabolism. Aust. J. Plant Physiol. 21:497–506.

Terashima, I. & Hikosaka, K. (1995) Comparative ecophysiology of leaf and canopy photosynthesis. Plant Cell Environ 18:1111–1128.

Terashima, I. & Inoue, Y. (1985) Vertical gradients in photosynthetic properties of spinach chloroplasts dependent on intra-leaf light environment. Plant Cell Physiol. 26:781–785.

Vogelmann, T.C. (1993) Plant tissue optics. Annu. Rev. Plant Physiol. Plant Mol. Biol. 44:231–251.

expense of photosynthetic capacity. In a comparison of understory, midcanopy, and canopy species of the Costa Rican rainforest trees, Poorter et al. (1995) found small differences in leaf absorptance (400 to 700 nm) among species. Shade species had a higher leaf absorptance in the lower-light environments, as opposed to the sun species that showed greater absorptance in high-light environments. Chlorophyll concentrations per unit leaf area were the same in all three environments because a greater chlorophyll concentration per unit leaf fresh mass and per unit leaf protein in understory species compensates for the greater leaf thickness of sun species. Some highly shade-adapted species [e.g., *Hedera helix* (ivy)] may also have higher chlorophyll levels per unit leaf area in shade in comparison to sun. This might be due to the fact that their leaves do not get much thinner in the shade; however, there may also be some photodestruction of chlorophyll in high light in such species. In most species, however, higher levels of chlorophyll per unit fresh mass are compensated for by the smaller number and size of the mesophyll cells, so that the chlorophyll level per unit area is largely unaffected.

The **ratio** between **chlorophyll a** and **chlorophyll b** is less in shade-acclimated leaves. These leaves have relatively more chlorophyll associated with the **LHC** (which has a lower chlorophyll a to b ratio) than with the photosystems. The decreased chl a/chl b ratio is therefore a reflection of the greater investment in the LHC (Evans 1988). The larger proportion of LHC is located in the **larger grana** of the shade-acclimated chloroplast (Fig. 12).

Sun leaves have larger amounts of Calvin-cycle enzymes per unit leaf area as compared with shade leaves due to more cell layers, a larger number of chloroplasts per cell, and a larger volume of stroma, where these enzymes are located, per chloroplast (Fig. 13). Sun leaves also have more stroma-exposed thylakoid membranes, which contain the b_6f cytochromes and ATPase. All these components

determine a leaf's photosynthetic capacity. Hence, sun leaves, when compared with shade leaves, have more of the components that determine **photosynthetic capacity** per unit leaf area. Because the amount of chlorophyll per unit area is more or less equal among leaf types, they also have a higher photosynthetic capacity per unit chlorophyll. The biochemical gradients across a leaf are similar to

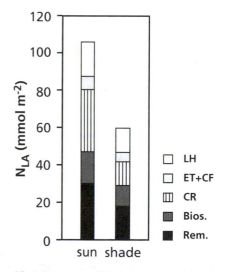

FIGURE 13. Nitrogen partitioning among various components in shade- and sun-acclimated leaves. Most of the leaf's nitrogen (here expressed per unit leaf area; NLA) in herbaceous plants is involved in the photosynthetic apparatus. Some of the fraction labeled Biosynthesis and Remainder is indirectly involved in synthesis and maintenance processes associated with the photosynthetic apparatus. LH = light harvesting (LHC, PSI, PSII), ET + CF = electron transport components and coupling factor (ATPase), CR = enzymes associated with carbon reduction (Calvin cycle, mainly Rubisco), Bios = Biosynthesis (nucleic acids and ribosomes), Rem = remainder, other proteins and nitrogen-containing compounds (mitochondrial enzymes, amino acids, cell-wall proteins, alkaloids, etc.) (after Evans & Seemann 1989).

those observed within a canopy, with adaxial (upper) cells having more nitrogen, Rubisco, and electron carriers, but less chlorophyll than abaxial (lower) cells (Terashima & Hikosaka 1995). The high photosynthetic capacity of sun leaves (or in adaxial cells) is disadvantageous in low light because it is achieved at the expense of high rates of dark respiration (see chapter on respiration) and a large investment of resources such as nitrogen (see chapter on growth and allocation).

3.2.4 The Light-Response Curve of Sun and Shade Leaves Revisited

Table 2 summarizes the differences in characteristics between shade-acclimated and sun-acclimated leaves. The higher A_{max} of sun leaves as compared

with shade leaves is associated with greater amounts of compounds that determine photosynthetic capacity that are located in the greater number of chloroplasts per area and the larger stroma volume and stroma-exposed thylakoids in chloroplast. The increase of A_{max} with increasing amount of these compounds is almost linear (Evans & Seemann 1989). Hence, investment in compounds that determine photosynthetic capacity is proportionally translated into photosynthetic rate at high irradiance levels.

The higher rate of dark respiration in sun leaves is probably due to a greater demand for respiratory energy, for the maintenance of the larger number of leaf cells, and for greater protein content per cell and/or for the export of the products of photosynthesis from the leaf (see Sect. 4.4 in the chapter on respiration).

TABLE 2. Overview of generalized differences in characteristics between shade- and sun-acclimated leaves.

	Sun	Shade
Structural		
Leaf dry mass per area	high	low
Leaf thickness	thick	thin
Palisade parenchyma thickness	thick	thin
Spongy parenchyma thickness	similar	similar
Stomatal density	high	low
Chloroplast per area	many	few
Thylakoids per stroma volume	low	high
Thylakoids per granum	few	many
Biochemical		
Chlorophyll per chloroplast	low	high
Chlorophyll per area	similar	similar
Chlorophyll per dry mass	low	high
Chlorophyll a/b ratio	high	low
Light-harvesting Complex per area	low	high
Electron transport components per area	high	low
Coupling factor (ATPase) per area	high	low
Rubisco per area	high	low
Nitrogen per area	high	low
Xanthophylls per area	high	low
Gas exchange		
Photosynthetic capacity per area	high	low
Dark respiration per area	high	low
Photosynthetic capacity per dry mass	similar	similar
Dark respiration per dry mass	similar	similar
Carboxylation capacity per area	high	low
Electron transport capacity per area	high	low
Quantum yield	similar	similar
Curvature of light-response curve	gradual	acute

The absorption of photons by a leaf is not a simple function of its irradiance and chlorophyll concentration. A relationship with a negative exponent would be expected as described for monochromatic light and pigments in solution (Lambert-Beer's law). The situation in a leaf is more complicated, however, because preferential absorption of red and blue light by chlorophyll causes changes in the spectral distribution of light through the leaf. Moreover, the path length of light is complicated because of reflection inside the leaf and changes in the proportions of direct and diffuse light. Empirical equations, such as a hyperbole, can be used to describe light absorption by chlorophyll. For a healthy leaf, the quantum yield based on incident light is directly proportional to the amount of photons absorbed.

The cause of the decrease in convexity (cf. Eq. 7) of the light-response curve with increasing growth irradiance (Fig. 9) is not well understood. It is probably partly associated with the level of light-acclimation of the chloroplast in the cross-section of a leaf in relation to the distribution of light within the leaf (Leverenz 1987).

As argued earlier, a high A_{max} is beneficial in high-light conditions because the prevailing high irradiance can be efficiently exploited. Such a high A_{max}, however, would not be of much use in the shade because the high irradiance required to utilize the capacity occurs only infrequently, and a high A_{max} is associated with high rates of respiration and a large investment of nitrogen. By contrast, high chlorophyll concentrations that allow efficient light capture per unit biomass invested in leaves should be advantageous at low irradiance. There is apparently, a "trade-off" between investment of resources in carbon-assimilating capacity and in light harvesting. An important aspect of shade acclimation and shade adaptation in leaves is the ratio of photosynthetic capacity to chlorophyll concentration.

Plant species appear to respond differently to variation in growth irradiance (Fig. 14) Four functional groups can be discerned:

1. Fast-growing herbaceous species from habitats with a dense but short vegetation have high A_{max}/chlorophyll ratios that decrease strongly with decrease in light availability. *Plantago lanceolata* (plantain) is an example.
2. Pioneer tree species such as *Betula pendula* (birch) have a high A_{max}/chlorophyll ratio; however, this ratio does not change much with growth irradiance.

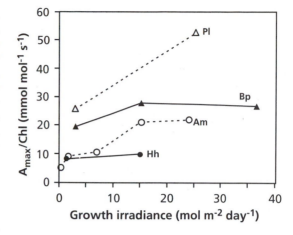

FIGURE 14. Light-saturated rate of CO_2 assimilation (A_{max}) per unit chlorophyll in relation to growth irradiance for four species. *Plantago lanceolata* (Pl) (Poot et al. 1996; copyright Physiologia Plantarum), *Betula pendula* (Bp) (Öquist et al. 1982; copyright Blackwell Science Ltd), *Alocasia macrorrhiza* (Am) (Sims & Pearcy 1991), *Hedera helix* (Hh) (T.L. Pons, unpublished data).

3. A plastic response is also found in shade-adapted plants such as herbaceous understory species (*Alocasia macrorrhiza*) and forest trees that tolerate shade as seedlings. The A_{max}/chlorophyll ratio, however, is much lower over the entire range of irradiance levels.
4. A low A_{max}/chlorophyll ratio that changes little with growth irradiance is found in woody shade-adapted species such as *Hedera helix* (ivy).

3.2.5 The Environmental Signal for Shade Acclimation in Chloroplasts

Is the acclimation of the biochemistry of the chloroplast a response to the level of irradiance or to the spectral composition of the radiation? To answer this question the *aurea* mutant of *Lycopersicon esculentum* (tomato) which lacks phytochrome A has been used (cf. Box 7 for details on this light-sensing pigment in plants). As expected, the *aurea* mutant does not show the typical enhanced stem elongation in response to radiation filtered by a vegetation cover ("green shade," as opposed to "neutral shade") (cf. Sect. 5.1.1 in the chapter on growth and allocation for the effect of phytochrome on stem elongation under shade). The *aurea* mutant shows a much greater reduction in its maximum rate of photosynthesis when exposed to "green shade," whereas the response to "neutral shade" is the same as in the wild-type. This indicates that phytochrome A plays a role in the acclimation of this component of CO_2 assimilation to shade. The *aurea* mutant, however, does show the typical response to shade light with respect to the chl a/chl b ratio and the investment in light-harvesting complex. Hence, the level of irradiance, rather than its spectral composition as detected by phytochrome, appears to trigger acclimation of light-harvesting aspects of photosynthesis; however, the exact mechanism accounting for the effect of irradiance level on shade acclimation of chloroplasts is unknown (Smith et al. 1995).

3.3 Effects of Excess Irradiance

At irradiance levels that are beyond the linear, light-limited, region of the light-response curve, some of the photons absorbed by chlorophyll can-

not be used in photochemistry. Plants have mechanisms to dispose of this excess excitation energy safely. When these mechanisms are at work, the quantum yield of photosynthesis is **temporarily** reduced, which often occurs at high irradiance in many plants. The excess excitation energy, however, may also cause damage to the photosynthetic membranes if the dissipation mechanisms are inadequate. This leads to **longer-lasting** (days) **photoinhibition**. In extreme cases this high-light stress may lead to breakdown of chlorophyll (bleaching).

A technique used for the quantification of photoinhibition is in vivo chlorophyll fluorescence. This technique has also been employed to quantify the rate of electron transport and other characteristics of the photosynthetic apparatus. The technique and some applications of chlorophyll fluorescence are described in Box 4.

Box 4
Chlorophyll Fluorescence

When chlorophyllous tissue is irradiated, with photosynthetically active radiation (400 to 700 nm) or wavelengths shorter than 400 nm, it emits radiation of longer wavelengths (approx. 680 to 760 nm). This fluorescence originates mainly from chlorophyll a associated with photosystem II (PSII). The measurement of the kinetics of chlorophyll fluorescence has been developed into a sensitive tool for probing state variables of the photosynthetic apparatus *in vivo*. In an ecophysiological context, this is a useful technique to quantify effects of stress on photosynthetic performance that is also applicable under field conditions.

Photons absorbed by chlorophyll give rise to (1) an excited state of the pigment which is channeled to the reaction center and may give rise to photochemical charge separation. The quantum yield of this process is given by ϕ_P. Alternative routes for the excitation energy are (2) dissipation as heat (ϕ_D) and (3) fluorescent emission (ϕ_F). These three processes are competitive. This leads to the assumption that the sum of the quantum yields of the three processes is unity:

$$\phi_P + \phi_D + \phi_F = 1 \qquad (1)$$

Since only the first two processes are subject to regulation, the magnitude of fluorescence depends on the added rates of photochemistry and heat dissipation. Measurement of fluorescence, therefore, provides a tool for quantification of these processes.

Basic Fluorescence Kinetics

When a leaf is subjected to strong white light after incubation in darkness, a characteristic pattern of fluorescence follows which is known as the Kautsky curve (e.g., Bolhar-Nordenkampf & Ögren 1993, Schreiber et al. 1994). It rises immediately to a low value (F_0), which is maintained only briefly in strong light, but can be monitored for a longer period in weak intermittent light (Fig. 1, left). This level of fluorescence (F_0) is indicative of open reaction centers due to a fully oxidized state of the primary electron acceptor Q_A. In strong saturating irradiance, fluorescence rises quickly to a maximum value (F_m, Fig. 1, left), which indicates closure of all reaction

continued

Box 4. *Continued*

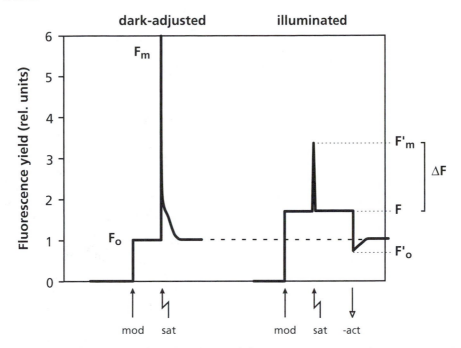

FIGURE 1. Fluorescence kinetics in dark-incubated (adjusted) and illuminated leaves in response to a saturating pulse of white light. mod = modulated measuring light on; sat = saturating pulse on; -act = actinic light off. For explanation of fluorescence symbols see text (after Schreiber et al. 1995).

centers as a result of fully reduced Q_A. When light is maintained, fluorescence decreases gradually (quenching) to a stable value as a result of induction of photosynthetic electron transport and dissipation processes.

After a period of illumination at a sub-saturating irradiance, fluorescence stabilizes at a value F, somewhat above F_0 (Fig. 1, right). When a saturating pulse is given under these conditions, fluorescence does not rise to F_m, but to a lower value called F_m'. Although reaction centers are closed at saturating light, dissipation processes compete now with fluorescence, which causes the quenching of F_m to F_m'.

Since all reaction centers are closed during the saturating pulse, the photochemical quantum yield (ϕ_P) is zero and, therefore, the quantum yields of dissipation at saturating light (ϕ_{Dm}) and fluorescence at saturating light (ϕ_{Fm}) are unity:

$$\phi_{Dm} + \phi_{Fm} = 1 \qquad (2)$$

It is further assumed that there is no change in the relative quantum yields of dissipation and fluorescence during the saturating pulse:

$$\frac{\phi_{Dm}}{\phi_{Fm}} = \frac{\phi_D}{\phi_F} \qquad (3)$$

Photochemical quantum yield (ϕ_P) is normally referred to as ϕ_{II} because it originates mainly from PSII. It can now be expressed in fluorescence parameters only, on the basis of equations (1) and (3).

$$\phi_{II} = \frac{\phi_{Fm} - \phi_F}{\phi_{Fm}} \qquad (4)$$

The fluorescence parameters F_0 and F_m can be measured with time-resolving equipment, where the sample is irradiated in darkness with $\lambda < 680$ nm and fluorescence is detected as emitted radiation $\lambda > 680$ nm. White light sources, however, typically also have radiation in the wavelength region of chlorophyll fluorescence. For measurements in any light condition, systems have been developed that use a weak modulated light source in conjunction with a detector that monitors only the fluorescence emitted at the frequency and phase of the source. A strong white light source for generat-

continued

Box 4. *Continued*

ing saturating pulses ($>5000\,\mu mol\,m^{-2}\,s^{-1}$) and an actinic light source typically complete such systems. The modulated beam is sufficiently weak for measurement of F_0. This is the method used in the example given in Figure 1. The constancy of the measuring light means that any change in fluorescence signal is proportional to ϕ_F. This means that the maximum quantum yield (ϕ_{IIm}) as measured in dark-incubated leaves is:

$$\phi_{IIm} = (F_m - F_0)/F_m = F_v/F_m \quad (5)$$

where F_v is the variable fluorescence, the difference between maximal and minimal fluorescence. In illuminated samples the expression becomes:

$$\phi_{II} = (F_m' - F)/F_m' = \Delta F/F_m' \quad (6)$$

where ΔF is the increase in fluorescence due to a saturating pulse superimposed on the illumination irradiance. $\Delta F/F_m'$ has values equal to or lower than F_v/F_m, the difference increases with increasing irradiance.

The partitioning of fluorescence quenching due to photochemical (qP) and nonphotochemical (qN) processes can be determined. These are defined as:

$$qP = (F_m' - F)/(F_m' - F_0') = \Delta F/F_v' \quad (7)$$

$$qN = 1 - (F_m' - F_0')/(F_m - F_0) = 1 - F_v'/F_v \quad (8)$$

F_0 may be quenched in light, and is then called F_0' (Fig. 1 right). The measurement of this parameter may be complicated, particularly under field conditions. We can also use another term for nonphotochemical quenching (NPQ) which does not require the determination of F_0':

$$NPQ = (F_m - F_m')/F_m' \quad (9)$$

The theoretical derivation of the fluorescence parameters as based on the assumptions described earlier is supported by substantial empirical evidence. The biophysical background of the processes, however, is not yet fully understood.

Relationships with Photosynthetic Performance

Maximum quantum yield after dark incubation (F_v/F_m) is typically very stable at values around 0.8 in healthy leaves. It appears that F_v/F_m correlates well with the quantum yield of photosynthesis measured as O_2 production or CO_2 uptake at low irradiance (Fig. 2). In particular, the reduction of the quantum yield by photoinhibition can be evaluated with this fluorescence parameter. A decrease in F_v/F_m can be due to a decrease in F_m and/or an increase in F_0. A fast- and a slow-relaxing component can be distinguished. The fast component is alleviated within

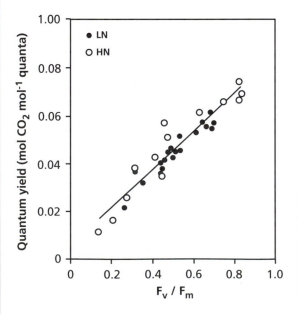

FIGURE 2. The relationship between quantum yield, as determined from the rate of oxygen evolution at different levels of low irradiance, and the maximum quantum yield of PSII determined with chlorophyll fluorescence (F_vF_m, ϕ_{IIm}). Measurements were made on *Glycine max* (soybean) grown at high (open symbols) and low (filled symbols) nitrogen supply (Kao & Forseth 1992). Copyright Blackwell Science Ltd.

continued

Box 4. *Continued*

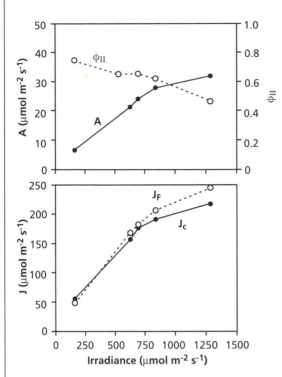

FIGURE 3. Relationship of chlorophyll fluorescence parameters and rates of CO_2 assimilation in the C_3 plant *Flaveria pringlei*. A = rate of CO_2 assimilation; ϕ_{II} = quantum yield of PSII in light ($\Delta F/F_m'$); J_F = electron-transport rate calculated from ϕ_{II} and irradiance; J_c = electron-transport rate calculated from gas-exchange parameters (after Krall & Edwards 1992). Copyright Physiologia Plantarum.

a few hours of low light or darkness and is therefore only evident during daytime; it is supposed to be involved in protection of PSII against overexcitation. The slow-relaxing component remains several days and is considered as an indication of (longer-lasting) damage to PSII. Such damage can be the result of sudden exposure of shade leaves to full sun light, or a combination of high irradiance and extreme (high or low) temperature. The way plants cope with this combination of stress factors determines their performance in particular habitats where such conditions occur.

Quantum yield in light ($\Delta F/F_m'$) can be used to derive the rate of electron transport (J_F).

$$J_F = I \; \Delta F/F_m' \; \text{abs} \; 0.5 \qquad (10)$$

where I is the irradiance and abs is the photon absorption by the leaf and 0.5 refers to the equal partitioning of photons between the two

photosystems (Genty et al. 1989). For comparison of J_F with photosynthetic gas-exchange rates, the rate of the carboxylation (V_c) and oxygenation (V_o) reaction of Rubisco must be known. In C_4 plants and in C_3 plants at low O_2 and/or high CO_2, V_o is low and can be ignored. Hence, J_c can simply be derived from the rate of O_2 production or CO_2 uptake. For a comparison of J_F with J_c in normal air in C_3 plants, V_o must be estimated from the intercellular partial pressure of CO_2 (see Box 1). Photosynthetic rates generally show good correlations with J_F (Fig. 3). J_F may be somewhat higher than the rate of electron transport calculated from the rate of O_2 release or CO_2 uptake (Fig. 3). This can be ascribed to electron flow associated with nonassimilatory processes, or with assimilatory processes that do not result in CO_2 absorption, such as nitrate reduction. A point of concern is whether the chloroplast population monitored for fluorescence is representative for the functioning of all chloroplasts across the whole leaf depth. The good correlation of gas exchange and fluorescence data in many cases indicates that this is indeed the case, at least in a relative sense. Hence, J_F is also referred to as the relative rate of electron transport.

References and Further Reading

Bolhar-Nordenkampf, H.R. & Öquist, G. (1993) Chlorophyll fluorescence as a tool in photosynthesis research. In: Photosynthesis and production in a changing environment, D.O. Hall, J.M.O. Scurlock, H.R. Bolhar-Nordenkampf, R.C. Leegood, & S.P. Long (eds). Chapman & Hall, London, pp. 193–206.

Genty, B., Briantais, J.-M., & Baker, N.R. (1989) The relationship between the quantum yield of photosynthetic electron transport and quenching of chlorophyll fluorescence. Biochim. Biophys. Acta 990:87–92.

Kao, W.Y. & Forseth, I.N. (1992) Diurnal leaf movement, chlorophyll fluorescence and carbon assimilation in soybean grown under different nitrogen and water availabilities. Plant Cell Environ. 15:703–710.

Krall, J.P. & Edwards, G.E. (1992) Relationship between photosystem II activity and CO_2 fixation. Physiol. Plant. 86:180–187.

Krause, G.H. & Weis, E. (1991) Chlorophyll fluorescence and photosynthesis: The basics. Annu. Rev. Plant Physiol. Plant Mol. Biol. 42:313–349.

Schreiber, U, Bilger, W., & Neubauer, C. (1995) Chlorophyll fluorescence as a non-intrusive indicator for rapid assessment of in vivo photosynthesis. In: Ecophysiology of photosynthesis, E.-D. Schools & M.M. Caldwell (eds). Springer-Verlag, Berlin, pp. 49–70.

3.3.1 Photoinhibition—Protection by Carotenoids of the Xanthophyll Cycle

Plants acclimated to high-light dissipate excess energy through reactions mediated by a particular group of **carotenoids** (Fig. 15). This dissipation process is induced by accumulation of protons in the thylakoid lumen, which is triggered by excess light. The strong acidification of the lumen induces an enzymatic conversion of the carotenoid violaxanthin into zeaxanthin (Demmig-Adams 1990, Gilmore 1997). Excess energy is transferred to the **zeaxanthin**, which acts as a "lightning rod," receiving the energy from the same energized form of chlorophyll that is normally used in photosynthesis, and then dissipating the energy harmlessly as heat (Demmig et al. 1987, Johnson et al. 1993a, 1993b) (Fig. 16). This energy dissipation can be measured by chlorophyll fluorescence (cf. Box 4) and is termed *high energy-dependent* or *pH-dependent fluorescence quenching*. In the absence of the **xanthophyll cycle**, excess energy would be passed on to oxygen via chlorophyll, which leads to

photooxidative damage. The first signs of such photooxidative damage appear in the D1 protein of PSII. More excessive damage may lead to destruction of membranes and oxidation of chlorophyll (bleaching). This can be seen in some ornamental plants with yellowish leaves, which have a defect in the dissipation and repair mechanism, as well as in shade plants transferred to high light.

In sun-exposed sites, diurnal changes in irradiance are closely tracked by the level of antheraxanthin and zeaxanthin. In shade conditions, sunflecks lead to the rapid appearance of antheraxanthin and zeaxanthin, and high levels are maintained subsequent to the first sunfleck. This regulation mechanism ensures that no competing dissipation of energy occurs when light is limiting for photosynthesis, whereas damage is prevented when light is absorbed in excess. The xanthophyll cycle is critical to understanding the basic light response curve of photosynthesis (Fig. 8). At irradiances above those where photosynthesis responds linearly to light, the xanthophyll cycle processes most energy that is not funneled into

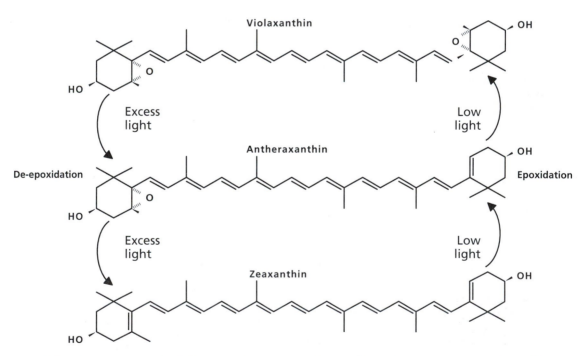

FIGURE 15. Scheme of the xanthophyll cycle and its regulation by excess or limiting light. Upon exposure to excess light, a rapid stepwise removal (de-epoxidation) of two oxygen functions (the epoxy groups) in violaxanthin takes place. This results in a lengthening of the conjugated system of double bonds from 9 in violaxanthin to 10 and 11 in antheraxanthin and zeaxanthin, respectively. The de-epoxidation step occurs in minutes. Under low-light conditions, the opposite process, epoxidation, takes place. It may take minutes, hours, or days, depending on environmental conditions (Demmig-Adams & Adams 1996). Copyright Elsevier Science Ltd.

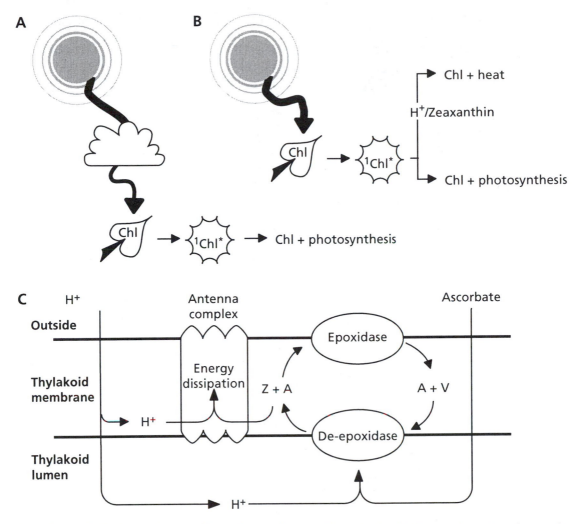

FIGURE 16. (A, B) Depiction of the conditions where A all or B only part of the sunlight absorbed by chlorophyll within a leaf is used for photosynthesis. Safe dissipation of excess energy requires the presence of zeaxanthin as well as a low pH in the photosynthetic membranes. The same energized form of chlorophyll is used either for photosynthesis or loses its energy as heat. (C) Depiction of the regulation of the biochemistry of the xanthophyll cycle, as well as the induction of xanthophyll cycle-dependent en-ergy dissipation by pH. De-epoxidation to antheraxanthin (A) and zeaxanthin (Z) from violaxanthin (V) requires a low pH in the lumen of the thylakoid as well as reduced ascorbate. In addition, a low pH of certain domains within the membrane, together with the presence of zeaxanthin or antheraxanthin, is required to induce the actual energy dissipation. This dissipation takes place within the light-collecting antenna complexes (Demmig-Adams & Adams 1996). Copyright Elsevier Science Ltd.

photochemistry. Sun-grown plants typically, not only contain a larger fraction of the carotenoids as zeaxanthin in high light, but their total pool of carotenoids is also larger (Fig. 17). The pool of reduced ascorbate, which also plays a role in the xanthophyll cycle (Fig. 16), is also severalfold greater in plants acclimated to high light (Logan et al. 1996).

3.3.2 Chloroplast Movement in Response to Changes in Irradiance

The leaf's absorptance is affected by the concentration of chlorophyll in the leaf and the path length of light in the leaf, as well as by the location of the chloroplasts. **Light-induced movements** of chloroplasts are affected only by wavelengths

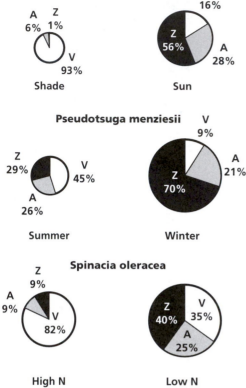

3.4 Responses to Variable Irradiance

We have so far discussed only steady-state responses to light, meaning that a particular environmental condition is maintained until a constant response is achieved. Conditions in the real world, however, are typically not constant, with irradiance being the most rapidly varying environmental factor. Because photosynthesis primarily depends on irradiance, the dynamic response to variation in irradiance deserves particular attention.

The irradiance level above a leaf canopy changes with time of day and with cloud cover, often by more than an order of magnitude within seconds. In a leaf canopy, irradiance, particularly direct radiation, changes even more. In a forest, direct sunlight may penetrate through holes in the overlying leaf canopy, casting sunflecks on the forest floor. These move with wind action and position of the sun, thus exposing both leaves in the canopy and shade plants in the understory to short periods of bright light. Sunflecks typically account for 40 to 60% of total irradiance in understory canopies of dense tropical and temperate forests and are quite variable in duration and intensity (Chazdon & Pearcy 1991).

3.4.1 Photosynthetic Induction

When a leaf that has been in darkness or low light for some time (hours) is transferred to a high saturating level of irradiance, the photosynthetic rate increases only gradually over a period of up to 1 hour to a new steady-state rate (Fig. 18), with stomatal conductance increasing more or less in parallel. The initial interpretation, however, that limitation of photosynthesis during induction is due to stomatal opening appeared to be wrong. If it were correct, then the intercellular partial pressure of CO_2 (p_i) should drop immediately upon transfer to high irradiance. As can be seen from Figure 18, there is a more gradual decline over the first minutes and then a slow increase until full induction. There are apparently, additional limitations at the chloroplast level. The demand for CO_2 increases faster than the supply during the initial decline in p_i. During its subsequent rise, the supply increases faster than the demand. This is illustrated in Figure 19, where A is plotted as a function of p_i during photosynthetic induction. During the first 1 or 2 minutes there is a fast increase in demand for CO_2 (fast **induction** component) that appears to be related with fast light induction of some Calvin-cycle enzymes and build-up of metabolite pools (Sassenrath-Cole et al. 1994). The slower phase of

FIGURE 17. Differences in zeaxanthin (Z), violaxanthin (V), and antheraxanthin (A) contents of leaves upon acclimation to the light level (*Vinca minor*, periwinkle), season (*Pseudotsuga menziesii*, Douglas fir), and nitrogen supply (*Spinacia oleracea*, spinach). The total areas reflect the concentration of the three carotenoids relative to that of chlorophyll (Demmig-Adams & Adams 1996). Copyright Elsevier Science Ltd.

below 500 nm. High intensities in this wavelength region cause the chloroplasts to line up along the vertical walls, parallel to the light direction, rather than along the lower cell walls, perpendicular to the direction of the radiation, as in control leaves. Chloroplast movements are pronounced in such shade-tolerant understory species as *Oxalis oregana*, where they may decrease the leaf's absorptance by as much as 20%, thereby increasing both transmittance and reflectance. Other species [e.g., the shade-avoiding *Helianthus annuus* (sunflower)], show no blue light–induced chloroplast movement or change in absorptance. Chloroplast movements in shade plants exposed to high light avoids photoinhibition (Brugnoli & Björkman 1992).

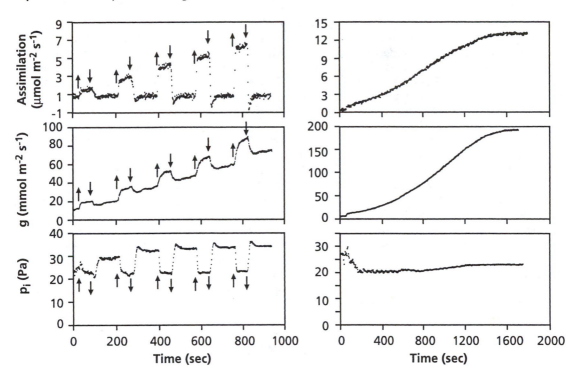

FIGURE 18. Photosynthetic induction in *Toona australis*, which is an understory species from the tropical rainforest in Australia. (Left panels) Leaves were exposed to five "sunflecks", indicated by arrows, with a low background level of irradiance in between. (Right panels) Time course of the rate of CO_2 assimilation (top), stomatal conductance (middle), and the intercellular CO_2 partial pressure (bottom) of plants that were first exposed to a low irradiance level and then transferred to high saturating irradiance (Chazdon & Pearcy 1986).

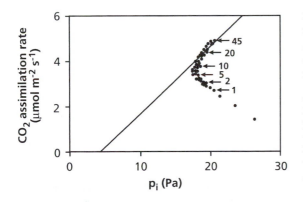

FIGURE 19. Photosynthetic response of *Alocasia macrorrhiza* to intercellular partial pressure of CO_2 (p_i) during the induction phase after a transition from an irradiance level of 10 to $500\,\mu mol\,m^{-2}s^{-1}$ (light saturation). The solid line represents the A–p_i relationship of a fully induced leaf calculated as Rubisco-limited rates. Numbers indicate minutes after transition (Kirschbaum & Pearcy 1988). Copyright American Society of Plant Physiology.

increase in demand until approximately 10 minutes is dominated by the light-activation of Rubisco. After that, p_i increases, which indicates that a decrease in stomatal limitation dominates further rise in photosynthetic rate.

Loss of photosynthetic induction occurs in low light, but at a lower rate than induction in high light, particularly in forest understory species. Hence, in a sequence of sunflecks, photosynthetic induction increases from one sunfleck to the next, until a high induction state is reached. Sunflecks can now be used efficiently (Fig. 18).

3.4.2 Light Activation of Rubisco

Rubisco, as well as other enzymes of the Calvin cycle, must be **activated by light**, before they have a high catalytic activity (Pons et al. 1992, Fig. 20). The increase in Rubisco activity, due to its activation by light, closely matches the increased photosynthetic rate at a high irradiance (Salvucci 1989). Two mechanisms are involved in the activation of Rubisco. First, the covalent binding of CO_2 to a lysine residue (**carbamylation**) activates Rubisco in all species investigated so far. This activation is catalyzed by **Rubisco activase**, whose activity increases with increasing rate of electron transport. Second, a natural inhibitor of Rubisco has been found in some, but not all, species. This inhibitor, 2-**carboxy-D-arabinitol 1-phosphate** (CA1P), is an analogue of the extremely short-lived product of the carboxylation reaction (Fig. 20).

Light activation of Rubisco, which is a natural process that occurs at the beginning of the light period in all plants, is likely to be an important aspect of the regulation (fine-tuning) of photosynthesis. In the absence of such light activation, the three phases of the Calvin cycle (i.e., carboxylation, reduction, and regeneration of RuBP) would likely compete for substrates, which would lead to oscillation of the rate of CO_2 fixation upon the beginning of the light period. It may also protect active sites of Rubisco during inactivity in darkness. The regulation mechanism occurs at the expense of low rates of CO_2 assimilation during periods of low induction.

3.4.3 Postillumination CO_2 Assimilation and Sunfleck Utilization Efficiency

After a sunfleck, the rate of O_2 evolution stops immediately, whereas CO_2 assimilation continues for a brief period. This is called **post-illumination CO_2 fixation** (Fig. 21). CO_2 assimilation in the Calvin cycle requires both NADPH and ATP, which are generated during the light reactions. This post-illumination CO_2 fixation is important, particularly in short sunflecks, relative to photosynthesis during the sunfleck, thus increasing total CO_2 assimilation due to the sunfleck above what would be expected from steady state measurements (Fig. 22).

CO_2 assimilation due to a sunfleck also depends on induction state, as mentioned before. Leaves become increasingly induced with longer sunflecks of up to a few minutes. At low induction states, sunfleck utilization efficiency decreases below what would be expected from steady-state rates (Fig. 22). Forest understory plants tend to utilize sunflecks more efficiently than do plants from short vegetation, particularly flecks of a few seconds to a few minutes. Accumulation of larger Calvin-cycle metabolite pools and longer maintenance of photosynthetic induction are possible reasons. Efficient utilization of sunflecks is crucial for understory plants because most radiation comes in the form of relatively long-lasting sunflecks, and half the plant's assimilation may depend on these short periods of high irradiance.

3.4.4 Metabolite Pools in Sun and Shade Leaves

As explained in Section 2.1.3 of this chapter, the photophosphorylation of ADP depends on the

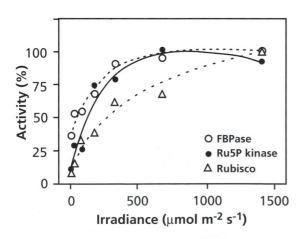

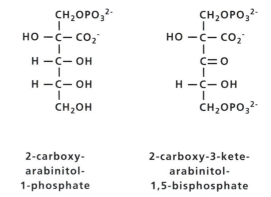

FIGURE 20. (Left) Light activation of Rubisco and other Calvin cycle enzymes. The increase in Rubisco activity with increasing irradiance closely matches that of the photosynthetic rate (Salvucci 1989; copyright Physiologia Plantarum). Activation of Rubisco occurs via two mechanisms. First, the covalent binding of CO_2 to a lysine residue (carbamylation) occurs in all species investigated. This activation is catalyzed by Rubisco activase, whose activity increases with increasing rate of electron transport. Second, a natural inhibitor of Rubisco has been found in some, but not all, species. This inhibitor, 2-carboxy-D-arabinitol 1-phosphate (CA1P), is an analogue of the extremely short-lived product of the carboxylation reaction; the structure of both molecules is depicted in the right-side part of the figure.

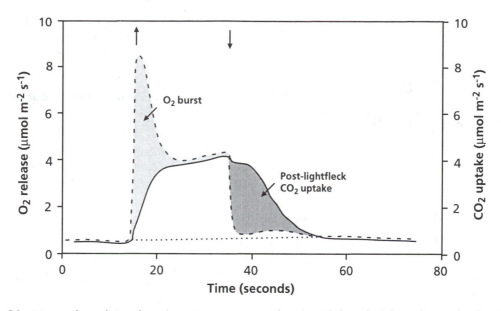

FIGURE 21. CO_2 uptake and O_2 release in response to a "sunfleck." Arrows indicate the beginning and end of the "sunfleck" (Pearcy 1990). With permission, from the An-nual Review of Plant Physiology Plant Molecular Biology, Vol. 41, copyright 1990, by Annual Reviews Inc.

proton gradient across the thylakoid membrane. This gradient is still present immediately following a sunfleck; therefore, ATP can still be generated for a brief period. The formation of NADPH, however,

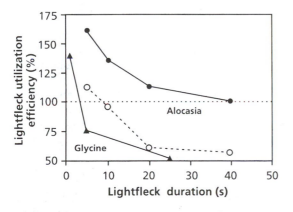

FIGURE 22. Efficiency of "sunfleck" utilization as dependent on duration of the "lightfleck" and induction state in two species. Efficiency is defined here as the increase in assimilation due to the lightfleck compared to the increase derived from steady state rates. *Alocasia macrorrhiza* (an understory species) was studied at high (closed symbols) and low induction state (open symbols). Induction state of *Glycine max* (soybean, a sun species) was approximately 50% of its maximum. Efficiencies were calculated as total CO_2 assimilation due to the sunfleck relative to that calculated from the steady state rates at the high irradiance (sunfleck) and the low (background) irradiance (Pearcy 1988; copyright CSIRO, Australia, Pons & Pearcy 1992; copyright Blackwell Science Ltd).

directly depends on the flux of electrons from water, via the photosystems and the photosynthetic electron-transport chain, and therefore comes to an immediate halt after the sunfleck. In addition, the concentration of NADPH in the cell is too low to sustain Calvin-cycle activity. Storage of the reducing equivalents takes place in triose-phosphates (Fig. 23), which are intermediates of the Calvin cycle.

To allow the storage of reducing power in intermediates of the Calvin cycle, the phosphorylating step leading to the substrate for the reduction reaction must proceed. This can be realized by regulating the activity of two enzymes of the Calvin cycle that both utilize ATP: phosphoglycerate kinase and ribulose-phosphate kinase (Fig. 4). When competing for ATP in vitro, the second kinase tends to dominate, leaving little ATP for phosphoglycerate kinase. If this were to happen in vivo as well, no storage of reducing equivalents in triose-phosphate would be possible and CO_2 assimilation would not continue beyond the sunfleck. The concentration of triose-phosphate at the end of a sunfleck is relatively greater in shade leaves than it is in sun leaves, whereas the opposite is found for ribulose-1,5-bisphosphate. This indicates that the steps in the Calvin cycle that lead to ribulose-monophosphate are somehow suppressed. Competition for ATP between the kinase is therefore prevented, and the reducing power from NADPH can be transferred to 1,3-bisphosphoglycerate, which leads to the formation of triose-phosphate.

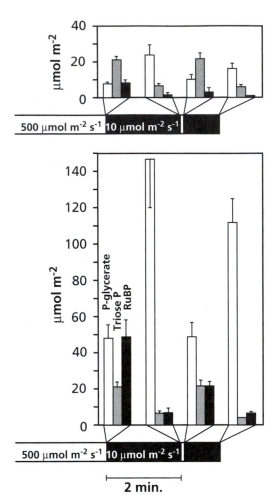

FIGURE 23. Pool sizes of a number of Calvin-cycle intermediates, expressed per unit leaf area: phosphoglycerate, triose-phosphate (dihydroxy-acetone phosphate plus glyceraldehyde 3-phosphate) and RuBP in shade-grown (top) and light-grown (bottom) *Alocasia macrorrhiza*, which is a species from the understory of a tropical rain forest in Australia. The leaves were exposed to high irradiance (500 µmol m⁻² s⁻¹, white bars), followed by a 2-min exposure to low irradiance 10 µmol m⁻² s⁻¹, black bars), a brief period of high irradiance, and, finally, a period of low irradiance. Pool sizes were determined at the end of each period of exposure to high or low irradiance (Sharkey et al. 1986a). Copyright American Society of Plant Physiology.

Contrary to the situation for sun leaves, the contribution of ribulose-bisphosphate to the post-illumination CO_2 fixation is relatively small in shade leaves. The opposite is found for triose-phosphate, which is not of major importance in sun leaves, but is significant in shade leaves (Table 3). The situation illustrated for *Alocasia*

macrorrhiza is not unique for species of the understory because it has also been found for the shade leaves of common bean plants (*Phaseolus vulgaris*, common bean).

3.4.5 Net Effect of Sunflecks on Carbon Gain and Growth

Although most understory plants can maintain a positive carbon balance with diffuse light in the absence of sunflecks, daily carbon assimilation and growth rate in moist forests correlate closely with irradiance received in sunflecks (Fig. 24). Sunflecks also account for an increasing proportion of total carbon gain (9 to 46%) as their size and frequency increase. In dry forests, where understory plants experience both light and water limitation, sunflecks may reduce daily carbon gain on cloud-free days (Young & Smith 1983). Thus, the net impact of sunflecks on carbon gain depends on both cumulative irradiance and other potentially limiting factors.

4. Partitioning of the Products of Photosynthesis and Regulation by "Feedback"

4.1 Partitioning Within the Cell

Most of the products of photosynthesis are exported out of the chloroplast to the cytosol as **triose-phosphate** in exchange for **inorganic phosphate** (P_i). Triose-phosphate is the substrate for the synthesis of sucrose in the cytosol (Fig. 25) and for the formation of cellular components in the source leaf. Sucrose is largely exported to other parts (sinks) of the plant via the phloem. Some species primarily export oligosaccharides or sugar-alcohols rather than sucrose (see Sect. 2.1 in the chapter on long-distance transport).

Partitioning of the products of the Calvin cycle within the cell is largely controlled by the concentration of P_i in the cytosol. If this concentration is high, then rapid **exchange** for triose-phosphate allows export of most of the products of the Calvin cycle. If the concentration of P_i drops, then the exchange rate will decline, and the concentration of triose-phosphate in the chloroplast increases. Inside the chloroplasts, the triose-phosphates are used for the synthesis of **starch**, releasing P_i within the chloroplast. The partitioning, therefore, of the products of photosynthesis between **export** to the cytosol versus **storage** in the chloroplasts is largely

TABLE 3. The potential contribution of triose-phosphates and ribulose-1,5-bisphosphate to the post-illumination CO_2 assimilation of *Alocasia macrorrhiza* and *Phaseolus vulgaris* (common bean).

	Alocasia		Phaseolus	
	Shade	Sun	Shade	Sun
RuBP (μmol m^{-2})	2.0	14.5	2.9	5.3
Triose-phosphates (μmol m^{-2})	16.3	18.0	19.8	10.5
Total potential CO_2 fixation (μmol m^{-2})	12.0	25.0	15.0	12.0
Potential efficiency (%)	190.0	204.0	154.0	120.0
Triose-P/RuBP	4.9	0.7	4.1	1.2
Postillumination ATP required (μmol g^{-1} Chl)	13.0	22.0	63.0	29.0

Source: Sharkey et al. 1986a.

* The values for the intermediates give the difference in their pool size at the end of the sunfleck and 1 min later. The total potential CO_2 assimilation is RuBP + 3/5 triose-P pool size. The potential efficiency was calculated on the assumption that the rate of photosynthesis during the 5 second sunfleck was equal to the steady state value measured after 20 min in high light.

determined by the availability of P_i in the cytosol. This regulation can be demonstrated by experiments in which the concentration of cytosolic P_i has been manipulated.

The concentration of cytosolic P_i can be decreased by incubating leaf discs in a solution that contains mannose. Mannose is readily taken up and enzymatically converted into mannose phosphate, thus sequestering some of the P_i originally present in the cytosol. Starch accumulates in the chloroplasts under these conditions. At extremely low cytosolic P_i concentrations, the rate of photosynthesis is also reduced.

In intact plants the rate of photosynthesis is also reduced when the plant demand for carbohydrate (**reduced sink strength**) is decreased; for example, by the removal of part of the fruits or "girdling" of the petiole. (Girdling involves damaging the phloem tissue of the stem, either by a temperature treatment or mechanically.) When leaf discs are taken from plants with reduced sink capacity, the addition of mannose has very little effect, presumably because the cytosolic P_i concentration is already very low before mannose addition (Table 4). Restricting the sink capacity therefore appears to sequester the cytosolic P_i, and sink limitation leads to a "**feedback inhibition**" of photosynthesis. When the level of P_i in the cytosol is increased, by floating the leaf discs on a phosphate buffer, the rate of photosynthesis also drops, but there is no accumulation of starch (Table 4). This is likely to be due to the very rapid export of triose-phosphate from the chloroplasts, in exchange for P_i, depleting the Calvin cycle of intermediates.

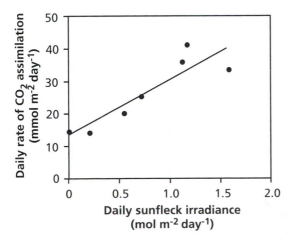

FIGURE 24. Total carbon gain of *Adenocaulon bicolor* as a function of daily photon flux contributed by sunflecks in the understory of a temperate redwood forest (Chazdon & Pearcy 1991). Copyright American Institute of Biological Sciences.

4.2 Regulation of the Rate of Photosynthesis by Feedback

As described in Section 4.1, when the rate of incorporation and/or export of the products of photosynthesis is low, the rate of photosynthesis may become restricted by "feedback inhibition." Under such conditions, phosphorylated intermediates of

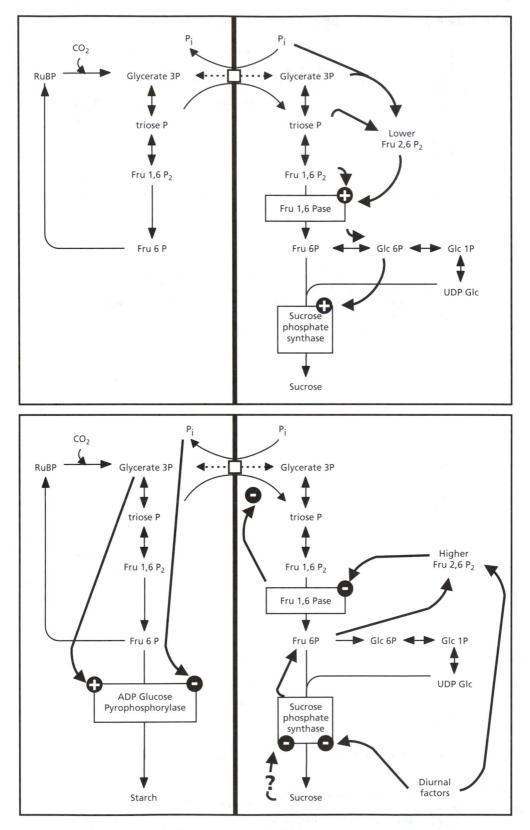

FIGURE 25. The formation of triose-phosphate in the Calvin cycle. Triose-P is exported to the cytosol, in exchange for P_i (upper part), or used as a substrate for the synthesis of starch in the chloroplast (lower graph), when export of triose phosphate stagnates. Fructose 2,6 bis phosphate is a regulatory molecule, and not an intermediate of glycolysis. The − and + signs refer to inhibition and stimulation of enzyme activities, respectively (after Stitt 1997).

TABLE 4. Rates of CO_2 assimilation ($\mu mol\,m^{-2}\,s^{-1}$) and the accumulation of ^{14}C in soluble sugars ("ethanol-soluble") and starch ("$HClO_4$-soluble") (^{14}C as % of total ^{14}C recovered) in leaf discs of *Gossypium hirsutum* (cotton) and *Cucumis sativus* (cucumber) floating on a Tris-maleate buffer, a phosphate buffer, or a mannose solution.

	Control			Girdled		
	CO_2 fixation	Ethanol-soluble	$HClO_4$-soluble	CO_2 fixation	Ethanol-soluble	$HClO_4$-soluble
Cotton						
Tris-maleate	18	83	17	12	76	24
Phosphate	18	87	13	10	83	17
Mannose	12	54	46	10	76	24
Cucumber						
Tris-maleate	13	76	24	6	40	60
Phosphate	9	82	18	5	76	24
Mannose	9	55	45	4	40	60

Source: Plaut et al. 1987.

*Leaves were taken from control plants ("control") or from plants whose petioles had been treated in such a way as to restrict phloem transport ("girdled").

the pathway leading to sucrose accumulate, inexorably decreasing the cytosolic P_i concentration. In the absence of sufficient P_i in the chloroplast, the formation of ATP is reduced and the activity of the Calvin cycle declines. That is, less intermediates are available and less RuBP is regenerated, so that the carboxylating activity of Rubisco and hence the rate of photosynthesis drops.

How important is feedback inhibition in plants whose sink has not been manipulated? To answer this question, the **oxygen sensitivity** of photosynthesis must be determined. The rate of net CO_2 assimilation normally increases when the oxygen concentration is lowered from a normal 21% to 1 or 2% because of the suppression of the oxygenation reaction (see Sect. 2.1.5 of this chapter). When the activity of Rubisco is restricted by the regeneration of RuBP, lowering of the oxygen concentration enhances the net rate of CO_2 assimilation to a lesser extent. This feedback inhibition is found at a high irradiance and also at a suboptimal temperature, which restricts phloem loading. Under these conditions the capacity to assimilate CO_2 exceeds the capacity to export and further metabolize the products of photosynthesis. As a result, phosphorylated intermediates of the pathway from triose-phosphate to sucrose accumulate, which sequesters phosphate. P_i therefore starts to limit photosynthesis and the rate of photosynthesis declines, as soon as the capacity to channel triose-phosphate to starch is saturated. Figure 26, which shows the response of the net rate of CO_2 assimilation to N_2 at four levels of irradiance, illustrates this point.

The assessment of feedback inhibition of photosynthesis using the oxygen sensitivity of this process is complicated by the fact that the relative

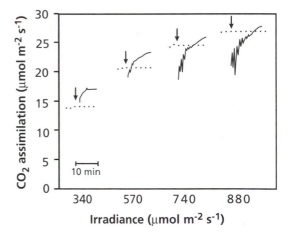

FIGURE 26. The response of the CO_2 assimilation rate to a change in oxygen concentration at four levels of irradiance. The gas phase was changed from air to N_2 at the times indicated by the arrows. The CO_2 partial pressure in the atmosphere surrounding the leaf was maintained at 55 Pa and the leaf temperature was 15°C. At a relatively low irradiance ($340\,\mu mol\,m^{-2}\,s^{-1}$) the rate of CO_2 assimilation is rapidly enhanced when the oxygen concentration is decreased, whereas at high irradiance ($880\,\mu mol\,m^{-2}\,s^{-1}$), CO_2 assimilation first decreases and is only marginally enhanced after several minutes. Intermediate results are found at irradiance in between these extremes (Sharkey et al. 1986b). Copyright American Society of Plant Physiology.

activities of the carboxylating and oxygenating re-actions of Rubisco also depend on temperature (cf. Sect. 7.1 of this chapter). To resolve this problem, a mathematical model of photosynthesis was used (See Box 1). This model incorporates biochemical information on the photosynthetic reactions and simulates the effect of lowering the oxygen concentration at a range of temperatures. Comparison of the observed effect of the decrease in oxygen concentration (Fig. 27, lower middle and right panels) with the model results in which the feedback inhibition of photosynthesis has been omitted (Fig. 27, lower left panel) shows the extent of feedback inhibition in normal plants under normal conditions. In the presence of feedback inhibition (lower middle and right panels, i.e., real plants in natural conditions), photosynthesis is not strongly inhibited by oxygen because CO_2 assimilation is inhibited by feedback whether oxygenase activity occurs or not. This is especially true at low temperature where feedback inhibition is most strongly developed. If, however, feedback inhibition is eliminated in the model (lower left panel), CO_2 assimilation becomes more sensitive to oxygen (i.e., to oxygenase activity) at low temperature; this occurs whether the carboxylation is limited by RuBP or by Rubisco. Feedback inhibition is predominantly associated with species that accumulate starch in their chloroplasts, rather than sucrose and hexoses in the cytosol and vacuoles. Because genetically transformed plants of the same species, which lack the ability to store starch, behave like the starch-accumulating wild-type, the reason for this difference remains obscure (Goldschmidt & Huber 1992). It could reflect the mode of phloem loading (i.e., either symplasmic or apoplasmic) (see the chapter on long-distance transport).

4.3 Glucose Repression of Genes Encoding Calvin-Cycle Enzymes

Although the feedback mechanism outlined in Section 4.2 of this chapter is likely to be important in the initial phase when carbohydrate production exceeds its export, mechanisms at the level of **gene transcription** probably play a more important role in the long term. Leaves of *Triticum aestivum* (wheat) that have been fed with 1% glucose have a lower photosynthetic capacity, as well as lower levels of mRNA encoding several Calvin-cycle enzymes, including the small subunit of Rubisco (Jones et al. 1996). This effect is specific in that glucose feeding to leaves enhances the activity of

several glycolytic enzymes in these leaves, again due to regulation of gene expression (Krapp & Stitt 1994).

It is likely that regulation of gene expression by carbohydrates plays an important role in the regulation of the activity of the "source" (leaves) by the demand in the "sink" (e.g., fruits) (Krapp et al. 1993). Such regulation at the level of gene transcription is similarly likely to play a role in the acclimation of the photosynthetic apparatus to elevated concentrations of atmospheric CO_2 (Sect. 12 of this chapter).

4.4 Ecological Impacts Mediated by Source-Sink Interactions

Many ecological processes affect photosynthesis through their impact on plant demand for carbohydrate (Sect. 4.2). In general, processes that increase carbohydrate demand cause an increase in photosynthetic rate, whereas factors that reduce carbohydrate demand reduce photosynthesis.

Although defoliation generally reduces carbon assimilation by the defoliated plant by reducing the biomass of photosynthetic tissue, it causes a compensatory increase in photosynthetic rate of remaining leaves through several mechanisms. The increased sink demand for carbohydrate generally leads to an increase in A_{max} in the remaining leaves. Defoliation also reduces environmental constraints on photosynthesis by increasing light penetration through the canopy and by increasing the biomass of roots available to support each remaining leaf. The resulting increases in light and water availability can enhance photosynthesis under shaded and dry conditions, respectively.

5. Responses to Availability of Water

The inevitable loss of water, when the stomates open to allow photosynthesis, may lead to a decrease in leaf relative water content (RWC), if the water supply from roots does not match the loss from leaves. The decline in RWC may directly or indirectly affect photosynthesis. Effects of the water supply on photosynthesis will be described in this section. In addition, we will discuss genetic adaptation and phenotypic acclimation to water shortage.

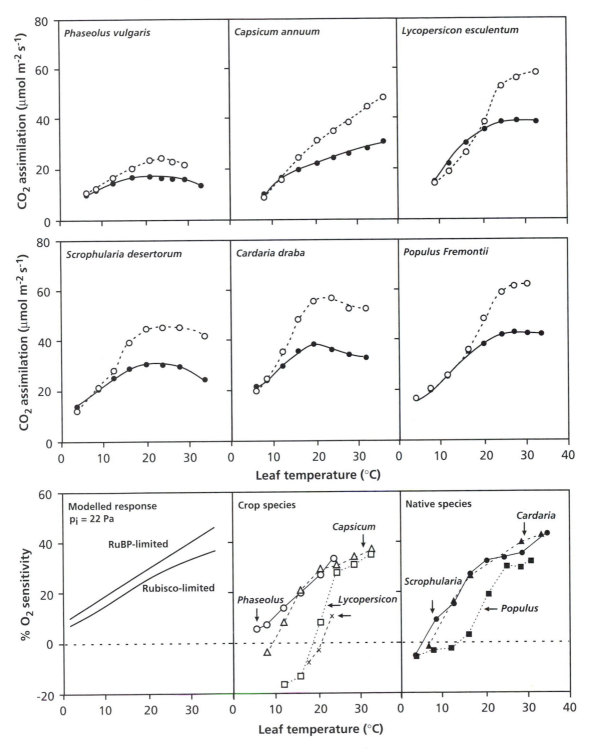

FIGURE 27. The effect of temperature on the net rate of CO_2 assimilation at 18 (filled symbols) and 3 (open symbols) kPa O_2 (top, middle), and the oxygen sensitivity of photosynthesis (bottom) for a number of species. The model (lower left panel) simulated oxygen sensitivity in the absence of feedback inhibition of CO_2 assimilation. The middle and lower right panels show distinct feedback inhibition for *Lycopersicon* (tomato) and *Populus* (poplar) at low temperatures, and less feedback inhibition for *Phaseolus* (bean), *Capsicum* (pepper), *Scrophularia*, and *Cardaria*. All plants were grown outdoors (Sage & Sharkey 1987). Copyright American Society of Plant Physiology.

5.1 Regulation of Stomatal Opening

Stomatal opening tends to be regulated such that photosynthesis is approximately colimited by CO_2 diffusion through stomates and light-driven electron transport. This is seen in Figure 6 as the intersection between the line describing the leaf's conductance for CO_2 transport (supply function) and the A–p_i curve (demand function). A higher conductance and higher p_i would only marginally increase CO_2 assimilation, but they would significantly increase transpiration because transpiration increases linearly with g_s (Fig. 28). At lower conductance, water loss declines linearly with g_s because the vapor pressure difference between the leaf and the air (VPD) remains the same (see Sect. 2.2.2); However, p_i also declines because the demand for CO_2 remains the same, and the difference with p_a increases. This increased CO_2 concentration gradient across the stomata counteracts the decrease in g_s. Photosynthesis, therefore, does not decline as much as transpiration does with decreasing p_i. The result is an increasing **water-use efficiency** (WUE) (carbon gain per water lost) with decreasing g_s. Less of the total photosynthetic capacity is used, however, which leads to a reduced **photosynthetic nitrogen-use efficiency** (PNUE) (carbon gain per unit leaf nitrogen; see Sect. 6.1 in this chapter). Intersection in the region where carboxylation and electron transport are co-limiting, therefore, is a compromise between an efficient use of water and an efficient use of the photosynthetic machinery.

Plants tend to reduce stomatal opening under water stress so that WUE is maximized at the expense of PNUE. Under limited availability of nitrogen, stomata may open further, causing a high PNUE at the expense of WUE (Table 5). In Section 3.2.2 of the chapter on life cycles, an example is given of a change from a high PNUE and

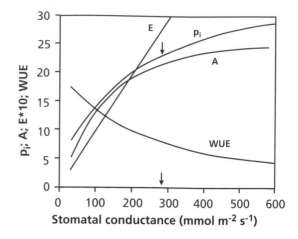

FIGURE 28. The effect of stomatal conductance (g_s) on the transpiration rate (E, mmol m^{-2} s^{-1}), rate of CO_2 assimilation (A, μmol m^{-2} s^{-1}), intercellular CO_2 partial pressure (p_i, Pa), and photosynthetic water-use efficiency [WUE, mmol CO_2 (mol $H_2O)^{-1}$]. Calculations were made for a constant leaf temperature of 25°C and without any boundary layer resistance. The arrow indicates g_s at the colimitation point of carboxylation and electron transport. For the calculations, equations as described in Box 1 have been used.

low WUE in compound bipinnate leaves to a high WUE and low PNUE in phyllodes of an *Acacia* species.

When a plant is subjected to water stress, the stomata tend to close. This closure response is regulated initially by **abscisic acid** (ABA), a phytohormone which is produced by roots in contact with dry soil and is transported to the leaves (Sect. 5.4.1 in the chapter on plant water relations; Box 8 in the chapter on growth and allocation). Secondly, there is also an effect of decreasing turgor potential in the leaf on stomatal opening, possibly mediated via ABA produced in the leaf. Stomatal conductance may also decline in response

TABLE 5. Water-use efficiency (WUE, A/g_s) and nitrogen-use efficiency of photosynthesis (PNUE, A/N_{LA}) of leaves of Helianthus annuus (sunflower), growing in a field in the middle of a hot, dry summer day in California.*

	N_{LA} mmol m^{-2}	A μmol m^{-2} s^{-1}	g_s mol m^{-2} s^{-1}	p_i Pa	WUE mmol mol^{-1}	PNUE μmol mol^{-1} s^{-1}
High N + W	190	37	1.2	24	31	195
Low W	180	25	0.4	20	63	139
Low N	130	27	1.0	26	27	208

Source: Fredeen et al. 1991.

* Plants were irrigated and fertilized (high N + W), only irrigated but not fertilized (low N), or only fertilized but not irrigated (low W). Because transpiration is approximately linearly related to g_s, A/g_s is used as an approximation of WUE.

to increasing VPD (Sect. 5.4.3 in the chapter on plant water relations). For example, epidermal peels of a leaf that are floated on water will close stomates in response to dry air. The result of these regulatory mechanisms is that, in many cases, transpiration is kept constant over a range of VPDs, and leaf water potential is kept constant over a range of soil water potentials. Water loss is therefore restricted when dry air is likely to impose water stress (a feedforward response) or when the plant experiences incipient water stress (a feedback response) (Schulze 1986). In dry environments these two regulatory mechanisms often cause midday stomatal closure and therefore declines in photosynthesis (Fig. 34 in plant water relations).

It was long assumed that stomata respond homogeneously over the entire leaf. However, leaves of droughted plants exposed to $^{14}CO_2$ show a heterogeneous distribution of fixed ^{14}C. This shows that some stomates close completely (there is no radioactivity close to these stomates), whereas others hardly change their aperture (label is located near these stomates) (Downton et al. 1988, Terashima et al. 1988). When stomatal closure is "**patchy**", the p_i cannot be determined in the manner as outlined in section 2.2.2, and the original conclusion that p_i is constant and controlled by photosynthesis, rather than the other way around, is not valid.

Leaves of plants that reduce stomatal conductance during the middle of the day may only close some of their stomates, while others remain open (Beyschlag & Pfanz 1990). This nonuniform reaction of stomata may occur only when plants are rapidly exposed to drought, whereas stomata may respond in a more uniform manner when the drought occurs more slowly (Gunasekera & Berkowitz 1992). Stomatal patchiness can also occur in dark-acclimated leaves upon exposure to bright light (Eckstein et al. 1996).

5.2 The A–p_i Curve as Affected by Water Stress

Water stress alters both the supply and the demand functions of photosynthesis (Fig. 29). The decline in stomatal conductance causes the reduced slope of the supply function with drought. The major change in the demand function is a decrease in A_{max}. The mechanism of this **down-regulation** of photosynthetic capacity in response to water stress is not fully understood. Because high irradiance and high temperature often coincide with drought, however,

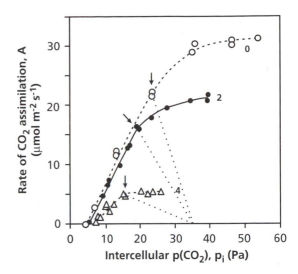

FIGURE 29. Photosynthesis (A) versus intercellular CO_2 (p_i) relationships for leaves of *Phaseolus vulgaris* (bean). Plants were grown in pots from which water was withheld for the indicated number of days. Arrows indicate p_i at ambient CO_2. The broken lines indicate the supply function (Von Caemmerer & Farquhar 1984).

photoinhibition may be involved. Similarly, because growth is inhibited more strongly than photosynthesis by drought, feedback inhibition may play an additional role. The net effect of the down-regulation of A_{max} under water stress is that the intercellular CO_2 partial pressure (p_i) is higher than would be expected if stomatal closure were the only factor causing a reduction in assimilation in droughted plants. The reduction in A_{max} allows photosynthesis to continue operating near the break-point between the RuBP-limited and the CO_2-limited regions of the A–p_i curve. Thus, drought-acclimated plants maximize the effectiveness of both light and dark reactions of photosynthesis under dry conditions at the cost of reduced photosynthetic capacity under favorable conditions. The decline in A_{max} in droughted plants is associated with declines in all biochemical components of the photosynthetic process.

The changes in photosynthesis in species and cultivars that are genetically adapted to drought are similar to those described earlier for drought acclimation. Drought-adapted wheat (*Triticum aestivum*) cultivars have a lower stomatal conductance and p_i than do less-adapted cultivars. In addition, stomatal conductance and photosynthesis in desert shrubs is lower than it is in less-drought-adapted plants and it declines less in response to water stress, largely due to osmotic adjustment (see the chapter on plant water relations).

5.3 Carbon Isotope Discrimination in Relation to Water-Use Efficiency

Carbon isotope composition of plant tissues provides an integrated measure of the photosynthetic water-use efficiency (WUE = A/E) during the time when the carbon comprising these tissues was being assimilated (Ehleringer 1993). As explained in Box 2, air has a $\delta^{13}C$ of -8‰, and the major steps in C_3 photosynthesis that discriminate are diffusion (4.4‰) and carboxylation (30‰, including dissolution of CO_2). The isotopic composition of a leaf will approach that of the process that most strongly limits photosynthesis. If stomates were almost closed and diffusion were the rate-limiting step, then $\delta^{13}C$ of leaves would be about -12‰ (= $-8 - 4.4$); if carboxylation were the only limiting factor, then we would expect a $\delta^{13}C$ of -38‰ ($=-8 - 30$). A typical range of $\delta^{13}C$ in C_3 plants is -25 to -29‰, which indicates colimitation by diffusion and carboxylation (O'Leary 1993). $\delta^{13}C$, however, varies among plant species and environment depending on the rate of CO_2 assimilation and stomatal conductance, which together determine p_i/p_a. The discrimination, Δ, is defined as (Box 2)

$$\Delta = \left(4.4 + 22.6(p_i/p_a)\right] \times 10^{-3}, \text{ or}:$$
$$\delta^{13}C_{air} - \delta^{13}C_{leaf} = 4.4 + 22.6(p_i/p_a), \quad (8)$$

which indicates that a high p_i/p_a (due to high stomatal conductance or low rate of CO_2 assimilation) result in a large discrimination (strongly negative $\delta^{13}C$). We can now use this information to estimate an integrated WUE for the plant. Following Equations 1 and 2 for A (Sect. 2.2.2), the water-use efficiency (WUE = A/E) is given by:

$$WUE = A/E = g_c(p_a - p_i)/g_w(e_i - e_a)$$
$$= p_a(1 - p_i/p_a)/1.6(e_i - e_a) \quad (9)$$

given that $g_c/g_w = 1.6$ (the molar ratio of diffusion of water vapor and CO_2 in air). Equation 9 tells us that the WUE is high, if the stomatal conductance is low in comparison with the capacity to assimilate CO_2 in the mesophyll. Under these circumstances p_i (and p_i/p_a) will be small. The right-hand part of Equation 9 then approximates [$p_a/(1.6(e_i - e_a)$], and **diffusion** is the predominant component that determines the fractionation of carbon isotopes and approaches a value of 1.0044. On the other hand if the stomatal conductance is large, then WUE is small, p_i approximates p_a and the right-hand part of Equation 9 approaches 1.027. Fractionation of the carbon isotopes is now largely due to the **biochemical discrimination by Rubisco.**

As expected from this analysis, there is a good correlation between WUE and the carbon isotope discrimination. *Triticum aestivum* (wheat) grown under dry conditions has a higher WUE and a lower $\delta^{13}C$ than do plants that are well supplied with water (Fig. 30). In addition, those genotypes that perform best under drought (greatest WUE) have the lowest $\delta^{13}C$, so that isotopic composition can be used to select for genotypes with improved performance under drought. A similar correlation between WUE and $\delta^{13}C$ has been found for cultivars of other species (e.g., *Hordeum vulgare*, barley; Hubick & Farquhar 1989, and *Lycopersicon*, tomato; Martin & Thorstenson 1988).

5.4 Other Sources of Variation in Carbon Isotope Ratios in C_3 Plants

Given the close relationship between WUE and $\delta^{13}C$, isotopic composition can be used to infer average WUE during growth (Sect. 6 in water relations). For example, $\delta^{13}C$ is higher (less negative) in desert than mesic plants, and it is higher in tissue produced during dry seasons (Smedley et al. 1991) or in dry years (Fig. 31). This indicates that plants growing in dry conditions have a lower p_i than do those in moist conditions; however, VPD is also higher in dry conditions, so WUE need not be as high for plants in dry conditions as the isotopic composition implies. On the other hand, other factors can alter isotopic composition without altering WUE. For example, $\delta^{13}C$ of plant tissue increases with altitude due to altitudinal gradients in CO_2 and water vapor concentration from the atmosphere to the leaf (Marshall & Zhang 1993). In addition, $\delta^{13}C$ is higher at the bottom than at the top of the canopy due to the contribution of ^{13}C-depleted CO_2 from soil respiration (Medina & Klinge 1983). Thick leaves may be enriched in ^{13}C, which could reflect a higher internal resistance to diffusion (Vitousek et al. 1990). Thick leaves, however, tend to have a high ratio of internal surface area to leaf area (Field & Mooney 1986), so the reason for a higher internal resistance in thick leaves is not immediately obvious (see Sect. 2.2.3).

Annuals discriminate more strongly against ^{13}C than do perennials; additionally, herbs discriminate more than grasses, and a root parasite (*Comandra umbellata*) was found to discriminate more than any of the surrounding species (see also the chapter on parasitic plants). These patterns suggest a high stomatal conductance and low WUE in the annuals and herbs. The low WUE of parasitic plants is

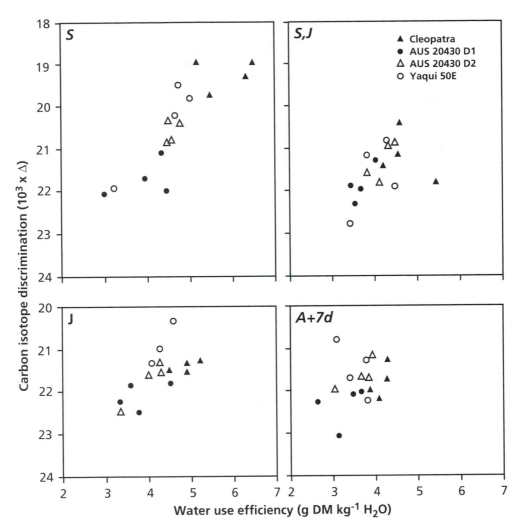

FIGURE 30. A comparison of the water-use efficiency (WUE) of four *Triticum aestivum* (wheat) cultivars, grown under different conditions. S, tubes watered to field capacity at sowing, and not watered again; S, J, same, except that tubes were rewatered to about one-half of field capacity at jointing; J, tubes were maintained at field capacity until plants were at the jointing stage; A + 7d, soil in tubes was maintained just below field capacity by replacing the water used each week, returned to field capacity at anthesis and no longer watered 7 days after anthesis. In this experiment, WUE was derived from the aboveground biomass produced and the cumulative water use during the production of that biomass, not from gas exchange experiments (cf. Sect. 6 in the chapter on plant water relations) (Farquhar & Richards 1984). Copyright CSIRO, Australia.

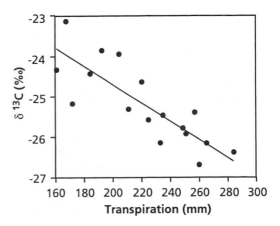

FIGURE 31. Carbon isotope composition of *Pseudotsuga menziesii* (Douglas fir) needles as a function of cumulative transpiration in trees growing with different moisture availabilities (Livingston & Spittlehouse 1993).

important in nutrient acquisition (see Sect. 3 in the chapter on parasitic associations).

6. Effects of Soil Nutrient Supply on Photosynthesis

6.1 The Photosynthesis–Nitrogen Relationship

Because the photosynthetic machinery accounts more than half of the nitrogen in a leaf (Fig. 13), it is not surprising that photosynthesis is strongly affected by nitrogen availability. A_{max} increases linearly with leaf nitrogen concentration (Fig. 32), regardless of whether the variation in leaf nitrogen is caused by differences in soil nitrogen availability, leaf age, or species composition (Field & Mooney 1986). The slope of this relationship is much steeper for C_4 plants than it is for C_3 plants (Sect. 9.5), and it also differs among C_3 species (Evans 1989; Sect. 4.2.1 in mineral nutrition). When leaves of different age are compared (e.g., leaves in a canopy profile), the light-saturated photosynthetic rate per unit nitrogen (**maximum photosynthetic nitrogen-use efficiency**; $PNUE_{max}$) is highest in leaves with high nitrogen concentrations (Pons et al. 1989). This is primarily due to the non-zero intercept for the A_{max} versus nitrogen relationship, which extrapolates to zero photosynthesis at a leaf nitrogen concentration of about $0.5 mmol N g^{-1}$. When variation in photosynthesis and leaf nitrogen concentration is due to variation in soil nitrogen supply, however, the A_{max} versus nitrogen relationship extrapolates to the origin (Fig. 32A)

The strong A_{max} versus nitrogen relationship cannot be due to any simple direct nitrogen limitation of photosynthesis because both carbon isotope studies and A–p_i curves generally show that photosynthesis is colimited by CO_2 diffusion and photosynthetic capacity. The entire photosynthetic process is instead down-regulated under conditions of nitrogen limitation, with declines in Rubisco, chlorophyll, and stomatal conductance (Sect. 5.1, Table 5). Photosynthesis operates near the point of transition between diffusion limitation and biochemical limitation in both high- and low-nitrogen plants. The net effect of this coordinated response of all photosynthetic components is that p_i/p_a and $\delta^{13}C$ show no consistent relationship with leaf nitrogen (Rundel & Sharifi 1994).

Why does photosynthesis correlate so consistently with nitrogen when other nutrients, like **phosphorus**, play a clear regulatory role in photosynthesis (see Sect. 4.1)? The close relationship between photosynthesis and leaf nitrogen reflects a

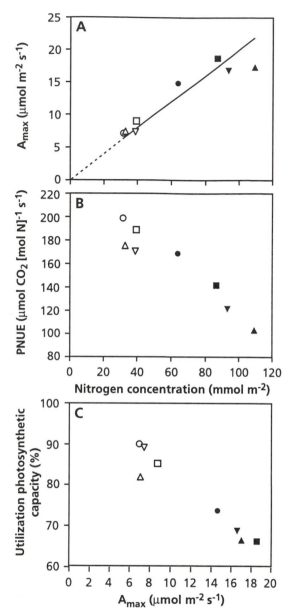

FIGURE 32. (A) The light-saturated rate of CO_2 assimilation (A_{max}) of four grasses grown at high (filled symbols) and low (open symbols) nitrogen supply. (B) Photosynthetic nitrogen-use efficiency (PNUE) determined at growth irradiance for the same grasses. Note the higher PNUE for plants grown at a low supply of nitrogen. (C) The proportional utilization of the total photosynthetic capacity at growth irradiance, calculated as the ratio of the rate at growth irradiance and A_{max} (Pons et al. 1994). Copyright SPB Academic Publishing.

combination of the larger proportional investment of nitrogen than of phosphorus in photosynthetic machinery, widespread nitrogen limitation in natural ecosystems, and the close correlation between tissue nitrogen and phosphorus concentrations over a range of environmental conditions (see the chapter on mineral nutrition). In some field studies, especially in conifers that often grow on low-phosphorus soils, photosynthesis may show little correlation with tissue nitrogen, but a strong correlation with tissue phosphorus (Reich & Schoettle 1988). The low photosynthetic rate of plants grown at low phosphorus supply may reflect feedback inhibition due to slow growth and low concentrations of P_i in the cytosol (Sect. 4.1) or low concentrations of Rubisco and other photosynthetic enzymes.

6.2 Interactions of Nitrogen, Light, and Water

Because of the coordinated responses of all photosynthetic processes, any environmental stress that reduces photosynthesis will reduce both the diffusional and the biochemical components (Table 5). Nitrogen concentration per unit leaf area, therefore, is typically highest in sun leaves and declines toward the bottom of the canopy. In single-species canopies, this partially reflects higher rates of CO_2 assimilation of young, high-nitrogen leaves in high-light environments (Pons et al. 1989). In multispecies canopies, however, the low leaf nitrogen concentration in understory species clearly reflects the adjustment of photosynthetic capacity to light availability. In addition, in dry environments plants typically have a low photosynthetic capacity, which matches the low stomatal conductance in these environments (Table 5).

6.3 Photosynthesis, Nitrogen, and Leaf Lifespan

As will be discussed in the chapters on mineral nutrition and growth and allocation, plants acclimate and adapt to low soil nitrogen and low soil moisture by producing long-lived leaves that are thicker and have a high leaf-mass density [low specific leaf area, SLA (i.e., leaf area per unit leaf mass)] and a low leaf nitrogen concentration. Both broadleafed and conifer species show a single strong negative correlation between leaf lifespan and either leaf nitrogen or mass-based photosynthetic rate (Fig. 33). The low SLA in long-lived leaves relates to structural properties required to withstand unfavorable environmental conditions (see chapter on growth and allocation). There is a strong positive correlation between SLA and leaf nitrogen concentration with a slope of 114 mmol N m^{-2}, ranging between 70 and 290 mmol N m^{-2} for different data sets (Fig. 34A). Conifers tend to have high leaf N per area (210 mmol N m^{-2}), probably because projected leaf area is a poor approximation of true leaf area in these species. Together the greater leaf thickness and low concentrations of nitrogen per unit leaf mass result in low rates of photosynthesis on a leaf-mass basis in long-lived leaves (Fig. 33). There is a much weaker relationship between leaf N and photosynthetic rate on a leaf-area basis (in fact, no relationship in conifers) because photosynthesis and nitrogen vary much less among species on an area than on a mass basis (Field & Mooney 1986, Reich et al. 1995). Because of the association of A_{max} with leaf nitrogen and g_s with A_{max}, maximum stomatal

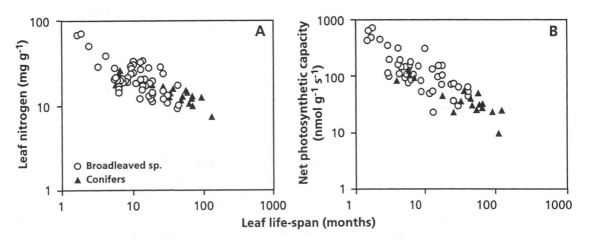

FIGURE 33. Mass-based leaf nitrogen (A) and maximum rate of CO_2 assimilation (B) as a function of leaf lifespan in broadleafed and confier trees (Reich et al. 1995). Copyright Academic Press.

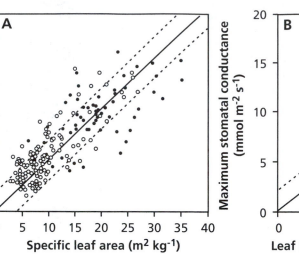

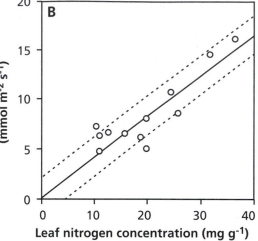

FIGURE 34. (A) Leaf nitrogen concentration as a function of specific leaf area of herbaceous (●) and woody (○) plants and (B) maximal stomatal conductance as a function of leaf nitrogen concentration in a broad survey of terrestrial plants. The solid line is the best-fit regression through the origin; dashed lines are 95% confidence intervals (Schulze et al. 1994). With permission, from the Annual Review of Ecology and Systematics, Vol. 25, copyright 1994, by Annual Reviews Inc.

conductance correlates strongly with leaf nitrogen (Fig. 34B).

In summary, we have seen that inadequate supplies of nitrogen leads to a decrease in the biochemical determinants of photosynthetic capacity and stomatal conductance with relatively small variation in p_i/p_a. This is at variance with the situation where water is limiting, when p_i/p_a tends to decrease, or where light is limiting the rate of CO_2 assimilation, when p_i/p_a tends to increase. Over short time scales (minutes) feedforward and feedback processes regulate stomatal conductance and activity of photosynthetic enzymes. During acclimation and adaptation, changes in morphology (largely leaf thickness) and photosynthetic capacity exert additional controls.

7. Photosynthesis and Leaf Temperature: Effects and Adaptations

Temperature has a major effect on enzymatically catalyzed reactions and membrane processes; therefore, it affects photosynthesis. Because the **activation energy** of different reactions often differs among plants acclimated or adapted to different temperature regimes, photosynthesis may be affected accordingly. In this section temperature

effects on photosynthesis will be explained in terms of underlying biochemical, biophysical, and molecular processes.

Differences among plants in their capacity to perform at extreme temperatures often correlate with the plant's capacity to photosynthesize at these temperatures. This may reflect both the adjustment of photosynthesis to the demand of the sinks (see Sect. 4 of this chapter) and changes in photosynthetic machinery during acclimation and adaptation.

7.1 Effects of High Temperatures on Photosynthesis

Many plants show an optimum temperature for photosynthesis close to their normal growth temperature. Below this optimum, enzymatic reaction rates, primarily associated with the **dark reactions**, are temperature-limited. At high temperatures the **oxygenating reaction** of Rubisco increases more than the carboxylating one so that photorespiration becomes proportionally more important. This is partly because the solubility of CO_2 declines with increasing temperature more strongly than does that of O_2. In addition, the effect of temperature on photosynthesis of C_3 plants is due to the effects of temperature on kinetic properties of Rubisco (i.e., the K_m-values for CO_2 and O_2) (Fig. 35). These combined effects cause a decline in net photosynthesis at high temperature.

Adaptation to high temperature typically causes a shift of the temperature optimum for net photosynthesis to higher temperatures (Fig. 36, Fig. 37). The temperature optimum for photosynthesis similarly shifts to higher temperatures when coastal and desert populations of *Atriplex lentiformis* are acclimated to high temperatures (Fig. 36). For the coastal population the rates of photosynthesis and growth are low at high temperatures because the fluidity of membranes increases strongly with temperature, which hampers membrane-associated function. In *Nerium oleander* (oleander) adaptation to high temperatures is associated with an increasing degree of saturation of the membrane lipids, which decreases the membranes' fluidity. Species

adapted to hot environments often show temperature optima for photosynthesis that are quite close to the temperature at which enzymes are inactivated.

7.2 Effects of Low Temperatures on Photosynthesis

Many (sub)tropical plants grow poorly or become damaged at temperatures between 10 and 20°C. Such damage is called **chilling injury** and differs from frost damage, which only occurs below 0°C. Part of the chilling injury is associated with the photosynthetic apparatus. The following aspects play a role:

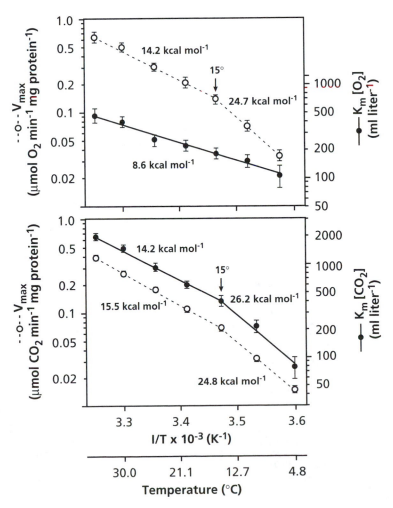

FIGURE 35. Temperature dependence of V_{max} and the K_m of the oxygenating (top) and the carboxylating (bottom) reaction of Rubisco. V_{max} is the rate of the carboxylating or oxygenating reaction at a saturating concentration of CO_2 and O_2, respectively. The K_m is the concentration of CO_2 and O_2 at which the carboxylating and oxygenating reaction, respectively, proceed at the rate which equals $1/2$ V_{max}. Note that a logarithmic scale is used for the y-axis and that the inverse of the absolute temperature is plotted on the x-axis ("Arrhenius-plot"). In such a graph, the slope gives the activation energy, which is a measure for the temperature dependence of the reaction (Berry & Raison 1981).

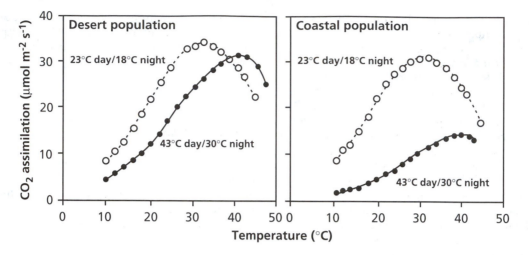

FIGURE 36. Photosynthesis in response to temperature of two populations of the C_4 species *Atriplex lentiformis* (Pearcy 1977, as cited in Berry & Raison 1981).

(1) Decrease in membrane fluidity
(2) Changes in the activity of membrane-associated enzymes and processes, such as the photosynthetic electron transport
(3) Loss of activity of cold-sensitive enzymes

Chilling resistance probably involves **reduced saturation** of membrane fatty acids, which increases membrane fluidity and so compensates for the effect of low temperature on membrane fluidity.

Chilling often leads to **photoinhibition** and **photooxidation** because the biophysical reactions of photosynthesis (photon capture and transfer of excitation energy) are far less affected by temperature than are the biochemical steps, including electron transport and activity of the Calvin cycle. Chlorophyll continues to absorb light at low temperatures, but the electrons cannot be transferred at a sufficiently high rate to the normal acceptors. The reversible reactions leading to reduced efficiency

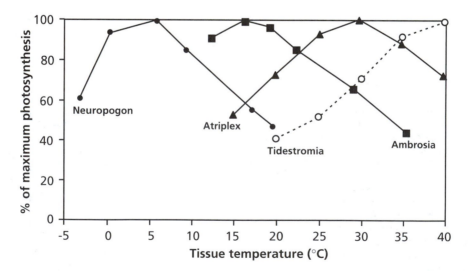

FIGURE 37. Photosynthetic response to temperature in plants from contrasting temperature regimes. Curves from left to right are for *Neuropogon acromelanus*, an antarctic lichen, *Ambrosia chamissonis*, a cool coastal dune plant, *Atriplex hymenelytra*, an evergreen desert shrub, and *Tidestromia obliongifolia*, a summer-active desert perennial (Mooney 1986). Copyright Blackwell Science Ltd.

(photoinhibition) may be adaptive because they avoid further damage. In the absence of such mechanisms, electrons are transferred in an aspecific manner to various compounds. This may lead to the bleaching of chlorophyll (photooxidation). Hardening (i.e., acclimation to low temperatures) of *Thuja plicata* (western red cedar) seedlings is associated with some loss of chlorophyll and with increased levels of carotenoids, giving the leaves a red-brown color. Exposure to low temperatures causes a decline in photosynthetic capacity and the quantum yield of photosynthesis, as evidenced from the decline in chlorophyll fluorescence (i.e., in the ratio F_v/F_m) (cf. Box 4). The carotenoids are likely to prevent damage that might otherwise occur as a result of photooxidation (cf. Sect. 3.3.1 of this chapter). Upon transfer of the seedlings to a normal temperature (dehardening) the carotenoids disappear within a few days (Weger et al. 1993).

One of the key enzymes of the C_4 pathway in *Zea mays* (maize), pyruvate P_i-dikinase, readily loses its activity at low temperature; hence, the leaves' photosynthetic capacity declines. This accounts for part of the chilling sensitivity of most C_4 plants. Loss of activity of pyruvate P_i-dikinase at low temperatures can be prevented by protective ("compatible") compounds (cf. Sect. 3.4.5 in the chapter on plant water relations), but it remains to be investigated if this plays a major role in intact C_4 plants (Krall et al. 1989).

Maximum rates of photosynthesis by arctic and alpine plants measured in the field are similar to those of temperate-zone species, but they are reached at lower temperatures—often 10 to 15°C (Fig. 37). These substantial photosynthetic rates at low temperatures are achieved in part by high concentrations of Rubisco, which require a high tissue nitrogen concentration. This may account for the high tissue nitrogen concentration of arctic and alpine plants despite low nitrogen availability in soils (Chapin & Shaver 1985, Körner & Larcher 1988). Although temperature optima of arctic and alpine plants are 10 to 30°C lower than those of temperate plants, they are still 5 to 10°C higher than average summer leaf temperatures in the field.

8. Effects of Air Pollutants on Photosynthesis

Many pollutants reduce plant growth, partly through their negative effects on photosynthesis. Those pollutants like SO_2 and ozone, which enter the leaf through stomates, directly damage the photosynthetic cells of the leaf. In general, any factor that increases stomatal conductance (e.g., high supply of water, high light intensity, high supply of nitrogen) increases the movement of pollutants into the plant and, therefore, their impact on photosynthesis. The negative effect of SO_2 on the net rate of CO_2 assimilation is only partly accounted for by an effect on the photosynthetic capacity; the rate of dark respiration is also enhanced. Although stomatal conductance declines upon exposure to SO_2, this is not the cause of the effect of SO_2 on photosynthesis because p_i/p_a is not affected (Table 6). The increased rate of dark respiration may be associated with repair of damage due to exposure to SO_2 (Black 1982, Black & Unsworth 1997, Winner & Mooney 1980a,b).

The major effect of SO_2 on growth and yield of *Vicia faba* (broad bean) is due to leaf injury (necrosis and abscision of leaves), rather than direct effects on gas exchange characteristics (photosynthesis and respiration) (Kropf 1989).

TABLE 6. Effects of exposure to SO_2 ($400\,\mu g\,m^{-3}$) for 2 hours on the light-saturated rate of CO_2 assimilation (A_{max}), carboxylation efficiency, quantum yield (ϕ, at saturating p_i), p_i/p_a, rate of dark respiration (R_d), and the CO_2-compensation point (Γ) of *Vicia faba* (broad bean).

	Before exposure	After exposure
A_{max}, $\mu mol\,m^{-2}\,s^{-1}$	16.5	14.0
ϕ, mol (CO_2) mol^{-1} (photons)	0.07	0.07
Carboxylation efficiency ($mm\,s^{-1}$)	2.1	1.5
p_i/p_a	0.7	0.7
R_d, $\mu mol\,m^{-2}\,s^{-1}$	0.9	1.1
Γ (Pa)	4.1	5.3

Source: Kropf 1989.

9. C_4 Plants

9.1 Introduction

The first sections of this chapter dealt primarily with the characteristics of photosynthesis of C_3 species. There are also species with photosynthetic characteristics quite different from these C_3 plants. These so-called C_4 species, which were first discovered by Hatch and Slack (1966), belong to widely different taxonomic groups; the C_4 syndrome is very rare among tree species (Table 7). Although their different anatomy has been well-documented for more than a century, the biochemistry and physiology of C_4 species has been elucidated only in the last few decades.

None of the reactions or anatomical features of C_4 plants are really unique to these species; however, they are all linked in a manner quite different from that in C_3 species. Based on differences in biochemistry, physiology, and anatomy, three subtypes of C_4 species are discerned (Table 8). In addition, there are intermediate forms between C_3 and C_4 metabolism (Sect. 9.6 of this chapter).

9.2 Biochemical and Anatomical Aspects

The anatomy of C_4 plants differs strikingly from that of C_3 plants (Fig. 38). C_4 plants are characterized by their **Kranz anatomy**, which is a sheath of thick-walled cells that surround the vascular bundle. (*Kranz* is the German word for *wreath*.) These thick walls of the bundle sheath cells may be impregnated with suberin, but this does not appear to be essential to reduce the gas diffusion between the bundle sheath and the mesophyll. In some C_4 species (NADP-ME* types), the cells of the bundle sheath contain large chloroplasts with mainly stroma thylakoids and very little stacking, which suggests a poorly developed PSII. The bundle sheath cells are connected via **plasmodesmata** with the adjacent thin-walled mesophyll cells, with large intercellular spaces.

TABLE 7. Taxonomic survey of families with C_4 photosynthesis and some examples of genera known to contain both C_3 and C_4.

Family	Genera containing both C_3 and C_4 species
Dicotyledonae	
Acanthaceae	
Aizoaceae	
Amaranthaceae	*Alternanthera*
Asteraceae	*Flaveria*
Boraginaceae	*Heliotropium*
Capparidaceae	
Caryophyllaceae	
Chenopodiacea	*Atriplex*
	Bassia
	Suaeda
Euphorbiaceae	*Euphorbia*
Molluginaceae	*Mollugo*
Nyctaginaceae	*Boerhavia*
Portulacaceae	
Scrophulariaceae	
Zygophyllaceae	*Kallstroemia*
	Zygophyllum
Monocotyledonae	
Cyperaceae	*Cyperus*
	Scirpus
Poaceae	*Alloteropsis*
	Panicum

Source: **Osmond et al. 1982.**

CO_2 is first assimilated in the mesophyll cells, which is catalyzed by **PEP-carboxylase**, a light-activated enzyme that is located in the cytosol. PEP-carboxylase uses phosphoenolpyruvate (PEP) and HCO_3^- as substrates. HCO_3^- is formed by hydratation of CO_2, which is catalyzed by **carbonic anhydrase.** The high affinity of PEP-carboxylase for HCO_3^- reduces p_i to about 10 Pa, which is less than half the p_i of C_3 plants. PEP is produced in the light from pyruvate and ATP and catalyzed by pyruvate-P_i-dikinase, which is a light-activated enzyme that is located in the chloroplast. The product of the reaction that is catalyzed by PEP-

TABLE 8. **Main differences between the three subtypes of C_4 species.**

Major decarboxylase in BSC	Decarboxylation occurs in	Major substrate moving from		Photosystems
		MC to BSC	BSC to MC	
NADP-malic enzyme	chloroplast	malate	pyruvate	I
NAD-malic enzyme	cytosol	aspartate	alanine	I and II
PEP carboxy-kinase	mitochondria	aspartate	PEP	I and II

*MC, mesophyll cells; BSC, vascular bundle sheath cells.

carboxylase is oxaloacetate, which is reduced to malate. On the other hand, oxaloacetate may be transaminated in a reaction with alanine to form aspartate. It depends on the subtype of the C$_4$ species (Table 8) whether malate or aspartate, or a mixture of the two, is formed. Malate (or aspartate) diffuses via plasmodesmata to the vascular bundle sheath cells, where it is decarboxylated, producing CO$_2$ and pyruvate (or alanine). CO$_2$ is then fixed by Rubisco in the chloroplasts of the **bundle sheath cells**, which have a normal Calvin cycle, as in C$_3$ plants. Rubisco is not present in the mesophyll cells, which do not have a complete Calvin cycle and do not accumulate starch.

Fixation of CO$_2$ by PEP-carboxylase and the subsequent decarboxylation occur relatively quickly, allowing the build-up of a high partial pressure of CO$_2$ in the vascular bundle sheath. When the outside partial pressure of CO$_2$ is 35 Pa, that at the site of Rubisco in the chloroplasts of the vascular bundle is 100 to 200 Pa. The p$_i$, which is the partial pressure in the intercellular spaces in the mesophyll, is only about 10 Pa. With such a steep gradient in the partial pressure of CO$_2$ it is inevitable that some CO$_2$ diffuses back from the bundle sheath to the mesophyll, but this is only about 20%. In other words, C$_4$ plants have a mechanism to enhance the partial pressure of CO$_2$ at the site of Rubisco to an extent that the **oxygenation reaction of Rubisco** is virtually fully **inhibited**. As a result, C$_4$ plants have negligible rates of photorespiration.

Based on the enzyme involved in the decarboxylation of the C$_4$ compounds transported to the vascular bundle sheath, three groups of C$_4$ species are discerned: NADP-malic enzyme-, NAD-malic enzyme-, and PEP-carboxykinase-types (Table 8, Fig. 38). In NAD-ME-subtypes, which decarboxylate malate (produced from imported aspartate) in the bundle sheath mitochondria, the mitochondrial frequency is severalfold higher than that in NADP-ME-subtypes. The specific activity of the mitochondrial enzymes involved in C$_4$ photosynthesis is also greatly enhanced (Hatch & Carnal 1992). The NAD-ME group of C$_4$ species tends to occupy the driest habitats, although the reason for this is unclear (Ehleringer & Monson 1993).

Decarboxylation of malate occurs only during assimilation of CO$_2$, and vice versa. The explanation for this is that the NADPH needed to decarboxylate malate is produced in the Calvin cycle, during the assimilation of CO$_2$. At least in the more "sophisticated" NADP-ME C$_4$ plants, such as *Zea mays* (maize) and *Saccharum officinale* (sugar cane), the NADPH required for the photosynthetic reduction of CO$_2$ originates from the activity of NADP-malic

enzyme. Because two molecules of NADPH are required per molecule of CO$_2$ fixed by Rubisco, this amount of NADPH is not sufficient for the assimilation of all CO$_2$. Additional NADPH is required to an even larger extent if aspartate, or a combination of malate and aspartate, diffuses to the bundle sheath. It is assumed that this additional NADPH can be imported via a "shuttle," which involves PGA and dihydroxyacetone phosphate (DHAP). Part of the PGA that originates in the bundle-sheath chloroplasts returns to the mesophyll. Here it is reduced, producing DHAP, which diffuses to the bundle sheath. On the other hand, NADPH that is required in the bundle sheath cells might originate from the removal of electrons from water. This reaction requires the activity of PSII, next to PSI. As already suggested by the lack of grana thylakoids in their chloroplast, PSII is only poorly developed in the bundle sheath cells, at least in the "more sophisticated" C$_4$ species. The poor development of PSII activity in the bundle sheath indicates that very little oxygen is evolved in the cells that contain Rubisco, which greatly favors the carboxylation reaction over the oxygenation.

The greatly suppressed oxygenation activity of Rubisco in C$_4$ species tends to increase their **quantum yield**. The formation of PEP from pyruvate, however, which is catalyzed by pyruvate-P$_i$-dikinase, requires the equivalent of two molecules of ATP per molecule of PEP. This is the cause for the lower quantum yield of photosynthesis of C$_4$ plants, when compared with C$_3$ plants, when measured under conditions where photorespiration is suppressed (2% O$_2$) (Fig. 39). In summary, C$_4$ photosynthesis concentrates CO$_2$ at the site of carboxylation by Rubisco in the bundle sheath. This is accomplished at the cost of two additional molecules of ATP per CO$_2$ transported to the bundle sheath.

9.3 Physiology of C$_4$ Photosynthesis

These differences in anatomy and biochemistry result in strikingly different A–p_i curves between C$_3$ and C$_4$ plants. First, the **CO$_2$-compensation point** of C$_4$ plants is only 0 to 0.5 Pa CO$_2$, as compared with 4 to 5 in C$_3$ plants. Second, this compensation point is not affected by the oxygen concentration, as opposed to that of C$_3$ plants, which is considerably less at a low oxygen concentration (i.e., when photorespiration is suppressed). Third, the p$_i$ (the **internal partial pressure of CO$_2$** in the mesophyll) at an ambient partial pressure of 35 Pa is only about 10 Pa, as compared with approximately 25 Pa in C$_3$ plants.

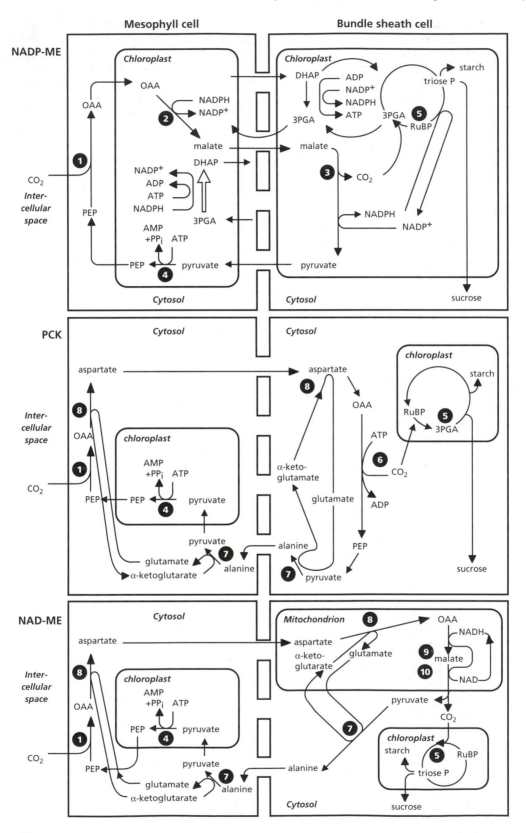

FIGURE 38.

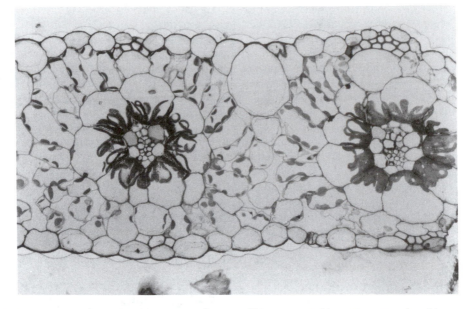

FIGURE 38. (Facing page) Schematic representation photosynthetic metabolism in the three C$_4$ types distinguished according to the decarboxylating enzyme. NADP-ME, NADP-requiring malic enzyme; PCK, PEP carboxykinase; NAD-ME, NAD-requring malic enzyme. Numbers refer to enzymes: (1) PEP carboxylase, (2) NADP-malate dehydrogenase, (3) NADP-malic enzyme, (4) pyruvate-P$_i$- dikinase, (5) Rubisco, (6) PEP carboxykinase, (7) alanine aminotransferase, (8) aspartate amino transferase, (9) NAD-malate dehydrogenase, (10) NAD-malic enzyme (after Lawlor 1993). (Above) Cross section of a leaf of the monocotyledonous C$_4$ species *Eleusine coracana* (courtesy J.R. Evans, Research School of Biologial Sciences, Australian National University, Canberra, Australia).

There are also major differences in the characteristics of the light-response curves of C$_3$ and C$_4$ species. When measured at 30°C or higher, the slope of the light-response curve is considerably steeper for C$_4$ plants. It is also independent of the oxygen concentration, in contrast to that of C$_3$ plants. At relatively high temperatures, therefore, the **quantum yield** of photosynthesis is higher for C$_4$ plants and is not affected by temperature. By contrast, the quantum yield of C$_3$ plants declines with increasing temperature, due to the proportionally increasing oxygenating activity of Rubisco (cf. Fig. 39 in Sect.

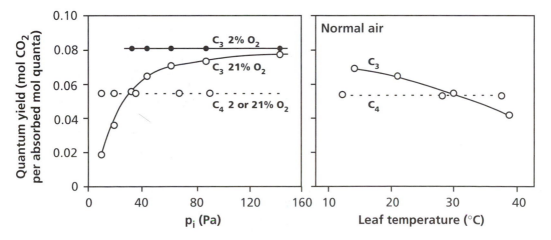

FIGURE 39. The effect of temperature and the intercellular partial pressure of CO$_2$ on the quantum yield of the photosynthetic CO$_2$ assimilation in a C$_3$ and a C$_4$ plant (Björkman 1981).

9.2 of this chapter). At a normal oxygen concentration and a p_a of 35 Pa, the quantum yield is higher for C_4 plants at high temperatures due to photorespiration in C_3 species, but lower at low temperatures due to the additional ATP required to regenerate PEP in C_4 species. When measured at a low oxygen concentration (to eliminate photorespiration) and a p_a of 35 Pa, the quantum yield is invariably higher for C_3 plants. The rate of CO_2 assimilation of C_4 plants typically saturates at a higher irradiance than does that of C_3 plants because of the higher CO_2 concentration at the site of Rubisco in C_4 plants. In C_3 plants, the light-response curve levels off because CO_2 becomes the limiting factor for the net CO_2 assimilation. Increasing the p_i shifts the irradiance at which light saturation is reached to higher levels in C_3 plants.

9.4 Intercellular and Intracellular Transport of Metabolites of the C_4 Pathway

Transport of the metabolites that move between the two cell types occurs by **diffusion** through **plasmodesmata**. The concentration gradient between the mesophyll and bundle sheath cells is sufficiently high to allow diffusion at a rate that readily sustains photosynthesis, with the exception of that of pyruvate. How can we account for rapid transport of pyruvate from the bundle sheath to the mesophyll if there is no concentration gradient?

Uptake of pyruvate in the chloroplast of the mesophyll cells is a light-dependent process, which requires a specific energy-dependent carrier. Active uptake of pyruvate into the chloroplast reduces the pyruvate concentration in the cytosol to a low level. Hence, the pyruvate concentration in the cytosol of the mesophyll cells, to which diffusion from the bundle sheath takes place, is considerably less than suggested by the data on the average concentration in the mesophyll cell. This explains why pyruvate can readily diffuse to the mesophyll cells (Flügge et al. 1985).

In the chloroplasts of the mesophyll cells, pyruvate is converted into PEP, which is exported to the cytosol in exchange for P_i. The same translocator is probably also used to export triose-phosphate in exchange for PGA. This translocator operates in the reverse direction in mesophyll and bundle sheath chloroplasts in that PGA is imported and triose-phosphate is exported in the mesophyll chloroplasts, whereas the chloroplasts in the bundle sheath export PGA and import triose-phosphate.

The chloroplast envelope of the mesophyll cells also contains a translocator for the transport of dicarboxylates (i.e., malate, oxaloacetate, aspartate, and glutamate). Transport of these acids occurs by exchange. The uptake of oxaloacetate, in exchange for other dicarboxylates, is competitively inhibited by these other dicarboxylates, with the values for K_i being in the same range as those for K_m. [K_i is the inhibitor (dicarboxylate) concentration at which the inhibition of the transport process is half that of the maximum inhibition by that inhibitor; K_m is the substrate (oxaloacetate) concentration at which the transport process occurs at half the maximum rate.] Such a system does not allow rapid import of oxaloacetate. A special transport system, which carries oxaloacetate without exchange against other dicarboxylates, takes care of rapid import of oxaloacetate into the mesophyll chloroplasts.

9.5 Photosynthetic Nitrogen-Use Efficiency, Water-Use Efficiency, and Tolerance of High Temperatures

The high partial pressure of CO_2 in the vascular bundle sheath of C_4-plants, which is the site of Rubisco, allows different kinetic properties of Rubisco. Table 9 shows that the $K_m(CO_2)$ of Rubisco from terrestrial C_3 plants is indeed lower than that from C_4 plants.

A high K_m (i.e., a low affinity) for CO_2 of Rubisco is not a disadvantage for the photosynthesis of C_4 plants if the high partial pressure of CO_2 in the bundle sheath is considered. For C_3 plants a low K_m for CO_2 is vital because the p_i is far from saturating for Rubisco in their mesophyll cells.

The advantage of the high K_m of the C_4 Rubisco may be indirect in that it allows a high **maximum rate** per unit protein of the enzyme. That is, the tighter CO_2 is bound to Rubisco, the longer it takes for the carboxylation to be completed. In C_3 plants, a high affinity is essential; therefore the maximum activity cannot be high. C_4 plants, which do not require a high affinity, do indeed have an enzyme with a high maximum catalytic activity, which allows more moles of CO_2 to be fixed per unit Rubisco and time (Table 10). It is interesting that the alga *Chlamydomonas reinhardtii*, which has a CO_2-concentrating mechanism (cf. Sect. 11.3 of this chapter), also has a Rubisco enzyme with a high K_m (low affinity) for CO_2 and a high V_{max} and k_{cat} (Table 10).

The biochemical and physiological differences between C_4 and C_3 plants have important ecological implications. The abundance of C_4 monocots in re-

TABLE 9. The K$_m$ for CO$_2$ of the enzyme Rubisco from a large number of terrestrial and aquatic C$_3$ and C$_4$ species.

Order	Number of species tested	Photosynthetic pathway	K$_m$(CO$_2$)	
			mM	Pa
Terrestrial plants				
Bryophyta	1	C$_3$	23	69
Pteridophyta	5	C$_3$	19	55
Gymnospermae	3	C$_3$	20	60
Monocotyledonae	3	C$_3$	17	51
	1	C$_4$	34	99
Dicotyledonae				
• "Crassinucelli"	16	C$_3$	18	54
	3	C$_4$	30	90
• "Tenuinucelli"	8	C$_3$	19	57
Aquatic plants				
Chlorophyta	4	C$_3$	60	180
Bryophyta	1	C$_3$	40	120
Monocotyledonae	5	C$_3$	41	123
Dicotyledonae	4	C$_3$	40	116

Source: Yeoh et al. 1981.
*To convert values from concentration to partial pressure, a solubility for CO$_2$ of 334 mmol Pa^{-1} was used and an atmospheric pressure of 100 Pa (Von Caemmerer et al. 1994).

gional floras correlates most strongly with growing season temperature, whereas C$_4$ dicot abundance correlates more strongly with aridity (Ehleringer & Monson 1993). At regional and local scales, areas with **warm-season rainfall** have greater C$_4$ abundance than do regions with cool season precipitation. Along local gradients, C$_4$ species occupy microsites that are the warmest or have the driest soils. In communities with both C$_3$ and C$_4$ species, C$_3$ species are most active early in the growing season when conditions are cool and moist, whereas C$_4$ activity increases as conditions become warmer and drier. When taken together these

patterns suggest that high photosynthetic rates at high temperature (due to **lack of photorespiration**) and high WUE (due to the **low p$_i$,** which enables C$_4$ plants to have a **lower stomatal conductance** for the same CO$_2$ assimilation rate) are the major factors governing the ecological distribution of the C$_4$ photosynthetic pathway. Any competitive advantage of the high WUE of C$_4$ plants, however, has been difficult to document experimentally (Ehleringer & Monson 1993). This may well be due to the fact that, in a competitive situation, any water that is left in the soil by a plant with a high WUE is available for a competitor with lower WUE.

C$_4$ plants generally have lower tissue nitrogen concentrations because they have three to six times less Rubisco than do C$_3$ plants as well as very low levels of the photorespiratory enzymes, although some of the advantage is lost by the investment of nitrogen in the enzymes of the C$_4$ pathway. C$_4$ plants also have equivalent or higher photosynthetic rates than C$_3$ do plants, which results in a higher rate of photosynthesis per unit of leaf nitrogen (**PNUE**), especially at high temperatures (Fig. 40). The higher PNUE of C$_4$ plants is accounted for by: (1) suppression of the oxygenase activity of Rubisco, so that the enzyme is only used for the carboxylation reaction; (2) the lack of photorespiratory enzymes; (3) the higher catalytic activity of Rubisco (Table 10). Just as in a comparison of C$_3$ species, which differ in PNUE (Sect. 4.2.1

TABLE 10. Kinetic parameters of the enzyme Rubisco, extracted from five C$_3$ species, eleven C$_4$ species, and one alga.*

Photosynthetic pathway	Specific activity	k$_{cat}$	Number of species tested
C$_3$	3.1	29	5
C$_4$-NADP-ME	6.3	58	4
C$_4$-NAD-ME	5.8	53	3
C$_4$-PCK	5.5	51	4
C$_3$-alga	6.7	61	1

Source: Seemann et al. 1984.
*The specific activity is expressed as μmol CO$_2$ fixed per minute and mg protein; k$_{cat}$ is expressed as mol CO$_2$ per mol Rubisco enzyme per second.

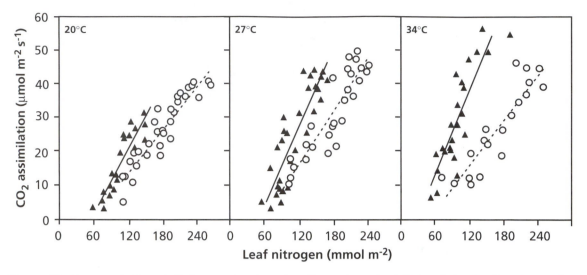

FIGURE 40. The rate of CO_2 assimilation as a function of the organic nitrogen concentration in the leaf and the temperature, as measured for the C_3 plant *Chenopodium album* (pigweed, circles) and the C_4 plant *Amaranthus retroflexus* (triangles) (Sage & Pearcy 1987b). Copyright American Society of Plant Physiology.

in mineral nutrition), there is no consistent tendency of C_4 species to have increased abundance or a competitive advantage in low-nitrogen soils (Christie & Detling 1982, Sage & Pearcy 1987a). This suggests that the high PNUE of C_4 species may be less important than their high WUE and high optimum temperature of photosynthesis in explaining patterns of distribution.

9.6 C_3–C_4 Intermediates

More than 20 plant species exhibit photosynthetic traits that are intermediate between C_3 and C_4 plants (e.g., species in the genera *Alternanthera, Flaveria, Neurachne, Moricandia, Panicum,* and *Parthenium*). These show **reduced rates of photorespiration** and **CO_2-compensation points** in the range of 0.8 to 3.5 Pa, compared with to 4 to 5 in C_3 and 0 to 0.5 in C_4 plants (Table 11). They have a weakly developed Kranz anatomy, compared with the true C_4 species, but Rubisco is located both in the mesophyll and the bundle sheath cells (Brown & Bouton 1993).

Two main types of intermediates are distinguished. In the first type (e.g., *Alternanthera ficoides, A. enella, Moricandia arvensis, Panicum milioides*) the activity of key enzymes of the C_4 pathway (i.e., pyruvate-P_i-dikinase, PEP carboxylase, NAD-malic enzyme, NADP-malic enzyme, and PEP carboxykinase) is very low, and they do not have a functional C_4-acid cycle. Their low CO_2-compensation point is due to the light-dependent recapture by mesophyll cells of CO_2 released in photorespiration in the bundle sheath cells. The bundle sheath cells contain a large fraction of the organelles involved in photorespiration, compared with that in C_3 species (Table 11). The CO_2-compensation point is negatively correlated with the percentage of photorespiratory organelles that occur in bundle sheath cells (Brown & Hattersley 1989). These C_3–C_4 intermediates have evolved a system to scavenge CO_2 that escapes from the bundle sheath cells but do not have the CO_2-concentrating mechanism of true C_4 species (Ehleringer & Monson 1993). In the leaves of these intermediate species, **glycine decarboxylase**, which is a key enzyme in photorespiration that releases the photorespiratory CO_2, occurs exclusively in the cells that surround the vascular bundle sheath. The mitochondria of the mesophyll cells, like those of true C_4 plants, have a low activity of glycine decarboxylase. Immunogold labeling shows a four times greater amount of one of the subunits of glycine decarboxylase in the bundle sheath cells than in the mesophyll cells (Morgan et al. 1992). It is likely that products of the oxygenation reaction, including glycine, move to the bundle sheath cells. The products are presumably metabolized in the bundle sheath; therefore, serine can move back to the mesophyll (Fig. 41). Due to the exclusive location of glycine decarboxylase in the bundle sheath cells, the release of CO_2 in photorespiration occurs close to the vascular tissue, with chloroplasts occur-

TABLE 11. The number of chloroplasts and of mitochondria plus peroxisomes in bundle sheath cells compared to mesophyll cells (BSC/MC) and the CO$_2$-compensation point (Γ, Pa), of C$_3$, C$_4$ and C$_3$–C$_4$ intermediates belonging to the genera *Panicum, Neurachne, Flaveria,* and *Moricandia*.

Species	Photosynthetic pathway	BSC/MC chloroplasts	BSC/MC mitochondria + peroxysomes	Γ
P. milioides	C$_3$–C$_4$	0.9	2.4	1.9
P. miliaceum	C$_4$	1.1	8.4	0.1
N. minor	C$_3$–C$_4$	3.1	20.0	0.4
N. munroi	C$_4$	0.8	4.9	0.1
N. tenuifolia	C$_3$	0.6	1.2	4.3
F. anomala	C$_3$–C$_4$	0.9	2.3	0.9
F. floridana	C$_3$–C$_4$	1.4	5.0	1.3
F. linearis	C$_3$–C$_4$	2.0	3.6	1.2
F. oppositifolia	C$_3$–C$_4$	1.4	3.6	1.4
F. brownii	C$_4$-like	4.2	7.9	0.2
F. trinerva	C$_4$	2.2	2.4	0
F. pringlei	C$_3$	0.5	1.0	4.3
M. arvensis	C$_3$–C$_4$	1.4	5.2	3.2
M. spinosa	C$_3$–C$_4$	1.6	6.0	2.5
M. foleyi	C$_3$	1.5	3.3	5.1
M. moricandioides	C$_3$	2.0	2.8	5.2

Source: Brown & Hattersley 1989.

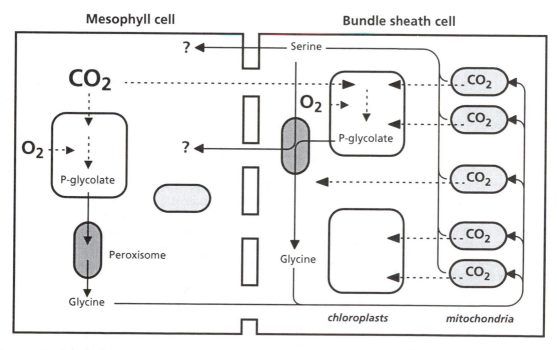

FIGURE 41. A model of the photorespiratory metabolism in leaves of the C$_3$–C$_4$ intermediate *Moricandia arvensis,* showing the recapture of CO$_2$ released by glycine decarboxylase. The model accounts for the low CO$_2$-compensation point and the low apparent rate of photorespiration in this type of intermediate (Morgan et al. 1992). Copyright SPB Academic Publishing.

ring between these mitochondria and the intercellular spaces. Glycine decarboxylase is only found in the enlarged mitochondria arranged along the cell walls adjacent to the vascular tissue and overlain by chloroplasts. This location of glycine decarboxylase increases the diffusion path for CO_2 between the site of release and the atmosphere, and allows the recapture of a large fraction of the photorespiratory CO_2, which is released by glycine decarboxylase, by Rubisco located in the bundle sheath. The location of glycine decarboxylase in the bundle sheath allows some build-up of CO_2, although not to the same extent as in the true C_4 plants. The advantage in terms of the net rate of CO_2 assimilation is rather small as compared with that in a true C_4 plant (Von Caemmerer 1989).

In the second type of intermediate species (e.g., *Flaveria anomala*, *Neurachne minor*), the activity of key enzymes of the C_4 pathway is considerable. Rapid fixation of $^{14}CO_2$ into C_4 acids, followed by transfer of the label to Calvin-cycle intermediates, has been demonstrated. These species have a limited capacity for C_4 photosynthesis, but lower quantum yields than either C_3 or C_4, presumably because the operation of the C_4 cycle in these plants does not really lead to a concentration of CO_2 to the extent it does in true C_4 species.

In addition to the C_3–C_4 intermediate species, there is at least one species (*Eleocharis vivipara*) that is capable of either C_3 or C_4 photosynthesis in different tissues, depending on environmental conditions (Ehleringer et al. 1991). C_4 species can shift to a CAM mode, as in the genus *Portulaca* (discussed in Section 10.4).

In the beginning of the 1970s, when the C_4 pathway was only partially understood, there were attempts to cross C_3 and C_4 species of *Atriplex*. One of the reasons to do this was that this was considered to be a useful approach to enhance the rate or efficiency of photosynthesis and yield of C_3 parents. The complexity of anatomy and biochemistry of the C_4 plants is such, however, that these crosses have not produced any useful progeny (Brown & Bouton 1993). Because molecular techniques have become available to allow silencing and overexpression of specific genes in specific cells, attempts have been made to reduce the activity of glycine decarboxylase, which is the key enzyme in photorespiration, in mesophyll cells of C_3 plants, and to overexpress the gene in the bundle sheath. Although these attempts have been successful from a molecular point of view, in that the aim of selectively modifying the enzyme activity was achieved, no results have yet been obtained to show enhanced rates of photosynthesis. This is perhaps not unexpected in view of the rather small advantage the true C_3–C_4 intermediates are likely to have in comparison with C_3 relatives.

9.7 Evolution and Distribution of C_4 Species

C_4 species represent approximately 5% of all plant species, C_3 species accounting for about 85%, and CAM species (Sect. 10) for 10%. The C_4 pathway evolved nearly synchronously on several continents in the late Miocene 5 to 7 million years ago, probably in response to a reduction in atmospheric CO_2 concentration. This change in atmospheric CO_2 concentration resulted from both the **photosynthetic activity** of plants and tectonic and subsequent **geochemical events**. To put it briefly, the collision of the Indian subcontinent caused the uplift of the Tibetan Plateau, causing new earth crust that consumed CO_2 to become exposed over a vast area. The reaction $CaSiO_3 + CO_2$ $\leftarrow - \rightarrow CaCO_3 + SiO_2$ was responsible for the dramatic decline in atmospheric CO_2 concentration (Ehleringer & Monson 1993, Raymo & Ruddiman 1992).

Low atmospheric CO_2 concentrations would increase photorespiration and favor the CO_2-concentrating mechanisms and lack of photorespiration that characterize C_4 species. Considering the three subtypes of C_4 species and their occurrence in at least 18 different families of widely different taxonomic groups, C_4 plants must have evolved from C_3 ancestors independently a number of times. Morphological and ecogeographical information (Powell 1978) combined with molecular evidence (Kopriva et al. 1996) suggests that C_4 photosynthesis has evolved twice in different lineages within the genus *Flaveria*. The physiology of C_3–C_4 intermediates suggests that the mechanism to recapture CO_2 evolved before the CO_2-concentrating mechanism (cf. Sect. 9.5; Bauwe et al. 1987). The phylogeny of *Flaveria* species, as deduced from an analysis of the nucleotide sequences that encode a subunit of glycine decarboxylase, suggests that C_4 species derived from C_3–C_4 intermediates (Kopriva et al. 1996).

Apart from the decreasing atmospheric CO_2 concentration, high temperatures and water shortage may have been driving forces for the evolution of C_4 species. This makes it likely that C_4 species evolved primarily in tropical areas, which continue to be their centers of distribution. Tropical and temperate grasslands, with abundant warm-season precipitation, are dominated by C_4 species.

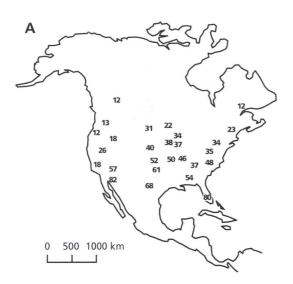

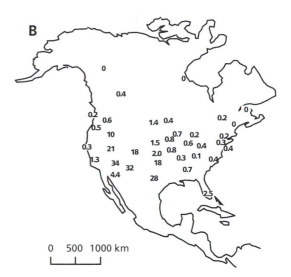

FIGURE 42. Geographic distribution of C₄ species in North America. (A) percentage of grass taxa that are C₄ plants. (B) percentage of dicotyledon taxa that are C₄ plants in regional floras of North America (Teeri & Stowe 1976, and Stowe & Teeri 1978, as cited in Osmond et al. 1982).

The high concentration of CO_2 at the site of Rubisco allows net CO_2 assimilation at relatively high temperatures, when photorespiration results in low net photosynthesis of C_3 species due to the increased oxygenating activity of Rubisco. This explains why C_4 species naturally occur in warm, open ecosystems, where C_3 species are less successful (Figs. 42 and 43). There is no a priori reason, however, why C_4 photosynthesis could not function in cooler climates. The lower quantum yield of C_4 species at low temperature may be important in dense canopies where light limits photosynthe-

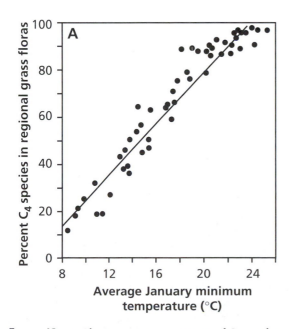

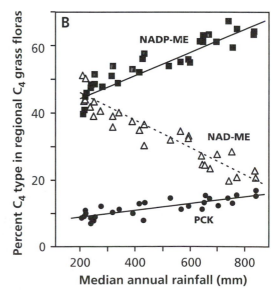

FIGURE 43. (A) The percentage occurrence of C₄ metabolism in grass floras of Australia in relation to temperature in the growing season (January). (B) The percentage occurrence of C₄ grass species of the three metabolic types in regional floras in Australia in relation to median annual rainfall (Henderson et al. 1995).

sis (and where quantum yield is therefore important). Quantum yield, however, is less important at higher levels of irradiance, and there is quite a wide temperature range where the quantum yield is still high compared with that of C_3 plants (Fig. 39). The high sensitivity to low temperature of pyruvate-P_i-dikinase, which is a key enzyme in the C_4 pathway, may be the main reason why C_4 species have rarely expanded to cooler places (cf. Sect. 7.2 of this chapter). Compatible solutes can decrease the low-temperature sensitivity of this enzyme; this could allow the expansion of C_4 species into more temperate regions in the future. On the other hand, rising atmospheric CO_2 concentration may offset the advantages of the CO_2-concentrating mechanism of C_4 photosynthesis (cf. Sect. 12 in this chapter).

9.8 Carbon Isotope Composition of C_4 Species

Although the Rubisco enzyme of C_4 plants discriminates between $^{12}CO_2$ and $^{13}CO_2$, just like that of C_3 plants, the fractionation of C_4 species is considerably less than that in C_3 plants. This is explained by the small extent to which inorganic carbon **diffuses back** from the vascular bundle to the mesophyll, as outlined in Section 9.4 of this chapter. In addition, the inorganic carbon that does diffuse back to the mesophyll cells will be **refixed** by PEP-carboxylase, which has a very high affinity for bicarbonate and discriminates very little between carbon isotopes (cf. Box 2). The discrimination against $^{13}CO_2$, therefore, is very small in C_4 species (Fig. 44). Consistent with this reasoning, the $\delta^{13}C$ **values** of C_4 plants (-8 to -16%; modal value -12.6%) are similar to values that would be expected if diffusion, rather than biochemistry, were the only factor limiting assimilation [i.e., -12.4% ($= -8 -4.4\%$)].

The isotopic differences between C_3 and C_4 plants are large compared with isotopic changes that occur during digestion by herbivores or decomposition by soil microbes. This makes it possible to determine the relative abundance of C_3 and C_4 species in the diets of animals by analyzing tissue samples of animals ("You are what you eat") or as sources of soil organic matter in paleosols (old soils). These studies have shown that many generalist herbivores show a preference for C_3 rather than C_4 plants (Ehleringer & Monson 1993). C_3 species, however, also tend to have more toxic secondary metabolites, which cause other herbivores to show exactly the opposite preference. Examination of paleosols documented the decrease

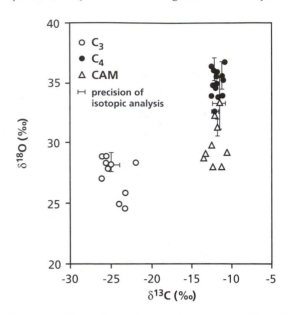

FIGURE 44. The carbon and oxygen isotope composition of C_3, C_4, and CAM species. Differences in oxygen isotope composition are due to differences in the source of water and fractionation during transpiration, which occurs more slowly for the heavier isotope (Sternberg et al. 1984). Copyright American Society of Plant Physiology.

in C_4 tropical grasslands at the Pleistocene-Holocene boundary, coincident with the rise in atmospheric CO_2. This pattern is consistent with the hypothesized importance of declines in atmospheric CO_2 as an important driving force for the evolution of C_4 photosynthesis.

10. CAM Plants

10.1 Introduction

In addition to C_3 and C_4 species, there are many succulent plants with another photosynthetic pathway: **Crassulacean Acid Metabolism** (CAM). This pathway is named after the Crassulaceae, which is a family in which many species show this type of metabolism. CAM, however, also commonly occurs in other families, such as the Cactaceae, Euphorbiaceae, Orchidaceae, and Bromeliaceae (e.g., *Ananas comosus*, pineapple). There are about 10,000 CAM species from 25 to 30 families (Table 12), all Angiospermae, with the exception of a few fern species, which also have CAM characteristics.

The unusual capacity of CAM plants to fix CO_2 into organic acids in the dark, which causes **nocturnal acidification**, with deacidification during the

TABLE 12. **Taxonomic survey of flowering plant families known to have species showing crassulacean acid metabolism.**

Agavaceae	Geraniaceae
Aizoaceae	Gesneriaceae
Asclepidiaceae	Labiatae
Asteraceae	Liliaceae
Bromeliaceae	Oxalidaceae
Cactaceae	Orchidaceae
Clusiaceae	Piperaceae
Crassulaceae	Polypodiaceae
Cucurbitaceae	Portulacaceae
Didieraceae	Rubiaceae
Euphorbiaceae	Vitaceae

Sources: Kluge & Ting 1978, Medina 1996.

day, has been known for almost two centuries. A full appreciation of CAM as a photosynthetic process was greatly stimulated by analogies with C_4 species.

The productivity of most CAM plants is fairly low; however, this is not an inherent trait of CAM species because *Agave mapisaga* and *A. salmiana*, for example, may achieve an average productivity of $40 Mg ha^{-1} year^{-1}$. An even higher productivity has been observed for both irrigated and fertilized and carefully pruned *Opuntia amyclea* and *O. ficus-indica* ($46 Mg ha^{-1} year^{-1}$; Nobel et al. 1992). These are among the highest productivities reported for any species. In a comparison of two succulent species with similar growth forms, *Cotyledon orbiculata* (CAM) and *Othonna optima* (C_3), during the transition from the rainy season to subsequent drought, the daily net rate of CO_2 assimilation is similar for the two species. This shows that rates of photosynthesis of CAM plants may be as high as those of C_3, plants, if morphologically similar plants adapted to the same habitat are compared (Eller & Ferrari 1997).

As with C_4 plants, none of the reactions of CAM is really unique to these species; however, they proceed at different times of the day from those in C_3 and C_4 species. Two subtypes of CAM species are discerned based on differences in the major decarboxylating enzyme. In addition, there are intermediate forms between C_3 only and CAM, as well as facultative CAM plants.

10.2 Physiological, Biochemical, and Anatomical Aspects

CAM plants are characterized by their **succulence** (but this is not pronounced in epiphytic CAM

plants; Sect. 10.5), the capacity to fix CO_2 at night via **PEP-carboxylase**, the accumulation of **malic acid** in the vacuole, and the subsequent deacidification during the day, when CO_2 is released from malic acid and fixed in the Calvin cycle, using Rubisco. (*Succulence* is defined as the volume of water in the leaf at a relative water content of 100% per unit leaf area.)

CAM plants show a strong fluctuation in pH of the cell sap because of the synthesis and breakdown of malic acid. The concentration of this acid may increase to 100 mM. By isolating vacuoles of the CAM plant *Kalanchoe daigremontiana*, it has been demonstrated that at least 90% of all the acid in the cells is in the vacuole. In addition, the kinetics of malic acid efflux from the leaves of *K. daigremontiana* provide evidence for the predominant location of malic acid in the vacuole.

At night, CO_2 is fixed in the cytosol which is catalyzed by PEP-carboxylase and produces oxaloacetate (Fig. 45). PEP originates from the breakdown of glucose in glycolysis; glucose is formed from starch. Oxaloacetate is immediately reduced to malate which is catalyzed by malate dehydrogenase. Malate is transported to the large vacuoles in an energy-dependent manner, which involves a proton pump. An H^+-ATPase and a pyrophosphatase pump protons into the vacuole so that malate can follow passively. It becomes malic acid in the vacuole.

The release of malic acid from the vacuole during the day is supposedly passive. Malate is then decarboxylated by malic enzyme (NAD- or NADP-dependent) or by PEP-carboxykinase. CAM species are subdivided like C_4 species, depending on the decarboxylating enzyme. The malic enzyme subtypes (ME-CAM) have a cytosolic NADP-malic enzyme, as well as a mitochondrial NAD-malic enzyme; they use a chloroplastic pyruvate-P_i-dikinase to convert the C_3 fragment that originates from the decarboxylation reaction into carbohydrate. Unlike C_4 species, ME-CAM species are not further subdivided into NADP- and NAD-malic enzyme subtypes. PEPCK-type CAM plants have very low malic enzyme activities (as opposed to PEPCK-C_4 plants) and no pyruvate-P_i-dikinase activity, but high activities of PEP-carboxykinase.

The C_3 fragment (pyruvate or PEP) that is formed during the decarboxylation is converted into starch, and the CO_2 that is released is fixed by Rubisco, much the same as it is in C_3 plants. The stomata are closed during the decarboxylation of malic acid and the fixation of CO_2 by Rubisco in the Calvin cycle. They are open during the nocturnal fixation of CO_2.

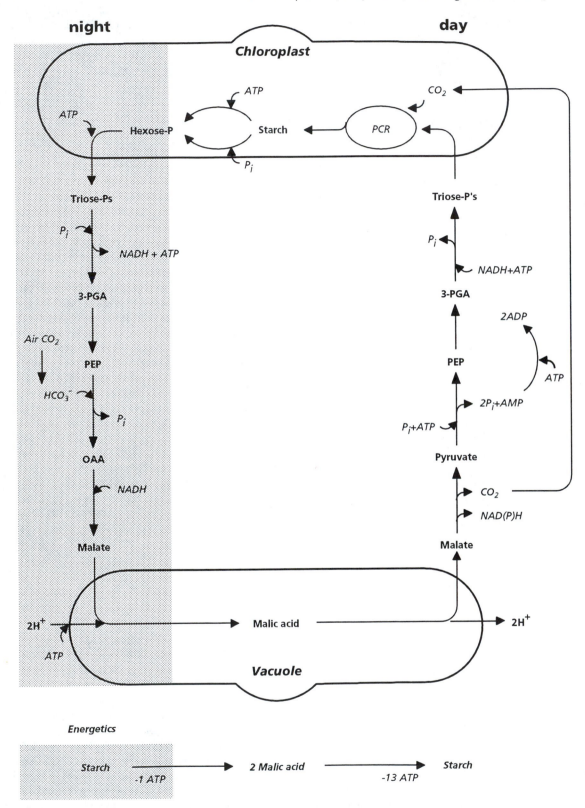

FIGURE 45. Metabolic pathway and cellular compartmentation of Crassulacean Acid Metabolism, showing the separation in night and day of carboxylation and decarboxylation (Black et al. 1996).

The CAM traits can be summarized as follows:

1. Fluctuation of organic acids, mainly of malic acid, during a diurnal cycle;
2. Fluctuation of the concentration of sugars and starch, opposite to the fluctuation of malic acid;
3. A high activity of PEP-carboxylase (at night) and of a decarboxylase (during the day);
4. Large vacuoles in cells that contain chloroplasts;
5. Some degree of succulence;
6. The gas exchange of the leaves occurs predominantly at night.

Four "phases" in the diurnal pattern of CAM are discerned (Fig. 46). **Phase I**, the carboxylation phase, starts at the beginning of the night. The rate of carboxylation declines toward the end of the night, and the malic acid concentration reaches its maximum. The stomatal conductance and the CO_2 fixation change more or less in parallel; carbohydrates are broken down. **Phase II**, at the beginning of the day, is characterized by a high rate of CO_2 fixation, which generally coincides with an increased stomatal conductance. CO_2 fixation by PEP carboxylase, along with malic acid formation, coincide with the fixation of CO_2 by Rubisco. Fixation by PEP carboxylase is gradually taken over by Rubisco. C_3 photosynthesis predominates in the last part of phase II, using exogenous CO_2 as the substrate. In **phase III** the stomata are fully closed and malic acid is decarboxylated. The p_i may then increase to values above 1000 Pa, which is when the normal C_3 photosynthesis takes place and when sugars and starch accumulate. When malic acid is depleted, the stomata open again, possibly because p_i drops to a low level; this is the beginning of **phase IV**. Gradually more exogenous and less endogenous CO_2 is fixed by Rubisco. In this last phase, CO_2 may be fixed by PEP-carboxylase again, as indicated by the photosynthetic quotient (PQ); that is, the ratio of O_2 release and of CO_2 uptake. The PQ is 1.0 over an entire day (Table 13), but deviations from this value occur, depending on the carboxylation process (Fig. 47).

In phase III, when the stomata are fully closed, malic acid is decarboxylated, and the p_i is very high, **photorespiration** is suppressed, as indicated by the relatively low rate of oxygen uptake (as measured using $^{18}O_2$; Fig. 47). In phase IV, when malic acid is depleted and the stomata open again, photorespiration does occur, as demonstrated by increased uptake of $^{18}O_2$.

How do CAM plants regulate the activity of the two carboxylating enzymes and decarboxylating enzymes in a coordinated way to avoid futile

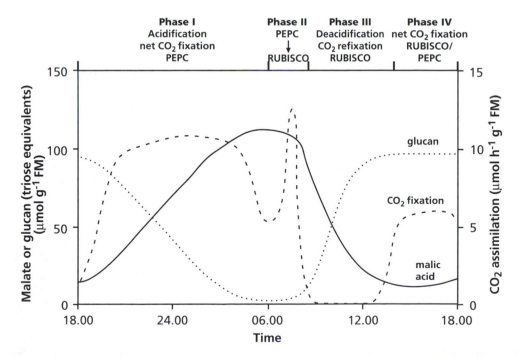

FIGURE 46. CO_2 fixation in CAM plants, showing diurnal patterns for net CO_2 assimilation, malic acid concentration, and carbohydrate concentrations; PEPC is PEP carboxylase (after Leegood & Osmond 1990, Osmond & Holtum 1981).

TABLE 13. Cumulative CO_2 and O_2 exchange and Photosynthetic Quotient (PQ) in the dark and in the light [in mmol plant^{-1} (12h)$^{-1}$] of one shoot of *Ananas comosus* (pineapple), measured over two consecutive days.*

| | Dark | | Light | | |
	CO$_2$ assimilation	O$_2$ consumption	CO$_2$ assimilation	O$_2$ release	PQ
Day 1	10.6	6.4	10.4	27.1	0.99
Day 2	11.1	6.3	10.7	27.5	0.98

Source: Coté et al. 1989.
*PQ is the ratio of the total net amount of O_2 evolved in the light to total amount of CO_2 fixed in the light plus dark period.

cycles? Rubisco is inactive at night for the same reason as in C_3 plants: This enzyme is light-activated and therefore cannot function in the dark (cf. Sect. 3.4.2 of this chapter). In addition, the kinetic properties of PEP-carboxylase are modulated. In *Mesembryanthemum crystallinum* and in *Crassula argentea*, PEP-carboxylase occurs in two configurations: a "day-configuration" and a "night-configuration." The night-configuration is relatively insensitive to malate (the K_i for malate is 0.06 to 0.9 mM, depending on pH) and has a high affinity for PEP (the K_m for PEP is 0.1 to 0.3 mM). The day-configuration is strongly inhibited by malate (the K_i for malate is 0.004 to 0.07 mM, again depending on the pH) and has a low affinity for PEP (the K_m for PEP has increased to 0.7 to 1.25 mM). When

malate is rapidly exported to the vacuole at night in phase I, therefore, the carboxylation of PEP readily takes place, whereas it is suppressed during the day in phase III.

The fluctuations in the kinetic properties of PEP-carboxylase, do not require protein synthesis. In *Crassula argentea* the change in configuration involves a change from a dimer, with subunits of 100 kDa, during the day, to a tetramer, with subunits of the same molecular mass, at night. When the enzyme is isolated from leaves collected during the day (dimer, sensitive to malate inhibition) a gradual transformation to the tetramer (insensitive to malate inhibition) takes place during storage during 49 days at −70°C. If malate is added to the enzyme, after 11 days at low temperature,

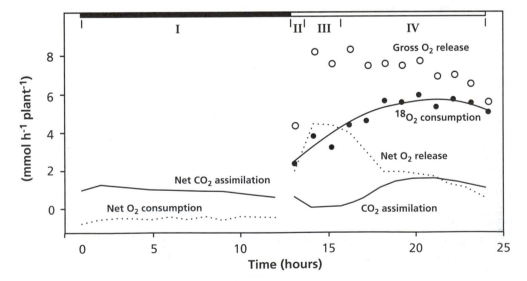

FIGURE 47. Gas exchange of *Ananas comosus* (pineapple) during the dark and light period. O_2 consumption during the day was measured using the stable isotope $^{18}O_2$. Gross O_2 production is the sum of the net O_2 production and $^{18}O_2$ consumption. The phases are explained in the legend to Figure 46 and the text of section 10.2 (Coté et al. 1989). Copyright American Society of Plant Physiology.

TABLE 14. Effects of malate and glucose-6-phosphate (G-6-P) on the kinetic parameters of PEP-carboxylase.

	V_{max} mmol mg^{-1}(Chl) min^{-1}	Ratio	K_m mM	Ratio
Control	0.42	1.0	0.13	1.0
+1 mM G-6-P	0.45	1.07	0.08	0.61
+2 mM G-6-P	0.47	1.12	0.05	0.39
+5 mM malate	0.31	0.74	0.21	1.60
+5 mM malate and 2 mM G-6-P	0.34	0.81	0.05	0.39

Source: Kluge & Ting 1978.
*The ratio gives the ratio of the values in the presence and absence of effectors.

then a large fraction of the enzyme is transformed into the dimer, which increases the sensitivity for malate inhibition. Addition of PEP in vitro has the opposite effect: formation of the tetramer.

The sensitivity for malate inhibition tends to be lost during storage of PEP-carboxylase from the CAM plant *Kalanchoe daigremontiana*. This effect is stronger in the presence of inorganic phosphate, PEP, 3-phosphoglycerate (PGA), and glucose-6-phosphate. The presence of malate slows down the rate of loss of sensitivity to malate inhibition. In this case, there is no change from dimer to tetramer; the enzyme remains a dimer with a molecular mass of 200 kDa. There is some evidence that the modification of the kinetic properties involves the phosphorylation and dephosphorylation of PEP-carboxylase.

Through the modification of its kinetic properties, the inhibition of PEP-carboxylase is likely to prevent a futile cycle of carboxylation and concomitant decarboxylation reactions. Further evidence that such a futile cycle does not occur comes from studies on the labeling with ^{13}C of the first or fourth carbon atom in malate. This pattern of labeling is constant during de-acidification. If a futile cycle were to occur, then doubly labeled malate should appear because fumarase in the mitochondria would randomize the label in the malate molecule. Such randomization does occur during the acidification phase, which indicates rapid exchange of the malate pools of the cytosol and the mitochondria, before malate enters the vacuole.

Next to malate, glucose 6-phosphate is also an effector of PEP-carboxylase: It enhances the V_{max} and decreases the K_m(PEP) (Table 14). The physiological significance of this effect is that glucose 6-phosphate, which is produced from glucose, during its conversion into PEP stimulates the carboxylation of PEP.

Temperature has exactly the opposite effect on the kinetic properties of PEP carboxylase

from a CAM plant and that from a C_4 plant (Fig. 48). These temperature effects help to explain why a low temperature at night enhances acidification.

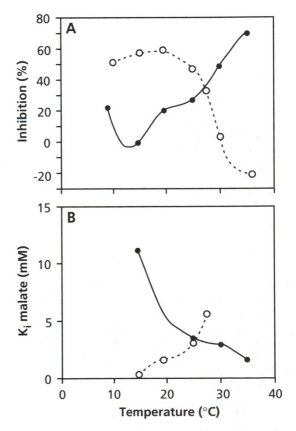

FIGURE 48. The effect of temperature on kinetic properties of PEP carboxylase from leaves of a *Crassula argenta* (a CAM plant, solid symbols) and *Zea mays* (a C_4 plant, open symbols). (A) Effect on percentage inhibition by 5 mM malate. (B) Effect on the inhibition constant (K_i) for malate (Wu & Wedding 1987). Copyright American Society of Plant Physiology.

10.3 Water-Use Efficiency

Because CAM plants keep their stomates closed during the day when the VPD between the leaves and the surrounding air is highest, and open at night when VPD is lowest, they have a very high water-use efficiency. As long as they are not severely stressed, which leads to complete closure of their stomata, the WUE of CAM plants tends to be considerably higher than that of both C_3 and C_4 plants (Kluge & Ting 1978) (Table 8 in plant water relations).

10.4 Incomplete and Facultative CAM Plants

Some CAM plants may not open their stomata during the night when they are exposed to severe water stress (Bastide et al. 1993). Even under these conditions, they may continue to show a diurnal fluctuation in malic acid concentration, as first found in *Opuntia basilaris*. The CO_2 they use to produce malic acid at night does not come from the air; rather, it is entirely derived from respiration. It is released again during the day, which allows some Rubisco activity. This metabolism is termed **CAM idling**. Fluorescence measurements have indicated that the photosystems remain intact during severe drought. This has led to the suggestion that CAM idling prevents photodamage, in addition to recapturing respiratory CO_2. In the absence of an electron acceptor, damage of PSII might occur at the high irradiance to which these plants are exposed. CAM idling can be considered as a modification of normal CAM. The plants remain "ready to move" as soon as the environmental conditions improve, but they keep their stomates closed during severe drought.

Some plants show a diurnal fluctuation in the concentration of malic acid without a net CO_2 uptake at night, but with normal rates of CO_2 assimilation during the day. These plants are capable of recapturing most of the CO_2 that is derived from respiration and to use this as a substrate for PEP-carboxylase. This is termed **CAM cycling** (Patel & Ting 1987). In *Peperomia camptotrichia*, 50% of the CO_2 that is released in respiration during the night is fixed by PEP-carboxylase. At the beginning of the day, some CO_2 is available for photosynthesis, even when the stomatal conductance is very small. In *Talinum calycinum*, which occurs naturally on dry rocks, CAM cycling may reduce water loss by 44%. In this way CAM cycling enhances a plant's water-use efficiency (Harris & Martin 1991).

CAM idling typically occurs in ordinary CAM plants that are exposed to severe water stress and have a very low stomatal conductance throughout the day and night. CAM cycling, conversely, occurs in plants that have a high stomatal conductance and normal C_3 photosynthesis during the day, but which refix the CO_2 produced in dark respiration at night, which ordinary C_3 plants lose to the atmosphere.

In a limited number of species, CAM only occurs upon exposure to drought stress: **facultative CAM plants**. For example, in plants of *Agave deserti*, *Clusia uvitana*, *Mesembryanthemum crystallinum*, and *Portulacaria afra*, irrigation with saline water or drought can change from a virtually normal C_3 photosynthesis to the CAM mode (Fig. 49; Winter et al. 1992). There is at least one genus that contains C_4 species that can shift from a normal C_4 mode under irrigated conditions to a CAM mode under water stress: *Portulaca grandiflora*, *P. mundula*, and *P. oleracea* (Koch & Kennedy 1982, Mazen 1996). ABA can also induce CAM photosynthesis at a comparable rate (Chu et al. 1990). The transition from the C_3 or C_4 to the CAM mode coincides with an enhanced PEP-carboxylase activity and of the mRNA coding for this enzyme. Upon removal of NaCl from the root environment of *Mesembryanthemum crystallinum*, the level of mRNA that encodes PEP-carboxylase declines by 77% in 2 to 3 hours. The amount of the PEP-carboxylase enzyme itself declines more slowly: The activity declines after 2 to 3 days to half its original level (Vernon et al. 1988).

10.5 Distribution and Evolution of CAM Species

CAM is undoubtedly an adaptation to drought because CAM plants close their stomates during most of the day. This is illustrated in a survey of epiphytic bromeliads in Trinidad (Fig. 50). There are two major ecological groupings of CAM plants: **succulents** from arid and semiarid regions and **epiphytes** from tropical and subtropical regions (Ehleringer & Monson 1993). Although CAM plants are uncommon in cold environments, this may reflect their evolutionary origin in warm climates rather than a temperature sensitivity of the CAM pathway (Nobel & Hartsock 1990). Roots of some Orchidaceae, which lack stomata, also show CAM. CAM species are well adapted to dry environments and naturally occur in semiarid and tropical environments. Some submerged aquatics also show CAM photosynthesis (see Sect. 11.5 of this chapter).

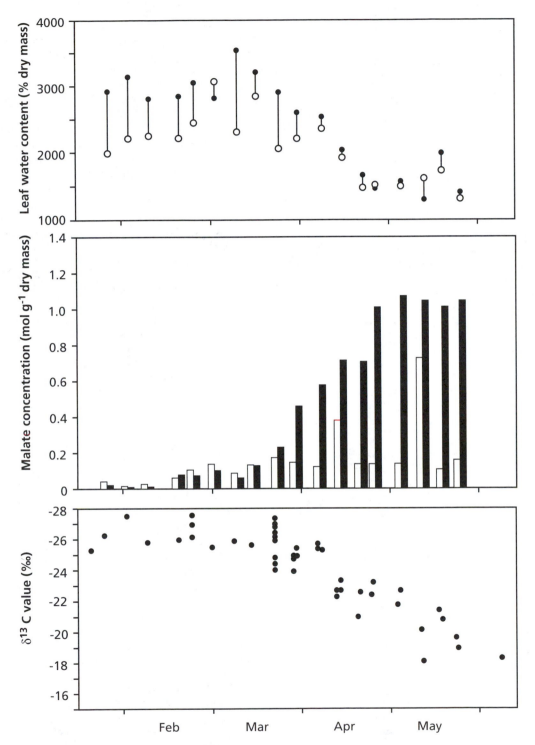

FIGURE 49. Induction of CAM in the facultative CAM species *Mesembryanthemum crystallinum*, growing in its natural habitat on rocky coastal cliffs of the Mediterranean Sea. There is a shift from the C₃ mode to CAM, coinciding with less discrimination against the heavy carbon isotope. Filled symbols and bars represent the end of the night; open symbols and bars represent end-of day values (Osmond et al. 1982).

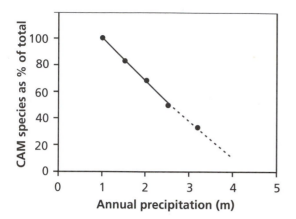

FIGURE 50. The relationship between percentage of epiphytic bromeliad species with CAM metabolism and mean annual rainfall in Trinidad (Winter & Smith 1996).

In temperate and alpine regions, (intermediate) CAM plants occur in niches that are rather dry. For example, *Sedum acre* (stonecrop), on roofs in the Netherlands, is an intermediate CAM plant.

10.6 Carbon Isotope Composition of CAM Species

Like Rubisco from C_3 and C_4 plants, the enzyme from CAM plants discriminates against $^{13}CO_2$. The fractionation is considerably less than that of C_3 plants and similar to that of C_4 species (Fig. 44). This is expected, as the stomata are closed during malate decarboxylation and fixation of CO_2 by Rubisco. Hence, only a small amount of inorganic carbon diffuses back from the leaves to the atmosphere, as discussed in Sections 9.3 and 9.4 of this chapter.

Upon a shift from C_3 to CAM photosynthesis in **facultative CAM plants**, the stomates are closed during most of the day and open at night. The **carbon isotope discrimination** consequently decreases, in accordance with the smaller extent of discrimination in CAM species (Fig. 49).

11. Specialized Mechanisms Associated with Photosynthetic Carbon Acquisition in Aquatic Plants

11.1 Introduction

In terrestrial plants, CO_2 diffuses from the air through the stomata to the mesophyll cells. In aquatic plants this diffusion process is often limited, because of the absence of stomata, a thick boundary layer around the leaves, the low rate of CO_2 transport in water, or a combination of these factors. How do aquatic plants cope with these problems? To achieve a high rate of photosynthesis and avoid high rates of photorespiration, special mechanisms are required to allow sufficient diffusion of CO_2 to match the requirement for photosynthesis. As discussed in this section different mechanisms have evolved in different species.

Another feature of the habitat of submerged aquatics is the low irradiance. Leaves of many aquatics have the traits typical of shade leaves, as discussed in Section 3.2.

11.2 The CO_2 Supply in Water

Molecular CO_2 is readily available in fresh water. Between 10 and 20°C, the partitioning coefficient (i.e., the ratio between the molar concentration of CO_2 in air and that in water) is about 1. The equilibrium concentration in water is then approximately 12.8 μM. Under these conditions, leaves of submerged aquatic macrophytes experience about the same CO_2 concentration as those in air. The **diffusion** of dissolved gasses in water, however, occurs approximately 10^4 times more slowly than that in air, which leads to rapid depletion of CO_2 around the leaf during CO_2 assimilation. To make things worse, the O_2 concentration inside the leaf increases, inexorably leading to conditions that restrict the **carboxylating** activity and favor the **oxygenating** activity of Rubisco.

The transport of CO_2 through the unstirred ("boundary") layer is only by diffusion. The thickness of the boundary layer is proportional to the square root of the leaf dimension, as measured in the direction of the streaming water, and inversely proportional to the flow of the streaming water. It ranges from 10 μm in well-stirred media to 500 μm in nonstirred media. The unstirred layer is often a major factor that limits an aquatic macrophyte's rate of photosynthesis.

CO_2 dissolved in water interacts as follows:

$$2\,H^+ + CO_3^{2-} \Leftrightarrow H^+ + HCO_3^- \Leftrightarrow CO_2 + H_2O$$
$$\Leftrightarrow H_2CO_3 \tag{10}$$

$$HCO_3^- \Leftrightarrow OH^- + CO_2 \tag{11}$$

Because the concentration of H_2CO_3 is very low in comparison with that of CO_2, these two species are commonly combined and indicated as $[CO_2]$.

The interconversion between carbon dioxide and bicarbonate is slow, at least in the absence of the enzyme **carbonic anhydrase**. The presence of the dissolved inorganic carbon compounds strongly depends on the pH of the water (Fig. 51). In ocean water, the contribution of dissolved inorganic carbon species shifts as follows: When the pH increases from 7.4 to 8.3 CO_2 as a fraction of the total inorganic carbon pool decreases from 4 to 1%, that of HCO_3^- from 96 to 89%, and that of CO_3^{2-} increases from 0.2 to 11%.

During darkness, the CO_2 concentration in ponds and streams is generally high, exceeding the concentration that is in equilibrium with air, due to respiration of aquatic organisms and the slow exchange of CO_2 between water and the air above it. The high CO_2 concentration coincides with a relatively low pH. During the day the CO_2 concentration declines rapidly, and the pH rises accordingly. The rise in pH, especially in the unstirred layer, is a crucial factor. Whereas the concentration of all dissolved inorganic carbon (i.e., CO_2, HCO_3^-, and CO_3^{2-}) may decline by a few percent only, the CO_2 concentration declines much more because the high pH shifts the equilibrium from CO_2 to HCO_3^- (Fig. 51). Because of this, carbon assimilation of submerged leaves that only use CO_2 as a carbon source, and not HCO_3^-, is often limited by the supply of CO_2.

11.3 The Use of Bicarbonate by Aquatic Macrophytes

Many, but by no means all, aquatic macrophytes can use bicarbonate, in addition to CO_2, as a carbon source for photosynthesis (Prins et al. 1982). This might be achieved either by **active uptake** of bicarbonate itself or by **proton extrusion**, commonly at the abaxial side of the leaf, thus lowering the pH in the extracellular space and shifting the equilibrium toward CO_2 (Elzenga & Prins 1988). In some species [e.g., *Elodea canadensis* (waterweed)] the conversion of HCO_3^- into CO_2 is also catalyzed by apoplasmic **carbonic anhydrase**. This enzyme has been demonstrated in the extracellular space of a number of aquatic macrophytes. In *Ranunculus penicillatus* spp. *pseudofluitans*, a freshwater aquatic buttercup, the enzyme is closely associated with the epidermal cell wall (Newman & Raven 1993). Active uptake of bicarbonate also requires proton extrusion, to provide a driving force. The two mechanisms, therefore, are hard to distinguish. To discriminate between the two mechanisms, isotopic disequilibrium experiments have been carried out. This method is based on the relatively slow equilibration between carbon dioxide and bicarbonate. (When one of the two carbon compounds is added, it takes approximately 80 seconds before isotopic equilibrium is established.) Such experiments were done with protoplasts (which lack an extracellular carbonic anhydrase), fixing either ^{14}C-labeled carbon dioxide or bicarbonate. These experiments led to the conclusion that bicarbonate use by *Potamogeton lucens* (ribbonweed) is due to the formation of carbon dioxide in the extracellular space and not to the uptake of bicarbonate (Staal et al. 1989).

Aquatic plants that use bicarbonate in addition to CO_2 have a mechanism to concentrate CO_2 in their chloroplasts. Although this **CO_2-concentrating mechanism** differs from that of C_4 plants (Sect. 9.2), its effect is rather similar: It suppresses the oxygenating activity of Rubisco and lowers the CO_2-compensation point. In *Elodea canadensis*, *Potamogeton lucens*, and other aquatic macrophytes, the capacity to acidify the lower side of the leaves, and thus to use bicarbonate, is expressed most at high irradiance and low dissolved inorganic carbon concentration in the water (Elzenga & Prins 1989). The capacity of the carbon-concentrating mechanism also depends on the nitrogen supply: The higher the supply of N, the greater the capacity of the photosynthetic apparatus as well as that of the carbon-concentrating mechanism (Madsen & Baattrup-Pedersen 1995). Acidification of the lower side of the leaves is accompanied by an increase in

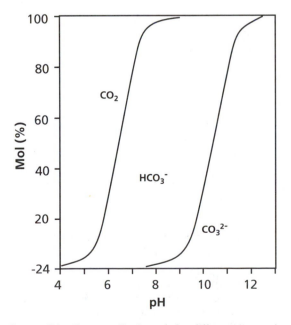

FIGURE 51. The contribution of the different inorganic carbon species strongly depends on the pH of the water (Osmond et al. 1982).

extracellular pH at the upper side of the leaves. The leaves become "polar" when the carbon supply from the water is less than the CO_2-assimilating capacity (Prins & Elzenga 1989). There are also anatomical differences between the upper and lower side of "polar" leaves: The lower epidermal cells are often **transfer cells** that are characterized by ingrowths of cell-wall material, which increases the surface area of the plasma membrane. They also contain numerous mitochondria and chloroplasts. At the upper side of the leaves, the pH increase leads to precipitation of calcium carbonates. This process plays a major role in the geological sedimentation of calcium carbonate (Sect. 11.7).

Due to the use of bicarbonate, the internal CO_2 concentration may become much higher than it is in terrestrial C_3 plants. This implies that they do not need a Rubisco enzyme with a high affinity for CO_2. It is interesting that just like C_4 plants, they have a Rubisco with a relatively high K_m for CO_2. The values are approximately twice as high as those of terrestrial C_3-plants (Table 9 in this chapter). This high K_m should allow a high maximum catalytic activity of Rubisco that is similar to that of an HCO_3^--using green alga, *Chlamydomonas reinhardtii*; this has a Rubisco with a relatively high specific activity, close to that of C_4 species and substantially higher than that of C_3 species (see Table 9 in this chapter).

Hydrilla verticillata has an inducible CO_2-concentrating mechanism, even when the pH of the medium is so low that there is no bicarbonate available. These plants have an inducible C_4-type photosynthetic cycle in that the CO_2 concentration in their chloroplasts is sufficiently high to suppress photorespiration and lower the CO_2-compensation point, but they lack the typical Kranz anatomy that is typical for C_4 species (Sect. 9.2). This mechanism is induced when the plants are grown in water that contains low concentrations of dissolved inorganic carbon (Reiskind et al. 1997). There appears to be a clear ecological benefit to this, possibly unique, CO_2-concentrating mechanism when the canopy becomes dense and the dissolved oxygen concentration is very high. Under these conditions photorespiration decreases photosynthesis of a C_3-type plant by at least 35%, whereas this decrease is only about 4% in *H. verticillata* (Bowes & Salvucci 1989).

11.4 The Use of CO_2 from the Sediment

Macrophytes like water lilies that have an internal ventilation system assimilate CO_2 arriving from the roots due to pressurized flow (cf. Sect. 4.1.4 in plant respiration). The use of CO_2 from the sediment is only minor for wetland species such as *Scirpus lacustris* and *Cyperus papyrus*, where it approximates 0.25% of the total CO_2 uptake in photosynthesis (Farmer 1996). For *Stratiotes aloides* (water soldier), the sediment is a major source of CO_2, although only after diffusion into the water column. The carbon in *Stratiotes*-dominated ponds has lower $\delta^{13}C$ values (-9.1 to $-13.1‰$) than would be expected if the CO_2 were in equilibrium with air because of CO_2 that originates from the decomposition of organic matter in the sediment (Prins & De Guia 1986). Discrimination by the plants is fairly small, with indicates that diffusion is limiting photosynthesis (cf. Box 2) and/or that bicarbonate is a significant carbon source (the $\delta^{13}C$ of bicarbonate is significantly less than that of CO_2).

Submerged macrophytes of the isoetid life form (quillworts) receive a very large portion of their carbon for photosynthesis directly from the sediment via their roots: 60 to 100% (Table 15). This capability may be an adaptation to growth in low pH, carbon-poor ("softwater") lakes, where these plants are common. None of the investigated species from "hardwater" lakes or marine systems show significant CO_2 uptake via their roots (Farmer 1996). In the quillworts, CO_2 diffuses from the sediment, via the lacunal air system to the submerged leaves. These leaves are thick with thick cuticles, have no functional stomata but large air spaces, so that gas exchange with the atmosphere is hampered; the emerged leaves have only very few stomata at the leaf base, and normal densities at the leaf tips (Fig. 52). The chloroplasts in isoetid leaves are concentrated around the lacunal system. The air spaces in the leaves are connected with those in stems and roots, with a relatively short distance in between. At night, only part of the CO_2 coming up from the sediment via the lacunal system is fixed, the rest being lost to the atmosphere.

CO_2 transport from the sediment is also of quantitative importance in the terrestrial isoetid, *Stylites andicola*. This is an unusual plant that has no stomata.

11.5 Crassulacean Acid Metabolism (CAM) in Water Plants

Though aquatic plants by no means face the same problems that are connected with water shortage as desert plants, some of them (*Isoetes* species) have a similar photosynthetic metabolism: the Crassulacean Acid Metabolism (CAM; Keeley 1990,

T<small>ABLE</small> 15. Assimilation of $^{14}CO_2$ derived from the air or from the rhizosphere by leaves and roots of *Littorella uniflora* (Isoetaceae).

| | $^{14}CO_2$ assimilation $\mu g\ C\ (g\ leaf\ or\ root\ DM)^{-1}h^{-1}$ | | | |
| | Leaves | | Roots | |
Source:	Air	Rhizosphere	Air	Rhizosphere
CO_2 concentration around the roots (mM)				
0.1	300	340	10	60
	(10)	(50)	(0.3)	(70)
0.5	350	1330	10	170
	(5)	(120)	(0.3)	(140)
2.5	370	8340	10	570
	(4)	(1430)	(0.3)	(300)

Source: Nielsen et al. 1991.
* $^{14}CO_2$ was added to the air around the leaves or to the water around the roots (rhizosphere). Measurements were made in the light and in the dark; values of the dark measurements are given in brackets.

Keeley & Busch 1984). They accumulate **malic acid** during the night, and have rates of CO_2 fixation during the night that are similar in magnitude to those during the day, when the CO_2 supply from the water is very low (Fig. 53). Contrary to the submerged leaves of the same plants, the aerial leaves of *Isoetes howellii* do not show a diurnal fluctuation in the concentration of malic acid.

Why would an aquatic plant have a photosynthetic pathway similar to that which is common in species from dry habitats? CAM in *Isoetes* may be an adaptation to very low levels of CO_2 in the water, especially during the day (Fig. 53), that allows the plants to assimilate additional CO_2 at night. This

nocturnal CO_2 fixation gives them access to a carbon source that is unavailable for other species. Even though some of the carbon fixed in malic acid comes from the surrounding water, where it accumulates due to the respiration of aquatic organisms, some is also derived from the plant's respiration during the night.

11.6 Variation in Carbon Isotope Composition Between Water Plants and Between Aquatic and Terrestrial Plants

There is a wide variation in carbon isotope composition among different aquatic plants, as well as a large difference between aquatic and terrestrial plants (Fig. 54). A low carbon isotope discrimination might reflect the employment of the C_4 pathway of photosynthesis, although the typical Kranz anatomy is lacking. A well-developed Kranz anatomy has been established only for *Neostapfia colusana* (Keeley & Sandquist 1992). A low carbon isotope discrimination might also reflect the CAM pathway of photosynthesis. This pathway, however, appears to be restricted to the Isoetaceae, which often have rather negative $\delta^{13}C$ values due to the isotope composition of the substrate (Table 16). Four factors account for the observed variation in isotope discrimination of freshwater aquatics (Keeley & Sandquist 1992):

1. The isotope composition of the carbon source: It ranges from a $\delta^{13}C$ value of +1‰, for HCO_3^-

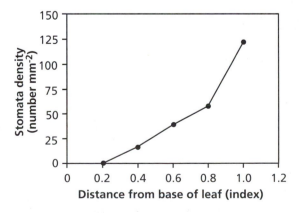

F<small>IGURE</small> 52. The stomatal density along mature leaves of *Littorella uniflora* increases from the base to the tip (Nielsen et al. 1991).

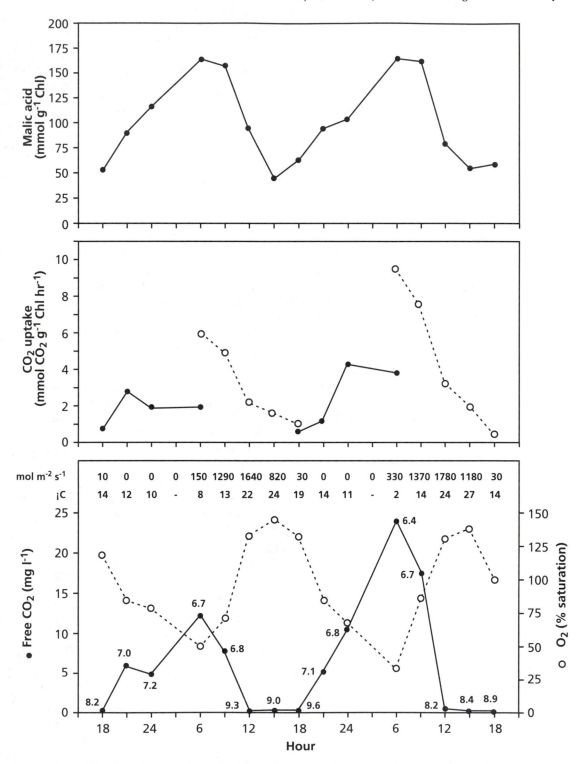

FIGURE 53. Malic acid levels, rates of CO_2 uptake, irradiance at the water surface, water temperatures, and concentrations of CO_2 and O_2 in the dark (filled symbol) and in the light (open symbols) for submerged *Isoetes howellii* leaves in a pool. The numbers near the symbols in the bottom figure give the pH values (Keeley & Busch 1984). Copyright American Society of Plant Physiology.

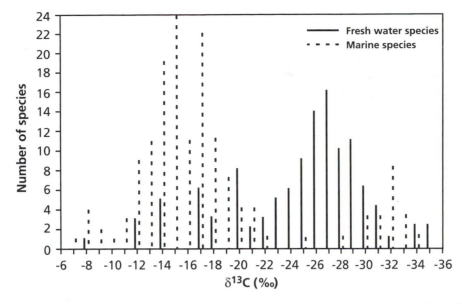

FIGURE 54. Variation in the carbon isotope composition ($\delta^{13}C$ values) of freshwater and marine aquatic species. The observed variation is due to variation in $\delta^{13}C$ values of the substrate and in the extent of diffusional limitation (Osmond et al. 1982).

derived from limestone, to $-30‰$, for CO_2 derived from respiration. The average $\delta^{13}C$ value of CO_2 in air is $-8‰$. The isotope composition also changes with the water depth (Table 17).

2. The species of carbon fixed: Some species assimilate HCO_3^-, which has $\delta^{13}C$ values of -1 to $+3‰$.

3. C_3, C_4 and CAM photosynthetic pathways are present in aquatic plants.

4. Diffusion through unstirred boundary layers and carbon uptake may be important factors limiting photosynthesis, thus decreasing the carbon isotope discrimination (cf. Box 2).

The isotope composition of the carbon source and diffusional processes are therefore more important factors that account for the variation as described in Figure 54 than are biochemical differences in the photosynthetic pathway (Osmond et al. 1982).

11.7 The Role of Aquatic Macrophytes in Carbonate Sedimentation

The capacity of photosynthetic organisms (e.g., *Chara*, *Potamogeton* and *Elodea*) to acidify part of the apoplast and use bicarbonate (Sect. 11.3) plays a major role in the formation of calcium precipitates in freshwater systems where macrophytes are abandant, on both an annual and a geological time scale. Many calcium-rich sediments are of plant origin, according to the overall equation:

TABLE 16. Carbon isotope composition ($\delta^{13}C$ in ‰) of submerged and emergent *Isoetes howellii* plants.*

Pondwater carbonate	-15.5 to -18.6
Submerged	
Leaves	-27.9 to -29.4
Roots	-25.8 to -28.8
Emergent	
Leaves	-29.4 to -30.1
Roots	-29.0 to -29.8

Source: Keeley & Busch 1984.
*Values are given for both leaves and roots and also for the pondwater carbonate.

TABLE 17. Changes in the dissolved carbon isotope composition ($\delta^{13}C$ in ‰) with depth as reflected in the composition of the organic matter at that depth.

Water depth (m)	$\delta^{13}C$ (‰)
1	-20.80
2	-20.75
5	-23.40
7	-24.72
9	-26.79
11	-29.91

Source: Osmond et al. 1982.

$$Ca^{2+} + 2HCO_3^- \rightarrow CO_2 + CaCO_3 \qquad (12)$$

This reaction occurs in the alkaline compartment that is provided at the upper side of the leaves of "polar" aquatic macrophytes. Similar amounts of carbon are assimilated in photosynthesis and precipitated as carbonate. If only part of the CO_2 released in this process is assimilated by the macrophyte, as may occur under nutrient-deficient conditions, then CO_2 is released to the atmosphere. On the other hand if the alkalinity of the compartment is relatively low, then there is a net transfer of atmospheric CO_2 to the water (McConnaughey et al. 1994).

Equation 12 shows that aquatic photosynthetic organisms play a major role in the global carbon cycle, even on a geological time scale. On the other hand, rising atmospheric CO_2 concentrations have an acidifying effect and thus dissolve part of the calcium carbonate precipitates in sediments; therefore, they contribute to a further rise in atmospheric CO_2 (Sect. 12).

12. Effects of the Rising CO₂ Concentration in the Atmosphere

Vast amounts of carbon are present in carbonates in the earth's crust. Some CO_2 enters the atmosphere when these carbonates are used for making cement, but apart from that carbonates are only biologically important on a geological time scale. The reduced carbon, which is present as coal, oil, and natural gas, is far more relevant in that the burning of these fossil fuels, together with land use change, increase the CO_2 input into the atmosphere by $7.1\ 10^{15}$ g of carbon per year. Compared with the total amount of carbon in the atmosphere, $750\ 10^{15}$ g, such inputs are substantial and inevitably affect the CO_2 concentration in the earth's atmosphere (Fig. 55).

Since the beginning of the industrial revolution in the late eighteenth century, the atmospheric CO_2 concentration has increased from about $290\,\mu mol\,mol^{-1}$ to the current level of more than $350\,\mu mol\,mol^{-1}$ (≈ 35 Pa at sea level). The concentration continues to rise by about $1.5\,\mu mol\,mol^{-1}$ per year (Fig. 56), mainly caused by fossil fuel burning and changes in land use (Houghton et al. 1991). Measurements of CO_2 concentrations in **ice cores** indicate a preindustrial value of about $280\,\mu mol\,mol^{-1}$ during the past 10,000 years, but about $205\,\mu mol\,mol^{-1}$ some 20,000 years ago (Lemon 1983). Considerable quantities of CO_2 have also been released into the atmosphere as a result of **defor-**estation, **ploughing of prairies**, and other landuse changes. This process still adds $1.6 \pm 1.0\ 10^{15}$ g of carbon per year to the atmosphere. **Combustion of fossil fuel** adds about $5.5 \pm 0.5\ 10^{15}$ g of carbon per year for a total input to the atmosphere of about $7.1 \pm 1.1\ 10^{15}$ g of carbon (Schimel 1995). The increase in the atmosphere, however, is only $3.2 \pm 0.2\ 10^{15}$ g of carbon per year. About $2.0 \pm 0.8\ 10^{15}$ g of the "missing" carbon is taken up in the oceans, and a similar amount ($2.1\ 10^{15}$ g) is fixed in terrestrial biomass (Schlesinger 1993). Analysis of atmospheric CO_2 concentrations and its isotopic composition shows that north-temperate and boreal forests are the most likely sinks for the missing carbon. This increased terrestrial uptake of CO_2 has many causes, including stimulation of photosynthesis by elevated $[CO_2]$ (about half of the increased terrestrial uptake) or by nitrogen deposition in nitrogen-limited ecosystems and regrowth of midlatitude forests following abandonment of agricultural lands (each about 25% of the enhanced terrestrial uptake) (Schimel 1995).

Because the rate of net CO_2 assimilation in C_3 plants is not CO_2-saturated at 35 Pa CO_2, the rise in CO_2 concentration is more likely to enhance photosynthesis in C_3 than it is in C_4 plants, where the rate of CO_2 assimilation is virtually saturated at a CO_2 concentration of $350\,\mu mol\,mol^{-1}$ (Bowes 1993). The consequences of an enhanced rate of photosynthesis for plant growth will be discussed in Section 5.8 of the chapter on growth and allocation.

12.1 Acclimation of Photosynthesis to Elevated CO₂ Concentrations

There is generally a reduction of the photosynthetic capacity upon long-term exposure to 70 Pa CO_2, which is associated with reduced levels of Rubisco and organic nitrogen per unit leaf area (Long et al. 1993). This **down-regulation** of photosynthesis increases with duration of the CO_2 exposure and is most pronounced in plants grown under low nutrient supplies (Curtis 1996). By contrast, water-stressed plants often increase net photosynthesis in response to elevated $[CO_2]$. Herbaceous plants consistently reduce stomatal conductance in response to elevated $[CO_2]$ so that p_i does not increase as much as would be expected from the increase in p_a. The trees studied to date, however, do not show any reduction in stomatal conductance in response to elevated $[CO_2]$. The decrease in stomatal conductance of herbaceous C_3 plants often indirectly stimulates photosynthesis in dry environments by reducing the rate of soil drying and therefore the water limitation of photosynthesis.

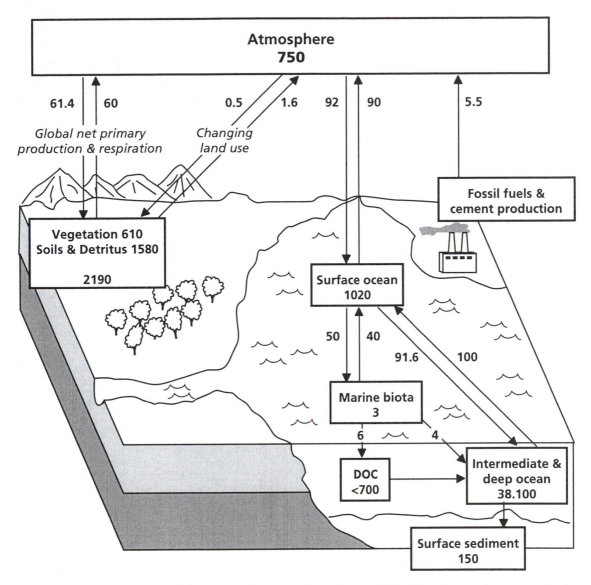

FIGURE 55. Global reservoirs and fluxes of carbon, expressed in 10^{15} g carbon per year. The atmosphere gains carbon from the burning of fossil fuel and loses carbon to the oceans. Carbon in plant biomass and soil carbon is lost through changes in land use, but may increase locally due to a "CO$_2$ fertilization effect". DOC is dissolved organic carbon that is dissolved in ocean water (Schimel 1995). Copyright Blackwell Science Ltd.

How do herbaceous plants sense that they are growing at an elevated CO$_2$ concentration and then down-regulate their photosynthetic capacity? Acclimation is not due to sensing the CO$_2$ concentration itself; rather, it is due to sensing the concentration of sugars in the leaf cells, more precisely the soluble hexose sugars. This sensing is mediated by a specific hexokinase, which is an enzyme that phosphorylates hexose by hydrolyzing ATP. In transgenic plants in which the level of hexokinase is greatly reduced, down-regulation of photosynthesis upon prolonged exposure to high [CO$_2$] is considerably less. The **sugar-sensing mechanism** affects the transcription of nuclear encoded photosynthesis-associated genes via an unknown transduction pathway. Among the first proteins to be affected are the small subunit of Rubisco and Rubisco activase. Upon longer exposure, the levels of thylakoid proteins and chlorophyll are also reduced (Table 18).

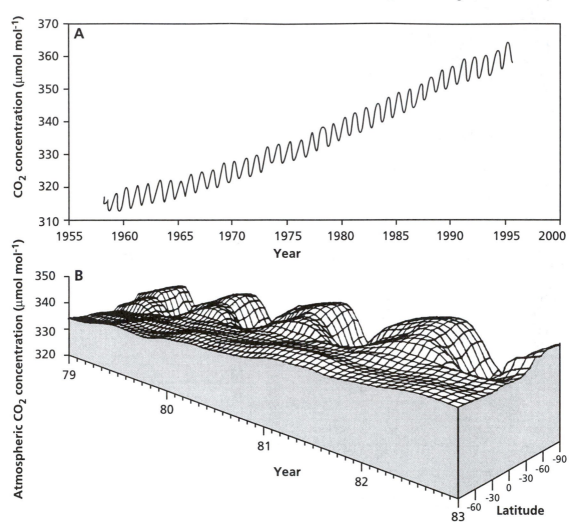

FIGURE 56. The steady rise in carbon dioxide concentration of approximately 1.5 µmol mol⁻¹, as measured at Mauna Loa (Hawaii) (A: Keeling & Whorf 1996). Superimposed on the steady rise is an annual fluctuation, especially for the northern hemisphere (B: Goudriaan 1987; copyright Netherlands Journal of Agricultural Science).

TABLE 18. Light-saturated rate of photosynthesis (A_{max}, measured at the CO_2 concentrations at which the plants were grown), *in vitro* Rubisco activity, chlorophyll concentration and the concentration of hexose sugars in the fifth leaf of *Lycopersicon esculentum* (tomato) at various stages of development.

Leaf expansion (% of full expansion)	Exposure time (days)	A_{max} (μmol m^{-2} s^{-1})		Rubisco activity (μmol m^{-2} s^{-1})		Chlorophyll (mg m^{-2})		Glucose (mg m^{-2})		Fructose (mg m^{-2})	
		Control	High	Control	High	Control	High	Control	High	Control	High
2	0	16.3	21.3	22.6	—	270	—	750	—	500	—
60	11	18.9	28.7	20.5	25.1	480	520	1000	1200	1400	1400
95	22	15.0	25.1	15.7	12.7	540	500	1100	1250	1800	2100
100	31	9.3	18.0	9.5	4.9	450	310	1100	2100	1800	4200

Source: Van Oosten & Besford 1995, Van Oosten et al. 1995.

*Plants were grown at different atmospheric CO_2 concentrations: control, 350 µmol CO_2 mol⁻¹; high, 700 µmol CO_2 mol⁻¹.

The sugar-sensing mechanism may well play a role in acclimation of photosynthesis to other environmental conditions, such as irradiance level, water stress, and nitrogen supply, but there is no information to corroborate this suggestion (Koch 1996).

12.2 Effects of Elevated CO$_2$ on Transpiration-Differential Effects on C$_3$, C$_4$, and CAM Plants

Photosynthesis in different types of plants responds to varying degrees to elevated CO$_2$. For example, C$_4$ plants, whose rate of photosynthesis is virtually saturated at 35 Pa, generally respond less to elevated CO$_2$ than do C$_3$ plants.

Also *Opuntia ficus-indica*, which is a CAM species cultivated worldwide for its fruits and cladodes, responds to the increase in CO$_2$ concentration in the atmosphere. The rate of CO$_2$ assimilation is initially enhanced, both at night and during the day, but this disappears upon prolonged exposure to elevated CO$_2$ (Cui & Nobel 1994).

13. Summary: What Can We Gain from Basic Principles and Rates of Single-Leaf Photosynthesis?

Numerous examples have been given on how differences in photosynthetic traits enhance a genotype's **survival** in a specific environment. These include specific biochemical pathways (C$_3$, C$_4$, and CAM) as well as more intricate differences between sun and shade plants, aquatic and terrestrial plants, and plants that differ in their photosynthetic nitrogen-use efficiency and water-use efficiency. Information on photosynthetic traits is also highly relevant when trying to understand effects of global environmental changes in temperature and atmospheric CO$_2$ concentrations. For a physiological ecologist, a full appreciation of the process of leaf photosynthesis is quintessential.

What we *cannot* derive from measurements on photosynthesis of **single leaves** is what the rate of photosynthesis of an **entire canopy** will be. To arrive at these rates, we need to take the approach discussed in a later chapter that deals with scaling-up principles. It is also quite clear that short-term measurements on the effect of atmospheric CO$_2$ concentrations are not going to tell us what will happen in the long term. Acclimation of the photosynthetic apparatus ("down-regulation") may take

place, which reduces the initial effect. Most importantly, we *cannot* derive growth rates from rates of photosynthesis of a single leaf. Growth rates are not simply determined by rates of single leaf photosynthesis per unit leaf area; rather, they are also determined by the total leaf area per plant and by the fraction of daily produced photosynthates required for plant respiration.

References and Further Reading

Anderson, J.W. & Beardall, J. (1991) Molecular Activities of Plant Cells. An Introduction to Plant Biochemistry. Blackwell Scientific Publications, Oxford.

Azcón-Bieto, J. (1983) Inhibition of photosynthesis by carbohydrates in wheat leaves. Plant Physiol. 173:681–686.

Bastide, B., Sipes, D., Hann, J., & Ting, I.P. (1993) Effect of severe water stress on aspects of crassulacean acid metabolism in *Xerosicyos*. Plant Physiol. 103:1089–1096.

Bauwe, H., Kerberg, O., Bassuner, R., Parnik, T., & Bassuner, B. (1987) Reassimilation of carbon dioxide by *Flaveria* (Asteraceae) species representing different types of photosynthesis. Planta 172:214–218.

Berry, J.A. & Raison, J.K. (1981) Responses of macrophytes to temperature. In: Encyclopedia of plant physiology, N.S., Vol. 12A, O.L. Lange, P.S. Nobel, C.B. Osmond, & H. Ziegler (eds). Springer-Verlag, Berlin, pp. 277–338.

Beyschlag, W. & Pfanz, H. (1990) A fast method to detect the occurrence of nonhomogeneous distribution of stomatal aperture in heterobaric plant leaves. Experiments with *Arbutus unedo* L. during the diurnal course. Oecologia 82:52–55.

Björkman, O. (1981) Responses to different quantum flux densities. In: Encyclopedia of plant physiology, N.S., Vol. 12A, O.L. Lange, P.S. Nobel, C.B. Osmond, & H. Ziegler (eds). Springer-Verlag, Berlin, pp. 57–107.

Black, C.C., Chen, J.Q., Doong, R.L., Angelov, M.N., & Sung, S.J.S. (1996) Alternative carbohydrate reserves used in the daily cycle of Crassulacean acid metabolism. In: Crassulacean acid metabolism, biochemistry, ecophysiology and evolution. Ecological Studies 114, K. Winter & J.A.C. Smith (eds). Springer-Verlag, Berlin, pp. 31–45.

Black, V.J. (1982) Effects of sulphur dioxide on physiological processes in plants. In: Effects of gaseous air pollution in agriculture and horticulture, M.H. Unsworth & D.P. Ormrod (eds). Butterworths, London, pp. 67–91.

Black, V.J. & Unsworth, M.H. (1997) A system for measuring effects of sulphur dioxide on gas exchange of plants. J. Exp. Bot. 30:81–88.

Bowes, G. (1993) Facing the inevitable: Plants and increasing atmospheric CO$_2$. Annu. Rev. Plant Physiol. Plant Mol. Biol. 44:309–332.

Bowes, G. & Salvucci, M.E. (1989) Plasticity in the photosynthetic carbon metabolism of submersed aquatic macrophytes. Aquat. Bot. 34:233–286.

Brown, R. & Bouton, J.H. (1993) Physiology and genetics of interspecific hybrids between photosynthetic types. Annu. Rev. Plant Physiol. Plant Mol. Biol. 44:435–456.

Brown, R.H. & Hattersley, P.W. (1989) Leaf anatomy of C_3–C_4 species as related to evolution of C_4 photosynthesis. Plant Physiol. 91:1543–1550.

Brugnoli, E. & Björkman, O. (1992) Chloroplast movements in leaves: Influence on chlorophyll fluorescence and measurements of light-induced absorbance changes related to ΔpH and zeaxanthin formation. Photosynth. Res. 32:23–35.

Brugnoli, E. & Lauteri, M. (1991) Effects of salinity on stomatal conductance, photosynthetic capacity, and carbon isotope discrimination of salt-tolerant (*Gossypium hirsutum* L.) and salt-sensitive (*Phaseolus vulgaris* L.) C_3 non-halophytes. Plant Physiol. 95:628–635.

Campbell, W.J. & Ogren, W.L. (1990) A novel role for light in the activation of ribulose-bisphosphate carboxylase/oxygenase. Plant Physiol. 92:110–115.

Chapin III, F.S. & Shaver, G.R. (1985) Individualistic growth response of tundra plant species to manipulation of light, temperature, and nutrients in a field experiment. Ecology 66:564–576.

Chazdon, R.L. & Pearcy, R.W. (1986) Photosynthetic responses to light variation in rainforest species. I. Induction under constant and fluctuating light conditions. Oecologia 69:517–523.

Chazdon, R.L. & Pearcy, R.W. (1991) The importance of sunflecks for forest understory plants. BioSciences 41:760–766.

Christie, E.K. & Detling, J.K. (1982) Analysis of interference between C_3 and C_4 grasses in relation to temperature and soil nitrogen supply. Ecology 63:1277–1284.

Chu, C., Dai, Z., Ku, M.S.B., & Edwards, G.E. (1990) Induction of crassulacean metabolism in the facultative halophyte *Mesembryanthemum crystallinum* by abscisic acid. Plant Physiol. 93:1253–1260.

Coté, F.X., André, M., Folliot, M., Massimino, D., & Daguenet, A. (1989) CO_2 and O_2 exchanges in the CAM plant *Ananas comosus* (L.) Merr. determination or total and malate-decarboxylation-dependent CO_2-assimilation rates; study of light O_2-uptake. Plant Physiol. 89:61–68.

Cui, M. & Nobel, P.S. (1994) Gas exchange and growth responses to elevated CO_2 and light levels in the CAM species *Opuntia ficus-indica*. Plant Cell Environ. 17:935–944.

Curtis, P.S. (1996) A meta-analysis of leaf gas exchange and nitrogen in trees grown under elevated carbon dioxide. Plant Cell Environ. 19:127–137.

DeLucia, E.H., Nelson, K., Vogelmann, T.C., & Smith, W.K. (1996) Contribution of intercellular reflectance to photosynthesis in shade leaves. Plant Cell Environ. 19:159–170.

Demmig, B., Winter, K., Kriger, A., & Czygan, F.-C. (1987) Photoinhibition and zeaxanthin formation in intact leaves. A possible role of the xanthophyll cycle in the dissipation of excess light energy. Plant Physiol. 84:218–224.

Demmig-Adams, B. (1990) Carotenoids and photoprotection: A role for the xanthophyll zeaxanthin. Biochim. Biophys. Acta. 1020:1–24.

Demmig-Adams, B. & Adams, W.W. (1992) Photoprotection and other responses of plants to high light stress. Annu. Rev. Plant Physiol. Plant Molec. Biol. 43:599–626.

Demmig-Adams, B. & Adams, W.W. (1996) The role of xanthophyll cycle carotenoids in the protection of photosynthesis. Trends in Plant Sciences 1:21–26.

Downton, W.J.S., Loveys, B.R., & Grant, W.J.R. (1988) Stomatal closure fully accounts for the inhibition of photosynthesis by abscisic acid. New Phytol. 108:263–266.

Eckstein, J., Beyschlag, W., Mott, K.A., & Ryell, R.J. (1996) Changes in photon flux can induce stomatal patchiness. Plant Cell Environ. 19:1066–1074.

Ehleringer, J.R. (1993) Gas exchange implications of isotopic variation in arid-land plants. In: Water deficits: Plant responses from cell to community, J.A.C. Smith & H. Griffiths (eds). Bios Scientific Publishers, Oxford, pp. 265–284.

Ehleringer, J.R. & Monson, R.K. (1993) Evolutionary and ecological aspects of photosynthetic pathway variation. Annu. Rev. Ecol. Syst. 24:411–439.

Ehleringer, J.R., Sage, R.F., Flanagan, L.B., & Pearcy, R.W. (1991) Climate change and the evolution of C_4 photosynthesis. Trends Ecol. Evol. 6:95–99.

Ehleringer, J., Björkman, O., & Mooney, H.A. (1976) Leaf pubescence: Effects on absorptance and photosynthesis in a desert shrub. Science 192:376–377.

Eller, B.M. & Ferrari, S. (1997) Water use efficiency of two succulents with contrasting CO_2 fixation pathways. Plant Cell Environ. 20:93–100.

Elzenga, J.T.M. & Prins, H.B.A. (1988) Adaptation of *Elodea* and *Potamogeton* to different inorganic carbon levels and the mechanism for photosynthetic bicarbonate utilisation. Aust. J. Plant Physiol. 15:727–735.

Elzenga, J.T.M. & Prins, H.B.A. (1989) Light-induced polar pH changes in leaves of *Elodea canadensis*. I. Effects of carbon concentration and light intensity. Plant Physiol. 91:62–67.

Evans, J.R. (1988) Acclimation by the thylakoid membranes to growth irradiance and the partitioning of nitrogen between soluble and thylakoid proteins. Aust. J. Plant Physiol. 15:93–106.

Evans, J.R. (1989) Photosynthesis and nitrogen relationships in leaves of C_3 plants. Oecologia 78:9–19.

Evans, J.R. & Seemann, J.R. (1989) The allocation of protein nitrogen in the photosynthetic apparatus: Costs, consequences, and control. In: Photosynthesis, W.R. Briggs (ed). Alan Liss Inc., New York.

Evans, J.R. & Von Caemmerer, S. (1996) Carbon dioxide diffusion inside leaves. Plant Physiol. 110:339–346.

Farmer, A.M. (1996) Carbon uptake by roots. In: Plant Roots: The hidden half, Y. Waisel, A. Eshel, & U. Kafkaki (eds). Marcel Dekker, Inc., New York, pp. 679–687.

Farquhar, G.D. & Richards, R.A. (1984) Isotopic composition of plant carbon correlates with water-use efficiency of wheat genotypes. Aust. J. Plant Physiol. 11:539–552.

Farquhar, G.D. & Sharkey, T.D. (1982) Stomatal conductance and photosynthesis. Annu. Rev. Plant Physiol. 33:317–345.

Field, C.B. & Mooney, H.A. (1986) The photosynthesis-nitrogen relationship in wild plants. In: On the economy of plant form and function, T.J. Givnish (ed). Cambridge University Press, Cambridge, pp. 25–55.

Flanagan, L.B. & Jefferies, R.L. (1989) Photosynthetic and stomatal responses of the halophyte, *Plantago maritima* L. to fluctuations in salinity. Plant Cell Environ. 12:559–568.

Flügge, U.I., Stitt, M., & Heldt, H.W. (1985) Light-driven uptake or pyruvate into mesophyll chloroplasts from maize. FEBS Lett. 183:335–339.

Fredeen, A.L., Gamon, J.A., & Field, C.B. (1991) Responses of photosynthesis and carbohydrate partitioning to limitations in nitrogen and water availability in field grown sunflower. Plant Cell Environ. 14:969–970.

Gilmore, A.M. (1997) Mechanistic aspects of xanthophyll cycle-dependent photoprotection in higher plant chloroplasts and leaves. Physiol. Plant. 99:197–209.

Goldschmidt, E.E. & Huber, S.C. (1992) Regulation of photosynthesis by end-product accumulation in leaves of plants storing starch, sucrose, and hexose sugars. Plant Physiol. 99:1443–1448.

Gornic, G., Le Gouallec, J.-L., Briantais, J.M., & Hodges, M. (1989) Effect of dehydration and high light on photosynthesis of two C_3 plants (*Phaseolus vulgaris* L. and *Elatostoma repens* (Lour) Hall f.). Planta 177:84–90.

Goudriaan, J. (1987) The biosphere as a driving force in the global carbon cycle. Neth. J. Agric. Sci. 35:177–187.

Goudriaan, J. (1993) Interaction of ocean and biosphere in their transient response to increasing atmospheric CO_2. Vegetatio 104/105:329–337.

Goudriaan, J. (1987) The biosphere as a driving force in the global carbon cycle. Neth. J. Agric. Sci. 35:177–187.

Gunasekera, D. & Berkowitz, G.A. (1992) Heterogenous stomatal closure in response to leaf water deficits is not a universal phenomenon. Plant Physiol. 98:660–665.

Houghton, J.T., Jenkins, G.J., & Ephraums, J.J. (1990) Climate change, The IPCC scientific assessment. Cambridge University Press, Cambridge.

Harris, F.S. & Martin, C.E. (1991) Correlation between CAM-cycling and photosynthetic gas exchange in five species of (*Talinum*) (Portulacaceae). Plant Physiol. 96:1118–1124.

Hubick, K. & Farquhar, G.D. (1989) Carbon isotope discrimination and the ratio of carbon gained to water lost in barley cultivars. Plant Cell Environ. 12:795–804.

Hanson, H.C. (1917) Leaf structure asrelated to environment. Am. J. Bot. 4:533–560.

Hatch, M.D. & Carnal, N.W. (1992) The role of mitochondria in C_4 photosynthesis. In: Molecular, biochemical and physiological aspects of plant respiration, H. Lambers & L.H.W. van der Plas (eds). SPB Academic Publishing, The Hague, pp. 135–148.

Hatch, M.D. & Slack, C.R. (1966) Photosynthesis by sugar cane leaves—a new carboxylation reaction and the pathway of sugar formation. Biochem. J. 101:103–111.

Hattersley, P.W. (1983) The distribution or C_3 and C_4 grasses in Australia in relation to climate. Oecologia 57:113–128.

Henderson, S., Hattersley, P., Von Caemmerer, S., & Osmond, C.B. (1995) Are C_4 pathway plants threatened by global climatic change? In: Ecophysiology of photosynthesis, E.-D. Schulze & M.M. Caldwell (eds). Springer-Verlag, Berlin, pp. 529–549.

Hougthon, J.T., Jenkins, G.J., & Ephraums, J.J. (1990) Climate Change, The IPCC Scientific Assesment. Cambridge University Press, Cambridge.

Johnson, G.N., Young, A.J., Scholes, J.D., & Horton, P. (1993a) The dissipation of excess excitation energy in British plant species. Plant Cell Environ. 16:673–679.

Johnson, G.N., Scholes, J.D., Horton, P., & Young, A.J. (1993b) Relationship between carotenoid composition and growth habit in British plant species. Plant Cell Environ. 16:681–686.

Johnson, H.B., Polley, H.W., & Mayeux, H.S. (1993) Increasing CO_2 and plant-plant interactions: Effects on natural vegetation. Vegetatio 104/105:157–170.

Jones, P.G., Lloyd, J.C., & Raines, C.A. (1996) Glucose feeding of intact wheat plants represses the expression of a number of Calvin cycle genes. Plant Cell Environ. 19:231–236.

Kao, W.-Y. & Forseth, I.N. (1992) Diurnal leaf movement, chlorophyll fluorescence and carbon assimilation in soybean grown under different nitrogen and water availabilities. Plant Cell Environ. 15:703–710.

Keeley, J.E. (1990) Photosynthetic pathways in freshwater aquatic plants. Trends Ecol. Evol. 5:330–333.

Keeley, J.E. & Busch, G. (1984) Carbon assimilation characteristics of the aquatic CAM plant, *Isoetes howellii*. Plant Physiol. 76:525–530.

Keeley, J.E. & Sandquist, D.R. (1991) Diurnal photosynthesis cycle in CAM and non-CAM seasonal-pool aquatic macrophytes. Ecology 72:716–727.

Keeley, J.E. & Sandquist, D.R. (1992) Carbon: Freshwater aquatics. Plant Cell Environ. 15:1021–1035.

Keeling, C.D. & Whorf, T.P. (1996) Atmospheric CO_2 records from sites in the SIO air sampling network. In: Trends: A compendium of data on global change. Carbon Dioxide Information Center, Oak Ridge National Laboratory, Oak Ridge.

Kirschbaum, M.U.F. & Pearcy, R.W. (1988) Gas exchange analysis of the relative importance of stomatal and biochemical factors in photosynthetic induction in *Alocasia macrorrhiza*. Plant Physiol. 86:782–785.

Kluge, M. & Ting, I.P. (1978) Crassulacean acid metabolism. Analysis of an ecological adaptation. Springer-Verlag, Berlin.

Koch, K.E. (1996) Carbohydrate-modulated gene expression in plants. Annu. Rev. Plant Physiol. Plant Mol. Biol. 47:509–540.

Koch, K.E. & Kennedy, R.A. (1982) Crassulacean acid metabolism in the succulent C_4 dicot, *Portulaca oleracea*

L. under natural environmental conditions. Plant Physiol. 69:757–761.

Kopriva, S., Chu, C.-C., & Bauwe, H. (1996) Molecular phylogeny of *Flaveria* as deduced from the analysis of nucleotide sequences encoding the H-protein of the glycine cleavage system. Plant Cell Environ. 19:1028–1036.

Körner, C. (1989) Bedeutung der Wälder in Naturhaushalt einer vom Menschen ver Šnderten Welt. In: Die Bedrohung der Wälder, H. Franz (ed). Verlag der Oesterreichischen Akademie der Wissenschaften, Vienna, pp. 7–40.

Korner, C. & Larcher, W. (1988) Plant life in cold climates. Symposium of the Society of Experimental Biology 42:25–57.

Knight, J.D., Livingston, N.J., & Van Kessel, C. (1994) Carbon isotope discrimination and water-use efficiency of six crops grown under wet and dryland conditions. Plant Cell Environ. 17:173–179.

Krall, J.P., Edwards, G.E., & Andrea, C.S. (1989) Protection of pyruvate, P_i dikinase from maize against cold lability by compatible solutes. Plant Physiol. 89:280–285.

Krapp, A. & Stitt, M. (1994) Influence of high carbohydrate content on the activity of plastidic and cytosolic isozyme pairs in photosynthetic tissues. Plant Cell Environ. 17:861–866.

Krapp, A., Hofmann, B., Schafer, C., & Stitt, M. (1993) Regulation of the expression of rbcS and other photosynthetic genes by carbohydrates: A mechanism for the "sink-regulation" of photosynthesis? Plant J. 3:817–828.

Krause, G.H. & Weis, E. (1991) Chlorophyll fluorescence and photosynthesis: The basics. Annu Rev. Plant Physiol. Plant Mol. Biol. 42:313–349.

Kropf, M. (1989) Quantification of SO_2 effects on physiological processes, plant growth and crop production. PhD Thesis, Wageningen Agricultural University, the Netherlands.

Kruger, I. & Kluge, M. (1987) Diurnal changes in the regulatory properties of phosphoenolpyruvate carboxylase in plants: Are alterations in the quaternary structure involved? Bot. Acta 101:24–27.

Lawlor, D.W. (1993) Photosynthesis; Molecular, physiological and environmental processes. Longman, London.

Leegood, R.C. & Osmond, C.B. (1990) The flux of metabolites in C4 and CAM plants. In: Plant physiology, biochemistry and molecular biology, D.T. Dennis & D.H. Turpin (eds). Longman Scientific & Technical, Singapore, pp. 274–298.

Lemon, E.R. (ed) (1983) CO_2 and Plants. The response of plants to rising atmospheric carbon dioxide. Westview Press, Boulder.

Leverenz, J.W. (1987) Chlorophyll content and the light response curve of shade adapted conifer needles. Physiol. Plant. 71:20–29.

Livingston, N.J. & Spittlehouse, D.L. (1993) Carbon isotope fractionation in tree rings in relation to the growing season water balance In: Stable isotopes and plant carbon-water relations, J.R. Ehleringer, A.E. Hall, &

G.D. Farquhar (eds). Academic Press, San Diego, pp. 141–153.

Logan, B.A., Barker, D.H., Demmig-Adams, B., & Adams, W.W. III (1996) Acclimation of leaf carotenoid composition and ascorbate levels to gradients in the light envoironment within an Australian rainforest. Plant Cell Environ. 19:1083–1090.

Long, S.P., Baker, N.R., & Raines, C.A. (1993) Analysing the responses of photosynthetic CO_2 assimilation to long-term elevation of atmospheric CO_2 concentration. Vegetatio 104/105:33–45.

Madsen, T.V. & Baattrup-Pedersen, A. (1995) Regulation of growth and photosynthetic performance in *Elodea canadensis* in response to inorganic nitrogen. Func. Ecol. 9:239–247.

Mansfield, T.A., Hetherington, A.M., & Atkinson, C.J. (1990) Some current aspects of stomatal physiology. Annu. Rev. Plant Physiol. Plant Mol. Biol. 41:55–75.

Marshall, J.D. & Zhang, J. (1993) Altitudinal variation in carbon isotope discrimination by conifers. In: Stable isotopes and plant carbon-water relations, Ehleringer, J.R., Hall, A.E., & Farquhar, G.D. (eds). Academic Press, San Diego, pp. 187–199.

Martin, B. & Thorstenson, Y.R. (1988) Stable carbon isotope composition ($\delta^{13}C$), water use efficiency, and biomass productivity of *Lycopersicon esculentum*, *Lycopersicon pennellii*, and the F_1 hybrid. Plant Physiol. 88:213–217.

Mazen, A.M.A. (1996) Changes in levels of phosphoenolpyruvate carboxylase with induction of Crassulacean acid metabolism (CAM)-like behavior in the C_4 plant *Portulaca oleracea*. Physiol. Plant. 98:111–116.

McConnaughey, T.A., LaBaugh, J.W., Rosenberry, D.O., Striegl, R.G., Reddy, M.M., & Schuster, P.F. (1994) Carbon budget for a groundwater-fed lake: Calcification supports summer photosynthesis. Limnol. Oceanogr. 39:1319–1332.

Medina, E. & Klinge, H. (1983) Productivity of tropical woodlands. In: Encyclopedia or plant physiology, N.S. Vol. 12D, O.L. Lange, P.S. Nobel, C.B. Osmond, & H. Ziegler (eds). Springer-Verlag, Berlin, pp. 281–303.

Medina, E. (1996) CAM and C_4 plants in the humid tropics. In: Tropical forest plant ecophysiology, S.D. Mulkey, R.L. Chazdon, & A.P. Smith (eds). Chapman & Hall, New York, pp. 56–88.

Meinzer, F., Goldstein, G., & Grantz, D.A. (1990) Carbon isotope discrimination in coffee genotypes grown under limited water supply. Plant Physiol. 92:130–135.

Monsi, M. & Saeki, T. (1953) Über den Lichtfaktor in den Pflanzengesellschaften und sein Bedeutung für die Stoffproduktion. Jap. J. Bot. 14:22–52.

Mooney, H.A. (1986) Photosynthesis. In: Plant ecology, M.J. Crawley (ed). Blackwell Scientific Publications, Oxford, pp. 345–373.

Morgan, C.L., Turner, S.R., & Rawsthorne, S. (1992) Cell-specific distribution of glycine decarboxylase in leaves of C_3, C_4 and C_3–C_4 intermediate species. In: Molecular, biochemical and physiological aspects of plant respiration, H. Lambers & L.H.W. Van der Plas (eds). SPB Academic Publishing, The Hague, pp. 339–343.

Morison, J.I.L. (1987) Intercellular CO_2 concentration and stomatal response to CO_2. In: Stomatal function, E. Zeiger, G.D. Farquhar, & I.R. Cowan (eds). Stanford University Press, Stanford, pp. 229–251.

Nakano, Y. & Edwards, G.E. (1987) Hill reaction, hydrogen peroxide scavenging, and ascorbate peroxidase activity or mesophyll and bundle sheath chloroplasts or NADP-malic enzyme type C4 species. Plant Physiol. 85:294–298.

Newman, J.R. & Raven, J.R. (1993) Carbonic anhydrase in *Ranunculus penicillatus* spp. *pseudofluitans*: Activity, location and implications for carbon assimilation. Plant Cell Environ. 16:491–500.

Nielsen, S.L., Gacia, E., & Sand-Jensen, K. (1991) Land plants or amphibious *Littorella* uniflora (L.) Aschers. maintain utilization of CO_2 from sediment. Oecologia 88:258–262.

Nobel, P.S. & Hartsock, T.L. (1990) Diel patterns of CO_2 exchange for epiphytic cacti differing in succulence. Physiol. Plant. 78:628–634.

Nobel, P.S., Garcia-Moya, E., & Quero, E. (1992) High annual productivity of certain agaves and cacti under cultivation. Plant Cell Environ. 15:329–335.

Ögren, E. (1993) Convexity of the photosynthetic light-response curve in relation to intensity and direction of light during growth. Plant Physiol. 101:1013–1019.

Ogren, W.L. (1984) Photorespiration: Pathways, regulation, and modification. Annu. Rev. Plant Physiol. 35:415–442.

O'Leary, M.H. (1993) Biochemical basis of carbon isotope fractionation. In: Stable isotopes and plant carbon-water relations, J.R. Ehleringer, A.E. Hall, & G.D. Farquhar (eds). Academic Press, San Diego, pp. 19–28.

Öquist, G., Brunes, L., & Hällgren, J.E. (1982) Photosynthetic efficiency of Betula pendula acclimated to diferent quantum flux densities. Plant Cell Environ. 5:9–15.

Osmond, C.B. & Holtum, J.A.M. (1981) Crassulacean acid metabolism. In: The biochemistry of plants. A comprehensive treatise, Vol. 8, P.K. Stumpf & E.E. Conn (eds). Academic Press, New York.

Osmond, C.B., Björkman, O., & Anderson, D.J. (1980) Physiological processes in plant ecology, ecological studies, Vol. 36. Springer-Verlag, Berlin.

Osmond, C.B., Winter, K., & Ziegler, H. (1982) Functional significance or different pathways or CO_2 fixation in photosynthesis. In: Encyclopedia or plant physiology, N.S. Vol. 12B, O.L. Lange, P.S. Nobel, C.B. Osmond, & H. Ziegler (eds). Springer-Verlag, Berlin, pp. 479–548.

Patel, A. & Ting, I.P. (1987) Relationship between respiration and CAM-cycling in *Peperomia camptotricha*. Plant Physiol. 84:640–642.

Pearcy, R.W. (1988) Photosynthetic utilisation of lightflecks by understorey plants. Aust. J. Plant Physiol. 15: 223–238.

Pearcy, R.W. (1990) Sunflecks and photosynthesis in plant canopies. Annu. Rev. Plant Physiol. Plant Mol. Biol. 41:421–453.

Pearcy, R.W., Osteryoung, K., & Calkin, H.W. (1985) Photosynthetic responses to dynamic light environments by Hawaiian trees. Time course of CO_2 uptake and carbon gain during sunflecks. Plant Physiol. 79:896–902.

Peisker, M. & Henderson, S.A. (1992) Carbon: Terrestrial C_4 plants. Plant Cell Environ. 15:987–1004.

Plaut, Z., Mayoral, M.L., & Reinhold, L. (1987) Effect of altered sink: Source ratio on photosynthetic metabolism of source leaves. Plant Physiol. 85:786–791.

Pons, T.L. & Pearcy, R.W. (1992) Photosynthesis in flashing light in soybean leaves grown in different conditions. II. Lightfleck utilization efficiency. Plant Cell Environ. 15:577–584.

Pons, T.L., Schieving, F., Hirose, T., & Werger, M.J.A. (1989) Optimization of leaf nitrogen allocation for canopy photosynthesis in *Lysimachia vulgaris*. In: Causes and consequences of variation in growth rate and productivity of higher plants, H. Lambers, M.L. Cambridge, H. Konings, & T.L. Pons (eds). SPB Academic Publishing, The Hague, pp. 175–186.

Pons, T.L., Pearcy, R.W., & Seemann, J.R. (1992) Photosynthesis in flashing light in soybean leaves grown in different conditions. I. Photosynthetic induction state and regulation of ribulose-1,5-bisphosphate carboxylase activity. Plant Cell Environ. 15:569–576.

Pons, T.L., Van der Werf, A., & Lambers, H. (1994) Photosynthetic nitrogen use efficiency of inherently slow- and fast-growing species: Possible explanations for observed differences. In: A Whole-plant perspective of carbon-nitrogen interactions, J. Roy & E. Garnier (eds). SPB Academic Publishing, pp. 61–77.

Poorter, L., Oberbauer, S.F., & Clark, D.B. (1995) Leaf optical properties along a vertical gradient in a tropical rain forest canopy in Costa Rica. Am. J. Bot. 82:1257–1263.

Poot, P., Pilon, J., & Pons, T.L. (1996) Photosynthetic characteristics of leaves of male sterile and hermaphroditic sex types of Plantago lanceolata grown under conditions of contrasting nitrogen and light availabilities. Physiol. Plant. 98:780–790.

Potvin, C. (1986) Differences in photosynthetic characteristics among northern and southern C_4 plants. Physiol. Plant. 69:659–664.

Powel, A.M. (1978) Systematics of *Flaveria* (Flaveriinae-Asteraceae). Ann. Missouri Bot. Garden 65:590–636.

Prins, H.B.A. & Elzenga, J.T.M. (1989) Bicarbonate utilization: function and mechanism. Aquat. Bot. 34:59–83.

Prins, H.B.A., Snel, J.F.H., Zanstra, P.E., & Helder, R.J. (1982) The mechanism of bicarbonate assimilation by the polar leaves of *Potamogeton* and *Elodea*. CO_2 concentrations at the leaf surface. Plant Cell Environ. 5:207–214.

Pyankov, V.I. & Kondratchuk, A.V. (1995) The specific features of structure of photosynthetic apparatus of plants of the East Pamirs. Proc. Russian Acad. Sci. 344:712–716 (in Russian).

Pyankov, V.I. & Kondratchuk, A.V. (1998) Meso-structure of photosynthetic apparatus of tree plants of The East Pamirs of different ecological and altitudinal groups. Russian J. Plant Physiol., in press.

Prins, H.B.A. & de Guia, M.B. (1986) Carbon source of the water soldier, *Stratiotes aloides* L. Aquat. Bot. 26:225–234.

Quick, W.P., Chaves, M.M., Wendler, R., David, M., Rodrigues, M.L., Passaharinho, J.A., Pereira, J.S., Adcock, M.D., Leegood, R.C., & Stitt, M. (1992) The effect of water stress on photosynthetic carbon metabolism in four species grown under field conditions. Plant Cell Environ. 15:25–35.

Rajendrudu, G., Prasad, J.S.R., & Das, V.S.R. (1986) C_3 C_4-intermediate species in *Alternanthera* (Amaranthaceae). Leaf anatomy, CO_2 compensation point, net CO_2 exchange and activities or photosynthetic enzymes. Plant Physiol. 80:409–414.

Ray, T.B. & Black, C.C. (1979) The C_4 and crassulacean acid metabolism pathways. In: Encyclopedia of plant physiology, N.S. Vol. 6, M. Gibbs & E. Latzko (eds). Springer-Verlag, Berlin, pp. 77–101.

Raymo, M.E. & Ruddiman, W.F. (1992) Tectonic forcing of late Cenozoic climate. Nature 359:117–122.

Reich, P.B. & Schoettle, A.W. (1988) Role of phosphorus and nitrogen in photosynthetic and whole plant carbon gain and nutrient use efficiency in eastern white pine. Oecologia 77:25–33.

Reich, P.B., Koike, T., Gower, S.T., & Schoettle, A.W. (1995) Causes and consequences of variation in conifer leaf life-span, In: Ecophysiology of coniferous forests, W.K. Smith & T.M. Hinckley (eds). Academic Press, San Diego, pp. 225–254.

Reiskind, J.B., Madsen, T.V., Van Ginkel, L.C., & Bowes, G. (1997) Evidence that inducible C_4-type photosynthesis is a chloroplastic CO_2-concentrating mechanism in *Hydrilla*, a submersed monocot. Plant Cell Environ. 20:211–220.

Rozema, J., Lambers, H., Van de Geijn, S.C., & Cambridge, M.L. (1992) CO_2 and Biosphere. Kluwer, Dordrecht.

Rundel, P.W. & Sharifi, M.R. (1993) Carbon isotope discrimination and resource availability in the desert shrub *Larrea tridentata*, In: Stable isotopes and plant carbon-water relations, J.R. Ehleringer, A.E. Hall, & G.D. Farquhar (eds). Academic Press, San Diego, pp. 173–185.

Sage, R.F. & Sharkey, T.D. (1987) The effect of temperature on the occurrence of O_2 and CO_2 insensitive photosynthesis in field grown plants. Plant Physiol. 84:658–664.

Sage, R.F. & Pearcy, R.W. (1987a) The nitrogen use efficiency or C_3 and C_4 plants. I. Leaf nitrogen, growth, and biomass partitioning in *Chenopodium album* (L.) and *Amaranthus retroflexus*. Plant Physiol. 84:954–958.

Sage, R.F. & Pearcy, R.W. (1987b) The nitrogen use efficiency or C_3 and C_4 plants. II. Leaf nitrogen effects on the gas exchange characteristics or *Chenopodium album* (L.) and *Amaranthus retroflexus*. Plant Physiol. 84:959–963.

Sage, R.F., Sharkey, T.D., & Seemann, J.R. (1989) Acclimation of photosynthesis to elevated CO_2 in five C_3 species. Plant Physiol. 89:590–596.

Salvucci, M.E. (1989) Regulation of Rubisco activity *in vivo*. Physiol. Plant. 77:164–171.

Sassenrath-Cole, G.F., Pearcy, R.W., & Steinmaus, S. (1994) The role of enzyme activation state in limiting carbon assimilation under variable light conditions. Photosynth. Res. 41:295–302.

Scheller, H.V. & Moller, B.L. (1990) Photosystem I polypeptides. Physiol. Plant. 78:484–494.

Schlesinger, W.H. (1993) Response of the terrestrial biosphere to global climate change and human perturbation. Vegetatio 104/105:295–305.

Schimel, D.S. (1995) Terrestrial ecosystems and the carbon cycle. Global Change Biol. 1:77–91.

Schulze, E.-D. (1986) Carbon dioxide and water vapor exchange in response to drought in the atmosphere and in the soil. Annu. Rev. Plant Physiol. 37:247–274.

Schulze, E.D. & Hall, A.E. (1982) Stomatal responses, water loss and CO_2 assimilation rates of plants from contrasting environments. In: Encyclopedia of plant physiology new series (Physiological plants ecology II, Water relations and carbon assimilation) V. 12B. Springer Verlag, Berlin, pp. 181–230.

Schulze, E.-D., Kelliher, F.M., Körner, C., Lloyd, J., & Leuning, R. (1994) Relationships among maximum stomatal conductance, ecosystem surface conductance, carbon assimilation rate, and plant nitrogen nutrition: A global ecology scaling exercise. Annu. Rev. Ecol. Syst. 25:629–660.

Seemann, J.R. (1989) Light adaptation/acclimation of photosynthesis and the regulation of ribulose-1,5-bisphosphate carboxylase activity in sun and shade plants. Plant Physiol. 91:379–386.

Servaites, J.C. (1990) Inhibition of ribulose 1,5-bisphosphate carboxylase/oxygenase by 2-carboxyarabinitol-1-phosphate. Plant Physiol. 92:867–870.

Seemann, J.R., Badger, M.R., & Berry, J.A. (1984) Variations in the specific activity of ribulose-1,5-bisphosphate carboxylase between species utilizing differing photosynthetic pathways. Plant Physiol. 74:791–794.

Sharkey, T.D., Seemann, J.R., & Pearcy, R.W. (1986a) Contribution of metabolites of photosynthesis to postillumination CO_2 assimilation in response to lightflecks. Plant Physiol. 82:1063–1068.

Sharkey, T.D., Stitt, M., Heineke, D., Gerhardt, R., Raschke, K., & Heldt, H.W. (1986b) Limitation of photosynthesis by carbon metabolism. II. CO_2-insensitive CO_2 uptake results from limitation of triose phosphate utilization. Plant Physiol. 81:1123–1129.

Sims, D.A. & Pearcy, R.W. (1989) Photosynthetic characteristics of a tropical forest understorey herb, *Alocasia macrorrhiza*, and a related crop species, *Colocasia esculenta*, grown in contrasting light environments. Oecologia 79:53–59.

Smedley, M.P., Dawson, T.E., Comstock, J.P., Donovan, L.A., Sherrill, D.E., Cook, C.S., & Ehleringer, J.R. (1991) Seasonal carbon isotope discrimination in a grassland community. Oecologia 85:314–320.

Smith, H., Samson, G., & Fork, D.C. (1993) Photosynthetic acclimation to shade: Probing the role of phytochromes using photomorphogenetic mutants of tomato. Plant Cell Environ. 16:929–937.

Staal, M., Elzenga, J.T.M., & Prins, H.B.A. (1989) $^{14}CO_2$ fixation by leaves and leaf cell protoplasts of the submerged aquatic angiosperm *Potamogeton lucens*: Carbon dioxide or bicarbonate? Plant Physiol. 90:1035–1040.

Sternberg, L.O., DeNiro, M.J., & Johnson, H.B. (1984) Isotope ratios of cellulose from plants having different photosynthetic pathways. Plant Physiol. 74:557–561.

Stitt, M. (1997) The flux of carbon between the chloroplast and the cytoplasm. In: Plant metabolism, D.T. Dennis, D.H. Turpin, D.D. Lefebvre, & D.B. Layzell (eds). Longman Scientific and Technical, Singapore, pp. 382–400.

Terashima, I. & Hikosaka, K. (1995) Comparative ecophysiology of leaf and canopy photosynthesis. Plant Cell Environ. 18:1111–1128.

Terashima, I., Wong, S.C., Osmond, C.B., & Farquhar, G.D. (1988) Characterisation of non-uniform photosynthesis induced by abscisic acid in leaves having different mesophyll anatomies. Plant Cell Physiol. 29:385–394.

Ting, I.T. (1985) Crassulacean acid metabolism. Annu. Rev. Plant Physiol. 36:595–622.

Van Oosten, J.-J. & Besford, R.T. (1995) Some relationships between the gas exchange, biochemistry and molecular biology of photosynthesis during leaf development of tomato plants after transfer to different carbon dioxide concentrations. Plant Cell Environ. 18:1253–1266.

Van Oosten, J.J., Wilkins, D., & Besford, R.T. (1995) Acclimation of tomato to different carbon dioxide concentrations. Relationships between biochemistry and gas exchange during leaf development. New Phytol. 130:357–367.

Vernon, D.M., Ostrem, J.A., Schmitt, J.M., & Bohnert, H. (1988) PEPCase transcript levels in *Mesembryanthemum crystallinum* decline rapidly upon relief from salt stress. Plant Physiol. 86:1002–1004.

Vitousek, P.M., Field, C.B., & Matson, P.A. (1990) Variation in foliar $\delta^{13}C$ in Hawaiian *Metrosideros polymorpha*: A case of internal resistance? Oecologia 84:362–370.

Vogelmann, T.C., Nishio, J.N., & Smith, W.K. (1996) Leaves and light capture: Light propagation and gradients of carbon fixation within leaves. Trends in Plant Science 1:65–70.

Von Caemmerer, S. (1989) A model of photosynthetic CO_2 assimilation and carbon-isotope discrimination in leaves of certain C_3–C_4 intermediates. Planta 178:463–474.

Von Caemmerer, S. & Farquhar, G.D. (1984) Effects of partial defoliation, changes of irradiance during growth, short-term water stress and growth at enhanced $p(CO_2)$ on photosynthetic capacity of leaves of *Phaseolus vulgaris* L. Planta 160:320–329.

Von Caemmerer, S., Evans, J.R., Hudson, G.S., & Andrews, T.J. (1994) The kinetics of ribulose-1,5-bisphosphate carboxylase/oxygenase in vivo inferred from measurements of photosynthesis in leaves of transgenic tobacco. Planta 195:88–97.

Walker, D.A. (1980) Regulation of starch synthesis in leaves. The role of orthophosphate. In: Physiological aspects of crop productivity. International Potash Institute, Bern, pp. 195–207.

Weger, H.G., Silim, S.N., & Guy, R.D. (1993) Photosynthetic acclimation to low temperature by western red cedar seedlings. Plant Cell Environ. 16:711–717.

Willeford, K.O. & Wedding, R.T. (1987) pH effects on the activity and regulation of the NAD malic enzyme. Plant Physiol. 84:1080–1083.

Winner, W.E. & Mooney, H.A. (1980a) Ecology of SO_2 resistance. I. Effects of fumigations on gas exchange of deciduous and evergreen shrubs. Oecologia 44:290–295

Winner, W.E. & Mooney, H.A. (1980b) Ecology of SO_2 resistance. II. Photosynthetic changes of shrubs in relation to SO_2 absorption and stomatal behavior. Oecologia 44:296–302.

Winter, K. & Smith, J.A.C. (1996) An introduction to crassulaceaen acid metabolism. Biochemical principles and ecological diversity. In: Crassulacean acid metabolism, biochemistry, ecophysiology and evolution. Ecological Studies 114, K. Winter & J.A.C. Smith (eds). Springer-Verlag, Berlin, pp. 1–13.

Winter, K., Zotz, G., Baur, B., & Dietz, K.-J. (1992) Light and dark CO_2 fixation in *Clusia uvitana* and the effects of plant water status and CO_2 availability. Oecologia 91:47–51.

Wu, M.-X. & Wedding, R.T. (1985) Regulation of phosphoenolpyruvate carboxylase from *Crassula* by interconversion of oligomeric forms. Arch. Biochem. Biophys. 240:655–662.

Wu, M.-X. & Wedding, R.T. (1987) Regulation of phosphoenolpyruvate carboxylase from *Crassula argentea*. Further evidence on the dimer-tetramer interconversion. Plant Physiol. 84:1084–1087.

Yeoh, H.-H., Badger, M.R., & Watson, L. (1981) Variations in kinetic properties of ribulose-1,5-bisphosphate carboxylase among plants. Plant Physiol. 67:1151–1155.

Young, D.R. & Smith, W.K. (1983) Effect of cloudcover on photosynthesis and transpiration in the subalpine understory species *Arnica latifolia*. Ecology 64:681–687.

2B.   Respiration

1. Introduction

A large portion of the carbohydrates that a plant assimilates each day are expended in respiration in the same period (Table 1). Dark respiration is needed to produce the energy and carbon skeletons to sustain plant growth. A significant part of respiration, however, may proceed via a nonphosphorylating pathway that is cyanide-resistant and generates less ATP than the cytochrome pathway, which occurs in both plants and animals. We have no satisfactory answer to date to the question why plants have a respiratory pathway that is not linked to ATP production; however, several hypotheses will be discussed in this chapter. It we seek to explain the carbon balance of a plant and to understand plant performance and growth in different environments, then it is imperative first to try to obtain a good understanding of respiration.

The types and rates of plant respiration are controlled by a combination of energy demand, substrate availability, and oxygen supply. At low levels of oxygen, respiration cannot proceed by normal aerobic pathways, and fermentation starts to take place, with ethanol and lactate as major end-products. The ATP yield of fermentation is considerably less than that of normal aerobic respiration. In this chapter, we will discuss the control over respiratory processes, the demand for respiratory energy and the significance of respiration for the plant's carbon balance, as these are influenced by species and environment.

2. General Characteristics of the Respiratory System

2.1 The Respiratory Quotient

The respiratory pathways in plant tissues include glycolysis, which is located both in the cytosol and in the plastids, the pentose phosphate pathways, which is located in the plastids, the tricarboxylic acid (TCA) or Krebs cycle, which is in the matrix of mitochondria, and the electron-transport pathways, which are in the inner mitochondrial membrane.

The **respiratory quotient** (RQ, the ratio between the number of moles of CO_2 released and that of O_2 consumed) is a useful index of the types of substrates used in respiration and the subsequent use of respiratory energy to support biosynthesis. In the absence of biosynthetic processes, the RQ of respiration in nonphotosynthetic tissues is expected to be 1.0, if sucrose is the only substrate for respiration and is fully oxidized to CO_2 and H_2O. For **roots** of young seedlings, measured in the absence of a nitrogen source, values close to 1.0 have indeed been found, but most experimental RQ values differ from unity (Table 2). RQ values for **leaves** tend to be close to 1.0, whereas values for germinating **seeds** depend on the storage compounds in the seeds. For seeds of *Triticum* (wheat), in which carbohydrates are major storage compounds, RQ is close to unity, whereas for the

TABLE 1. Utilization of photosynthates in plants, as dependent on the nutrient supply.

Item	Utilization of photosynthates % of C fixed	
	Free nutrient availability	Limiting nutrient supply
Shoot growth	40*–57	15–27*
Root growth	17–18*	33*–35
Shoot respiration	17–24*	19–20*
Root respiration	8–19*	38*–52
• growth	3.5–4.6*	6*–9
• maintenance	0.6–2.6*	?
• ion acquisition	4–13*	?
Volatile losses	0–8	0–8
Exudation	<5	<23
N_2-fixation	negligible	5–24
Mycorrhiza	negligible	7–20

Source: Van der Werf et al. 1994.

*, inherently slow-growing species; ?, no information for nutrient-limited conditions.

fat-storing seeds of *Linum* (flax) RQ values as low as 0.4 are found (Stiles & Leach 1936).

The nature of the respiratory substrate strongly influences RQ. The RQ can be greater than 1, if **organic acids** are an important substrate because organic acids are more oxidized than sucrose and, therefore, produce more CO_2 per unit O_2. On the other hand, RQ will be less than 1, if compounds that are more reduced than sucrose (e.g., **lipids** and **protein**) are a major substrate, as occurs during starvation of excised root tips (Table 2). Substrates available to support root respiration depend on processes that occur throughout the plant. For example, organic acids (malate) produced during the reduction of nitrate in leaves can be transported and decarboxylated in the roots, releasing CO_2 and increasing RQ (Ben Zioni et al. 1971). If **nitrate reduction** proceeds in the roots, then the RQ is also expected to be greater than 1 because an additional two molecules of CO_2 are produced per molecule of nitrate reduced to ammonium. Values of RQ, therefore, are lower in plants that use NH_4^+ as an N source than they are in plants grown with NO_3^- or, symbiotically, with N_2 (Table 2).

Biosynthesis influences RQ in several ways. Carboxylating reactions consume CO_2, reducing RQ, whereas decarboxylating reactions produce CO_2 and, therefore, increase RQ. In addition, synthesis of oxidized compounds, such as organic acids, decreases RQ, whereas the production of reduced compounds, such as lipids, leads to higher RQ values. The average molecular formula of the biochemical compounds typical of plant biomass is

TABLE 2. The respiratory quotient (RQ) of root respiration of several herbaceous species.

Species	RQ	Special remarks
Allium cepa	1.0	Root tips
	1.3	Basal parts
Dactylis glomerata	1.2	
Festuca ovina	1.0	
Galinsoga parviflora	1.6	
Helianthus annuus	1.5	
Holcus lanatus	1.3	
Hordeum distichum	1.0	
Lupinus albus	1.4	
	1.6	N_2-fixing
Oryza sativa	1.0	NH_4^+-fed
	1.1	
Pisum sativum	0.8	NH_4^+-fed
	1.0	
	1.4	N_2-fixing
Zea mays	1.0	Fresh tips
	0.8	Starved tips

Source: Various authors as summarized in Lambers et al. 1996.

* All plants were grown in nutrient solution, with nitrate as the N-source, unless stated otherwise. The *Pisum sativum* (pea) plants were grown with a limiting supply of combined N, so that their growth matched that of the symbiotically grown plants.

more reduced than sucrose, so RQ values influenced by biosynthesis should be greater than 1, as generally observed (Table 2; for further information, see Table 11 in Sect. 5.2.2).

RQ values of root respiration increase with increasing potential growth rate of a species (Fig. 1).

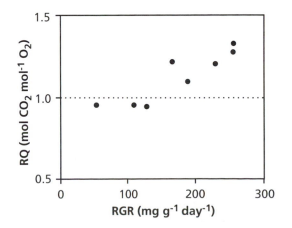

FIGURE 1. The respiratory quotient of several fast- and slow-growing grass species, grown with free access to nutrients and with nitrate as the source of nitrogen (I. Scheurwater, unpublished data).

Faster-growing species could have higher rates of nitrate reduction in their roots than slow-growing ones, which raises RQ. If the fast-growing species were to make greater use of the malate shuttle (i.e., decarboxylate relatively more malate and exchange the resultant HCO_3^- for nitrate and take up relatively less K^+) (cf. Ben Zioni et al. 1971), then this might also partly account for their higher RQ values; however, there is currently no evidence to confirm this hypothesis.

In summary, the patterns of RQ in plants clearly demonstrate that in roots it depends on the plant's growth rate. For all organs, it depends on the predominant respiratory substrate, integrated whole-plant processes, and ecological differences among species.

2.2 Glycolysis, the Pentose Phosphate Pathway, and the Tricarboxylic (TCA) Cycle

The first step in the production of energy for respiration occurs when glucose (or starch or other storage carbohydrates) is metabolized in glycolysis or in the oxidative pentose phosphate pathway (Fig. 2). **Glycolysis** involves the conversion of glucose, via phospho*enol*pyruvate (PEP), into malate and pyruvate. In contrast to mammalian cells, where virtually all PEP is converted into pyruvate, malate is the major end-product of glycolysis in plant cells and is thus the major substrate for the mitochondria. Key enzymes in glycolysis are controlled by adenylates (AMP, ADP, and ATP) in such a way as to speed up the rate of glycolysis when the demand for metabolic energy (ATP) increases.

Oxidation of one glucose molecule in glycolysis produces two malate molecules, without a net production of ATP. When pyruvate is the end-product, there is a net production of two ATP molecules in glycolysis. Despite the production of NADH in one step in glycolysis, there is no net production of NADH when malate is the end-product because of the need for NADH in the reduction of oxaloacetate, which is catalyzed by malate dehydrogenase.

Unlike glycolysis, which is predominantly involved in the breakdown of sugars and ultimately in the production of ATP, the **oxidative pentose phosphate** pathway plays a more important role in producing intermediates (e.g., amino acids, nucleotides) and NADPH. There is no evidence for a control of this pathway by the demand for energy.

The malate and pyruvate formed in glycolysis in the cytosol are imported into the mitochondria, where they are oxidized in the **TCA cycle**. Complete oxidation of one molecule of malate yields four molecules of CO_2, five molecules of NADH, and one molecule of $FADH_2$, as well as one molecule of ATP (Fig. 2). NADH and $FADH_2$ subsequently donate their electrons to the electron-transport chain (Sect. 2.3.1).

2.3 Mitochondrial Metabolism

The malate formed in glycolysis in the cytosol is imported into the mitochondria and oxidized partly via **malic enzyme**, producing pyruvate and CO_2, and partly via **malate dehydrogenase**, producing oxaloacetate. Pyruvate is then oxidized in the **TCA cycle**, so that malate is regenerated. Oxidation of malate and other NAD-linked substrates of the TCA cycle is associated with complex I (Sect. 2.3.1). In mitochondria there are four major complexes associated with electron transfer and one associated with oxidative phosphorylation, all located in the inner mitochondrial membrane. In addition, there are two small redox molecules, ubiquinone (Q) and cytochrome c, that play a role in electron transfer. Finally, in plant mitochondria there is the cyanide-resistant alternative oxidase, which is also located in the inner membrane (Fig. 3).

2.3.1 The Complexes of the Electron-Transport Chain

Complex I is the main entry point of electrons from NADH that is produced in the **TCA cycle** or in **photorespiration** (glycine oxidation). Complex I is the **first coupling site** or **site 1** of proton extrusion, and this is linked to ATP production. Succinate is the only intermediate of the TCA cycle that is oxidized by a membrane-bound enzyme: succinate dehydrogenase (Fig. 3). Electrons enter the respiratory chain via complex II and are transferred to ubiquinone. NAD(P)H that is produced outside the mitochondria, also feeds its electrons into the chain at the level of ubiquinone (Fig. 3). As with complex II, the external dehydrogenases are not connected with the translocation of H^+ across the inner mitochondrial membrane. Less ATP, therefore, is produced per oxygen atom when succinate or NAD(P)H are oxidized in comparison with that when malate, citrate, or oxoglutarate are oxidized. Complex III transfers electrons from ubiquinone to

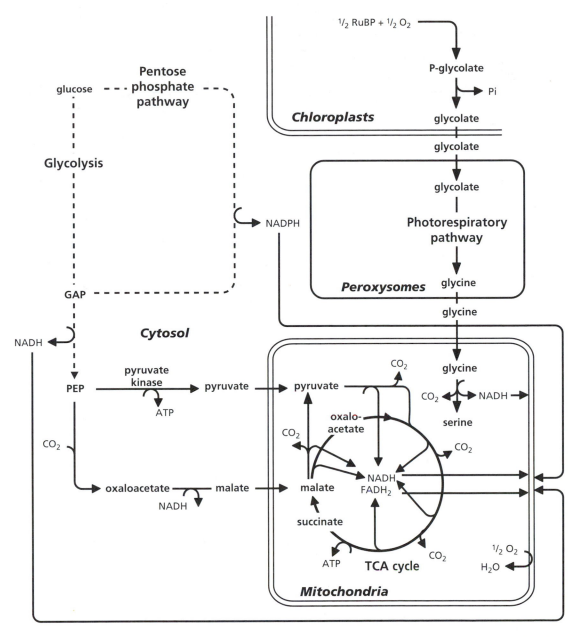

FIGURE 2. The major substrates for the electron transport pathways. Glycine is only a major substrate in photosynthetically active cells of C_3 plants when photorespiration plays a role (after Lambers 1997a). Reprinted with permission from Nature © copyright 1997 Macmillan Magazines Ltd.

cytochrome c, which is coupled with the extrusion of four protons per electron pair to the intermembrane space and is therefore **site 2** of proton extrusion. Complex IV is the terminal oxidase of the cytochrome pathway, accepting electrons from cytochrome c and donating these to O_2. It also generates a proton-motive force, which makes complex IV the **third coupling site**.

2.3.2 A Cyanide-Resistant Terminal Oxidase

Mitochondrial respiration of many tissues from higher plants is not fully inhibited by inhibitors of the cytochrome path (e.g., KCN). This is due to the presence of a cyanide-resistant, alternative electron-transport pathway, which consists of one enzyme, the **alternative oxidase**, that is firmly embedded in

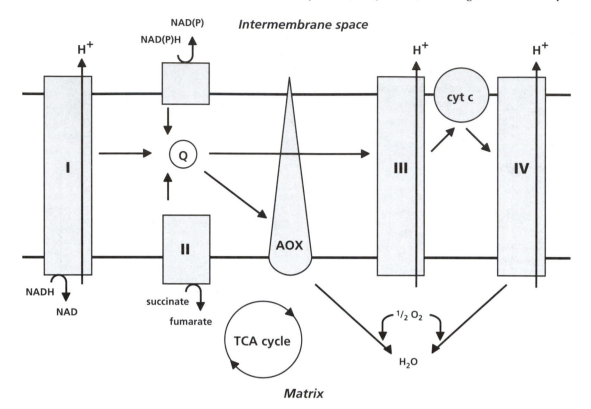

Figure 3. The organization of the electron-transporting complexes of the respiratory chain in higher plant mitochondria. All components are located in the inner mitochondrial membrane. Some of the components are membrane-spanning; others face the mitochondrial matrix or the space between the inner and the outer mitochondrial membrane. Q (ubiquinone) is a mobile pool of quinone and quinol molecules (after McIntosh 1994). Copyright American Society of Plant Physiologists.

the inner mitochondrial membrane. The branching point of the alternative path from the cytochrome path is at the level of **ubiquinone**, which is a component common to both pathways. Transfer of electrons from ubiquinone to oxygen via the alternative path is not coupled to the extrusion of protons from the matrix to the intermembrane space. Hence, the transfer of electrons from NADH produced inside the mitochondria to O_2 via the alternative path yields only one third of the amount of ATP that is produced when the cytochrome path is used.

2.3.3 Substrates, Inhibitors, and Uncouplers

Figure 2 summarizes the major substrates for mitochondrial oxygen uptake as well as their origin. Oxidation of glycine is of quantitative importance only in tissues that exhibit **photorespiration**. Glycolysis may start with glucose, as depicted here, or with starch, sucrose, or any major transport carbohydrate or sugar alcohol imported via the phloem (cf. Sect. 2 of the chapter on long-distance transport).

A range of respiratory inhibitors have helped to elucidate the organization of the respiratory pathways. To give just one example, **cyanide** effectively blocks complex IV and has been used to demonstrate the existence of the alternative path. **Uncouplers** make membranes, including the inner mitochondrial membrane, permeable to protons and, therefore, prevent oxidative phosphorylation. Many compounds that inhibit components of the respiratory chain or have an uncoupling activity naturally occur as **secondary compounds** in plant and fungal tissues (cf. Sects. 2 and 3.1 of ecological biochemistry) which protects these tissues from being grazed or infected by other organisms. In addition, concentrations of CO_2, in a range which is expected to occur within the twenty-first century, inhibit leaf respiration, due to inhibition of, for example, cytochrome oxidase and succinate dehydrogenase (Sect. 3.6). Do the high CO_2 concentrations which normally occur in soil also inhibit root respiration? There is a remarkable lack of information in the literature to answer this obvious question in a satisfactory manner (Sect. 4.7).

2.3.4 Respiratory Control

To learn more about the manner in which plant respiration responds to the demand for metabolic energy, we will first describe some experiments with **isolated mitochondria**. Freshly isolated intact mitochondria in an appropriate buffer, which is a condition referred to as "state 1," do not consume an appreciable amount of oxygen; in vivo they rely on a continuous import of respiratory substrate from the cytosol (Fig. 4). Upon addition of a respiratory substrate ("state 2") there is some oxygen uptake, but still not much; for rapid rates of respiration to occur in vivo, import of additional metabolites is required. As soon as ADP is added, a

rapid consumption of oxygen can be measured. This "state" of the mitochondria is called "state 3." In vivo, rapid supply of ADP will occur when a large amount of ATP is required to drive biosynthetic or transport processes. Upon conversion of all ADP into ATP ("state 4"), the respiration rate of the mitochondria declines again to the rate found before addition of ADP (Fig. 4). Upon addition of more ADP, the mitochondria go into state 3 again, followed by state 4 upon depletion of ADP. This can be repeated until all oxygen in the curette is consumed. Thus, the respiratory activity of isolated mitochondria is effectively controlled by the availability of ADP: **respiratory control**, quantified in the "respiratory control ratio" (the ratio of the rate at substrate saturation in the presence of ADP and that under the same conditions, but after ADP has been depleted; Fig. 4). The same respiratory control occurs in intact tissues and is one of the mechanisms ensuring that the rate of respiration is enhanced when the demand for ATP increases.

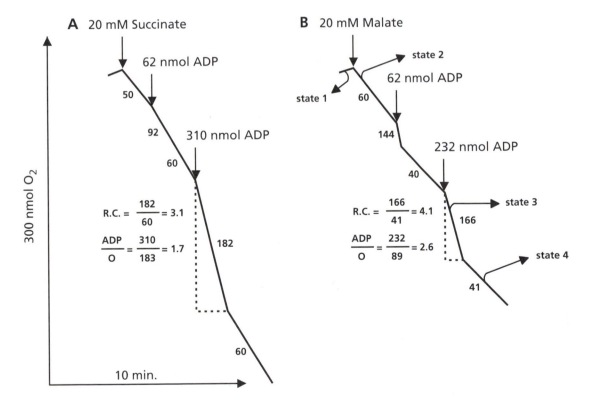

FIGURE 4. The "states" of isolated mitochondria. The ADP:O ratio is calculated from the oxygen consumption during the phosphorylation of a known amount of added ADP (state 3). The respiratory control ratio (R.C.) is the ratio of the rate of oxygen uptake in state 3 and state 4. State 1 refers to the respiration in the absence of respiratory substrate and state 2 is the respiration after addition of respiratory substrate, but before addition of ADP. The numbers next to lines give the rates of O_2 consumption (based on unpublished data from A.M. Wagner, Free University of Amsterdam).

2.4 A Summary of the Major Points of Control of Plant Respiration

We briefly discussed the control of glycolysis by "energy demand" (Sect. 2.2) and a similar control by "energy demand" of mitochondrial electron-transport, termed **respiratory control** (Sect. 2.3.4). The effects of **energy demand** on dark respiration are a function of metabolic energy required for **growth, maintenance**, and **transport** processes; therefore, when tissues grow fast, take up ions rapidly, and/or have a fast turnover of proteins, they generally have a high rate of respiration. At low levels of **respiratory substrate** (carbohydrates, organic acids), however, the activity of respiratory pathways may be substrate-limited. When substrate levels increase, respiration may exceed a rate required to satisfy the demand for metabolic energy. Under such conditions, the activity of the cyanide-resistant, alternative respiratory path increases. In the long run, the respiratory capacity is enhanced and adjusted to the high substrate input, through the transcription of specific genes that code for respiratory enzymes. Figure 5 summarizes these and several other points of control. Plant respiration is clearly quite flexible and responds rapidly to the demand for respiratory energy as well as the supply of respiratory substrate. The production of ATP coupled to the oxidation of substrate may also vary widely due to the presence of both nonphosphorylating and phosphorylating paths.

2.5 ATP Production in Isolated Mitochondria and In Vivo

The rate of oxygen consumption during the phosphorylation of ADP can be related to the total ADP that must be added to consume this oxygen. This allows calculation of the ADP:O ratio in vitro. This ratio is around 3 for NAD-linked substrates and around 1.5 for succinate and external NAD(P)H. Nuclear magnetic resonance (NMR) spectroscopy has been used to estimate ATP production and oxygen consumption in intact tissues, as will be outlined in Section 2.5.2.

2.5.1 Oxidative Phosphorylation: The Chemiosmotic Model

During the transfer of electrons from various substrates to oxygen via the cytochrome path, protons are extruded into the space between the inner and outer mitochondrial membranes. This generates a **proton-motive force** across the inner mitochondrial membrane that drives the synthesis of ATP. The basic features of this chemiosmotic model are (Elthon & Stewart 1983):

1. Protons are transported outward, coupled to the transfer of electrons, thus giving rise to both a **proton gradient** (ΔpH) and a **membrane potential** ($\Delta\Psi$)
2. The inner membrane is impermeable to protons and other ions, except by special transport systems
3. There is an **ATP synthetase** (also called **ATPase**), which transforms the energy of the electrochemical gradient generated by the proton-extruding system into ATP

The pH gradient, ΔpH, and the membrane potential, $\Delta\Psi$, are interconvertible. It is the combination of the two that forms the **proton-motive force** (Δp), which is the driving force for ATP synthesis, catalyzed by an ATPase:

$$\Delta p = \Delta\Psi - 2.3 RT/F.\Delta pH \qquad (1)$$

where R is the gas constant ($J\,mol^{-1}$), T is the absolute temperature (K), and F is Faraday's number (Coulomb). Both components in the equation are expressed in millivolts. Approximately one ATP is produced per three protons transported.

2.5.2 ATP Production In Vivo

As mentioned earlier, ATP production in vivo can be measured using **NMR spectroscopy**. This technique relies on the fact that certain nuclei, including ^{31}P, possess a permanent magnetic moment because of nuclear spin. Such nuclei can be made "visible" in a strong external magnetic field, in which they orient their nuclear spins in the same direction. It is just like the orientation of a small magnet in response to the presence of a strong one. Nuclear magnetic resonance spectroscopy allows one to monitor the absorption of radiofrequency by the oriented spin population in the strong magnetic field. The location of the peaks in an NMR spectrum depends on the molecule in which the nucleus is present as well as on the "environment" of the molecule (e.g., pH). Figure 6 illustrates this point for a range of phosphate-containing molecules (Roberts 1984).

The resonance of specific P-containing compounds can be altered by irradiation with radiofrequency power. If this irradiation is sufficiently strong ("saturating"), then it disorientates the nuclear spins of that P-containing compound so that its peak disappears from the spectrum.

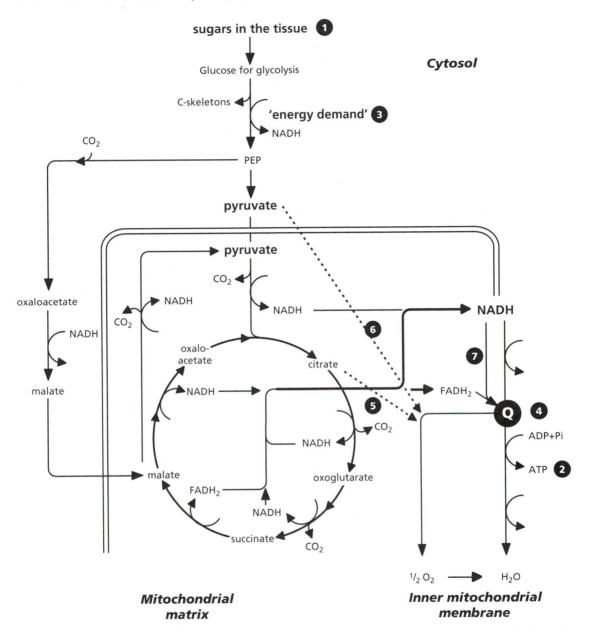

sugars in the tissue ❶

Cytosol

Glucose for glycolysis

C-skeletons ◄

'energy demand' ❸

NADH

CO_2

PEP

pyruvate

pyruvate

CO_2

NADH

oxaloacetate

NADH

CO_2

NADH

NADH

malate

oxalo-acetate

citrate

❻

❼

$FADH_2$

Q ❹

❺

ADP+Pi

CO_2

NADH

ATP ❷

malate

oxoglutarate

$FADH_2$

NADH

succinate ◄

CO_2

$^1/_2 O_2$ ⟶ H_2O

**Mitochondrial
matrix**

**Inner mitochondrial
membrane**

FIGURE 5. A simplified scheme of respiration and its major control points. Controlling factors include the concentration of respiratory substrate [e.g., glucose (1), and adenylates (2, 3)]. Adenylates may exert control on electron transport via a constraint on the rate of oxidative phosphorylation (2) as well as on glycolysis, via modulation of the activity of key enzymes in glycolysis, phosphofructokinase, and pyruvate kinase ("energy demand", 3). When the input of electrons into the respiratory chain is very high, a large fraction of ubiquinone becomes reduced and the alternative path becomes more active (4). When the rate of glycolysis is very high, relative to the activity of the cytochrome path, organic acids may accumulate (5, 6). The accumulation of citric acid may reduce the sulfide bonds of the alternative oxidase and thus enhance the capacity of the alternative path (5). Pyruvate accumulation may activate the alternative oxidase and allow it to function at a low level of reduced ubiquinone (6). There is increasing evidence that the nonphosphorylating rotenone-insensitive bypass (7) operates only when the concentration of NADH is very high.

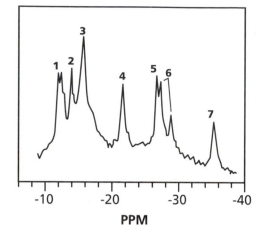

FIGURE 6. NMR spectrum of root tips of *Pisum sativum* (pea), showing peaks of, for example, glucose-6-phosphate (1), (P$_i$) (2, 3), and ATP (4, 5) in a living plant cell. The exact radiofrequency at which a phosphate-containing compound absorbs depends on the pH. This explains why there are two peaks for P$_i$: a small one for the cytosol (2), where the pH is approximately 7, and a larger one for the vacuole (3), where the pH is lower (Roberts 1984). Copyright American Society of Plant Physiologists.

Figure 7A illustrates this for the γ-ATP phosphate atom, which is the phosphate atom that is absent in ADP. Upon hydrolysis of ATP, the γ-ATP phosphate atom becomes part of the cytoplasmic P$_i$ pool. For a brief period, therefore, some of the P$_i$ molecules also contain disoriented nuclear spins; specific radiation of the γ-ATP peak decreases the P$_i$ peak. This phenomenon is called "saturation transfer" (Fig. 7). Saturation transfer has been used to estimate the rate of ATP hydrolysis to ADP and P$_i$ in vivo.

If the rate of disappearance of the saturation in the absence of biochemical exchange of phosphate between γ-ATP and P$_i$ is known, then the rate of ATP hydrolysis can be derived from the rate of loss of saturation. This has been done for root tips for which the oxygen uptake was measured in parallel experiments. In this manner ADP:O ratios in *Zea mays* (maize) root tips exposed to a range of conditions have been determined (Table 3).

The ADP:O ratios for the root tips supplied with 50 mM glucose are remarkably close to those expected when glycolysis plus the TCA cycle are responsible for the complete oxidation of exogenous glucose, as long as the alternative path does not contribute to the oxygen uptake (Table 3). KCN decreases the ADP:O ratio of glucose oxidation in a manner to be expected from mitochondrial studies.

SHAM, which is an inhibitor of the alternative path, has no effect on the rate of ATP production. So far, maize root tips are the only intact plant material used for the determination of ADP:O ratios in vivo. We cannot assume, therefore, that the ADP:O ratio in vivo is invariably 3. In fact, the ratio under most circumstances is likely to be far less than 3 (Sect. 2.6.2).

2.6 Regulation of Electron Transport Via the Cytochrome and the Alternative Paths

The existence of two mitochondrial respiratory pathways both of which transport electrons to oxygen, raises the question if and how the **partitioning of electrons** between the two paths is regulated. This is important because the cytochrome path is coupled to proton extrusion and the production of ATP, whereas transport of electrons via the alternative path is not, at least not from the

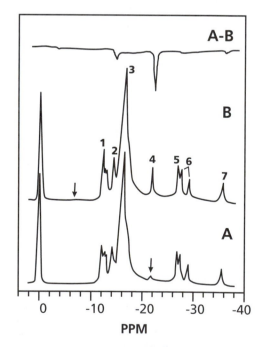

FIGURE 7. Saturation transfer from γ-ATP phosphate to cytosolic P$_i$ in root tips of *Zea mays* (maize). Spectrum A was obtained with selective presaturation of the γ-ATP peak. Spectrum B was obtained with selective presaturation of a point equidistant from the cytosolic P$_i$ peak. Spectrum A-B gives the difference between the two spectra, showing the transfer of saturation from γ-ATP to cytosolic P$_i$ (Roberts et al. 1984a). Copyright American Society of Plant Physiologists.

TABLE 3. The in vivo ADP:O ratios in root tips of *Zea mays* (maize) determined with the saturation transfer ^{31}P NMR technique and oxygen uptake measurements.*

Exogenous substrate	Oxygen concentration	Inhibitor	Rate of O_2 uptake	Rate of ATP production	ADP:O ratio
Glucose	100	None	90	143	3.2
Glucose	0	None	0	<20	—
None	100	None	60	93	3.0
Glucose	100	KCN	55	26	1.0
Glucose	100	KCN + SHAM	15	<20	—
Glucose	100	SHAM	84	137	3.2

Source: Roberts et al. 1984a.

*The oxygen concentration was either that in air (100) or zero. Rates of ATP production and O_2 consumption are expressed as $nmol\,g^{-1}\,FM\,s^{-1}$. Exogenous glucose was supplied at 50 mM. The concentration of KCN was 0.5 mM and that of SHAM 2 mM; this is sufficiently high to fully block the alternative path in maize root tips.

point where both pathways branch to oxygen (ubiquinone).

2.6.1 Competition or Overflow?

There was initially a substantial amount of experimental data collected with isolated mitochondria that suggested that inhibition of the alternative path had no effect on the activity of the cytochrome path. It was widely believed, therefore, that the alternative path did *not* compete for electrons with the cytochrome path, and that it served as an overflow when the cytochrome path was (virtually) saturated with electrons (Bahr & Bonner 1973). It was found much later that the activity of the cytochrome path increases linearly with the fraction of ubiquinone (the common substrate with the alternative path) that is in its reduced state (Q_r/Q_t). By contrast, the alternative path showed no appreciable activity until a substantial (30 to 40%) fraction of the ubiquinone was in its reduced state; the activity then increased rapidly (Dry et al. 1989). By that stage a sound biochemical explanation seemed available for the "**energy overflow**" model.

Research with "titration experiments" that use intact tissues provided additional evidence that the alternative pathway was primarily an overflow for electrons when the cytochrome pathway became saturated. In these experiments, oxygen uptake was measured over a range of concentrations of an inhibitor of the cytochrome path in the absence and presence of an inhibitor that fully blocks the alternative path (Fig. 8). If the alternative path competed with the cytochrome path for electrons, then we would expect the inhibition of the cytochrome path to be greater in the presence than in the absence of an inhibitor of the alternative path. Because this pattern was not observed (Fig. 8), it was concluded

that the alternative path functioned largely as an overflow for electrons. More recent findings, however, led to a thorough revision of this model, as outlined in the following section.

2.6.2 The Intricate Regulation of the Alternative Oxidase

Evidence has indicated that the alternative pathway can change its level of activity, so that it sometimes competes with the cytochrome pathway from electrons. When embedded in the inner mitochondrial membrane, the alternative oxidase exists as a dimer, with the two subunits linked by **disulfide bridges**. These sulfide bridges may be oxidized or reduced. If they are reduced, then the alternative oxidase switches to a **higher-activity state**, as opposed to the **lower-activity state** when the sulfide bridges are oxidized. In vitro the change from the oxidized to the reduced state can be brought about by isocitrate and other organic acids. Note that this is not an effect of the organic acids being used as a respiratory substrate. In addition, citrate accumulation enhances expression of the gene that encodes the alternative oxidase (Vanlerberghe & McIntosh 1996).

The alternative oxidase's capacity to oxidize its substrate (Q_r) also increases in the presence of **pyruvate** and some other **organic acids** (Millar et al. 1996). When that happens, the alternative path becomes engaged even when less than 30% of ubiquinone is in its reduced state. Note again that this is not because pyruvate is used as a respiratory substrate. In other words, the alternative pathway becomes active even when the cytochrome pathway is not fully saturated (Fig. 9).

Pyruvate, isocitrate, and other organic acids probably accumulate when their production ex-

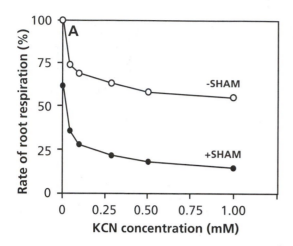
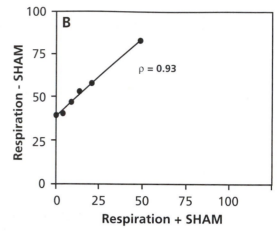

FIGURE 8. (A). Oxygen uptake by intact roots of *Carex acutiformis* at a range of concentrations of a specific inhibitor of the cytochrome path (KCN) in the absence and presence of a concentration of an inhibitor which fully blocks the alternative path (SHAM). (B) The rates obtained in the absence of an inhibitor of the cytochrome path are plotted against those obtained in the presence of an inhibitor blocking the cytochrome path. A straight line with a slope of 1 indicates that the alternative path is not employed until the cytochrome path reaches saturation. The slope of the line ($\rho = 0.93$) was not significantly different from 1.0. The applied inhibitor of the alternative path was salicylhydroxamic acid, SHAM (after Van der Werf et al. 1988). Copyright Physiologia Plantarum.

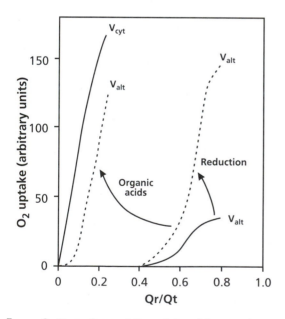

FIGURE 9. Dependence of the activity of the cytochrome path and of the alternative path on the fraction of ubiquinone that is in its reduced state (Q_r/Q_t). When the alternative oxidase is in its "reduced" (higher-activity) configuration, it has a greater capacity to accept electrons. In its reduced state, the alternative oxidase can be affected by organic acids, which enhance its activity at low levels of Q_r. [Based on Day et al. (1995) Dry et al. (1989). Hoefnagel & Wiskich 1996, and Umbach et al. (1994), as summarized in Lambers (1997b).]

ceeds their rate of oxidation by the mitochondria. When organic acids induce the higher-activity state of the alternative oxidase and increase its activity at low levels of Q_r/Q_t, competition between the two pathways may occur. Indeed, both in vitro (Hoefnagel et al. 1995, Ribas-Carbo et al. 1995) and in vivo (Atkin et al. 1995), the two respiratory pathways can compete for electrons.

3. The Ecophysiological Function of the Alternative Path

Why should plants produce and maintain a pathway that supports nonphosphorylating electron transport in mitochondria? It could merely be a relic or an "error" in the biochemical machinery that has not yet been eliminated by natural selection. On the other hand, there may be situations where respiration in the absence of ATP production could serve important physiological functions. This section discusses the merits of hypotheses put forward to explain the presence of the alternative path in higher plants. Testing of these hypotheses will require the use of mutants lacking alternative path activity, some of which have now been produced with molecular techniques.

3.1 Heat Production

An important consequence of the lack of coupling to ATP production in the alternative pathway is that the energy produced by oxidation is released as **heat**. More than 200 years have passed since Lamarck described heat production in *Arum*, and more than 60 years since thermogenesis was linked to cyanide-resistant respiration (Laties 1998). This heat production is ecologically important in the spadix (inflorescence) of *Arum* lilies as well as in some other flowers (Knutson 1974). These inflorescences may expand in early spring when air temperatures are low and can "melt" their way through late-lying snow. Preceding the upsurge in respiration, which is generally referred to as the "respiratory crisis," salicylic acid accumulates, and this triggers the increase in respiration in some parts of the flower (Raskin et al. 1987). During its respiratory crisis the respiration rate of this reproductive organ increases to very high levels. As a result, the temperature rises to approximately 10°C

above ambient so that odoriferous amines are volatilized, pollinators are attracted, and rates of ovule and pollen development increase (Meeuse 1975). Just before the respiratory crisis the alternative oxidase is transformed from its oxidized **lower-activity configuration** into its reduced, **higher-activity state** (Umbach & Siedow 1993). During heat production the respiration of the spadix is largely cyanide-resistant. This contributes to the heat production, as the lack of proton extrusion coupled with electron flow allows a large fraction of the energy in the substrate to be released as heat. This regulated thermogenic activity in inflorescences is analogous, in effect but not in mechanism, to the uncoupled respiration that occurs in thermogenic tissues (brown fat) of some mammals under cold conditions.

Heat production also occurs in the flowers of *Nelumbo nucifera* (lotus), which is presumably also linked to activity of the alternative path. These flowers regulate their temperature with remarkable precision (Fig. 10). When the air temperature varies

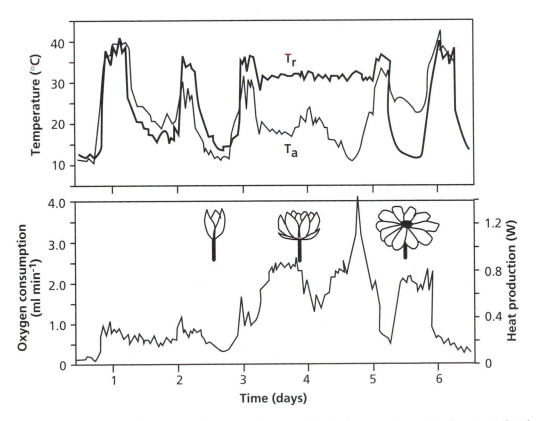

FIGURE **10.** Temperature of the receptacle (T$_r$) and ambient air (T$_a$) and rates of oxygen consumption throughout the thermogenic phase in *Nelumbo nucifera* (the sacred lotus). Oxygen consumption is converted to heat production, assuming 21.1 J per ml of oxygen (Seymour & Schultze-Motel 1996). Copyright Nature.

between 10 and 30°C, the flowers remain between 30 and 35°C. The stable temperature is a consequence of increasing respiration rates in proportion to decreasing temperatures. Such a phenomenon of thermoregulation in plants is known for only two other species: *Philodendron selloum* and *Symplocarpus foetidus* (skunk cabbage) (Knutson 1974). It has been suggested that the heat production in lotus is an energetic reward for pollinating beetles. These are trapped overnight, when they feed and copulate, and then carry the pollen away (Seymour & Schultze-Motel 1996).

Can the alternative oxidase also play a significant role in increasing the temperature of leaves, for example during exposure to low temperature? There is indeed some evidence for increased heat production (7 to 22% increase) in low-temperature–resistant plants (Moynihan et al. 1996). Using an approach outlined in the chapter on the plant's energy balance, however, it can readily be calculated that such an increase in heat production *cannot* lead to a significant temperature rise in leaves (less that 0.1°C); hence, it is unlikely to play a role in any cold-resistance mechanism. Other ecophysiological roles must be invoked to explain the contribution of the alternative path in respiration of nonthermogenic organs.

3.2 Can We Really Measure the Activity of the Alternative Path?

Does the alternative path also play a role in the respiration of "ordinary" tissues, such as roots and leaves? The application of specific inhibitors of the alternative path suggests that the alternative path does contribute to the respiration of roots and leaves of several species (Tables 4, 5). The decline in respiration, however, upon addition of an inhibitor of the alternative path may underestimate the actual activity of the alternative path. If the two pathways competed for electrons, as we now know they often do, then the inhibition is less than the activity of the alternative path (Table 5). Thus, any observed inhibition of respiration following the addition of an alternative pathway inhibitor indicates that some alternative pathway activity was present prior to inhibition, but provides no quantitative estimate of its activity (Day et al. 1996).

Stable isotopes can be used to estimate alternative path activity without the complications caused by use of inhibitors. The alternative oxidase and cytochrome oxidase discriminate to a different extent against the heavy isotope of oxygen (^{18}O) when reducing O_2 to produce water (Guy et al. 1992, Robinson et al. 1995). This allows calculation of the

TABLE 4. A comparison of the KCN-resistance of respiration of intact tissues of a number of species and of O_2 uptake by mitochondria isolated from these tissues.

Species	Tissue	Cyanide-resistance (%)	
		Whole tissue	Mitochondria
Gossypium hirsutum	Roots	36	22
Phaseolus vulgaris	Roots	61	41
Spinacia oleracea	Roots	40	34
Triticum aestivum	Roots	38	35
Zea mays	Roots	47	32
Pisum sativum	Leaves	39	30
Spinacia oleracea	Leaves	40	27

Source: Lambers 1997b.

*The percentage KCN-resistance of intact tissue respiration was calculated from the rate measured in the presence of 0.2 mM KCN and that measured in the presence of 0.1 µM FCCP, an uncoupler of the oxidative phosphorylation; this was done to obtain a rate of electron transfer through the cytochrome path similar to the state 3 rate (cf. Fig. 4). KCN-resistance of isolated mitochondria was calculated from the rate in the presence and absence of 0.2 mM KCN. Mitochondrial substrates were 10 mM malate plus 10 mM succinate and a saturating amount of ADP. KCN-resistant oxygen uptake by isolated mitochondria was fully inhibited by inhibitors of the alternative path; in the presence of both KCN and SHAM approximately 10% of the control respiration proceeded in some of the tissues ("residual respiration").

TABLE 5. KCN-resistance, expressing the total respiratory electron flow through the alternative path under the conditions of measurement, and SHAM-inhibition of root respiration.*

Species	KCN-resistance	SHAM-inhibition
Cytochrome path saturated:		
Carex diandra	66	29
Hordeum distichum	34	0
Pisum sativum	40	11
Plantago lanceolata	53	45
Cytochrome path not saturated:		
Festuca ovina	53	1
Phaseolus vulgaris	57	4
Poa alpina	41	1
Poa costiniana	61	0

Source: Atkin et al. 1995.
* Values are expressed in percentage of the control rate of respiration. KCN and SHAM (salicylhydroxamic acid) are specific inhibitors of the cytochrome path and the alternative path, respectively. SHAM inhibition equals the activity of the alternative path only when the cytochrome path is saturated. If the cytochrome path is not saturated, then SHAM-inhibition is less than the activity of the alternative path; in fact, its activity may be as high as the KCN-resistant component of root respiration.

partitioning of electron flow between the two pathways in the absence of added inhibitors, even in intact tissues. The discrimination technique showed that the alternative pathway may account for more than 40% of all respiration. The role of the alternative path in roots and leaves cannot be that of heat production. What might be its role in these tissues?

3.3 The Alternative Path as an Energy Overflow

As pointed out earlier the quantitative significance of the alternative path increases when the production of organic acids is not matched by their oxidation; therefore, they accumulate. This observation led to the "energy overflow hypothesis" (Lambers 1997b). It states that respiration via the alternative path only proceeds in the presence of high concentrations of respiratory substrate. It considers the alternative path as a sort of coarse control of carbohydrate metabolism, but not as an alternative to the finer control by adenylates (Sects. 2.1 and 2.2).

The continuous employment of the alternative oxidase under normal "nonstress" conditions may ensure a rate of carbon delivery to the root that

enables the plant to cope with "stress." The carbon demand of a tissue probably increases suddenly when a stress is imposed. For example, a decrease in soil water potential increases the roots' carbon demand for synthesis of compatible solutes for osmotic adjustment. Attack by parasites and pathogens similarly, may suddenly increase carbon demands for tissue repair and the mobilization of plant defenses. The alternative oxidase activity may also prevent the production of superoxide and/or hydrogen peroxide. Superoxide is produced when electron transport through the cytochrome path is impaired (e.g., due to low temperature or desiccation injury), and this is partly due to a reaction of ubisemiquinone with molecular oxygen (Purvis & Shewfelt 1993). Superoxide, like other toxic free radicals, can cause severe metabolic disturbances. So far, the various interpretations of the physiological function of an "energy overflow" remain speculative.

3.4 Using the Alternative Path in Emergency Cases

The activity of the cytochrome path may increase upon increased availability of ADP, at the expense of the activity of the alternative path. Addition of nitrate to 2-week old Pisum sativum (pea) roots, grown without nitrate, showed this effect (De Visser et al. 1986). The increased energy demand for nitrate uptake should increase the concentration of ADP in the cell, thus increasing glycolytic activity. This then leads to a greater input of electrons into the respiratory chain than can be accommodated by the cytochrome path, which already operates at its maximum activity. As a result, the alternative path becomes engaged. Because more electrons are fed into the mitochondrial electron-transport chain than can be accepted by the cytochrome path, this model is termed the energy overcharge model. It could apply to nonsteady-state conditions, as in the previous experiments, and/or to conditions when the activity of the cytochrome path is controlled more by substrate supply than by adenylates.

3.5 NADH-Oxidation in the Presence of a High Energy Charge

If cells require a large amount of carbon skeletons (e.g., oxoglutarate or succinate) but do not have a high demand for ATP, then the operation of the alternative path could prove useful; however, can

TABLE 6. Total respiratory activity in the absence of inhibitors and in the presence of SHAM (leaving cytochrome path activity only) or KCN (leaving only alternative path activity) in protoplasts of guard cells and mesophyll cells from *Pisum sativum* (pea).

	Guard cells	Mesophyll cells
Respiratory O_2 uptake		
$\mu mol\,g^{-1}$ (chlorophyll) s^{-1}	78.8	2.7
$\mu mol\,g^{-1}$ protein s^{-1}	1.5	0.5
$nmol\,mm^{-3}\,s^{-1}$	0.07	0.02
Respiratory O_2 uptake (as % of total respiration)		
when SHAM present	67	34
when KCN present	38	66

Source: Vani & Reghavendra 1994.

we envisage such a situation in vivo? Whenever the rate of carbon skeleton production is high, there tends to be a great need for ATP to further metabolize and incorporate these skeletons. When plants are infected by pathogenic microorganisms, however, they tend to produce **phytoalexins** (cf. Sect. 3 of the chapter on effects of microbial pathogens), which might well require engagement of the alternative path (cf. Sect. 4.8).

The example of the synthesis of phytoalexins, however, which requires the production of carbon skeletons without concomitant need for ATP, is probably exceptional. Other examples include the infected cells in root nodules of legumes; however, these have less alternative path capacity than any other cells from the same nodules or other tissues of the same plants (cf. Sect. 3.5 of the chapter on symbiotic associations). Leaf guard cells also have the capacity to synthesize malate rapidly, which plays a role in stomatal opening (cf. Sect. 5.4.2 of the chapter on plant water relations). Guard cells have a very high respiratory capacity when compared with adjacent mesophyll cells. Inhibitor studies suggest that the alternative path is only partly operative, however, and that it constitutes only one third of total respiration in the guard cells, as opposed to full engagement and accounting for two thirds of respiration in the mesophyll cells (Table 6). Hence, rapid synthesis of organic acids is probably not invariably associated with greater activity of the alternative path.

There may be a need for a nonphosphorylating path to allow rapid oxidation of malate in **CAM plants** during the day (Lance et al. 1985). There are, unfortunately, no techniques available to assess alternative path activity in the light. If measurements are made in the dark during the normal light period, however, then malate decarboxylation in

CAM plants is indeed associated with increased engagement of the alternative path (Table 7). Malate decarboxylation, however, naturally occurs in the light (Sect. 10.2 of the chapter on photosynthesis). It therefore remains to be confirmed that the alternative path plays a vital role in crassulacean acid metabolism.

3.6 Continuation of Respiration When the Activity of the Cytochrome Path Is Restricted

Naturally occurring **inhibitors** of the cytochrome path (e.g., cyanide, sulfide, carbon dioxide, and nitric oxide) may reach such high concentrations in the tissue that respiration via the cytochrome path is partially or fully inhibited (Millar & Day 1997, Palet et al. 1991).

TABLE 7. Respiration, oxygen isotope discrimination, and partitioning of electrons to the cytochrome and the alternative pathway in leaves of *Kalanchoe daigremontiana*.*

Parameter	Acidification	Deacidification
Respiration $\mu mol\ O_2\ m^{-2}\,s^{-1}$	1.8	2.6
Discrimination ‰	22.4	25.0
Cytochrome path $\mu mol\ O_2\ m^{-2}\,s^{-1}$	1.3	1.4
Alternative path $\mu mol\ O_2\ m^{-2}\,s^{-1}$	0.5	1.2

Source: Robinson et al. 1992.

*Measurements were made in the dark, during the normal dark period (acidification phase) and the normal light period (deacidification phase, when rapid decarboxylation occurs).

Dry seeds, including those of species such as *Cucumis sativus* (cucumber), *Hordeum vulgare* (barley), *Oryza sativa* (rice), and *Xanthium pennsylvanicum* (cocklebur) contain **cyanogenic compounds**, such as cyanohydrin, cyanogenic glycosides and cyanogenic lipids. Such compounds liberate free HCN after hydrolysis during imbibition. Upon imbibition and triggered by ethylene, seeds producing these cyanogenic compounds produce a mitochondrial β-cyanoalanine synthase that detoxifies HCN (Hagesawa et al. 1995). Despite this detoxifying mechanism, some HCN is likely to be present in the mitochondria of germinating seeds.

Some plants produce **sulfide** (e.g., species that belong to the Cucurbitaceae) (Rennenberg & Filner 1983). Sulfide is also produced by anaerobic sulfate-reducing microorganisms. It may occur in high concentrations in the phyllosphere of aquatic plants or the rhizosphere of flooded plants. Carbon dioxide levels also increase in flooded soils.

It has repeatedly been found that leaf respiration in the dark is inhibited at elevated concentrations of CO_2 in the atmosphere, such as predicted to occur in the twenty first century (e.g., El Kohen et al. 1991, Wullschleger et al. 1994). Such inhibition is probably partly due to inhibition of the cytochrome path because in vitro cytochrome oxidase is inhibited by CO_2 concentrations in the range known to inhibit respiration of intact leaves (Gonzalez-Meler et al. 1996). The presence of an alternative path, which is unaffected by inhibitors that block the cytochrome path, may allow continued respiration and ATP production, albeit with low efficiency, under such conditions.

In addition, when the activity of the cytochrome path is restricted by **low temperature**, the alternative path might increase in activity. In fact, exposure to low temperature enhances the amount of alternative oxidase in mitochondria (Vanlerberghe & McIntosh 1992). Such an induction is also achieved when the activity of the cytochrome path is restricted in other ways (e.g., by application of inhibitors of mitochondrial protein synthesis) (Day et al. 1995), or of the cytochrome path (Wagner et al. 1992). It is interesting that only those inhibitors of the cytochrome path that enhance superoxide production lead to induction of the alternative oxidase. In addition, superoxide itself can also induce expression of the alternative oxidase. This has led to the suggestion that **active oxygen species**, including H_2O_2, are part of the signal(s) communicating cytochrome path restriction in the mitochondria to the nucleus, thus inducing alternative oxidase synthesis (Wagner & Krab 1995).

In the absence of an alternative oxidase, inhibition or restriction of the activity of the cytochrome path would inexorably lead to the accumulation of fermentation products, as found in transgenic plants lacking the alternative oxidase (Vanlerberghe et al. 1995). In addition, it might cause the ubiquinone pool to become highly reduced, which might well lead to the formation of free radicals and concomitant damage to the cell (Purvis & Shewfelt 1993). Further work with genotypes lacking the alternative path is an essential avenue of future research on the ecophysiological role of the alternative path in plant functioning.

4. Effects of Environmental Conditions on Respiratory Processes

4.1 Flooded, Hypoxic, and Anoxic Soils

Plants growing in flooded soil are exposed to **hypoxic** (low-oxygen) or **anoxic** (no-oxygen) conditions in the root environment, and they experience a number of problems, including an insufficient supply of O_2 and accumulation of CO_2 (Sect. 4.7) and changes in plant water relations (Sect. 3 of the chapter on plant water relations).

4.1.1 Inhibition of Aerobic Root Respiration

The most immediate effect of soil flooding on plants is a decline in the partial pressure of O_2 in the soil. In water-saturated soils the air that is normally present in the soil pores is almost completely replaced by water. The **diffusion** of gases in water is approximately 10,000 times slower than that in air. In addition, the **solubility** of oxygen in water is less than in air (at 25°C approximately 0.5 mmol O_2 dissolves per liter of water, whereas air contains approximately 10 mmol). The oxygen supply from the soil, therefore, is decreased to the extent that aerobic root respiration, and hence ATP production, is restricted. Under these conditions the synthesis of RNA and proteins is strongly suppressed, but that of specific m-RNAs and **anaerobic polypeptides** is induced. Among these "anaerobic polypeptides" is the fermentative enzyme **alcohol dehydrogenase** (Andrews et al. 1993).

4.1.2 Fermentation

When insufficient oxygen reaches the site of respiration, such as in seeds germinating under water and submerged rhizomes, ATP may be produced through **fermentative processes**. These tissues generate energy in glycolysis, producing ethanol, and sometimes lactate. Lactate tends to be the product of fermentation immediately after the cells are deprived of oxygen. Lactate accumulation causes decreased pH in the cytosol (Sect. 4.1.3), which inhibits lactate dehydrogenase and activates the first enzyme of ethanol fermentation: pyruvate decarboxylase. When lactate accumulation does not stop, cytosolic acidification may lead to cell death (Rivoal & Hanson 1994).

It was initially believed that root metabolism cannot continue in flooded conditions because of the production of toxic levels of **ethanol**. Ethanol, however, does not really inhibit plant growth until concentrations are reached that far exceed those found in flooded plants (Table 8). Despite a widespread belief to the contrary, ethanol plays only a minor role in flooding injury to roots and shoots (Alpi et al. 1985, Jackson et al. 1982). As long as there is no accumulation of acetaldehyde, which is the product of pyruvate decarboxylase and the substrate for alcohol dehydrogenase that reduces acetaldehyde to ethanol, alcoholic fermentation is unlikely to lead to plant injuries (Perata & Alpi 1991). If **acetaldehyde** does accumulate, however, for example upon re-aeration, then this may cause injury, because acetaldehyde is a potential toxin (Armstrong et al. 1994). It is the low potential of ATP production and its metabolic consequences, rather than the toxicity of the products of fermentative metabolism, which constrain the functioning of plants under anoxia (Sect. 4.1.3).

Continued fermentation requires the mobilization of a large amount of reserves, such as starch. Seeds of most species fail to germinate under anoxia, but those of *Oryza sativa* (rice) are an exception (Perata & Alpi 1993). As opposed to seeds of cereals like *Triticum aestivum* (wheat) and *Hordeum vulgare* (barley), rice seeds produce α-amylase under anoxia; this enzyme allows the degradation of starch, and therefore sustains a rapid fermentative metabolism (Guglielminetti et al. 1995, Perata et al. 1993). Plants that occur naturally in flooded soils have a well-developed **aerenchyma**, which is a system of air spaces for gas transport that allows diffusion of O_2 to the roots (Konings & Lambers 1990). Such an aerenchyma avoids inhibition of respiration due to lack of O_2, which is inevitable for plants that are not adapted to wet soils (Perata & Alpi 1993).

The **energetic efficiency** of ethanol formation, which only produces two molecules of ATP in glycolysis ("substrate phosphorylation") per molecule of glucose, is considerably less than that of aerobic respiration, which produces around 36 ATP per molecule of glucose, if the most efficient mitochondrial electron-transport path is used exclusively. In addition, a large fraction of the lactate may be secreted into the rhizosphere (e.g., in some *Limonium* species). Although such secretion prevents acidification of the cytosol, it also represents a substantial carbon loss to the plant (Rivoal & Hanson 1993).

4.1.3 Cytosol Acidosis

A secondary effect of the decline in root respiration and ATP production in the absence of oxygen is a decrease in the pH of the cytosol (**acidosis**), due in part to accumulation of organic acids in fermentation and the TCA cycle. In addition, in the absence of O_2 as a terminal electron acceptor, ATP production decreases, so there is less energy available to maintain ion gradients within the cell. Acidification of the cytosol reduces the activity of many cytosolic enzymes, whose pH optimum is around 7; therefore, it severely disturbs the cell's metabolism so that protons leak from the vacuole to the cytosol. The extent of this cytosolic acidification is less in the presence of nitrate (Figure 11). Nitrate reduction leads to the formation of hydroxyl ions, which partly neutralize the protons and prevent severe acidosis. In addition, nitrate reduction requires the

TABLE 8. The effect of supplying ethanol in aerobic and anaerobic nutrient solutions to the roots of *Pisum sativum* (garden pea) at a concentration close to that found in flooded soil (i.e., 3.9 mM) or greater than that.

	Aerobic control	Aerobic + ethanol	Anaerobic control	Anaerobic + ethanol
Ethanol in xylem sap (mM)	37	540	90	970
Stem extension (mm)	118	108	94	74
Final fresh mass (g)				
shoot	11.9	11.9	10.7	11.4
roots	7.8	9.7	5.7	6.1

Source: Jackson et al. 1982.

FIGURE 11. The effect of hypoxia on root tips of *Zea mays* (maize), in the presence (triangles), and absence (circles) of nitrate. (A) The effect on the pH of the cytosol, as measured in experiments using ^{31}P-NMR spectroscopy; (B) the increase in fresh mass during 48 hours in air, after the indicated period of hypoxia. The location of the inorganic phosphate (P_i) peaks in an NMR spectrum depend on the "environment" of the molecule (e.g., pH) (Fig. 6 in this chapter). NMR spectroscopy can therefore, be used to determine the peak wavelength at which P_i absorbs the magnetic radiation and hence the pH in the cytosol (Roberts 1985). Copyright American Society of Plant Physiologists.

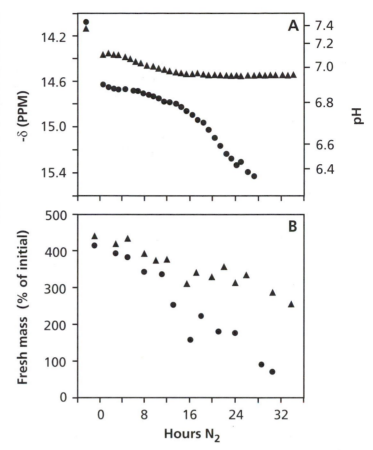

oxidation of NADH, which produces NAD. This allows the continued oxidation of organic acids in the TCA cycle, thus preventing their accumulation and associated drop in pH.

4.1.4 Avoiding Hypoxia: Aerenchyma Formation

Wetland plants, including crop species such as *Oryza sativa* (rice), have evolved mechanisms to prevent the problems associated with flooded soils. The most important adaptation to flooded soils is the development of a functional **aerenchyma**, which is a continuous system of air spaces in the plant that allows the transport of oxygen from the shoot or the air to the roots. In many species, special structures allow the diffusion of oxygen from the air into the plant: the pneumatophores of mangroves, the knee roots of *Taxodium distichum* (bald cypress), and lenticels in the bark of many wetland trees.

Because there is a gradient in partial pressure within the aerenchyma, oxygen will move by **dif-** **fusion** to the roots. In aquatic plants like *Nuphar lutea* (yellow water lily) and *Nelumbo nucifera* (lotus), however, there is also a **pressurized flow-through** system, which forces O_2 from young emergent leaves to the roots and rhizomes buried in the anaerobic sediment (Dacey 1980, 1987). Such a mass flow requires a difference in atmospheric pressure between leaves and roots. The diurnal pattern of the mass flow of air to the roots suggests that the energy to generate the pressure comes from the sun. Rather than being the photosynthetically active component of radiation, however, it is the long-wave region (heat) that increases the atmospheric pressure inside young leaves by as much as 300 Pa. How can these young leaves draw in air against a pressure gradient? To understand this we have to realize that the atmosphere inside the leaf is saturated with water vapor and that movement of gases occurs by **diffusion**, along a gradient in partial pressure, and by **mass flow**, depending on the **porosity** of the pathway. The porosity of the young emergent leaves is such that gas flux by diffusion (i.e., down a concentration gradient) is more important than a mass flux due to a difference in atmospheric pressure. The concentration gradient is due to the evaporation from the cells inside the leaf, which dilutes the other gases in the intercellular spaces, thus creating a diffusion gradient for air between the atmosphere and the intercellular spaces. The slightly higher atmospheric pressure inside young leaves forces air, which has been enriched in oxygen by photosynthesis, to move along a pressure gradient from young leaves to roots and rhizomes. Some of the air from roots and rhizomes, which is enriched with CO_2 from respiration, is then forced to older leaves. Isotope studies show that much of this CO_2 is subsequently assimilated in photosynthesis. The reason that only young leaves cause this internal ventilation is the higher porosity of the older leaves, which does not allow them to draw in more air through diffusion than is lost via mass flow. The quantity of air flow through a single petiole is enormous: as much as $221 day^{-1}$, with peak values as high as $1 ml s^{-1}$ and rates of $8 mm s^{-1}$. The transport of oxygen from the shoot by convective gas flow is also likely to contribute to the flow of oxygen to roots of other species growing in an anaerobic soil (Armstrong et al. 1991). Pressurized flow of oxygen is important in water lilies, and it also plays a role in the oxygen supply to the roots and rhizosphere of many **emergent macrophytes**. The vital element is that a compartment exists surrounded by walls with sufficiently small pores to allow diffusion to occur at greater rates than mass flow (Armstrong et al. 1994).

Aerenchymatous plants often transport more O_2 to the roots than is consumed by root respiration. The **outward diffusion of O_2** into the rhizosphere allows the oxidation of potentially harmful compounds. This can readily be seen when excavating a plant from a reduced substrate. The bulk substrate itself is black, due to the presence of FeS, but the soil in the immediate vicinity of the roots of such a plant will be brown or red, which indicates the presence of oxidized iron (Fe^{3+}, "rust"), which is less soluble than the reduced Fe^{2+}.

Aerenchyma is not an unmitigated benefit to plants. Aerenchymatous roots characteristically have a large diameter, which may well reduce the surface area per unit biomass. Because plant nutrient uptake is strongly affected by root diameter and surface area, one cost associated with the development of aerenchyma may be a decline in rates of nutrient uptake per unit root biomass.

Aerenchyma also serves as a conduit of soil gases to the atmosphere. **Methane** (CH_4) is a bacterial product commonly produced in anaerobic soils. In rice paddies and natural wetlands most CH_4 is

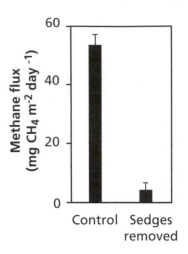

FIGURE 12. Methane flux in a tundra wetland in which sedges are present (control) or have been experimentally removed (data from Torn & Chapin 1993).

transported to the atmosphere through plant aerenchyma. Experimental removal of sedges from wetland substantially reduces CH_4 flux and causes CH_4 to accumulate in soils (Fig. 12). CH_4 production and transport to the atmosphere is a topic of current concern because CH_4 is a "greenhouse gas" that absorbs infrared radiation 20 times more effectively than does CO_2. Increases in atmospheric CH_4 have contributed approximately 17% of the warming potential of the atmosphere that has caused recent global warming (Houghton et al. 1996). Although the relative importance of the multiple causes of this increased atmospheric CH_4 are uncertain, the expansion of rice agriculture and associated CH_4 transport via aerenchyma from the soil to the atmosphere are undoubtedly an important contributor.

4.2 Salinity and Water Stress

Sudden exposure of sensitive plants to salinity or water stress often enhances their respiration. For example, the root respiration of *Hordeum vulgare* (barley) increases upon exposure to 10mM NaCl. This may either reflect an increased **demand for respiratory energy** or an increased activity of the alternative path, when carbon use for growth is decreased more than carbon gain in photosynthesis (see Sect. 5.3 of the chapter on growth and allocation).

Long-term exposure of sensitive plants to salinity or drought gradually decreases respiration, which is part of the general decline in carbon assimilation and overall metabolism associated with slow growth under these conditions (Sect. 5.3 of the chapter on growth and allocation). Additional declines in root respiration of *Triticum aestivum* (wheat) plants upon exposure to dry soil may reflect a specific decline in the alternative path. The decline correlates with the accumulation of **osmotic solutes**, reducing the availability of sugars, hence providing less "grist for the mill" of the alternative path.

Leaves also show a decline in respiration, as leaf water potential declines. The decline is most likely associated with a decrease in the energy requirement for growth or the export of photoassimilates (Collier & Cummins 1996).

Species differ in their respiratory response to drought, primarily due to differences in sensitivity of growth to drought. When salt-adapted plants are exposed to mild salinity stress, they accumulate **compatible solutes**, such as sorbitol (cf. Sect. 3 of the chapter on plant water relations). Accumulation of these sugar alcohols requires glucose as a substrate, but it does not directly affect the concentration of carbohydrates or interfere with growth. Studies of root respiration, using an inhibitor of the alternative path, suggested that sorbitol accumulation is associated with a reduction in activity of the nonphosphorylating alternative respiratory pathway (Fig. 13). It is interesting that the amount of sugars that are "saved" by the decline in respiration in the alternative pathway is the same as that used as the substrate for the synthesis of sorbitol, which suggests that accumulation of compatible solutes by drought-adapted plants may have a minimal respiratory cost.

Prolonged exposure of salinity-adapted species (**halophytes**) to salt concentrations that are sufficiently low not to affect their growth has no effect on the rate of root respiration. This similarity in growth and respiratory pattern under saline and nonsaline conditions confirms the conclusion from inhibitor studies that the respiratory costs of coping with mild salinity levels are negligible in salt-adapted species. The respiratory costs of functioning in a saline environment for adapted species that accumulate NaCl are also likely to be relatively small because of the low respiratory costs of absorbing and compartmentalizing salt when grown in saline soils. For salt-excluding **glycophytes**, however, there may be a large respiratory cost associated with salt exclusion. If one were to generalize broadly, then it appears that root respiration of salt-resistant species is stimulated if NaCl stimulates growth, but inhibited if growth is reduced (Lambers et al. 1996).

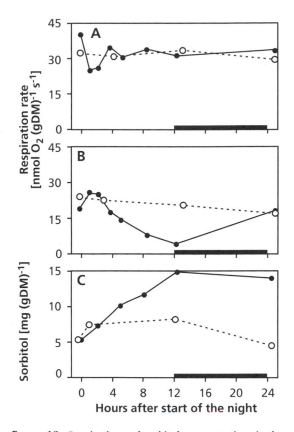

FIGURE 13. Respiration and sorbitol concentrations in the roots of *Plantago coronopus* grown in the absence of NaCl (open symbols) and upon exposure to 50 mM NaCl at the beginning of the experiment (filled symbols). (A) Respiration in the presence of an inhibitor of the alternative path (putatively the activity of the cytochrome path). (B) Respiration that was sensitive to the inhibitor of the alternative path (putatively the activity of the alternative path). (C) The concentration of sorbitol (after Lambers et al. 1981). Copyright Physiologia Plantarum.

4.3 Nutrient Supply

Root respiration generally increases when roots are suddenly exposed to increased ion concentrations in their environment, which is a phenomenon known as **salt respiration**. The stimulation of respiration is at least partly due to the increased **demand for respiratory energy** for ion transport. The added respiration may also reflect a replacement of osmotically active sugars by inorganic ions, thereby leaving a large amount of sugars to be respired via the **alternative path**.

When plants are grown at a low supply of nutrients, their rate of root respiration is lower than that of plants that are well-supplied with mineral nutrients (Fig. 14). This is expected, because their rates of

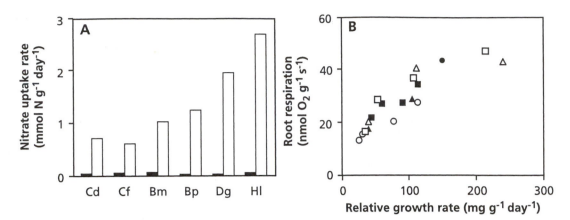

FIGURE 14. (A) Rates of net inflow of nitrate of six grass species grown at two nitrogen addition rates, allowing a near-maximum relative growth rate (open columns) or an RGR well below RGR$_{max}$ (black colums). (B) Root respiration of the same inherently fast- and slow-growing grasses as shown in A, now compared at a range of nitrogen addition rates that allow a near-maximum relative growth rate or a relative growth rate below RGR$_{max}$, with the lowest RGR being $40\,mg\,g^{-1}\,day^{-1}$. Cd, *Carex diandra* (open circles); Cf, *Carex flacca* (filled triangle); Bm, *Briza media* (filled squares); Bp, *Brachypodium pinnatum* (filled circles); Dg, *Dactylis glomerata* (open squares); Hl, *Holcus lanatus* (open triangles) (Van der Werf et al. 1992a). Copyright SPB Academic Publishing.

growth and ion uptake are greatly reduced; however, rates of root respiration per ion absorbed or per unit root biomass produced at the low nitrate supply are relatively high, if we compare these rates with those of plants which grow and take up ions at a *much* higher rate. This suggests that **specific costs** of growth (i.e., cost per unit biomass produced), maintenance (cost per unit biomass to be maintained), or ion transport (cost per unit nutrient absorbed) must increase in plants grown at a limiting nutrient supply (cf. Sect. 5.2.4). The effect of nutrient supply on shoot respiration is not as pronounced nor as consistent (Van der Werf et al. 1994).

4.4 Irradiance

The respiratory response of plants to light and assimilate supply depends strongly on time scale. The immediate effect of low light is to reduce the **carbohydrate status** of the plant and, therefore, the supply of substrates available to support respiration (Fig. 15). As expected from the overflow model of respiration, it is primarily the alternative path that declines in response to short-term declines in carbohydrate status in both roots (Fig. 15) and leaves (Azcón-Bieto et al. 1983, Noguchi & Terashima 1997). Thus, CO_2 production and O_2 consumption via respiration probably decline at night and in cloudy weather more strongly than does ATP production. Upon prolonged exposure to low irradi-

ance, the relative contribution of the alternative path to root respiration tends to be similar to that at high irradiance (Gloser et al. 1996).

Roots and leaves that are subjected to an increased carbohydrate supply gradually **acclimate** over several hours by increasing their respiratory activity. The respiration of excised roots or leaves that are depleted of sugars is initially stimulated by exogenous sugars simply because these sugars provide the **substrate** for respiration (cf. Fig. 15). Such a stimulatory effect of an increase in carbohydrate supply, however, often increases with increasing time of exposure due to an increased **respiratory capacity** that involves **gene expression**. It is also found in roots that are still attached to the shoot; for example, when shaded plants are exposed to increased irradiance. In addition, after the removal of all but one seminal root of *Hordeum distichum* (barley) seedlings, the soluble sugar concentration and respiration of the remaining seminal root increases (Farrar & Jones 1986). On the other hand, after pruning of the shoot to one leaf blade, both the soluble sugar concentration and the respiration of the seminal roots decrease. These effects on respiration reflect the **coarse control** of the respiratory capacity upon pruning or sucrose feeding (Bingham & Farrar 1988, Williams & Farrar 1990). This illustrates the adjustment of the respiratory capacity to the root's carbohydrate level. The respiratory apparatus can be modified in response to the internal and/or the external environment of the plant. These responses are an important component of the plant's response to environmental factors such as light and nutrients.

Changes in respiratory capacity induced by pruning and long-term changes in carbohydrate supply reflect acclimation of the respiratory machinery. The protein pattern of the roots of pruned plants is affected within 24 hours (Williams et al. 1992). Mitochondria isolated from such roots show changes in respiratory properties and activities of cytochrome oxidase (McDonnel & Farrar 1992). Glucose feeding to leaves enhances the activity of several glycolytic enzymes in these leaves due to regulation of **gene expression** by carbohydrate levels (Krapp & Stitt 1994). The capacity to use carbohydrates in respiration is clearly enhanced when the respiratory substrate supply increases and declines with decreasing substrate supply.

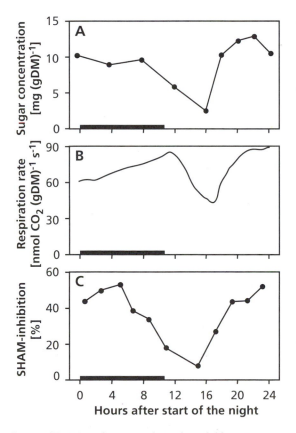

FIGURE 15. Diurnal course of (A) the soluble sugar concentration, (B) the rate of CO_2 production, and (C) the component of respiration that is sensitive to an inhibitor of the alternative path (putatively the contribution of the alternative path to O_2 uptake) in the roots of *Cucumis sativus* (cucumber) (after Lambers et al. 1996). By courtesy of Marcel Dekker, Inc.

The plant's potential to adjust its respiratory capacity to environmental conditions is ecologically significant. Individual plants acclimated to low light generally have low leaf respiration rates. Thus, acclimation accentuates the short-term declines in respiration due to substrate depletion. After transfer of the understory species *Alocasia odora* from full sun to 20% of that light intensity or vice versa, the leaves adjust to the new environment within a week. This is much faster than the change in photosynthetic characteristics (Noguchi & Terashima 1997).

As with acclimation, understory species that are **adapted** to low light generally exhibit lower respiration rates than high-light adapted species. For example, a species from the understory of Australian rain forests (*Alocasia odora*) has lower rates of both photosynthesis and respiration than does a sun species (*Spinacia oleracea*, spinach), when the two species are compared under the same growth conditions (Table 9). The net daily carbon gain of the leaves (photosynthesis minus respiration) is rather similar for the two species, when expressed as a proportion of photosynthesis. Understory species of *Piper* also have lower respiration rates than do species from shaded and exposed habitats when both are grown in the same environment (Fredeen & Field 1991). Because rates of photosynthesis and respiration show parallel differences between sun and shade species (both lower in the shade species), differences in the carbon balance between sun and shade species probably reflect different patterns of biomass allocation rather than differences in photosynthesis and respiration.

4.5 Temperature

Respiration increases as a function of temperature, with the degree of increase being dependent on the **temperature coefficient** (Q_{10}) of respiration (Criddle et al. 1994). This temperature effect on respiration is characteristic of most heterothermic organisms and is a logical consequence of the temperature sensitivity of the enzymatically catalyzed reactions involved in respiration and of the increased ATP requirements as metabolic rates increase. The temperature stimulation of respiration also reflects the increased demand for energy to support the increased rates of biosynthesis, transport, and protein turnover that occur at high temperatures (cf. Sect. 5.2).

The rate of respiration at any given measurement temperature also depends on the growth tem-

TABLE 9. The daily carbon budget (mmol g^{-1} day^{-1}) of the leaves of *Spinacia oleracea*, a sun species, and *Alocasia odora*, a shade species, when grown in different light environments.*

Irradiance	Photosynthesis		Leaf respiration		Net leaf carbon gain	
	Spinacia	*Alocasia*	*Spinacia*	*Alocasia*	*Spinacia*	*Alocasia*
500	26	nd	3.4 (13)	nd	23 (87)	nd
320	21	11	2.4 (12)	1.1 (10)	18 (88)	9.4 (90)
160	15	9	1.7 (11)	0.82 (9)	14 (89)	8.2 (91)
40	nd	4.5	nd	0.76 (17)	nd	3.7 (83)

Source: Noguchi et al. 1996; K. Noguchi, personal communication.
*Irradiance is expressed in μmol m^{-2} s^{-1}. Percentages of the photosynthetic carbon gains have been indicated in brackets; nd is not determined; in the original paper the species name is erroneously given as *A. macrorrhiza*.

perature to which a plant is acclimated. Temperature acclimation results in **homeostasis** of respiration, such that warm-acclimated (temperate, lowland) and cold-acclimated alpine (Larigauderie & Körner 1995) or high-arctic (Semikhatova et al. 1992) plants display similar rates of respiration when measured at their respective growth temperatures (Fig. 16).

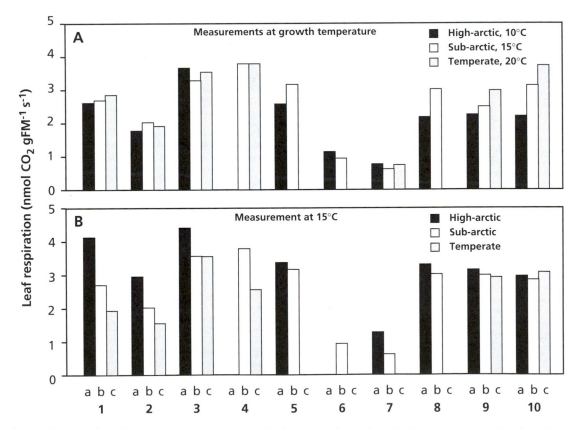

FIGURE 16. The effect of measurement temperature on leaf dark respiration (nmol CO_2 g^{-1} FM s^{-1}) of the same or closely related species growing in their natural high-arctic, subarctic, or temperate environment. (A) Respiration at the temperature in the natural environment, i.e., 10, 15 and 20°C for the high-arctic, subarctic and temperate region, respectively. (B) Respiration at the same temperature (15°C), irrespective of the temperature in the growth environment. Numbers and letters below each set of vertical bars refer to the following species: 1a: *Primula tschuktschorum*; 1b,c: *P. elatior* 2a: *Ranunculus sulphureus* 2b,c: *R. acris*; 3a: *Polygonum bistorta*; 3b: *P. viviparum*; 3c: *P. oviculare*; 4b,c: *Trolleus europaeus*; 5a,b *Salix reticulata*; 6a: *Claytonia arctica*; 6b: *C. asarifolia*; 7: *Rhodiola rosea*; 8a: *Oxygraphis glacialis*; 8b: *O. vulgaris*; 9: *Cardamine pratensis*; 10: *Oxyria digyna* (Semikhatova et al. 1992). Copyright Academic Press Ltd.

Acclimation of root respiration to temperature occurs in *Plantago lanceolata* (plantain) and *Zostera marina* (eelgrass), but not in *Picea glauca* (white spruce), *Picea engelmannii* (spruce), or *Abies lasiocarpa* (fir) (Lambers et al. 1996). In those species where no thermal acclimation takes place, root respiration rates will depend entirely on ambient temperature, with low temperatures severely restricting root metabolism (Weger & Guy 1991). In general, the extent of root and leaf acclimation is often greatest in plants from more thermally variable environments (Billings et al. 1971). Cold-acclimated plants generally have more mitochondria than do warm-acclimated plants, but they have similar rates of respiration per mitochondrion (Klikoff 1966). The mechanism of temperature acclimation of respiration is not really understood. It may well involve a change in carbohydrate level in the tissue, which affects the expression of the genes that encode respiratory enzymes (cf. Sect. 4.4).

In addition to the acclimation potential of total root respiration, acclimation may also change the partitioning of electrons between the cytochrome and alternative pathways. Transfer from 20 to 13°C tends to lead to a transient decline in SHAM-resistant root respiration of *Plantago lanceolata* (plantain) (Smakman & Hofstra 1982). We presume that some change in the rate of energy-requiring processes or in the mitochondria causes this shift in electron transport between pathways. Such mitochondrial adjustments have been found in callus-forming potato tuber discs (Hemrika-Wagner et al. 1982). In contrast to the data on *Plantago lanceolata*, Weger and Guy (1991) did not find any acclimation of root mitochondrial electron transport to the growth temperature for *Pinus glauca* (pine), as was observed in respiration of intact conifer roots. At any one temperature, SHAM-sensitive and SHAM-resistant respiration rates are independent of the growth temperature (4, 11, or 18°C) (Lambers et al. 1996). Thermal acclimation of root respiration, therefore, appears to be species-dependent. Needles of cold-hardened plants that maintain relatively low rates of respiration when exposed to higher temperatures maintain higher concentrations of soluble sugars, which in turn confers greater **frost tolerance**. This suggests that plants that show a large capacity to adjust their respiration may avoid problems associated with excessive accumulation of sugars, while being more frost sensitive than plants that lack the capacity to adjust their respiration (cf. Sect. 3.5 of the chapter on effects of radiation and temperature).

When grown and measured under the same conditions species or populations that have evolved in

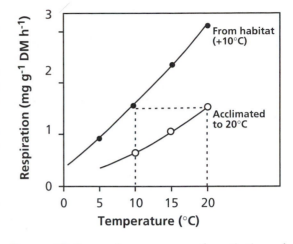

FIGURE 17. Temperature response of respiration of *Vaccinium myrtillus* (bilberry) shoots of populations acclimated to 10°C and 20°C. The dashed line shows the respiration rate of each plant at its acclimation temperature (Körner & Larcher 1988). Copyright The Company of Biologists.

cold climates typically have higher respiration rates than do warm-adapted populations (Fig. 17). For example, arctic and alpine populations typically have higher respiration rates than do their temperate counterparts when grown in a common environment (Mooney & Billings 1961). As with acclimation, this homeostatic adjustment in respiration causes the two populations to have similar respiration rates when each is grown in its natural environment. Due to their high respiratory capacity, arctic plants are able to grow just as rapidly as their temperate counterparts, despite much lower ambient temperatures (Chapin 1983). The high respiratory capacity of arctic and alpine species has been implicated in the low carbohydrate status and high mortality rate of alpine species transplanted to low elevation.

4.6 Low pH and High Aluminum Concentrations

Root respiration rate increases as the **pH** in the rhizosphere decreases to a level below that at which growth is no longer possible (Fig. 18). Net H^+ release from roots by H^+-ATPase activity is a prerequisite for continued root growth and limits root growth at very low pH values (Schubert et al. 1990). One way of coping with excess H^+ uptake at a low pH is to increase active H^+ pumping by plasma-membrane ATPases. This increases the **demand for respiratory energy** (Fig. 18). Increased respiration rates, therefore, can allow plants to maintain root growth at noncritically low pH values by increasing the supply of ATP for H^+ pumping by plasma-membrane ATPases.

At very low pH values, root growth, net H^+ release, and respiration rates decline (relative to rates at pH 7.0). The increased entry of H^+ into the roots under these circumstances appears to be responsible for these effects (Yan et al. 1992). Such increased uptake of H^+ tends to disturb cytosolic pH and ultimately root growth. The decrease in root respiration at very low pH, therefore, might result from the decreased respiratory demand for growth.

A second avenue by which low pH increases respiration involves the increased solubility of **aluminum** at low pH (see Sect. 3.1 of the chapter on mineral nutrition). Respiration of intact roots increases in response to aluminum in both aluminum-resistant and sensitive cultivars of *Triticum aestivum* (wheat) (Collier et al. 1993). This effect is due to aluminum, and not to a low pH, but it would normally occur when plants are exposed to a low pH that enhances the solubility of aluminum. For the wheat roots, the increase is entirely due to an enhanced level of SHAM-resistant respiration (putatively cytochrome path activity). At higher aluminum concentrations, respiration rates gradually decline to the control rates due to inhibition of growth. These declines in root growth and respiration occur at much higher aluminum concentrations in the resistant than in the sensitive cultivar. A similar stimulation of root respiration by aluminum was observed for a sensitive (Tan & Keltjens 1990a) and a resistant (Tan & Keltjens 1990b) cultivar of *Sorghum bicolor*. Respiration of excised root tips of a sensitive cultivar of *Triticum aestivum* (wheat) is already inhibited at $75\mu M$ aluminum (De Lima & Copeland 1994). The difference in result between whole root systems and those on root tips suggests that the latter are more sensitive to aluminum.

The increase in respiration of the intact roots suggests that root functioning in the presence of aluminum imposes a demand for additional respiratory energy. These increased costs have little to do with the mechanism explaining resistance (i.e., excretion of chelating organic acids or phosphate) because such excretion does not occur to any major extent in the sensitive cultivar (cf. Sect. 3.1 of the chapter on mineral nutrition).

4.7 Partial Pressures of CO_2

Partial pressures of CO_2 in the air pockets in the soil are up to 30-fold higher than those in the atmosphere. Although respiration rates are highest in superficial layers of soil where root biomass is concentrated, the CO_2 concentration increases with increasing profile depth, due to the restricted diffusion of gases in soil pores (Fig. 19).

The partial pressures of CO_2 in the soil may increase substantially upon **flooding** of the soil. Values of 2.4 and 4.2mmol CO_2 mol^{-1} (0.24 and 0.42%, respectively) occur in flooded soils that support the growth of desert succulents, as opposed to 0.54 and 1.1mmol mol^{-1} in the same soils when well-drained (Nobel & Palta 1989). Good & Patrick (1987) found CO_2 concentrations of 56 and 38% in silt loam, supporting the growth of *Fraxinus pennsylvanica* (ash) and *Quercus nigra* (oak), respectively. In the air spaces of these roots, CO_2 concentrations are as high as 10% (*Fraxinus*) and 15%

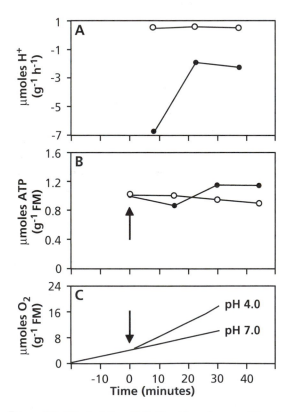

FIGURE 18. Effect of the pH in the rhizosphere on (A) net H^+ release, (B) ATP concentration, and (C) respiration of *Zea mays* (maize) roots. Seedlings were grown at pH 7.0, and either kept at pH 7.0 (open symbols) or exposed to a pH of 4.0 (filled symbols) at the time indicated by the arrow. Note that the slopes in A and C give the *rate* of H^+ release and respiration (Yan et al. 1992). Copyright American Society of Plant Physiologists.

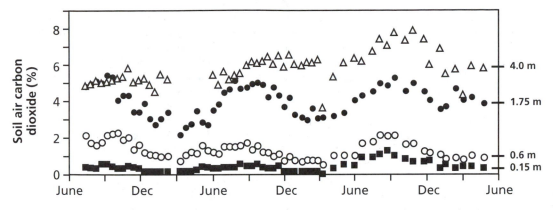

FIGURE 19. Seasonal fluctuations in carbon dioxide concentration in soil at four depths. Measurements were made at the Calhoun Experimental Forest, southern Piedmont, South Carolina, USA (Richter & Markewitz 1995). Copyright American Institute of Biological Sciences.

(*Quercus*). Do such high CO_2 concentrations have any direct or indirect effect on root respiration?

Compared with the vast number of papers on the inhibition of leaf respiration immediately upon exposure to an elevated atmospheric CO_2 concentration (0.7 vs. $0.35 \, mmol \, CO_2 \, mol^{-1}$ in normal air), very little is known on effects of $[CO_2]$ on root respiration. Root respiration is reversibly inhibited by $5 \, mmol \, CO_2 \, mol^{-1}$ in two cacti (*Opuntia ficus-indica* and *Ferocactus acanthodes*) (Nobel & Palta 1989). Full inhibition occurs at $20 \, mmol \, CO_2 \, mol^{-1}$ (2%), which is irreversible if it lasts for 4 hours, leading to the death of cortical cells. Very similar effects occur for another desert succulent: *Agave deserti* (Palta & Nobel 1989). The death of the cortical cells, however, is unlikely to be due to inhibition of respiration because exposure to a root atmosphere without O_2 for a similar period is fully reversible for the same plants. Qi et al. (1994) also found inhibition of root respiration by soil CO_2 levels in a range normally found in soil for a C_3 plant (*Pseudotsuga menziessii*), whereas no such inhibition is found for a range of other species studied by other authors (e.g., Bouma et al. 1997). Because respiration is only affected by **CO_2**, and *not* by **bicarbonate** (Palet et al. 1992), the pH of the root environment will greatly affect experimental results (cf. Fig. 51 in the chapter on photosynthesis).

How can we account for effects of CO_2 concentration on respiration? The effects of soil CO_2 concentrations on root respiration might be *indirect*, due to inhibition of energy-requiring processes. There may also be *direct* effects of a high concentration of CO_2 on respiration (e.g., inhibition of **cytochrome oxidase**) (see Sect. 3.6). Other mitochondrial enzymes are also affected by high concentrations of

inorganic carbon (Gonzalez-Meler et al. 1996, Shipway & Bramlage 1973). Malic enzyme, which oxidizes malate to form pyruvate and CO_2, is rather strongly inhibited by HCO_3^- in a range that might well account for inhibition of respiration by CO_2 as found for some tissues (Chapman & Hatch 1977, Neuburger & Douce 1980). Some of the effects as found in vitro for several mitochondrial enzymes, however, do not appear until very high CO_2 concentrations—much higher than expected to occur in intact roots.

The information in the literature is still too scanty to draw the robust conclusion that CO_2 levels normally occurring in well-drained soil inhibit respiration to a major extent (Lambers et al. 1996).

There are also effects of long-term exposure of plants to elevated $[CO_2]$, such as expected some time during the twenty-first century, on plant respiration. Both a stimulation and an inhibition can be found; these effects are *indirect*, due to changes in allocation, plant growth rate, chemical composition of the biomass, and so on, rather than accounted for by the *direct* effects as described in this section (Poorter et al. 1992).

4.8 Effects of Plant Pathogens

Pathogen attack on roots or leaves causes an increase in respiration, but the pattern of this respiratory response may differ between sensitive and resistant varieties of plants. For example, **nematode** infection of roots of a susceptible variety of *Lycopersicon esculentum* (tomato) causes root respiration first to increase, but then to return to the level of uninfested plants. By contrast, the resistant variety shows no initial change in root respiration in response to nematode attack, but after 8 days the respiration rate exceeds that of control plants. There is some evidence, based on inhibitor studies, that the enhanced respiration in the susceptible variety is largely due to enhanced activity of the alternative path (Zacheo & Molinari 1987).

Just as with tomato roots, leaves of a susceptible variety of *Hordeum distichum* (barley) show a large increase in respiration when infected with the **fungus** causing powdery mildew. This is not unexpected because both fungus and host have high demands for energy (the fungus for growth; the host for defense). In the case of barley, most of the respiration is accounted for by the host. Inhibitor studies suggest that the alternative path may account for half of the increase in leaf respiration. More resistant cultivars of the same species do not appear to show alternative path activity following infection with powdery mildew (but this might reflect the problems with assessing the alternative path activity; Sect. 3.2) (Table 10).

The conclusion that there is an increase in activity of the alternative pathway, rather than that of the cytochrome path only, seems counterintuitive. Could it be due to experimental error, as hinted at in Section 3.2? It is most unlikely that all the results can be dismissed as an experimental error because **mRNA levels** encoding the alternative oxidase, the amount of **alternative oxidase protein**, and the concentration of **pyruvate** strongly increase in leaves of *Arabidopsis thaliana* upon infection with the leaf-spotting bacterium *Pseudomonas syringae* (Simons et al. 1998). What could be the functional significance for an increase of this pathway? Pathogenic fungi may produce **ethylene** and enhance the concentra-

tion of **salicylic acid** in the plant. These compounds may trigger the increased activity of the alternative path. In ripening fruits ethylene enhances alternative respiration; salicylic acid induces the large increase in respiration in the spadix of thermogenic *Arum* species (Sect. 3.1) and in vegetative organs of nonthermogenic plants. It is quite likely that the enhanced synthesis of defense-related compounds (phytoalexins and other phenolics; cf. Sect. 3 in effects of microbial pathogens) requires a large production of NADPH in the **oxidative pentose phosphate** pathway (Fig. 2) (Shaw & Samborski 1957). This pathway, unlike glycolysis (Fig. 3), is not regulated by the demand for metabolic energy. Products of the oxidative pentose phosphate pathway can enter glycolysis, bypassing the steps controlled by energy demand. Additional NADPH can be produced by cytosolic **NADP-malic enzyme**, which oxidizes malate, producing pyruvate and CO_2. This enzyme is induced upon the addition of "elicitors" (i.e., chemical components of a microorganism that induce the synthesis of defense compounds in plant cells) (cf. Sect. 3 of the chapter on effects of microbial pathogens) (Schaaf et al. 1995). The increased activity of the oxidative pentose pathway and of NADP-malic enzyme probably lead to the delivery of a large amount of pyruvate and malate to the mitochondria, without there being a large need for ATP. As a result the cytochrome path becomes saturated with electrons, the alternative oxidase is activated (Sect. 2.6.2), and much of the electrons are transported via the alternative pathway (cf. Sect. 3.3) (Simons & Lambers 1998).

5. The Role of Respiration in Plant Carbon Balance

5.1 Carbon Balance

As discussed earlier, approximately half of all the photosynthates produced per day are respired in the same period, with the exact fraction depending on species and environmental conditions (Table 1). The level of irradiance and the photoperiod appear to affect the carbon balance of acclimated plants to a relatively small extent, but factors such as inadequate nutrient supply and water stress may greatly increase the proportion of photosynthates used in respiration. This is accounted for by a much stronger effect of nutrients on biomass allocation, when compared with that of irradiance and photoperiod (chapter on growth and allocation).

TABLE 10. Leaf respiration (μmol O_2 m^{-2}s^{-1}) of a mildew-susceptible and a resistant variety of *Hordeum distichum*, 9 days after infection with powdery mildew.*

	Susceptible		Resistant	
	Control	Infected	Control	Infected
No inhibitors	1.2	2.1	1.3	1.6
+SHAM	1.0	0.8	1.1	1.4
+KCN	0.9	0.8	0.7	0.7

Source: Farrar & Rayns 1987.
*SHAM and KCN were used as specific inhibitors of the alternative and cytochrome path, respectively. KCN-resistance gives the maximum possible activity of the alternative path. SHAM-inhibition gives the minimum flux through the alternative path (cf. Sect. 3.2).

Root temperature is also likely to affect plant carbon balance because this has a major effect on biomass allocation (Sect. 5.2.2 of the chapter on growth and allocation).

5.1.1 Root Respiration

Root respiration accounts for approximately 10 to 50% of the total carbon assimilated each day in photosynthesis (Table 1) and is a major proportion of the plant's carbon budget (Fig. 20). This percentage is much higher in slow-growing plants than it is in fast-growing plants. This is true for a comparison of species that vary in their **potential growth rate** (Poorter et al. 1991) and for plants of the same species that vary in growth rate, due to variation in the **nutrient supply** (Van der Werf et al. 1992a). Root temperatures that enhance biomass allocation to roots (Sect. 5.2.2 of growth and allocation) probably also increase the proportion of carbon required for root respiration. When slow growth is due to exposure to low light levels, however, no greater respiratory burden is incurred (cf. Sect. 4.4). To some extent the proportionally greater carbon use in slow-growing plants is accounted for by their relatively low carbon gain per unit plant mass (Poorter et al. 1995). This does not, however,

explain the entire difference; variation in respiratory efficiency and/or respiratory costs for processes like ion transport may play an additional role (Sect. 5.2.3).

Root respiration provides the driving force for root growth and maintenance and for ion absorption and transport into the xylem. The percentage of total assimilates used in root respiration tends to decrease as plants age, perhaps reflecting an increased efficiency of respiration, if the nonphosphorylating alternative path contributes to a smaller extent to root respiration in older plants. On the other hand, such a decrease may be due to a decrease in the demand for respiratory energy, when the energy required for root growth and ion uptake decreases with increasing age. Furthermore, the root mass ratio tends to decrease with increasing age, thus decreasing the respiratory burden of roots.

The fraction of carbohydrates used in root respiration, including the respiration of symbionts, if present, is affected by both abiotic and biotic environmental factors (Table 1). Root respiration is higher in the presence of an **N_2-fixing symbiont** than it is when nonnodulated roots are supplied with nitrate as a nitrogen source. This reflects the greater energy requirement for N-assimilation dur-

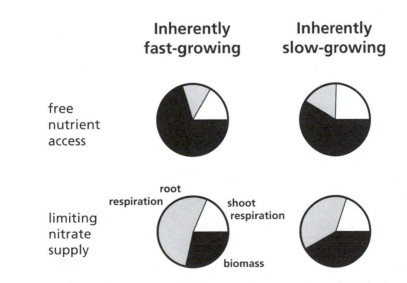

FIGURE 20. The fraction of all carbohydrates produced in photosynthesis per day that is consumed in respiration as dependent on species and the nitrogen supply. Measurements were made on inherently fast-growing (pies on the left) and slow-growing (pies on the right) grass species grown with free nutrient availability (pies at the top), and at a nitrogen supply that allowed a relative growth rate of approximately $40\,mg\,g^{-1}\,day^{-1}$ (pies at the bottom). The black section of the pie refers to carbon invested in growth; the other two sections refer to carbon used in shoot respiration (white sector) and root respiration (shaded sector) (Van der Werf et al. 1992a). Copyright SPB Academic Publishing.

ing N_2-fixation compared with that during NO_3^--assimilation (Sect. 3 of the chapter on symbiotic associations). The fraction of carbohydrates used in root respiration is also greater in the presence of a symbiotic **mycorrhizal fungus** than it is in nonsymbiotic plants (Table 1).

The *proportion* of the carbohydrates translocated to roots that is used in respiration, rather than root biomass accumulation, increases with plant age. This is primarily due to the increasing role of maintenance respiration, as root growth slows down and as the *quantity* of assimilates translocated to roots declines (cf. Sect. 5.2). Low nutrient supply also increases the proportion of carbohydrates respired in the roots. At a high supply of nutrients, plants respire approximately 40% of the carbon imported into the roots. This fraction increases to 60% at very low nutrient supply (Van der Werf 1996). This increase is largely accounted for by a relatively high carbon requirement for maintenance processes, as compared with that in growth processes. An additional factor is the proportionally low requirement for root growth (relative to that for maintenance) under these low-nutrient conditions. Finally, specific costs for maintenance or ion uptake might increase when nutrients are in short supply (Van der Werf et al. 1994; Sect. 5.3).

5.1.2 Respiration of Other Plant Parts

Leaf respiration provides some of the metabolic energy for leaf growth and maintenance, for ion transport from the xylem and export of solutes to the phloem. Leaf respiration, however, when expressed as a fraction of the carbon gain in photosynthesis, varies much less than root respiration because photosynthesis, leaf respiration, and biomass allocation are affected similarly by changes in nutrient supply. This is at variance with the situation for roots, where a major cause of the large variation found for root respiration (Table 1) is the effect of nutrient supply and genotype on **biomass allocation** to roots.

Rates of photosynthesis and leaf respiration often vary in a similar manner with changes in environment (e.g., nitrogen supply and growth irradiance). To some extent this may reflect the respiratory costs of the export of photosynthates from leaves, which vary with the carbon gained in photosynthesis. There may be greater maintenance costs, however, in leaves with high rates of photosynthesis and high protein concentrations or lower specific costs for major energy-requiring processes (e.g., for transport of assimilates from the mesophyll to the sieve

tubes) (Van der Werf 1995). These ecophysiological aspects of leaf respiration have remained largely unexplored.

The respiration of other plant parts (e.g., fruits) is largely accounted for by their growth rate and the respiratory costs per unit of growth. The maintenance component also plays a role. In green fruits, a substantial proportion of this energetic requirement may be met by photosynthesis in the fruit (Blanke & Whiley 1995, De Jong & Walton 1989). Respiration of the petals of the arctic herb *Saxifraga cernua* (saxifrage) shows a distinct peak when the petals are about 5 days old. A major part of the increased respiration appears to be accounted for by the alternative path (Collier et al. 1991).

5.2 Respiration Associated with Growth, Maintenance, and Ion Uptake

The rate of respiration depends on three major energy-requiring processes: **maintenance** of biomass, **growth**, and (ion) **transport**. This can be summarized in the following overall equation:

$$r = r_m + c_g \times RGR + c_t \times TR \qquad (1)$$

where r is the rate of respiration (normally expressed as nmol O_2 or $CO_2 g^{-1} s^{-1}$; however, to comply with the units in which RGR is expressed, we use $\mu mol\, g^{-1} day^{-1}$ here); r_m is the rate of respiration to produce ATP for the maintenance of biomass; c_g (mmol O_2 or $CO_2 g^{-1}$) is the respiration to produce ATP for the synthesis of cell material; RGR is the relative growth rate of the roots ($mg\, g^{-1} day^{-1}$); c_t (mol O_2 or $CO_2 mol^{-1}$) is the rate of respiration required to support TR, the transport rate ($\mu mol\, g^{-1} day^{-1}$). In roots TR is equal to the net ion uptake rate (NIR) and the rate of xylem loading; in photosynthesizing leaves TR is equal to the rate of export of the products of photosynthesis (from mesophyll to sieve tubes). Although respiration can be measured as either O_2 uptake or CO_2 release, the measurements do not yield exactly the same values. First, RQ may not equal 1.0 (Sect. 2.1); second, the rate of CO_2 release varies with the rate of nitrate reduction, whereas rates of O_2 consumption do not. For this reason **O_2 consumption** is preferred as a basis to compare plants when we are interested in **respiratory efficiency**, whereas **CO_2 release** is preferred when comparing the **carbon budgets** of different plants.

By examining these three requirements for respiratory energy, we can estimate how the ATP produced in respiration is used for major plant functions. Note that this equation assumes a tight correlation between the rate of respiration and the rates of major energy-requiring processes; there is no implicit assumption that respiration controls the rate of the energy-requiring processes, or vice versa.

5.2.1 Maintenance Respiration

Once biomass is produced, energy must be expended for repair and maintenance. Estimates of the costs of maintaining biomass range from 20 to 60% of the photosynthates produced per day in both herbaceous (De Visser et al. 1992) and woody species (Ryan et al. 1994), with the high values pertaining to plants that grow very slowly. The energy demands of the individual maintenance processes in vivo are not well known, and reliable estimates of individual maintenance costs are scarce. A major part of the maintenance energy costs is supposed to be associated with **protein turnover** and with the maintenance of **ion gradients** across membranes. These costs of maintenance have been estimated from basic biochemical principles (Penning de Vries 1975).

In higher plants approximately 2 to 5% of all the proteins are replaced daily, with extreme estimates being as high as 20% (Bouma et al. 1994, Van der Werf et al. 1992b). It is quite likely that protein turnover rates vary among plant organs, species, and with growth conditions, but the data are too scanty to make firm statements. The cost of synthesizing proteins from amino acids is at least 4 to 5 ATP, and possibly double that, per peptide bond, or approximately 0.26 (possibly 0.52) g glucose g^{-1} protein (De Visser et al. 1992). Approximately 75% of amino acids from degraded proteins are recycled (Davies 1979). The remaining 25%, however, must be synthesized from basic carbon skeletons, which is a cost of 0.43 g glucose g^{-1} protein. The total cost of protein turnover is about 28 to 53 mg glucose $g^{-1} day^{-1}$, or 3 to 5% of dry mass per day. Similar calculations for lipids suggest that membrane turnover constitutes a much lower energy requirement—approximately 1.7 mg glucose $g^{-1} day^{-1}$, or approximately 0.2% of dry mass per day. Based on an experimentally determined protein half-life of 5 days, the respiratory energy requirement to sustain protein turnover is approximately 1 mmol ATP g^{-1} (DM) day^{-1} (i.e., 7% of the total respiratory energy produced) in roots of *Dactylis glomerata*. Expressed as a fraction of the total maintenance requirement as derived from a multiple regression analysis (Sect. 5.2) (i.e., 2.7 mmol ATP g^{-1} (DM) day^{-1} for *Carex* species), the maintenance requirement for protein turnover is quite substantial (Van der Werf et al. 1992b).

Maintenance of ion gradients is also an important maintenance process. Some estimates suggest that costs of maintaining ion gradients are up to 30% of the respiratory costs involved in ion uptake, or approximately 20% of the total respiratory costs of young roots (Bouma & De Visser 1993).

Other processes (e.g., cytoplasmic streaming and turnover of other cellular constituents) are generally assumed to have a relatively small cost. Based on these many (largely unproven) assumptions, the total estimated maintenance respiration is approximately 30 to 60 mg glucose $g^{-1} day^{-1}$ (about 3 to 6% of dry mass day^{-1}). Measured values of maintenance respiration (8 to 60 mg glucose $g^{-1} day^{-1}$) suggest that these rough estimates are reasonable.

These experimental values for maintenance respiration suggest that protein turnover and the maintenance of solute gradients are by far the largest costs of maintenance in plant tissues. If true, then this conclusion has important implications for plant carbon balance because it suggests that any factor that increases protein concentration or turnover or the leakiness of membranes will increase maintenance respiration.

The often observed positive correlation of respiration rate with nitrogen concentration (e.g., De Visser et al. 1992, Ryan 1995) would seem consistent with the prediction that maintenance respiration depends on protein concentration. Thus, leaves that have a high nitrogen investment in Rubisco and other photosynthetic enzymes have a correspondingly high maintenance respiration. Whether this is a general phenomena remains to be investigated (Van der Werf et al. 1992b). Higher respiration rates might also reflect greater costs for the loading of photosynthates in the phloem, which is an active process (Sect. 3.3 of the chapter on long-distance transport). Whatever the explanation for the higher leaf respiration rates, they do contribute to their higher light-compensation point (Sect. 3.2.1 of the chapter on photosynthesis) and, therefore, place an upper limit on the irradiance level at which these leaves can maintain a positive carbon balance. Thus, there is a trade-off between high metabolic activity (requiring high protein concentrations and rapid loading of the phloem) and the associated increase in cost of maintenance and transport.

The stimulation of maintenance respiration by temperature is a logical consequence of the increased leakage and of protein turnover that occurs at high temperature. This provides a conceptual framework for studies that seek to explain why different tissues and species differ in their Q_{10} of respiration. This could reflect differences in membrane properties upon prolonged exposure to higher temperatures or of thermal stability of proteins, with corresponding differences in protein turnover (Criddle et al. 1994). It might also reflect a difference in contribution of cytochrome oxidase and the alternative oxidase.

5.2.2 Growth Respiration

Production of biomass (**biosynthesis**) requires the input of carbohydrates, partly to generate ATP and NAD(P)H for biosynthetic reactions and partly to provide the carbon skeletons present in biomass (Fig. 21; Table 11). Plant tissue is generally more reduced than the carbohydrates from which it is produced, therefore, the cost of biosynthesis from primary substrates must include the carbohydrates necessary to supply the extra reducing power (e.g., for the reduction of nitrate). If a more reduced source of nitrogen is absorbed instead (e.g., ammonium or amino acids) (cf. Sect. 2.2 of mineral nutrition), then biosynthetic costs are less. When a tissue senesces, most of the chemical constituents are lost to the plant; however, some are resorbed and can be used in the production of new tissues. The **final cost** of producing a tissue is the **initial cost** minus **resorption** (Fig. 22).

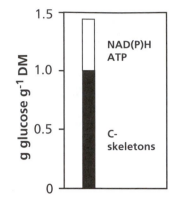

FIGURE 21. Construction costs of leaf biomass. Most of the glucose required for biomass production ends up in the carbon compounds in the biomass. Because the average carbon compound in biomass is more reduced than the carbohydrates from which it is produced, some glucose is required to produce NADPH. Some of the glucose is required to produce ATP, that drives many energy-requiring biosynthetic reactions in the cell. The data are for an "average" leaf.

In photosynthetically active leaves, some of the metabolic energy [ATP and NAD(P)H] may come directly from photosynthesis. In heterotrophic tissues such as roots, and in leaves in the dark,

TABLE 11. Values for characterizing the conversion of substrates to products during biosynthesis, excluding costs of substrate uptake from the environment.*

Compound	PV′	ORF′	CPF′	RQ′	HRF	ERF
Amino acids with NH_4^+	700	169	5,772	34	−11.2	−1.4
Amino acids with NO_3^-	700	169	5,772	34	26.7	39.0
Protein with NH_4^+	604	163	5,727	35	−12.9	34.9
Protein with NO_3^-	604	163	5,727	35	31.4	82.0
Carbohydrates	853	0	1,295	—	−3.6	12.2
Lipids	351	0	10,705	—	−10.1	51.0
Lignin	483	1388	5,545	4	−4.3	18.7
Organic acids	1104	0	−1,136	—	16.9	−4.5

* Production Value, PV′: milligrams of the end product per gram of substrate required for carbon skeletons and energy production, without taking into account the fate of excess or shortage of NAD(P)H and ATP (the term *production value*, PV, is used when PV′ is corrected for this component); Oxygen Requirement Factor, ORF′: μmol of oxygen consumed per gram of substrate required for carbon skeletons and energy production, without taking into account the fate of excess or shortage of NAD(P)H and ATP; Carbon dioxide production factor, CPF′: μmol of CO_2 produced per gram of substrate required for carbon skeletons and energy production, without taking into account the fate of excess or shortage of NAD(P)H and ATP (the term *carbon dioxide production factor*, CPF, is used when CPF′ is corrected for this component); RQ′ is the ratio of CPF′ and ORF′; Hydrogen Requirement Factor, HRF: mmoles of NAD(P)H required (−) or produced (+) per gram of end product; Energy Requirement Factor, ERF: mmoles of ATP required (−) or produced (+) per gram of end product (Penning de Vries et al. 1974). More recent findings, for example on the importance of targeting sequences of proteins that are required to "direct" the synthesized proteins to a specific compartment in the cell, indicate that the costs for protein synthesis are likely to be substantial, possibly even double the value as presented in this table.

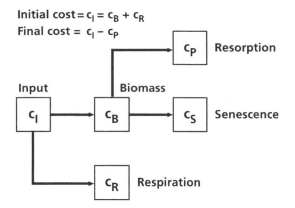

Initial cost $= c_I = c_B + c_R$
Final cost $= c_I - c_P$

FIGURE 22. Fate of carbon that is initially invested (C_I) in synthesizing a structure. Some of the carbon is retained in the biomass (C_B), the remainder is required for respiration (C_R). Of the carbon in the biomass (C_B), most is lost or respired when a plant part is shed (C_S) but some is resorbed (C_P) for subsequent use (Chapin 1989). Copyright by The University of Chicago.

respiration provides the required energy. The amount of respiratory energy that is required for biosynthesis can be calculated from the composition of the biomass in a number of ways, as will be discussed in this section.

First, costs for biosynthesis can be derived from detailed information on the **biochemical composition**, combined with biochemical data on the costs of synthesis of all the major compounds: protein, total nonstructural carbohydrates (i.e., sucrose, starch, fructans), total structural carbohydrates (i.e., cellulose, hemicellulose), lignin, lipid, organic acids, minerals. This can be extended to include various other compounds like soluble amino acids,

nucleic acids, tannins, lipophilic defense compounds, alkaloids, and so on, but these are mostly ignored and they are generally combined with the major ones. Taking glucose as the standard substrate for biosynthesis, one can estimate the amount of glucose required to provide the carbon skeletons, reducing equivalents, and ATP for the biosynthesis of plant compounds in tissues (Table 11) (Poorter & Villar 1997).

Note that the amount of product produced per unit carbon substrate (production value, PV') varies nearly threefold among chemical constituents (Table 11), with lipid and lignin being "most expensive" (i.e., requiring greatest glucose investment per gram of product) and organic acids "least expensive". Compounds like proteins and lipids are very costly in terms of ATP required for their biosynthesis (ERF), whereas carbohydrates and lignin are not. There are both expensive and cheap ways to produce structure in plants (lignin and cellulose/hemicellulose, respectively) and to store energy (lipids and carbohydrates, respectively) (Chapin 1989). Plants generally use energetically cheap structural components (cellulose/hemicellulose) and energy stores (carbohydrates). By contrast, mobile animals and small seeds, where mass is an important issue, often use lipids as their energy store (Levin 1974). Immobile animals, like plants, use carbohydrate (glycogen) as their primary energy store. We can calculate the costs for a gram of biomass once we know the costs and concentrations of the major compounds in plant biomass. As for individual compounds, these costs can be expressed in terms of glucose, oxygen requirement, carbon dioxide release, requirement for reducing power, and ATP (Table 12).

The major underlying assumption in the approach based on the biochemical composition of the

TABLE 12. An example of a simplified calculation of the variables that characterize biosynthesis of biomass from glucose, nitrate, and minerals.

Compound	Concentration in biomass (mg g^{-1} DM)	Glucose required for synthesis (mg)	O$_2$ required for synthesis (μmol)	CO$_2$ production during synthesis (μmol)	NAD(P)H required for synthesis (mmol)	ATP required for synthesis (mmol)
N-compounds	230	371	65	2,100	7.14	17.83
Carbohydrates	565	662	0	857	−2.03	6.92
Lipids	25	71	0	807	0.25	1.27
Lignin	80	166	230	918	−0.34	1.50
Organic acids	50	45	0	−52	−0.84	−0.23
Minerals	50	0	0	0	0	0
Total	1000	1315	295	4630	3.68	27.29

Source: Penning de Vries et al. 1974.

biomass is that glucose is the sole substrate for all ATP, reductant, and carbon skeletons. When some of these resources are derived directly from photosynthesis, costs may be lower. Costs may be higher when the alternative path, rather than the cytochrome path, plays a predominant role in respiration. If we restrict this approach to nonphotosynthetic tissues in which the contribution of the cytochrome and alternative respiratory pathway is known, then there is still a source of error, if these tissues import compounds other than glucose (e.g., amino acids) as a substrate for biosynthesis.

A second method for estimating the construction cost is based on information on the **elemental composition** of tissues: C, H, O, N, and S (McDermitt &

Loomis 1981). The constructions costs that are not covered by this equation include costs of mineral uptake and transport of various compounds in the plant, costs for providing ATP for biosynthetic reactions, and reductant required to reduce molecular oxygen in some biosynthetic reactions. This method is less laborious than the first method, which requires detailed chemical analysis, however, it is based on the observations of the first method (i.e., that expensive compounds are generally more reduced than glucose, whereas cheap compounds are more oxidized) (Poorter 1994). Although this method, which is based on elemental analysis of plant biomass, may seem to be a crude approach, it is surprisingly effective. First, this is because two thirds of the construction costs are costs to provide

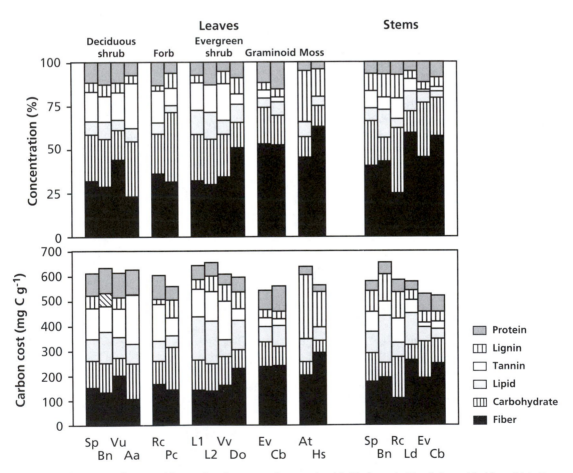

FIGURE 23. The chemical composition and carbon cost of producing leaves and stems of thirteen species of tundra plants. Species shown are *Salix pulchra* (Sp), *Betula nana* (Bn), *Vaccinium uliginosum* (Vu), *Arctostaphylos alpina* (Aa), *Rubus chamaemorus* (Rc), *Pedicularis capitata* (Pc), *Ledum decumbens* (Ld, including 1-year-old, L1, and 2-year-old, L2, leaves), *Vaccinium vitis-idaea* (Vv), *Dryas octopetala* (Do), *Eriophorum vaginatum* (Ev), *Carex bigelowii* (Cb), *Aulocomnium turgidum* (At), and *Hylocomium splendens* (Hs) (Chapin 1989). Copyright by The University of Chicago.

carbon skeletons rather than for respiration (Fig. 21). Second, most of the carbon that does not end up in the carbon skeletons of biomass is required to reduce carbon skeletons, and not for the production of ATP. Even in the absence of detailed information on respiratory pathways, therefore, construction costs can be estimated rather accurately. In fact, the second method can be simplified even further, taking into account only the **carbon and ash content** of biomass and ignoring minor constituents that have only a small effect on the production value (Vertregt & Penning de Vries 1987).

The level of reduction of plant biomass is approximately linearly related to its **heat of combustion** as well as its costs of construction (McDermitt & Loomis 1981). For example, lipids are highly reduced compounds and have a high heat of combustion. A third method, therefore, uses this approximation to arrive at costs for providing carbon skeletons and reductant for biosynthesis (Williams et al. 1987).

Given the threefold range in the cost of producing different organic constituents in plants and the large range in concentrations of these constituents among plant parts and species (often two- to tenfold; Fig. 23), we might expect large differences in costs of synthesizing tissues of differing chemical composition. A given tissue, however, tends to have *either* a high concentration of proteins and tannins (allowing high metabolic activity and chemical defense of these tissues) *or* a high concentration of lignin and lipophilic secondary metabolites (Chapin 1989). The negative correlation between the concentrations of these two groups of expensive constituents is seen in the comparison of leaves versus stems or in the comparison of leaves of rapidly growing species (e.g., forbs) and slowly growing species (evergreen shrubs) (Fig. 23). The net result of this trade-off between expensive components that allow rapid metabolic activity (proteins and tannins) versus those that allow persistence (lignin and lipophilic defensive compounds) is that the cost of all plant species and plant parts are remarkably similar—approximately 1.5 g glucose per gram of biomass (Figs. 23 and 24). Another important correlation that explains the similarity of construction costs across species and tissues is that tissues of fast-growing species that have high protein concentrations (an expensive constituent) also have high concentrations of minerals (a cheap constituent) (Poorter 1994). This explains why extremely simple relationships are excellent predictors of costs of synthesis.

The similarity of cost of synthesis across species, plant parts, and environments (Chapin 1989,

Poorter 1994) is surprising and differs from early conclusions that emphasized the high costs associated with lipids and lignin in evergreen leaves (Miller & Stoner 1979). Small seeds are an exception to this generalization because seed lipids are primarily an energy store (rather than an antiherbivore defense) and are positively associated with protein concentration (Fig. 24), which leads to a high carbon cost. The similarity among species and tissues in carbon cost of synthesis has the practical consequence that biomass is an excellent predictor of carbon cost. One possible ecological explanation for this pattern is that carbon is such a valuable resource that natural selection has led to the same minimal carbon cost for the construction of most plant parts. An alternative, and far more probable, explanation is that the negative correlations among expensive constituents and the positive correlation between protein and minerals have a basic physiological significance that coincidentally leads to a similar carbon cost of synthesis in most structures. For example, lignin and protein concentrations may be negatively correlated because young expanding cells have a high protein concentration; however, cell expansion would be prevented by lignin or heavy lignification might render cell walls less permeable to water and solutes, which would be disadvantageous in tissues with high metabolic activity (as gauged by high protein concentration). In general, currently available data suggest that costs of synthesis differ much less within (10 to 20%) and among (25%) ecosystems than do other

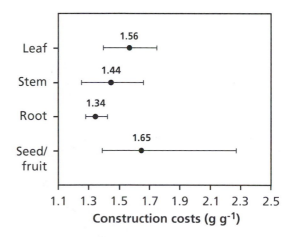

FIGURE 24. Range of construction costs for a survey of leaves (*n* = 123), stems (*n* = 38), roots (*n* = 35), and fruits/seeds (*n* = 31). Values are means and 10th and 90th percentiles (Poorter 1994). Copyright SPB Academic Publishing.

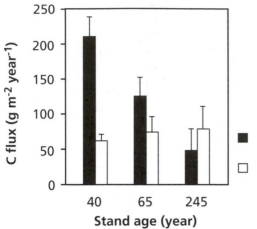

FIGURE 25. Annual carbon use for stem and branch growth (growth costs) and for stem and branch maintenance respiration in a lodgepole pine (*Pinus contorta*) successional sequence. Error bars show 95% confidence intervals (Ryan & Waring 1992). Copyright Ecological Society of America.

causes of variation in carbon balance, such as respiration and allocation (Chapin 1989, Poorter 1994).

Decreases in growth respiration may be the cause of declines in forest productivity in late succession, although increased maintenance respiration is often assumed to be involved (e.g., Waring & Schlesinger 1985). Maintenance respiration remains relatively constant through succession, whereas growth respiration declines (Fig. 25). The likely cause of reduced growth in old forest stands is a reduced carbon gain caused by loss of leaf area and loss of photosynthetic capacity associated with reduced hydraulic conductance and in some cases with reduced nutrient availability (Ryan et al. 1997).

5.2.3 Respiration Associated with Ion Transport

Ion transport across membranes may occur via ion channels, if transport is down an electrochemical potential gradient, or via ion carriers, which allow transport against an electrochemical potential gradient (Sect. 2.2.1 of the chapter on mineral nutrition). Because cation **transport from the rhizosphere** into the symplast mostly occurs down an electrochemical potential gradient, cation channels are often involved in this transport. Even cation uptake via channels, however, requires respiratory energy to extrude protons into the apoplast, thus creating an **electrochemical potential gradient**. Transport of anions from the rhizosphere into the symplast almost invariably occurs against an electrochemical potential gradient; hence, it requires respiratory energy, mostly because such anion transport is coupled to proton re-entry into the cells.

The situation is exactly the opposite for the transport of ions from the symplast to the xylem. (**xylem loading**) Anions might enter the xylem via channels, as this transport is mostly down an electrochemical potential gradient; however, we still know very little about such a mechanism (De Boer & Wegner 1997). The transport of most cations is against an electrochemical potential gradient and therefore must involve the expenditure of metabolic energy. In the context of the present section it is important to realize that the transport of cations to the xylem depends directly on metabolic energy. Release of anions into the xylem may be passive, but it still depends on the presence of an electrochemical potential gradient that can only be maintained by the expenditure of metabolic energy, predominantly by aerobic respiration. On the other hand, resorption of anions must be active (involving carriers), whereas that of cations may occur via channels (De Boer & Wegner 1997, Wegner & Raschke 1994).

When nitrate is the major source of nitrogen, this will be the major anion absorbed because only 10% and 1%, respectively, as much phosphate and sulfate compared with nitrate are required to produce biomass. Uptake of amino acids will also be against an electrochemical potential gradient and hence require a proton-cotransport mechanism similar to that described for nitrate. Like the uptake of nitrate and amino acids, phosphate uptake also occurs via a proton-symport mechanism. When phosphate availability is low, however, phosphate acquisition may require excretion of organic acids (cf. Sect. 2.2.5 of the chapter on mineral nutrition), which will incur additional carbon expenditure. Similarly, phosphate acquisition through a symbiotic association with mycorrhizal fungi requires additional carbon; this aspect is further discussed in Section 2.6 of the chapter on symbiotic associations.

As long as there is an electrochemical potential gradient, which is a prerequisite for the uptake of anions, cations can enter the symplast passively. In fact, plants may well need mechanisms to excrete cations that have entered the symplast passively to avoid excessive uptake of some cation (e.g., Na^+) (Sect. 3.4.2 of the chapter on mineral nutrition). When NH_4^+ is the predominant nitrogen source for the plant, such as in acid soils where rates of nitrification are low, this can enter the symplast via a cation channel. Rapid uptake of NH_4^+, however, must be balanced by excretion of H^+ in order to maintain a negative membrane potential. Hence, ammonium uptake also occurs with expenditure of respiratory energy.

When nitrate is the predominant N-source, rather than ammonium or amino acids, there are additional costs for its reduction. These show up with carbon costs and CO_2 release, but not in O_2 uptake (Table 11) because some of the NADH generated in respiration is used for the reduction of nitrate rather than for the reduction of O_2. As a result, the respiratory quotient strongly depends on the source of N (ammonium or nitrate; Table 2) and on the rate of nitrate reduction. Costs associated with nitrate acquisition are less when the reduction of nitrate occurs in leaves exposed to relatively high light intensities, as opposed to reduction in the roots, because the reducing power generated in the light reactions exceeds that needed for the reduction of CO_2 in the Calvin cycle under these conditions (cf. Sect. 3.2.1 of the chapter on photosynthesis).

Given that nitrogen is a major component of plant biomass, most of the respiratory energy associated with nutrient acquisition in plants with free access to nutrients will be required for the uptake of this nutrient.

5.2.4 Experimental Evidence

Measurements made with roots provide an opportunity to test the concepts of maintenance respiration, growth respiration, and uptake respiration. We assume that the rate of respiration for maintenance of root biomass is linearly related to the root biomass to be maintained. We also assume that the rate of respiration for ion transport is proportional to the amount of ions taken up, whereas that for root growth is proportional to the relative growth rate of the roots, as long as the chemical composition of the root biomass does not

change in a manner that affects the specific costs of biomass synthesis. Based on these assumptions, which are largely untested, the rate of ATP production per gram of roots and per day can be related to the relative growth rate of the roots and the rate of anion uptake by the roots. The maintenance respiration is superimposed. The costs of the three processes can then be estimated by multiple regression. The analysis is presented graphically in a three-dimensional graph (Fig. 26A). If a plant's relative growth rate and rate of anion uptake are very closely correlated, which is common, then a multiple regression analysis cannot separate the costs of growth from those of ion uptake (Fig. 26B).

Using the analysis as depicted in Figure 26A, respiratory costs for growth, maintenance, and ion uptake have been obtained for a limited number of species (Table 13A). The correlation between relative growth rate and nutrient uptake is quite often so tight that a linear regression analysis, as depicted in Figure 26B, is the only approach possible (Table 13B). There is quite a large variation in experimental values among species. This may reflect real differences between species; however, the variation may also indicate that the statistical analysis "explained" part of respiration by ion uptake in one experiment and by maintenance in another. For example, a costly process like ion leakage from roots, followed by reuptake, may show up in the slope or in the y-intercept in the graph, and suggest large costs for ion uptake and for maintenance, respectively. At the highest rates of growth and ion uptake (young plants, fast-growing species) these data suggest that respiration for growth and ion uptake together account for about 60% of root respiration and that maintenance respiration is relatively small. With increasing age, when growth and ion uptake slow down, maintenance respiration accounts for an increasing proportion of total respiration (more than 85%).

The specific costs for *Carex* (sedge) species (Table 13A) were used to calculate the rate of root respiration of 24 other herbaceous species of differing potential growth rate, whose rates of growth and ion uptake were known. These calculations greatly overestimated the root respiration of fast-growing species when compared with measured values (Fig. 27). This could indicate that either the efficiency of respiration is greater (relatively more cytochrome path and less alternative path) in fast-growing species, or that the specific costs for growth, maintenance, or ion uptake are lower for fast-growing species. Is there any evidence to support either hypothesis?

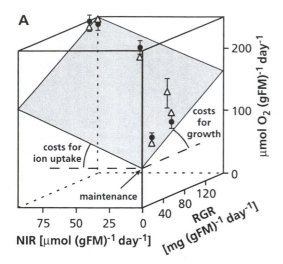

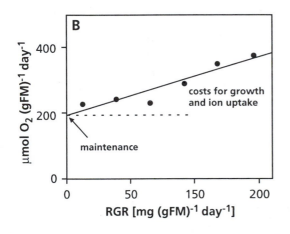

TABLE 13. (A) Specific respiratory energy costs for the maintenance of root biomass, for root growth, and for ion uptake. (B) Specific respiratory energy costs for the maintenance of root biomass and for root growth, including costs for ion uptake.

A.

	Carex	Solanum	Zea
Growth, mmol O_2 $(gDM)^{-1}$	6.3	10.9	9.9
Maintenance, nmol O_2 $(gDM)^{-1}s^{-1}$	5.7	4.0	12.5
Anion uptake, mol O_2 $(molions)^{-1}$	1.0	1.2	0.53

B.

	Dactylis	Festuca	Quercus	Triticum
Growth + ion uptake, mmol O_2 $(gDM)^{-1}$	11	19	12	18
Maintenance, nmol $(gDM)^{-1}s^{-1}$	26	21	6	22

Sources: (A) The values were obtained using a multiple regression analysis, as explained in Figure 25A [Van der Werf et al. 1988: average values for *Carex acutiformis* and *C. diandra* (sedges); Bouma et al. 1996: *Solanum tuberosum* (white potato); Veen 1980: *Zea mays* (maize)]. (B) The values were obtained using a linear regression analysis, as explained in Figure 25B [I. Scheurwater et al., unpublished: *Dactylis glomerata* and *Festuca ovina*; Mata et al. 1996: *Quercus suber* (cork oak); Van den Boogaard, as cited in Lambers et al. 1996: *Triticum aestivum* (wheat)].

FIGURE 26. (A) Rate of O_2 consumption [μmol g^{-1} (FM) day^{-1}] in roots as related to both the relative growth rate (RGR) of the roots and their net rate of anion uptake (NIR). (B) Rate of O_2 consumption (μmol g^{-1} (FM) day^{-1}) in roots as related to the relative growth rate of the roots. The plane in (A) and line in (B) give the predicted mean rate of O_2 consumption. The intercept of the plane in (A) and the line in (B) with the *y*-axis gives the rate of O_2 consumption in the roots that is required for maintenance. The slope of the projection of the line on the *y–z* plane gives the O_2 consumption required to produce 1 g of biomass. When projected on the *x–y* plane, the slope gives the specific respiratory costs for ion transport. In (B) the slope gives costs for growth including ion uptake (after Lambers et al. 1996). By courtesy of Marcel Dekker, Inc.

There is no convincing evidence for a more efficient respiration in roots of fast-growing species. Figure 28 shows that the specific respiratory costs for root growth are somewhat higher for fast-growing species. In addition, maintenance costs, if anything, are higher, rather than lower for roots of fast-growing species; this may well be connected with their higher protein concentrations and associated turnover costs. Most likely the discrepancy between the expected and measured rates of root respiration (Fig. 27) is based on higher **specific costs** for ion uptake in the inherently slow-growing species (Fig. 28) (Lambers et al. 1998). On the other hand, there might also be respiratory energy costs in roots of slow-growing species of which we are not aware.

The rate of root respiration of plants grown with a limiting nutrient supply is lower than that of plants grown with free access to nutrients, but not nearly as low as expected from their low rates of growth and nutrient acquisition (Sect. 4.3). This again suggests increased specific costs, possibly for ion uptake. Further experimental evidence is

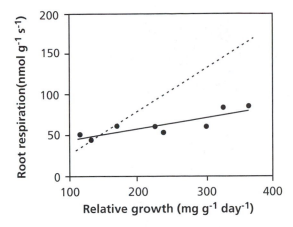

FIGURE 27. The rate of root respiration of fast-growing and slow-growing herbaceous C_3 species. The broken line gives the calculated respiration rate, assuming that specific costs for growth, maintenance, and ion uptake are the same as those given in Table 13 for *Carex* and identical for all investigated species (Poorter et al. 1991). Copyright Physiologia Plantarum.

Rates of photosynthesis per unit leaf area are poorly correlated with rates of growth, let alone final yield (Evans 1980). One of the reasons that has emerged in this chapter on plant respiration is that the fraction of all carbohydrates that are gained in photosynthesis and subsequently used in respiration varies considerably. First, slow-growing genotypes use relatively more of their photoassimilates for respiration. Second, many environmental variables affect respiration more than photosynthesis. This is true both because respiration rate is sensitive to environment and because the size of nonphotosynthetic plant parts, relative to that of

needed to address this important question concerning the carbon balance and growth of slow-growing plants.

In summary, these experimental data suggest that the concepts of respiration associated with growth, maintenance, and ion uptake are useful tools in understanding the carbon balance of plants and that the partitioning of respiration among these functions may differ substantially with environment and the type of plant species.

6. Plant Respiration: Why Should It Concern Us from an Ecological Point of View?

A large number of measurements have been made on the gas exchange (i.e., rates of photosynthesis, respiration, and transpiration) of different plants growing under contrasting conditions. Those measurements have yielded fascinating experimental results, some of which have been discussed in our chapter on photosynthesis; however, there is often the (implicit) assumption that rates of photosynthesis provide us with vital information on plant growth and productivity. Photosynthesis is certainly responsible for most of the gain in plant biomass; however, can we really derive essential information on growth rate and yield from measurements on photosynthesis only?

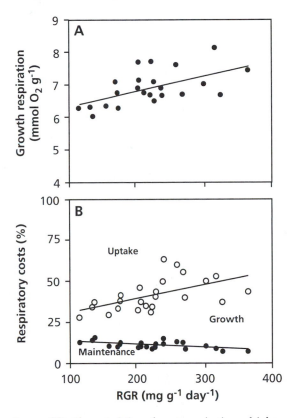

FIGURE 28. Characteristics of root respiration of inherently fast- and slow-growing herbaceous species, grown at free nutrient availability. (A) Respiratory cost for growth, as derived from an analysis of the roots' chemical composition and known cost for the synthesis of the various plant compounds. (B) Assuming similar respiratory efficiencies and maintenance costs for all species and using the costs for growth as given in (A), the specific costs for ion uptake were calculated. It is suggested that these costs are substantially higher for slow-growing herbaceous species than for fast-growing ones (Poorter et al. 1991). Copyright Physiologia Plantarum.

the photosynthetically active leaves depends on the environment, as will be discussed in the chapter on growth and allocation. An important message from this chapter on plant respiration should be that measurements of leaf photosynthesis by themselves cannot provide us with sound information on a plant's growth rate or productivity.

A second message worth emphasizing here is that respiration and the use of respiratory energy [NAD(P)H, ATP] are not as tightly linked as long believed. Respiration may proceed via a pathway that does not yield the respiratory products needed for growth, but produces heat instead. This component may be major, at least in some plants under some conditions. In specific tissues the production of heat may be of use to the plant, but the ecophysiological meaning of it in other tissues is still puzzling many plant scientists.

References and Further Reading

Alpi, A., Perata, P., & Beevers, H. (1985) Physiological responses of cereal seedlings to ethanol. J. Plant Physiol. 119:77–85.

Armstrong, W., Armstrong, J., Beckett, P.M., & Justin, S.H.F.W. (1991) Convective gas-flows in wetland plant aeration. In: Plant life under oxygen deprivation, M.B. Jackson, D.D. Davies, & H. Lambers, H. (eds). SPB Academic Publishing, The Hague, pp. 283–302.

Armstrong, W., Jackson, M.B., & Brändle, R. (1994) Mechanisms of flood tolerance in plants. Acta Bot. Neerl. 43:307–358.

Andrews, D.L., Cobb, B.G., Johnson, J.R., & Drew, M.C. (1993) Hypoxic and anoxic induction of alcohol dehydrogenase in roots and shoots of seedlings of Zea mays. Adh transcripts and enzyme activities. Plant Physiol. 101:407–414.

Atkin, O.K., Villar, R., & Lambers, H. (1995) Partitioning of electrons between the cytochrome and the alternative pathways in intact roots. Plant Physiol. 108:1179–1183.

Azcón-Bieto, J., Lambers, H., & Day, D.A. (1983) Effect of photosynthesis and carbohydrate status on respiratory rates and the involvement of the alternative pathway in leaf respiration. Plant Physiol. 72:598–603.

Bahr, J.T. & Bonner, W.D. (1973) Cyanide-insensitive respiration. II. Control of the alternate pathway. J. Biol. Chem. 248:3446–3450.

Ben Zioni, A., Vaadia, Y., & Lips, S.H. (1971) Nitrate uptake by roots as regulated by nitrate reduction products of the shoot. Physiol. Plant. 24:288–290.

Billings, W.D., Godfrey, P.J., Chabot, B.F., & Bourque, D.P. (1971) Metabolic acclimation to temperature in arctic and alpine ecotypes of Oxyria digyna. Arct. Alpi. Res. 4:227–289.

Bingham, I.J. & Farrar, J.F. (1988) Regulation of respiration in barley roots. Physiol. Plant. 73:278–285.

Blanke, M.M. & Whiley, A.W. (1995) Bioenergetics, respiration costs and water relations of developing avocado fruit. J. Plant Physiol. 145:87–92.

Bouma, T. & De Visser, R. (1993) Energy requirements for maintenance of ion concentrations in roots. Physiol. Plant. 89:133–142.

Bouma, T., De Visser, R., Janssen, J.H.J.A., De Kock, M.J., Van Leeuwen, P.H., & Lambers, H. (1994) Respiratory energy requirements and rate of protein turnover in vivo determined by the use of an inhibitor of protein synthesis and a probe to assess its effect. Physiol. Plant. 92:585–594.

Bouma, T., Broekhuysen, A.G.M., & Veen, B.W. (1996) Analysis of root respiration of Solanum tuberosum as related to growth, ion uptake and maintenance of biomass: A comparison of different methods. Plant Physiol. Biochem. 34:795–806.

Bouma, T., Nielsen, K.L., Eissenstat, D.M., & Lynch, J.P. (1997) Estimating respiration of roots in soil: Interactions with soil CO_2, soil temperature and soil water content. Plant Soil. 195:221–232.

Chapin III, F.S. (1989) The costs of tundra plant structures: Evaluation of concepts and currencies Am. Nat. 133:1–19.

Chapman, K.S.R. & Hatch, M.D. (1977) Regulation of mitochondrial NAD-malic enzyme involved in C_4 pathway photosynthesis. Arch. Biochem. Biophys. 184:298–306.

Collier, D.E. & Cummins, W.R. (1996) The rate of development of water deficits affects Saxifraga cernua leaf respiration. Physiol. Plant. 96:291–297.

Collier, D.E. & Cummins, W.R. (1991) Respiratory shifts in developing petals of Saxifraga cernua. Plant Physiol. 95:324–328.

Collier, D.E., Ackermann, F., Somers, D.J., Cummins, W.R., & Atkin, O.K. (1993) The effect of aluminium exposure on root respiration in an aluminium-sensitive and an aluminium-tolerant cultivar of Triticum aestivum. Physiol. Plant. 87:447–452.

Criddle, R.S., Hopkin, M.S. McArthur, E.D., & Hansen, L.D. (1994) Plant distribution and the temperature coefficient of metabolism. Plant Cell Environ. 17:233–243.

Dacey, J.W.A. (1980) Internal winds in water lilies: an adaptation for life in anaerobic sediments. Science 210:1017–1019.

Dacey, J.W.A. (1987) Knudsen-transitional flow and gas pressurization in leaves of Nelumbo. Plant Physiol. 85:199–203.

Davies, D.D. (1979) Factors affecting protein turnover in plants. In: Nitrogen assimilation of plants, E.J. Hewitt & C.V. Cutting (eds). Academic Press, London, pp. 369–396.

Day, D.A., Whelan, J., Millar, A.H., Siedow, J.N., & Wiskich J.T. (1995) Regulation of the alternative oxidase in plants and fungi. Aust. J. Plant Physiol. 22:497–509.

Day, D.A., Whelan, J., Millar, A.H., Siedow, J.N., & Wiskich, J.T. (1995) Regulation of the alternative oxidase in plants and fungi. Aust. J. Plant Physiol. 22:497–509.

Day, D.A., Krab, K., Lambers, H., Moore, A.L., Siedow, J.N., Wagner, A.M., & Wiskich, J.T. (1996) The cyanide-resistant oxidase: To inhibit or not to inhibit, that is the question. Plant Physiol. 110:1–2.

De Boer, A.H. & Wegner, L.H. (1997) Regulatory mechanisms of ion channels in xylem parenchyma cells. J. Exp. Bot. 48:441–449.

De Jong, T.M. & Walton, E.F. (1989) Carbohydrate requirements of peach fruits, growth and respiration. Tree Physiol. 5:329–335.

De Lima, M.L. & Copeland, L. (1994) The effect of aluminium on respiration of wheat roots. Physiol. Plant. 90:51–58.

De Visser, R., Spreen Brouwer, K., & Posthumus, F. (1986) Alternative path mediated ATP synthesis in roots of *Pisum sativum* upon nitrogen supply. Plant Physiol. 80:295–300.

De Visser, R., Spitters, C.J.T., & Bouma, T. (1992) Energy costs of protein turnover: Theoretical calculation and experimental extimation from regression of respiration on protein concentration of full-grown leaves. In: Molecular, biochemical and physiological aspects of plant respiration, H. Lambers & L.H.W. Van der Plas (eds). SPB Academic Publishing, The Hague, pp. 493–508.

Dry, I.B., Moore, A.L., Day. D.A., & Wiskich, J.T. (1989) Regulation of alternative pathway activity in plant mitochondria. Non-linear relationship between electron flux and the redox poise of the quinone pool. Arch. Biochem. Biophys. 273:148–157.

El Kohen, A., Pontailler, J.-Y., & Mousseau, M. (1991) Effect of doubling of atmospheric CO_2 concentration on dark respiration in aerial parts of young chestnut trees (*Castanea sativa* Mill.). C.R. Acad. Sci, Paris, t. 312, Series III: 477–481.

Elthon, T.E. & Stewart, C.R. (1983) A chemiosmotic model for plant mitochondria. BioScience 33:687–692.

Evans, L.T. (1980) The natural history of crop yield. Amer. Scientist 68:388–397.

Farrar, J.F. (1992) Beyond photosynthesis: the translocation and respiration of diseased leaves. In: Pests and pathogens, P.G. Ayres (ed). Bios Scientific Publishers, Oxford, pp. 107–127.

Farrar, J.F. & Jones, C.L. (1986) Modification of respiration and carbohydrate status of barley roots by selective pruning. New Phytol. 102:513–521.

Farrar, J.F. & Rayns, F.W. (1987) Respiration of leaves of barley infected with powdery mildew: Increased engagement of the alternative oxidase. New Phytol. 102:119–125.

Fredeen, A.L. & Field, C.B. (1991) Leaf respiration in *Piper* species native to a Mexican rainforest. Physiol. Plant. 82:85–92.

Gloser, V., Scheurwater, I., & Lambers, H. (1996) The interactive effect of irradiance and source of nitrogen on growth rate and root respiration of *Calamagrostis epigejos*. New Phytol. 134:407–412.

Gonzalez-Meler, Ribas-Carbo, M., Siedow, J.N., & Drake, B.G. (1996) Direct inhibition of plant respiration by elevated CO_2. Plant Physiol. 112:1349–1355.

Good, B.J. & Patrick, W.H. (1987) Gas composition and respiration of water oak (*Quercus nigra* L.) and green ash (*Fraxinus pennsylvanica* Marsh.) roots after prolonged flooding. Plant Soil 97:419–427.

Guglielminetti, L., Yamaguchi, J., Perata, P., & Alpi, A. (1995) Amylolytic activities in cereal seeds under aerobic and anaerobic conditions. Plant Physiol. 109:1069–1076.

Gonzalez-Meler, M.A. (1995) Effect of increasing concentration of atmospheric carbon dioxide on plant respiration. PhD Thesis, Universitat de Barcelona, Barcelona.

Guy, R.D., Berry, J.A., Fogel, M.L., Turpin, D.H., & Weger, H.G. (1992) Fractionation of the stable isotopes of oxygen during respiration by plants—the basis of a new technique to estimate partitioning to the alternative path. In: Plant respiration. molecular, biochemical and physiological aspects, H. Lambers & L.H.W. Van der Plas (eds). SPB Academic Publishing, The Hague, pp. 443–453.

Hagesawa, R., Muruyama, A., Nakaya, M., & Esashi, Y. (1995) The presence of two types of β-cyanoalanine synthase in germinating seeds and their response to ethylene. Physiol. Plant. 93:713–718.

Hemrika-Wagner, A.M., Kreuk, K.C.M., & Van der Plas, L.H.W. (1982) Influence of growth temperature on respiratory characteristics of mitochondria from callus-forming potato tuber discs. Plant Physiol. 70:602–605.

Hoefnagel, M.H.N., Millar, A.H., Wiskich, J.T., & Day, J.T. (1995) Cytochrome and alternative respiratory pathways compete for electrons in the presence of pyruvate in soybean mitochondria. Arch. Biochem. Biophys. 318:394–400.

Hoefnagel, M.H.N. & Wiskich, J.T. (1996) Alternative oxidase activity and the ubiquinone redox level in soybean cotyledon and *Arum* spadix mitochondria during NADH and succinate oxidation. Plant Physiol. 110:1329–1335.

Jackson, M.B., Hermann, B., & Goodenough, A. (1982) An examination of the importance of ethanol in causing injury to flooded plants. Plant Cell Environ. 5:163–172.

Klikoff, L.C. (1966) Temperature dependence of the oxidative rates of mitochondria in *Danthonia intermedia*, *Pentstemon davidsonii* and *Sitanion hystrix*. Nature 212:529–530.

Konings, H. & Lambers, H. (1990) Respiratory metabolism, oxygen transport and the induction of aerenchyma in roots. In: Plant life under oxygen deprivation, M.B. Jackson, D.D. Davies, & H. Lambers (eds). SPB Academic Publishing, The Hague, pp. 247–265.

Körner, C. & Larcher, W. (1988) Plant life in cold environments. In: Plants and temperature. Symposium of the Society of Experimental Biology, Vol. 42, S.P. Long & F.I. Woodward (eds). The Company of Biologists, Cambridge, pp. 25–57.

Knutson, R.M. (1974) Heat production and temperature regulation in eastern skunk cabbage. Science 186:746–747.

Krapp, A. & Stitt, M. (1994) Influence of high carbohydrate content on the activity of plastidic and cytosolic isozyme pairs in photosynthetic tissues. Plant Cell Environ. 17:861–866.

Lambers, H. (1985) Respiration in intact plants and tissue: Its regulation and dependence on environmental factors, metabolism and invaded organisms. In: Encyclopedia of plant physiology, N.S. Vol. 18, R. Douce and D.A. Day (eds). Springer-Verlag, Berlin, pp. 418–473.

Lambers, H. (1997a) Oxidation of mitochondrial NADH and the synthesis of ATP. In: Plant physiology, biochemistry and molecular biology, D.T. Dennis, D.H. Turpin, D. Lefebvre, & D.B Layzell (eds). Longmon, London, pp. 200–219.

Lambers, H. (1997b) Respiration and the alternative oxidase. In: A Molecular approach to primary metablism in plants, C.H. Foyer & P. Quick (eds). Taylor and Francis, London, pp. 295–309.

Lambers, H. & Van der Plas, L.H.W. (eds) (1992) Molecular, biochemical and physiological aspects of plant respiration. SPB Academic Publishing, The Hague.

Lambers, H. & Van der Werf, A. (1988) Variation in the rate of root respiration of two *Carex* species: A comparison of four related methods to determine the energy requirements for growth, mintenance and ion uptake. Plant Soil 111:207–211.

Lambers, H., Atkin, O.K., & Scheurwater, I. (1996) Respiratory patterns in roots in relation to their functioning. In: Plant roots: The hidden half, Y. Waisel, A. Eshel, & U. Kafkaki (eds). Marcel Dekker, Inc. New York, pp. 323–362.

Lambers, H., Scheurwater, I., Mata, C., & Nagel, O.W. (1998) Root respiration of fast- and slow-growing plants, as dependent on genotype and nitrogen supply: A major clue to the functioning of slow-growing plants. In: Inherent variation in plant growth. Physiological mechanisms and ecological consequences. H. Lambers, H. Poorter & M.M.I. Van Vuuren (eds). Backhuys, Leiden, in press.

Lance, C., Chauveau, M., & Dizengremel, P. (1985) The cyanide-resistant of plant mitochondria. In: Encyclopedia of plant physiology, R. Douce & D.A. Day (eds). Springer-Verlag, Berlin, pp. 202–247.

Larigauderie, A. & Körner, C. (1995) Acclimation of dark leaf respiration to temperature in alpine and lowland plant species. Ann. Bot. 76:245–252.

Laties, G.G. (1998) The discovery of the cyanide-resistant alternative path and its' aftermat. In: Discoveries in plant biology, Vol. 1, S.-D. Kung & S.F. Yang (eds). World Scientific Publishing Co., Hong Kong University of Science and Technology, Hong Kong, in press.

Levin, D.A. (1974) The oil content of seeds: an ecological perspective. Am. Nat. 108:193–206.

Mata, C., Scheurwater, I., Martins-Louçao, M.-A., & Lambers, H. 1996. Root respiration, growth and nitrogen uptake of *Quercus suber* L. seedlings. Plant Physiol. Biochem. 34:727–734.

McDermitt, D.K. & Loomis, R.S. (1981) Elemental composition of biomass and its relation to energy content, growth efficiency and growth yield. Ann. Bot. 48:275–290.

McDonnel, E. & Farrar, J.F. (1992) Substrate supply and its effect on mitochondrial and whole tissue respiration in barley roots. In: Plant respiration. molecular, biochemical and physiological aspects, H. Lambers & L.H.W. Van der Plas (eds). SPB Academic Publishing, The Hague, pp. 455–462.

McIntosh, L. (1994) Molecular biology of the alternative oxidase. Plant Physiol. 105:781–786.

Meeuse, B.J.D. (1975) Thermogenic respiration in aroids. Annu. Rev. Plant Physiol. 26:117–126.

Millar, A.H. & Day, D.A. (1997) Nitric oxide inhibits the cytochrome oxidase but not the alternative oxidase of plant mitochondria. FEBS Lett., in press

Millar, A.H., Hoefnagel, M.H.N., Day, D.A., & Wiskich, J.T. (1996) Specificity of the organic acid activation of the alternative oxidase in plant mitochondria. Plant Physiol. 111:613–618.

Miller, P.C. & Stoner, W.A. (1979) Canopy structure and environmental interactions. In: Topics in plant population biology, O.T. Solbrig, S. Jain, G.B. Johnson, & P.H. Raven (eds). Columbia University Press, New York, pp. 428–458.

Mitchell, P. (1966) Chemiosmotic coupling in oxidative and photosynthetic phosphorylation. Biol. Rev. 41:445–502.

Mooney, H.A. & Billings, W.D. (1961) Comparative physiological ecology of arctic and alpine populations of *Oxyria digyna*. Ecol. Monogr. 31:1–29.

Moynihan, M.R., Ordentlich, A., & Raskin, I. (1995) Chilling-induced heat evolution in plants. Plant Physiol. 108:995–999.

Neuberger, M. & Douce, R. (1980) Effect of bicarbonate and oxaloacetate on malate oxidation by spinach leaf mitochondria. Biochim. Biophys. Acta 589:176–189.

Nobel, P.S. & Palta, J.A. (1989) Soil O_2 and CO_2 effects on root respiration of cacti. Plant Soil 120:263–271.

Noguchi, K. & Terashima, I. (1997) Different regulation of leaf respiration between *Spinacia oleracea*, a sun species, and *Alocasia odora*, a shade species. Physiol. Plant. 101:1–7.

Noguchi, K., Sonoike, K., & Terashima, I. (1996) Acclimation of respiratory properties of leaves of *Spinacia oleracea* (L.), a sun species, and of *Alocasia macrorrhiza* (L.) G. Don., a shade species, to changes in growth irradiance. Plant Cell Physiol. 37:377–384.

Ordentlich, A., Linzer, R.A., & Raskin, I. (1991) Alternative respiration and heat evolution in plants. Plant Physiol. 97:1545–1550.

Palet, A., Ribas-Carbo, M., Argiles, J.M., & Azcón-Bieto, J. (1991) Short-term effects of carbon dioxide on carnation callus cell respiration. Plant Physiol. 96:467–472.

Palet, A., Ribas-Carbo, M., Gonzalez-Meler, M.A., Aranda, X., & Azcón-Bieto, J. (1992) Short-term effects of CO_2/bicarbonate on plant respiration. In: Molecular, biochemical and physiological aspects of plant respiration, H. Lambers & L.H.W. Van der Plas (eds). SPB Academic Publishing, The Hague, pp. 597–602.

Palta, J.A. & Nobel, P.S. (1989) Influence of soil O_2 and CO_2 on root respiration for *Agave deserti*. Physiol. Plant. 76:187–192.

Penning de Vries, F.W.T. (1975) The cost of maintenance processes in plant cells. Ann. Bot. 39:77–92.

Penning de Vries, F.W.T., Brunsting, A.H.M., & Van Laar, H.H. (1974) Products, requirements and efficiency of biosynthesis: A quantitative approach. J. Theor. Biol. 45:339–377.

Perata, P. & Alpi, A. (1991) Ethanol induces injuries to carrot cells. Plant Physiol. 95:748–752.

Perata, P. & Alpi, A. (1993) Plant responses to anaerobiosis. Plant Sci. 93:1–17.

Perata, P., Pozueta-Romero, J., Akazawa, T., & Yamaguchi, J. (1992) Effect of anoxia on the induction of a-amylase in cereal seeds. Planta 191:402–408.

Poorter, H. (1994) Construction costs and payback time of biomass: A whole plant perspective. In: A whole plant perspective on carbon-nitrogen interactions, J. Roy & E. Garnier (eds). The Hague, SPB Academic Publishing, pp. 111–127.

Poorter, H. & Villar, R. (1997) Chemical composition of plants: Causes and consequences of variation in allocation of C to different plant compounds. In: Resource allocation in plants, physiological ecology series, F. Bazzaz & J. Grace (eds). Academic Press, San Diego, pp. 39–72.

Poorter, H., Van der Werf, A., Atkin, O.K., & Lambers, H. (1991) Respiratory energy requirements of roots vary with the potential growth rate of a plant species. Physiol. Plant. 83:469–475.

Poorter, H., Gifford, R.M., Kriedemann, P.E., & Wong, S.C. (1992) A quantitative analysis of dark respiration and carbon content as factors in the growth response of plants to elevated CO_2. Aust. J. Bot. 40:501–513.

Poorter, H., Van de Vijver, C.A.D.M., Boot, R.G.A., & Lambers, H. (1995) Growth and carbon economy of a fast-growing and a slow-growing grass species as dependent on nitrate supply. Plant Soil 171:217–227.

Purvis, A.C. & Shewfelt, R.L. (1993) Does the alternative pathway ameliorate chilling injury in sensitive plant tissues? Physiol. Plant. 88:712–718.

Raskin, I., Ehmann, A., Melander, W.R., & Meeuse, B.J.D. (1987) Salicylic acid: A natural inducer of heat production in Arum lilies. Science 237:1601–1602.

Raskin, I., Turner, I.M., & Melander, W.R. (1989) Regulation of heat production in the inflorescence of an Arum lily by endogenous salicylic acid. Proc. Natl. Acad. Sci. 86:2214–2218.

Rennenberg, H. & Filner, P. (1983) evelopmental changes in the potential for H_2S emission in cucurbit plants. Plant Physiol. 71:269–275.

Ribas-Carbo, M., Berry, J.A., Yakir, D., Giles, L., Robinson, S.A., Lennon, A.M., & Siedow, J.N. (1995) Electron partitioning between the cytochrome and alternative pathways in plant mitochondria. Plant Physiol. 109:829–837.

Qi, J., Marshall, J.D., & Mattson, K.G. (1994) High soil carbon dioxide concentrations inhibit root respiration of Douglas fir. New Phytol. 128:435–442.

Richter, D.D. & Markewitz, D. (1995) How deep is soil? BioScience 45:600–609.

Rivoal, J. & Hanson, A.D. (1993) Evidence for a large and sustained glycolytic flux to lactate in anoxic roots of some members of the halophytic genus Limonium. Plant Physiol. 101:553–560.

Rivoal, J. & Hanson, A.D. (1994) Metabolic control of anaerobic glycolysis. Overexpression of lactate dehydrogenase in transgenic tomato roots supports the Davies-Roberts hypothesis and points to a critical role for lactate secretion. Plant Physiol. 106:1179–1185.

Roberts, J.K.M. (1984) Study of plant metabolism in vivo using NMR spectroscopy. Annu. Rev. Plant Physiol. 35:375–386.

Roberts, J.K.M. (1985) Further evidence that cytoplasmic acidosis is a determinant of flooding intolerance in plants. Plant Physiol. 77:492–494.

Roberts, J.K.M., Wemmer, D., & Jardetzky, O. (1984a) Measurements of mitochondrial ATP-ase activity in maize root tips by saturation transfer ^{31}P nuclear magnetic resonance. Plant Physiol. 74:632–639.

Roberts, J.K.M., Callis, J., Jardetzky, O., Walbot, V., & Freeling, M. (1984b) Cytoplasmic acidosis as a determinant of flooding intolerance in plants. Proc. Natl. Acad. Sci. USA 81:6029–6033.

Robinson, S.A., Yakir, D., Ribas-Carbo, M., Giles, L. Osmond, C.B., Siedow, J.N., & Berry, J.A. (1992) Measurements of the engagement of cyanide-resistant respiration in the crassulacean acid metabolism plant Kalanchoe daigremontiana with the use of on-line oxygen isotope discrimination. Plant Physiol. 100:1087–1091.

Robinson, S.A., Ribas-Carbo, M., Yakir, D., Giles, L., Reuveni, Y., & Berry, J.A. (1995) Beyond SHAM and cyanide: Opportunities for studying the alternative oxidase in plant respiration using oxygen isotope discrimination. Aust. J. Plant Physiol. 22:487–496.

Ryan, M.G. (1995) Foliar maintenance respiration of subalpine and boreal trees and shrubs in relation to nitrogen content. Plant Cell Environ 18:765–772.

Ryan, M.G. & Waring, R.H. (1992) Maintenance respiration and stand development in a subalpine lodgepole pine forest. Ecology 73:2100–2108.

Ryan, M.G., Linder, S., Vose, J.M., & Hubbard, R.M. (1994) Dark respiration of pines. Ecol. Bull. 43:50–63.

Ryan, M.G., Binkley, D., & Fownes, J.H. (1997) Age-related decline in forest productivity: Pattern and process. Adv. Ecol. Res. 27:213–262.

Schaaf, J., Walter, M.H., & Hess, D. (1995) Primary metabolism in plant defense. Regulation of bean malic enzyme gene promoter in transgenic tobacco by development and environmental cues. Plant Physiol. 108:949–960.

Schubert, S., Schubert, E., & Mengel, K. (1990) Effect of low pH of the root medium on proton release, growth, and nutrient uptake of field beans (Vicia faba). Plant Soil 124:239–244.

Semikhatova, O.A., Gerasimenko, T.V., & Ivanova, T.I. (1992) Photosynthesis, respiration, and growth of plants in the Soviet Arctic. In: Arctic Ecosystems in a Changing Climate, F.S. Chapin, R.L. Jefferies, J.F. Reynolds, G.R. Shaver, & J. Svoboda (eds). Academic Press, San Diego, pp. 169–192.

Seymour, R.S. & Schultze-Motel, P. (1996) Thermoregulating lotus flowers. Nature 383:305.

Shaw, M. & Samborski, D.J. (1957) The physiology of host-parasite relations. III The pattern of respiration in rusted and mildewed cereal leaves. Can. J. Bot. 35:389–407.

Shipway, M.R. & Bramlage, W.J. (1973) Effects of carbon dioxide on activity of apple mitochondria. Plant Physiol. 51:1095–1098.

Simons, B.H. & Lambers, H. (1998) The alternative oxidase: is it a respiratory pathway allowing a plant to cope with stress? In: Plant responses to environmental stresses: From phytohormones to genome reorganization. H.R. Lerner (ed). Plenum Press, New York, in press

Simons, B.H., Mulder, L., Van Loon, L.C., & Lambers, H. (1998) Enhanced expression and activation of the alternative oxidase in the interaction of *Arabidopsis* with pathogenic *Pseudomonas syringae* strains. Plant Physiol., submitted

Sims, D.A. & Pearcy, R.W. (1991) Photosynthesis and respiration in *Alocasia macrorrhiza* following transfers to high and low light. Oecologia 86:447–453.

Smakman, H. & Hofstra, R. (1982) Energy metabolism of *Plantago lanceolata* as affected by change in root temperature. Physiol. Plant. 56:33–37.

Stiles, W. & Leach, W. (1936) Respiration in plants. Methuen & Co., London.

Tan, K. & Keltjens, W.G. (1990a) Interaction between aluminium and phosphorus in sorghum plants. I. Studies with the aluminium sensitive genotype TAM428. Plant Soil 124:25–23.

Tan, K. & Keltjens, W.G. (1990b) Interaction between aluminium and phosphorus in sorghum plants. II. Studies with the aluminium tolerant genotype SC0 283. Plant Soil 124:25–32.

Torn, M.S. & Chapin III, F.S. (1993) Environmental and biotic controls over methane flux from arctic tundra. Chemosphere 26:357–368.

Skubatz, H., Nelson, T.A., Meeuse, B.J.D., & Bendich, A.J. (1991) Heat production in the voodoo lily (*Sauromatum guttatum*) as monitored by infrared thermography. Plant Physiol. 95:1084–1088.

Vertregt, N. & Penning de Vries, F.W.T. (1987) A rapid method for determining the efficiency of biosynthesis of plant biomass. J. Theor. Biol. 128:109–119.

Umbach, A.L. & Siedow, J.N. (1993) Covalent and noncovalent dimers of the cyanide-resistant alternative oxidase protein in higher plant mitochondria and their relationship to enzyme activity. Plant Physiol. 103:845–854.

Umbach, A.L., Wiskich, J.T., & Siedow, J.N. (1994) Regulation of alternative oxidase kinetics by pyruvate and intermolecular disulfide bond redox status in soybean seedling mitochondria. FEBS Lett. 348:181–184.

Van der Werf, A. (1995) Growth analysis and photoassimilate partitioning. In: Photoassimilate distribution in plants and crops: Source-sink relationships, E. Zamski & A.A. Schaffer (eds). Marcel Dekker, New York, pp. 1–20.

Van der Werf, A. (1996) Growth, carbon allocation, and respiration as affected by nitrogen supply: Aspects of the carbon balance. In: Dynamics of roots and nitrogen in intercropping systems of the semiarid tropics, O. Ito, C. Johansen, J.J. Adu-Gyamfi, K. Katayama, J.V.D.K. Kumar Rao, & T.J. Rego (eds). Japan International Research Center for Agricultural Sciences, Ibaraki.

Van der Werf, A., Kooijman, A., Welschen, R., & Lambers, H. (1988) Respiratory costs for the maintenance of biomass, for growth and for ion uptake in roots of *Carex diandra* and *Carex acutiformis*. Physiol. Plant. 72:483–491.

Van der Werf, A., Welschen, R., & Lambers, H. (1992a) Respiratory losses increase with decreasing inherent growth rate of a species and with decreasing nitrate supply: A search for explanations for these observations. In: Molecular, biochemical and physiological aspects of plant respiration, H. Lambers & L.H.W. Van der Plas (eds). SPB Academic Publishing, The Hague, pp. 421–432.

Van der Werf, A., Van den Berg, G., Ravenstein, H.J.L., Lambers, H., & Eising, R. (1992b) Protein turnover: A significant component of maintenance respiration in roots? In: Molecular, biochemical and physiological aspects of plant respiration, H. Lambers & L.H.W. Van der Plas (eds). SPB Academic Publishing, The Hague, pp. 483–492.

Van der Werf, A., Poorter, H., & Lambers, H. (1994) Respiration as dependent on a species' inherent growth rate and on the nitrogen supply to the plant. In: A whole-plant perspective of carbon–nitrogen interactions, J. Roy & E. Garnier (eds). SPB Academic Publishing, The Hague, pp. 61–77.

Vani, T. & Raghavendra, S. (1994) High mitochondrial activity but incomplete engagement of the cyanide-resistant alternative pathway in guard cell protoplasts of pea. Plant Physiol. 105:1263–1268.

Vanlerberghe, G.C. & McIntosh, L. (1992) Lower growth temperatures increase alternative oxidase protein in tobacco callus. Plant Physiol. 100:115–119.

Vanlerberghe, G.C. & McIntosh, L. (1996) Signals regulating the expression of the nuclear gene encoding alternative oxidase of plant mitochondria. Plant Physiol. 111:589–595.

Vanlerberghe, G.C., Day, D.A., Wiskich, J.T., Vanlerberghe, A.E., & McIntosh, L. (1995) Alternative oxidase activity in tobacco leaf mitochondria. Plant Physiol. 109:353–361.

Veen, B.W. (1980) Energy costs of ion transport. In: Genetic engeneering of osmoregulation. Impact on plant productivity for food, chemicals and energy, D.W. Rains, R.C. Valentine, & C. Holaender (eds), Plenum Press, New York, pp. 187–195.

Wagner, A.M. & Krab, K. (1995) The alternative respiration pathway in plants: Role and regulation. Physiol. Plant. 95:318–325.

Wagner, A.M., Van Emmerik, W.A.M., Zwiers, J.H., & Kaagman, H.M.C.M. (1992) Energy metabolism of *Petunia hybrida* cell suspensions growing in the presence of antimycin A. In: Plant respiration. Molecular, biochemical and physiological aspects, H. Lambers & L.H.W. Van der Plas (eds). SPB Academic Publishing, The Hague, pp. 609–614.

Waring R.H. & Schlesinger, W.H. (1985) Forest ecosystems: Concepts and management. Academic Press, Orlando.

Weger, H.G. & Guy, R.D. (1991) Cytochrome and alternative pathway respirtion in white spruce (*Picea glauca*) roots. Effects of growth and measurement temperature. Physiol. Plant 83:675–681.

Wegner, L. & Raschke, K. (1994) Ion channels in the xylem parenchyma of barley roots. A procedure to isolate protoplasts from this tissue and a patch-clamp exploration of salt passageways into xylem vessels. Plant Physiol. 105:799–813.

Williams, J.H.H. & Farrar, J.F. (1990) Control of barley root respirtion. Physiol. Plant. 79:259–266.

Williams, J.H.H., Winters, A.L., & Farrar, J.F. (1992) Sucrose: a novel plant growth regulator. In: Plant respiration. Molecular, biochemical and physiological aspects, H. Lambers & L.H.W. Van der Plas (eds).

SPB Academic Publishing, The Hague, pp. 463–469.

Yan, F., Schubert, S., & Mengel, K. (1992) Effect of low root medium pH on net proton release, root respiration, and root growth of corn (*Zea mays* L.) and broad bean (*Vicia faba* L.). Plant Physiol. 99:415–421.

Williams, K., Percival, F., Merino, J., & Mooney, H.A. (1987) Estimation of tissue construction cost from heat of combustion and organic nitrogen content. Plant Cell Environ. 10:725–734.

Williams, K., Field, C.B., & Mooney, H.A. (1989) Relationship among leaf construction costs, leaf longevity and light environment in rainforest plants of the genes *Piper*. Am. Nat. 133:198–211.

Wullschleger, S.D., Ziska, L.H., & Bunce. J.A. (1994) Respiratory responses of higher plants to atmospheric CO_2 enrichment. Physiol. Plant. 90:221–229.

Zacheo, G. & Molinari, S. (1987) Relationship between root respiration and seedling age in tomato cultivars infested by *Meloidogyne incognita*. Ann. Appl. Biol. 111:589–595.

2C. Long-Distance Transport of Assimilates

1. Introduction

The evolution of cell walls allowed plants to solve the problem of osmoregulation in freshwater environments; however, cell walls restrict motility and place constraints on the evolution of long-distance transport systems. Tissues are too rigid for a heart-pump mechanism; rather, higher plants have two systems for long-distance transport. The dead elements of the xylem allow transport of water and solutes between sites of different water potentials. This transport system will be dealt with in the chapter on plant water relations. The other transport system, the phloem, allows the mass flow of carbohydrates and other solutes from a **source** region, where the hydrostatic pressure in the phloem is relatively high, to a **sink** region with lower pressure (Fig. 1).

Plants differ markedly in the manner in which the products of photosynthesis pass from the mesophyll cells to the sieve tubes (**phloem loading**) through which they are then transported to a site where they are unloaded and metabolized (Fig. 1). Plants also differ with respect to the major carbon-containing compounds that occur in the sieve tubes, which is the complex that consists of sieve elements and companion cells. It is only fairly recently that we have become aware of the close association between the type of phloem loading (symplasmic or apoplasmic) and the type of major carbon compound (sucrose or oligosaccharides) transported in the phloem. Sucrose is a sugar composed of two hexose units, whereas an oligosaccharide is com-posed of more than two units. In addition, there appears to be a close association between the pattern of phloem loading and the **ecological distribution** of species. It is this close association between transport mechanism and ecological adaptation that we will explore in this chapter.

2. Major Transport Compounds in the Phloem: Why Not Glucose?

In animals, glucose is the predominant transport sugar, albeit at much lower concentrations than that of predominant sugars in the sieve tubes of higher plants. In plants, sucrose is a major constituent of phloem sap, whereas glucose and other monosaccharides are only found in trace concentrations (Table 1). Why not glucose?

A comparison of the physical properties of glucose and sucrose does not provide a compelling reason for the predominance of sucrose. A good long-distance transport compound, however, should be nonreducing to avoid a nonenzymatic reaction with proteins or other compounds during its transport. At the same time the transported compound should be protected from enzymatic attack until it arrives at its destination. In this way the flow of carbon in plants can be controlled by the presence of key sucrose-hydrolyzing enzymes in appropriate tissues. Thus, sucrose appears to be a preferred compound because it is "**protected**".

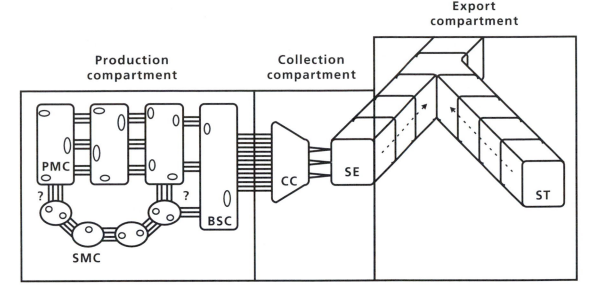

FIGURE 1. The phloem-loading pathway comprises three compartments with separate functions. The photosynthates move from the production compartment, comprising the palisade (PMC) and spongy (SMC) mesophyll and bundle sheath (BSC), to the collection compartment. The site of the collection compartment is the phloem in the minor vein ends; it includes companion cells (CC) and sieve elements (SE). Companion cells occur in all species, but they differ widely in their ultrastructure and biochemistry. The collection compartment is connected via plasmodesmata with the export compartment. In the pro- duction compartment the palisade mesophyll cells are connected via plasmodesmata, just like the spongy mesophyll cells. The absence of sound information on plasmodesmatal connections between the two types of mesophyll cells is indicated by the questions marks. The frequency of plasmodesmatal connections between the production compartment and the collection compartment differs enormously, as outlined in this chapter. The sieve tubes (ST) of the export compartment comprise both sieve elements and companion cells (Van Bel 1996).

Other "protected" sugars include the oligosaccharides of the raffinose family: raffinose, stachyose, verbascose. These sugars are formed by the addition of one, two, or three galactose molecules to a sucrose molecule (Fig. 2). They are major transport sugars in a range of species. Other transport compounds are the sugar alcohols (sorbitol, mannitol, dulcitol, myoinositol; Fig. 2) [e.g., in Apiaceae (e.g., *Apium graveolens*, celery) Rosaceae, Combretaceae, Celastraceae and Plantaginaceae] (Table 1). The composition of the phloem transport fluid differs widely between species, but there is one similarity: very low quantities of monosaccharides (glucose, fructose) (Turgeon 1995).

transport system: the sieve elements. In the gymnosperm redwood tree *Sequiodendron giganteum* (giant redwood) the source-sink distance of the phloem path may be as much as 110 m because of the enormous height of the tree. This example is somewhat extreme because sinks mostly receive assimilates from adjacent source leaves, but it emphasizes the point that transport may have to occur over vast distances, for example to growing root tips far removed from source leaves. Long-distance transport in the phloem is considered a **mass flow**, driven by a difference in hydrostatic pressure, created by phloem loading in source leaves and unloading processes in sink tissues. How are the products of photosynthesis in the mesophyll loaded into the sieve tubes?

3. Phloem Structure and Function

In the process of transport of assimilates from the site of their synthesis (the **source**) to the site where they are used (the **sink**), the products of photosynthesis have to move from the mesophyll cells to the

3.1 Symplasmic and Apoplasmic Pathways of Phloem Loading

As long as there are sufficient plasmodesmatal connections, movement of sugars from the mesophyll cells to the sieve tubes may occur symplasmically.

Table 1. List of sugars and sugar alcohols in sieve-tube exudates of a number of species from different families.*

	Sucrose	Raffinose	Stachyose	Verbascose	Mannitol	Sorbitol	Dulcitol	Myoinositol
Aceraceae								
Acer circinatum	++++	+						
A. saccharum	++++	+						
Anacardiaceae								
Anacardia excelsum	+++	tr	tr					
Annonaceae								
Annona bullata	+++	tr	tr					
Apocynaceae								
Nerium oleander	+++	++	+	+				+
Aquifoliaceae								
Ilex aquifolium	+++		+					
Asteraceae								
Artemisia canariensis	+	tr						
Berberidaceae								
Mahonia aquifolium	++	+						+
Bignoniaceae								
Catalpa bignonioides	+++	+	+++	+	tr			+
Budleiaceae								
Budleya albiflora	++	+++	++++					+
Buxaceae								
Buxus sempervirens	++++	+++	+++	+				++
Caesalpiniaceae								
Bauhinia candicans	+++							
Caprifoliaceae								
Lonicera fragrantissima	+	++	tr					+
Casuarinaceae								
Casuarina equisetifolia	+++							
Celastraceae								
Celastrus orbiculata	++	+	+	tr			+++	tr
Cornaceae								
Cornus mas	++++	+	+					tr
Coryleaceae								
Alnus glutinosa	++++	++		++				tr
Ericaceae								
Rhododendron houlstonii	+++	++	+					
Fabaceae								
Robinia pseudoacacia	+++							

TABLE 1. *Continued*

	Sucrose	Raffinose	Stachyose	Verbascose	Mannitol	Sorbitol	Dulcitol	Myoinositol
Fagaceae								
Castanea dentata	+++	tr						
Quercus robur	++++	tr		+				tr
Hyperiaceae								
Clusia rosea	+++							
Juglandaceae								
Juglans regia	++++	+	++				+	
Lamiaceae								
Rosmarinus officinalis	+++	++	tr	tr				+
Mimosaceae								
Acacia baileyana	+++		tr	tr				
Oleaceae								
Fraxinus americana	++	++	+++	tr	+++			
Prunoideae								
Prunus padus	+++	++	++			++++	+	
Rosaceae								
Cotoneaster hupehensis	+++	tr				++++	tr	
Salicaceae								
Populus candicans	++++	+	+					+
Scrophulariaceae								
Paulownia tomentosa	++++	+++	++++	++	tr			+
Tiliaceae								
Tilia americana	+++	+	+					+
Ulmaceae								
Ulmus americana	+++	tr						
Verbenaceae								
Gmelina arborea	++	++	+++	tr				
Vitaceae								
Parthenocissus henryana	++	tr	+					tr

Source: Zimmermann & Ziegler 1995.
*++++, approx. 20 to 30% (w/v, sucrose equivalent), +++, ca. 10 to 20%, ++, ca. 2 to 10%, +, ca. 0.5 to 2%, tr, trace (0.1 to 0.5%).

(The **symplast** is the entire space in a plant that is surrounded by the plasma membranes. The **apoplast** is the space outside the plasma membrane, including the cell walls as well as the xylem conduits.) That is, sucrose moves from the cytosol of the mesophyll cells via **plasmodesmata** to the cytosol of neighboring phloem cells. Plasmolysis studies on leaves of dicotyledonous species, however, have shown that the solute concentration in the sieve elements of minor veins is up to ten-fold higher than that of the surrounding cell types. Symplasmic phloem loading by diffusion through plasmodesmata therefore seems thermodynamically impossible, unless the sugar is somehow "trapped." We will discuss shortly that **intermediary cells** (Fig. 3A), companion cells located between the mesophyll cells and the sieve elements, play an important role in "trapping" the sugar. There is no evidence for carrier-mediated transport of either sucrose or stachyose in mature leaves of

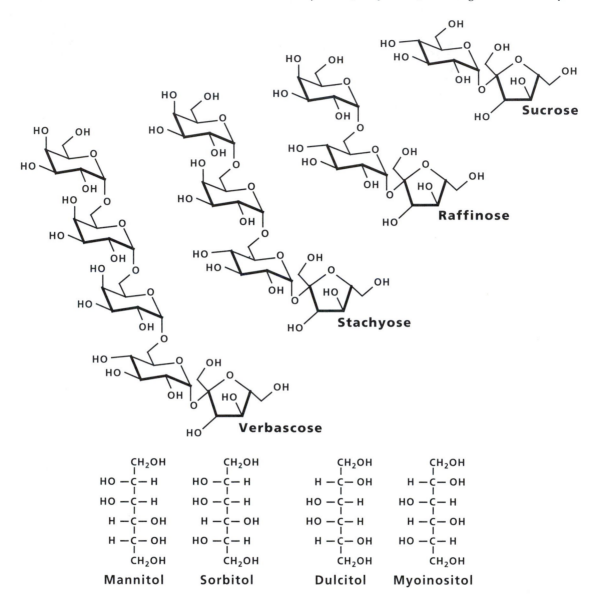

FIGURE 2. The chemical structure of the major sugars and some sugar alcohols transported in sieve-tube exudate. Note that not all of these compounds occur in the phloem sap of every species (Table 1).

species showing the symplasmic pathway (Turgeon 1996).

If plasmodesmata are absent between mesophyll cells and sieve-tube elements, transport from the mesophyll to the sieve tubes cannot occur symplasmically. It involves the release of sucrose from the cytosol of the mesophyll cells to the cell walls (the apoplast). Cells located between the mesophyll cells and the sieve elements, take up the sucrose from the apoplast against a concentration gradient. It is mediated via a sucrose-proton-cotransport carrier in their highly invaginated plasma membranes (these companion cells are indicated as **transfer cells** (Fig. 3B). This presence of the sucrose carrier explains both the selectivity for the exported sugar sucrose and the high sugar concentration in the sieve tubes.

Depending on species, either the symplasmic or the apoplasmic pathway of phloem loading is followed. Ultrastructural information on plasmo-

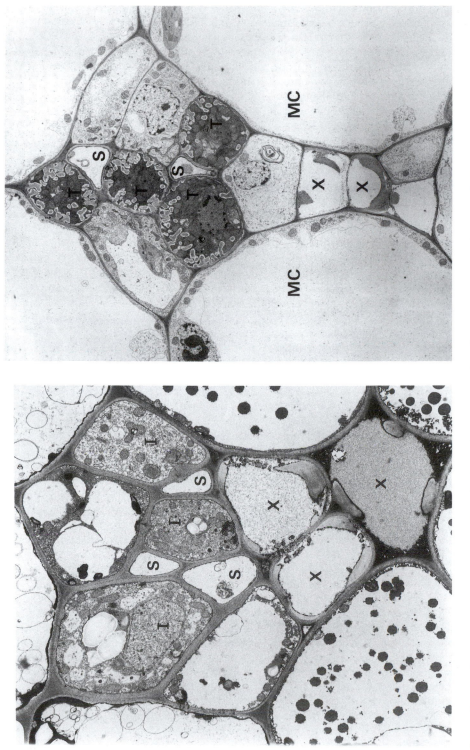

FIGURE 3. (Left) EM photograph of a cross-section of a leaf minor vein of *Lythrum salicaria*, showing three intermediary cells (I), adjacent to three sieve tube (S), and three xylem elements (courtesy A. Ammerlaan & R. Kempers, Utrecht University, the Netherlands). (Right) EM photograph of a cross-section of a leaf minor vein of *Senecio vulgaris* (groundsel), showing four transfer cells (T) adjacent to two sieve tubes (S); M indicates two mesophyll cells (MC) and two xylem elements (X) (courtesy B.E.S. Gunning, Australian National University, Canberra, Australia; Gunning & Steer 1996). Copyright Gustav Fischer Verlag.

desmatal connections is essential to decide if the symplasmic pathway is possible, even though it cannot prove that it is the predominant path. It should also be noted that in some species, including temperate trees like *Populus deltoides* (poplar), the solute concentration in the phloem is lower than that of the mesophyll, which is in striking contrast to all other plants examined. Plants with low solute concentrations in the phloem may not "load" the phloem in the sense that loading occurs against a concentration gradient (Turgeon 1996).

3.2 Minor Vein Anatomy

Gamalei (1989, 1991) studied minor vein anatomy of more than 1000 species. He distinguished several gradations of **plasmodesmatal connectivity** between the cells that are closely associated with the sieve elements (companion cells or intermediary cells, as detailed further on) and the mesophyll cells. These range from entirely symplasmic phloem-loading (Type 1) to entirely apoplasmic phloem-loading (Type 2b). In the herb *Senecio vernalis* the plasmodesmatal frequency is around 0.03 plasmodesmata per square micrometer interface area, against 60 in the evergreen tree *Fraxinus ornus* (manna ash). During the evolution from evergreen trees to annual herbs there was apparently a decrease in plasmodesmatal connectivity between the minor vein phloem and the mesophyll. Type 1 [families with many trees and shrubs (e.g., Hydrangeaceae) as well as others: Lamiaceae, Onagraceae] exhibits about three orders of magnitude more plasmodesmatal contacts than type 2b (many herbaceous species: Asteraceae, Brassicaceae, Chenopodiaceae, Fabaceae, Solanaceae). Types 1–2a and 2a are intermediate between the two extremes and differ by about one and two orders of magnitude in plasmodesmatal frequency, respectively. There is increasing evidence that the minor veins of grasses show low plasmodesmatal frequencies and that grasses therefore follow an apoplasmic loading pathway (Botha & Cross 1997, Botha & Van Bel 1992).

The **intermediary cells** of type 1 are especially large, with a dense cytoplasm, many small vacuoles, numerous mitochondria, rudimentary plastids, and numerous plasmodesmata. On the other hand, the companion cells of type 2b are smaller and have numerous invaginations (**transfer cells**). This points to their different function in the phloem-loading process: symplasmic in type 1 versus apoplasmic in type 2b; type 1–2a and 2a occupy

an intermediate position. Families with the type 1 configuration (symplasmic loading) translocate 20 to 80% of the sugars as oligosaccharides. Apoplasmic phloem loading coincides with sucrose being the major transport sugar, whereas the intermediate types are likely to load sucrose apoplasmically and the other compounds symplasmically (Turgeon 1995, 1996).

3.3 Sugar Transport Against a Concentration Gradient

It is easy to see how sucrose can reach a very high concentration in the sieve tubes of the species with an apoplasmic loading pathway. It first enters the apoplast from mesophyll cells and is then **actively** accumulated against a concentration gradient by the companion cells. In species with this type of apoplasmic phloem loading, the photosynthate concentration in the sieve tubes is higher than that in the mesophyll cells. Mass transfer rates tend to exceed those in species with a symplasmic phloem-loading pathway, at least under temperate conditions.

In species with a symplasmic pathway, photosynthate concentrations may be higher in the sieve tubes than they are in the cytosol of the mesophyll, which points to an energy-requiring step in the pathway. In these species, sucrose and some galactose first diffuse through plasmodesmata to the intermediary cells. There, some of the sucrose is metabolized to make galactose. Sucrose is then linked with one, two, or three molecules of galactose in an **ATP-requiring** process. The oligosaccharides thus formed (raffinose, stachyose, and verbascose, respectively) are too large to diffuse back to the mesophyll. This is referred to as the **polymerization trap mechanism** (Turgeon 1991, 1996). It is hypothesized that the oligosaccharides can diffuse to the sieve elements through wider, branched plasmodesmata between the intermediary cells and the sieve tubes. In this way, the concentration of photosynthate, but not that of sucrose, can be higher in the sieve tubes than it is in the mesophyll cells. The osmotic gradient is often not as large as that in species with the apoplasmic pathway.

Figure 4 summarizes the major aspects of the two pathways of phloem loading in minor veins, together with their different anatomy and biochemistry. The energy-requiring step in plants with the symplasmic pathway occurs inside the intermediary cells, which form oligosaccharides. In plants

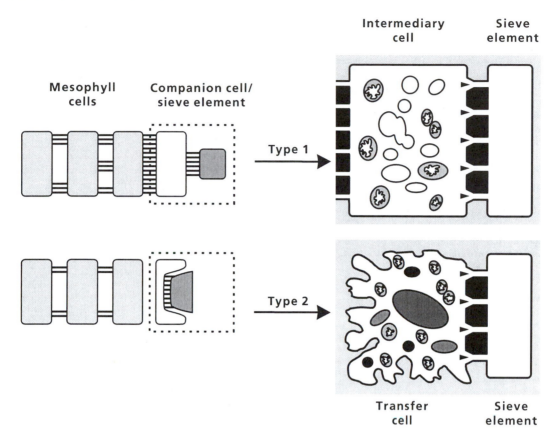

FIGURE 4. The symplasmic (top) and apoplasmic (bottom) pathway of phloem loading in minor veins. In plants with the symplasmic pathway (type 1), there are numerous plasmodesmatal connections between the mesophyll cells and the cells adjacent to the sieve tube in the minor veins. The adjacent cell is termed an *intermediary cell*. It has a number of small vacuoles and functions as a trapping mechanism. Smaller sugar molecules (monosaccharides and/or sucrose) move via relatively narrow plasmodesmata from the mesophyll to the intermediary cells. They are transformed here into larger ones (of the raffinose family).

The larger sugar molecules can move to the sieve tubes through relatively wide plasmodesmata; they are too large to move back to the mesophyll. In plants with the apoplasmic pathway (type 2), there are very few plasmodesmatal connections between the mesophyll cells and the cells adjacent to the sieve tube in the minor vein. The adjacent cell is termed a "companion cell." It has one large vacuole and is characterized by numerous invaginations in the plasma membrane (hence, it is also called a transfer cell). For further explanation see the text in Sections 3.1–3.4.

with the apoplasmic pathway, the energy-requiring step is the uptake of sucrose from the apoplast by the transfer cells.

3.4 Variation in Transport Capacity

Leaves with the **symplasmic pathway** of phloem loading may have a **lower capacity** to export assimilates from the leaves. As a result, when growing in well-watered conditions and at high irradiance, leaves with the symplasmic pathway accumulate more nonstructural carbohydrates than

do leaves with the apoplasmic pathway (Table 2). When exposed to elevated atmospheric CO_2 concentrations, which are expected to occur in the second half of the twenty-first century, carbohydrate levels in leaves increase. In absolute terms, the increase in nonstructural carbohydrates is the same, irrespective of the pathway of phloem loading. The concentrations in leaves, however, with a symplasmic pathway build up to a level that may well be close to that which becomes damaging for the chloroplast (Sasek et al. 1985). Temperate-zone tree species, which are supposed to have the symplasmic pathway of phloem loading in the mi-

TABLE 2. Concentrations of total nonstructural carbohydrates and starch (% of dry mass) in leaves of plant species grouped by the type of phloem loading.*

	Total nonstructural carbohydrates			Starch		
	Low CO_2	High CO_2	Difference	Low CO_2	High CO_2	Difference
Herbaceous species (apoplasmic pathway)						
Bellis perennis	14	18	+33	9	15	+62
Brassica nigra	12	17	+39	10	12	+22
Linum usitatissimum	21	19	−9	9	10	+1
Valeriana officinalis	16	27	+73	13	24	+84
average (9 species)	19	25	+29	15	20	+35
Herbaceous species (symplasmic pathway)						
Cucurbita pepo	45	53	+17	44	52	+17
Epilobium angustifolium	20	26	+27	11	17	+59
Lamium galeobdolon	29	36	+22	26	34	+31
Origanum vulgare	41	41	−1	38	38	−1
average (11 species)	36	41	+14	32	38	+16
Temperate zone tree species (symplasmic pathway)						
Fraxinus ornus	26	35	+36	24	35	+43
Ginkgo biloba	31	34	+10	26	29	+11
Quercus robur	27	33	+22	20	26	+29
Tilia cordata	22	27	+24	16	22	+32
average (8 species)	22	28	+27	18	23	+33

Source: Körner et al. 1995. Copyright Blackwell Science, Ltd.

*Plants were grown in normal air (low CO_2; 350 μmol mol^{-1}) or at an elevated atmospheric CO_2 concentration (high CO_2; 600 μmol mol^{-1}) and the difference between values (in %) is also indicated.

nor veins, do not fit the pattern found for the herbaceous species (Table 2). With the exception of *Ginkgo biloba*, which has a symplasmic pathway, they possibly have a "hidden" apoplasmic pathway of minor vein loading (Van Bel 1992).

The relatively low transport capacity of the species with a symplasmic loading system may be due to the limited capacity of the intermediary cells to make oligosaccharides. Mass transfer in the sieve tubes may also be slower, however, due to the relatively **low solubility** of raffinose compared with sucrose (Fig. 5). At low temperatures, therefore, raffinose is not an appropriate solute, which explains why species with the symplasmic pathway tend to transport other, more complex oligosaccharides of the raffinose family, whose solubility is considerably higher. The highest concentrations of raffinose in phloem sap are in the range of 400 mM, whereas sucrose is soluble at concentrations of 1 M. Because high solute concentrations are essential to drive phloem transport, raffinose cannot serve as the sole transport sugar solute, espe-

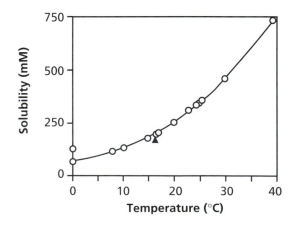

FIGURE 5. Solubility of raffinose at different temperatures. The solubility in artificial phloem sap is indicated by the triangle (Turgeon 1995). Copyright American Society of Plant Physiologists.

cially not at low temperatures. Verbascose is found in some species, but only as a minor component (Turgeon 1995).

4. Phloem Loading and Ecological Distribution

The apoplasmic pathway of phloem loading predominates in the temperate and arid zones, and the symplasmic pathway is typical for tropical rainforests (Gamalei 1991, Van Bel 1993). Translocation of raffinose sugars is especially common in tropical trees and vines. It may well be due to the limited solubility of raffinose that selection occurred against these transport compounds during the

progressive extension of species into temperate habitats.

Ancient families typically show the type 1 (symplasmic) configuration. With progressive evolution, types 1–2a, 2a and 2b emerged (Fig. 6). What might be the selective pressures that have led to the emergence of the more recent apoplasmic pathway of phloem loading?

Extreme temperatures and/or water stress may have been major environmental factors necessitating the apoplasmic pathway. At temperatures below 10°C, plasmodesmata may not function properly and the intermediary cells may have a restricted capacity to produce oligosaccharides. In addition, the viscosity of the oligosaccharide solution may become too high to allow transport of a sufficient magnitude (Fig. 5).

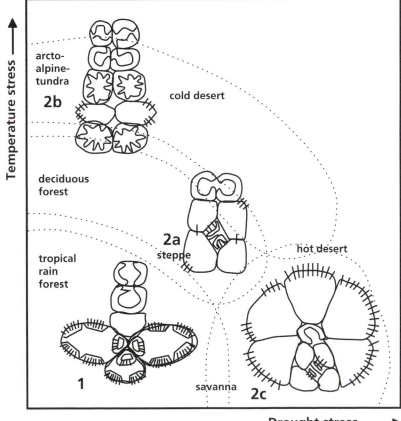

FIGURE 6. The distribution of the minor vein types over various terrestrial ecosystems. Type 2c is similar to type 2a with regard to the symplasmic connections between the sieve element/companion cell complex and the neighboring cells, but has many conspicuous plasmodesmata between the mesophyll and the bundle sheath (C_4-plants). Typical dicotyledonous 2c-families include the Portulacaceae and Amaranthaceae (Van Bel & Gamalei 1992). Copyright Blackwell Science Ltd.

A large number of gymnosperm (*Taxus, Pinus, Picea*) and angiosperm (*Acer, Betula, Fagus, Fraxinus, Prunus, Populus, Tilia, Ulmus*) tree species from cool environments in central Europe, however, do have high plasmodesmatal frequencies (>10 plasmodesmata per square micrometer interface area), which suggests that these species have a symplasmic phloem-loading system. In conifers, there is no evidence for a trapping mechanism, and sucrose is the major transport metabolite. It may simply move by diffusion down a concentration gradient between the mesophyll cells and the phloem. This may restrict their transport capacity in winter and partly account for the winter depression of photosynthetic capacity of conifer needles. In the dicot tree *Fraxinus excelsior* (ash) stachyose, and to a lesser extent raffinose and sucrose, are major transport metabolites in the phloem. Considering the likely mechanistic link between transport sugars of the raffinose family and symplasmic phloem loading, this tree presumably also has a symplasmic pathway of phloem loading (Blechschmidt-Schneider et al. 1997).

In *Sorbus aucuparia* (Rosaceae) the sugar alcohol sorbitol is the main carbohydrate in the phloem, next to sucrose. It is not yet clear how sorbitol or other polyols enter the sieve tubes, but it may well be via a proton-cotransport mechanism as described for mannitol in *Apium graveolens* (celery) (Keller 1991). This would point to an apoplasmic pathway, which is in agreement with ultrastructural evidence (Gamalei 1989). In *Petroselinum crispum* (parsley), however, the loading of mannitol is insensitive to an inhibitor of the system responsible for active uptake, which suggests symplasmic loading. This is puzzling because mannitol is a smaller molecule than sucrose and cannot be "trapped" (Turgeon 1996). Trees that belong to the Fabaceae appear to employ the apoplasmic pathway, with sucrose being the major phloem constituent.

5. Phloem Unloading

In the sink, the solutes imported via the phloem are unloaded. Depending on the kind of sink (roots, shoot meristems, developing seeds, storage organs, or parasitic organism), unloading may be **symplasmic** or **apoplasmic**. The unloading may be passive, as in developing kernels of *Zea mays* (maize), or they may involve a carrier-mediated energy-dependent mechanism (Thorne 1985). Active phloem unloading allows control of solute concentration in the apoplast of the sink, and, thus, control

over the rate of movement of solutes toward the sink. As long as the solute concentration is rather high, water will move osmotically into the apoplast and decrease the hydrostatic pressure in the sieve tubes. This will speed up the mass flow in the sieve tubes (Wolswinkel & Ammerlaan 1986). Seeds of legumes maintain a high solute level in the seed coat and hence maintain a rapid import of solutes via the phloem. Root tips show a symplasmic pathway of phloem unloading (Dick & ap Rees 1975, Giaquinta et al. 1983) and hence possibly exert little control over the amount of solutes moving towards them (Lambers & Atkin 1995). The lack of fine control over solute import by roots is consistent with the well-developed alternative pathway of respiration as an overflow control over sugar supply to this organ (cf. Sect. 3.3 in the chapter on respiration). Mature parts of the roots are symplasmically isolated from the sieve tubes and therefore import of solutes from the sieve tubes to mature root cells must involve an apoplasmic step (Oparka et al. 1994, Pritchard 1996).

Rapid phloem unloading occurs when a **phloem-feeding organism** (e.g., an aphid) injects its stylet into a sieve tube. The hydrostatic pressure in the sieve tube pushes the contents of the sieve tube into the aphid. The aphid predominantly absorbs nitrogenous compounds and excretes much of the carbohydrates as "honeydew." Another special site where phloem unloading occurs is the **haustoria of parasites**, if these depend on the host for their carbon supply. The release of solutes from the phloem of the host is strongly stimulated by the presence of such a parasite, by an as yet unidentified mechanism (see Sect. 4 in the chapter on parasitic associations; Wolswinkel et al. 1984).

Phloem unloading is affected in a rather special manner by **root nematodes** (e.g., the parasitic nematodes *Meloidogyne incognita* and *Heteroderma schachtii*), which can act as major sinks (Dorhout et al. 1993). Unloading from the sieve element companion cell complexes occurs specifically into the "syncytium," the nematode-induced feeding structure within the vascular cylinder of the root. The infective juvenile nematode selects a procambial or cambial cell as an initial syncytial cell, from which a syncytium develops by integration of neighboring cells. The developing nematode depends entirely on the expanding syncytium, withdrawing nutrients from it through a feeding tube. Unlike in the root tip, the transport of sugars from the phloem to the syncytium in this host–pathogen relationship is apoplasmic. The syncytium is not connected via plasmodesmata with the normal root cells. In an as yet unidentified manner, the nematode induces massive leakage from the phloem, thus reducing the transport of phloem solutes to the rest of the roots (Böckenhoff et al. 1996).

6. The Transport Problems of Climbing Plants

Vines can be considered as "mechanical parasites." They invest too little in wood to support themselves and thus depend on other plants for mechanical support. Xylem (wood) tissue has both a transport and a mechanical support function. As will be discussed in Section 5.3.5 of the chapter on plant water relations, vines have fewer but longer and wider xylem vessels in their stem per unit stem cross-

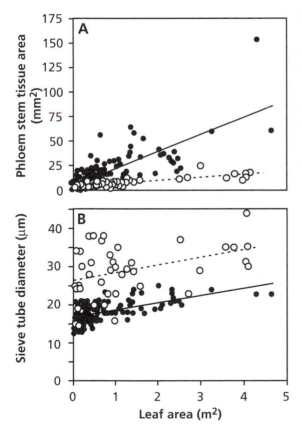

FIGURE 7. Phloem area (A) and maximum diameters of sieve tubes (B) of contrasting *Bauhinia* species. Values are plotted as a function of the leaf area distal to the investigated stem section for stems of lianas (dashed line, open symbols) and congeneric trees and shrubs (solid line, closed symbols) (Ewers & Fisher 1991).

sectional area. They also have less lignified phloem fibers than trees and shrubs have and less phloem tissue per unit of distal area. Their stems are thin relative to the distal leaf area, when compared with plants that support themselves (Fig. 7). How do vines manage to have sufficient phloem transport capacity?

Compared with trees and shrubs, vines have **wider sieve tubes** (Fig. 7). Because the hydraulic conductance, by Hagen-Poiseuille's law for ideal capillaries, is proportional to the fourth power of the conduit diameter (cf. Sect. 5.3.1 in the chapter on plant water relations), the larger diameter compensates for the smaller total area. The obvious advantages of fewer sieve tubes with a larger diameter are that relatively few resources need to be allocated to the phloem in the stem, which is therefore light, preventing the supporting plant from toppling over. For a similar investment in stem, the climbing plant will reach a greater height than a nonclimbing plant. If few sieve tubes with large diameters are so advantageous for climbing plants, why do not *all* plants have such wide tubes in their phloem? There is likely a disadvantage in having large-diameter sieve tubes, in that physical damage to a small number of sieve tubes causes a larger proportional loss of transport capacity. Such damage may be mechanical or due to phloem-sucking arthropods or pathogens. As in the xylem of plants with contrasting strategy (cf. Sect. 5.3.5 in the chapter on plant water relations), there may be a trade-off between transport **capacity** and **safety**.

7. Summary: Phloem Loading, Plant Performance, and Plant Distribution

After several years of debate on whether phloem loading occurs via an apoplasmic or a symplasmic pathway, there is now general agreement that both pathways occur, depending on species. The symplasmic pathway may have appeared first in evolution, since it occurs in more ancestral families; however, molecular evidence has cast doubt on some traditional views of taxonomic relationships within the angiosperms. The polymer trap mechanism of loading might not be as ancestral as originally thought (Turgeon 1996). Could it be derived from a more primitive symplasmic phloem-loading system? As long as plant life occurred under tropical conditions, the symplasmic pathway with its trapping mechanism was effective. As soon as plants faced cooler conditions, however, the symplasmic pathway probably became a major bottleneck for plant functioning. The need for a pathway that is less affected by low temperatures gradually emerged, and plants evolved with the apoplasmic pathway of phloem loading.

The apoplasmic pathway of phloem loading is probably an important ecophysiological attribute that allows a species to occupy cold habitats and perform well under cool conditions. At low temperature the viscosity of the sugar solutions in the sieve tubes, which in species with the symplasmic pathway typically includes sugars that are poorly soluble at low temperature, is very high. The high viscosity may restrict the transport of sugars in the phloem, and it possibly could inhibit photosynthesis. Inhibition of photosynthesis at low temperature may well be a secondary effect. That is, it is not necessarily the primary process affected by low temperature. Rather, feedback inhibition due to slow export of sugars may reduce the rate of carbon assimilation. Understanding of transport mechanisms is therefore critical to interpretation of photosynthesis measurements in the field or the design of programs to improve plant performance under cold conditions, through breeding or molecular approaches.

Are there also disadvantages associated with the apoplasmic pathway, which appears to have a greater capacity and function better under extreme temperature conditions? The proton-pumping activity of the transfer cells will require a substantial amount of metabolic energy. It remains to be demonstrated, however, that this energy requirement is greater than that for the polymerization that occurs in the intermediary cells. Disadvantages associated with the apoplasmic pathway are not immediately obvious. The symplasmic pathway could be a mere relic of the past, with no great penalty as long as the climate is not too harsh and rates of photosynthate production are modest, so that there is no great strain on the transport system. These penalties may increase, however, with rising CO_2 concentrations in the atmosphere and increasing concentrations of carbohydrates in the mesophyll.

References and Further Reading

Blechschmidt-Schneider, S., Eschrich, W., & Jahnke, S. (1997) Phloem loading, translocation and unloading processes. In: Trees—contributions to modern tree physiology, H. Rennenberg, W. Eschrich, & H. Ziegler (eds). Backhuys, Leiden, pp. 139–163.

Böckenhoff, A., Prior, D.A.M., Gruddler, F.M.W., & Oparka, K.J. (1996) Induction of phloem unloading in *Arabidopsis thaliana* roots by the parasitic nematode *Heterodera schachtii*. Plant Physiol. 112:1421–1427.

Botha, C.E.J. & Cross, R.H.M. (1997) Plasmodesmatal frequency in relation to short-distance transport and phloem loading in leaves of barley (*Hordeum vulgare*). Phloem is not loaded directly from the symplast. Physiol. Plant. 99:355–362.

Botha, C.E.J. & Van Bel, A.J.E. (1992) Quantification of symplastic continuity as visualised by plasmodesmograms: Diagnostic value for phloem-loading pathways. Planta 187:359–366.

Dick, P.S. & ap Rees, T. (1975) The pathway of sugar transport in roots of *Pisum sativum*. J Exp Bot 26:305–314.

Dorhout, R., Gommers, F.J., & Köllöffel, C. (1993) Phloem transport of carboxyfluorescein through tomato roots infected with *Meloidogyne incognita*. Physiol. Mol. Plant Pathol. 43:1–10.

Ewers, F.W. & Fisher, J.B. (1991) Why vines have narrow stems: Histological trends in *Bauhinia fassoglensis* (Fabaceae). Oecologia 88:233–237.

Gamalei, Y.V. (1989) Structure and function of leaf minor veins in trees and herbs. A taxonomic review. Trees 3:96–110.

Gamalei, Y.V. (1991) Phloem loading and its development related to plant evolution from trees to herbs. Trees 5:50–64.

Giaquinta, R.T., Lin, W., Sadler, N.L., & Franceschi, V.R. (1983) Pathway of phloem unloading of sucrose in corn roots. Plant Physiol. 72:362–367.

Gunning, B.E.S. & Steer, M.W. (1996) Plant cell biology, structure and function. Jones and Bartlett Publishers, London.

Keller, F. (1991) Carbohydrate transport in discs of storage parenchyma of celery petioles. 2. Uptake of mannitol. New Phytol. 117:423–429.

Körner, C., Pelaez-Riedl, S., & Van Bel, A.J.E. (1995) CO_2 responsiveness of plants: A possible link to phloem loading. Plant Cell Environ. 18:595–600.

Lambers, H. & Atkin, O.K. (1995) Regulation of carbon metabolism in roots. In: Carbon partitioning and source-sink interactions in plants, M.A. Madore & W.J. Lucas (eds). American Society of Plant Physiologists, Rockville, pp. 226–238.

Oparka, K.J., Duckett, C.M., Prior, D.A.M., & Fisher, D.B. (1994) Real time imaging of phloem unloading in the root tip of *Arabidopsis*. Plant J. 6:759–766.

Pritchard, J. (1996) Aphid stylectomy reveals an osmotic step between sieve tube and cortical cells in barley roots. J. Exp. Bot. 47:1519–1524.

Sasek, T.W., DeLucia, E.H., & Strain, B.R. (1985) Reversibility of photosynthetic inhibition in cotton after long-term exposure to elevated CO_2 concentrations. Plant Physiol. 78:619–622.

Thomas, R.B. & Strain, B.R. (1985) Root restriction as a factor in photosynthetic acclimation of cotton seedlings grown in elevated carbon dioxide. Plant Physiol. 96:627–634.

Thorne, J.H. (1985) Phloem unloading of C and N assimilates in developing seeds. Annu. Rev. Plant Physiol. 36:317–343.

Turgeon, R. (1991) Symplasmic phloem loading and the sink-source transition in leaves: a model. In: Recent advances in phloem transport and assimilate compartmentation. Bonnemain, J.L., Delrot, S. Lucas, W.J., & Dainty, J. eds. Ouest Edition, Nantes pp. 18–22.

Turgeon, R. (1995) The selection of raffinose family oligosaccharides as translocates in higher plants. In: Carbon Partitioning and Source-Sink Interactions in Plants, M.A. Madore & W.J. Lucas (eds). American Society of Plant Physiologists, Rockville, pp. 195–203.

Turgeon, R. (1996) Phloem loading and plasmodesmata. Trends Plant Sci. 1:418–423.

Van Bel, A.J.E. (1992) Different phloem loading machineries correlated with the climate. Acta Bot. Neerl. 41:121–141.

Van Bel, A.J.E. (1993) Strategies of phloem loading. Annu. Rev. Plant Physiol. Plant Mol. Biol. 44:253–281.

Van Bel, A.J.E. (1996) Carbohydrate processing in the mesophyll trajectory in symplasmic and apoplasmic phloem loading. Prog. Botany 57:140–167.

Van Bel, A.J.E. & Gamalei, Y.V. (1992) Ecophysiology of phloem loading in source leaves. Plant Cell Environ. 15:265–270.

Wolswinkel, P. & Ammerlaan, A. (1986) Turgor-sensitive transport in developing seeds of legumes: The role of the stage of development and the use of excised vs. attached seed coats. Plant Cell Environ 9:133–140.

Wolswinkel, P., Ammerlaan, A., & Peters, H.F.C. (1984) Phloem unloading of amino acids at the site of Cuscuta europaea. Plant Physiol. 75:13–20.

Zimmermann, M.H. & Ziegler, H. (1975) List of sugars and sugar alcohols in sieve-tube exudates. In: Encyclopedia of Plant Physiology, Vol. 1, M.H. Zimmermann & J.A. Milburn (eds). Springer-Verlag, Berlin, pp. 480–503.

3
Plant Water Relations

1. Introduction

Although water is the most abundant molecule on the earth's surface, the availability of water is the factor that most strongly restricts terrestrial plant production on a global scale. Low water availability limits the productivity of many natural ecosystems, particularly in dry climates (Fig. 1). In addition, losses in crop yield due to water stress exceed losses due to all other biotic and environmental factors combined (Boyer 1985). Regions where rainfall is abundant and fairly evenly distributed over the growing season, such as in the wet tropics, have lush vegetation. Where summer droughts are frequent and severe, forests are replaced by grasslands, as in the Asian steppes and North American prairies. Further decrease in rainfall results in semidesert, with scattered shrubs, and finally deserts. Even the effects of temperature are partly exerted through water relations because rates of evaporation and transpiration are correlated with temperature. Thus, if we want to explain natural patterns of productivity or to increase productivity of agriculture or forestry, then it is crucial that we understand the controls over plant water relations and the consequences for plant growth of an inadequate water supply.

1.1 The Role of Water in Plant Functioning

Water is important to the physiology of plants because of the crucial role that water plays in all physiological processes and because of the large quantities that are required. Water typically comprises 80 to 95% of the biomass of nonwoody tissues such as leaves and roots. At the cellular level, water is the major medium for transporting metabolites through the cell. Because of its highly polar structure, water readily dissolves large quantities of ions and polar organic metabolites like sugars, amino acids, and proteins, which are critical to metabolism and life. At the whole-plant level, water is the medium that transports the raw materials (carbohydrates and nutrients) as well as the phytohormones that are required for growth and development from one plant organ to another. Unlike most animals, plants lack a well-developed skeletal system and depend largely on water for their overall structure and support. Due to their high concentrations of solutes, plant cells exert a positive pressure (turgor) against their cell walls which is the basic support mechanism in plants. Large plants gain additional structural support from the lignified cell walls of woody tissues. When plants lose turgor (wilt), they no longer carry out physiological functions such as cell expansion and photosynthesis. Prolonged periods of wilting usually kill the plant.

A second general reason for the importance of water relations to the physiological ecology of plants is that plants require vast quantities of water. Whereas plants incorporate more than 90% of absorbed nitrogen, phosphorus, and potassium, and about 10 to 70% of photosynthetically fixed carbon into new tissues (depending on respiratory demands for carbon), less than 1% of the water ab-

154

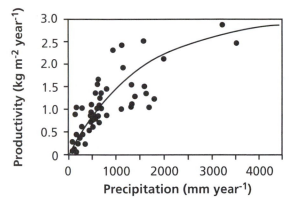

FIGURE 1. Correlation between net primary production and precipitation for the major world ecosystems (Lieth 1975).

TABLE 2. The ratio of the surface area of mesophyll cells and that of the leaf (A_{mes}/A) as dependent on species and growing conditions.*

Leaf morphology/habitat	A_{mes}/A
Shade leaves	7
Mesomorphic leaves	12–19
Xeromorphic sun leaves	17–31
Low altitude (600 m)	37
High altitude (3000 m)	47

Species	A_{mes}/A
Plectranthus parviflorus	
High light	39
Low light	11
Alternanthera philoxeroides	
High light	78
Low light	50

*The data on leaves of species with different morphologies are from Turrel (1936), those on low-altitude and high-altitude species from Körner et al. (1989), those on *Plectranthus parviflorus* from Nobel et al. (1975), and those on *Alternanthera philoxeroides* (alligator weed) from Longstreth et al. (1985).

sorbed by plants is retained in biomass (Table 1). The remainder is lost by **transpiration**, which is the evaporation of water from plants. The inefficient use of water by terrestrial plants is an unavoidable consequence of photosynthesis. The stomates, which allow CO_2 to enter the leaf, also provide a pathway for water loss. CO_2 that enters the leaf must first cross the wet walls of the mesophyll cells, before diffusing to the site of carboxylation. The surface area of mesophyll cells exposed to the internal air spaces of the leaf is about 7 to 80 times the external leaf area, depending on species and plant growth conditions (Table 2). This causes the air inside the leaf to be saturated with water vapor (100% relative humidity), which creates a strong gradient in water vapor concentration from the inside to the outside of the leaf.

TABLE 1. Concentration of major constituents in a hypothetical herbaceous plant and the amount of each constituent that must be absorbed to produce a gram of dry biomass.*

Resource	Concentration (% of fresh mass)	Quantity required ($mg\,g^{-1}$)
Water	90	2500
Carbon	4	40
Nitrogen	0.3	3
Potassium	0.2	2
Phosphorus	0.02	0.2

*The values only give a rough approximation and vary widely between species and with growing conditions, indicated in Section 4.3 of the chapter on mineral nutrition for nutrients and in Section 6 of this chapter for water.

1.2 Transpiration as an Inevitable Consequence of Photosynthesis

Transpiration is an inevitable consequence of photosynthesis; however, it does have important effects on the plant because it is a major component of the leaf's energy balance. As water evaporates from mesophyll cell surfaces, it cools the leaf. In the absence of transpiration, the temperature of large leaves can rapidly rise to lethal levels. We further discuss this effect of transpiration in the chapter on the plant's energy balance. The transpiration stream also allows transport of nutrients from the bulk soil to the root surface and of solutes, such as inorganic nutrients, amino acids, and phytohormones, from the root to transpiring organs. As will be discussed later, however, such transport in the xylem also occurs in the absence of transpiration, so that the movement of materials in the transpiration stream is not strongly affected by transpiration rate.

In this chapter, we will describe the environmental factors that govern water availability and loss, the movement of water into and through the plant, and the physiological adjustments that plants make to variation in water supply over diverse time scales. We will emphasize the mechanisms by which individual plants adjust water relations in response to variation in water supply and the adaptations that have evolved in dry environments.

2. Water Potential

The status of water in soils, plants, and the atmosphere is commonly described in terms of **water potential** (ψ_w) (i.e., the chemical potential of water in a specified part of the system compared with the chemical potential of pure water at the same temperature and atmospheric pressure; it is measured in units of pressure (MPa). The water potential of pure, free water at atmospheric pressure and at a temperature of 298K is 0MPa (by definition) (Box 5).

In an isothermal two-compartment system, in which the two compartments are separated by a **semipermeable membrane**, water will move from a high to a low water potential. If we know the water potential in the two compartments, therefore, we can predict the direction of water movement;

Box 5
The Water Potential of Osmotic Solutes and the Air

We are quite familiar with the fact that water can have a potential: We know that the water at the top of a falls or in a tap has a higher potential than that at the bottom of the falls or outside the tap. Transport of water, however, occurs not invariably as a result of differences in hydrostatic pressure, but also due to differences in vapor pressure (cf. Sect. 2.2.2 of the chapter on photosynthesis) or to differences in the amount of dissolved osmotic solutes in two compartments separated by a semipermeable membrane. In fact, in all these cases there is a difference in water potential, which drives the transport of water. For a full appreciation of many aspects of plant water relations, we will first introduce the concept of the **chemical potential of water**, for which we use the symbol μ_w.

By definition, the chemical potential of pure water under standard conditions (298K and standard pressure), for which the symbol μ_w° is used, is zero. We can also calculate the chemical potential of water under pressure, water that contains osmotic solutes, or water in air. This can best be explained using a simple example, comparing the chemical potential of water in two sealed containers of similar size, 1L (Fig. 1). One of these containers (A) contains pure water under standard conditions: $\mu_w = \mu_w^\circ = 0$. Of course the gas phase is in equilibrium with the liquid pure water, and the vapor pressure is p_0. The second container (B) contains a 1M sucrose solution in water. The gas phase will again be in equilibrium with the liquid phase; the vapor pressure is p. The vapor pressure, however, will be less than p_0 because the sucrose

molecules interact with the water molecules via hydrogen bonds, so that the water molecules cannot move in the gas phase as readily as in the situation of pure water. How large is the difference between p and p_0?

To answer this question we use Raoult's law, which states:

$$p/p_0 = N_w \qquad (1)$$

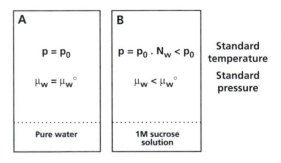

FIGURE 1. The difference in water potential between two systems. The system at the left is a sealed container with pure water at standard temperature and pressure; the partial water vapor pressure in this container is p_0 and the chemical potential of water in this system is μ_w°. The system at the right is a container with a solution of 1M sucrose at the same temperature and pressure; the water vapor pressure can be calculated according to Raoult's law ($p = p_0.N_w$) and the chemical potential of water in this system is μ_w. The difference in chemical potential between the two systems can be calculated as explained in the text.

continued

Box 5. *Continued*

where N_w is the mol fraction (i.e., the number of moles of water divided by the total number of moles in container B; in our case of 1 mole of sucrose in 1 L water (= 55.6 moles of water) N_w equals 55.6/56.6 = 0.982); p_o is the vapor pressure (in Pa) above pure water, at standard pressure and temperature. We can calculate the difference in potential between the two containers ($\mu_w - \mu_w^o$) by considering the amount of work needed to obtain the same (higher) pressure in container B as in container A. To achieve this, we need to compress the gas in container B until the pressure equals p_o:

$$\mu_w - \mu_w^0 = \int_{p_0}^{p} V \, dp = RT \ln\left(\frac{p}{p_0}\right) \qquad (2)$$

where V is the volume (m^3) of container B, which is compressed until p_o is reached, R is the gas constant ($J \, mol^{-1} K^{-1}$), and T is the absolute temperature (K).

Combination of Equations 1 and 2 yields:

$$\mu_w - \mu_w^o = RT \ln N_w \qquad (3)$$

Because N_w is the mole fraction of water and N_s is the mole fraction of the solute (in our example, 1/56.6 = 0.018), we can write Equation 3 as:

$$\mu_w - \mu_w^o = RT \ln(1 - N_s) \qquad (4)$$

As long as we consider solutions in a physiologically relevant range (i.e., not exceeding a few molar) Equation 4 approximates:

$$\mu_w - \mu_w^o = -RT N_s \qquad (5)$$

(as can readily be calculated for our example of a 1 M solution of sucrose, N_s is 0.018 en $\ln(1 - N_s)$ = −0.018).

Dividing N_s by the molar volume of pure water (V_w^o, $m^3 mol^{-1}$) we arrive at the concentration of the solute, c_s (in $mol \, m^{-3}$):

$$N_s/V_w^o = c_s \qquad (6)$$

We make one further change, by introducing the molar volume of pure water ($m^3 mol^{-1}$; at 273 K) in Equation 5:

$$\frac{\mu_w - \mu_w^o}{V_w^o} = -RT c_s = \Psi \qquad (7)$$

Ψ is the water potential. Because we are dealing with the water potential of a solution in this example, we refer to this potential as the osmotic potential of water (Ψ_π). The dimension is Pascal (Pa). It is often more convenient, however, to use

megapascals (MPa = 10^6 Pa) instead (one MPa equals 10 bars, a unit used in the literature, or 10 atmospheres, a unit that is no longer used).

We can therefore calculate that our 1 M sucrose solution has an osmotic potential of −2.4 MPa, which approximates a pressure of a water column of about 250 m!!! In equilibrium, the water potential of the gas phase above the 1 M sucrose solution also equals −2.4 MPa. In the case of electrolytes, the calculation is slightly more complicated in that the dissociation of the solute has to be taken into account.

By modifying Equation 7 we can also calculate the water potential of air that is not in equilibrium with pure water [i.e., with a relative humidity (RH) of less than 100%]:

$$\frac{\mu_w - \mu_w^o}{V_w^o} = \frac{RT}{V_w^o} \ln\left(\frac{p}{p_o}\right) \qquad (8)$$

For air of 293 K and a RH of 75% Ψ equals −39 MPa (to calculate this you need to know that the molar volume of water (molecular mass = 18) at 293 K is 18 $10^{-6} m^3 mol^{-1}$). Values for Ψ of air of different RH are presented in Table 1. Note that even when the water vapor pressure is only marginally lower than the saturated water vapor pressure, the water potential is rather negative.

TABLE 1. The water potential (MPa) of air at a range of relative humidities and temperatures.*

Relative humidity (%)	$-\Psi$ (MPa) at different temperature (°C)				
	10	15	20	25	30
100	0	0	0	0	0
99.5	0.65	0.67	0.68	0.69	0.70
99	1.31	1.33	1.36	1.38	1.40
98	2.64	2.68	2.73	2.77	2.81
95	6.69	6.81	6.92	7.04	7.14
90	13.75	13.99	14.22	14.45	14.66
80	29.13	29.63	30.11	30.61	31.06
70	46.56	47.36	48.14	48.94	49.65
50	90.50	92.04	93.55	95.11	96.50
30	157.2	159.9	162.5	165.2	167.6
10	300.6	305.8	310.8	316.0	320.6
RT/V_w	130.6	132.8	135.0	137.3	139.2

Note: The values were calculated using the formula: $\Psi = -RT/V_w \ln(\%$ relative humidity/100).
*Note that all values for Ψ are *negative* and that the effect of the temperature is exclusively due to the appearance of temperature in the equation given in the last line of this table, rather than to any effect of temperature on p_o.

however, it is certainly *not* true that water invariably moves down a gradient in water potential. For example, in the phloem of a source leaf the water potential is typically more negative than it is in the phloem of the sink. In this case water transport is driven by a difference in hydrostatic pressure and water moves up a gradient in water potential. When dealing with a nonisothermal system, such as a warm atmosphere and a cold leaf, water vapor may condense on the leaf, despite the fact that the water potential of the air is more negative than that of the leaf.

Water potential in any part of the system is the algebraic sum of the **osmotic potential**, ψ_π, and the **hydrostatic pressure**, ψ_p (the component of the water potential determined by gravity is mostly ignored):

$$\psi_w = \psi_\pi + \psi_p \tag{1}$$

where water potential is the overall pressure on water in the system. The **osmotic potential** is the chemical potential of water in a solution due to the presence of dissolved materials. The osmotic potential always has a negative value because water tends to move across a semipermeable membrane from pure water (the standard against which water potential is defined) into water-containing solutes (Box 5). The higher the concentration of solutes, the lower (more negative) is the osmotic potential. The **hydrostatic pressure**, which can be positive or negative, refers to the physical pressure exerted on water in the system. For example, water in the turgid root cortical cells or leaf mesophyll cells is under positive **turgor pressure** exerted against the cell walls, whereas water in the dead xylem vessels of a rapidly transpiring plant is typically under **suction tension** (negative pressure). Large negative hydrostatic pressures arise because of capillary effects—the attraction between water and hydrophilic surfaces at an air–water interface (Box 6). Total water potential can have a positive or negative value, depending on the algebraic sum of its components. When dealing with the water potential in soils, an additional term is used: the **matric potential**, ψ_m. The matric potential refers to the force with which water is adsorbed onto surfaces such as cell walls, soil particles, or colloids. As such it is actually a convenient *alternative* to hydrostatic pressure for characterizing the water status of a porous solid. The hydrostatic pressure and the matric potential should therefore never be added! (Passioura 1988c). The matric potential always has a negative value, because the forces tend to hold water in place, relative to pure water in the absence of adsorptive surfaces. The matric potential becomes more negative as the water film becomes thinner (smaller cells or thinner water film in soil).

Now that we have defined the components of water potential, we will show how these components vary along the gradient from soil to plant to atmosphere.

3. Water Availability in Soil

The availability of soil water to plants depends primarily on the quantity of water stored in the soil and its relationship to soil water potential. Clay and organic soils, which have small soil particles, have more small soil pores; these small capillaries generate very negative pressures (large suction tensions) (Box 6). Pores larger than 30 μm hold the water only rather loosely, so the water drains out following a rain. Pores smaller that 0.2 μm hold water so tightly to surrounding soil particles that the drainage rate often becomes very small once the large pores have been drained. As a result, most plants cannot extract water from these pores at sufficiently high rates. It is thus the intermediate-sized pores (0.2 to 30 μm diameter) that hold water that can be tapped by plants.

In friable soil, roots can explore a large fraction of the soil volume; hence, the volume of water that is available to the roots is relatively large. Upon soil compaction, roots are unable to explore as large a fraction of the soil volume; the roots then tend to be clumped into sparse pores and water uptake is restricted. Compacted soils, however, are not uniformly hard and usually contain structural cracks and biopores (i.e., continuous large pores formed by soil fauna and roots). Roots grow best in soil with an intermediate density, which is soft enough to allow good root growth but sufficiently compact to give good root-soil contact (Stirzaker et al. 1996).

Water movement between root and soil can be limited by incomplete root–soil contact, such as that caused by air gaps due to root shrinkage during drought. It can also be influenced by a **rhizosheath** that is composed of soil particles bound together by root exudates and root hairs (McCully & Canny 1988). Rhizosheaths are limited to distal root regions, which generally have a higher water content than do the more proximal regions (Huang et al. 1993), which is partly accounted for by the immaturity of the xylem in the distal region (Wang et al. 1991). The rhizosheath virtually eliminates root–soil air gaps, thus facilitating water uptake in moist soil. On the other hand, bare roots restrict

Box 6
Positive and Negative Hydrostatic Pressures

Positive values of hydrostatic pressure in plants are typically found in living cells and are accounted for by high concentrations of osmotic solutes. Large negative values arise because of **capillary effects** (i.e., the attraction between water and hydrophilic surfaces at an air–water interface). It is this attraction that explains the negative matric potential in soil and the negative hydrostatic pressure in the xylem of a transpiring plant.

The impact of the attraction between water and hydrophilic surfaces on the pressure in the adjacent water can be understood by imagining a glass capillary tube, with radius a (m), placed vertically with one end immersed in water. Water will rise in the tube, against the gravitational force, until the mass of the water in the tube equals the force of attraction between the water and the glass wall. A fully developed meniscus will exist (i.e., one with a radius of curvature equal to that of the tube). The meniscus of the water in the glass tube is curved, because it supports the mass of the water.

The upward acting force in the water column equals the perimeter of contact between water and glass ($2\pi a$) multiplied by the surface tension, γ ($N\,m^{-1}$), of water; namely, $2\pi a\gamma$ (provided the glass is perfectly hydrophilic, when the contact angle between the glass and the water is zero; otherwise, this expression has to be multiplied by the cosine of the angle of contact). When in equilibrium, there must be a difference in pressure, ΔP (Pa) across the meniscus, equal to the force of attraction between the water and the capillary wall (i.e., the pressure in the water is less than that of the air). The downward acting force (N) on the meniscus is the difference in pressure multiplied by the cross-sectional area

of the capillary tube (i.e., $\pi.a^2.\gamma.P$). Thus, because these forces are equal in equilibrium, we have:

$$\pi.a^2.\Delta P = 2\pi.a.\gamma \qquad (1)$$

and:

$$\Delta P = 2\pi.a.\gamma/\pi.a^2 = 2\gamma/a \qquad (2)$$

The surface tension of water is $0.075\,N\,m^{-1}$ at about 20°C, so $\Delta P = 0.15/a$ (Pa). Thus a fully developed meniscus in a cylindrical pore of radius, say 1.5 μm, would have a pressure drop across it of 1.0 MPa; the pressure, P, in the water would therefore be −0.1 MPa if referenced to normal atmospheric pressure, or −0.9 MPa absolute pressure (given that standard atmospheric pressure is approximately 0.1 MPa).

This reasoning also pertains to pores that are not cylindrical. It is the radius of curvature of the meniscus that determines the pressure difference across the meniscus, and this curvature is uniform over a meniscus that occupies a pore of any arbitrary shape. It is such capillary action that generates the large negative pressures (large suction tension) in the cell walls of leaves that induce the long-distance transport of water from the soil through a plant to sites of evaporation. The pores in cell walls are especially small (approximately 5 nm) and are therefore able to develop very large suction tensions, as they do in severely water-stressed plants.

Reference and Further Reading

Passioura, J.B. (1998) Chapter 3.2 In: Plants in action. B.J. Atwell, R.L. Bieleski, D. Eamus, M.H. Turnbull, & P.E. Kriedemann (eds), Macmillan Australia, Melbourne, in press.

water loss from roots to a drier soil (North & Nobel 1997).

3.1 The Field Capacity of Different Soils

Field capacity is defined as the water content after the soil becomes saturated, followed by complete

gravitational drainage. The water potential of nonsaline soils at field capacity is close to zero (−0.01 to −0.03 MPa). There is a higher soil water content at field capacity in fine-textured soils with a high clay or organic matter content (Fig. 2). The lowest water potential at which a plant can access water from soil is the **permanent wilting point**. Although species differ in the extent to which they

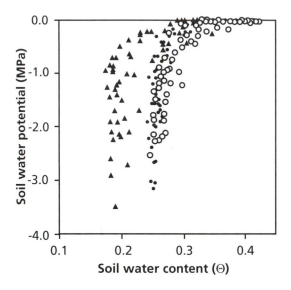

FIGURE 2. Relationship between soil water potential and volumetric soil water content (ratio of volume taken up by water and total soil volume, θ) at different soil depths: 25 cm, solid triangles; 50 to 80 cm, open circles; 110 to 140 cm, filled circles. The top horizon was a silty clay loam, the middle layer was enriched with clay and in the deepest soil layer the clay content decreased again. Soil water potential was measured with tensiometers and micropsychrometers and soil water content with a neutron probe. Data were obtained over 1 year while water content fell during drought (Bréda et al. 1995). Copyright Kluwer Academic Publishers.

then the soil water potential is −0.48 MPa. As the soil dries out, the salts become more concentrated and further add to the negative value of the soil water potential. When half of the water available at field capacity has been absorbed, the osmotic component of the soil water potential will have dropped to almost −1 MPa. Under such situations the osmotic component of the soil water potential clearly cannot be ignored.

Soil organic matter affects water retention because of its hydrophilic character and its influence on soil structure. Increasing the organic matter content from 0.2 to 5.4% more than doubles the water-holding capacity of a sandy soil; from 0.05 to 0.12 (vol/vol). In silty soils, which have a larger water-holding capacity, the absolute effect of organic matter is similar, but less dramatic when expressed as a percentage; it increases from about 0.20 to less than 0.30%. Effects on plant-available water content are smaller because the water content at field capacity as well as that at the permanent wilting point is enhanced (Kern 1995). Roots may promote the development of soil aggregates, through the release of organic matter, and thus affect soil hydraulic properties; however, organic matter may also have the effect of repelling water, if it is highly hydrophobic. Such situations may arise when plant-derived waxy compounds accumulate on the soil surface. These reduce the rate at which water penetrates the soil so that much of the precipitation from a small shower may be lost through run-off or evaporation, rather than become available for the plant.

3.2 Water Movement Toward the Roots

Water moves relatively easily through soil to the roots of a transpiring plant by flowing down a gra-

can draw down soil water (e.g., from −1.0 to −8.0 MPa), as discussed later, a permanent wilting point of −1.5 MPa is common for many herbaceous species. The **available water** is the difference in the amount of soil water between field capacity and permanent wilting point, −1.5 MPa (by definition). The amount of available water is higher in clay than it is in sandy soils (Fig. 2, Table 3).

In a moist soil, the smallest soil pores are completely filled with water and only the largest pores have air spaces. As soil moisture declines, the thickness of the water film surrounding soil particles declines, and remaining water is held more tightly to soil particles, giving a low (negative) matric potential. Finally, the hydrostatic pressure (reflecting gravity or the mass of the water column) is generally negligible in soils. In nonsaline soils, the matric potential is the most important component of soil water potential.

In saline soils, the osmotic potential adds an additional important component. If plants are well-watered with a saline solution of 100 mM NaCl,

TABLE 3. Typical pore-size distribution and soil water contents of different soil types.

Parameter	Soil type		
	Sand	Loam	Clay
Pore space (% of total)			
>30 μm particles	75	18	6
0.2–30 μm	22	48	40
<0.2 μm	3	34	53
Water content (% of volume)			
Field capacity	10	20	40
Permanent wilting point	5	10	20

dient in hydrostatic pressure. If the soil is especially dry (with a water potential less than −1.5 MPa), then there may be significant movement as water vapor. Under those conditions, however, transpiration rates will be very low. Gradients in osmotic potential move little water because the transport coefficients for diffusion are typically orders of magnitude smaller than for flow down an hydrostatic gradient. Movement across the interface between root and soil is more complicated. There may be a mucilaginous layer that contains pores so small that the flow of water across it is greatly hindered. There may also be a lack of hydraulic continuity between root and soil if the root is growing in a pore wider than itself or if the root has shrunk. A root generally has access to all available water within 6 mm of the root. As the soil dries and the matric forces holding water to soil particles increases, movement of liquid water through soils declines (Fig. 3).

In a situation where the soil is relatively dry and the flow of water through it is limiting water uptake by the roots, the following equation approximates water uptake by the roots:

$$d\theta'/dt = D\,(\theta' - \theta_a)/2b^2 \qquad (2)$$

where $d\theta'/dt$ is the rate of fall of mean soil water content, θ', with time, t; D is the diffusivity of soil water, which is approximately constant with a value of $2\cdot10^{-4}\,m^2 day^{-1}$ ($0.2\cdot10^{-8}\,m^2 s^{-1}$), during the extraction of about the last third of the available water in the soil (Fig. 3), when the flow is likely to be limiting the rate of water uptake; θ_a is the soil water content at the surface of the root; and b is the radius of a putative cylinder of soil surrounding the root, to which that root effectively has sole access, and can be calculated as $b = (\pi L)^{-1/2}$, where L (m·m^{-3}), the rooting density,

is the length of root per unit volume of soil (m^3) (Passioura 1991).

Under the reasonable assumption that θ_a is constant, as it would be if the root were maintaining a constant water potential of say −1.5 MPa at its surface (Fig. 3), the equation can be integrated to give:

$$(\theta' - \theta_a)_d = (\theta' - \theta_a)_0 \exp(-Dt/2b^2)$$
$$= \theta_{d0} \exp(-t/t^*) \qquad (3)$$

where $(\theta' - \theta_a)_0$ is $(\theta' - \theta_a)$ when t = 0, and t^* (equal to $2b^2/D$) is the time constant for the system: the time taken for the mean soil water content to fall to $1/e$ (i.e., 0.37) of its initial value. If D is $2.10^{-4}\,m^2 day^{-1}$, then t^* is $b^2 \times 10^4$ days. If the roots are evenly distributed in the soil, then, even at a low rooting density, L, of 0.1 m.m^{-3}, t^* (calculated from $b^2 = 1/[\pi.L)]$) is only about 3 days. Roots, therefore, should readily be able to extract all the available water from the soil. When the soil is compacted, roots are not distributed so evenly through the soil (cf. section 5.5 of the chapter on growth and allocation), and Equations 2 and 3 are no longer applicable. Under those conditions t^* could become of the order of weeks. The parameter t^* changes with soil type and soil depth, but is not strongly affected by the nature of the plant extracting the water (Passioura 1991).

If a plant does not absorb all the ions arriving at the surface of its roots, the osmotic potential will drop locally, either only in the apoplast of the roots or possibly in the rhizosphere as well. This is more pronounced in fertilized or saline soils than it is in nutrient-poor, nonsaline soils. The result is that plants have greater difficulty in extracting water from soil than expected from the average soil water potential (Stirzaker & Passioura 1996).

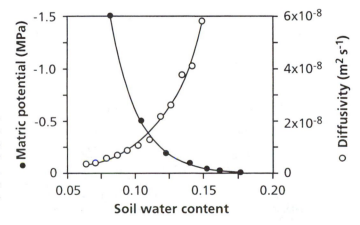

FIGURE 3. The matric potential and diffusivity of soil water as a function of the volumetric water content (ratio of volume taken up by water and total soil volume) of a sandy loam soil (55% coarse sand, 19% fine sand, 12% silt, and 14% clay) (Stirzaker & Passioura 1996). Copyright Blackwell Science Ltd.

3.3 Rooting Profiles as Dependent on Soil Moisture Content

As long as the upper soil is fairly moist, plants tend to absorb most of their water from shallower soil regions, which is where roots are concentrated. As the soil dries out, relatively more water is absorbed from deeper layers. For a crop of *Triticum aestivum* (wheat) growing under Mediterranean rain-fed conditions, very little water tends to be available in the top 45cm; most of the water is absorbed from the layer 45 to 135cm. When irrigated, the zone from 45 to 90cm is the main source for water (Van den Boogaard et al. 1996). Water from the deepest layers, even from those where no roots penetrate, may become available through capillary rise (Fig. 4). The actual rooting depth varies greatly among species, with some species reaching depths much greater than the example presented in Figure 4.

The root-trench method, in combination with measurements of volumetric soil water content, is not only a laborious and expensive method to obtain information on where most of the water comes from that a tree transpires, it is also an indirect one. If the **isotope signature** of water differs among soil layers, then this value can be used to obtain information on which soil layers the water comes from

and which roots are functional in the uptake of water. This technique has shown that perennial groundwater sources are more important than anticipated (Thorburn & Ehleringer 1995). For example, in a Utah desert scrub community, most plants use a water source derived from winter storm recharge for their early spring growth (Ehleringer et al. 1991). As this water source is depleted, however, only the deep-rooted woody perennials continue to tap this source, and more shallow-rooted species such as annuals, herbaceous perennials, and succulent perennials depend on summer rains (Fig. 5). Plants that have an isotopic composition representative of deep water are less water-stressed and have higher transpiration rates and lower water-use efficiency (Sect. 6) than do species with a shallow-water isotopic signature.

3.4 Roots Sense Moisture Gradients and Grow Toward Moist Patches

As with so many other fascinating phenomena in plants, Darwin (1880) already noticed that roots have the amazing ability to grow away from dry sites and toward wetter pockets in the soil: They are **hydrotropic**. Positive hydrotropism occurs due to inhibition of root cell elongation at the humid side

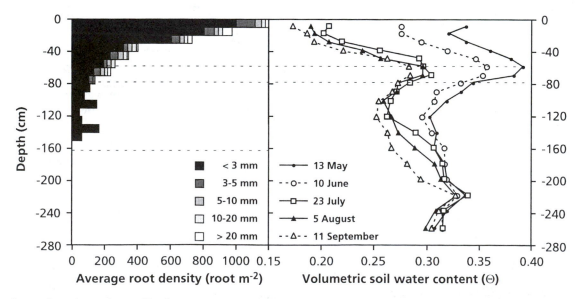

FIGURE 4. (Left) Rooting profile of *Quercus petraea* (sessile oak) as dependent on soil depth. Roots were divided in different diameter classes, indicated in the box. (Right) Volumetric water content of the soil in which the oak tree was growing, as dependent on depth and time of the year (indicated in the box). A clay-enriched horizon at around 50cm depth is indicated by the two broken lines. The third broken line at 160cm depth indicates the depth of the trench that was dug to make the measurements (Bréda et al. 1995). Copyright Blackwell Science Ltd.

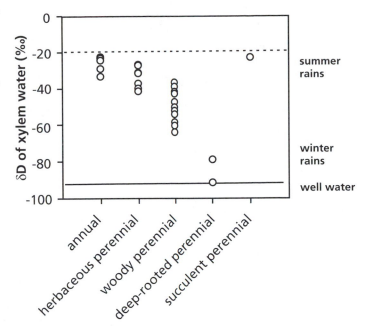

FIGURE 5. Hydrogen isotope ratios (δD) of xylem water during the summer from different growth forms in a Utah desert scrub community. The mean winter precipitation δD was −88‰, whereas summer precipitation δD ranged from −22 to −80‰ (Ehleringer et al. 1991).

of the root. The elongation at the dry side is either unaffected or slightly stimulated, resulting in a curvature of the root and growth toward a moist patch (Takahashi 1994). The root cap is most likely the site of **hydrosensing** (Takahashi & Scott 1993), but the exact mechanism of hydrotropism is not known. It involves an increase in cell-wall extensibility of the root cells that face the dry side (see Sect. 2.2 of the chapter on growth and allocation) (Hirasawa et al. 1997). The hydrotropic response is stronger in roots of *Zea mays* (maize), which is a species that tolerates relatively dry soils, than it is in those of *Pisum sativum* (pea), and it shows a strong interaction with the root's gravitropic response (Fig. 6).

Within a tissue, the water relations of individual cells may differ widely. This is both the basis of stomatal opening and closure (Sect. 5.4.2) and leaf rolling and movements (Sect. 5.4.6), as well as of **tissue tension**. The presence of tissue tension can readily be demonstrated by cutting a stem of a dandelion (*Taraxacum officinale*) parallel to its axis. Upon cutting, the stem halves curl outward, illustrating that the inner cells were restrained by outer cells and unable to reach full turgor before the cut. Tissue tension plays a major role in the closing mechanism of the carnivorous plant *Dionaea muscipula* (Venus' fly trap) (Sect. 3.1 of the chapter on carnivory).

4. Water Relations of Cells

There are major constraints that limit the mechanisms by which plants can adjust cellular water potential. Adjustment of the water potential must come through variation in hydrostatic pressure or osmotic potential. Live cells must maintain a positive hydrostatic pressure (i.e., remain turgid) to be physiologically active, so osmotic potential is the only component that live cells can adjust to modify water potential within hours. In the long term, plants can adjust by changing the elasticity of their cell walls. By contrast with living cells, dead xylem cells have very dilute solutes, so their water potential can change only through changes in hydrostatic pressure.

4.1 Osmotic Adjustment

As the soil dries, causing soil water potential to decline, live cells adjust their water status by accumulating osmotically active compounds, which reduces the osmotic potential and, therefore, helps maintain turgor. As a result of the increased concentration of osmotic solutes, cells have a higher turgor when fully hydrated. In addition, they lose their turgor at a more negative water potential compared with the turgor loss point of nonacclimated plants (Nabil & Coudret 1995, Rodriguez et al. 1993), thereby enabling the plant to continue to acquire water from soil at low soil water potentials. The osmotic solutes in the vacuole, which constitutes most of the volume of the plant cell, are often

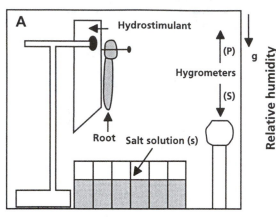

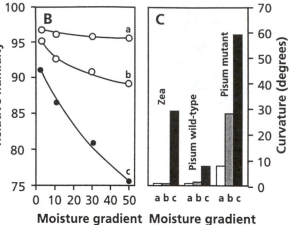

FIGURE 6. Hydrotropism in roots of *Zea mays* (maize) and of the wildtype, and the ageotropic mutant *ageotropum*) of *Pisum sativum* (pea). (A) Diagram showing the humidity-controlled chamber. Roots were placed 2 to 3 mm from the "hydrostimulant" (wet cheesecloth). Saturated solutions of salts create the humidity gradient. Different salts (KCl, K_2CO_3) give different gradients. The relative humidity and temperature was measured with a thermohygrometer (P). A stationary hygrometer (S) measured the relative humidity in the chamber. The arrow and letter g indicate the direction of gravitational force. (B) Moisture gradients, between 0 and 50 mm from the hydrostimulant, created by using no salt (a), KCl (b), or K_2CO_3 (c). (C) Root curvature 10 h after the beginning of hydrostimulation by the three moisture gradients shown in (B) (Takahashi & Scott 1993). Copyright Blackwell Science Ltd.

inorganic ions and organic acids. Such compounds reduce the activity of cytoplasmic enzymes, so plants tend to synthesize other **compatible** solutes in the cytoplasm (i.e., solutes that do not have a negative effect on cell metabolism). Such compatible solutes include glycinebetaine, sorbitol, and proline. These compounds are not highly charged, and they are polar, highly soluble, and have a larger hydration shell (the layer of water molecules surrounding each molecule) than denaturing mol-ecules, like NaCl. Compatible solutes do not interfere with the activity of enzymes at a concentration where NaCl strongly inhibits them (Fig. 7). Some of these compatible solutes (e.g., sorbitol, mannitol, and proline) are effective as hydroxyl radical scavengers in vitro, but this is not the case for glycinebetaine. A role as radical scavenger in vivo still needs to be established (Smirnoff & Cumbes 1989).

Some plants accumulate fructans (i.e., one glucose molecule linked to two or more fructose molecules) when exposed to drought. Fructan accumulation confers greater drought resistance, partly because these solutes play a role in osmotic adjustment, but presumably also because fructans protect membranes. Transgenic tobacco plants (*Nicotiana tabacum*) that contain the genetic information that enables them to accumulate fructans show greater drought resistance than wildtype plants (Pilon-Smits et al. 1995).

4.2 Cell-Wall Elasticity

When cells lose water, they decrease in volume until the turgor is completely lost. The extent to which the cells can decrease in volume and hence the extent to which their water potential can decrease until the turgor-loss point is reached depends on the **elasticity** of their cell walls. Cells with highly elastic walls, such as those of the CAM plant

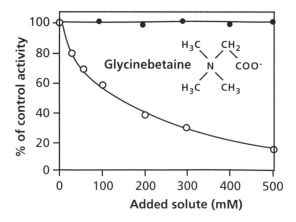

FIGURE 7. The effect of NaCl (open symbols) and glycinebetaine (filled symbols) on the activity of malate dehydrogenase from barley leaves (Pollard & Wyn Jones 1979). The chemical structure of glycinebetaine, a compatible solute in many higher plant species, is also given.

Kalanchoe daigremontiana, contain more water at full turgor; hence, their volume can decrease more before the turgor-loss point is reached. The elasticity of the cell walls depends on chemical interactions between the various cell-wall components. Cells with elastic walls can therefore store water that they accumulate during the night and gradually lose again during the day due to the leaf's transpiration. In this way they can afford to lose more water temporarily than is imported from the root environment.

A greater elasticity of cell walls is expressed as a smaller **elastic modulus, ε** (MPa), which describes the amount by which a small change in volume (ΔV, m^3) brings about a change in turgor, ΔP (MPa) at a certain initial cell volume (Tyree & Jarvis 1982).

$$\Delta P = \varepsilon \cdot \Delta V / V, \text{ or } \varepsilon = dP/dV \cdot V \qquad (4)$$

Thick-walled cells generally have greater values for ε than do thin-walled cells. For the Mediterranean, evergreen, sclerophyllous trees, the **bulk elastic modulus** has been derived from **Höfler diagrams**, such as given in Figure 8, and hence refers to an entire leaf. At full turgor, the change in turgor for a change in volume is much greater for *Laurus nobilis* than for *Olea oleaster* (i.e., ε is greatest for *L. nobilis*) (Table 4). The elastic modulus can also be determined for individual cells by using a pressure probe (cf. Sect. 5.3.1 of this chapter). The greater

elasticity of the leaf cell walls of species from the drier sites, in comparison with species from moister sites, implies that its cells can lose more water before they reach the turgor-loss point (Table 4). This does *not* imply that cells from plants that grow in arid conditions have larger cells; rather, it shows that they have cells that can shrink more during periods of water shortage without damage to the cytoplasm. In other words, they have a greater capacity to store water. A small elastic modulus (low rigidity) thus contributes to turgor maintenance in much the same way as a decrease in osmotic potential (Robichaux et al. 1986).

At a given value for the water potential, more elastic cells, like those of *Olea oleaster* (olive) may have a more negative osmotic potential and a more positive turgor, in comparison with the values for less elastic cells like those of *Laurus nobilis* (laurel). The water potential of the leaf cells of *Olea oleaster* (with relatively elastic cell walls), however, will inexorably decline to more negative values than those of *Laurus nobilis* (with more rigid cell walls) because the concentration of osmotic solutes in the cells rises (Fig. 8). In nonadapted species, leaves lose turgor ($\psi_p = 0$) at higher relative water content and higher leaf water potential (Table 4; Nabil & Coudret 1995). The protoplasm of the leaf cells of adapted species (e.g., *O. oleaster*) must have the capacity to tolerate more negative water potentials to survive

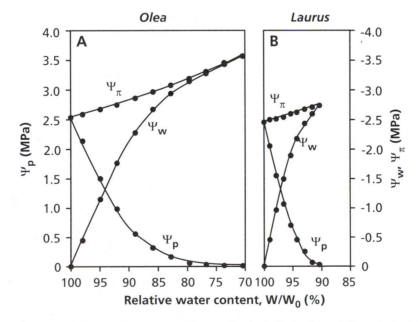

FIGURE 8. Höfler diagrams, relating turgor pressure (Ψ_p), osmotic potential (Ψ_π), and water potential (Ψ_w), to relative water content for leaves of two Mediterranean tree species. (A) *Olea oleaster* (olive); (B) *Laurus nobilis* (lau-rel). The bulk elastic modulus, ε, is the initial slope of Ψ_p with relative water content (Lo Gullo & Salleo 1988). Copyright Trustees of The New Phytologist.

TABLE 4. The elastic modulus of 1-year-old leaves of three Mediterranean evergreen, sclerophyllous trees, growing in the same Mediterranean environment, but at locations differing in water availability.*

Species	Elastic modulus, ε at full turgor (MPa)	
	Wet season	Dry season
Olea oleaster	19.5	19.3
Ceratonia siliqua	20.5	24.5
Laurus nobilis	28.1	40.7

Source: Lo Gullo & Selleo 1988.
* *Olea oleaster* (olive) is the most desiccation-tolerant, followed by *Ceratonia siliqua* (carob); *Laurus nobilis* (laurel) grows at somewhat wetter locations, near river banks. The elastic modulus was determined at full turgor, in both May (wet season) and September (dry season).

the greater loss of water from their cells with more elastic cell walls (Lo Gullo & Salleo 1988).

Comparing hemiepiphytic *Ficus* (fig) species, which start their life as epiphyte and subsequently establish root connections with the ground, leaf cells have a less negative osmotic potential at full turgor and lower bulk elastic modulus (more elastic cells) in the epiphytic stage than they do as terrestrial trees. The combination of the difference in osmotic potential and elastic modulus between epiphytic and terrestrially rooted *Ficus* species results in a less negative turgor pressure but similar relative water content at the turgor-loss point in the epiphytic stage. Lower osmotic potentials (in the tree stage) should allow leaves to withstand greater evaporative demand without wilting in order to mobilize water from deeper and/or drier soil layers. This strategy, however, requires that there be some substrate moisture in the first place. Given the substrate of the epiphyte, which dries rapidly, frequently, and uniformly, a more favorable strategy appears to gather water from the rooting medium when it is readily available for storage within leaf cells (Holbrook & Putz 1996).

The advantage of having elastic cell walls is related to the storage of water, which is especially important when water is intermittently in short supply, as in epiphytic *Ficus* species. Is there also, however, an advantage in having less elastic cell walls when water is in short supply? As will be pointed out in the chapter on growth and allocation, rigid cell walls limit the growth of expanding cells and hence the leaf expansion. The changes in cell wall properties that increase the elastic modulus may prevent the leaves from expanding

the transpiring area too much (cf. Sect. 5.3.2 of the chapter on growth and allocation). Such limitations in leaf expansion reflect a different strategy. Rather than closing their stomata and allowing their leaves to expand at a similar rate, as in the wetter season, they maintain a high stomatal conductance, but reduce their rate of leaf area expansion. The rigidity of the walls is possibly enhanced by an increase in the amount of pectin or phenolics.

4.3 Evolutionary Aspects

Both the capacity to adjust the concentration of osmotic solutes and the elasticity of the leaves' cell walls are under genetic control. The genus *Dubautia* (Asteraceae) consists of a wide range of species, some of which are restricted to dry habitats and others to moister sites. They are therefore ideally suited to an analysis of the survival value of specific traits related to plant water relations. Individuals of the species *Dubautia scabra*, which is restricted to a relatively moist 1935 lava flow in Hawaii, have less negative osmotic potentials, lower turgor and a higher elastic modulus than those of *Dubautia ciliolata*, which is restricted to an older drier lava flow. Hybrids of the two species are common and show values for osmotic potential, turgor, and cell-wall elasticity intermediate between the two parents (Robichaux et al. 1986).

A wider comparison of six other *Dubautia* species from Hawaii confirms the results obtained with *D. scabra* and *D. ciliolata*. The species from a wet forest (12,300 mm rainfall per year) has larger leaves with a lower cell-wall elasticity (higher elastic modulus) and less negative osmotic potential than the one from a dry scrub habitat (400 mm per year). The species in between these extremes showed values for leaf size, wall elasticity, and osmotic potential that were intermediate. Just like *D. scabra* and *D. ciliolata*, the species also differ in that the ones from dry sites have 13 chromosome pairs, as opposed to 14 pairs in the ones from mesic and wet habitats. The loss of one chromosome pair may have allowed species to exploit drier habitats (Robichaux et al. 1986). What kind of genetic information would therefore have to be lost to improve the species' performance in drier habitats remains to be established.

As discussed in Section 4.1, **fructan accumulation** confers greater drought resistance. Some prominent fructan-accumulating families include Poaceae (*Triticum*, wheat, and *Hordeum*, barley), Liliaceae (*Allium*, onion, *Tulipa*, tulip), and Asteraceae (*Helianthus tuberosus*, Jerusalem artichoke,

Cichorium intybus, cichory). In plants, the synthesis of fructans involves at least two enzymes. The first catalyzes the formation of a trisaccharide (one molecule of glucose and two fructose molecules); the second extends this trisaccharide with fructose residues (Pollock & Cairns 1991). Fructan-accumulating taxa increased some 30 to 15 million years ago, when the climate shifted toward seasonal droughts. The distribution of present-day fructan-accumulating species corresponds with regions of seasonal droughts. The appearance of the genes coding for fructan-synthesizing enzymes probably allowed the fructan flora to cope with seasonal droughts (Hendry 1993). The deduced amino acid sequence of key enzymes in the formation of fructans shows a high homology with plant **invertases**, which are ubiquitous enzymes that hydrolyze sucrose, producing glucose and fructose (Sprenger et al. 1995, Vijn et al. 1997). It is quite likely that the genes in fructan-producing taxa

emerged as a result of duplication of the invertase gene, followed by slight modification.

5. Water Movement Through Plants

5.1 General

Water transport from the soil, through the plant, to the atmosphere, takes place in a soil–plant–air continuum that is interconnected by a continuous film of liquid water (Fig. 9). Water moves through the plant along a **gradient**, either from high to low **water potential** (if transport occurs across a semipermeable membrane), from high to low **hydrostatic pressure** (if no such membrane is involved), or from a high to a low **partial water vapor pressure**. The low partial pressure of water vapor in the

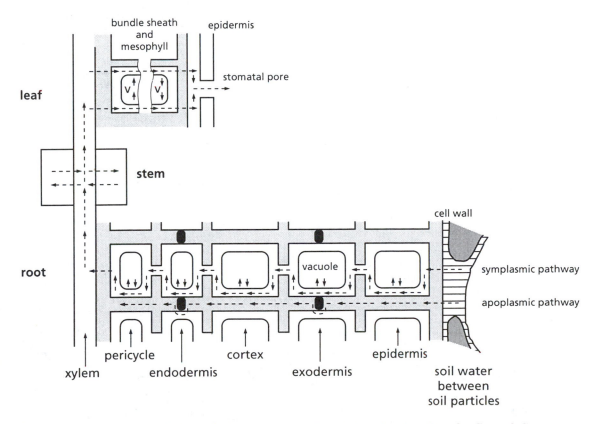

FIGURE 9. Water transport in the soil–plant–air continuum. Water can move through the cell walls (apoplast), or cross the plasma membrane and move through the cytoplasm and plasmodesmata (symplast). Water cannot move through the suberized Casparian bands in the wall of all endodermal and exodermal cells, including passage cells. Note that the exodermis is absent in some species, in which case water can move from the soil through the apoplast as far as the endodermis.

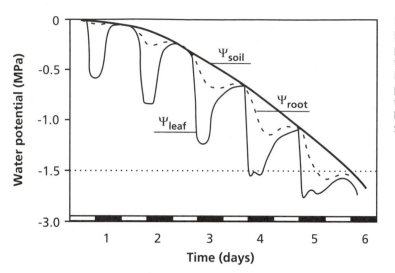

FIGURE 10. Typical diurnal changes in leaf water potential, Ψ_{leaf}, root water potential, Ψ_{root}, and soil water potential, Ψ_{soil}, of a transpiring plant rooted in soil allowed to dry from a water potential near zero to a water potential at which wilting occurs. The dark bars indicate the night period (after Slatyer 1967).

air, compared with that inside the leaves, is the major driving force for water loss from leaves, which, in turn, drives water transport along the gradient in hydrostatic pressure between the xylem in roots and leaves, and down a gradient in water potential between the soil and the cells in the roots (Fig. 10). As soils dry out, there are parallel decreases in soil water potential and plant water potential, both immediately before dawn (when water stress is minimal) and at midday (when water stress is maximal) (Fig. 11). The passive movement of water along a gradient differs strikingly from plant acquisition of carbon and nutrients, which occurs through the expenditure of metabolic energy. The steepest gradient in the soil–plant–atmosphere continuum occurs at the leaf surface, which indicates that the stomata are the major control point for plant water relations. There are substantial resistances, however, to water movement in soil, roots, and stems, so short-term stomatal controls are constrained by supply from the soil and resistances to transfer through the plant. An appreciation of these controls that operate at different time scales is essential to a solid understanding of plant water relations.

Water flux, J (mm^{3}s^{-1}) (i.e., the rate of water movement between two points in the soil–plant–atmosphere system) is determined by both the gradient between two points and the resistance to flow between these points. The conductance, L_p (mm^{3}s^{-1}MPa^{-1}) (i.e., the inverse of resistance) is often a more convenient property to measure. As pointed out earlier, the gradient along which water moves is *not* invariably a gradient in water potential ($\Delta\psi_w$, MPa), but it may be a gradient in hydrostatic pressure ($\Delta\psi_p$, MPa) or in water vapor pressure (Δe;

see Eq. 2 in the chapter on photosynthesis). In the case of a gradient in water potential, we can write:

$$J = L_p.\Delta\psi_w \qquad (5)$$

During the day, the water potential of leaves often declines, when the conductance of the roots or

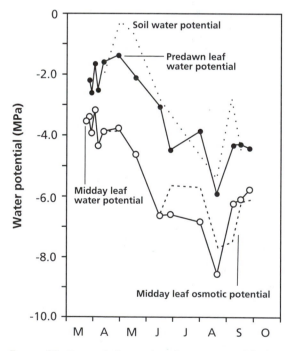

FIGURE 11. Seasonal changes in soil water potential (dotted line), predawn leaf water potential (solid line with solid symbols), midday leaf water potential (solid line with open symbols), and midday leaf osmotic potential (dashed line) in the C$_4$ plant *Hammada scoparia* (Schulze 1991). Copyright Academic Press, Ltd.

stems is too low to allow a supply of water to the leaves which is sufficient to meet the loss of water by transpiring leaves. This is not invariably found, however, because roots in drying soil send signals to the leaves, which reduce the stomatal conductance and hence water loss (cf Sect. 5.4.1 of this chapter).

5.2 Water in Roots

When growing in moist soils, cell membranes are the major resistance to water flow through roots. Water travels along two pathways from the outside to the inside of the root. If there is no **exodermis**, then water may move through the **apoplast** (i.e., the cell walls and other spaces outside of living cells) or through the **symplast** (i.e., the space comprising all the cells of a plant's tissues connected by plasmodesmata and surrounded by a plasma membrane) (Fig. 9). Water must eventually enter the symplast at the **endodermis**, which is the innermost cortical layer of suberized cells. The radial and transverse walls of the endodermal cells are thick, frequently lignified, and impregnated with waxy suberin, which forms a **Casparian band** (Fig. 12A). These hydrophobic bands completely encircle each endodermal cell and prevent further transport of water through the apoplast. Even when the neighboring cortical and pericycle cells plasmolyze, the plasma membrane of the endodermal cells remains attached to the Casparian band. Plasmodesmata, which connect the endodermis with the central cortex and pericycle, remain intact and functional during the deposition of suberin lamellae. **Passage cells** frequently occur in both the endodermis and the exodermis; in the endodermis they are typically located in close proximity with the xylem (Fig. 12B, C). Passage cells appear as short cells with a Casparian band, but the suberin lamellae and thick cellulosic walls that characterize other endodermal and exodermal cells are either absent or are formed at a much later stage of development. The passage cells become the only cells that present a plasma membrane surface to the soil solution once the epidermal and cortex cells die, which naturally occurs in some herbaceous and woody species. Passage cells then provide areas of low resistance to water flow (Peterson & Enstone 1996).

In most plants water entry into the symplast must already occur at the **exodermis**, which is a cell layer just inside the epidermal cells with properties similar to the endodermis. The exodermis is the outermost layer of cortical cells, adjacent to the epidermis. It is a type of hypodermis and occurs in roots of most species. Only 9% of all investigated species have either no hypodermis or have a hypodermis without suberized walls (Peterson 1989). At the endodermis or exodermis, water has to enter the cells, passing the plasma membrane, before it can arrive in the xylem vessels. For a long time it was thought that the water had to pass through the lipid bilayer itself. Like other organisms, however, plants have a family of **water-channel proteins**, also termed **aquaporins**, which are inserted into membranes and allow passage of water in a single file (Chrispeels & Agre 1994). The first water-channel proteins in plants were found in the tonoplast. They were also demonstrated more recently in the plasma membrane, where they play a role in the normal water uptake by plants (Daniels et al. 1994, Maggio & Joly 1995). Some of the water-channel proteins are blocked by mercury, which binds to cysteine in the protein. The amount of water-channel proteins decreases during the night and starts to increase again just before the beginning of the light period, which suggests rapid turnover. Environmental factors that affect the roots' hydraulic conductance may well affect the number or status of the water channels, but little is known about it so far.

At a **low temperature**, when membrane lipids are less fluid and membrane proteins are somewhat immobilized, the resistance of the plasma membrane to the water flow is high. Adaptation and acclimation to low temperature generally involves a shift to more unsaturated fatty acids, which increases the fluidity of these membranes at low temperature. The resistance to water flow is also high in plants exposed to soil **flooding**, which results in a low oxygen concentration in the soil, followed by inhibition of the normal aerobic respiration. This quite possibly results in a smaller amount of the water-channel proteins and hence a lower hydraulic conductance (see Sect. 5.6 of the chapter on growth and allocation). An excess of water in the soil may paradoxically cause symptoms that also occur in water-stressed plants: wilting, accumulation of ABA, and stomatal closure (cf. Sect. 5.6.2 of the chapter on growth and allocation).

As the soil dries, roots and soils shrink, which reduces the contact between roots and the soil water films, as well as conductance of water into the root. In **dry environments**, the contact between roots and soil is the greatest resistance to water flux from soil to leaves. Plants increase root conductance primarily by increasing allocation to production of new roots. Root hairs may be important in that they maintain contact between roots and soil. The role of

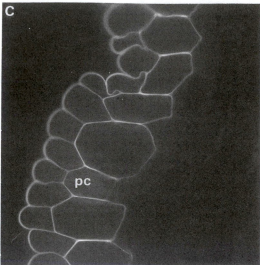

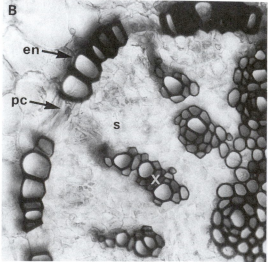

FIGURE 12. (A) Transverse section of a lateral root of *Zea mays* (maize). The endodermis (indicated by hollow arrows) and exodermis (indicated by black arrows) both have a Casparian band, suberized lamellae and a suberized and lignified tertiary wall. The section has been observed under fluorescent light, and the suberin and lignin molecules around the endodermal and exodermal cells autofluoresce very brightly. The section was prestained with periodic-acid Schiff's reaction to minimize autofluorescence from the walls of the other cell types (courtesy X.L. Wang and M.E. McCully, Biology Department, Carleton University, Ottawa, Canada). (B) Cross-section of an orchid (*Phalenopsis* sp.) root, showing the endodermis (en), a passage cell (pc), and the xylem (x) within the stele (s). Bar is 100 μm. (C) Cross-section of an onion root (*Allium cepa*), stained and viewed with UV, to show the walls of the epidermis and exodermis. Note the shorter passage cells (pc) in the exodermis. Bar is 800 μm (courtesy D.E. Enstone, Biology Department, University of Waterloo, Canada). Copyright Physiologia Plantarum.

mycorrhizal associations in water transport is discussed in Section 2.7 of the chapter on symbiotic associations. A high root mass ratio (RMR, root mass as a fraction of total plant mass) is typical of any plant grown under dry conditions, and drought-adapted C_3 species typically have higher root mass ratios when compared with nonadapted C_3 species. There are very few comparisons of the root mass ratio of C_3 and C_4 species, but the few data there indicate that RMR is lower in C_4, which correlates with their higher water-use efficiency (cf. Sect. 9.5 of the chapter on photosynthesis) (e.g., Kalapos et al. 1996). When these adaptations and acclimation responses to low temperature or low water supply are combined in natural vegetation patterns, aboveground biomass declines dramatically along a gradient of increasing aridity, but root biomass often remains relatively constant (Table 5), resulting in an increased root mass ratio.

In extremely dry soils, where the soil water potential is lower than that of plants, and plants can no longer extract water from soil, it may be advantageous to increase root resistance. For example, cactuses shed fine roots in summer and so prevent water loss to soil. Cactuses quickly produce new roots (**rain roots**) within 24 h after a rain to exploit new sources of soil moisture (Nobel 1996). Some plants have **contractile roots**. These decrease in length and increase in width, and so maintain hydraulic contact with the surrounding soil. During root contraction in *Hyacinthus* (hyacinth), mature cortical cells increase in diameter while decreasing in length, suggesting a change in wall extensibility in one or more directions (cf. Sect. 2.2 of the chapter on growth and allocation; Pritchard 1994). [Contractile roots also offer the explanation for why geophytes tend to pull themselves into the ground over the years (Pütz 1996).]

Many of the dominant woody species growing in arid and semi-arid conditions have **dual root systems**. Shallow, superficial roots operate during spring, and the deep-penetrating part of the root, which is usually located in relatively unweathered bedrock, operates during dry summers. Because most of the nutrients tend to be in the superficial soil layers, rather than the unweathered deep bedrock, most of the nutrients will be taken up by the shallow roots. Plants that grow on shallow soil or even bare rock in the Israeli maquis continue to

transpire during the entire summer by growing roots in rock fissures. On such sites with shallow soil in semi-arid climate conditions, roots of some plants (e.g., *Arctostaphyllos viscida*, whiteleaf manzanita, and *Arbutus menziesii*, Pacific madrone, of the Pacific Northwest in the United States) can utilize water from the bedrock. Roots of such plants occupy rock fissures as small as 100 μm. The cortex of such roots may become flat, with winglike structures on the sides of the stele (Zwieniecki & Newton 1995).

Water in the xylem vessels of the roots is normally under tension (negative hydrostatic pressure). Under moist conditions and low transpiration, however, the loading of solutes into the xylem may be rapid enough to produce a very negative osmotic potential in the xylem. Water may then move osmotically into the xylem vessels and create a positive hydrostatic pressure, forcing water up through the xylem into the stem. This phenomenon is known as **root pressure**. Root pressure can push xylem sap out through the leaf tips of short-statured plants (guttation), which is a phenomenon that contributes to the formation of "dew" on leaves. Root pressure is important in reestablishing continuous water columns in stems, after these columns break (see later). Using Equation 7 in Box 5, we can calculate that xylem sap containing 10 to 100 mM solutes can be "pushed" up the stem as high as 2.6 to 26 m. The liquid exuding from tree stumps and wounds in stems may also result from root pressure (e.g., in *Vitis vinifera*, grape) and *Betula nigra* (river birch); however, the xylem sap

TABLE 5. **Above-ground and below-ground biomass and the root mass ratio in various forest ecosystems, partly grouped by climate, climatic forest type, and species.**

Ecosystem	Above-ground biomass ($g m^{-2}$)	Below-ground biomass ($g m^{-2}$)	RMR* ($g g^{-1}$)
Boreal			
Broadleaf deciduous	50	25	0.32
Needle-leaf evergreen	30–140	7–33	0.20–0.30
Cold-temperate			
Broadleaf deciduous	175–220	25–50	0.13–0.19
Needle-leaf deciduous	170	40	0.18
Needle-leaf evergreen	210–550	50–110	0.14–0.28
Warm-temperate			
Broadleaf deciduous	140–200	40	0.21
Needle-leaf evergreen	60–230	30–35	0.15

Source: Vogt et al. 1996.
*RMR, ratio of root mass to total plant mass.

exuded by palms and several maples (e.g., *Acer saccharum*, sugar maple, and *A. nigrum*, black sugar maple, which is often tapped commercially to make sugar or syrup) results from stem pressure, and *not* from root pressure (Kramer 1969).

Hydraulic lift is the movement of water from deep moist soils to drier surface soils through the root system. This occurs primarily at night, when stomates are closed, so that the plant is at equilibrium with root water potential. Under these circumstances, water will move from deep moist soils with a high water potential into the root and out into dry surface soils of low water potential. Although hydraulic lift was first observed in dry grasslands (Caldwell & Richards 1989), it also occurs during dry periods in temperate forests, when high leaf area and high transpiration rates deplete water from upper soil horizons. For example, adult sugar maples (*Acer saccharum*) derive all transpirational water from deep roots; however, 3 to 60% of water transpired by shallow-rooted species without direct access to deep water comes from water that is hydraulically lifted by sugar maple (Dawson 1993). Deep ground-water often has a different isotopic signature than does surface water, making it possible to determine the original source of water transpired by plants. Hydraulic lift can modify competitive interactions among plants in unexpected ways by resupplying water to shallow-rooted species during dry periods, thereby modifying both water supply and the conditions for nitrogen mineralization and diffusion in dry soils. For example, 20 to 50% of the water used by shallow-rooted *Agropyron desertorum* comes from water that is hydraulically lifted by neighboring sage brush (*Artemisia tridentata*) in the Great Basin desert of western North America (Richards & Caldwell 1987). We will discuss hydraulic lift in this context in the chapter on interactions among plants.

Water can occasionally move from leaves through the stem and roots to soil. This has been documented in the Atacama Desert in coastal Peru, where coastal fog can wet leaves of *Prosopis tamarugo* (mesquite) sufficiently to generate a water-potential gradient from the leaf to the soil. The soil becomes wet and the water absorbed from the air is later available for uptake by the roots (Mooney et al. 1980).

5.3 Water in Stems

Ever since the phenomenon of atmospheric pressure was recognized, it has been evident that even a perfect vacuum pump cannot lift water any higher

than 10 m. In addition, even a relatively small xylem vessel with a radius of 20 μm only accounts for about 0.75 m of sap ascent by capillary action; however, plants appear to be able to pull water well beyond this limit. Some of them, like the giant redwood (*Sequoia gigantea*) in California or karri (*Eucalyptus diversicolor*) in Western Australia, lift substantial quantities of water close to 100 m daily. If a vacuum pump cannot lift water higher than 10 m, then the pressure in the xylem must be lower than that delivered by such a pump (i.e., it must be negative)!

Water in the xylem of the stem, in contrast to that in the live cells of the roots, is under tension (negative hydrostatic pressure) in transpiring plants. As explained in Box 6, these suction tensions are due to interactions of water molecules with the capillaries in the cell walls of transport vessels. In fact, the water column in a 100 m tall tree is held in place by the enormous **capillary forces** in the xylem at the top of the tree. Due to the **cohesion** among water molecules from hydrogen bonding, the water column in the stem is "sucked upward" to replace water that is transpired from leaves (**cohesion theory**; Dixon 1914, Steudle 1995). What is our evidence for such **suction tensions** or negative hydrostatic pressures and what exactly do they mean?

5.3.1 Can We Measure Negative Xylem Pressures?

Evidence for negative pressure in the xylem has been obtained using the **pressure chamber** (Scholander 1965). A cut stem is placed in the apparatus and sealed from the atmosphere, with the cut stem extending out (Fig. 13). A pressure is then applied just high enough to make the xylem sap in the stem appear at the cut surface. The positive pressure applied ("balancing pressure") is equal to the negative pressure in the xylem when the plant was still intact. Although there are problems using this technique when using plants with a low relative water content, the pressure chamber is widely used to assess the water potential in plants.

For a full appreciation of the ascent of sap in plants, we need to consider carefully the exact site the water is coming from that is pushed back into the xylem when pressure is applied to the pressure chamber (Box 6). This water is pushed out of the many capillaries in the **walls** of the xylem vessels and adjacent cells, where it was held in place by strong **capillary forces** between the water molecules and the cell walls (Fig. 14). In a transpiring plant water continuously moves from these capil-

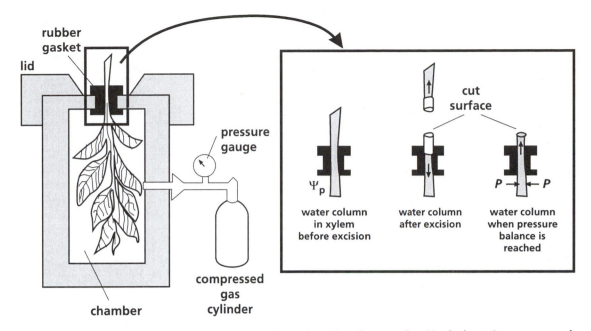

FIGURE 13. Schematic representation of the Scholander pressure chamber that allows measurement of negative hydrostatic pressures in the xylem. A cut shoot or twig is sealed around the stem and placed upside down in the chamber and the chamber is hermetically sealed. Positive pressure is exerted on the shoot or twig, using a gas cylinder. When the exerted positive hydrostatic pressure equals the negative water potential (negative osmotic potential and negative pressure) in the xylem, the xylem fluid will appear at the cut surface. After determination of the osmotic potential of the xylem fluid, the negative hydrostatic pressure is calculated.

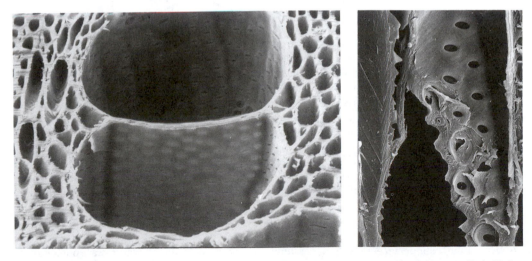

FIGURE 14. (Left) A scanning electron micrograph of two adjacent xylem vessels and surrounding cells in the sapwood of *Eucalyptus marginata* (jarrah). Note the thick walls of the xylem vessels and the pits in the walls of these vessels (courtesy J. Kuo, The University of Western Australia, Australia). (Right) A scanning electron micrograph of the sap-wood of *Populus nigra* hand-sectioned tangentially and slightly oblique. Bordered pits are shown between two adjacent vessels. Ray parenchyma cells are cut transversely in a vertical row to the right and left. Several pit apertures are seen (top central) in face view, with pit membranes visible through them. Lower down, the blade has removed the top borders of four pits, showing the interiors of the lower bordered pit chambers without any pit membranes. A pit can be seen centrally half-closed by a torn pit membrane: The blade has folded the upper chamber above it, revealing the inner cavity (indicated by arrow). The delicate pit membrane capillaries are sufficiently fine to filter fine carbon suspensions from indian ink. They are normally supported from mechanical disruption by the borders if subjected to powerful pressure flow from liquids or gases. Gases can pass when the membranes are physically torn; alternatively, the suctions may be so great that air bubbles are pulled through the capillaries in the membranes initiating cavitation in the conduits in which the air bubble enters. (courtesy J.A. Milburn, University of New England, Australia).

laries to the intercellular spaces in leaves where the water potential is more negative as long, as the water vapor pressure is not saturated (Box 5). Due to the strong capillary forces this water is replaced by water in the lumen of the xylem vessels. These strong capillary forces keep the entire water column in the xylem vessels in place and prevent it from retreating, such as happens when the stem is cut. At physiological temperatures, the **cohesive forces** between the water molecules are so strong that the water column in the xylem is unlikely to break (but see Sect. 5.3.3).

A **pressure probe** has been used to obtain further evidence for negative xylem pressures. It involves the insertion in a cell of a small glass microcapillary. The fluid in the capillary will then be pushed back by the turgor pressure. The force to push back the meniscus of the fluid to its position before insertion is then measured, using a sensitive pressure transducer. In this way the pressure in individual cells, such as stomatal guard cells, can be measured very accurately. When using this technique for xylem elements, little or no hydrostatic tension was found under a variety of conditions (Zimmermann et al. 1994). Rather than rejecting the cohesion theory, however, the method itself has been questioned because gaining access to the xylem requires impaling the stem with a glass microcapillary, starting from the outside.

The most compelling recent evidence in favor of the cohesion theory for the ascent of sap comes from measurements using a device that involves spinning a length of branch about its center to create a known tension based on centrifugal forces. Results of such experiments agree perfectly with those obtained with the pressure chamber (Holbrook et al. 1995). Tensions can therefore be created in xylem vessels and measured accurately by the pressure chamber, but what do these tensions really represent? When stating that the xylem is under tension, we do not actually mean the xylem conduit itself. Rather, the suction tension, or negative hydrostatic pressure, refers to the **adhesive forces** that tightly hold the water in the small capillaries in the wall of the xylem conduits (Shackel 1996).

5.3.2 The Flow of Water in the Xylem

Hydraulic resistance in the shoot xylem accounts for 20 to 60% of the total pressure difference between the soil and the air in transpiring trees and crop plants (Sperry 1995). In woody plants most of this pressure difference occurs in small twigs and branches, where the cross-sectional area of xylem is small (Gartner 1995). The water flow (J_v, $mm^3 mm^{-2} s^{-1}$ = $mm s^{-1}$) in xylem vessels is approximated by the Hagen-Poiseuille equation, which describes transport of fluids in ideal capillaries:

$$J_v = \left(\pi.R^4.\Delta\psi_p\right)/8.\eta.L \tag{6}$$

where $\Delta\psi_p$ (MPa) is the difference in hydrostatic pressure, R (mm) is the radius of the single element with length L (mm) through which transport takes place, and η ($mm^{-2} MPa\,s$) is the viscosity constant. This equation shows that the **hydraulic conductance** is proportional to the fourth power of the vessel diameter. The hydraulic conductance of a stem, with only a few xylem vessels with a large diameter is therefore much higher than that of a stem with many more vessels with a small diameter, but the same total xylem area. In addition, the pits in the connecting walls of the tracheids impose a substantial resistance to water flow (Nobel 1991).

Plants differ widely with respect to the diameter and length of their xylem vessels (Table 6). Vessel length in trees varies from less than 0.1 m to well over 10 m or as long as the whole stem. There is no obvious advantage associated with either short or

TABLE 6. Hydraulic conductance of xylem conduits, maximum velocity of water transport through the conduits and xylem diameter for stems of different types of plants.

	Hydraulic conductivity of xylem lumina ($m^2 s^{-1} MPa^{-1}$)	Maximum velocity ($mm s^{-1}$)	Vessel diameter μm
Evergreen conifers	5–10	0.3–0.6	<30
Mediterranean sclerophylls	2–10	0.1–0.4	5–70
Deciduous diffuse porous	5–50	0.2–1.7	5–60
Deciduous ring porous	50–300	1.1–12.1	5–150
Herbs	30–60	3–17	
Lianas	300–500	42	200–300

Source: Milburn 1979; Zimmerman & Milburn 1982.

long vessels; it might be the accidental result of other variables of tree growth, such as mechanical requirement for fiber length (Zimmermann & Milburn 1982), or that small vessels are less prone to **freezing-induced cavitation**, the breakage of the water column in a transport vessel (cf. Sect. 5.3.3 of this chapter). Vessel length tends to correlate with vessel diameter. In deciduous trees, xylem vessels produced early in the season tend to be longer and wider than the ones produced later in the year. The difference in xylem diameter between early and late wood shows up as the "year rings" of the trunk of these "ring-porous" trees. "Diffuse-porous" trees, on the other hand, with a random distribution of wide and narrow vessels throughout the year, do not show such distinct year rings (Zimmermann 1983).

Vines, which have relatively narrow stems, have long vessels with a large diameter, compared with related species or to the species in which they climb (cf. Sect. 5.3.6 of this chapter). Because the hydraulic conductance is proportional to the fourth power of the vessel diameter, the larger diameter compensates for the smaller total area. For example, the stem of the liana *Bauhinia fassoglensis* has a conductance equal to the tree *Thuja occidentalis* with a tenfold greater sapwood area (Ewers & Fisher 1991). Xylem vessels with a narrow diameter have the disadvantage of a low **hydraulic conductance**. Because more of the total xylem area is taken up by the xylem walls, they provide greater **mechanical strength**. The narrow xylem vessels, however, are also less vulnerable to freezing-induced cavitation (cf. Sect. 5.3.6 of this chapter).

5.3.3 Cavitation or Embolism: The Breakage of the Xylem Water Column

Cavitation is unlikely to be caused simply by breakage of capillary water columns, which, at moderate temperatures, does not occur except at considerably higher tensions than ever occur in the xylem (i.e., in excess of 100 MPa) (Tyree & Sperry 1989). Cavitation, or embolism, however, does occur and leads to the filling of the xylem with water vapor and/or air, rather than water. Under water stress, when the tensions in the xylem become very high, cavitation is nucleated by the **entry of air** through the largest pores in the walls of the transport vessels, located in primary walls of the interconduit pits, the **pit membrane** (Fig. 15). Water then begins to evaporate explosively into the air bubble. Short acoustic pulses are registered during cavitation induced by water stress, allowing sound recordings to document cavitation rate. The bubble expands and interrupts

the water column. Entry of air into the xylem conduit depends on the size of the pores in the pit membrane. The thin, porous areas in conduit walls allow passage of water between conduits, but not a gas–water meniscus. This minimizes the spread of air bubbles into neighboring conduits. The tension required to cause cavitation is a function of the permeability of the interconduit pits to an air–water interface, which depends on pore diameters (Pockman et al. 1995, Sperry 1995). Pore diameters range from less that 0.05 to more than 0.4 μm (as opposed to <0.01 μm in the cell wall proper), depending on species and location in the plant. Embolism reduces the ability to conduct water and, if severe enough, will limit growth.

Species differ considerably in their vulnerability to cavitation, with the less vulnerable species tending to be more **desiccation-tolerant** (Tyree & Sperry 1989). For example, stems of *Populus fremontii* (poplar) show complete cavitation at −1.6 MPa, whereas those of *Salix gooddingii* (willow), *Acer negundo* (boxelder), *Abies lasiocarpa* (fir), and *Juniperus monosperma* (juniper) have a threshold at −1.4, −1.9, −3.1, and less than −3.5 MPa, respectively (Fig. 16). Vulnerability of species to cavitation in response to water stress is determined by the diameter of pores in pit membranes, because this determines the xylem tension at which an air bubble is sucked into the xylem lumen. Because even the widest vessels in ring-porous trees are sufficiently narrow to prevent breaking of a water column other than by the mechanisms accounted for, conduit diameter has no relationship to the vulnerability to cavitation in response to drought stress (Fig. 17) (Sperry 1995).

Cavitation is also caused by **freezing and thawing** of the xylem sap when it is under tension (Fig. 17). In this case, cavitation occurs at much lower tension and is induced by a different mechanism: Dissolved gases in the sap are insoluble in ice and freeze out as bubbles. If these bubbles are large enough when tension develops during thawing, then they will grow and cause cavitation. When cavitation is caused by freeze–thaw cycles, we can predict that wide and long vessels will be more vulnerable than small ones. Differences in conduit diameter are the main factors that account for species differences in vulnerability to cavitation from freeze–thaw events (Fig. 17). If the air freezes out as one large bubble, rather than a number of smaller ones, then the greater dissolved air content in larger conduits will give rise to larger bubbles that cause embolism at lower tensions (Sperry & Sullivan 1992). Most embolism in temperate woody plants occurs in response to freeze–thaw events

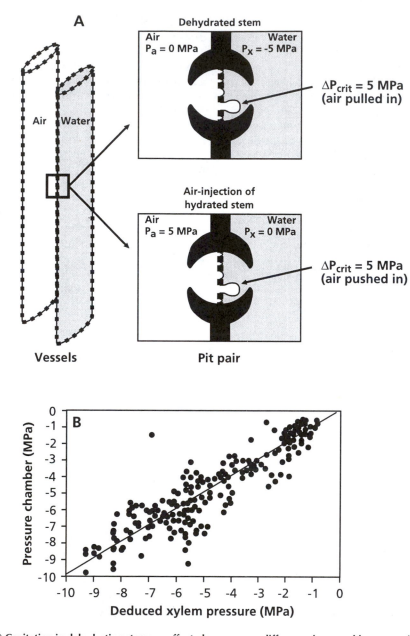

FIGURE 15. (A) Cavitation in dehydrating stems as affected by "air seeding." Two adjacent xylem vessels are shown. The right-hand one is filled with xylem fluid. The left-hand has cavitated and is therefore filled with water vapor at a very low pressure, near vacuum: 0 MPa. Pits between the vessels allow water flow and prevent passage of an air–water meniscus in the event that one vessel becomes air-filled. The top part of the illustration shows how a small air bubble is pulled in through the pit membrane pores when the pressure difference between the two vessels exceeds a critical threshold. In the example shown, this occurs at a xylem pressure of −5 MPa. In the experimental design shown by the lower part of the illustration, the critical pressure difference is exceed by pressurizing the air in the embolized vessel, while the fluid in the other vessel is at atmospheric pressure. Now an air bubble is pushed in. The top part illustrates what is happening in a real plant. (B) There is a very close agreement between pressure differences at which cavitation occurs in real plants, using the pressure chamber to determine the negative pressure in the xylem, and those achieved as illustrated by the bottom part of the top figure. This confirms that negative pressures must occur in the xylem, and that the failure of measuring these with a pressure probe reflects problems with that method (Sperry et al. 1996). Copyright Blackwell Science Ltd.

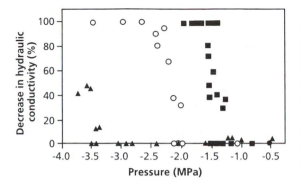

FIGURE 16. Decrease in hydraulic conductivity in the xylem of stems of *Populus fremontii* (poplar, squares), *Acer negundo* (box elder, circles), and *Juniperus monosperma* (juniper, triangles) against xylem hydrostatic pressure. The hydrostatic pressure in the xylem was varied by centrifuging stems which were cut and then recut under water to a length of 260–400mm (Pockman et al. 1995). Reprinted with permission from Nature 378:715–716 copyright (1995) Macmillan Magazines Ltd.

during winter and during the growing season rather than in response to drought. Embolism correlates more closely with the number of freeze–thaw episodes than it does with degree of frost (Sperry 1995).

The capacity of xylem to withstand freeze–thaw embolisms has important consequences for the evolution and distribution of woody plants. Evergreen trees from cold climates are more likely to be actively transpiring (and therefore developing negative xylem potentials) when freeze–thaw events occur. It is probable that for this reason they have small tracheids that are less likely to cavitate and easier to refill (see later). The disadvantage of narrow conduits is a low conductance; this low conductance may partly account for a low relative growth rate (Woodward 1995). Among deciduous woody plants there are ring-porous trees that produce large vessels during rapid growth in early spring and diffuse-porous species that have smaller diameter conduits. Ring-porous species cannot refill overwintering xylem, so their transpiration is entirely supported by current year's xylem, which therefore requires large-diameter vessels with high conductance. These species leaf out at least 2 weeks later than co-occurring diffuse-porous species, presumably because of their greater vulnerability to spring frosts (Sperry 1995). Some diffuse-porous species can refill embolized over-winter conduits and are particularly successful in cold climates, whereas other diffuse-porous species cannot.

Embolism can also be induced by **pathogens**. Although it has been known for quite some time that vascular diseases induce water stress in their host by reducing the hydraulic conductivity of the xylem, embolism as a cause for this has received very little attention. In the case of Dutch Elm disease, however, embolism precedes any occlusion of vessels by other means. The exact cause of embolism remains unclear. It might be due to a pathogen-induced increase in stomatal conductance or decrease of water uptake by the roots. Pathogens might also change the xylem sap chemistry. For example, millimolar concentrations of oxalic acid, which is produced by many pathogenic fungi, lower the surface tension of the xylem sap. In *Acer saccharum* (sugar maple) and *Abies balsamea* (balsam fir), oxalic acid reduces the tension at which air can enter the xylem (Tyree & Sperry 1989).

5.3.4 Can Embolized Conduits Resume Their Function?

Functioning of the embolized vessels can be resumed upon refilling the vessel with water. This can occur at night under moist conditions, when **root pressure** builds up a positive xylem pressure. In more extreme cases it may not occur until it rains (Fig. 18). Because a solute concentration of 100mM exerts just enough positive pressure to balance a water column of about 26m, it is unlikely that cavitated conduits in the top of a tall tree are ever refilled by root pressure. As mentioned in passing in Section 5.2, trees can also build up stem pressure.

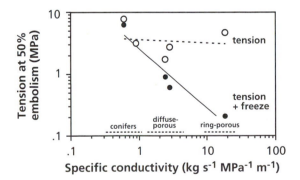

FIGURE 17. Tension at the time of 50% embolism as dependent on the size of the tracheids or vessels. The experiment as carried out with trees exposed to water stress (tension, open circles) and with trees exposed to a freeze–thaw cycle (tension + freeze, solid circles) (Sperry & Sullivan 1992). Copyright American Society of Plant Physiologists.

It is unknown if stem pressure can also restore cavitated xylem conduits.

Cavitated conduits can refill by **dissolution** of the bubble, which can occur at moderately negative xylem pressures. Dissolution of bubbles under tension may require narrow conduits (Sperry 1995), which perhaps explains the tendency of desert plants and plants from cold environments to have narrow conduits (see also Sect. 5.3.5). When such a moderately negative water potential cannot be reached, the xylem remains filled with water vapor and the conduit no longer functions in water transport (Yang & Tyree 1992). Failure to refill cavitated vessels can sometimes be advantageous. For example, conduits of cactus xylem cavitate when soil gets extremely dry, preventing water from being lost from the body of the plant to the soil.

5.3.5 Trade-Off Between Conductance and Safety

Species differences in xylem anatomy and function reflect the **trade-off** between a large xylem diameter, which maximizes **conductance**, and a small diameter, which increases the **strength** of the wood and minimizes the chances of **cavitation** due to freeze–thaw events. For example, **vines**, which have a small stem diameter, have large vessels with a high conductance and rapid water movement through the vessels, compared with other species (Table 6). Their stem, however, does not have the strength of that of a tree with similar leaf area. Many plants, including herbs and crop plants, function close to the water potential where cavitation occurs. This suggests that the investment in transport conduits is such that it is only just sufficient to allow the required rate of water transport during the growing season (Tyree & Sperry 1989).

Woody species function close to the theoretical limit of the hydraulic conductance of their xylem conduits and loss of xylem conductance due to embolism is a regular event. Species differ enormously in their vulnerability with respect to water-stress induced cavitation (Fig. 19). *Juniperus virginiana* (juniper) is less vulnerable than *Abies balsamea* (balsam fir). Among the hardwoods there are also major differences: *Cassipourea elliptica*, which seldom has a xylem water potential in its natural habitat of less than −1.5 MPa, is more vulnerable than *Rhizophora mangle*, which regularly experiences xylem water potentials below −4.0 MPa. In general, the vulnerability of a species correlates negatively with the xylem tensions it experiences in

the natural habitat. The risk of cavitation plays a major role in the differentiation between drought-adapted and mesic species. On the one hand smaller interconduit pores confer resistance to cavitation. On the other hand, they may reduce the hydraulic conductivity of the xylem. The **safer** the xylem, the **less efficient** it may be in water conduction.

Species differ in the safety margins for cavitation by water stress (Fig. 20). Species from mesic habitats, in which there are ample opportunities for nighttime recovery, operate quite close (ca. 0.4–0.6 MPa) to the xylem tensions that cause 100% cavitation. Species from dry habitats, however, which experience negative water potentials for months at a time, operate with a much larger safety margin (Fig. 20). For example, 100% cavitation in creosote bush (*Larrea tridentata*) occurs at < -16 MPa,

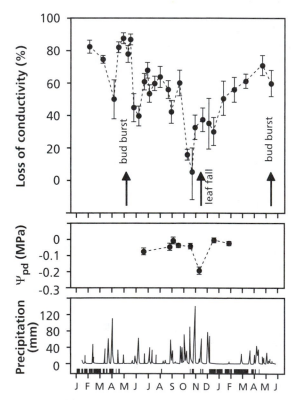

Figure 18. **Seasonal changes of xylem embolism in apical twigs, expressed as a percentage loss of hydraulic conductivity (top); predawn water potential (middle); precipitation at the site of the studied tree *Fagus sylvatica*. The occurrence of subzero temperatures is marked by black bars at the bottom of the lower graph. Arrows in the top figure indicate bud burst and leaf fall (Magnani & Borghetti 1995). Copyright Blackwell Science Ltd.**

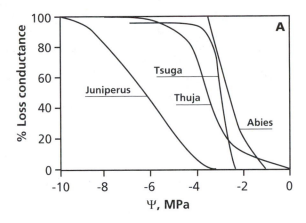

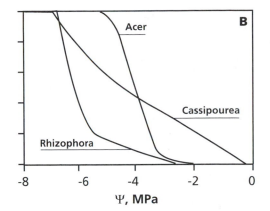

FIGURE 19. Vulnerability of various species to embolism, measured as the loss of hydraulic conductance versus water potential in the xylem. (Left) coniferous species (*Juniperus virginiana, Thuja occidentalis, Tsuga canadensis*); (right) hardwood species (*Rhizophora mangle,* *Acer saccharum, Cassipourea elliptica*) (Tyree & Sperry 1989). With permission, from the Annual Review of Plant Physiology and Plant Molecular Biology, Vol. 40, copyright 1989, by Annual Reviews Inc.

whereas minimum water potentials observed in the field are closer to −9 MPa (Sperry 1995).

Cavitation induced by freezing stress occurs at less negative water potential in wide and long xylem conduits than it does in shorter and narrower ones (Fig. 17). This may explain why xylem diameters are less in species from high latitude or altitude (Baas 1986). It may also account for the rarity of woody vines at high altitude (Ewers et al. 1990). If, contrary to a widely held belief, however, the breaking of the water column in the xylem at moderate temperatures is *not* related to conduit size (Fig. 17), then why do **desert plants** tend to have narrow vessels? Conduits with a small diameter also tend to have smaller **pit membrane pores** than do wide ones, and this may well explain why desert plants have small xylem diameters. That is, the correlation does *not* reflect a direct causal relation. Pits may differ widely in different species, however, and the correlation between **pit membrane** pore size and xylem diameter is not very strict, which accounts for the generally poor correlation between xylem diameter and vulnerability to cavitation in different taxa (Sperry 1995). The length of the xylem conduit is also important, and this is often correlated with conduit diameter. Many short and narrow xylem conduits (such as those concentrated in the nodes or junctions of a stem segment) may be of ecological significance in that they prevent emboli from traveling from one internode to the next or from a young twig to an older one, thus acting as "safety zones" (Lo Gullo et al. 1995).

5.3.6 Transport Capacity of the Xylem and Leaf Area

In a given stand of trees, there is a strong linear relationship between the cross-sectional area of sapwood (A_s), that part of the xylem that functions in water transport, and the foliage area (A_f) supported by that xylem. Given that hydraulic

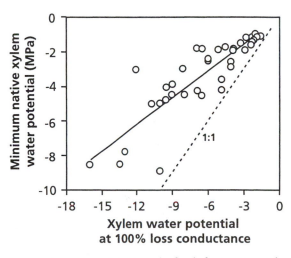

FIGURE 20. Xylem pressure required to induce 100% cavitation versus minimum pressure observe in the field. Dashed line indicates 1:1 relationship. Each point is a different species (Sperry 1995). Copyright Academic Press, Ltd.

conductance of stems differs among species and environments, however, it is not surprising that the ratio of foliage area to sapwood area $(A_f : A_s)$ differs substantially among species and environments (Table 7). Desiccation-resistant species generally support much less leaf area per unit of sapwood than desiccation-sensitive species (Table 7). This is logical because vessels are narrower in species from dry habitats; hence, more sapwood is needed for a similar transport capacity (Zimmermann & Milburn 1982). Any factor that speeds the growth of a stand (i.e., higher "site quality") generally increases $A_f : A_s$ because it increases vessel diameter (Fig. 21A). For example, nutrient addition and favorable moisture status enhance $A_f : A_s$, and dominant trees have greater $A_f : A_s$ than do subdominants (Margolis et al. 1995). When conductance per unit sapwood is also considered, there is a much more consistent relationship between foliage area and sapwood area (Fig. 21B).

Vines have less xylem tissue area per unit of distal leaf area (i.e., per unit leaf area for which they provide water). Their stems are thin relative to the distal leaf area, when compared with plants that support themselves. Vines compensate for this by having vessels with a large diameter (Fig. 22). It is interesting that the correlation between sapwood area and distal leaf area also holds when the leaf area is that of a mistletoe tapping the xylem, even when there is no host foliage on the branch (see Sect. 2.3 of the chapter on parasitic associations). Because there are no phloem connections between the xylem-tapping mistletoe and its host tree, the correlation cannot be accounted for by signals leaving the leaves and traveling through the phloem. This raises the unanswered question on how leaf area controls sapwood area. Or is it the other way around?

5.3.7 Storage of Water in Stems

Plants store some water in stems, which can temporarily supply the water for transpiration. For example, water uptake in many trees lags behind transpirational water loss by about 2 hours (Fig. 23) because the water initially supplied to leaves comes from parenchyma cells in the stem. Withdrawal of stem water during the day causes stem diameter to fluctuate diurnally, being greatest in the early morning and smallest in late afternoon. Most **stem**

TABLE 7. Typical ratios of foliage area (A_f) to sapwood area (A_s) of conifers.

Species	Common name	$A_f : A_s$ $(m^2 m^{-2})$
Mesic environments		
Abies balsamea	Balsam fir	6700–7100
A. amabilis	Pacific silver fir	6300
A. grandis	Grand fir	5100
A. lasiocarpa	Subalpine fir	7500
Larix occidentalis	Western larch	5000
Picea abies	Norway spruce	4600
P. engelmanni	Engelmann spruce	2900–3400
P. sitchensis	Sitka spruce	4500
Pseudotsuga menziesii	Douglas fir	3800–7000
Tsuga heterophylla	Western hemlock	4600
T. mertensiana	Mountain hemlock	1600
Average		5000 ± 500
Xeric environments		
Juniperus monosperma	One-seeded juniper	800
J. occidentalis	Western juniper	1800
Pinus contorta	Lodgepole pine	1100–3000
P. edulis	Pinyon pine	2500
P. nigra	Austrian pine	1500
P. ponderosa	Ponderosa pine	1900
P. sylvestris	Scotch pine	1400
P. taeda	Loblolly pine	1300–3000
Average		1800 ± 200

Source: Margolis et al. 1995.

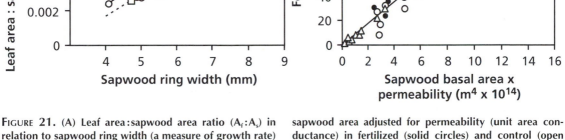

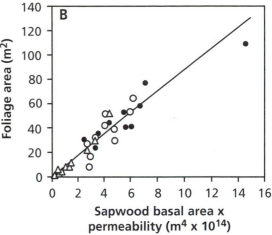

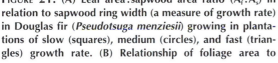

FIGURE 21. (A) Leaf area:sapwood area ratio (A_f:A_s) in relation to sapwood ring width (a measure of growth rate) in Douglas fir (*Pseudotsuga menziesii*) growing in plantations of slow (squares), medium (circles), and fast (triangles) growth rate. (B) Relationship of foliage area to sapwood area adjusted for permeability (unit area conductance) in fertilized (solid circles) and control (open circles) trees of Sitka spruce (*Picea sitchensis*) and control trees of lodgepole pine (*Pinus contorta*) (open triangles) (Margolis et al. 1995). Copyright Academic Press, Ltd.

shrinkage occurs in living tissues external to the xylem, where cells have more elastic walls and cells decrease in volume when water is withdrawn. In trees the stem water provides less than 10 to 20% of the daily water transpired in most plants, so it forms an extremely small buffer. In addition, if this water becomes available as a result of cavitation, which stops the functioning of the cavitated conduit, the benefit of such a store is questionable.

Under some circumstances, however, stem water storage is clearly important. For example, in tropical dry forests, the loss of leaves in the dry season by drought-deciduous trees eliminates transpirational water loss. Stem water storage makes an important contribution to the water required for flowering and leaf flushing by these species during the dry season (Borchert 1994). Water storage in these trees is inversely related to wood density (Fig. 24). Early successional shade-intolerant species grow rapidly, have low wood density, and, therefore, high water storage that enables them to flower during the dry season and to reflush leaves late in the dry season. By contrast, slow-growing deciduous trees with high wood density and low water storage remain bare to the end of the dry season (Borchert 1994). Storage of water in stems is also important in reducing winter desiccation [e.g., of the needles of *Picea engelmannii* (Engelmann spruce) that grow at the timberline]. Water in the stem may become available when the

soil is frozen and air temperatures are above −4°C (Sowell et al. 1996).

In herbaceous plants and succulents, which have more elastic cell walls than those of the sapwood in trees, storage in the stem is more important. Small herbaceous plants also transpire water made available by cavitation of some of the conduits in the stem. They refill the xylem by root pressure during the following night. Water storage is most evident in the concertinalike stems of the giant saguaro cactus (*Carnegia gigantea*), which can hold as much as 5000 kg of water (Zimmermann & Milburn 1982). Stored water of succulents may allow transpiration to continue for several weeks after uptake from soil has ceased.

5.4 Water in Leaves and Water Loss from Leaves

The earliest known measurements of stomata were made in 1660 by Mariotte, a French mathematician and physicist who earned his living as a clergyman in Dijon. Fifteen years later, Malpighi, who was a professor of medicine at Bologna and Pisa, mentioned porelike structures on leaf surfaces (Meidner 1987). Leaves inexorably lose water through their stomatal pores, as a consequence of the photosynthetic activity of the mesophyll leaf cells. Stomates exert the greatest short-term control over plant wa-

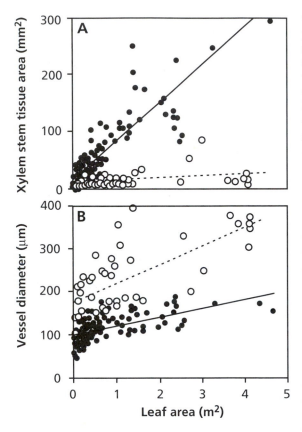

FIGURE 22. Xylem area (A) and maximum diameters of vessels (B) of contrasting *Bauhinia* species. Values are plotted as a function of the leaf area distal to the investigated stem section for stems of lianas (dashed line, open symbols) and congeneric trees and shrubs (solid line, closed symbols) (Ewers & Fisher 1991).

ter relations because of the steep gradient in water potential between leaf and air. There are two major interacting determinants of plant water potential: soil moisture, which governs water supply, and transpiration, which governs water loss. Both of these factors exert their control primarily by regulating stomatal conductance. Stomatal conductance depends both on the availability of moisture in the soil and on vapour pressure in the air, as will be outlined shortly.

5.4.1 Effects of Soil Drying on Leaf Conductance

Leaves of "isohydric" species, which control gas exchange in such a way that daytime leaf water status is unaffected by soil water deficits, must control stomatal conductance by messages arriving from the root. This is an example of **feedforward control**. That is, stomatal conductance declines before any adverse effects of water shortage arise in the leaves. Isohydric species include *Zea mays* (maize) and *Vigna sinensis* (cowpea). The phytohormone abscisic acid (ABA) is the predominant message arriving from roots in contact with drying soil (Davies et al. 1994). Soil drying enhances the concentration of this hormone in the xylem sap as well as in the leaves (Tardieu et al. 1992, Correia et al. 1995). In addition, injection of ABA in the stem of maize plants has fairly similar effects on the ABA-concentration in the xylem sap and on stomatal conductance as exposure to a drying soil. The stomata of desiccated plants, however, become more "sensitized" to the ABA signal, possibly by a

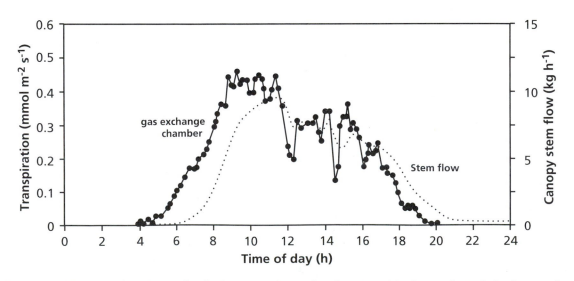

FIGURE 23. Diurnal pattern of water flow in the stem and water loss from transpiring leaves of a *Larix* (larch) tree. The difference between the two lines represents stem storage (Schulze et al. 1985).

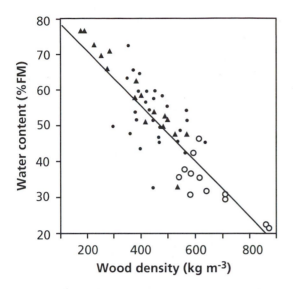

FIGURE 24. Relationship between stem water storage and wood density in 32 species of deciduous trees from a dry tropical forest in Costa Rica (Borchert 1994). Deciduous trees with high (triangles) or low (open circles) water storage or evergreen trees (closed circles). Copyright Ecological Society of America.

The mechanism by which roots sense dry soil is not clear. ABA, like any other acid, crosses membranes in its undissociated form. It therefore, accumulates in soil, especially when the rhizosphere is alkaline. Because the concentration of ABA in soil increases when water is limiting for plant growth, it has been speculated that roots may sense drying soils through ABA. Because the presence of NaCl inhibits the microbial degradation of ABA, the concentration of ABA also tends to be higher in saline soils. This might offer a mechanism for the roots to

combination of other chemical signals transported in the xylem and the low water potential of the leaf itself (Fig. 25). Among trees, isohydric species are those that generally occur in mesic habitats and operate at water potentials extremely close to potentials causing complete cavitation (Sect. 5.3.3; Sperry 1995). These species seldom experience more than 10% loss in conductance due to cavitation because their effective control of stomatal conductance minimizes diurnal variation in leaf water potential.

In "anisohydric" species, such as *Helianthus annuus* (sunflower), both the leaf water potential and leaf conductance decline with decreasing soil water potential. In these species, both root-derived ABA and leaf water status regulate stomatal conductance. A controlling influence of leaf water status on stomatal conductance need not be invoked. Rather, leaf water status is likely to vary as a consequence of water flux through the plant, which is controlled by stomatal conductance. The spectrum in stomatal "strategy" between isohydric and anisohydric plants is determined by the degree of influence of leaf water status on stomatal control for a given concentration of ABA in the xylem. Correlations between leaf conductance and leaf water status are only observed in plants where leaf water status has no controlling action on the stomata (Tardieu et al. 1996).

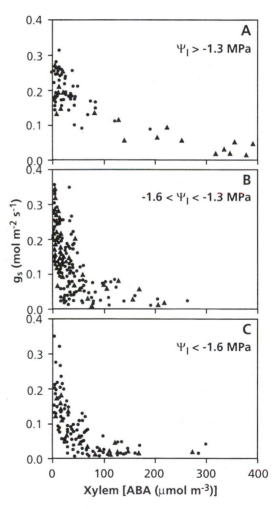

FIGURE 25. Leaf conductance (g_s) as a function of the concentration of ABA in the xylem sap of field-grown *Zea mays* (maize) plants. Measurements were made over three ranges of leaf water potential (ψ_l). Relationships found by feeding the plants artificial ABA are depicted as triangles. Circles refer to leaf conductance and endogenously produced ABA (Davies et al. 1994). Copyright American Society of Plant Physiologists.

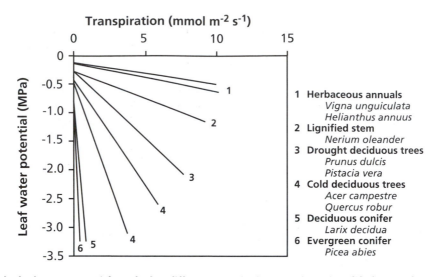

FIGURE 26. The leaf water potential reached at different transpiration rates in various life forms, when exposed to the same water supply (Schulze 1991). Copyright Academic Press, Ltd.

sense a low osmotic potential in the soil (Hartung et al. 1996).

The relationship between leaf conductance (and hence transpiration) and leaf water potential differs strikingly among growth forms (Fig. 26). Because the difference in leaf water potential and soil water potential is the driving force for water transport in the plant, the slope of the relationship in Figure 26 gives the conductance for water transport of the entire system. This conductance is greatest in herbaceous annuals and smallest in evergreen conifers. Herbaceous species like *Helianthus annuus* change stomatal conductance and transpiration dramatically in response to small changes in leaf water potential. By contrast, stomatal conductance and transpiration are insensitive to progressively larger changes in water potential as we go from herbaceous annuals to woody shrubs to deciduous trees to conifers. There is a corresponding decrease in conduit diameter and increase in the margin of safety against cavitation (Sect. 5.3.3). These patterns demonstrate the close integration of various parameters that determine plant "strategies" of water relations.

Among trees, anisohydric species maintain a larger safety margin against cavitation, but they also experience more cavitation (up to 50% loss of conductance), compared with isohydric species. In anisohydric species, the closure of stomates in response to declines in leaf water potential is essential; otherwise, the effects of soil drying would be augmented by cavitation that would cause further declines in leaf water potential and lead to runaway cavitation (Sperry 1995).

How do we know that the decreased stomatal conductance is really accounted for by signals from the roots in contact with drying soil, rather than by the low water potential in the leaf itself? To address this question, Passioura (1988b) used a pressure vessel placed around the roots of a *Triticum aestivum* (wheat) seedling growing in drying soil. As the soil dried out, the hydrostatic pressure on the roots was increased so as to maintain shoot water potential similar to that of well-watered plants. Despite having the same leaf water status as the control plants, the treated wheat plants showed reductions in leaf conductance similar to those of plants in drying soil outside a pressure chamber. Additional evidence has come from experiments with small apple trees (*Malus* x *domestica*) growing in two containers. Soil drying in one container restricts leaf expansion and initiation, with no obvious effect on shoot water relations. These effects on leaves of wheat seedlings and apple trees must therefore be attributed to effects of soil drying that do not require a change in shoot water status (Davies et al. 1994). They are a clear example of **feedforward** control; however, in other species [e.g., *Pseudotsuga menziesii* (Douglas fir) and *Alnus rubra* (alder)], stomata do not respond to soil drying according to a feedforward model. When their leaf water status is manipulated in a pressure chamber, stomatal conductance responds to turgor in the leaves within minutes. In these species stomatal control is hydraulic and no chemical signal from the roots appears to be involved (Fuchs & Livingston 1996).

How do chemical messengers and hydraulic signals affect leaf conductance? To answer this question we first need to explore the mechanism of opening and closing of the stomata.

5.4.2 The Control of Stomatal Movements and Stomatal Conductance

Although the anatomy of stomata differs among species, they have a number of traits in common. First, there are two **guard cells** above a **stomatal cavity**. Because the cell walls of these adjacent cells are only linked at their distal end, they form a pore whose aperture can vary because of the swelling or shrinking of the guard cells. Next to the guard cells, there are often a number of lateral and distal **subsidiary cells** (Sharpe et al. 1987). Stomatal closure occurs when solutes are transported from the guard cells, via the apoplast, to the subsidiary cells, followed by water movement along an osmotic gradient. Stomatal opening occurs by the transport of solutes and water in the opposite direction, from subsidiary cells, via the apoplast, to the guard cells (Fig. 27).

The stomatal pore becomes wider when the guard cells take up solutes and water due to the special structure of the cells, which are attached at their distal ends, and the ultrastructure of their cell walls. The **ultrastructural features** include the radial orientation of rigid microfibrils in the walls, which allow the cells to increase in volume only in a longitudinal direction, especially toward their end. In addition, the guard cells of some species show some thickening of the cell wall bordering the pore. This may help to explain the movement of the guard cells, but the radial orientation of the microfibrils is the most important feature. The combination of the structural and ultrastructural characteristics force the stomata to open when the guard cells increase in volume. This can happen in minutes and requires rapid and massive transport of solutes across the plasma membrane of the guard cells (Raschke 1987). Which solutes are transported and how is such transport brought about?

The major ion that is transported is K^+, whose concentration may rise by $0.5 M$. K^+ may be accompanied, immediately or with some delay, by Cl^-. On the other hand, the charge may be (temporarily) balanced by negative charges produced in the guard cells, with the major one being malate produced from carbohydrate inside the guard cell (MacRobbie 1987). Guard cells also accumulate sucrose during stomatal opening, but its role is not yet

fully understood (Outlaw 1995). **Ion-selective channels** play a role in the transport of both K^+ and Cl^-, in the opening as well as the closing reaction. The channels responsible for the entry of K^+ are open only when the membrane potential is very negative. A very negative membrane potential results from the activation of an H^+-pumping ATPase in the plasma membrane of the guard cells. Activation may be due to light, involving a blue-light receptor (Sect. 5.4.4). The ion-selective channels responsible for the release of K^+ open when the membrane potential becomes less negative (Hedrich & Schroeder 1989).

We now know that ABA affects some of these K^+-selective channels. The decrease in stomatal conductance as affected by ABA involves both an inhibition of the opening response and a stimulation of the closing reaction. This is a consequence of inhibition by ABA of the channel that allows K^+ entry and activation of the channel that determines K^+ release. Calcium plays a role as "second messenger" in the inhibition of the inward channel. ABA enhances the calcium concentration in the cytosol, which in its turn inhibits the inward channel. The outward channel is unaffected by $[Ca^{2+}]$ (Mansfield & McAinsh 1995).

Irradiance, the CO_2 concentration, and humidity of the air as well as water stress affect stomatal movements. There are photoreceptors in stomatal cells, which perceive a certain wavelength, thus affecting stomatal movements. We know very little about the exact mechanisms and the transduction pathways between perception of the environmental signal and the ultimate effect: stomatal movement and diurnal variation in leaf conductance. Even within one species, there can be drastically different diurnal courses of stomatal conductance at different times of year (associated with very different relative water contents and leaf water potentials). In addition, the leaf water potential at which leaf cells start to plasmolyze (the turgor loss point) can change through a season (Fig. 28). Such a change in turgor loss point must be associated with changes in elastic modulus (Table 4, Fig. 8) (see also Sect. 5.4.7 of this chapter).

5.4.3 Effects of Vapor Pressure Difference or Transpiration Rate on Leaf Conductance

Exposure of a single leaf or a whole plant to dry air is expected to increase transpiration because of the greater vapor pressure difference between the leaf and the air. Such a treatment, however, may also decrease stomatal conductance and hence affect transpiration (Lange et al. 1971). These effects

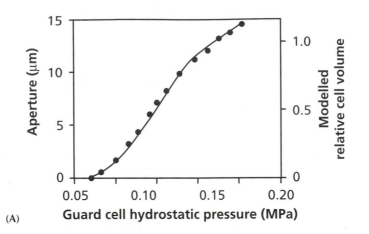

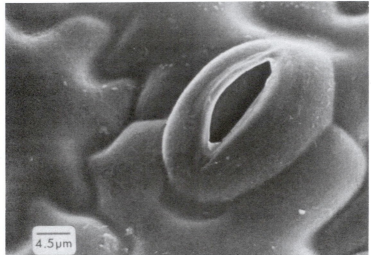

(A)

(B1)

(B2)

FIGURE 27. (A) Stomatal aperture and cell volume as a function of the guard cell hydrostatic pressure. The pressure in the cells of *Tradescantia virginiana* was controlled with a pressure probe apparatus after the guard cells had been filled with silicon oil (Franks et al. 1995). Copyright Blackwell Science Ltd. (B) Electron micrographs of an open and a closed stoma of the abaxial surface of a leaf of *Vicia faba* (broad bean). Leaves were plunge-frozen in liquid nitrogen and their surfaces were observed still frozen in a cryoscanning electron microscope (courtesy M.E. McCully, Biology Department, Carleton University, Ottawa, Canada).

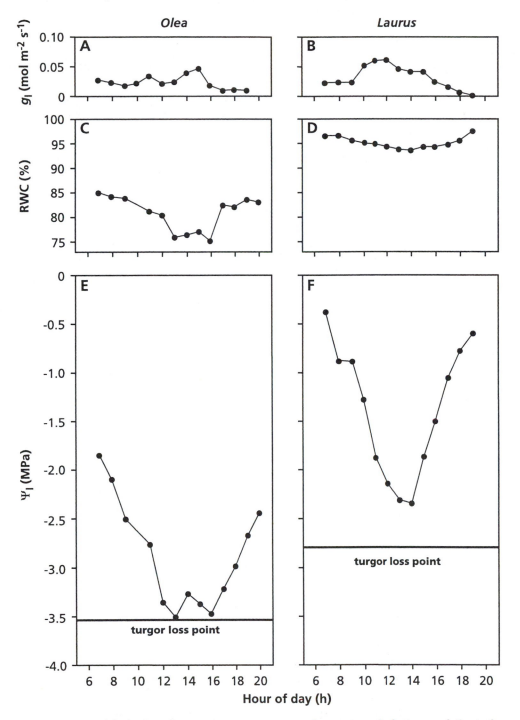

FIGURE 28. Time course of the leaf conductance to water vapor, the relative water content (RWC) of the leaves and the leaf water potential for two Mediterranean tree species; the relatively drought-tolerant *Olea oleaster* and and the less tolerant *Laurus nobilis*. RWC is defined as the amount of water per unit plant mass relative to the amount when the tissue is fully hydrated. Measurements were made in September (dry season) (Lo Gullo & Salleo 1988). Copyright Trustees of The New Phytologist.

on transpiration are readily appreciated when considering the equation introduced in Section 2.3.2 of the chapter on photosynthesis:

$$E = g_w(e_i - e_a)/P \qquad (7)$$

where g_w is the leaf conductance for water vapor transport; e_i and e_a are the water vapor partial pressures in the leaf and air, respectively, and P is the atmospheric pressure. Which environmental factors affect the vapor pressure difference between leaf and air and how can stomata respond to humidity?

The vapor pressure inside the leaf changes with leaf temperature. As temperature rises, the air can contain more water vapor, and evaporation from the wet surfaces of the leaf cells raises the vapor pressure to saturation. This is true for leaves of both well-watered and water-stressed plants. The air that surrounds the plant can also contain more humidity with rising temperature, but water vapor content of the air typically rises less rapidly than that of the leaf. If the vapor pressure outside the leaf remains the same, then the vapor pressure difference between leaf and air increases. This enhances

the leaves' transpiration in proportion to the increased vapor pressure difference (VPD), unless stomatal conductance declines. Such a decrease in stomatal conductance does indeed occur in some species [e.g., *Vigna unguiculata* (cowpea) and *Helianthus nuttallii*] (Fig. 29). In other species [e.g., *Prunus dulcis* (almond)] the stomata show very little response to air humidity (Schulze et al. 1987). Even in those species where the stomata do respond to humidity, however, this response is not always observed and much remains to be learned about the mechanism that accounts for this interesting phenomenon.

Note that as in Section 2.3 of the chapter on photosynthesis we use absolute values of water vapor in the air rather than relative humidity or water potential. The RH of the air is the absolute amount of water vapor (p) in the air as a proportion of the maximum amount of water vapor that can be held at that temperature (p_o). The water potential of the air relates to the relative humidity, as out lined in Box 5:

$$\psi_{air} = RT/V_w{}^o \ln p/p_o \qquad (8)$$

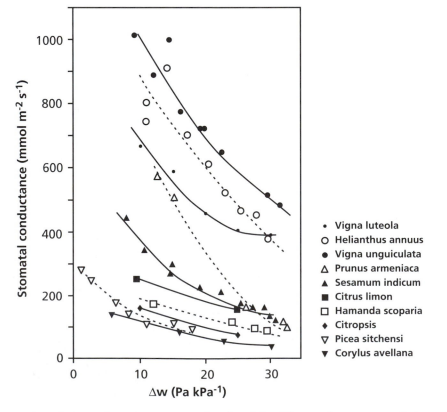

FIGURE 29. Stomatal conductance for water vapor as a function of the difference in water vapor between leaf and air (Schulze & Hall 1982).

where V_w° is the molar volume of water. For air with a temperature of 293 K and a relative humidity of 75%, the water potential $\psi_{air} = -39$ MPa (using the value for the molar volume of water at 293 K of $18 \ 10^{-6} \, m^3 \, mol^{-1}$). Air with a lower RH has an even more negative water potential. This shows that water potentials of air that contains less water vapor than the maximum amount are extremely negative (Box 5). This negative water potential of the air is the driving force for transpiration. When describing the transport of water in different parts of the soil–plant–atmosphere continuum, it is essential to use the concept of water potential. For an analysis of leaf gas exchange, however, it tends to be more appropriate to express the driving force for transpiration in terms of the vapor pressure difference between leaf and air, as is done for the diffusion of CO_2 from air to the intercellular spaces inside the leaf (Sect. 2.2.2 in the chapter on photosynthesis).

To further elucidate the mechanism that accounts for stomatal responses to humidity, transpiration was measured in several species using normal air and a helium:oxygen mixture (79:21 v/v, with CO_2 and water vapor added) (Mott & Parkhurst 1991). Because water vapor diffuses 2.33 times faster in the helium–oxygen mixture than it does in air, VPD between the leaf and the air at the leaf surface can be varied independently of the transpiration rate, and vice versa. The results of these experiments were consistent with a mechanism for stomatal responses to humidity that is based on the rate of water loss from the leaf. It suggests that stomata do not directly sense and respond to either the water vapor concentration at the leaf surface or the VPD between the leaf interior and the leaf surface.

The mechanism that accounts for the stomatal **response to humidity** of the air or **transpiration rate** is unknown (Bunce 1997, Monteith 1995). It can even be demonstrated in epidermal strips, isolated from the mesophyll. It is not universal, however, and it may even vary for one plant throughout a day (Franks et al. 1997). The consequence of this phenomenon is that a decrease in vapor pressure of the air has less effect on the leaf's water potential and relative water content than expected from the increase in vapor pressure difference between leaf and air. Stomatal response to humidity, therefore, allows an apparent **feedforward response** (Cowan 1977, Franks et al. 1997). It enables a plant to restrict excessive water loss before it develops severe water deficits and may enhance the ability of plants to use soil water supplies efficiently. The stomatal response to humidity inevitably reduces the

intercellular CO_2 pressure in the leaf, p_i, in response to low humidity and hence the rate of CO_2 assimilation. A compromise somehow, has to be reached, as will be discussed in Section 5.4.7 of this chapter.

5.4.4 Effects of Irradiance and CO_2 on Leaf Conductance

About a century ago, Francis Darwin (1898) already noted that the surface of a leaf facing a bright window had open stomata, whereas the surface away from the window had closed stomata. When he turned the leaf around, the stomata, which were closed before, opened. The ones that were open, then closed. Since Darwin's observation an overwhelming amount of evidence accumulated showing that stomata respond to light (Sharkey & Ogawa 1987). In Section 4.2 of the chapter on photosynthesis, we discussed the rapid response of stomata in plants exposed to sun flecks. The response to light ensures that stomata are only open when there is the possibility to assimilate CO_2. In this way water loss through transpiration is minimized. How, then, do stomata perceive the light and how is this subsequently translated into a change in stomatal aperture?

There are basically two mechanisms by which stomata respond to light. The *direct* response involves specific pigments in the guard cells. In addition, guard cells respond to the intercellular CO_2 concentration, which will be reduced by an increased rate of photosynthesis. This is the *indirect* response. A third mechanism might involve the transmission of some "factor" from the photosynthesizing mesophyll cells to the guard cells, but there is no direct evidence for this (Sharkey & Ogawa 1987).

The light response of guard cells is largely to blue light (with a peak at 436 nm) mediated by a blue-light photoreceptor, containing a flavin group (Zeiger et al. 1987). Stomata also open in response to red light (with a peak at 681 nm), which is perceived by chlorophyll. It is not certain if the red-light effects are mediated through photosynthesis because many guard cells lack the capacity to photophosphorylate and have little or no Calvin-cycle activity, due to the (virtual) absence of Rubisco and other Calvin-cycle enzymes (Sharkey & Ogawa 1987). Using chlorophyll fluorescence, however, it has been shown that guards cells do photophosphorylate and also have some Rubisco activity; this might be enough to have a regulatory role (Cardon & Berry 1992). It appears that the blue-light receptor has effects on biochemical events, such as an enhancement of PEP-carboxylase, which catalyzes malate formation. Blue light also affects K^+ channels in the plasma membrane of the guard cells, allowing massive and rapid entry of K^+ into the guard cells, which is the first step in the train of events that lead to stomatal opening (Assmann & Zeiger 1987).

Stomata can respond to CO_2, even when isolated or in epidermal peels, but the sensitivity varies greatly among species and depends on environmental conditions (Morison 1987). If stomata do respond, then the response is found in both light and dark conditions. Although the mechanism that accounts for the stomatal response remains unclear, it does play a major role in plant response to elevated atmospheric CO_2 concentrations. Under these conditions stomatal conductance is less than it is under present ambient conditions, enhancing the plant's photosynthetic water-use efficiency (cf. Sect. 10.2 of the chapter on photosynthesis).

5.4.5 The Cuticular Conductance and the Boundary Layer Conductance

We have so far only dealt with stomatal conductance. The **cuticular conductance** for carbon dioxide and water vapor is so low that it can be ignored in most cases, particularly when the stomatal conductance is not extremely low. It is widely believed that thick cuticles are better water barriers than thin ones, but all the experimental evidence shows this to be wrong. Cuticles are formed of three main constituents: waxes, polysaccharide microfibrils, and cutin, which is a three-dimensional polymer network of esterified fatty acids. The main barrier for diffusion is located within a waxy band, called the **skin**, whose thickness is much less than 1 μm (Kerstiens 1996).

In the continuum from the cell walls in the leaf, where evaporation takes place, to the atmosphere, there is one more step that cannot be ignored under many conditions. This is the leaf boundary layer conductance. We have already dealt with this in Section 2.3 of the chapter on photosynthesis and will come back to it in the chapter on the plant's energy balance.

5.4.6 Leaf Traits That Affect Leaf Temperature and Leaf Water Loss

As discussed in Section 5.4.3 of this chapter, leaf temperature affects the water vapor pressure inside the leaf; therefore, it is expected to affect transpira-

tion. At increasing irradiance, leaf temperatures might rise and enhance transpiration enormously. Plants, however, have mechanisms to minimize these effects. For example, water stress may cause **wilting** in large-leafed dicots even in moist soils (Chiariello et al. 1987) or **leaf rolling** in many Gramineae. The latter is associated with the presence of **bulliform** cells, which are large epidermal cells with thin anticlinal walls. A decline in relative water content reduces the volume of these cells to a greater extent than that of the surrounding cells, so that the leaves roll up. As a result, less radiation is absorbed, the boundary layer conductance of the adaxial surface is decreased, and further development of water stress symptoms is reduced (cf. Sect. 2 of the chapter on the plant's energy balance). Leaf rolling is probably a consequence of the relatively large elasticity of the cell walls and associated water relations of the bulliform cells compared with other epidermal leaf cells.

Leaf movements (heliotropisms) may also reduce the radiation load, as discussed in Section 2.2 of the chapter on the plant's energy balance. Such leaf movements require a leaf joint, or **pulvinus** at the base of the petiole or leaf sheath (Satter & Galston 1981). Solutes, especially K^+, are actively transported from one side of the pulvinus to the other (Fig. 30). Water follows passively and the turgor is increased, which causes movement of the petiole or leaf sheath.

Leaf movements have been studied in detail in *Glycine max* (soybean) (Oosterhuis et al. 1985) and in *Melilotus indicus* (Schwartz et al. 1987). In these plants, as in other Fabaceae and in *Mimosa* species, the (blue) light stimulus giving rise to leaf movement is perceived in the pulvinus itself (Vogelmann 1984). In species belonging to the Malvaceae, perception occurs in the leaf lamina (Schwartz et al. 1987). Both the adaxial (upper) and the abaxial (lower) side of the pulvinus of *Melilotus indicus* perceive the light stimulus. Light perception at the adaxial side causes the pulvinus to move upward, whereas perception of light at the abaxial side induces the pulvinus to cause a downward movement (Fig. 31).

Leaf movements of *Phaseolus vulgaris* (common bean) depend on air temperature (Fu & Ehleringer 1989). The effect of these leaf movements is that at a low air temperature the leaf is oriented in such a way as to enhance the incident radiation, whereas the opposite occurs at a high air temperature. As a result, the leaf temperature is closer to the optimum for photosynthesis (Fig. 32). The air temperature that induces the leaf movements in bean is perceived in the pulvinus, rather than in the leaf itself.

Other acclimations and adaptations that affect plant transpiration will be discussed in Section 2.2 of the chapter on the plant's energy balance.

5.4.7 Stomatal Control: A Compromise Between Carbon Gain and Water Loss

As first discussed in the chapter on photosynthesis, leaves are faced with the problem of a compromise between maximization of photosynthesis and minimization of transpiration. At a relatively

high leaf conductance when photosynthesis no longer increases linearly with p_i, transpiration increases more rapidly with increasing leaf conductance than does photosynthesis. The ratio of the change in E and the change in A (termed λ) also increases with increasing leaf conductance (Cowan 1977).

Figure 33 gives the rate of transpiration as a function of the rate of assimilation and the time of the day, assuming different values for leaf conductance or for λ. If we assume that stomata are regulated only to **maximize** carbon gain, then this produces a transpiration curve with one peak on the contour of the surface. The peak is due to the high water

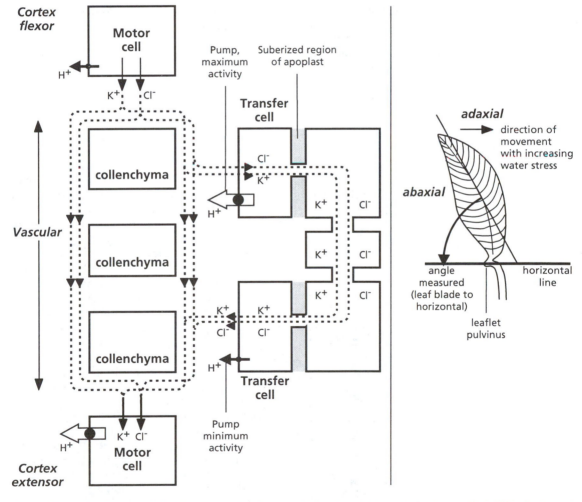

FIGURE 30. A flow diagram of the direction and pathways of net K⁺, Cl⁻, and H⁺ movements in a pulvinus during leaflet opening (after Satter & Galston 1981; with permission, from the Annual Review of Plant Physiology and Plant Molecular Biolology, Vol. 32, copyright 1981, by Annual Reviews Inc.) and the location of the pulvinus in *Glycine max* (soybean) (after Oosterhuis et al. 1985).

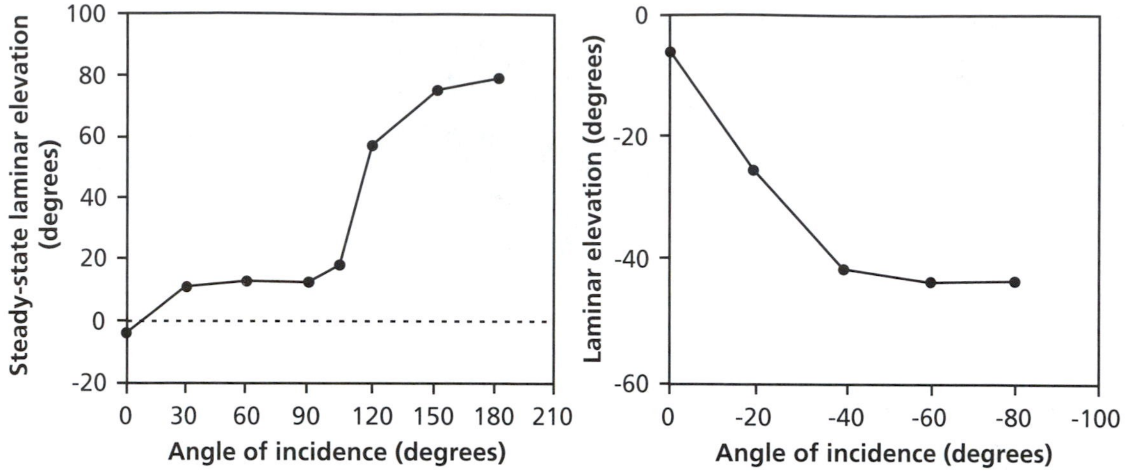

FIGURE 31. The orientation of the terminal leaf of a composite leaf of *Melilotus indicus*, as dependent on the angle of the incident radiation. An angle of 0° and +180° of the light refers to light in the horizontal plane, from the tip to the base of the leaf and from the base to the tip, respectively. An angle of +90° and −90° refers to light in the vertical plane, coming from above and below, respectively. For the leaf orientation the same terminology is followed. (Left) The pulvinus is irradiated from above. (Right) The pulvinus is irradiated from below (Schwartz et al. 1987). Copyright American Society of Plant Physiologists.

vapor pressure difference between the leaf and the atmosphere when the radiation level is high during the middle of the day. Assuming **optimization** of stomatal regulation gives a curve with two peaks, when λ is small (i.e., when carbon assimilation is an important criterion for optimization). When λ is large (larger than the example in Fig. 3, but stomatal conductance regulated to optimize carbon gain and water loss), a curve with only one peak is found. That is, the optimization model predicts that when plants function in an environment where water is scarce and/or the demand for transpiration is high, they are expected to operate according to the two-peaks curve. When they are well supplied with water and the transpiration demand is moderate, they are expected to operate according to the one-peak curve. The two-peak curve is achieved by (partial) closure of the stomata during that time of the day when the evaporative demand is highest, due to a large water vapor pressure difference between the leaf and the air.

How should stomata be regulated so as to maximize the fixation of CO_2 with a minimum loss of water? The optimization theory for stomatal action is based on the following assumption: Stomatal action is such that for each amount of CO_2 absorbed, the smallest possible amount of water is lost. The mathematics to solve such a problem requires a sophisticated approach, which will not be included here (Cowan 1977). The solution, however, can be presented very briefly: For each infinitesimally small change of E at a certain E, the change in A is constant, λ (Fig. 33).

The theoretical curves of Figure 33 agree with observations on both C_3 and C_4 plants in dry envi-

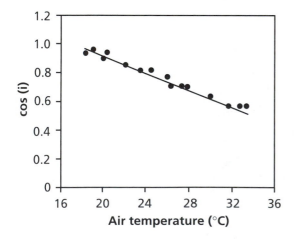

FIGURE 32. The correlation between the cosine of the angle between the incident light beam and the vector normal to the leaf lamina of *Phaseolus vulgaris* (common bean) as dependent on air temperature. Irradiance, atmospheric CO_2 concentration, and vapor pressure deficit were kept constant (Fu & Ehleringer 1989). Copyright American Society of Plant Physiologists.

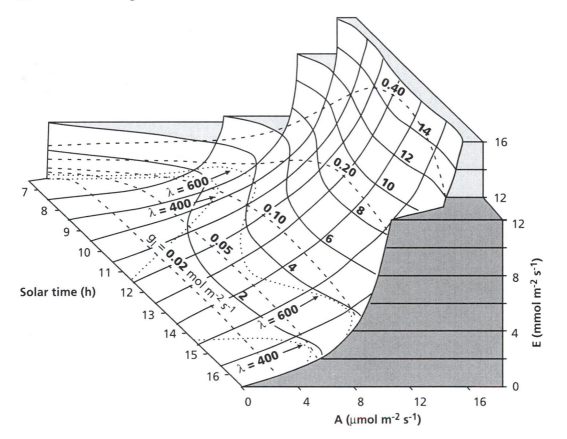

FIGURE 33. Calculated rates of evaporation (E), as a function of the rate of CO_2 assimilation (A) and time of the day, assuming certain characteristics of leaf metabolism and environment. The magnitudes for E are given on the contours of the surface. The broken lines are trajectories on the surface giving the diurnal variation in E and A for particular constant magnitudes of leaf conductance (g_l; 0.02, 0.05, 0.10, 0.20 or 0.40 mol m^{-2}s^{-1}). The dotted lines are trajectories for which λ is constant (400 or 600) (Cowan 1977). By permission of the Australian Academy of Science.

ronments, where curves with two peaks are quite common. When the water supply is favorable and the vapor pressure deficit is moderate, however, curves with only one peak are found (i.e., there is no partial midday stomatal closure) (Fig. 34). This has led to the conclusion that stomatal conductance is regulated so as to optimize carbon gain and water loss. It should be kept in mind, however, that this optimization approach, while very attractive for explanation of stomatal behavior, is teleological in nature; it has no mechanistic basis and is not easily used for predictive purposes.

Constancy of λ does not have to be the result of the action of stomata, but it may also be achieved by a specific leaf orientation. For example, vertical leaves absorb least radiation during the middle of the day as opposed to horizontal ones. A vertical orientation of leaves is typically associated with hot and dry places close to the equator. A horizontal leaf orientation is common in temperate regions, further away from the equator. Some leaves have the ability to orientate their leaves in response to environmental factors, including the angle of the incident radiation and leaf temperature. Such heliotropic leaf movements may also lead to the constancy of λ.

5.4.8 Water Storage in Leaves

Succulents store water in their leaves (Kluge & Ting 1978), often in specialized cells. For example, in the epiphytic *Peperomia magnoliaefolia*, water storage occurs in a multiple epidermis (**hydrenchyma**), just under the upper epidermis, which may account for 60% of the leaf volume (Fig. 35). The water-storage tissue of the epiphytic Bromeliad, *Guzmania monostachia*, may amount to as much as 67% of the total leaf volume on exposed sites; however, its vol-

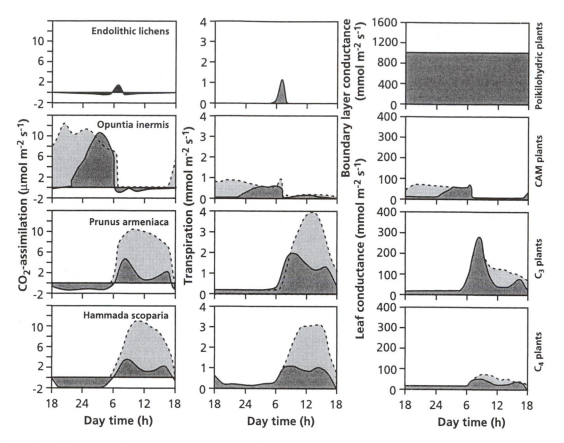

FIGURE 34. Diurnal variation in the rate of CO_2 assimilation (left), transpiration (middle), and leaf conductance (right) for four different plant types. Light shading (dashed line) shows wet season; dark shading (solid line) shows dry season (Schulze & Hall 1982).

ume is only 24% in the shade (Maxwell et al. 1992). The hydrenchyma in *Peperomia magnoliaefolia* consists of large cells with large vacuoles, but lacking chloroplasts. Their radial walls are thin and "collapse" when the cells lose water. Beneath the hydrenchyma is a layer of smaller cells that contain many chloroplasts: the **chlorenchyma**.

When the leaves lose water, the dehydration of the chlorenchyma is much less than that of the hydrenchyma. The hydrenchyma appears to function as a reservoir for water lost through transpiration. This allows the chlorenchyma to remain photosynthetically active. During water loss, both solutes and water move from the hydrenchyma to the chlorenchyma; therefore, they maintain a high water content in these cells. The total amount of water in the hydrenchyma of *P. magnoliaefolia* exceeds $1\,kg\,m^{-2}$ leaves. At an average transpiration rate of $0.2\,mmol\,H_2O\,m^{-2}s^{-1}$ during 12 hours of the day, this stored water allows the plant to continue

to transpire at the same rate for about 1 week. The stored water allows the plant to maintain a positive carbon balance in the absence of water uptake from the environment for several days.

5.5 Aquatic Angiosperms

Aquatic angiosperms are perhaps comparable to whales: They returned to the water, taking with them some features of terrestrial organisms. In perennially submerged angiosperms, where the pressure in the xylem is never negative, the xylem is somewhat "reduced." The structure is like that of resin ducts. The xylem ducts in submerged aquatics often have thin walls, whereas "conventionally" thick-walled xylem cells are found in aquatics whose tops are able to emerge from the water.

It is well-established that water transport from roots to leaves is possible in submerged aquatic angiosperms, and that it is important in the transport of nutrients and root-produced phytohormones to the stem and leaves. The roots of most aquatics serve the same role as those of terrestrial plants as the major site of nutrient uptake and in the synthesis of some phytohormones. In submerged angiosperms the driving force for xylem transport cannot be the transpiration, and

root pressure is the most likely mechanism (Pedersen & Sand-Jensen 1997, Zimmermann 1983).

6. Water-Use Efficiency

Water-use efficiency refers to the amount of water lost during the production of biomass or the fixation of CO_2 in photosynthesis. It is defined in two ways. First, the **water-use efficiency of productivity** is the ratio between (above-ground) gain in biomass and loss of water during the production of that biomass; the water loss may refer to total transpiration only, or include soil evaporation. Second, as explained in Section 5.2 of the chapter on photosynthesis, the **photosynthetic water-use efficiency** is the ratio between carbon gain in photosynthesis and water loss in transpiration, A/E. Instead of the ratio of the rates of photosynthesis and transpiration, the leaf conductances for CO_2 and water vapour can be used, g_c/g_w. As expected, there is generally a good correlation between the water-use efficiency of productivity and the photosynthetic water-use efficiency.

As explained in Box 2 in the chapter on photosynthesis, the carbon isotope composition of plant biomass is largely determined by the biochemical discrimination of Rubisco and by the fractionation during diffusion of CO_2 from the atmosphere to the intercellular spaces. The higher the stomatal conductance, relative to the activity of Rubisco, the less $^{13}CO_2$ ends up in the photosynthates and hence in plant biomass. This is the basis of the generally observed correlation between $\delta^{13}C$-values and both the intercellular partial pressure of CO_2 (p_i) and photosynthetic water-use efficiency (Fig. 36). As a result, **$\delta^{13}C$-values** can be used to assess a plant's water-use efficiency; however, differences in water-use efficiency determined at the leaf level may be reduced substantially at the canopy level, as further explained in the chapter on scaling-up (De Pury 1995).

A plant's water-use efficiency depends both on stomatal conductance and on the difference in vapor pressure in the leaf's intercellular air spaces and that in the air. Because temperature affects the vapor pressure in the leaf, temperature also has a pronounced effect on plant water-use efficiency, A/E.

There are major differences in photosynthetic water-use efficiency (A/E) between C_3, C_4 and CAM plants, as well as smaller differences among species of the same photosynthetic pathway (cf.

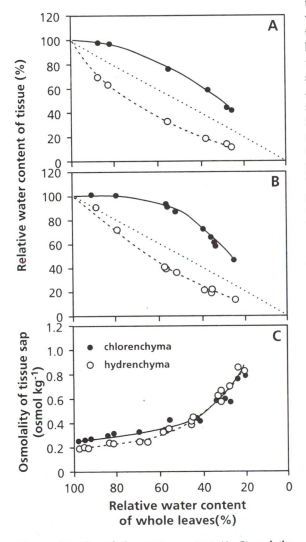

FIGURE 35. The relative water content (A, B) and the osmolality (C) of the hydrenchyma and chlorenchyma sap of *Peperomia magnoliaefolia* as dependent on the relative water content of whole leaves. The data in A and B refer to results obtained with detached and attached leaves, respectively; the broken line gives the relative water content if both tissues would lose water at the same rate (Schmidt & Kaiser 1987). Copyright American Society of Plant Physiologists.

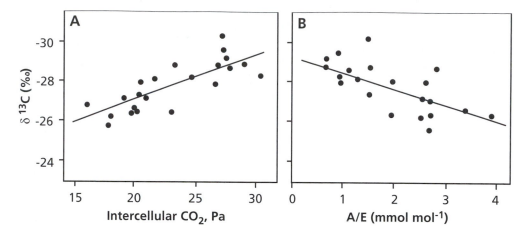

FIGURE 36. The relationship between carbon isotope composition ($\delta^{13}C$) and (A) average intercellular CO_2 concentration, and (B) daily photosynthetic water-use efficiency, assimilation/transpiration (A/E). The data points refer to mistletoes and host plants in central Australia (Reprinted with permission from Ehleringer et al. 1985). Copyright 1985 American Association for the Advancement of Science.

Sect. 5.2 of the chapter on photosynthesis). Xylem-tapping hemiparasitic plants have the lowest water-use efficiency, as discussed in Section 3 of the chapter on parasitic associations (Table 8).

7. Water Availability and Growth

During incipient water-stress, specific genes are induced (Fig. 37). Some water-stress–induced gene products are thought to protect cellular structures from the effects of water loss. They are predominantly hydrophilic and are probably located in the

TABLE 8. The photosynthetic water-use efficiency of plants with different photosynthetic pathway* and belong to different functional groups.[†]

Functional type	Water-use efficiency (mmol mol^{-1})
CAM-plants	4–20
C$_4$ plants	4–12
Woody C$_3$ plants	2–11
Herbaceous C$_3$ plants	2–5
Hemiparasitic C$_3$ plants	0.3–2.5

Source: Kluge & Ting 1978; Morrison 1993; Osmond et al. 1982; Shah et al. 1987.
*C_3, C_4 and CAM; for CAM-plants the high values refer to gas exchange during the night and the low values to the light period.
[†]All species are nonparasitic, unless stated otherwise, grown at an ambient CO_2 partial pressure of around 35 Pa and not exposed to severe water stress.

cytoplasm, likely involved in the **sequestration of ions**, which are concentrated during cellular dehydration. They are amphiphilic α-helices (i.e., they contain both hydrophilic and hydrophobic parts). The hydrophilic part binds ions, thus preventing damage, whereas the hydrophobic part is associated with membranes. Other proteins have many charged amino acids and are thought to have a large water-binding capacity. Some of the proteins may **protect other proteins**, by replacing water, be involved in **renaturation** of unfolded proteins, or have a **chaperon** function (i.e., allow the transport of proteins across a membrane, on their way to a target organelle) (Bray 1993).

At a low soil water potential, the rate of photosynthesis decreases, at least partly due to a decline in stomatal conductance, as discussed in Section 5.1 of the chapter on photosynthesis. As pointed out in Section 5.3 of the chapter on growth and allocation, however, effects of water stress on growth are largely accounted for by physiological processes other than photosynthesis. Many processes in the plant are far more sensitive to a low water potential than are stomatal conductance and photosynthesis. The growth reduction at a low soil water potential is therefore more likely due to inhibition of more sensitive processes such as **cell elongation** and **protein synthesis**; these processes are, at least partly, also controlled by **ABA** (Box 8 in the chapter on growth and alloeation) (Bradford & Hsiao 1982).

Above-ground plant parts respond more strongly to a decreased soil water potential than do roots. Is this perhaps due to a much greater effect of the low water potential on growth of leaves, as compared with that of the roots, simply because they are closer to the source of water? Do roots and leaves, on the other hand, have a different sensitivity for the water potential? In *Zea mays* (maize), the soil water potential causing growth reduction is indeed lower (more negative) for roots than it is for leaves, but this does not provide a conclusive answer to our question. In Section 5.3 of the chapter on growth and allocation, this problem will be addressed more elaborately. Lowering the water potential enhances the transport of assimilates to the roots, which is probably due to the growth reduction of the leaves. Because photosynthesis is less affected than leaf growth, sugar import as well as

root growth may be enhanced, with the overall effect that the leaf area ratio decreases upon a decrease in soil water potential. That is, the evaporative surface is reduced, relative to the water-absorbing surface.

8. Adaptations to Drought

Plants have adapted to a lack of water in the environment either by avoiding drought or by tolerating it. Desert annuals and drought-deciduous species **avoid** drought by remaining dormant until

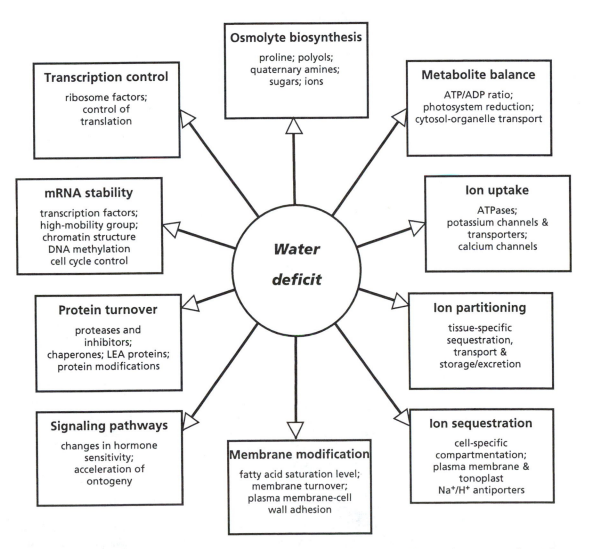

FIGURE 37. A plant's response to water deficit. When water is in short supply, many cellular responses change. These changes allow for continued growth under water stress, although possibly at a lower rate (after Bohnert et al. 1995). Copyright American Society of Plant Physiologists.

water arrives. Other plants in dry environments avoid drought by producing roots with access to deep ground-water. The alternative strategy is to **tolerate** drought. Tolerance mechanisms are found in evergreen shrubs and in plants that can dry out in the absence of water and "resurrect" upon exposure to water. Many plants in dry habitats exhibit intermediate strategies. For example, succulents, especially those with the CAM pathway (Sect. 10 of the chapter on photosynthesis), minimize effects of drought by opening their stomates at night and concentrating their activity in wet seasons, but they also have many characteristics typical of drought-tolerant species.

8.1 Desiccation-Avoidance: Annuals and Drought-Deciduous Species

A large proportion of the plant species in deserts are annuals with little or no physiological tolerance of drought. As will be further discussed in Section 2.2 of the chapter on life cycles, seeds of these species have water-soluble **germination inhibitors** so that germination occurs only after rains that are large enough both to leach out the inhibitors and to support growth. These species grow quickly following germination, often completing their life cycle in 6 weeks or less. These plants typically have high rates of photosynthesis and a high leaf area ratio to support their rapid growth and have correspondingly high stomatal conductances and transpiration rates (Mooney et al. 1976). Following the rain, ungerminated seeds synthesize more germination inhibitors, enabling them to "measure" the size of the next rain event.

The most obvious mechanism of acclimation to drought is perhaps a decrease in canopy leaf area. This can be rapid, through **leaf shedding**, or more slowly, through adjustments in allocation pattern (see Sect. 5.3 of the chapter on growth and allocation). In general, drought-deciduous species have high stomatal conductance and high rates of photosynthesis and transpiration when water is available but lose their leaves and enter **dormancy** under conditions of low water potential. As with desert annuals, their leaves exhibit no physiological adaptations for drought tolerance or water conservation. The advantages of a drought-voiding strategy (high rates of photosynthesis and growth under favorable conditions) are offset by the cost of producing new leaves in each new growth period. Some species (e.g., *Fouqueria splendens* in the deserts of North America) produce and lose leaves as many as six times per year. There is typically a 2 to 4 week lag

between onset of rains and full canopy development of drought-deciduous species. It is, therefore, not surprising that drought-tolerant evergreens displace drought-deciduous species as rains become more frequent and water availability increases (Fig. 38; Mooney & Dunn 1970).

Some desert plants, known as **phreatophytes**, produce extremely deep roots that tap the water table. Like the desert annuals and drought-deciduous shrubs, these plants generally have high rates of photosynthesis and transpiration with little capacity to restrict water loss or withstand drought. For example, mesquite (*Prosopis*) com-

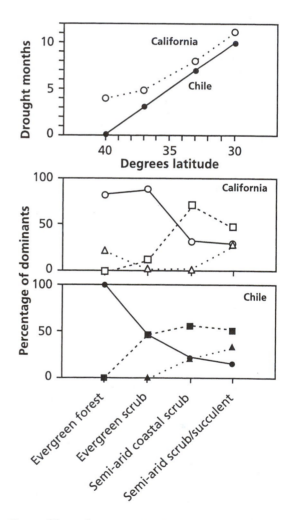

FIGURE 38. Leaf types of the dominant plants in major vegetation types along a latitudinal drought gradient in California and Chile. Leaf types are evergreen (circles), deciduous (squares), and succulent (triangles) (Mooney & Dunn 1970). Copyright by The University of Chicago.

monly occupies desert washes in the southeastern United States, where there is little surface water but where groundwater is close enough to the surface that seedlings can occasionally produce deep enough roots to reach this groundwater in wet years. In the same area, *Tamarix*, which is an exotic phreatophyte, has lowered the water table sufficiently through its high transpiration rate that other species of intermediate rooting depth are being eliminated (Van Hylckama 1974).

8.2 Dessication-Tolerance: Evergreen Shrubs

Most evergreen shrubs are exposed to water stress during part of the year, be it during the summer in a Mediterranean-type climate, or in winter in cooler climates. The evergreen and scleromorphic leaves of Mediterranean shrubs tend to have low rates of photosynthesis, compared to the leaves of deciduous shrubs. There is no a priori reason, however, for scleromorphic leaves to have a low photosynthetic capacity-and some *Eucalyptus* and *Banksia* species in Australia have a particularly high A_{max}.

Relatively drought-tolerant species (for example *Olea oleaster* in Fig. 28) typically have **a low maximum stomatal conductance** (and therefore low rates of transpiration and photosynthesis) and a low relative water content. They withstand lower water potentials, however, before stomatal closure and before loss of turgor because they have high solute concentrations in live cells (low osmotic potential) and a high resistance to cavitation of xylem. Natural selection leading to scleromorphic and evergreen growth habits are complex. A low nutrient availability can also promote the evolution of evergreen, scleromorphic leaves (Loveless 1961, 1962).

Mediterranean shrubs are also characterized by "**dual root systems**" that have both deep tap roots and shallow feeder roots. This architecture allows access to semi-permanent groundwater supplies as well as to surface precipitation. Other special features include the presence of vasicentric tracheids. These are tracheory elements with bordered pits about as large and frequent as those in vessel elements. Vasicentric tracheids may act as subsidiary conducting systems for water transport in the xylem. They lie adjacent to many vessels, providing a safety factor to allow continued conductance when xylem vessels have cavitated. Another adaptive xylem structure is the widespread occurrence of vascular tracheids, which is thought to provide a

safe conductive tissue at the time of maximum drought stress (Rundel 1995).

8.3 "Resurrection Plants"

An extreme case of desiccation tolerance of whole plants is that of the **resurrection plants** or "poikilohydric plants." Even after their protoplasm has dried out to the extent that the water potential of the cells is in equilibrium with dry air (with a relative humidity of 20 to 40% or even less), they can almost fully restore their physiological activity (Gaff 1981). Their dry, shriveled, and seemingly dead leaves regain turgor in less than 1 day after a shower, which makes the term **resurrection** most appropriate. Many mosses and ferns and some Angiospermae, including woody species (e.g., *Myrothamnus flabellifolius*) are characterized as resurrection plants. They are found in South Africa, America, and Australia in environments where droughts occur regularly (e.g., on rocky substrates).

There are two strategies among resurrection angiosperms:

1. Those that lose chlorophyll and break down their chloroplasts upon drying (**poikilochlorophyllous**)
2. Those that retain some or all of their chlorophyll and chloroplast ultrastructure (**homoiochlorophyllous**)

The poikilochlorophyllous species tend to take longer to recover than do the homoiochlorophyllous ones because the poikilochlorophyllous ones have to reconstitute their chloroplasts (Sherwin & Farrant 1996). The poikilochlorophyllous plants are all monocotyledonous, but some of the grasses are homoiochlorophyllous. The two strategies may have evolved in response to light stress, which is exacerbated during dehydration and rehydration. While the leaf tissue is dehydrating, dry, or rehydrating, light absorption should be minimal and the energy that is absorbed must be dissipated. The leaves of homoiochlorophyllous plants tend to roll or curl and produce protective pigments (e.g., anthocyanins) that act as screens. The poikilochlorophyllous plants tend to have elongate leaves that can only fold, thus leaving a greater surface exposed to light (Sherwin & Farrant 1998).

The exact nature of the reactivation of the physiological processes is not yet fully understood. The following must generally hold:

1. Any damage incurred during the drying phase is not lethal
2. Some of the metabolic functions are maintained in the dry state, to an extent that they can be deployed upon rewetting
3. Any damage incurred is repaired during rehydration

Even though the dehydrated homoiochlorophyllous resurrection plants may have lost most of their green color, their thylakoid membranes, chlorophyll complexes, mitochondria, and other membrane systems remain intact. Elements of the protein-synthesizing machinery, including mRNA, tRNA, and ribosomes, also remain functional. Using inhibitors of transcription and translation shows that membrane protection and repair does not require transcription of new gene products or translation of existing transcripts. Full recovery of the photosynthetic apparatus in the homoiochlorophyllous *Craterostigma wilmsii* requires protein synthesis, but not gene transcription. On the other hand, for the poikilochlorophyllous *Xerophyta humilis* both transcription and translation are required for full recovery (Dace et al. 1998). Recovery of woody resurrection plants, which need to restore their embolized xylem vessels before rewetting the leaves, is relatively slow. Bubbles appear to dissolve, enabling rapid refilling of the xylem conduits, whereas root pressure appears to play a minor role in this process (H.W. Sherwin, pers. comm.).

A large number of enzymes associated with carbon metabolism remain intact in the dry state, as found for *Selaginella lepidophylla* from the Chihuahuan Desert in Texas (Table 9). About 24 hours after rewetting, the plants have regained their green appearance, and rates of photosynthesis and respiration are again close to those of normal wet plants. At that time, the activity of nine out of the ten measured enzymes has increased, compared with that in plants in a dehydrated state. On average, 74% of the enzyme activity remains in the dry phase; however, this value is only 27% for the NADPH-dependent triose phosphate dehydrogenase in *S. lepidophylla*. In addition in a bryophyte, *Acrocladium cuspidatum*, the activity of this photosynthetic enzyme was reduced more than that of all other enzymes tested. It appears that enzymes involved in respiratory metabolism are conserved better than are those associated with photosynthesis.

The increase in activity of the enzymes that were not fully conserved in the dry phase may involve de novo protein synthesis (NADP-dependent triose phosphate dehydrogenase, Rubisco), but this is not invariably the case (pyruvate kinase) (Table 10). Rapid de novo synthesis, in addition to the maintenance of functional enzymes, is clearly important in the reactivation phase after rewetting. Maintenance of the protein-synthesizing machinery clearly therefore appears to be of vital importance.

During dehydration of the resurrection plants, as in "ordinary" plants, the phytohormone **ABA** accu-

TABLE 9. **The activity of three enzymes associated with photosynthesis and three involved in respiration.***

Enzyme	Enzyme activity (enzyme units g^{-1} DM)		Conservation (%)
	Desiccated	Hydrated	
Photosynthetic enzymes:			
Ribose-5-phosphate isomerase	7.56	9.24	82
Rubisco	0.60	0.96	62
(NADPH)Triose-phosphate dehydrogenase	0.48	1.80	27
Respiratory enzymes:			
Citrate synthase	1.76	2.05	86
Malate dehydrogenase	2.89	2.97	97
(NADH)Triose-phosphate dehydrogenase	1.13	1.40	81

Source: Harten & Eickmeier 1986.
* They were isolated from the resurrection plant *Selaginella lepidophylla*, both from dehydrated plants and 24 hours after rehydration.

TABLE 10. The effect of rehydration of the resurrection plant *Selaginella lepidophylla* on the activity of three enzymes, in the presence or absence of the different inhibitors of protein synthesis.*

Treatment	Enzyme activity (enzyme units g^{-1} DM)		
	Triose-phosphate dehydrogenase	Rubisco	Pyruvate kinase
Desiccated	1.67	0.60	0.92
24h Hydrated	4.43	0.98	1.35
24h Hydrated + CAP	2.25	0.75	1.51
24h Hydrated + CHI	4.09	0.96	1.54

Source: Harten & Eickmeier 1986.

*Chloramphenicol (CAP) inhibits protein synthesis in organelles; cycloheximide (CHI) inhibits protein synthesis in the cytosol. The effect of CAP on the activity of Rubisco agrees with the large subunit being synthesized in the chloroplast. The effect of CAP on NADP-dependent triose phosphate dehydrogenase is unexpected in that this enzyme is supposed to be synthesized in the cytosol only.

mulates. In resurrection plants, ABA induces the **transcription** of a number of genes (Bartels et al. 1990, Piatkowski et al. 1990), which encode proteins that are closely related to those that are abundantly induced during embryo maturation in the seeds of many higher plants or to some extent in water-stressed seedlings (Bartels & Nelson 1994, Ingram & Bartels 1996). The function of the proteins is not yet fully understood. They may be enzymes associated with the synthesis of compatible solutes, such as sucrose, trehalose, arbutin, and gluco-pyranosyl-glycerol (Bianchi et al. 1993, Ingram & Bartels 1996). In the small, herbaceous, homoio-chlorophyllous *Craterostigma plantagineum*, sucrose accumulates to high concentrations (up to 40% of the dry mass) while the concentration of the C8-sugar octulose declines. Sucrose and other solutes play a major role in stabilizing subcellular components, including membranes and proteins (Bartels & Nelson 1994, Ingram & Bartels 1996). The sugars ensure that the little amount of water left in the tissue occurs in a "glassy" state, like the glass in our windows, which is actually a fluid. Some of the gene products are proteins with both hydrophobic and hydrophilic zones; they may bind ions and be membrane-associated (Piatkowski et al. 1990). These may have an "osmoprotective" function, reducing potential damage by high solute concentrations. Other gene products are likely involved in carotene biosynthesis (Alamillo & Bartels 1996).

The genes expressed upon dehydration of resurrection plants are similar to those expressed at the end of the ripening of the embryo in **ripening seeds**, described as **late embryogenesis abundant**

genes, or *lea* genes. It appears that proteins involved in the survival of dehydrated embryos in dry seeds are similar to those that protect resurrection plants in their dehydrated state. Some of the genes that are expressed in resurrection plants during dehydration are also expressed in water-stressed poplar leaves, and more so in the more drought-resistant *Populus popularis* than in the less resistant *P. tomentosa* (Pelah et al. 1997).

9. Winter Water Relations and Freezing Tolerance

As discussed in Section 5.3.2 of this chapter, sub-zero temperatures may lead to the formation of air bubbles in xylem conduits, hence to **embolism**. The water in the xylem generally freezes between 0 and $-2°C$. Some water transport may still continue after embolism has occurred, although at a very low rate (around 3% of normal rates). This slow movement probably occurs either through late-wood tracheids or through cell-wall cavities (Tranquillini 1982).

Frost damage is also associated with the formation of extracellular **ice crystals** that cause dehydration of the cytoplasm and the formation of crystals inside the cells, both being associated with damage to membranes and organelles. The cells become leaky and their water potential declines sharply. Resistance mechanisms predominantly involve the prevention of the formation of intracellular ice crystals, either by restricting freezing to the extracellular compartment or by **supercooling** of

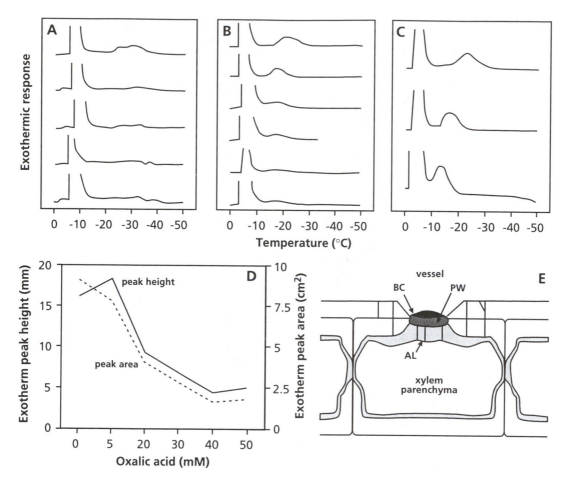

FIGURE 39. (A–C) The effect of macerase (an enzyme which hydrolyzes pectin), oxalic acid, and EGTA (which bind calcium, responsible for "cross-linking" in pectin) on the exothermal response. The left peak (which is not relevant in the present context) is due to freezing of extracellular water. The peak to the right decreases, or shifts to lower temperatures, upon removal of pectin. The data in B have been replotted in D, both as peak height and as peak area versus the concentration of oxalic acid.

(E) The structure of a pit between the xylem and a xylem-ray parenchyma cell of *Prunus persica* (peach). The pit membrane consists of three layers: an outermost black cap (BC) or toruslike layer, a primary wall (PW), and an amorphous layer (AL). The channels are meant to diagrammatically illustrate how pore size or continuity would affect the ability of a cell to exhibit deep supercooling (Wisniewski et al. 1991). Copyright American Society of Plant Physiologists.

the cellular solutions, in addition to biochemical mechanisms to withstand dehydration (Steponkus 1981).

Changes in the composition of cell walls play a major role in preventing ice formation. For example, deposition of **pectin** in the cell wall reduces the size of the microcapillaries in the walls, allowing a more negative water potential. Pectin formation in the pits between xylem and xylem-parenchyma cells closes these pores, so that water remains in the cells (Fig. 39). In spring, pectin is enzymatically removed again, coinciding with the loss of the capac-

ity to tolerate deep supercooling (Wisniewski et al. 1991). Deep supercooling is only possible to temperatures around −40°C; below that temperature ice formation occurs in the absence of crystallization nuclei.

In subarctic trees, which tolerate temperatures below −40°C, supercooling does not play a role. Ice formation starts around −2 to −5°C, but only in the cell wall. The cold acclimation that occurs in autumn is triggered by photoperiod and exposure to cool temperatures. It involves synthesis of membrane lipids with less saturated fatty acids, so they remain flexible at low temperatures, and the production of osmotically active solutes. Cells that would freeze at -3 to $-5°C$ in summer remain unfrozen to $-40°C$ in winter. At subfreezing temperatures, ice forms first in cell walls, reducing the concentration of extracellular liquid water. Water migrates out of cells along this water-potential gradient, increasing the intracellular solute concentration, which prevents intracellular freezing. The biochemical mechanisms to withstand this winter desiccation are identical to those caused by lack of water in deserts. It is therefore not surprising that species that tolerate extremely low temperatures are also highly desiccation-tolerant.

10. Salt Tolerance

Halophytes are species that typically grow in soils with high levels of NaCl and, hence, a low water potential. They accumulate NaCl in their vacuoles. By contrast, **glycophytes** have a limited capacity to transport NaCl into their vacuoles and are unable to tolerate high salinity levels. Cytoplasmic enzymes of glycophytes and halophytes are very similar with respect to their sensitivity to high concentrations of inorganic solutes (Fig. 6 in this chapter). Tolerance mechanisms of halophytes will be discussed in the chapter on mineral nutrition (Sect. 3.4).

In salt-sensitive plants, salinity affects water uptake by its effect on water potential, and it reduces the hydraulic conductance of the root cells. This effect can be ameliorated by calcium through an as-yet-unknown mechanism (Steudle 1994).

11. Final Remarks: The Message That Transpires

What have we finally learned from this chapter on water relations? First, that water is a major factor limiting plant growth in many ecosystems, but also that different species have evolved fascinating mechanisms to cope with this limiting factor, ranging from **avoidance** to **tolerance**. Tolerance at one level (e.g., of the roots) may allow drought avoidance at another (e.g., of the leaves). Plants have adapted to a limiting supply of water in their environment, but all plants, to varying degrees, can also acclimate to an environment where water is scarce.

The characteristics that enable plants to tolerate drought are highly interdependent (Table 11). To appreciate these mechanisms, a full understanding of the biophysical, physiological, and molecular aspects of plant water relations is essential. Such an appreciation is pivotal, if we aim to improve the performance of crop species in dry environments. This is not to say that other ecophysiological aspects are not of equal, or even greater, importance. In fact, vigorous early growth and early flowering may also greatly contribute to a greater water-use efficiency over the entire season.

Resurrection plants offer one of the most remarkable examples of how plants cope with a shortage of water in their environment. At one stage it may have been considered esoteric to study these peculiar plants, which would seem useless from an economic point of view. It now becomes increasingly clear, however, that resurrection plants show many similarities to ripening seeds and leaves that are

TABLE 11. Summary of characteristics of dessication-sensitive and dessication-tolerant evergreen species.

Characteristic	Dessication-sensitive species	Dessication-tolerant species
Maximum transpiration rate	high	low
Maximum photosynthetic rate	high	low
Maximum stomatal conductance	high	low
Specific leaf area	high	low
Leaf size	large	small
Leaf longevity	low	high
Potential growth rate	high	low
Root mass ratio	low	high
Leaf compatible solute concentration	low	high
Water potential at turgor loss	high	low
Stomatal regulation	iso/anisohydric	anisohydric
Safety margin for cavitation	small	large

able to cope with water stress. As such, resurrection plants both offer a model system to study water-stress resistance and they may also be the source of genes to be used to improve the performance of new crop varieties in dry environments. As so often in science, possibilities for applications emerge long after fascinating discoveries are being made on fundamental aspects of plant biology.

References and Further Reading

Alamillo, J.M. & Bartels, D. (1996) Light and stage of development influence the expression of desiccation-induced genes in the resurrection plant *Craterostigma plantagineum*. Plant Cell Environ. 19:300–310.

Assmann, S.M. & Zeiger, E. (1987) Guard cell bioenergetics. In: Stomatal function, E. Zeiger, G.D. Zeiger, & I.R. Cowan (eds). Stanford University Press, Stanford, pp. 125–162.

Baas, P. (1986) Ecological patterns in xylem anatomy. In: On the economy of plant form and function, T.J. Givnish (ed.). Cambridge University Press, Cambridge, pp. 327–352.

Bartels, D. & Nelson, D. (1994) Approaches to improve stress tolerance using molecular genetics. Plant Cell Environ. 17:659–667.

Bartels, D., Schneider, K., Terstappen, G., Piatkowski, D., & Salamini, F. (1990) Molecular cloning of abscisic acid-modulated genes which are induced during desiccation of the resurrection plant *Craterostigma plantagineum*. Planta 181:27–34.

Bewley, J.D. & Krochko, J.E. (1982) Desiccation-tolerance. In: Encyclopedia of plant physiology, N.S., Vol. 12B, O.L. Lange, P.S. Nobel, C.B. Osmond, & H. Ziegler (eds). Springer-Verlag, Berlin, pp. 325–400.

Bianchi, G., Gamba, A., Limiroli, R., Pozzi, N., Elster, R., Salamini, F., & Bartels, D. (1993) The unusual sugar composition in leaves of the resurrection plant *Myrothamnus flabellifolia*. Physiol. Plant. 87:223–226.

Bohnert, H.J., Nelson, D.E., & Jensen, R.G. (1995) Adaptations to environmental stresses. Plant Cell 7:1099–1111.

Boyer, J.S. (1985) Water transport. Annu. Rev. Plant Physiol. 36:473–516.

Borchert, R. (1994) Soil and stem water storage determine phenology and distribution of tropical dry forest trees. Ecology 75:1437–1449.

Bradford, K.J. & Hsiao, T.C. (1982) Physiological responses to moderate water stress. In: Encyclopedia of plant physiology, N.S., Vol. 12B, O.L. Lange, P.S. Nobel, C.B. Osmond, & H. Ziegler (eds). Springer-Verlag, Berlin, pp. 263–324.

Bray, E.A. (1993) Molecular responses to water deficit. Plant Physiol. 103:1035–1040.

Bréda, N., Granier, A., Barataud, F., & Moyne, C. (1995) Soil water dynamics in an oak stand. I. Soil moisture, water potential and water uptake by roots. Plant Soil 172:17–27.

Bunce, J.A. (1997) Does transpiration control stomatal responses to water vapour pressure deficit? Plant Cell Environ. 19:131–135.

Caldwell, M.M. & Richards, J.H. (1986) Competing root systems: Morphology and models of absorption. In: On the economy of plant form and function, T.J. Givnish (ed.). Cambridge University Press, Cambridge, pp. 251–273.

Caldwell, M.M. & Richards, J.H. (1989) Hydraulic lift: Water efflux from upper roots improves effectiveness of water uptake by deep roots. Oecologia 79:1–5.

Cardon, Z.G. & Berry, J. (1992) Effects of O_2 and CO_2 concentration on the steady-state fluorescence yield of single guard cell pairs in intact leaf discs of *Tradescantia albiflora*. Plant Physiol. 99:1238–1244.

Chiariello, N.R., Field, C.B., & Mooney, H.A. (1987) Midday wilting in a tropical pioneer tree. Funct. Ecol. 1:3–11.

Chrispeels, M.J. & Agre, P. (1994) Aquaporins: Water channel proteins of plant and animal cells. Trends Biochem. Sci. 19:421–425.

Correia, M.J., Pereira, J.S., Chaves, M.M., Rodrigues, M.L., & Pacheo, C.A. (1995) ABA xylem concentrations determine maximum daily leaf conductance of field grown *Vitis vinifera* L. plants. Plant Cell Environ. 18:511–521.

Cowan, I.R. (1977) Water use in higher plants. In: Water. Planets, plants and people, A.K. McIntyre (ed.). Australian Academy of Science, Canberra, pp. 71–107.

Dace, H., Sherwin, H.W., Illing, N., & Farrant, J.M. (1998) Use of metabolic inhibitors to elucidate mechanisms of recovery from desiccation stress in the resurrection plant *Xerophyta humilis*. Plant Growth Regul., in press.

Daniels, M.J., Mirkov, T.E., & Chrispeels, M.J. (1994) The plasma membrane of *Arabidopsis thaliana* contains a mercurey-insensitive aquaporin that is a homolog of the tonoplast water channel protein TIP. Plant Physiol. 106:1325–1333.

Darwin, C. (1880) The power of movement in plants. John Murray, London.

Darwin, F. (1898) Observations on stomata. Phil Trans. Royal Soc., Ser. B, 190:531–621.

Davies, W.J., Tardieu, F., & Trejo, C.L. (1994) How do chemical signals work in plants that grow in drying soil? Plant Physiol. 104:309–314.

Dawson, T.E. (1993) Hydraulic lift and water use by plants: Implications for water balance, performance and plant-plant interactions. Oecologia 95:565–574.

de Pury, D.G.G. (1995) Scaling photosynthesis and water use from leaves to paddocks. PhD Thesis, Australian National University, Canberra, Australia (Chap. 3).

Dixon, H.H. (1914) Transpiration and the ascent of sap in plants. Macmillan, London.

Ehleringer, J.R., Phillips, S.L., Schuster, W.S.F., & Sandquist, D.R. (1991) Differential utilization of summer rains by desert plants Oecologia 75:1–7.

Ehleringer, J.R., Schulze, E.-D., Ziegler, H., Lange, O.L., Farquhar, G.D., & Cowan, I.R. (1985) Xylem-tapping mistletoes: Water or nutrient parasites? Science 227: 1479–1481.

Ewers, F.W. & Fisher, J.B. (1991) Why vines have narrow stems: Histological trends in *Bauhinia fassoglensis* (Fabaceae). Oecologia 88:233–237.

Ewers, F.W., Fisher, J.B., & Chiu, S.T. (1990) A survey of vessel dimensions in stems of tropical lianas and other growth forms. Oecologia 84:544–552.

Fichtner, K. & Schulze, E.-D. (1990) Xylem water flow in tropical vines as measured by a steady state heating method. Oecologia 82:355–361.

Flowers, T.J., Troke, P.F., & Yeo, A.R. (1977) The mechanism of salt tolerance in halophytes. Annu. Rev. Plant Physiol. 28:89–121.

Franks, P.J., Cowan, I.R., Tyerman, D., Cleary, A.L., Lloyd, J., & Farquhar, G.D. (1995) Guard cell pressure/aperture characteristics measured with the pressure probe. Plant Cell Environ. 18:795–800.

Franks, P.J., Cowan, I.R., & Farquhar, G.D. (1997) The apparent feedforward response of stomata to air vapour pressure deficit: Information revealed by different experimental procedures with two rainforest species. Plant Cell Environ. 20:142–145.

Fu, Q.A. & Ehleringer, J.R. (1989) Heliotropic leaf movements in common beans controlled by air temperature. Plant Physiol. 91:1162–1167.

Fuchs, E.E. & Livingston, N.J. (1996) Hydraulic control of stomatal conductance in Douglas fir [*Pseudotsuga menziesii* (Mirb.) Franco] and alder [*Alnus rubra* (Bong)] seedlings. Plant Cell Environ. 19:1091–1098.

Gaff, D.F. (1981) The biology of resurrection plants. In: The biology of Australian plants, J.S. Pate & A.J. McComb (eds). University of Western Australia Press, pp. 115–146.

Gamon, J.A. & Pearcy, R.W. (1989) Leaf movement, stress avoidance and photosynthesis in *Vitis californica*. Oecologia 79:475–481.

Gartner, B.L. (1995) Patterns of xylem variation within a tree and their hydraulic and mechanical consequences. In: Plant stems. Physiology and functional morphology, B.L. Gartner (ed.). Academic Press, San Diego, pp. 125–149.

Hanson, A.D. & Hitz, W.D. (1982) Metabolic responses of mesophytes to plant water deficits. Annu. Rev. Plant Physiol. 33:163–203.

Harten, J.B. & Eickmeier, W.G. (1986) Enzyme dynamics of the resurrection plant *Selaginella lepidophylla* (Hook. & Grev.) spring during rehydration. Plant Physiol. 82:61–64.

Hartung, W., Sauter, A., Turner, N.C., Fillery, I., & Heilmeier, H. (1996) Abscisic acid in soils: What is its function and which mechanisms influence its concentration? Plant Soil 184:105–110.

Hedrich, R. & Schroeder, J.I. (1989) The physiology of ion channels and electrogenic pumps in higher plants. Annu. Rev. Plant Physiol. 40:539–569.

Hendrey, G.A.F. (1993) Evolutionary origins and natural functions of fructans—a climatological, biogeographic and mechanistic appraisal. New Phytol. 123:3–14.

Hirasawa, T., Takahashi, H., Suge, H., & Ishihara, K. (1997) Water potential, turgor and cell wall properties in elongating tissues of the hydrotropically bending roots of pea (*Pisum sativum* L.). Plant Cell Environ. 20:381–386.

Holbrook, N.M. & Putz, F.E. (1996) From epiphyte to tree: differences in leaf structure and leaf water relations associated with the transition in growth form in eight species of hemiepiphytes. Plant Cell Environ. 19:631–642.

Holbrook, N.M., Burns, M.J., & Field, C.B. (1995) Negative xylem pressures in plants: A test of the balancing-pressure technique. Science 270:1193–1194.

Huang, B., North, G.B., & Nobel, P.S. (1993) Soil sheath, photosynthate distribution to roots, and rhizosphere water relations of *Opuntia ficus-indica*. Int. J. Plant Sci. 154:425–431.

Ingram, J. & Bartels, D. (1996) The molecular basis of dehydration tolerance in plants. Annu. Rev. Plant Physiol. Plant Mol. Biol. 47:377–403.

Kalapos, T., Van den Boogaard, R., & Lambers, H. (1996) Effect of soil drying on growth, biomass allocation and leaf gas exchange of two annual grass species. Plant Soil 185:137–149.

Kao, W.-Y. & Forseth, I.N. (192) Diurnal leaf movement, chlorophyll fluorescence and carbon assimilation in soybean grown under different nitrogen and water availabilities. Plant Cell Environ. 15:703–710.

Kern, J.S. (1995) Evaluation of soil water retention models based on basic soil physical properties. Soil Sci. Soc. Am. J. 59:1134–1141.

Kerstiens, G. (1996) Signalling across the divide: A wider perspective of cuticular structure-function relationships. Trends Plant Sci. 1:125–129.

Körner, C., Neumayer M., Pelaez Menendez-Riedl, S., & Smeets-Scheel, A. (1989) Functional morphology of mountain plants. Flora 182:353–383.

Kramer, P.J. (1969) Plant & soil water relationships. McGraw-Hill, New York.

Lange, O.L., Lösch, R., Schulze, E.-D., & Kappen, L. (1971) Responses of stomata to changes in humidity. Planta 100:76–86.

Lieth, H. (1975) Modelling the primary productivity of the world. In: Primary productivity of the biosphere, H. Lieth & R.H. Whittaker (eds). Springer-Verlag, Heidelberg, pp. 237–283.

Lo Gullo, M.A. & Salleo, S. (1988) Different strategies of drought resistance in three Mediterranean sclerophyllous trees growing in the same environmental conditions. New Phytol. 108:267–276.

Lo Gullo, M.A., Salleo, S. Piaceri, E.C., & Rosso, R. (1995) Relations between vulnerability to xylem embolism and xylem conduit dimensions in young trees of *Quercus cerris*. Plant Cell Environ. 18:661–669.

Longstreth, D.J., Bolanos, J.A., & Goddard, R.H. (1985) Photosynthetic rate and mesophyll surface area in expanding leaves of *Alternanthera philoxeroides* grown at two light intensities. Am. J. Bot. 72:14–19.

Loveless, A.R. (1961) A nutritional interpretation of sclerophyllous and mesophytic leaves. Ann. Bot. 25:169–184.

Loveless, A.R. (1962) Further evidence to support a nutritional interpretation of sclerophylly. Ann. Bot. 26:551–561.

MacRobbie, E.A.C. (1987) Ionic relations of guard cells. In: Stomatal Function, E. Zeiger, G.D. Zeiger, & I.R. Cowan (eds). Stanford University Press, Stanford, pp. 125–162.

Magnani, F. & Borghetti, M. (1995) Interpretation of seasonal changes of xylem embolism and plant hydraulic resistance in *Fagus sylvatica*. Plant Cell Environ. 18:689–696.

Mansfield, T.A. & McAinsh, M.R. (1995) Hormones as regulators of water balance. In: Plant hormones, P.J. Davies (ed). Kluwer Academic Publishers, Dordrecht.

Margolis, H., Oren, R., Whitehead, D., & Kaufmann, M.R. (1995) Leaf area dynamics of conifer forests. In: Ecophysiology of coniferous forests, W.K. Smith & T.M. Hinckley (eds). Academic Press, San Diego, pp. 181–223.

Maxwell, C., Griffiths, H., Borland, A.M., Broadmeadow, M.S.J., & McDavid, C.R. (1992) Photoinhibitory responses of the epiphytic bromeliad *Guzmania monostachia* during the dry season in Trinidad maintain photochemical integrity under adverse conditions. Plant Cell Environ. 15:37–47.

Maggio, A. & Joly, R.J. (1995) Effects of mercuric chloride on the hydraulic conductivity of tomato root systems. Evidence for a channel-mediated water pathway. Plant Physiol. 109:331–335.

McCain, D.C., Croxdale, J., & Markley, J.L. (1993) The spatial distribution of chloroplast water in *Acer platanoides* sun and shade leaves. Plant Cell Environ. 16:727–733.

McCully, M.E. & Canny, M.J. (1988) Pathways and processes of water and nutrient movement in roots. Plant Soil 111:159–170.

Meidner, H. (1987) Three hundred years of research into stomata. In: Stomatal function, E. Zeiger, G.D. Zeiger, & I.R. Cowan (eds). Stanford University Press, Stanford, pp. 7–27.

Meidner, H. & Sheriff, D.W. (1976) Water and plants. Blackie, Glasgow.

Milburn, J.A. (1997) Water flow in plants. Longman, London.

Monteith, J.L. (1995) A reinterpretation of stomatal responses to humidity. Plant Cell Environ. 18:357–364.

Mooney, H.A. & Dunn, E.L. (1970) Photosynthetic systems of Mediterranean climate shrubs and trees of California and Chile. Am. Nat. 194:447–453.

Mooney, H.A., Ehleringer, J., & Berry, J.A. (1976) High photosynthetic capacity of a winter annual in Death Valley. Science 194:322–324.

Mooney, H.A., Gulmon, S.L., Rundel, P.W., & Ehleringer, J. (1980) Further observations on the water relations of Prosopis tamarugo of the northern Atacama desert. Oecologia 44:177–180.

Morison, J.I.L. (1987) Intercellular CO_2 concentration and stomatal response to CO_2. In: Stomatal function, E. Zeiger, G.D. Farquhar, & I.R. Cowan (eds). Stanford University Press, Stanford, pp. 229–251.

Morison, J.I.L. (1993) Response of plants to CO_2 under water limited conditions. Vegetatio 104/105:193–209.

Morison, J.I.L. & Gifford, R.M. (1983) Stomatal sensitivity of carbon dioxide and humidity. A comparison of two C_3 and two C_4 grass species. Plant Physiol. 71:789–796.

Mott, K.A. & Parkhurst, D.F. (1991) Stomatal responses to humidity in air and helox. Plant Cell Environ. 14:509–516.

Nabil, M. & Coudret, A. (1995) Effects of sodium chloride on growth, tissue elasticity and solute adjustments in two *Acacia nilotica* subspecies. Physiol. Plant. 93:217–224.

Nobel, P.S. (1991) Physicochemical and environmental plant physiology. Academic Press, San Diego.

Nobel, P.S. (1996) Ecophysiology of roots of desert plants, with special emphasis on agaves and cacti. In: Plant roots: The hidden half, Y. Waisel, A. Eshel, & U. Kafkaki (eds). Marcel Dekker, Inc., New York, pp. 823–858.

Nobel, P.S., Zaragoza, L.J., & Smith, W.K. (1975) Relationship between mesophyll surface area, photosynthetic rate, and illumination level during development for leaves of *Plectranthus parviflorus*. 55:1067–1070.

North, G.B. & Nobel, P.S. (1997) Drought-induced changes in soil contact and hydraulic conductivity for roots of *Opuntia ficus-indica* with and without rhizosheaths. Plant Soil 191:249–258.

Oertli, J.J. (1996) Transport of water in the rhizosphere and in roots. In: Plant roots: The hidden half. Y. Waisel, A. Eshel, & U. Kafkaki (eds). Marcel Decker, Inc., New York, pp. 607–633.

Oliver, M.J. (1991) Influence of protoplastic water loss on the control of protein synthesis in the desiccation-tolerant moss *Tortula ruralis*. Ramifications for a repair-based mechanism of desiccation tolerance. Plant Physiol. 97:1501–1511.

Oosterhuis, D.M., Walker, S., & Eastman, J. (1985) Soybean leaflet movement as an indicator of crop water stress. Crop. Sci. 25:1101–1106.

Osmond, C.B., Winter, K., & Ziegler, H. (1982) Functional significance of different pathways of CO_2 fixation in photosynthesis. In: Encyclopedia of plant physiology, N.S. Vol. 12B, O.L. Lange, P.S. Nobel, C.B. Osmond, & H. Ziegler (eds). Springer-Verlag, Berlin, pp. 479–547.

Outlaw, W.H. Jr. (1995) Stomata and sucrose: A full circle. In: Carbon partitioning and source-sink interactions in plants, M.A. Madore & W.J. Lucas (eds). American Society of Plant Physiologists, Rockville, MD, pp. 56–67.

Passioura, J.B. (1982) Water in the soil–plant–atmosphere continuum. In: Encyclopedia of plant physiology, N.S., Vol. 12B, O.L. Lange, P.S. Nobel, C.B. Osmond, & H. Ziegler (eds). Springer-Verlag, Berlin, pp. 5–33.

Passioura, J.B. (1988a) Water transport in and to roots. Annu. Rev. Plant Physiol. Plant Mol. Biol. 39:245–265.

Passioura, J.B. (1988b) Root signals control leaf expansion in wheat seedlings growing in drying soil. Aust. J. Plant Physiol. 15:687–693.

Passioura, J.B. (1988c) Responses to Dr P.J. Kramer's article, "Changing concepts regarding plant water relations." Plant Cell Environ. 11:569–571.

Passioura, J.B. (1991) Soil structure and plant growth. Aust. J. Soil Res. 29:717–728.

Pedersen, O. & Sand-Jensen, K. (1997) Transpiration does not control growth and nutrient supply in the amphibious plant *Mentha aquatica*. Plant Cell Environ. 20:117–123.

Pelah, D., Wang, W., Altman, A., Shoseyov, O., & Bartels, D. (1997) Differential accumulation of water stress-related proteins, sucrose synthase and soluble sugars in *Populus* species that differ in their water stress response. Physiol. Plant. 99:153–159.

Peterson, C.A. (1989) Significance of the exodermis in root function. In: Structural and functional aspects of transport in roots, B.C. Loughman, O. Gasparikova, & J. Kolek (eds). Kluwer Academic Publishers, Dordrecht, pp. 35–40.

Peterson, C.A. & Enstone, D.E. (1996) Functions of passage cells in the endodermis and exodermis of roots. Physiol. Plant. 97:592–598.

Piatkowski, D., Schneider, K., Salamini, F., & Bartels, D. (1990) Characterization of five abscisic acid-responsive cDNA clones isolated from the desiccation-tolerant plant *Craterostigma plantagineum* and their relationship to other water-stress genes. Plant Physiol. 94:1682–1688.

Pilon-Smits, E.A.H., Ebskamp, M.J.M., Paul, M.J., Jeuken, M.J.W., Weisbeek, P.J., & Smeekens, S.J.M. (1995) Improved performance of transgenic fructan-accumulating tobacco under drought stress. Plant Physiol. 107:125–130.

Pockman, W.T., Sperry, J.S., & O'Leary, J.W. (1995) Sustained and significant negative water pressure in xylem. Nature 378:715–716.

Pollard, A. & Wyn Jones, R.G. (1979) Enzyme activities in concentrated solutions of glycinebetaine and other solutes. Planta 144:291–298.

Pollock, C.J. & Cairns, A.J. (1991) Fructan metabolism in grasses and cereals. Annu. Rev. Plant Physiol. Plant. Mol. Biol. 42:77–101.

Pritchard, J. (1994) The control of cell expansion in roots. New Phytol. 127:3–26.

Pütz, N. (1996) Development and function of contractile roots. In: Plant roots: The hidden half. Y. Waisel, A. Eshel, & U. Kafkaki (eds). Marcel Decker, Inc., New York, pp. 859–894.

Richards, J.H. & Caldwell, M.M. (1987) Hydraulic lift: Substantial nocturnal water transport between soil layers by *Artemisia tridentata* roots. Oecologia 73:486–489.

Rundel, P.W. (1995) Adaptive significance of some morphological and physiological characteristics in Mediterranean plants: Facts and fallacies. In: Time scales of biological responses to water constraints. The case of Mediterranean Biota, J. Roy, J. Aronson, & F. di Castri (eds). SPB Academic Publishing, Amsterdam, pp. 119–139.

Pyankov, V.I. (1993) The role of the photosynthetic apparatus in adaptation of plants to environment. PhD Thesis, Moscow, Institute of Plant Physiology.

Raschke, K. (1987) Action of abscisic acid on guard cells. In: Stomatal function, E. Zeiger, G.D. Zeiger, & I.R. Cowan (ed). Stanford University Press, Stanford, pp. 253–279.

Robichaux, R.H. & Canfield, J.E. (1985) Tissue elastic properties of eight Hawaiian *Dubautia* species that differ in habitat and diploid chromosome number. Oecologia 66:77–80.

Robichaux, R.H., Holsinger, K.E., & Morse, S.R. (1986) Turgor maintenance in Hawaiian *Dubautia* species: The role of variation in tissue osmotic and elastic properties. In: On the economy of plant form and function, T.J. Givnish (ed). Cambridge University Press, Cambridge, pp. 353–380.

Rodriguez, M.L., Chaves, M.M., Wendler, R., David, M.M., Quick, W.P., Leegood, R.C., Stitt, M., & Pereira, J.S. (1993) Osmotic adjustment in water stressed grapevine leaves in relation to carbon assimilation. Aust. J. Plant Physiol. 20:309–321.

Satter, R.L. & Galston, A.W. (1981) Mechanism of control of leaf movements. Annu. Rev. Plant Physiol. 32:83–110.

Schmidt, J.E. & Kaiser, W.M. (1987) Response of the succulent leaves of *Peperomia magnoliaefolia* to dehydration. Plant Physiol. 83:190–194.

Scholander, P.F., Bradstreet, E.D., & Hemmingsen, E.A. (1965) Sap pressures in vascular plants. Science 148:339–346.

Schulze, E.-D. (1991) Water and nutrient interactions with plant water stress. In: Response of plants to multiple stresses, H.A. Mooney, W.E. Winner, & E.J. Pell (eds). Academic Press, San Diego, pp. 89–101.

Schulze, E.-D. & Hall, A.E. (1982) Stomatal responses, water loss, and CO_2 assimilation rates of plants in contrasting environments. In: Encyclopedia of plant physiology, N.S., Vol. 12B, O.L. Lange, P.S. Nobel, C.B. Osmond, & H. Ziegler (eds). Springer-Verlag, Berlin, pp. 181–230.

Schulze, E.-D., Cermak, J., Matyssek, R., Penka, M., Zimmermann, R., Vasicek, F., Gries, W., & Kucera, J. (1985) Canopy transpiration and water fluxes in the xylem of the trunk of *Larix* and *Picea* trees—a comparison of xylem flow, porometer and cuvette measurements. Oecologia 66:475–483.

Schulze, E.-D., Turner, N.C., Gollan, T., & Shakel, K.A. (1987) Stomatal responses to air humidity and soil drought. In: Stomatal function, E. Zeiger, G.D. Farquhar, & I.R. Cowan (eds). Stanford University Press, Stanford, pp. 311–321.

Schwartz, A., Gilboa, S. & Koller, D. (1987) Photonastic control of leaflet orientation in *Melilotus indicus* (Fabaceae). Plant Physiol. 84:318–323.

Shackel, K. (1996) To tense, or not too tense: reopening the debate about water ascent in plants. Trends Plant Sci. 1:105–106.

Shah, N., Smirnoff, N., & Stewart, G.R. (1987) Photosynthesis and stomatal characteristics of *Striga hermonthica* in relation to its parasitic habit. Physiol. Plant. 69:699–703.

Sharkey, T.D. & Ogawa, T. (1987) Stomatal responses to light. In: Stomatal function, E. Zeiger, G.D. Zeiger, &

I.R. Cowan (eds). Stanford University Press, Stanford, pp. 195–208.

Sharpe, P.J.H., Wu, H & Spence, R.D. (1987) Stomatal mechanics. In: Stomatal function, E. Zeiger, G.D. Zeiger, & I.R. Cowan (eds). Stanford University Press, Stanford, pp. 91–114.

Sherwin, H.W. & Farrant, H.W. (1996) Differences in rehydration of three desiccation-tolerant angiosperm species. Ann. Bot. 78:703–710.

Sherwin, H.W. & Farrant, H.W. (1998) Protection mechanisms against excess light in the resurrection plants *Craterostigma wilmsii* and *Xerophyta viscosa*. Plant Growth Regul., in press.

Slatyer, R.O. (1967) Plant–water relationships. Academic Press, London.

Smirnoff, N. & Cumbes, Q.J. (1989) Hydroxyl radical scavenging activity of compatible solutes. Phytochemistry 28:1057–1060.

Sowell, J.B., McNulty, S.P., & Schilling, B.K. (1996) The role of stem recharge in reducing the winter desiccation of *Picea engelmannii* (Pinaceae) needles at alpine timberline. Am. J. Bot. 83:1351–1355.

Sperry, J.S. (1995) Limitations on stem water transport and their consequences. In: Plant stems. Physiology and functional morphology, B.L. Gartner (ed). Academic Press, San Diego, pp. 105–124.

Sperry, J.S. & Sullivan, J.E. (1992) Xylem embolism in response to freeze-thaw cycles and water stress in ring-porous, diffuse-porous, and conifer species. Plant Physiol. 100:605–613.

Sperry, J.S. & Tyree, M.T. (1988) Mechanism of water stress-induced xylem embolism. Plant Physiol. 88:581–587.

Sperry, J.S. & Tyree, M.T. (1990) Water-stress–induced xylem embolism in three species of conifers. Plant Physiol. 88:581–587.

Sperry, J.S., Saliendra, N.Z., Pockman, W.T., Cochard, H., Cuizat, P., Davis, S.D., Ewers, F.W., & Tyree, M.T. (1996) New evidence for large negative xylem pressures and their measurement by the pressure chamber technique. Plant Cell Environ. 19:427–436.

Sprenger, N., Bortlik, K., Brandt, A., Boller, T., & Wiemken, A. (1995) Purification, cloning, and functional expression of scucrose:fructan 6-fructosyltransferase, a key enzyme of fructan synthesis in barley. Proc. Natl. Acad. Sci. USA 92:11652–11656.

Steponkus, P.L. (1981) Responses to extreme temperatures. Cellular and sub-cellular bases. In: Encyclopedia of plant physiology, N.S., Vol. 12A, O.L. Lange, P.S. Nobel, C.B. Osmond, & H. Ziegler (eds). Springer-Verlag, Berlin, pp. 371–402.

Steudle, E. (1994) Water transport across roots. Plant Soil 167:79–90.

Steudle, E. (1995) Trees under tension. Nature 378:663–664.

Stirzaker, R.J. & Passioura, J.B. (1996) The water relations of the root–soil interface. Plant Cell Environ. 19:201–208.

Stirzaker, R.J., Passioura, J.B., & Wilms, Y. (1996) Soil structure and plant growth: Impact of bulk density and biopores. Plant Soil 185:151–162.

Takahashi, H. (1994) Hydrotropism and its interaction with gravitropism in roots. Plant Soil 165:301–308.

Takahashi, H. & Scott, T.K. (1993) Intensity of hydrostimulation for the induction of root hydrotropism and its sensing by the root cap. Plant Cell Environ. 16:99–103.

Tardieu, F., Zhang, J., Katerji, N., Bethenod, O., Palmer, S., & Davies, W.J. (1992) Xylem ABA controls the stomatal conductance of field-grown maize subjected to soil compaction or soil drying. Plant Cell Environ. 15:193–197.

Tardieu, F., Lafarge, T., & Simonneau, T. (1996) Stomatal control by fed or endogenous xylem ABA in sunflower: interpretation of correlations between leaf water potential and stomatal conductance in anisohydric species. Plant Cell Environ. 19:75–84.

Thorburn, P.J. & Ehleringer, J.R. (1995) Root water uptake of field-growing plants indicated by measurements of natural-abundance deuterium. Plant Soil 177:225–233.

Tranquillini, W. (1982) Frost-drought and its ecological significance. In: Encyclopedia of plant physiology, N.S., Vol. 12B, O.L. Lange, P.S. Nobel, C.B. Osmond, & H. Ziegler (eds). Springer-Verlag, Berlin, pp. 379–400.

Turrel, F.M. (1936) The area of the internal exposed surface of dicotyledon leaves. Am. J. Bot. 23:255–264.

Tyree, M.T. & Jarvis, P.G. (1982) Water in tissues and cells. In: Encyclopedia of plant physiology, N.S., Vol. 12B, O.L. Lange, P.S. Nobel, C.B. Osmond, & H. Ziegler (eds). Springer-Verlag, Springer-Verlag, Berlin, pp. 36–77.

Tyree, M.T. & Sperry, J.S. (1989) Vulnerability of xylem to cavitation and embolism. Annu. Rev. Plant Physiol. Mol. Biol. 40:19–38.

Van den Boogaard, R., Veneklaas, E.J., Peacock, J., & Lambers, H. (1996) Yield and water use of wheat (*Triticum aestivum* L.) cultivars in a Mediterranean environment: Effects of water availability and sowing density. Plant Soil 181:251–262.

Van Hylckama, T.E.A. (1974) Water use by salt cedar as measured by the water budget method. U.S. geological survey papers, 491-E.

Vijn, I., Van Dijken, A., Sprenger, N., Van Dun, K., Weisbeek, P., Wiemken, A., & Smeekens, S. (1997) Fructan of the inulin neoseries is synthesized in transgenic chicory plants (*Cichorium intybus* L.) harbouring onion (*Allium cepa* L.) fructan:fructan 6G-fructosyltransferase. Plant J. 11:387–398

Vogelmann, T.C. (1984) Site of light perception and motor cells in a sun-tracking lupine (*Lupinus succulentus*). Physiol. Plant. 62:335–340.

Vogt, K.A., Vogt, D.A., Palmiotto, P.A., Boon, P., O'Hara, J., & Asbjornson, H. (1996) Review of root dynamics in forest ecosystems grouped by climate, climatic forest type and species. Plant Soil, 187:159–219.

Wang, X.-L., Canny, M.J., & McCully, M.E. (1991) The water status of the roots of soil-grown maize in relation to the maturity of their xylem. Physiol. Plant. 82:157–162.

Wisniewski, M., Davis, G., & Arora, R. (1991) Effect of macerase, oxalic acid, and EGTA on deep supercooling and pit membrane structure of xylem parenchyma of peach. Plant Physiol. 96:1354–1359.

Woodward, F.I. (1995) Ecophysiological controls of conifer distributions. In: Ecophysiology of coniferous forests, W.K. Smith & T.M. Hinckley (eds). Academic Press, San Diego, pp. 79–94.

Wyn Jones, R.G. & Gorham, J. (1983) Osmoregulation. In: Encyclopedia of plant physiology, N.S., Vol. 12C, O.L. Lange, P.S. Nobel, C.B. Osmond, & H. Ziegler (eds). Springer-Verlag, Berlin, pp. 35–58.

Yang, S. & Tyree, M.T. (1992) A theoretical model of hydraulic conductivity recovery from embolism with comparison to experimental data on *Acer saccharum*. Plant Cell Environ. 15:633–643.

Zeiger, E., Iino, M., Shimazaki, K.-I., & Ogawa, T. (1987) The blue-light response of stomata: Mechanism and function. In: Stomatal function, E. Zeiger, G.D. Zeiger, & I.R. Cowan (eds). Stanford University Press, Stanford, pp. 209–227.

Zimmermann, M.H. (1983) Xylem structure and the ascent of sap. Springer-Verlag, Berlin.

Zimmermann, M.H. & Milburn, J.A. (1982) Transport and storage of water. In: Encyclopedia of plant physiology, N.S., Vol. 12B, O.L. Lange, P.S. Nobel, C.B. Osmond, & H. Ziegler (eds). Springer-Verlag, Berlin, pp. 135–151.

Zimmermann, U., Meinzer, F.C., Benkert, R., Zhu, J.J., Schneider, H., Goldstein, G., Kuchenbrod, E., & Haase, A. (1994) Xylem water transport: Is the available evidence consistent with the cohesion theory? Plant Cell Environ. 17:1169–1181.

Zwieniecki, M.A. & Newton, M. (1995) Roots growing in rock fissures: Their morphological adaptation. Plant Soil 172:181–187.

4
Leaf Energy Budgets: Effects of Radiation and Temperature

4A. The Plant's Energy Balance

1. Introduction

Temperature is a major environmental factor that determines plant distribution. Temperature affects virtually all plant processes, ranging from enzymatically catalyzed reactions and membrane transport to physical processes such as transpiration and the volatilization of specific compounds. Species differ in the activation energy of particular reactions and, consequently, in the temperature responses of most physiological processes (e.g., photosynthesis, respiration, biosynthesis). Given the pivotal role of temperature in the ecophysiology of plants, it is critical to understand the factors that determine plant temperature. Air temperature in the habitat provides a gross approximation of plant temperature. Air temperature in a plant's **microclimate**, however, may differ from standard air temperature, and the actual temperature of a plant organ often deviates substantially from that of the surrounding air. We can only understand the temperature regime of plants and, therefore, the physiological responses of plants to their thermal environment through study of microclimate and plant energy budgets.

2. Energy Inputs and Outputs

2.1 A Short Overview of a Leaf's Energy Balance

Most leaves effectively absorb **short-wave solar radiation** (SR). A relatively small fraction of incident solar radiation is **reflected**, **transmitted**, or utilized for processes other than just heating. In bright sunlight the net absorption of solar radiation (SR_{net}) is the main energy input to a leaf. If such a leaf had no means to dissipate this energy, then its temperature would reach 100°C in less than 1 minute (Jones 1985). Thus, processes that govern heat loss by a plant are critical for maintaining a suitable temperature for physiological functioning.

Heat loss occurs by several processes (Fig. 1). A leaf emits long-wave infrared radiation (LR). At the same time, however, it absorbs LR emitted by surrounding objects and by the sky. The net effect of **emission** and **absorption** (LR_{net}) may be negative or positive. When there is a temperature difference between leaf and air, **convective heat transfer** (C) takes place in the direction of the temperature gradient. Another major component of the energy balance is cooling caused by **transpiration** (λE; where λ is the energy required for evaporation and E is transpiration). In addition, **metabolic processes** (M) generate heat, although this is typically small compared with the other components of the energy balance, and it is usually ignored. When the temperature rises in response to sunlight, most components of the energy balance that contribute to cooling increase in magnitude until energy gain and loss are in balance. At this point, the leaf has reached an equilibrium temperature (steady state), and the sum of all components of the energy balance must equal zero:

$$SR_{net} + LR_{net} + C + \lambda E + M = 0 \qquad (1)$$

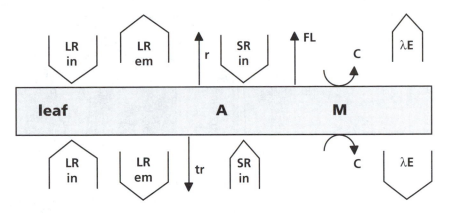

FIGURE 1. Schematic representation of the components of the energy balance of a leaf consisting of short-wave radiation (SR), long-wave radiation (LR), both incident (in) and emitted (em), convective heat transfer (C), and evaporative heat loss (λE). Reflection (r), transmission (tr), and fluorescent emission (FL) are only given for SR incident on the upper side of the leaf. A and M are CO_2-assimilation and heat-producing metabolic processes, respectively.

Changing conditions are more the rule than the exception in natural evironments. Any change in the components of the energy balance will alter leaf temperature. For a correct description of the time course of change, a **heat storage** term must be included; however, storage capacity is low in most leaves, except for succulents, and response times of leaf temperature to changing conditions are typically on the order of minutes or less.

2.2 Short-Wave Solar Radiation

Absorption of solar radiation normally dominates the input side of the energy balance of sunlit leaves during daytime. About 98% of the radiation emitted by the sun is in the range of 300 to 3000 nm (SR). Ultraviolet radiation (UV; 300 to 400 nm) has the highest energy content per quantum (shortest wavelength); it constitutes approximately 7% of solar radiation and is potentially damaging to a plant (cf. Sect. 2.2 of the chapter on effects of radiation and temperature). Plants absorb about 97% of incoming UV radiation (Fig. 2). About half of the energy content of solar radiation is in the waveband of 400 to 700 nm (photosynthetically active radiation, PAR), which can be used to drive photochemical processes (SR_A); most green leaves absorb around 85% of the incident radiation in this region (Fig. 2). Short-wave (solar) infrared radiation (IR_s; 700 to 3000 nm) is absorbed to a much lesser extent. This wavelength region can be divided in two parts: 700 to 1200 nm, which is largely **reflected** or **transmitted** by a leaf and which represents the largest part of IR_s in terms of energy content, and 1200 to 3000 nm, which is largely absorbed by water

in the leaf. The result is that about 50% of IR_s is absorbed.

Leaves have mechanisms that can modify the magnitude of the components that make up the amount of solar radiation absorbed (SR_{abs}): **incident radiation** (SR_{in}), **reflection** (SR_r) and **transmission** (SR_{tr}). Under conditions of high radiation load and water shortage, many shrubs exhibit steep leaf angles relative to the sun. This reduces midday SR_{in}, when temperatures are warmest, and increases SR_{in} in mornings and afternoons, when irradiance is less and temperatures are cooler. Angles can become progressively more horizontal in wetter communities, which increases SR_{in} at midday (Ehleringer 1988). Barrel cactuses similarly reduce SR_{in} by leaning toward the south in the United States and toward the north in Chile (Ehleringer et al. 1980). This increases the temperature of floral meristems in winter by 15 to 20°C and reduces heat load on the side of the cactus. Active leaf movements can also affect SR_{in} (Sect. 5.4.6 of plant water relations; Jurik et al. 1990). Such leaf movements (heliotropisms or **solar tracking**) may orient the leaf perpendicularly to the incident radiation (**diaheliotropism**), thus maximizing SR_{in} under conditions of low temperature and adequate soil moisture (Fig. 3). The most dramatic example of this could be heliotropic flowers of many arctic and alpine plants (e.g., *Dryas octopetala*, mountain avens) that move diurnally to continually face the sun, thus maximizing radiation gain. The parabolic shape of these heliotropic flowers reflects radiation toward the ovary. Thus, the shape and orientation of flowers maximizes rate of ovule development and attracts pollinators (Sect. 3.3.5 of life cycles; Kjellberg et al. 1982).

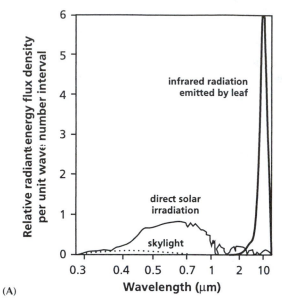

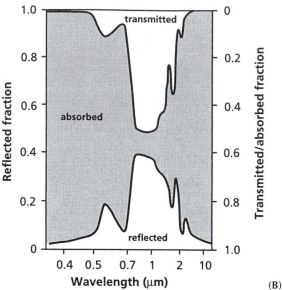

(A)

(B)

FIGURE 2. (A) Wavelength spectrum of skylight ($60\,W\,m^{-2}$), direct solar irradiance ($840\,W\,m^{-2}$), and infrared radiation emitted by a leaf ($900\,W\,m^{-2}$) at steady state at 25°C.

(B) Wavelength spectrum of absorbed, transmitted, and reflected irradiance (% of total) by a leaf (Gates 1965). Copyright Ecological Society of America.

When exposed to high temperature and water stress, leaves may become **paraheliotropic** (i.e., the leaf orientation is parallel to the incident radiation, thus minimizing SR_{in}) (Kao & Forseth 1992) (Fig. 3). Such movements keep SR_{abs} and thus leaf temperatures within limits, thus reducing transpiration and photoinhibition and maximizing the rate of CO_2 assimilation (Gamon & Pearcy 1989). Wilting and leaf rolling are additititional mechanisms by which leaves reduce incident radiation under conditions of water stress.

The **reflection** component of incident radiation (SR_r) is typically small (ca. 5 to 10%; Fig. 2) and is composed of reflection from the surface, which is largely independent of wavelength, and internal reflection, which is wavelength-specific because of absorption by pigments along the internal pathway (Sect. 2.1.1 of photosynthesis). In some plants, surface reflection can be high due to the presence of reflecting **wax** layers, short **white hairs**, or **salt crystals**. Reflection can change seasonally. The desert shrub, *Encelia farinosa*, produces new leaves during winter with sparse hairs that absorb 80% of incident radiation and raise leaf temperature several degrees above ambient (Fig. 4). Leaves produced in summer, however, when water is scarce, have dense reflective hairs that reduce absorptance to 30 to 40% of incident radiation and reduce leaf temperature below ambient. Shaving of these hairs from the leaf increases absorptance and raises leaf temperature by 5 to 10°C (Fig. 4; Fig. 2 in the chap-

ter on photosynthesis). The variation in leaf hair formation throughout the year might be controlled by photoperiod and gibberellins, as it is in *Arabidopsis thaliana* (Chien & Sussex 1996). Because *Encelia farinosa* has an optimum temperature for

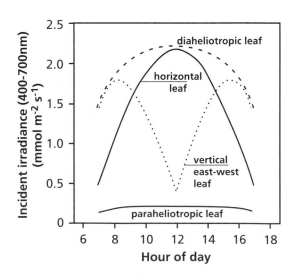

FIGURE 3. Photosynthetically active radiation incident on three leaf types over the course of a midsummer day: a diahelotropic leaf (cosine of incidence = 1.0); a vertical east–west facing leaf; a horizontal leaf, and a paraheliotropic leaf (cosine of incidence = 0.1) (Reprinted with permission from Ehleringer & Forseth 1980). Copyright 1980 American Association for the Advancement of Science.

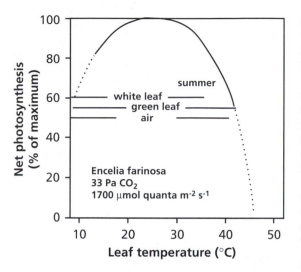

FIGURE 4. Daily ranges of leaf temperatures for green and white leaves of *Encelia farinosa* in the winter and summer, and the temperature dependence of photosynthesis (winter and summer) (Ehleringer & Mooney 1978).

photosynthesis that is below summer daytime temperature, this reduction in absorptance is critical to leaf carbon balance (Ehleringer & Björkman 1978, Ehleringer & Mooney 1978). Presence of leaf hairs in summer increases carbon gain and reduces water loss by 20 to 25% through amelioration of leaf temperature (Fig. 4). A white layer of salt on the leaves of *Atriplex hymenelytra*, which is excreted by salt glands, similarly reduces the absorptance and leaf

temperature, and enhances CO_2 assimilation and water-use efficiency because of a more favorable leaf temperature for photosynthesis (Mooney et al. 1977).

A small part (a few percentages) of SR_{abs} is emitted as fluorescence (SR_{FL}) (see Box 4). For simplicity's sake, SR_A and SR_{FL} are generally ignored in energy balance calculations, however, this may represent a significant error under conditions that are favorable for photosynthesis or where light intensity is low.

The components of solar radiation relevant for the energy balance of a leaf are summarized below, with values typical of a green leaf exposed to full sun at sea level on a cloudless day presented in Table 1 (Nobel 1983).

SR Short-wave (solar) radiation (300 to 3000 nm)
    SR_{in} incident radiation (PAR + IR_s)
    PAR photosynthetically active radiation (400 to 700 nm)
    IR_s short-wave infrared radiation (700 to 3000 nm)

$$SR_{in} = PAR + IR_s \qquad (2)$$
    SR_r reflected
    SR_{tr} transmitted
    SR_{abs} absorbed

$$SR_{abs} = SR_{in} - SR_r - SR_{tr} \qquad (3)$$
    SR_A used in photosynthesis
    SR_{FL} emitted as fluorescence

$$SR_{net} = SR_{abs} - SR_A - SR_{FL} \qquad (4)$$

TABLE 1. Components of energy gain (Wm^{-2}) by a leaf at the top, middle, and bottom of a forest canopy on sunny and cloudy days.*

Radiation component	Canopy					
	Cloudless day			Cloudy day		
	Top	Middle	Bottom	Top	Middle	Bottom
SR_{in}	1000	200	40	200	80	10
SR_{abs}	600	100	20	100	40	5
SR_A	7	2	0.4	2	0.8	0.1
SR_{net}	593	198	19.6	98	39.2	4.9
LR_{in}	650	750	833	792	800	802
LR_{abs}	624	720	800	760	768	770
LR_{em}	859	800	800	770	770	770
LR_{net}	−235	−80	0	−10	−2	0
TR_{abs}	1224	820	820	860	808	775
TR_{net}	358	118	19.6	88	37.2	4.9

Source: Data modified from Nobel 1983.
*Assume global incident irradiance of 833 Wm^{-2}, reflectance by the surroundings of 20%, leaf absorbance of 0.6, photosynthetic rate of 8 $\mu mol\,m^{-2}s^{-1}$ at the top of the canopy on a sunny day, leaf temperature of 25°C (top), 20°C (middle and bottom) on the sunny day, and of 17°C on the cloudy day. Total incoming radiation (SR_{in}) exceeds the global irradiance due to reflectance from surrounding leaves and twigs. Abbreviations as defined in text.

2.3 Long-Wave Radiation

A leaf loses heat by **emission** of long-wave infrared (LR_{em}). All objects above $0°K$ emit energy by radiation. LR_{em} is proportional to the fourth power of the absolute temperature (T) and the **emissivity** of the leaf (ε). The proportionality constant (σ, Stefan Bolzman contant) is $5.57 \ 10^{-8} W m^{-2} K^{-4}$):

$$LR_{em} = \varepsilon.\sigma.T^4 \qquad (5)$$

For a perfect radiator ("black body") ε is one. Real objects have lower values. Values for ε range from 0 to 1, with leaves having long-wave emissivities typically between 0.94 and 0.99. The leaf's emissivity equals 1 minus the reflection coefficient and is determined by such traits as leaf "roughness" (e.g., presence of hairs) and leaf color. Heat loss through long-wave infrared radiation is a major component of a leaf's energy balance; however, this may be balanced by an equally large absorption of LR emitted by surrounding objects and the sky. The high ε of leaves causes almost total absorbtion of incident LR. LR_{in} is low under a clear sky, which may have an effective radiation temperature around $-30°C$, but is high when plants grow widely spaced in dry sand or rocks that can reach temperatures of more than $70°C$ in full sun (Stoutjesdijk & Barkman 1987). LR_{abs} often accounts for half the total energy gain of a leaf in a canopy exposed to the sun (see later). The negative radiation balance of a leaf under a clear sky at night causes its temperature to drop below air temperature (negative ΔT), causing condensation of water (dew) on the leaf; however, not all water on a leaf in early morning is dew; it may also originate from guttation (Sect. 5.2 of plant water relations). A negative ΔT as a result of a negative radiation balance can also be found during daytime in leaves shaded from the sun but exposed to a blue sky, as on the north side of objects (rocks, walls, trees) at higher latitudes of the northern hemisphere (Nobel 1983, Stoutjesdijk & Barkman 1987).

The components of the long-wave radiation balance are summarized in what follows, with values typical of a green leaf exposed to full sun at sea level on a cloudless day presented in Table 1 (Nobel 1983):

LR   Long-wave (terrestrial) radiation (>3000 μm)
    LR_{in} incident radiation
    LR_r reflected
    LR_{abs} absorbed

$$LR_{abs} = LR_{in} - LR_r \qquad (6)$$
    LR_{em} emitted

$$LR_{net} = LR_{abs} - LR_{em} \qquad (7)$$

TR   Total radiation
$$TR_{abs} = SR_{abs} + LR_{abs} \qquad (8)$$
$$TR_{net} = SR_{net} + LR_{net} \qquad (9)$$

The energy gained by a leaf changes dramatically through the canopy, and the extent of this change depends on cloud conditions (Table 1, Fig. 5). In full sun, the short-wave radiation incident on a leaf exceeds that of incoming solar short-wave because of reflection from surrounding leaves and other surfaces, and the total energy absorbed by a leaf ($1224 W m^{-2}$ in Table 1) approaches that of the solar constant ($1360 W m^{-2}$) (i.e., the solar energy input above the atmosphere) (Nobel 1983). The total energy absorbed by a leaf declines dramatically in absence of direct solar irradiance, whether this is due to clouds or to canopy shading. Even leaves in full sun absorb more than half their energy as long-wave radiation, and leaves in cloudy or shaded conditions receive most energy as long-wave radiation emitted by objects in their surroundings. There is a sharp decline in total net radiation gained (and therefore energy that must be dissipated) through the canopy under both sunny and cloudy conditions.

2.4 Convective Heat Transfer

Leaf temperature is further determined by **convective (sensible) heat exchange** (C) which is proportional to the temperature difference (ΔT) between leaf (T_l) and air (T_a). The contribution of C to the energy balance of the leaf can be negative or positive when leaf temperature is higher or lower than air temperature, respectively, which is typically the case during daytime in the sun and during nighttime or shaded conditions, respectively. The magnitude of C depends on boundary layer conductance for heat transport (g_{ah}), which is proportional to the conductance for diffusion of CO_2 and H_2O (Fig. 6):

$$C = g_{ah}(T_a - T_l) \qquad (10)$$

The **boundary layer** is the layer of air close to a leaf (or any other surface) whose gas concentrations (e.g., CO_2 and H_2O), temperature, and pattern of air flow are modified by the leaf (see Section 2.2.2 in the chapter on pholosynthesis). As air moves across a leaf, it becomes increasingly influenced by the leaf surface, primarily by diffusion of heat and gases close to the leaf surface (conductive transfer) and by turbulent movement of air (convective transfer) at greater distance from the leaf surface (Fig. 7). Heat transport across the boundary layer is inversely

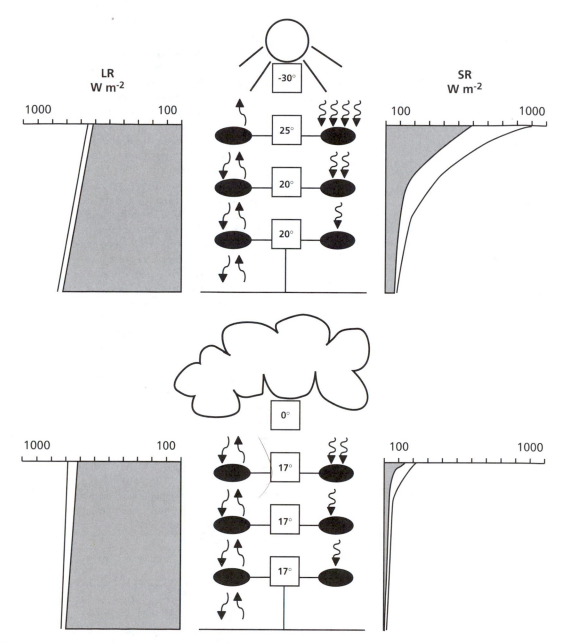

FIGURE 5. Schematic representation of the long-wave radiation (LR) and short-wave radiation (SR) inputs and outputs described in Table 1 for a sunny (upper) and cloudy (lower) day. Graphs show the vertical profile of incident radiation (solid line) and net radiation (shaded) (Nobel 1983). Copyright by W.H. Freeman and Company. Used with permission.

related to the **boundary layer thickness** (δ), which in turn depends on **wind speed** (u) and **leaf dimension** (leaf width measured in the direction of the wind; d) at the leading edge of the leaf, and by convection at the downwind side of the leaf.

$$\delta = 4\sqrt{d/u} \qquad (11)$$

From the preceding relationship it is clear that small leaves have temperatures closer to air temperature than do large leaves. Compound or highly dissected leaves are functionally similar to small leaves in this respect. Under hot dry conditions, most plants have small leaves because they cannot support high transpiration rates and must rely

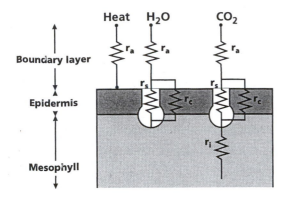

FIGURE 6. Resistances for the exchange of heat and gas between the leaf and the atmosphere. For CO_2, apart from the cuticular resistance (r_c) three resistances play a role (internal, r_i, stomatal, r_s, boundary layer, r_a), whereas there are only two and one for H_2O and heat exchange, respectively. Note that conductance is the inverse of resistance, $g = 1/r$ (Jones 1985). Copyright Pergamon Press.

largely on convective cooling to dissipate absorbed short-wave and long-wave radiation. Other mechanisms that reduce boundary layer thickness include an increase of effective windspeed across leaves with thin flattened petioles that cause leaves to flutter at low wind speeds, as in some *Populus* (poplar) species (*P. tremula* and *P. tremuloides*).

Some plants do not "obey the rules". *Welwitschia mirabilis* is an African perennial desert plant with extremely large leaves (0.5 to 1.0 m wide, and 1 to 2 m long) (Schulze et al. 1980). Those parts of the leaves not in contact with the ground, however, are only 4 to 6°C above air temperature. A high reflectivity of leaves (56%) minimizes SR_{abs}, and relatively cool shaded soils beneath leaves minimize LR_{in} (Fig. 8). There is negligible transpiration in summer, so it is primarily through these two mechanisms of minimizing energy gain that the plant avoids serious overheating.

It may be advantageous in cold environments to increase boundary layer thickness in order to raise leaf temperature at high irradiance. Some prostrate growth forms such as **cushion plants** in alpine habitats maximize boundary layer thickness by keeping leaves closely packed. They are also close to the ground, where they are in the boundary layer of the ground surface where temperatures are higher during daytime and wind speeds are lower. The plant's sensible heat loss is reduced, as is possibly its transpiration. Thus, it is possible that some prostrate alpine plants uncouple their microenvironment from ambient to an extent that temperatures of 27°C, relative humidities of 99%, and calm conditions occur around leaves, while the hiker may experience 10°C, a relative humidity of 50%, and a windspeed of $4 \, \text{m s}^{-1}$. This has been measured under clear sky conditions for *Celmisia longifolia*, *Verbascum thapsus*, and other rosette plants with sessile leaves (Körner 1983, Salisbury & Spomer 1964). The resulting higher plant temperatures are more favorable for many physiological processes. Leaf hairs are another mechanism by which plants in cold environments can increase boundary layer thickness and therefore leaf temperature. Plants from high altitudes, for example *Espeletia timotensis* in the Andes, may have **long hairs**. These increase reflection to a minor extent, but they do reduce the boundary layer conductance considerably so that the leaf temperature becomes higher than that of the air when the incident radiation is high (Fig. 9). Hence, leaf hairs have different functions with respect to the energy balance. Highly reflective hairs reduce absorption, and nonreflective hairs reduce boundary layer conductance.

2.5 Evaporative Energy Exchange

Heat loss associated with evaporation of water is the result of the energy demand for that process. Its

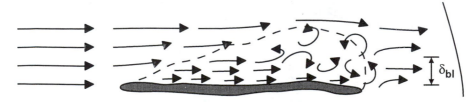

FIGURE 7. Schematic representation of the flow of nonturbulent air across a leaf. The arrows indicate the relative speed and direction of air movement. As air moves across a leaf, there is a laminar sublayer (short straight arrows), followed by a turbulent region. The effective boundary layer thickness (δ_{bl}) averages across these regions (Nobel 1983). Copyright by W.H. Freeman and Company. Used with permission.

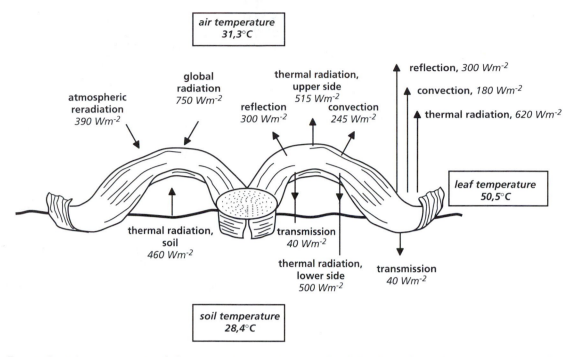

FIGURE 8. Major components of the energy budget of leaves of *Welwitschia mirabilis* in a coastal desert area of Namibia. Budgets are shown for leaves in contact with the ground and for those above the ground (Schulze et al. 1980).

contribution to the energy balance is typically negative during daytime when a leaf transpires, but it can be positive during nighttime when water condenses on the leaf. The rate of transpiration depends on leaf conductance for diffusion of water vapor (g_w), which consists of the **stomatal conductance** (g_s) and the **boundary layer conductance** (g_a) (Fig. 6) and the difference in vapor pressure between leaf and air ($e_i - e_a$). The latter is determined by leaf temperature and absolute air humidity (Sect. 2.3 of the chapter on photosynthesis). Heat loss through transpiration (E) is the product of the rate of transpiration and the latent heat of the vaporization of water (λ, $2450\,J\,g^{-1}$ at 20°C):

$$\lambda.E = -\lambda.g_w(e_i - e_a) \qquad (12)$$

A popular but incorrect idea is that plants control their leaf temperature by regulating transpiration. There is no evidence, however, for such a regulatory mechanism. Leaf cooling through transpiration is only beneficial at high temperatures, but then high transpiration rates may create problems for the plant if water loss is not matched by water uptake; limited water supply often coincides with high temperatures (cf. the chapter on plant water relations). Leaf cooling by transpiration also occurs at suboptimal temperatures because of stomatal

opening during photosynthesis, which creates an even less favorable temperature. Hence, leaf cooling must be considered a consequence of transpirational water loss that is inexorably associated with the stomatal opening required to sustain photosynthesis rather than a mechanism to control leaf temperature, as is the case for some animals.

In many situations, there is an inverse relationship between convective and evaporative heat exchange at any given irradiance. When stomatal conductance and evaporative heat loss decline, leaf temperature increases, causing an increase in both evaporative heat loss (by increasing water vapor concentration inside the leaf), and convective heat exchange (by increasing the temperature gradient between the leaf and the air.

2.6 Metabolic Heat Generation

The metabolic component (M) refers to heat production in **biochemical** reactions. Its contribution in leaves is very small under most circumstances and is generally ignored in calculation of the energy balance. In some plant organs, however, the contribution of metabolism to the energy balance may be substantial. In inflorescences such as the spadix

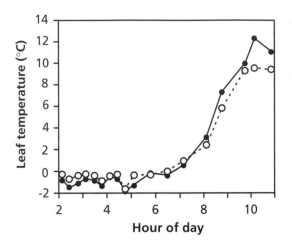

FIGURE 9. Leaf temperature of *Espeletia timotensis*, a giant rosette plant that occurs in the paramo zone of the Venezuelan Andes at elevations up to 4500 m. Measurements were made on plants growing in their natural environment, at different times of the day, both on intact hairy leaves (solid symbols), and on adjacent leaves of which the hairs were partly removed. Sunrise occurred around 6:45 a.m. The temperature of the intact leaf became higher than that of the shaved leaf when global radiation exceeded 300 W m⁻²; the thick leaf pubescence (up to 30 mm) increases boundary layer thickness and resistance to convective and latent heat transfer; effects of the pubescence on solar radiation absorption are minor. At night, the temperature of the intact leaves is somewhat lower than that of shaved leaves, due to reduced convective heat transfer from air to leaf (Meinzer & Goldstein 1985). Copyright Ecological Society of America.

of Araceae and flowers of Annonaceae and Nymphaceae, temperatures may rise several degrees above air temperature, due to their extremely high **respiration**, which may largely proceed through the alternative path (Sect. 2.6.1 of the chapter on respiration).

3. Modeling the Effect of Components of the Energy Balance on Leaf Temperature—A Summary of Hot Topics

The analysis of the exact contribution of the different components of the energy balance to leaf temperature is difficult when carried out with measurements only. The physical relationships as described earlier can be used to calculate leaf temperature from input parameters relevant for the energy balance of a leaf. By varying parameter values, the influence of a single parameter or combination of parameters on the final leaf temperature and components of the energy balance can be analyzed in a model. Although such a model uses symplifying assumptions, the outcome of these calculations appears to describe the real situation in a satisfactory manner (Campbell 1981).

Table 2 and Figure 10 illustrate the result of such model calculations. Leaf-to-air temperature difference (ΔT) is calculated for different conditions. Realistic parameter values for a clear day under moist conditions, an overcast day, a clear night, and a clear day in a desert were used. Components of the energy balance and ΔT are plotted against leaf dimension (d), stomatal conductance (g_s), and wind speed (u).

The calculations show that the difference in leaf–air temperature increases with increasing leaf width because the boundary layer conductance decreases, which results in a decrease in convective heat exchange (C). On a clear day, net long-wave radiation (LR) and evaporative cooling are quantitatively more important for the energy balance as a whole. At night, condensation causes the evaporative heat exchange (E) to be positive, which results in a lesser-than-expected cooling of the leaf below air temperature, as a result of the negative radiation balance. In a desert environment, the radiation load has increased enormously due to high

TABLE 2. Parameter values used for the calculations shown in figure 10.*

	Clear day (d, g_s, u)	Clear night (d)	Desert (d)	Overcast day (g_s)
Air temperature, °C	20	10	30	20
Soil temperature, °C	20	20	60	20
Sky temperature, °C	−20	−20	−20	+20
Short-wave radiation (SR_in), W m⁻²	800	0	800	100
Leaf dimension, mm	100	—	—	100
Wind speed, m s⁻¹	1	1	1	1
Relative humidity, %	65	100	30	80
Stomatal conductance, mmol m⁻² s⁻¹	400	0	30	—

*Abbreviations in brackets refer to the independent variable (*x*-axis) of the relationship.

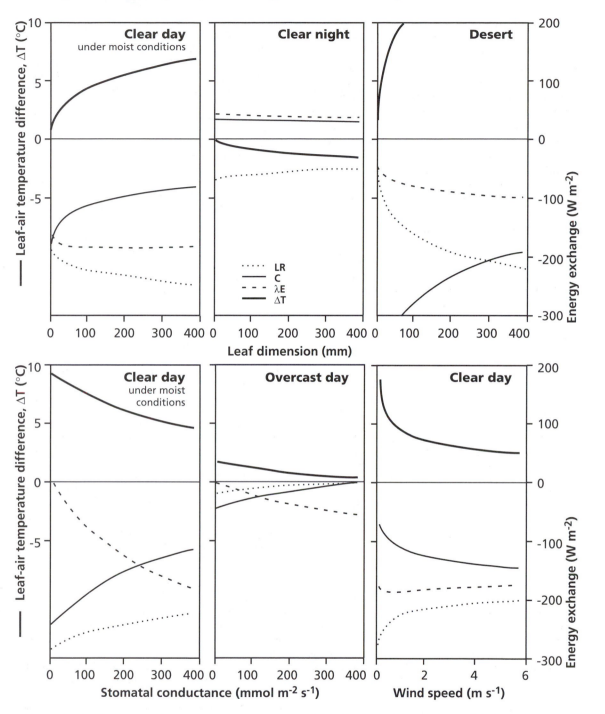

FIGURE 10. Results of energy balance model calculations for leaves in different conditions (model provided by F. Schieving, Utrecht University, the Netherlands). The difference in temperature between the leaf and the air (ΔT) and the components of the energy balance long-wave radiation (LR), convective heat exchange (C), and evaporative heat exchange (E) are plotted as a function of the leaf dimension (d), stomatal conductance (g_s), and wind speed (u) for conditions pertaining to a clear day in moist conditions, clear night, an overcast day, and for a clear day in a desert environment. The parameter values used for the calculations are shown in Table 2.

soil surface temperatures, and evaporative cooling is restricted due to stomatal closure. Leaf temperatures remain within tolerable limits only in small leaves that have large convective heat exchange. Such conditions occur both in deserts and, for example, in sand dunes in temperate climates on a sunny day. An increasing stomatal conductance causes a decrease in leaf temperature as a result of increasing evaporative cooling. On a cloudy day, this compensates for the small influx of short-wave radiation and brings the leaf temperature back to around air temperature at higher stomatal conductances. Wind reduces leaf temperature due to an increase in convective cooling, but, strangely enough, transpiration is hardly affected. The increase in boundary layer conductance with wind speed is counterbalanced by a decrease in leaf-to-air vapor pressure difference due to the decrease in leaf temperature. Largest effects on leaf temperature are found at the lower ranges of wind speeds.

References and Further Reading

Campbell, G.S. (1981) Fundamentals of radiation and temperature relations. In: Encyclopedia of plant physiology, Vol 12A, O.L. Lange, P.S. Nobel, C.B. Osmond, & H. Ziegler (eds). Springer-Verlag, Berlin, pp. 11–40.

Chien, J.C. & Sussex, I.M. (1996) Differential regulation of trichome formation on the adaxial and abaxial leaf surfaces by gibberellins and photoperiod in *Arabidopsis thaliana* (L.) Heynh. Plant Physiol. 111:1321–1328.

Ehleringer, J. (1983) Characterization of a glabrate *Encelia farinosa* mutant: Morphology, ecophysiology, and field observations. Oecologia 57:303–310.

Ehleringer, J. (1984) Ecology and ecophysiology of leaf pubescence in North American desert plants. In: Biology and chemistry of plant trichomes, E. Rodrigues, P.L. Healy, & I. Mehta (eds). Plenum Press, New York, pp. 113–132.

Ehleringer, J.R. (1988) Changes in leaf characteristics of species along elevational gradients on the wasatch front, Utah. Am. J. Bot. 75:680–689.

Ehleringer, J.R. & Björkman, O. (1978) Pubescence and leaf spectral characteristics in a desert shrub, *Encelia farinosa*. Oecologia 36:151–162.

Ehleringer, J.R. & Cook, C.S. (1990) Characteristics of *Encelia* species differing in leaf reflectance and transpiration rate under common garden conditions. Oecologia 82:484–489.

Ehleringer, J.R. & Forseth, I. (1980) Solar tracking by plants. Science 210:1094–1098.

Ehleringer, J.R. & Mooney, H.A. (1978) Leaf hairs: Effects on physiological activity and adaptive value to a desert shrub. Oecologia 37:183–200.

Ehleringer, J., Mooney, H.A., Gulmon, S.L., & Rundel, P. (1980) Orientation and its consequences for *Copiapoa* (Cactaceae) in the Atacama desert. Oecologia 46:63–67.

Gamon, J.A. & Pearcy, R.W. (1989) Leaf movement, stress avaiodance and photosynthesis in *Vitis californica*. Oecologia 79:475–481.

Gates, D.M. (1965) Energy, plants, and ecology. Ecology 46:1–13.

Grace, J.B. (1983) Plant-atmosphere relationships. Chapman & Hall, London.

Jones, M.B. (1985) Plant microclimate. In: Techniques in bioproductivity and photosynthesis, 2nd edition, J. Coombs, D.O. Hall, S.P. Long, & J.M.O. Scurlock (eds). Pergamon Press, Oxford, pp. 26–40.

Jurik, T.W., Zhang, H., & Pleasants, J.M. (1990) Ecophysiological consequences of non-random leaf orientation in the prairie compass plant, *Silphium laciniatum*. Oecologia 82:180–186.

Kao, W.-Y. & Forseth, I.N. (1992) Diurnal leaf movement, chlorophyll fluorescence and carbon assimilation in soybean grown under different nitrogen and water availabilities. Plant Cell Environ. 15:703–710.

Kjellberg, B., Karlsson, S., & Kerstensson, I. (1982) Effects of heliotropic movements of flowers of *Dryas octopetala* L. on gynoecium temperature and seed development. Oecologia 54:10–13.

Körner, C. (1983) Influence of plant physiognomie on leaf temperature on clear midsummer days in the Snowy Mountains, south-eastern Australia. Acta Oecologica 4:117–124.

Meinzer, F. & Goldstein, G. (1985) Some consequences of leaf pubescence in the andean giant rosett plant *Espeletia timotensis*. Ecology 66:512–520.

Mooney, H.A. Ehleringer, J.R., & Björkman, O. (1977) The energy balance of leaves of the evergreen shrub *Atriplex hymenelytra*. Oecologia 29:301–310.

Nobel, P.S. (1983) Biophysical plant physiology and ecology. W.H. Freeman and Co., San Francisco.

Salisbury, F.B. & Spomer, G.G. (1964) Leaf temperatures of alpine plants in the field. Planta 60:497–505.

Schulze, E.-D., Eller, B.M., Thomas, D.A., Von Willert, D.J., & Brinckmann, E. (1980) Leaf temperatures and energy balance of *Welwitschia mirabilis* in its natural habitat. Oecologia 44:258–262.

Schwartz, A., Gilboa, S., & Koller, D. (1987) Photonastic control of leaflet orientation in *Melilotus indicus* (Fabaceae). Plant Physiol. 84:318–323.

Smith, W.K. & Geller, G.N. (1980) Leaf and environmental parameters influencing transpiration: Theory and field measurements. Oecologia 46:308–313.

Stoutjesdijk, P. & Barkman, J.J. (1987) Microclimate, vegetation and fauna. Opulus Press, Upsala.

4B. Effects of Radiation and Temperature

1. Introduction

In the chapter on the plant's energy balance, we discussed plant traits that avoid deleterious effects of high levels of radiation. These mechanisms may not always suffice to prevent negative effects; some additional mechanisms and deleterious effects will be discussed in Section 2.1 of this chapter. Effects of ultraviolet radiation and a plant's mechanisms to avoid or repair damage will be treated in Section 2.2. Finally, some effects of both high and low temperatures that were not discussed earlier will be addressed in Section 3.

2. Radiation

2.1 Effects of Excess Irradiance

Species that are genetically well-adapted to shade often have a very restricted capacity to acclimate to a high irradiance. Unacclimated plants tend to be damaged by high irradiance levels, when the energy absorbed by the photosystems exceeds the energy that can be used in photochemistry (**photodamage**). Acclimated species have a protective mechanism that avoids photodamage: The energy absorbed by the light-harvesting complex is lost as heat; this mechanism is induced by acidification of the thylakoid lumen that results from the formation of a proton-motive force. Energy is dissipated in the light-harvesting systems of PSII, which

involves the **xanthophyll cycle** (Sect. 3.3.1 of the chaper on photosynthesis). The strong acidification of the lumen induces an enzymatic conversion of the carotenoid violaxanthin into zeaxanthin. Excess energy is transferred to zeaxanthin, which loses the energy as heat; this energy dissipation can be measured by chlorophyll fluorescence (cf. Box 4 in the chapter on photosynthesis) (see Sect. 3.3.1 in the chapter on photosynthesis).

2.2 Effects of Ultraviolet Radiation

Effects of ultraviolet (UV) radiation on plants have been studied for more than a century. The finding that the stratospheric UV-screening ozone layer is rapidly depleted due to human activities, however, has increased interest in this topic. Ozone in the Earth's atmosphere prevents all of the UV-C (<280 nm) and most of the UV-B (280 to 320 nm) radiation from reaching the Earth's surface (Fig. 1). Due to differences in optical density of the atmosphere, the ultraviolet radiation reaching the Earth is least at sea level in the arctic region (e.g., Alaska) and most at high altitude and low latitude (e.g., the Andes). Because most of the ozone is in the stratosphere above 15 km, however, the effect of altitude is much less pronounced than that of latitude. Cloud cover greatly reduces solar UV irradiance.

2.2.1 Damage by UV

Many compounds in plant cells absorb photons in the ultraviolet region (Fig. 2); the most destructive

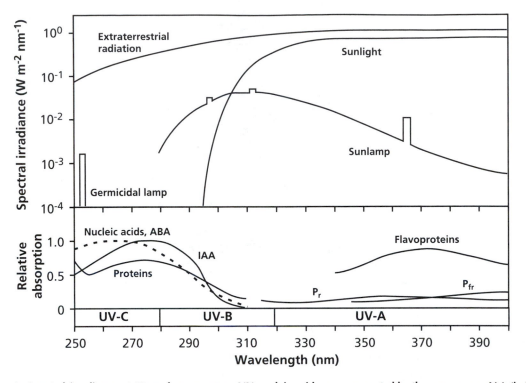

FIGURE 1. Spectral irradiance at 30 cm from common UV lamps, solar spectral irradiance before attenuation by the Earth's atmosphere (extraterrestrial), and as would be received at sea level at midday in the summer at temperate latitudes. The absorption spectra of a number of plant compounds are also shown; ABA (abscisic acid) and nu-cleic acids are represented by the same curve; IAA (indole acetic acid) and the two forms of phytochrome (P$_r$ and P$_{fr}$) are represented by the same curve as protein. Major subdivisions of the UV spectrum are indicated at the bottom; UV-B is ultraviolet light in the region 280 to 320 nm (Caldwell 1981).

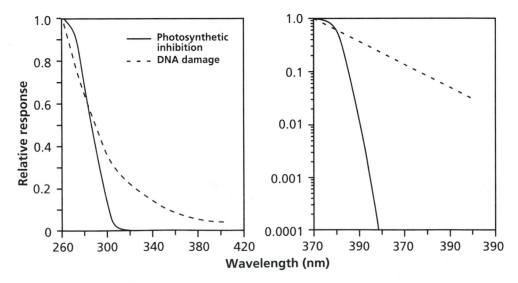

FIGURE 2. The damaging effect of UV on the dichlorophenol-indophenol reduction of isolated spinach chloroplasts ("photosynthetic inhibition") and on DNA in microorganisms ("DNA damage"), plotted on a linear (left) and an exponential (right) scale (Sewtlow 1974, and Jones & Kok 1966, as cited in Caldwell 1981).

actions of UV include effects on nucleic acids. DNA is by far the most sensitive nucleic acid. Upon absorption of UV radiation, polymers of pyrimidine bases, termed **dimers**, are formed, which lead to loss of biological activity. Although RNA and proteins also absorb UV radiation, much higher doses are required for inactivation to occur, possibly due to their higher concentration in the cell compared with DNA.

Algae and bacteria are considerably more sensitive to UV-B radiation than are leaves of higher plants, probably due to a difference in shielding of their DNA. Higher plants that are sensitive to solar UV show a reduction in photosynthetic capacity and in leaf expansion. Although part of the reduced leaf expansion may be the result of reduced photosynthesis, it also involves direct effects on cell division (Fig. 2), with both effects leading to reductions in plant growth and productivity (Van de Staaij et al. 1993). There may be additional effects on plant development (e.g., apical dominance) flower development, specific leaf area, and leaf abscission.

2.2.2 Protection Against UV: Repair or Prevention

Damage incurred by nucleic acids due to UV absorption can be repaired at the molecular level by splitting the pyrimidine dimers. Identification, followed by excision of the lesions from a DNA molecule and replacement by an undamaged patch that uses the other strand as a template, has also been demonstrated. Genotypes of *Oryza sativa* (rice) that lack the capacity to repair the damaged DNA show enhanced sensitivity to UV (Hidema et al. 1997). Plants have effective mechanisms to repair damage in all cells and organelles that contain DNA (Stapleton et al. 1997)

Plants can minimize UV exposure by having **steeply inclined leaves**, especially at lower latitudes and by **reflecting** or **absorbing UV** in the **epidermis**. Epidermal cells may selectively absorb UV because of the presence of **phenolic compounds** (flavonoids and flavones) (Stapleton & Walbot 1994). Next to flavonoids, sinapate esters of phenolics provide some protection against UV in Brassicaceae (e.g., *Arabidopsis thaliana*) (Sheahan 1996). The phenolic compounds sometimes occur in leaf hairs (Karabourniotis et al. 1992). There is evidence that both adaptation and acclimation to UV occur via the production of epidermal phenolic compounds (Ormrod et al. 1995). The most effective location for phenolics to screen UV is in the cell walls of epidermal cells rather than in their vacuoles, where phenolics may also accumulate.

The epidermis of evergreens transmits, on average, approximately 4% of the incident UV, and it does not allow penetration beyond 32 µm, as opposed to, on average, 28% and 75 µm, respectively, for leaves of deciduous plants (Day 1993). Conifer needles screen UV-B far more effectively because the absorbing compounds are located in the cell walls as well as inside their epidermal cells. The epidermis of herbaceous species is relatively ineffective at UV-B screening because UV-B may still penetrate through the epidermal cell walls, even if their vacuoles contain large amounts of UV-absorbing phenolics (Fig. 3 and Day et al. 1994).

Alkaloids also absorb UV radiation. The negative correlation between the frequency of species that produce alkaloids and the UV load in their natural environment suggests that alkaloids may represent a protective mechanism. **Cuticular waxes** may reflect UV radiation, although not to a greater extent than other wavelengths (Caldwell 1981).

3. Effects of Extreme Temperatures

3.1 How Do Plants Avoid Damage by Free Radicals at Low Temperature?

Variation in growth potential at different temperatures may reflect the rate of photosynthesis per unit leaf area, as discussed in Section 7 of the chapter on photosynthesis. A frequently observed effect of chilling is **photooxidation**, which occurs because the biophysical reactions of photosynthesis are far less temperature-sensitive than are the biochemical ones. Chlorophyll continues to absorb light at low temperatures, but the energy cannot be transferred to the normal electron-accepting components with sufficient speed to avoid **photoinhibition**. One mechanism by which cold-acclimated plants avoid photooxidation is to increase the components of the **xanthophyll cycle** (see Sect. 3.3.1 in the chapter on photosynthesis), thus preventing the formation of free radicals (i.e., toxic reactive molecules that rapidly lose an electron); radicals may form when oxygen is reduced to superoxide (Jabs et al. 1996). The xanthophyll cycle, however, is widespread among plants, which suggests that other mechanisms probably also protect the photosynthetic apparatus of cold-adapted species, especially if low temperatures coincide with high levels of irradiance, such as at high altitude.

There are several mechanisms that avoid damage by free radicals that may form at low temperature, especially in combination with a high irradiance.

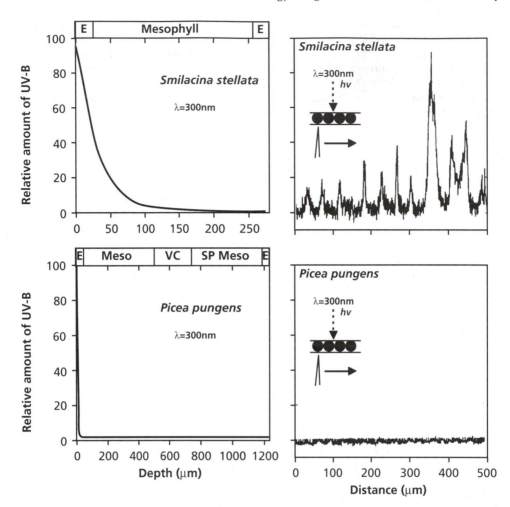

FIGURE 3. (Left) Relative amount of UV-B (300 nm) as a function of depth in intact foliage of a herbaceous species (*Smilacina stellata*) and a conifer (*Picea pungens*). Measurements were made with a fiberoptic microprobe. E, epidermis; Meso, mesophyll; VC, vascular cylinder; SP, Meso, spongy mesophyll. Note that UV-B penetrates into the mesophyll of the herbaceous leaf, whereas it is quickly attenuated in the epidermis of the conifer needle. (Right:) Pattern of UV-B transmission under an epidermal peel, removed from the rest of the leaf, of *Smilacina stellata* and

Picea pungens. Measurements were made by running a microscopic fiberoptic sensor along the underside of irradiated peels, parallel to the leaf axis, as illustrated by the image in the figures. UV-B penetrates (the spikes) through cell walls between cells of the epidermis in the herbaceous species, where UV-absorbing compounds are located in the vacuole. In the conifer species minimal transmission occurs because UV-absorbing compounds are present in the cell wall (Day et al. 1993). Copyright Blackwell Science Ltd.

Upon exposure to oxidative stress some free radicals will arise in nonacclimated plants. This induces the expression of genes coding for enzymes like chalcone synthase, which is involved in the synthesis of phenolic antioxidants (Henkow et al. 1996). High-alpine species contain **antioxidants**, such as ascorbic acid (vitamin C), α-tocopherol (vitamin E), and the tripeptide glutathione. Their concentration increases with increasing altitude (Fig. 4). The

alpine site and the lowland environments from which the plants shown in Figure 4 were collected are likely to receive a similar daily quantum input (Körner & Diemer 1987). The higher level of antioxidants in the high-altitude plants is therefore likely to be associated with the lower early-morning temperature or with the higher level of irradiance at peak times. The concentrations of antioxidants also show a diurnal pattern, with highest values at

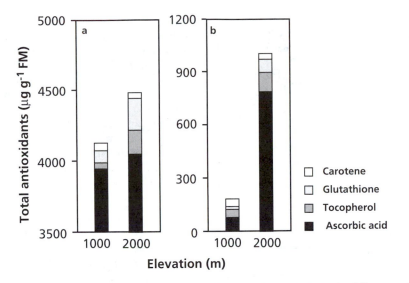

FIGURE 4. The concentration of various antioxidants in leaves of (a) *Homogyne alpina* and (b) *Soldanella pusilla* measured in plants growing at 1000 m (Wank) and at 2000 m (Obergurgl). Note the different scales on the y-axis (Wildi & Luetz 1996). Copyright Blackwell Science Ltd.

midday and lower ones during the night (Wildi & Luetz 1996).

Acclimation to low temperature in *Zea mays* (maize) is enhanced by exposure to a low soil water potential (Irigoyen et al. 1996). Both stresses could enhance the level of the phytohormone ABA (Box 8 in the chapter on growth and allocation), which is likely to be involved in acclimation to both a low soil water potential and a low temperature.

3.2 Heat-Shock Proteins

A sudden rise in temperature, close to the lethal temperature, induces the formation of mRNAs coding for **heat-shock proteins**. Some of the genes coding for heat-shock proteins are homologous with those from animals; in fact, heat-shock proteins were first discovered in *Drosophila*. Although the precise role of heat-shock proteins is not yet known, they do increase the plant's heat tolerance. Some of these proteins are only produced after exposure to high temperatures; others are also found after exposure to other extreme environmental conditions (e.g., drought). Heat-shock proteins may be involved in the protection of the photosynthetic apparatus and prevent photooxidation. Other heat-shock proteins belong to the class of the **chaperones**, which normally occur in plant cells, although in smaller quantities. Chaperones are involved in arranging the tertiary structure of

proteins. Heat-shock proteins are formed both after a sudden increase in temperature, and upon a more gradual and moderate rise in temperature, although not to the same extent. This class of proteins is, therefore, probably also involved in the tolerance of milder degrees of heat stress.

3.3 Is Isoprene Emission an Adaptation to High Temperatures?

There is increasing evidence that plants, especially some tree species and ferns, can cope with rapidly changing leaf temperatures through the production of the low-molecular-mass hydrocarbon: isoprene. Around Sydney in Australia, these hydrocarbons account for the haze in the Blue Mountains. The reason for high emission rates, however, have puzzled scientists for a long time. Many isoprene-emitting species lose about 15% of fixed carbon as isoprene emissions, with extreme values up to 50%. There should be sufficient evolutionary pressure to eliminate this process, if it serves no function. The finding that emissions increase at high temperature and under water stress has stimulated research into a role in coping with high leaf temperatures. The change in **isoprene emission** capacity through the canopy is similar to the change in **xanthophyll cycle** intermediates, which suggests that isoprene emission may be the plant's protection against excess heat, just as the xanthophyll cycle protects

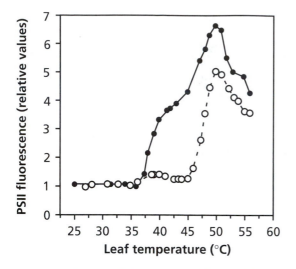

FIGURE 5. Photosystem II fluorescence of leaves of *Peraria lobata* (kudzu) as affected by temperature, measured both in the absence (filled circles) and presence (open circles) of isoprene. The increased fluorescence at higher temperature indicates that less energy is used for carbon reduction. In the presence of isoprene, the temperature at which carbon assimilation is affected shifts to higher values (Sharkey 1996). Copyright Elsevier Science Ltd.

against excess light (Sharkey 1997). Indeed, in the presence of realistic concentrations of isoprene, leaves are protected against high-temperature damage of photosynthesis (Fig. 5).

How hot do leaves normally get? Leaves of *Quercus rubra* (red oak) in the sun can be as much as 14°C above air temperature. The leaf temperature may drop by 8°C within minutes. Using isoprene may be an effective way of changing membrane properties rapidly enough to track leaf temperature. In plants that are not subject to such high temperatures or changes in leaf temperature, slower and less wasteful methods may be more effective.

3.4 Chilling Injury and Chilling Tolerance

Many (sub)tropical plants grow poorly at or are damaged by temperatures between 10 and 20°C. This type of damage is quite different from frost damage, which occurs at subzero temperatures, and is generally indicated as "**chilling injury.**" Different parts of the plant may well differ in their sensitivity to low temperatures. This sensitivity may also vary with age. For example, germinating

seeds and young seedlings of *Gossypium herbaceum* (cotton) and *Glycine max* (soybean) are far more chilling-sensitive than are mature plants. For *Oryza sativa* (rice) and *Sorghum bicolor* (sorghum), processes that occur in the phase just prior to flower initiation are most sensitive. Low temperatures may disturb the formation of pollen mother cells and thus cause sterility. Ripening fruits of (sub)tropical crops are also rapidly damaged by low temperatures.

The physiological cause of low-temperature damage varies among species and plant organs. The following factors play a role:

1. Changes in membrane fluidity
2. Changes in the activity of membrane-bound enzymes and processes, such as electron transport in chloroplasts and mitochondria, and in compartmentation
3. Loss of activity of low-temperature sensitive enzymes

Chilling resistance may involve **membrane properties**, which are affected by the composition of the membranes. Both the proteins and the lipids in the membrane may play a role. Chilling tolerance correlates with a high proportion of cis-unsaturated fatty acids in the phosphatidyl-glycerol molecules of chloroplast membranes. Evidence for this comes from work with *Nicotiana tabacum* (tobacco) plants transformed with glycerol-3-phosphate acyltransferase from either a cold-tolerant species or a cold-sensitive one. Overexpression of the enzyme from the cold-tolerant species increases cold-tolerance, whereas the tobacco plants become more sensitive to cold stress when overexpressing the enzyme from cold-sensitive plants. Cold sensitivity of the transgenic tobacco plants correlates with the extent of fatty acid unsaturation in phosphatidylglycerol, which is due to different selectivities for the saturated and cis-unsaturated fatty acids of the enzyme from contrasting sources (Bartels & Nelson 1994).

The degree of saturation of the fatty acids affects the membrane's fluidity, as shown for mitochondria: The ratio between unsaturated and saturated fatty acids is about 2 for chilling-sensitive species and about 4 for resistant species. It is likely that the mitochondrial membranes of sensitive species tend to "solidify" at a relatively high temperature, hampering membrane-associated processes and causing "leakage" of solutes out of various compartments or out of the cells. A reduction in membrane fluidity may also hamper the roots' rate of water uptake (cf. Sect. 5.2.1 in the chapter on growth and allocation).

3.5 Carbohydrates and Proteins Conferring Frost Tolerance

As outlined in Section 9 of the chapter on plant water relations, frost damage only occurs at sub-zero temperatures, when the formation of ice crystals causes damage to membranes and organelles and dehydration of cells. Cold tolerance is correlated with the concentration of **soluble carbohydrates** in the cells (Fig. 6; Sakai & Larcher 1987). These carbohydrates play a role in **cryoprotection** (Crowe et al. 1990, Sakai & Yoshida 1968). Differences in cold tolerance between *Picea abies* (Norway spruce), *Pinus contorta* (lodgepole pine), and *Pinus sylvestris* (Scots pine), following exposure of hardened needles to 5.5°C, are closely correlated with their carbohydrate concentration. *Picea abies* maintains high sugar concentrations by having larger reserves to start with and lower rates of respiration, which decline more rapidly when sugars are depleted (Ögren et al. 1997).

Cold stress leads to differential **gene expression**, and a wide range of cold-inducible genes have been isolated. Several of these genes occur in a wide range of plant species and contain conserved structural elements, which suggests that these elements are vital for functional reasons. Their role in low-temperature acclimation, however, is not yet clear. A group of low-temperature-induced genes is homologous to the genes preferentially expressed during embryo maturation and encode mainly hydrophilic proteins. These genes may be involved in the osmotic stress response that is common to cold, water, and salt stress (Bartels & Nelson 1994).

Many plants that naturally occur in temperate climates go though an annual cycle of frost **hardening** and **dehardening**, with maximum freezing tolerance occurring during winter. In herbaceous plants frost hardening occurs by exposure to low, nonfreezing temperatures. Upon exposure to 5/2 (day/night) °C, specific **proteins** accumulate in the apoplast of *Secale cereale* (winter rye) and other frost-resistant monocotyledonous species; remarkably, these proteins are similar to the pathogenesis-related proteins that are induced by microbial pathogens (Sect. 3 of the chapter on effects of microbial pathogens) (Antikainen & Griffith 1997). These proteins confer greater frost tolerance, as evidenced by less ion leakage from the leaves when exposed to subzero temperatures. When these proteins are experimentally removed from the apoplast, the plant's cold tolerance is lost (Marentes et al. 1993). Hence, it is likely that the accumulation of these proteins is causally linked to the increased frost tolerance; they may well have an effect on the growth of ice in the cell walls (Hon et al. 1994). Upon cold-acclimation, a specific glycoprotein ("cryoprotectin") accumulates in leaves of *Brassica oleracea* (cabbage) that protects thylakoids from nonacclimated leaves, both of cabbage and of other species such as *Spinacia oleracea* (spinach) (Sieg et al. 1996). Exposure to low temperature induces a specific class of proteins: lipid-transfer proteins. Although the name of these proteins suggests otherwise, they are unlikely to be involved in lipid transfer in vivo. The relationship between the putative protective role of lipid-transfer proteins and cold tolerance still needs to be determined (Kader 1997).

4. Global Change and Future Crops

Plants are frequently exposed to potential harmful radiation and adverse temperatures. Some of the protective mechanisms in plants are universal (e.g., the carotenoids of the xanthophyll cycle that protect against excess radiation). All plants also have mechanisms to avoid effects of UV radiation and repair UV damage. There is a wide variation among species, however, in the extent of the avoidance and

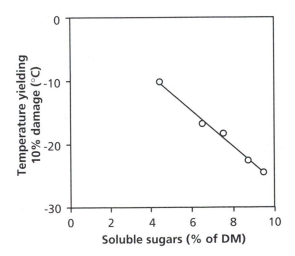

FIGURE 6. Temperature causing 10% damage of the needles of *Pinus sylvestris* (Scots pine) as dependent on the concentration of soluble carbohydrates in the needles. Variation in sugar concentration was obtained by exposure of intact plants to temperatures ranging from −8.5°C to 5.5°C for 16 weeks in midwinter (Ögren 1997). Copyright Heron Publishing.

probably also in the capacity to repair the damage. The rapid depletion of the stratospheric UV-screening ozone layer, due to human activities, imposes a selective force on plants to cope with UV.

Toxic reactive oxygen species (free radicals) are produced when the dark reactions of photosynthesis cannot cope with the high activity of the light reactions. This may occur under high-light conditions in combination with extreme temperatures. The xanthophyll cycle can prevent some of the potential damage by funneling off excess energy, acting as a lightning rod, at both high and low temperatures. Isoprene production possibly provides additional protection of leaves at high temperatures. Specific proteins and carbohydrates offer protection against temperature extremes. Further ecophysiological research on these compounds and the regulation of the genes that encode their production may help us to develop crop varieties that have a greater capacity to cope with extreme temperatures. Such plants will be highly desirable for agriculture in those parts of the world where extreme temperatures are a major factor limiting crop productivity.

References and Further Reading

Antikainen, M. & Griffith, M. (1997) Antifreeze protein accumulation in freezing-tolerant cereals. Physiol. Plant. 99:423–432.

Bartels, D. & Nelson, D. (1994) Approaches to improve stress tolerance using molecular genetics. Plant Cell Environ. 17:659–667.

Caldwell, M.M. (1981) Plant responses to solar ultraviolet radiation. In: Encyclopedia of plant physiology, N.S., Vol. 12A, O.L. Lange, P.S. Nobel, C.B. Osmond, & H. Ziegler (eds). Springer-Verlag, Berlin, pp. 169–197.

Crowe, J.H., Carpenter, J.F., Crowe, L.M., & Anchordoguy, T.J. (1990) Are freezing and dehydration similar stress factors? A comparison of modes of interaction of different biomolecules. Cryobiol. 27:219–231.

Day, T.A. (1993) Relating UV-B radiation screening effectiveness of foliage to absorbing-compound concentration and anatomical characteristics in a diverse group of plants. Oecologia 95:542–550.

Day, T.A., Martin, G., & Vogelmann, T.C. (1993) Penetration of UV-B radiation in foliage: evidence that he epidermis behaves as a non-uniform filter. Plant Cell Environ. 16:735–741.

Day, T.A., Howells, B.W., & Rice, W.J. (1994) Ultraviolet absorption and epidermal-transmittance in foliage. Physiol. Plant. 92:207–218.

Henkov, L., Strid, A., Berglund, T., Rydstrom, J., & Ohlsson, A.B. (1996) Alteration of gene expression in

Pisum sativum tissue cultures caused by the free radicalgenerating agent 2,2'-azobis (2-aminopropane) dihydrochloride. Physiol. Plant. 96:6–12.

Hidema, J., Kumagai, T., Sutherland, J.C., & Sutherland, B.M. (1997) Ultraviolet B-sensitive rice cultivar deficient in cyclobutyl pyrimidine dimer repair. Plant Physiol. 113:39–44.

Hon, W.-C., Griffith, M., Chong, P., & Yang, D.S.C. (1994) Extraction and isolation of antifreeze proteins from winter rye (Secale cereale L.) Leaves. Plant Physiol. 104:971–980.

Irigoyen, J.J., Perez de Juan, J., & Sanchez-Diaz, M. (1996) Drought enhances chilling tolerance in a chilling-sensitive maize (Zea mays) variety. New Phytol. 134;53–59.

Jabs, T., Dietrich, R.A., & Dangl, J.L. (1996) Initiation of runaway cell death in an Arabidopsis mutant by extracellular superoxide. Science 273:1853–1856.

Kader, J.-C. (1997) Lipid-transfer proteins: A puzzling family of plant proteins. Trends Plant Sci 2:66–70.

Karabourniotis, G., Papadopoulos, K., Papamarkou, M., & Manetas, Y. (1992) Ultraviolet-B radiation absorbing capacity of leaf hairs. Physiol. Plant. 86:414–418.

Körner, C. & Diemer, M. (1987) In situ photosynthetic responses to light, temperature and carbon dioxide in herbaceous plants from low and high altitude. Funct. Ecol. 1:179–194.

Marentes, E., Griffiths, M., Mlynarz, A., & Brush, R.A. (1993) Proteins accumulate in the apoplast of winter rye leaves during cold acclimation. Physiol. Plant. 87:499–507.

Ögren, E. (1997) Relationship between temperature, respiratory loss of sugar and premature dehardening in dormant Scots pine seedlings. Tree Physiol. 17:47–51.

Ögren, E., Nilsson, T., & Sundblad, L.-G. (1997) Relationships between respiratory depletion of sugars and loss of cold hardiness in coniferous seedlings overwintering at raised temperatures: Indications of different sensitivities of spruce and pine. Plant Cell Environ. 20:247–253.

Ormrod, D.P., Landry, L.G., & Conklin, P.L. (1995) Short-term UV-B radiation and ozone exposure effects on aromatic secondary metabolite accumulation and shoot growth of flavonoid-deficient Arabidopsis mutants. Physiol. Plant. 93:602–610.

Sakai, A. & Larcher, W. (1987) Frost survival of plants. Responses and adaptation to freezing stress. Springer-Verlag, Berlin.

Sakai, A. & Yoshida, S. (1968) The role of sugar and related compounds in variation of freezing resistance. Cryobiol. 5:160–174.

Sharkey, T.D. (1996) Emission of low molecular mass hydrocarbons from plants. Trends Plant Sci. 1:78–82.

Sharkey, T.D. (1997) Isoprene production in trees. In: Trees—contributions to modern tree physiology, H. Rennenberg, W. Eschrich, & H. Ziegler (eds). Backhuys, Leiden, pp. 111–120.

Sheahan, J.J. (1996) Sinapate esters provide greater UV-B

attenuation than flavonoids in *Arabidopsis thaliana* (Brassicaceae). Am. J. Bot. 83:679–686.

Sieg. F., Schröder, W., Schmitt, J.M., & Hincha, D.K. (1996) Purification and characterization of a cryoprotective protein (cryoprotectin) from the leaves of cold-acclimated cabbage. Plant Physiol. 111:215–221.

Stapleton, A.E. & Walbot, V. (1994) Flavonoids protect maize DNA from the induction of ultraviolet radiation damage. Plant Physiol. 105:881–889.

Stapleton, A.E., Thornber, C.S., & Walbot, V. (1997) UV-B component of sunlight causes measurable damage in field-grown maize (*Zea mays* L.): Developmental and cellular heterogeneity of damage and repair. Plant Cell Environ. 20:279–290.

Van de Staaij, J.W.M., Lenssen, G.M., Stroetenga, M., & Rozema, J. (1993) The combined effects of elevated CO_2 and UV-B radiation on growth characteristics of *Elymus athericus* (=*E. pycnathus*). Vegetatio 104/105:433–439.

Wildi, B. & Luetz, C. (1996) Antioxidant composition of selected high alpine plant species from different altitudes. Plant Cell Environ. 19:138–146.

5
Scaling-Up Gas Exchange and Energy Balance from the Leaf to the Canopy Level

1. Introduction

Having discussed the gas exchange and energy balance of individual leaves in previous chapters, we are now in a position to "scale up" to the canopy level. In moving between scales, it is important to determine which interactions should be considered and which can either be ignored or taken as independent variables. The water relations of plant canopies are distinctly different than would be predicted from the study of individual leaves because each leaf modifies the environment of adjacent leaves through reduced irradiance, wind speed, and vapor pressure deficit. These changes within the canopy reduce transpiration from each leaf more than would be predicted from an individual leaf model, using the atmospheric conditions above the canopy. For example, **irradiance** declines exponentially with leaf area index within the canopy (Box 7), reducing the energy that each leaf dissipates. Friction from the canopy causes **wind speed** to decline close to the canopy, just as it declines close to the ground surface. Wind speed then declines exponentially within the canopy and individual leaves within a canopy have lower boundary layer conductance than expected from meteorological conditions of the bulk air and leaf dimensions. Finally, transpiration by each leaf increases the **water vapor concentration** around adjacent leaves, as does evaporation from a wet soil surface. As stomatal conductance increases, the increasing water vapor concentration within the canopy re-

duces the driving force for transpiration, so that transpiration increases less than expected from the increase in stomatal conductance alone (Jarvis & McNaughton 1986).

Mathematical functions can be used to describe the effects of variables and their interactions in a model of the system. A good model for scaling will be based on mechanistic processes at a lower scale. Can we treat the canopy simply as one big leaf to arrive at the gas exchange and energy balance of a canopy? Do we need to sum up the gas exchange and energy balance of individual leaves and include many intricate details about microclimate and plant physiology? These questions will be addressed in the following sections.

2. Canopy Water Use

In Section 2.2 of the chapter on photosynthesis we discussed leaf transpiration as measured in gas exchange cuvettes. These systems involve enclosing a single leaf in a well-ventilated and environmentally controlled cuvette. In such cuvettes, the boundary layer is minimal and the transpiration has little effect on the conditions inside and around the leaf. For leaves in a canopy, however, boundary layers have a significant effect on the transpiration rate and the air in the boundary layer contains more water vapor than the ambient air. In a canopy, more so than in leaves measured in a leaf cuvette, transpiration is affected by both stomatal and **boundary**

Box 7
Optimization of Nitrogen Allocation to Leaves in Plants Growing in Dense Canopies

A theoretical optimum distribution of nitrogen over the leaves of a plant that maximizes whole plant photosynthesis per unit leaf N can be calculated (Anten et al. 1995, Evans 1993, Field 1983, Hirose & Werger 1987, Pons et al. 1989). Such an optimum distribution pattern depends on the distribution of light over the leaves of a plant growing in a dense canopy. The approach chosen here is for plant stands consisting of one species of even-sized individuals. Hence, the performance of the stand is identical to the performance of individual plants growing in the stand. The calculations consist of five parts that describe mathematically: (1) the distribution of irradiance in the leaf canopy where the plant is growing, (2) the dependence of photosynthetic rate on irradiance of leaves, (3) the relationships with leaf N of the parameters of the photosynthesis–irradiance relationship, (4) canopy photosynthesis by summation of photosynthetic rates in different canopy layers, and (5) the distribution of leaf N at maximum canopy photosynthesis per unit leaf N.

Following the approach discussed in Box 3 on gradients in leaves (attached to the chapter on photosynthesis), we can use the Lambert-Beer law to calculate the light absorption profile in the canopy. An extension of that formula gives the mean irradiance (I_L, $\mu mol\, m^{-2} s^{-1}$) incident on a leaf at a certain depth in the canopy expressed as cumulative leaf area index from the top of the canopy [F, m^2 (leaf area) m^{-2} (ground surface)]:

$$I_L = \frac{I_o \cdot K_L}{1-t} \exp^{(-K_L \cdot F)} \quad (1)$$

where I_o ($\mu mol\, m^{-2} s^{-1}$) is the irradiance above the canopy, and K_L and t are the dimensionless canopy extinction and leaf transmission coefficients for light, respectively (Hirose & Werger 1987). I_o is multiplied by K_L to account for the deviation of leaf angle from horizontal transmission of light by leaves is also accounted for.

Again following the approach in Box 3 in the chapter on photosynthesis, we calculate the photosynthetic rate in each canopy layer by using the light-response curve. For this purpose we use the equation introduced in Section 3.2.1 of the chapter on photosynthesis:

$$A = \frac{\phi \cdot I + A_{max} - \sqrt{\{(\phi \cdot I + A_{max})^2 - 4\Theta \cdot I \cdot \phi \cdot A_{max}\}}}{2\Theta} - R_{day} \quad (2)$$

where A ($\mu mol\, CO_2\, m^{-2} s^{-1}$) is the actual rate of photosynthesis, ϕ is the apparent quantum yield at low irradiance [mol CO_2 mol^{-1} (quanta)], I is irradiance ($\mu mol\, quanta\, m^{-2} s^{-1}$), A_{max} ($\mu mol\, CO_2\, m^{-2} s^{-1}$) is the light-saturated rate of photosynthesis, and Θ (dimensionless) describes the curvature on the A-I relationship.

Parameters of the light-response curve (Eq. 2) can be related to leaf nitrogen per unit leaf area (N_{LA}). Linear relationships give a satisfactory description of the increase of A_{max} and R_{day} with N_{LA}:

$$A_{max} = a_a(N_{LA} - N_b) \quad (3)$$

$$R_{day} = a_r(N_{LA} - N_b) + R_b \quad (4)$$

where a_a and N_b are the slope and intercept of the A_{max}-N_{LA} relation. N_b is the amount of N still present in leaves that have no photosynthetic capacity left. R_b is R_{day} in leaves with $N_{LA} = N_b$. The quantum yield, ϕ, and the curvature, Θ, may also depend on the leaf nitrogen concentration, N_{LA}, for which mathematical relationships can be formulated.

Canopy photosynthesis can now be calculated by using Equations 1–4, using the leaf N distribution in the canopy. For that purpose distribution functions may be used (Hirose & Werger 1987). Photosynthetic rates are summed over the different canopy layers and over a day or other time interval with varying irradiance. Daily course of irradiance may be described by a sinusoidal curve, or in any other way.

Maximum canopy photosynthesis at constant total leaf N of the plant is reached when at every depth in the canopy a change in leaf N (δN_{LA}) will result in the same change in daily photosynthesis (δA_{day}) (Field 1983):

$$\frac{\delta A_{day}}{\delta N_{LA}} = \lambda \quad (5)$$

continued

Box 9. *Continued*

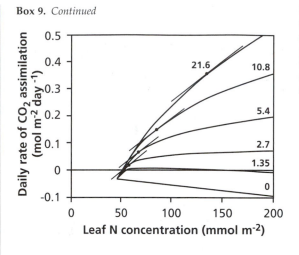

FIGURE 1. Calculated daily photosynthesis as a function of leaf N for different depths in a canopy with concomitantly different levels of irradiance (expressed as mol m^{-2} day^{-1}). The points of contact of the parallel tangents to the curves represent the optimal distribution pattern of N at a given total amount of leaf N (after Hirose & Werger 1987).

The constant λ is called the *Lagrange multiplier*. This is illustrated in Figure 1, where the points of contact of the tangents to the lines for daily photosynthesis at different canopy depths as a function of N_{LA} represent the optimal distribution of leaf N. Different total amounts of leaf N will result in different values for λ. In this way optimal leaf N distribution for maximum canopy photosynthesis of a plant per unit leaf N (Photosynthetic nitrogen-use efficiency, PNUE) can be calculated. Photosynthetic rates at actual distribution of leaf N in plants growing in leaf canopies has been compared with theoretically derived ones as described earlier, and with plants that have a uniform distribution. For instance, in the study of Pons et al. (1989) the performance of *Lysimachia vulgaris* at uniform and optimal distribution was 73% and 112%, respectively, of that at actual distribution. Hence, plants distribute leaf N close to the optimal distribution pattern.

A submodel of this model is a canopy photosynthesis model. This is a simplified one because both the light distribution and leaf photosynthesis use simplifications that are valid for the purpose of the preceding calculations, but not when we would be interested in the quantitative outcome of the canopy photosynthesis itself. The distribution of light as described here gives the average irradiance incident on leaves at a particular depth in a canopy with unidirectional light coming from straight overhead. It provides a reasonable approximation for diffuse light, but not for directional sunlight because spatial variation due to sunflecks and varying angle of inci-

dent sunlight are not accounted for. For the leaf photosynthesis module, the Farquhar and Von Caemmerer (1982) model could be used that not only accounts for varying conditions of irradiance as in this model, but also for variation in temperature and stomatal conductance. This model is described in Box 1 of the chapter on photosynthesis.

References and Further Reading

Anten, N.P.R., Schieving, F., & Werger, M.J.A. (1995) Patterns of light and nitrogen distribution in relation to whole canopy gain in C$_3$ and C$_4$ mono- and dicotyledonous species. Oecologia 101:504–513.

Evans, J.R. (1993) Photosynthetic acclimation and nitrogen partitioning within a lucerne canopy. II. Stability through time and comparison with a theoretical optimum. Aust. J. Plant Physiol. 20:69–82.

Field, C. (1983) Allocating leaf nitrogen for the maximization of carbon gain: Leaf age as a control on the allocation programme. Oecologia 56:341–347.

Hirose, T. & Werger, M.J.A. (1987) Maximising daily canopy photosynthesis with respect to the leaf nitrogen allocation pattern in the canopy. Oecologia 72:520–526.

Pons, T.L., Schieving, F., Hirose, T., & Werger, M.J.A. (1989) Optimization of leaf nitrogen allocation for canopy photosynthesis in *Lysimachia vulgaris*. In: Causes and consequences of variation in growth rate and productivity of higher plants, H. Lambers, M.L. Cambridge, H. Konings, & T.L. Pons (eds). SPB Academic Publishing, The Hague, pp. 175–186.

layer conductance. In effect, the boundary layer provides a **negative feedback** for transpiration. As a result, stomatal conductance has much less effect on canopy water loss than would be expected from study of single leaves (Jarvis & McNaughton 1986).

While transpiration from individual leaves in a leaf cuvette can be adequately described by the diffusion equation, transpiration from leaves in a canopy requires consideration of both diffusion and the leaf energy balance. The dual processes of **vaporization** and **diffusion** were first considered in an evaporation model by Penman (1948). This work was extended to include evaporation from vegetation by incorporation of a canopy conductance (Monteith 1963, 1965). This line of thinking, which leads to **single-layer** models, is to determine the evaporation if the plant canopy were no more than a partly wet plane at the lower boundary of the atmosphere. This conceptual plane, which is often referred to as a "big-leaf", is ascribed a physiological and aerodynamic resistance to water vapor transfer (Fig. 1). In an analogy with an individual leaf, a **canopy conductance** is introduced, which implicitly assumes that the conductances of individual leaves act in parallel, so that this canopy conductance can be determined by the leaf-area–weighted sum of leaf conductances (Monteith 1973). This approach ignores details of the canopy profile and simplifies the canopy to one single layer ("big leaf"; Field 1991). The "big-leaf" models are applicable only in circumstances where the detailed and complete spatial structure of the actual canopy microclimate and difference in response types of individual leaves that make up the canopy are irrelevant (Fig. 1).

Whenever details within the canopy (e.g., the interaction between microclimate and physiology) are important to estimate canopy gas exchange, big-leaf models are often not satisfactory. **Multilayer** models have therefore been developed

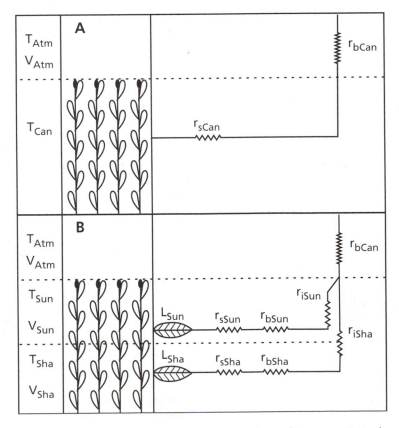

FIGURE 1. Schematic representation of (a) a single-layer ("big-leaf") model and (b) a multilayer model used for calculation of canopy evapotranspiration (modified after Raupach & Finnigan (1988). The model includes temperatures (T), vapor pressures (V), and resistances (r). Subscripts refer to the canopy (Can), the atmosphere (Atm), sunlit leaves (Sun), and shaded leaves (Sha). Resistances are stomatal (s), boundary layer (b), and in the air within the canopy (i).

(Cowan 1968, 1988). These models aim to describe both the evaporation of the entire canopy, and the partitioning of evaporation between the various components (e.g., soil, understory, and crown), together with other aspects of the canopy microclimate, such as profiles of leaf and air temperature and humidity of the air (Fig. 1).

Single-layer models are appropriate when one is concerned with vegetation essentially as a permeable lower boundary of the atmosphere or upper boundary of the soil, in systems with a length scale much larger than that of the vegetation itself. They are useful in hydrological modeling of large-scale or medium-scale catchments (e.g., areas where water is collected for urban use). On the other hand, multilayer models are appropriate when necessary to resolve details within the canopy, either because the detail is important in its own right or because the height scale is comparable to that of the system under investigation. They are relevant when dealing with interactions between microclimate and plant physiology or with hydrology in small catchments (Raupach & Finnigan 1988).

Water loss from communities includes both transpiration of leaves and evaporation directly from the soil. Evaporation from the soil may account for some 40% of the water used by a wheat crop in a mediterranean environment (Siddique et al. 1990). The soil component is affected by the level of radiation that penetrates through the canopy to the soil surface and hence by the canopy leaf area index. Evaporation from the soil is also affected by wetness of the soil surface, hydraulic conductivity of the soil, and wind speed beneath the canopy. The rate of soil evaporation is high when the surface is wet. As the soil dries out, the point of evaporation moves deeper into the soil and the surface layer offers a greater impedance, thus reducing soil evaporation. Measurements of soil evaporation beneath canopies can be obtained using minilysimeters. When the canopy intercepts most of the incident radiation, soil evaporation is likely to be a minor component of the total evaporation. If rain is infrequent and the soil surface dry, then soil evaporation tends to be insignificant. When the canopy is sparse, soil evaporation cannot be ignored and should be included in a bilayer evaporation model.

Canopies differ in the extent to which the behavior of individual leaves is "coupled" to the atmosphere. In **rough canopies**, such as those of forest trees or of small plants in complex terrain, the complex surface structure creates large eddies of air that penetrate the canopies. As a result, the air that surrounds each leaf has a temperature and humid-ity similar to that of bulk air, so that single-leaf models predict the behavior of leaves in canopies. On the other hand, individual leaves in **smooth canopies**, such as in crops or grasslands, are poorly coupled to the atmosphere. Leaf resistances are in series with the canopy boundary layer resistance (Fig. 1), which can be large relative to the combined effect of the leaf resistances smooth canopies, particularly when wind speeds are low. Under such conditions variation in leaf resistance does not play an important role in determining canopy evaporation.

3. Canopy CO_2 Fluxes

Carbon accumulation in communities involves exchanges of carbon with both the atmosphere and the soil (i.e., photosynthesis, plant respiration, and microbial respiration). **Photosynthesis** of the entire canopy can be approached as discussed in Section 2 for water use: with single-layer or multilayer models. It has proven consistently more difficult to approach **respiration** in an equally satisfactory manner. This offers a serious problem as the growth of the canopy is not a simple function of the canopy's photosynthesis. Respiration is a major and rather variable component of the plant's carbon budget (Sects. 1 and 4 of the chapter on respiration). Complications arise from the fact that respiration depends on metabolic activity as well as on carbohydrate status, in a manner that is not readily modeled. This remains one of those areas of the plant's physiology where more information is needed to allow scaling from the leaf's CO_2 flux to that of the canopy, especially if this approach is going to be used to address global issues.

There is currently considerable interest in appropriate means by which to scale gas exchange models up from leaf to the canopy level. Several models of canopy gas exchange based on equations developed for single leaves are already in use (**big-leaf** approach). Such models are relatively easy to handle, but they have the intrinsic disadvantage of major errors introduced when averaging gradients of light and photosynthetic capacity. It can also be modeled in a "multilayer" approach (Boxes 1 and 7). In big-leaf models of canopy photosynthesis, the Rubisco activity and electron-transport capacity per unit ground area are taken as the sums of activities per unit leaf area within the canopy. Such models overestimate rates of photosynthesis and require empirical curvature factors in the response to irradiance (cf. Sect. 3.2.1 of the chapter on photosynthesis). These curvature factors, however, are not constant; rather, they vary with leaf area index and nitrogen concentration (De Pury & Farquhar 1997).

Canopy photosynthesis can also be measured using large cuvettes that enclose entire plants or several plants in the canopy, or by eddy covariance, which is a micrometeorological approach that compares the concentrations of water vapor, CO_2 and heat in upward-moving versus downward-moving parcels of air. Figure 2 shows the rate of canopy CO_2 assimilation and total stomatal conductance of an entire macadamia tree (*Macadamia integrifolia*). Net CO_2 assimilation and stomatal conductance are related to photon irradiance, but the relationships differ for overcast conditions and clear sky.

The heterogeneity of the canopy may offer a complication in that the light environment is highly variable (cf. Sect. 3 of the chapter on photosynthesis) as are leaf physiological properties; moreover, photosynthesis and transpiration modify the air within the canopy, creating gradients in both humidity and temperature and in CO_2 concentration. Such gradients, however, tend to be much smaller than the variation that arises from irradiance and associated leaf traits. Errors associated with big-leaf models are avoided in models that treat the canopy in terms of a number of layers: the **multilayer** models. Thus, by combining a model of **leaf photosynthesis** with a model on **penetration of light** and on transport processes within the canopy, the flux from each canopy layer can de determined. The summation gives the canopy photosynthesis according to a **multilayer model**. Such models are essential for the analysis of the significance of variation in leaf traits along a depth gradient for plant performance. An example is the **allocation of nitrogen** to different leaves within a canopy that appears to be parallel with the light gradient in a canopy (Field 1983, Hirose & Werger 1987, Pons et al. 1989) (Box 7). Instead of complicated multilayer models, a simplification can be made by separately integrating the sunlit and shaded leaf fractions of the canopy. Such a **single-layer sun/shade model** is just as accurate and is also simpler (De Pury & Farquhar 1997).

4. Canopy Water-Use Efficiency

If canopies affect the gas exchange properties of individual leaves, then the water-use efficiency (WUE) of the canopy cannot simply be deduced from that of individual leaves measured under the prevailing bulk air conditions. Does this imply that genotypic differences in WUE at the leaf level (Sect. 6 of the chapter on plant water relations), disappear when studied in an ecologically more relevant context? When dealing with a **rough canopy** (Sect. 2), the differences certainly persist. In a **smooth canopy**, however, such as that of a wheat crop, the differences in conductance are diminished when scaling from the leaf to the canopy level (Fig. 3). Whereas the difference in photosynthetic water-use efficiency is as large as 24% at the leaf level, it is only 5% at the canopy level. Although part of this decrease in the genotypic difference in WUE, when scaling from a single leaf to the entire canopy, can be ascribed to the dominance of the **canopy boundary layer resistance** in the total canopy resistance (Sect. 2), the greater leaf area of the cultivar with lower WUE may also contribute to it. This greater leaf area reduces the canopy boundary layer conductance, which counteracts the greater stomatal conductance. In addition, much of the gain made by decreasing stomatal conductance and transpiration can be offset by greater evaporation from the soil when rate of leaf area development decreases simultaneously.

Wheat genotypes with a low WUE tend to develop their leaf area faster and have a higher leaf area ratio (LAR), in comparison with ones that have a higher WUE, with two important consequences. First, genotypes with a lower WUE transpire more of the available water early in the growing season, when the vapor pressure deficit of the air is relatively low due to low temperatures and, consequently, the WUE is high. Second, transpiration represents a greater fraction of the total crop water use of low-WUE genotypes due to the reduced evaporation from the soil, as mentioned earlier (Condon et al. 1993). A high LAR and vigorous early growth is clearly a major trait determining a crop's water use (Van den Boogaard et al. 1997). This calls for a line of plant breeding, which combines a high WUE (low δ-value) with vigorous early growth to reduce soil evaporation.

5. Canopy Effects on Microclimate: A Case Study

As pointed out earlier, individual leaves in smooth canopies, such as in crops or grasslands, are poorly coupled to the atmosphere. When stomatal conductance declines to low levels, leaves dissipate most heat through convective exchange, warming the air within the canopy. This creates turbulence

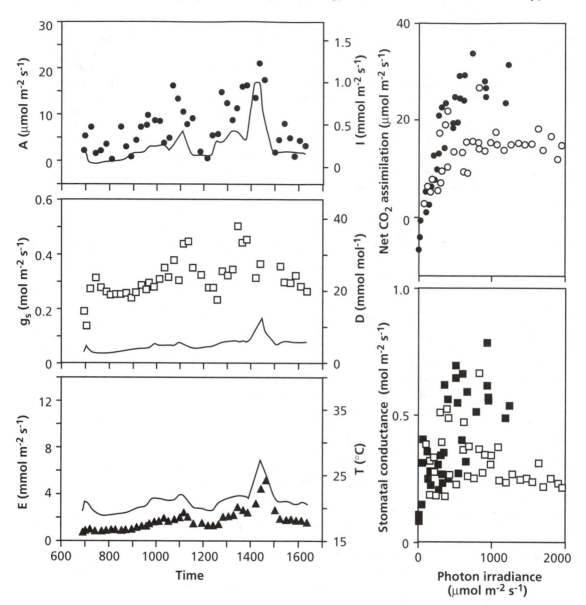

FIGURE 2. (Left) The rate of CO_2 assimilation, stomatal conductance, and transpiration (all expressed on a ground area basis) of an entire tree of *Macadamia integrifolia*, throughout an entire day. Diurnal changes in irradiance (I), leaf-to-air vapor pressure difference (D), and air temperature (T) are also shown (solid lines). (Right) The rate of net CO_2 assimilation and stomatal conductance (expressed on a ground area basis) of an entire tree of *Macadamia integrifolia*, as dependent on photon irradiance. The solid and open symbols refer to overcast and clear-sky conditions, respectively (Lloyd et al. 1995). Copyright CSIRO, Australia.

within the canopy, which brings new dry air into the canopy to increase transpiration.

The net loss of radiative energy from a surface exposed to the sky at night is balanced by the flow of heat from the overlying air and the underlying soil. During nights of radiation frost, temperatures of *Eucalyptus* leaves exposed to clear skies may be 1 to 3°C below those of the air. The resistance to heat transfer between air and grass is less than it is between air and soil because of the canopy's greater aerodynamic roughness. Because the thermal resistance of air within the grass sward is rather high, air temperatures immediately above the grass are lower than they are above bare soil. As a result, leaf

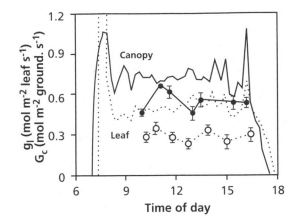

FIGURE 3. Comparison of the diurnal variation of leaf conductance, g_l (circles connected by lines) and canopy conductance, G_c (lines only) for two cultivars of *Triticum aestivum* (wheat), selected on the basis of their contrasting photosynthetic water-use efficiency at the leaf level (de Pury 1995). Reproduced with the author's permission.

temperatures of seedlings above grass tend to be lower than those above dry soil, which are lower than those above moist soil. Temperatures at the center of the larger leaves of *Eucalyptus pauciflora* (snow gum) become 0.5 to 1.5°C lower than those of the smaller leaves of *E. viminalis* (manna gum). Leaves of snow gum are 0.5 to 1.5°C cooler when horizontal than they are when vertical (Leuning & Cremer 1988). This affects the performance of plants growing above a grass canopy, as compared with those above bare patches (Sect. 2 of the chapter on interactions among plants).

6. Aiming for a Higher Level

When scaling processes from a single-leaf level to that of an entire canopy or community, we may be confronted with major complications. We may be able to acquire sufficient information about leaves in different layers of a canopy to use a "multilayer" approach and model the gas exchange of the entire canopy. It may also be difficult, costly, or even impossible, however, to acquire sufficient detail about all the leaves in a canopy to estimate rates of processes for an entire canopy. In these situations "big-leaf" models are particularly helpful.

When dealing with canopies, we have to be well aware of the fact that differences (e.g., in water-use efficiency), that are relatively large when studied at the leaf level may become smaller or disappear at the canopy level. Scaling from single leaves to communities will become increasingly important when ecophysiologists model effects of global change in temperature and atmospheric CO_2 concentration on primary productivity. Difficulties arise when dealing with the time factor; short-term effects of temperature on rates of processes may differ widely from those in acclimated plants. Here lies a major challenge for ecophysiologists wanting to deal with the future.

References and Further Reading

Balisky, A.C. & Burton, P.J. (1995) Root-zone temperature variation associated with microsite characteristics in high-elevation forest openings in the interior of British Columbia. Agric. For. Meteor. 7:31–54.

Condon, A.G., Richards, R.A., & Farquhar, G.D. (1993) Relationships between carbon isotope discrimination, water use efficiency and transpiration efficiency for dryland wheat. Aust. J. Agric. Res. 44:1693–1711.

Cowan, I.R. (1968) Mass, heat and momentum exchange between stands of plants and their atmospheric environment. Q.J.R. Meteor. Soc. 94:523–544.

Cowan, I.R. (1988) Stomatal physiology and gas exchange in the field. In: Flow and transport in the natural environment: Advances and applications, W.L. Steffen & O.T. Denmead (eds). Springer-Verlag, Berlin, pp. 160–172.

de Pury, D.G.G. (1995) Scaling photosynthesis and water use from leaves to paddocks. PhD Thesis, Australian National University, Canberra, Australia.

de Pury, D.G.G. & Farquhar, G.D. (1997) Simple scaling of photosynthesis from leaves to canopies without the errors of big-leaf models. Plant Cell Environ. 20:537–557.

Ehleringer, J.R. & Field, C.B. (eds) (1993) Scaling physiological processes: Leaf to globe. Academic Press, San Diego.

Field, C. (1983) Allocating leaf nitrogen for the maximization of carbon gain: Leaf age as a control on the allocation programme. Oecologia 56:341–347.

Field, C.B. (1991) Ecological scaling of carbon gain to stress and resource availability. In: Response of plants to multiple stress, H.A. Mooney, W.E. Winner & E.J. Pell (eds). Academic Press, San Diego, pp. 35–65.

Grace, J.B. (1983) Plant–atmosphere relationships. Chapman & Hall, London.

Hirose, T. & Werger, M.J.A. (1987) Maximising daily canopy photosynthesis with respect to the leaf nitrogen allocation pattern in the canopy. Oecologia 72:520–526.

Jarvis, P.G. & McNaughton, K.G. (1986) Stomatal control of transpiration: Scaling up from leaf to region. Adv. Ecol. Res. 15:1–49.

Jarvis, P.G., Miranda, H.S., & Muetzenfeldt, R.I. (1985) Modelling canopy exchanges of water vapour and carbon dioxide in coniferous forest plantations. In: The forest–atmosphere interaction, B.A. Hutchison & B.B. Hicks (eds). Reidel, Dordrecht, pp. 521–554.

Jones, H.G. (1983) Plant and microclimate. A quantiative approach to environmental physiology. Cambridge University Press, Cambridge.

Leuning, R. (1988) Leaf temperatures during radiation frost. II. Agric. For. Meteor. 42:135–155.

Leuning, R. & Cremer, K.W. (1988) Leaf temperatures during radiation frost. I. Agric. For. Meteor. 42:121–133.

Lloyd, J., Grace, J.B., Wong, S.C., Styles, J.M., Batten, D., Priddle, R., Turnbull, C., & McConchie, C.A. (1995a) Measuring and modelling whole-tree gas exchange. Aust. J. Plant Physiol. 22:987–1000.

Lloyd, J., Grace, J.B., Miranda, A.C., Meir, P., Wong, S.C., Miranda, H.S., Wright, I.R., Cash, J.H.C., & McIntyre, J. (1995b) A simple calibrated model of Amazonan rainforest productivity based on leaf biochemical properties. Plant Cell Environ. 18:1129–1145.

McMurtrie, R.E. (1993) Modelling of canopy carbon and water balance. In: Photosynthesis and production in a changing environment: A field and laboratory manual. D.A. Hall, J.M.O. Scurlock, H.R. Bolhar-Nordenkampf, R.C. Leegood, & S.P. Long (eds). Chapman & Hall, London, pp. 220–231.

Monteith, J.L. (1963) Gas exchange in plant communities. In: Environmental control of plant growth, L.T. Evans (ed). Academic Press, New York, pp. 95–112.

Monteith, J.L. (1965) Evaporation and environment. Symp. Soc. Exp. Biol. 19:205–234.

Monteith, J.L. (1973) Principles of Environmental Physics. Edward Arnold, London.

Penman, H.L. (1948) Natural evaporation from open water, bare soil and grass. Proc. R. Soc. London Series A. 193:120–145.

Pons, T.L., Schieving, F., Hirose, T., & Werger, M.J.A. (1989) Optimization of leaf nitrogen allocation for canopy photosynthesis in Lysimachia vulgaris. In: Causes and consequences of variation in growth rate and productivity of higher plants, H. Lambers, M.L. Cambridge, H. Konings, & T.L. Pons (eds). SPB Academic Publishing, The Hague, pp. 175–186.

Raupach, M.R. (1995) Vegetation-atmosphere interaction and surface conductance at leaf, canopy and regional scales. Agric. For. Meteor. 73:151–179.

Raupach, M.R. & Finnigan, J.J. (1988) "Single-layer" models of evaporation from plant canopies are incorrect but useful, whereas multilayer models are correct but useless. Aust. J. Plant Physiol. 15:705–716.

Siddique, K.H.M., Belford, R.K. & Tennant, D. (1990) Root:shoot ratios of old and modern, tall and semi-dwarf wheats in a mediterranean environment. Plant Soil 121:89–98.

Van den Boogaard, R., Alewijnse, D., Veneklaas, E.J., & Lambers, H. (1997) Growth and wateruse efficiency of ten Triticum aestivum L. cultivars at different water availability in relation to allocation of biomass. Plant Cell Environ. 20:200–210.

Williams, M., Rastetter, E.B., Fernandes, D.N., Goulden, M.L., Wofsy, S.C., Shaver, G.R., Melillo, J.M., Munger, J.W., Fan, S.-M., & Nadelhoffer, K.J. (1996) Modelling the soil-plant-atmosphere continuum in a Quercus-Acter stand at Harvest Forest: The regulation of stomatal conductance by light, nitrogen and soil/plant hydraulic properties. Plant Cell Environ. 19:911–927.

6
Mineral Nutrition

1. Introduction

If water is the environmental factor that most strongly constrains terrestrial productivity, then nutrients are an important additional factor. The productivity of virtually all natural ecosystems, even arid ecosystems, responds to addition of one or more nutrients, which indicates widespread nutrient limitation. Species differ widely in their ability to acquire nutrients from the soil. Some plants can take up iron, phosphate, or other ions from a calcareous soil from which other species cannot extract enough nutrients to persist. In other soils, the concentration of aluminum, heavy metals, or sodium chloride may reach toxic levels, but some species have genetic adaptations that enable them to survive in such environments. This does not mean that metallophytes need high concentrations of heavy metals or that halophytes require high salt concentrations to survive. These species perform well in the absence of these adverse conditions. Their distribution is restricted to these extreme habitats because, on one hand, these plants resist the adverse conditions, whereas most other plants do not. On the other hand, metallophytes and halophytes generally perform less well than most other plants in habitats without toxic levels of minerals or salts. Terms like **metallophytes**, **halophytes**, and others that we will encounter later in this chapter therefore refer to the **ecological amplitude** of the species rather than to their physiological requirements.

This chapter deals with both the acquisition and the use of nutrients by plants, concentrating on terrestrial plants that absorb nutrients predominantly via their roots from the soil. Leaves are also capable of acquiring nutrients. For example, volatile nitrogenous and sulfurous compounds, which may occur either naturally or as air pollutants in the atmosphere, can be taken up through the stomata. Nutrients in the water on wet leaves are also available for absorption by leaves. This may be of special importance for aquatic and epiphytic plants as well as for mosses. Other mechanisms to acquire nutrients include those found in carnivorous plants, which have an additional nutrient source from their prey, symbiotic associations with microorganisms, and parasitic associations with host plants. These will be treated in separate chapters.

2. Acquisition of Nutrients

Most terrestrial plants absorb the inorganic nutrients required for growth via their roots from the soil. For the uptake into the root cells transport proteins ("carriers" and "channels") are used (Sect. 2.2.1 in this chapter). Before describing mechanisms associated with transport across the plasma membrane, the movement of nutrients in the soil will be discussed.

2.1 Nutrients in the Soil

2.1.1 Nutrient Supply Rate

Nutrient supply rates in the **soil** ultimately govern the rates of nutrient acquisition by plants. Parent material, the rocks or sediments that give rise to soil, determines the proportions of minerals that are potentially available to plants. For example, granite is resistant to weathering and generally has lower concentrations of phosphorus and cations required by plants than does limestone. Other parent materials such as serpentine rock have high concentrations of heavy metals that are either not required by plants or are required in such low concentrations that their high concentrations in serpentine soils can cause toxic accumulations in plants. Various ecological factors (climate, vegetation, topography, and surface age) strongly influence weathering rates and rates of leaching loss and, therefore, the relationship between parent material and nutrient availability (Jenny 1980).

The **atmosphere** is the major source of nitrogen, through both biotic dinitrogen fixation (see chapter on symbiotic associations) and deposition of nitrate and ammonium in precipitation. There is also substantial input of cations from wet and dry deposition. Some of these cations (e.g., sodium) may come primarily from sea salt, particularly in coastal regions, but others (calcium, magnesium, and potassium) come from dust (from deserts, agricultural areas and unpaved roads) and from industrial pollution. These atmospheric inputs can be substantial. For example, atmospheric inputs of calcium are equivalent to 62%, 42%, and 154% of uptake by forests in the eastern United States, Sweden, and the Netherlands, respectively (Hedin et al. 1994), which is considerably higher than annual inputs by weathering. Thus, atmospheric inputs may determine external mineral supply to ecosystems much more than generally appreciated.

Soil pH is a major factor in determining the **availability** of nutrients in soils. High concentrations of hydrogen ions cause modest increases in nutrient input by increasing weathering rate (Johnson et al. 1972), but even greater loss of base cations by leaching. Acid rain is an increasing source of soil acidity caused by atmospheric deposition of nitric and sulfuric acid in precipitation. These acids first displace cations from the exchange complex on clay minerals and soil organic matter. Sulfate can then be leached below the root zone, carrying with it, to maintain charge balance, mobile mineral cations (e.g., potassium, calcium, and magnesium) and leaving behind a predominance of hydrogen and aluminum ions (Fig. 1). The availability of other

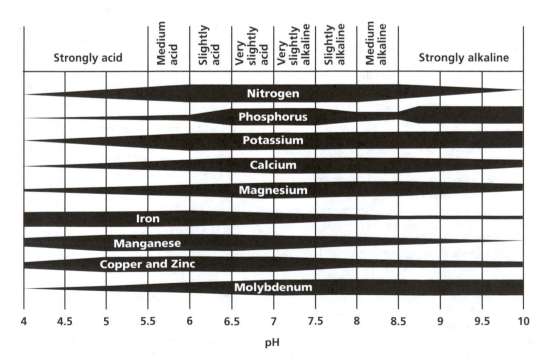

FIGURE 1. The availability of a number of ions in the soil as dependent on soil pH.

ions is strongly affected by pH because this affects their oxidation state and solubility (e.g., phosphorus, sulfur and aluminum) or affects the biological processes that control production and consumption (e.g., nitrogen, see later) (Fig. 1).

In the short term, recycling of nutrients from dead organic matter is the major direct source of soluble nutrients to soils (Table 1). Soluble cations like potassium and calcium are leached from dead organic matter, whereas organically bound nutrients like nitrogen and phosphorus must be released by **decomposition**. Phosphate is released by plant or microbial phosphatases, which cleave the phosphate ester bond. Nitrogen is released from dead organic matter yielding soluble organic nitrogen, which may be further decomposed to ammonium (nitrogen **mineralization**). Ammonium may then be oxidized, via nitrite, to nitrate (**nitrification**), and nitrate may be converted to gaseous dinitrogen (**denitrification**) (Fig. 2). The rates of these steps depend on temperature and soil conditions (e.g., pH and redox potential). At each step plants or soil microorganisms can take up soluble nitrogen, or nitrogen can be leached from the system, reducing the substrate available for the next nitrogen transformation. Therefore, the supply rates of the different forms of "available nitrogen" to plants and microbes must follow this same sequence: dissolved organic nitrogen ≥ ammonium ≥ nitrate (Eviner & Chapin 1997). If nitrogen supply rate always follows the same sequence in all soils, then why do the quantities and relative concentrations of these soluble forms of nitrogen differ among ecosystems?

First, microbes generally release phosphate or ammonium to the soil solution when their growth is more strongly limited by carbon than by nutri-

ents. As a result, plant litter with low nutrient concentrations and/or high concentrations of labile carbon (e.g., inputs of straw) immobilize nutrients, whereas microorganisms release nutrients when their growth is energy-limited. Second, environmental conditions further modify rates of specific nitrogen transformations. For example, cold anaerobic soils of the Arctic limit nitrogen mineralization and nitrification (an aerobic process) so that amino-acid nitrogen concentrations are quite high and nitrate concentrations low (Kielland 1994). On the other hand, in many arid and agricultural soils, high temperatures promote rapid mineralization and nitrification, and denitrification (an anaerobic process) occurs slowly, so nitrate is the most abundant form of soluble nitrogen. Finally, nitrogen uptake rates by plants and microorganisms modify availability of each nitrogen form to other organisms. For example, low concentrations of nitrate in acidic conifer forest soils are caused by rapid microbial nitrate uptake rather than by low nitrification rates (Stark & Hart 1997). These recent results raise questions about earlier generalizations that nitrification is inhibited in acid soils.

2.1.2 Nutrient Movement to the Root Surface

As roots grow through the soil, they **intercept** some nutrients. This amount, however, is often less than the amount contained in the growing root and therefore cannot serve as a net source of nutrients to the rest of the plant. That is, plants do not move toward the nutrients; rather, the nutrients must arrive by mass flow or diffusion (Table 2).

Rapid transpiration in plants results in substantial nutrient transport from the bulk soil to the root surface via **mass flow**. The extent to which mass flow is responsible for ion transport to the roots depends on the concentration of the different ions in the bulk solution relative to the requirement for plant growth (Table 2).

If less nutrients arrive at the root surface than required to sustain plant growth, then the concentration at the root surface drops, due to absorption by the roots. This creates a concentration gradient, which drives ion **diffusion** toward the root (e.g., for phosphate and potassium) (see Eq. 1 in Sect. 2.3 for the mathematical description of diffusion and mass flow). Other ions are delivered more rapidly by mass flow than they are required by the roots (e.g., calcium), which causes precipitation on the root surface (often as calcium sulfate). Diffusion from the bulk soil to the root surface depends both on the **concentration gradient** and on the **diffusion coefficient**. This coefficient, which varies among soil

TABLE 1. Major sources of available nutrients that enter the soil.

Nutrient	Source of plant nutrient (% of total)		
	Atmosphere	Weathering	Recycling
Temperate forest			
N	7	0	93
P	1	<10?	>89
K	2	10	88
Ca	4	31	65
Arctic tundra			
N	4	0	96
P	4	<1	96

Source: Chapin 1991.

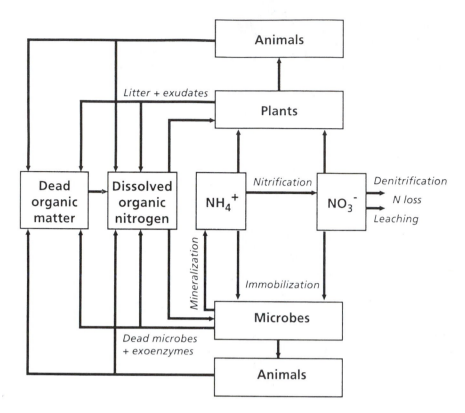

FIGURE 2. A simplified view of the terrestrial nitrogen cycle. All nitrogen pools (boxes) and transformations (arrows) are affected by both plants and microorganisms. Dead plants, animals, and microorganisms are decomposed, releasing dead organic matter and then dissolved organic nitrogen (e.g., amino acids, urea). Some of the dissolved organic nitrogen in soils originates from living organisms. Both plants and microorganisms are capable of using dissolved organic nitrogen. Microorganisms use the dissolved organic nitrogen as a carbon source, releasing N that is in excess of their requirement as NH_4^+. Both plants and microorganisms can use NH_4^+ as a source of N. Incorporation of NH_4^+ into soil microorganisms leads to N-immobilization; the reverse transformation is called *mineralization*. Immobilization predominates at high avail-ability of a carbon source, whereas mineralization is favored by a shortage of a source of carbon for microorganisms. Under aerobic conditions, some of the NH_4^+ is transformed into NO_3^- in a process called *nitrification*. In alkaline soil, nitrification predominantly results from autotrophic microorganisms, whereas in acid soil heterotrophic microorganisms are probably most important. NO_3^- is available for both plants and microorganisms; as with NH_4^+, some of the NO_3^- may be immobilized, or lost from the system through leaching or denitrification. This scheme ignores nitrogen deposition and dinitrogen fixation, pathways of nitrogen loss from the system, and fluxes between plant-microbial symbionts (Eviner & Chapin 1997). Reprinted with permission from Nature © copyright 1997 Macmillan Magazines Ltd.

types, differs by three orders of magnitude among common ions. It is large for nitrate, which therefore moves quickly to the root surface in moist soils, even when there is little water uptake. The diffusion coefficient is also fairly large for potassium, so that most plants can acquire sufficient potassium. Diffusion coefficients are very low for zinc and inorganic phosphate (Table 3), due to specific interactions with the clay minerals of the soil cation-exchange complex. Hence, variation in soil clay content is one of the factors that affects the diffusion coefficient. Nitrogen and phosphorus, which are the two nutrients that most frequently limit plant growth, are seldom supplied in sufficient quantities by mass flow to meet the plant requirement; therefore, diffusion generally limits their supply to the plant, particularly in natural ecosystems where soil solution concentrations are much lower than they are in agricultural soils (Table 4).

Most estimates of the importance of mass flow consider only water movement associated with transpiration. Bulk movement of soil solution, however, also occurs as a "wetting front" following rain. The wetting front carries ions with it and replen-

TABLE 2. The significance of root interception, mass flow, and diffusion in supplying *Zea mays* (maize) and a sedge tundra ecosystem with nutrients.[a]

Nutrient	Amount taken up by the crop	Approximate amounts supplied by		
		Root interception	Mass flow	Diffusion
Zea mays				
Nitrogen	190	2	150	38
Phosphorus	40	1	2	37
Potassium	195	4	35	156
Calcium*	40	60	165	0
Magnesium*	45	15	110	0
Sulfur	22	1	21	0
Copper*	0.1	—	0.4	—
Zinc	0.3	—	0.1	—
Boron*	0.2	—	0.7	—
Iron	1.9	—	1.0	—
Manganese*	0.3	—	0.4	—
Molybdenum*	0.01	—	0.02	—
Sedge tundra ecosystem				
Nitrogen	22	—	0.1	21.9
Phosphorus	1.4	—	0.01	1.4
Potassium	9.7	—	0.6	9.1
Calcium*	20.9	—	52	0
Magnesium	47.1	—	39.1	8.0

Source: Barber & Olson 1968, as cited in Clarkson 1981, and Jungk 1991; tundra data calculated from Shaver & Chapin 1991 and Chapin, unpublished.

[a] All data in $kg\,ha^{-1}$. The maize data pertain to a typical fertile silt loam and a crop yield of $9500\,kg\,ha^{-1}$ and the tundra data a wet sedge meadow with a low-nutrient peat soil. The amount supplied by mass flow was calculated from the concentration of the nutrients in the bulk soil solution and the rate of transpiration. The amount supplied by diffusion is calculated by difference; other forms of transport to the root (e.g., mycorrhizae) may also be important, but they are not included in these estimates. The elements marked * are potentially supplied in excess by mass flow; they may accumulate at the soil/root interface and diffuse back into the bulk soil.

ishes "diffusion shells" where plant uptake has reduced nutrient concentrations around individual roots. In Arctic tundra, where permafrost causes substantial lateral movement of water, bulk water flow accounts for 90% of the nutrient delivery to deep-rooted species (Chapin et al. 1988). Bulk water movement may have a large (but currently unknown) influence on nutrient supply in other wet ecosystems. Soil heterogeneity may influence the importance of bulk water flow for nutrient supply to roots. Roots and rain water both move preferentially through soil cracks created by worms or soil drying. For example, about 40% of the elongation of apple roots occurs along soil cracks made by worms. These effects of soil heterogeneity could increase the importance of bulk water movement as a mechanism of nutrient supply more than is currently appreciated.

TABLE 3. Typical values for diffusion coefficients for ions in moist soil.*

Ion	Diffusion coefficient $(m^2\,s^{-1})$
Cl^-	$2-9\,10^{-10}$
NO_3^-	$1\,10^{-10}$
SO_4^{2-}	$1-2\,10^{-10}$
$H_2PO_4^-$	$0.3-3.3\,10^{-13}$
K^+	$1-28\,10^{-12}$

Source: Clarkson 1981.

*The range of values represents values for different soil types.

Mass flow and diffusion cannot always account for the nutrient transport to the root surface. Mass flow delivers very little inorganic phosphate to the root and the diffusion coefficient for inorganic phosphate in soil is too low to allow much inorganic phosphate to move by diffusion. Some organic phosphate molecules may diffuse more rapidly and become available for the roots, but generally diffusion of organic phosphate is also slow (Sect. 2.2.5.1). If plants do not have access to this source of phosphorus, then special adaptations or

TABLE 4. Yields of seedlings of five species that differ in root surface area in phosphate-deficient soil with various additions of phosphate, relative to the yield obtained at the highest phosphate supply.*

| Species | Root diameter (mm) | Root hairs | | Mean mass (mg) at P-supply | | | |
		Frequency	Length (mm)	0	15	45	135
Podocarpus totara	>1.0	None	—	31	31	**40**	**100**
Coprosma robusta	0.2–0.3	Few	0.2	4	4	7	**100**
Leptospermum scoparium	0.15–0.2	Moderate	<1	16	26	**62**	**100**
Solanum nigrum	0.15–0.2	Abundant	1	1	**4**	**25**	100
Lolium perenne	>0.1	Abundant	1	101	—	—	**100**

Source: After Baylis 1970; 1972, as cited in Clarkson 1981.
*The amount of phosphate-fertilizer (in mg) per pot containing 200 g soil is indicated. All plants were grown without mycorrhiza. Values printed bold differ significantly from those to their immediate left.

acclimations are required to acquire phosphate when its concentration in the soil solution is low (Sect. 2.2.2). Mycorrhizae are an additional important mechanism of nutrient transport to the root (chapter on symbiotic associations).

Because nitrate moves more readily to the roots' surface, it would appear to be available in larger quantities than ammonium. Is nitrate really the predominant source of nitrogen for any plant? That depends to a large extent on environmental conditions. If the soil pH is low, then so is the rate of nitrification; that is, the oxidation of ammonium to nitrite and then to nitrate by ammonium-oxidizing and nitrite-oxidizing autotrophic bacteria, respectively. Under these conditions nitrate will not be a major source of N. The same is true for anaerobic soils because nitrification is an aerobic process. When soils are cold, such as in the Arctic, mineralization is slow, very little ammonium is made available, and a large fraction of the total pool of soil nitrogen is present as amino acids (Kielland 1994). Under such conditions amino acids tend to be a major source of N, rather than ammonium or nitrate (Chapin et al. 1993). That does not mean that arctic plants are unable to use nitrate or ammonium. If supplied in sufficient amounts, then they will absorb and assimilate it. Most plants from acid soils also appear to be capable of absorbing and assimilating nitrate and very few, if any, species appear to be incapable of using nitrate as a source of nitrogen (Atkin 1996).

Low water availability reduces diffusion rates below values found in moist soils because air replaces water in pores of dry soil, greatly lengthening the path from the bulk soil to the root surface (increased "tortuosity"). Ion mobility can decrease by two orders of magnitude between −0.01 and −1.0 MPa, which is a range in soil water potential that does not strongly restrict water uptake by most plants (Fig. 3). Because diffusion is the rate-limiting

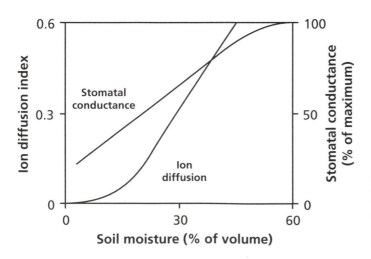

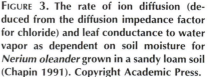

FIGURE 3. The rate of ion diffusion (deduced from the diffusion impedance factor for chloride) and leaf conductance to water vapor as dependent on soil moisture for *Nerium oleander* grown in a sandy loam soil (Chapin 1991). Copyright Academic Press.

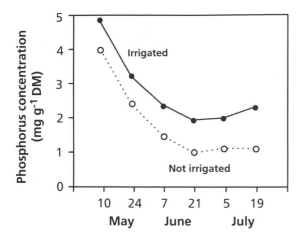

FIGURE 4. Phosphorus concentration in the shoots of *Hordeum vulgare* (barley) grown with or without irrigation (Chapin 1991). Copyright Academic Press.

step in uptake of the most strongly limiting nutrients (Table 2), reduction in water availability can greatly reduce plant growth. Two lines of evidence suggest that this may be a major causal mechanism by which low water supply restricts plant growth (Chapin 1991):

1. Tissue concentrations of growth-limiting nutrients often decline with water stress (Fig. 4), whereas one would expect tissue concentrations to increase if water restricted growth more than nutrient uptake.
2. Nutrient addition enhances growth of some desert annuals more than does water addition (Gutierrez & Whitford 1987).

For soil-mobile ions, such as nitrate, tissue concentrations vary with soil moisture availability in exactly the opposite manner as found for immobile ions. That is, in plants of Australian semi-arid mulga woodlands, the nitrate concentration in the tissue tends to be high and the rate of nitrate assimilation tends to be low, when the availability of soil moisture is low. After a shower, the nitrate concentration in the soil rises rapidly, the rate of nitrate assimilation in the tissue increases whereas the concentration of nitrate in the tissue declines (Erskine et al. 1996).

2.2 Root Traits That Determine Nutrient Acquisition

Rates of nutrient uptake depend on the quantity of root surface area and the uptake properties of this surface. Once nutrients arrive at the root surface, they must pass the plasma membrane of the root cells. As with carbon uptake by photosynthesis, the rate of nutrient uptake depends on both the concentration in the environment and the **demand** of the plant. The plant's demand is determined by its growth rate and the concentration of nitrogen in the tissues. At a high internal concentration, nutrient uptake is down-regulated. Despite this feedback mechanism, plants may show **luxury consumption** of specific nutrients (i.e., absorption at a higher rate than required to sustain growth), leading to the accumulation of that nutrient. Many species from nitrogen-rich sites [e.g., *Urtica dioica* (stinging nettle) *Spinacia oleracea* (spinach), and *Lactuca sativa* (lettuce)] show luxury consumption of nitrate and accumulate nitrate in their cells.

2.2.1 Increasing the Roots' Absorptive Surface

Because diffusion is the major process that delivers growth-limiting nutrients to plant roots (Table 2), the major way in which plants can augment nutrient acquisition is by increasing the size of the root system. The relative size, expressed as the **root mass ratio** (root mass as a fraction of total plant mass), is enhanced by growth at a low nutrient supply (**acclimation**) (Brouwer 1962). Similarly, plants **adapted** to low nutrient supply typically have a high root mass ratio. The high potential relative growth rate that characterizes plants on fertile soils requires that a large fraction of the plant's resources is allocated to leaves (chapter on growth and allocation). Species adapted to fertile soils, however, show no consistent difference in plasticity of root mass ratio with changes in soil fertility (Reynolds & D'Antonio 1996). Increased allocation of resources to root growth with acclimation or adaptation to infertile soils is particularly important for those ions that diffuse slowly in soil (e.g., phosphate; cf. Table 3). In a heterogeneous soil, roots tend to proliferate in those zones with highest availability of growth-limiting nutrients, rather than in the depleted zones, thus maximizing the effectiveness of each unit of root production (but, see Sect. 2.2.5).

The effective absorbing root surface can be enlarged by **root hairs** (Table 4). These root hairs vary in length from 0.2 to 2 mm, depending on species. Root hair length may increase from 0.1 to 0.8 mm by a reduction in the supply of nitrate or phosphate (Bates & Lynch 1996, Jungk 1996), but such a response is not invariably found (Powell 1974). The diameter of most roots involved in ion uptake is between 0.15 and 1.0 mm, so the presence of root hairs allows a considerably larger cylinder of the soil to be exploited by the root than could be achieved by a root without root hairs. This may enhance the phosphate-uptake capacity by a factor of 3 to 4 (Clarkson 1981). Root hairs have greatest effect on absorption of those ions that diffuse slowly in soil. Species with a high frequency of long root hairs yield relatively more when phosphate is limiting, in comparison with those with less frequent or shorter root hairs, which need a high phosphate supply for good growth (Table 4). Increasing the root mass ratio or production of root hairs must incur costs, in terms of investment of carbon, nitrogen, and other resources. To achieve a threefold expansion of the root surface by root hairs requires less than 2% of the costs associated with a similar increase realized by a greater investment in roots (Clarkson 1996). **Mycorrhizal associations** are even more effective in terms of enlarging the phosphate-absorbing surface per unit cost, also if we consider that the fungus requires additional plant-derived carbohydrates for its functioning (cf. Sect. 2.6 in the chapter on symbiotic associations).

2.2.2 Transport Proteins: Ion Channels and Carriers

Roots transport nutrients across their plasma membrane either by **diffusion** down an electrochemical potential gradient or by **active transport** against an electrochemical potential gradient. The electrochemical potential gradient is caused by the extrusion of protons by a **proton-pumping ATPase** that pumps H^+ from the cytosol across the plasma membrane, creating an electrical potential difference of 80 to 150 mV (negative inside) across the plasma membrane (Fig. 5A). Such a proton pump functions like the ATPase in the thylakoid membrane of the chloroplast (Sect. 2.1.3 in the chapter on photosynthesis) and the inner membrane of mitochondria (Sect. 2.5.1 in the chapter on respiration); however, now the ATPase acts in reverse: It uses ATP and extrudes protons. Cations tend to move inward and anions outward along this electrical gradient. The **Nernst equation** allows us to calculate that monovalent cations are at electrochemical equilibrium (no driving force for movement) if the concentration of the cation is 40- to 150-fold lower outside than inside the cell. For monovalent anions the reverse can be calculated: The concentration of an anion at electrochemical equilibrium is 40- to 150-fold lower inside than outside the cell. When concentration gradients are less than this, ions move in

the direction predicted by electrochemical gradient; when the concentration gradients exceed these values, ions move in the opposite direction (Fig. 5B).

For most ions, diffusion across the lipid bilayer of the plasma membranes is a very slow process, unless facilitated by special transport proteins. Such transport proteins include **ion-specific channels** (i.e., "pores" in the membrane through which ions can move single file) (cf. the water-channel proteins discussed in Sect. 5.2 of the chapter on plant water relations). The ion channels are either open or closed, depending on the membrane potential or the concentration of specific effectors (Fig. 5A). Ion channels have the advantage that they allow massive transport, albeit it only down an electrochemical potential gradient. If such a gradient does not exist or when the gradient is in the opposite direction, channels cannot be used for net transport. In that case, transport may require, first, the extrusion of protons via a proton-pumping ATPase (Fig. 5A). The proton gradient can then be used for uptake of ions, in a proton-cotransport mechanism via **carrier proteins** (Fig. 5A). Such carriers are like enzymes: they bind their substrates, followed by a specific reaction (release of the substrate at the other side of a membrane), and may be allosterically regulated. Carriers have a much lower transport capacity than channels. Both types of proteins are subject to turnover so that continuous protein synthesis is required to maintain ion transport.

Both channels and carriers are, in principle, ion-specific, but other ions with similar structure might occasionally enter the cell via these transport proteins. This may account for the entry of some heavy metals and aluminum in plant roots. Transport proteins are involved in the **influx** of nutrients from the rhizosphere as well as in the transport of some of the acquired nutrients into the **vacuoles** and the release into the **xylem vessels** (De Boer & Wegner 1997). Channels are also involved in ion **efflux**, sometimes spectacularly so, as during stomatal movements (Sect. 5.4.2 in the chapter on plant water relations), or they may be responsible for efflux of nutrients, which may occur simultaneously with nutrient influx. Uptake of sodium ions from a saline soil occurs down an electrochemical potential gradient, in which case the ions may be extruded with an energy-dependent carrier mechanism (Sect. 3.4.1 in this chapter).

Transport from the rhizosphere across the plasma membrane into the cytosol (influx) is mostly against an electrochemical potential gradient for all anions, and sometimes also for some cations. Such

A

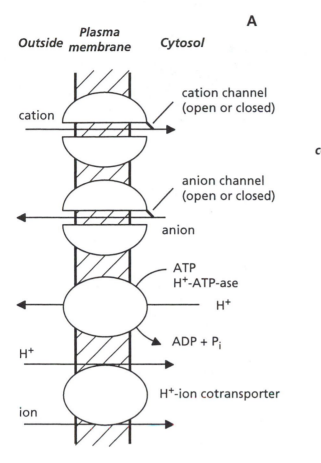

B

$$\Delta E = - \underbrace{\frac{RT}{zF}}_{-25.6} \ln \frac{[ion]_{in}}{[ion]_{out}} \ (mV)$$

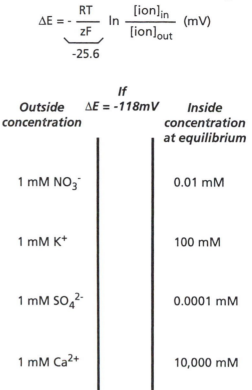

If

Outside concentration	$\Delta E = -118mV$	Inside concentration at equilibrium
1 mM NO_3^-		0.01 mM
1 mM K^+		100 mM
1 mM SO_4^{2-}		0.0001 mM
1 mM Ca^{2+}		10,000 mM

FIGURE 5. (A) Ion transport across the plasma membrane. The membrane potential is negative (i.e., there is a negative charge inside and a positive charge outside). Cations can enter via a cation channel, down an electrochemical potential gradient. Anions (e.g., nitrate) might leave the cytosol via an anion channel, down an electrochemical potential gradient. An H^+-ATPase ("proton pump") extrudes protons from the cytosol, thus creating a protonmotive force. Protons can be used to drive ion uptake against an electrochemical potential gradient. See the text for further explanation. (B) Schematic representation of the concentration of monovalent and divalent anions and cations that is expected if the plasma membrane is perfectly permeable for these ions in the absence of energy-requiring mechanisms at a membrane potential of 118 mV. The Nernst equation gives the relationship between the membrane potential ΔE and the outside and inside ion concentrations. R is the gas constant; T is the absolute temperature; z = the valency of the ion for which the equilibrium concentration is calculated; F is Faraday's number. See the text for further explanation.

transport must involve an active component (i.e., it requires **metabolic energy**). Transport mediated by channels, however, also requires metabolic energy, be it indirectly, to generate the electrochemical potential gradient. This requires respiratory energy: ATP is used to extrude protons, catalyzed by an H^+-ATPase, so that a membrane potential is created (inside negative). Efflux of ions, from the cytosol to the rhizosphere, is mostly down an electrochemical potential gradient for anions; the efflux of nitrate may be very low in some circumstances, but it may also be of similar magnitude as the influx (Oscarson et al. 1987, Ter Steege 1996). Like the nitrate-uptake system, the nitrate-efflux system is nitrate-inducible, and it strongly increases with increasing internal nitrate concentrations. The efflux system requires both RNA and protein synthesis, but has a much lower turnover rate than does the uptake system (Aslam et al. 1996). Nitrate efflux may contribute significantly to the respiratory costs associated with nutrient acquisition (cf. Sect. 5.2.3 in the chapter on plant respiration). Nitrate efflux may reflect a fine-control of net uptake, compared to the coarse-control of gene expression.

2.2.3 Acclimation and Adaptation of Uptake Kinetics

2.2.3.1 Response to Nutrient Supply

Nutrient uptake by roots increases in response to increasing nutrient supply up to some maximum uptake rate where a plateau is reached (Fig. 6A), which is very similar to the CO_2 or light-response curves of photosynthesis. If nutrient uptake is not limited by diffusion of the nutrient to the root surface, then the shape of this curve is also similar to that obtained with enzymes in solution (Michaelis-Menten kinetics). This leads to the suggestion that the **maximum inflow rate (I_{max})** may be determined largely by the abundance or specific activity of transport proteins in the plasma membrane; the K_m describes the **affinity** of the transport protein for its ion. This analogy, however, may not be entirely accurate because the access of ions to carriers and ion channels in plasma membranes of a structurally complex cortex is probably quite different from the access of substrates to an enzyme in a stirred solution. Nonetheless, I_{max} is a useful description of the capacity of the root for ion uptake, and K_m describes the capability of the root to utilize low concentrations of substrate (low K_m confers high affinity). C_{min} is the minimum ion concentration at which net uptake occurs (analogous to the light and CO_2-

compensation points of photosynthesis). C_{min} is determined by the balance of influx by ion-transport proteins and efflux along an electrochemical potential gradient. The experimental determination of C_{min} is difficult. For instance, in nonsterile conditions much of the labeled nutrient remaining in solution is in micoorganisms. If these are filtered out, then the C_{min} is often spectacularly lower than is usually determined in this critical experiment.

For many nutrients, roots have both a **high-affinity uptake system**, which functions well at low external concentration but has a low I_{max}, and a **low-affinity system**, which is slow at low external concentrations but has a high I_{max} (Borstlap 1983, Ullrich 1992). The high-affinity system is most probably carrier-mediated, whereas the low-affinity system may reflect the activity of a channel, at least for K^+. The ecophysiological significance of low-affinity systems for nitrate, which only allow significant uptake at nitrate concentrations well above that in most natural soils, still remains to be demonstrated.

When nutrients are in **short supply**, plants tend to show a **compensatory** response in that the I_{max} is increased and a high-affinity transport system is sometimes induced. For example, plants exhibit a high capacity (i.e., high I_{max}) to absorb phosphate when grown at a very low supply of phosphate, a high potential to absorb nitrate and ammonium under conditions when nitrogen is in short supply, a high potential to absorb potassium or sulfate when potassium or sulfate are limiting (Lee 1982,

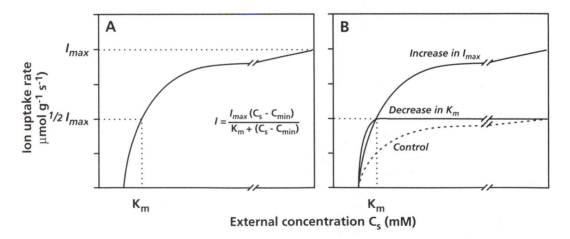

FIGURE 6. (A) The relationships between uptake rates (net inflow = I) of ions and their external concentrations (Cs). At C_{min} the net uptake is zero (influx = efflux) (after Marschner 1995). (B) Uptake kinetics in control plants and in plants grown with a shortage of nutrients. Note that both induction of a different high-affinity system and up-regulation of the same low-affinity system enhance the capacity for nutrient uptake at low external concentration.

TABLE 5. **Effect of a shortage of one nutrient or of water and exposure to a low irradiance on the maximum rate of nutrient uptake (I_{max}).***

Factor	Ion absorbed	Uptake rate by stressed plant (% of control)
Nitrogen	Ammonium	209
	Nitrate	206
	Phosphate	56
	Sulfate	56
Phosphorus	Phosphate	400
	Nitrate	35
	Sulfate	70
Sulfur	Sulfate	895
	Nitrate	69
	Phosphate	32
Water	Phosphate	13
Light	Nitrate	73

Source: Chapin 1991.
* Values for barley, except for water stress (tomato). Stress is due to low availability of the resource listed in the left-hand column.

Table 5). Information about other nutrients is sparse, but it suggests that there is little stimulation of the inflow of calcium, magnesium, and manganese (Robinson 1996). The compensatory increase in I_{max} for phosphate, nitrogen, and potassium in response to a shortage of these nutrients occurs over a 2- to 15-day period, and it is specific to the nutrient that limits growth: Nitrogen limitation increases the capacity to absorb nitrogen, but it decreases the capacity to absorb other nonlimiting nutrients (Table 5). The appearance of a high-affinity system (low K_m) is especially strong for K^+ and happens within an hour (Smart et al. 1996). The compensatory response is found in widely different taxa, with no evidence for any type of plant showing a greater increase than others (Robinson 1996).

The simplest explanation for these compensatory changes in I_{max} is that it involves synthesis of additional transport proteins for the growth-limiting nutrient. Indeed, there is an up-regulation of mRNA levels coding for a high-affinity sulfate uptake system and the capacity for sulfate uptake in roots of *Stylosanthes hamata* (Smith et al. 1995). A decrease in K_m could be due to induction of a high-affinity system or to allosteric effects on existing transporters (Smart et al. 1996). Both an increase in capacity (I_{max}) of a low-affinity system and induction of a high-affinity system could enhance the uptake capacity at a low nutrient supply (Fig. 6B).

The significance of the up-regulation of the uptake system for the plant is that the concentration of the limiting nutrient at the root surface is decreased, which increases the concentration gradient and the diffusion of the limiting nutrient from the bulk soil to the root surface. The significance of such up-regulation for plants growing in soil is probably relatively small for immobile ions such as phosphate, when compared with that for mobile ions such as nitrate. For immobile ions, it is the mobility in the soil, rather than the I_{max} of the roots, that determines the rate at which roots can acquire this nutrient from the rhizosphere (cf. Sects. 2.1.2 and 2.3).

2.2.3.2 Response to Nutrient Demand

Any factor that increases plant **demand** for nutrients appears to cause an increase in I_{max}. Up-regulation of the system for nitrate uptake upon an increased demand involves the nitrate concentration in the root itself as well as signals from the shoot, imported via the phloem (King et al. 1993). The signals that arrive via the phloem probably include a low concentration of amino acids and/or an increased concentration of organic acids (Touraine et al. 1994). In experiments on effects of the demand for phosphate, potassium or sulfate, effects of demand can be simulated by a period of starvation, as discussed in Section 2.2.3.1. The number of transport proteins in plasma membranes increases in the absence of any external ion; however, the influence of demand and of starvation on nitrate transport is more complex (Fig. 7).

For nitrate, as is the case for many other ions, there are two inducible uptake systems: a **high-affinity transport system** (HATS) and a **low-affinity transport system** (LATS) (Siddiqi et al. 1990). The high-affinity system involves both a constitutive and an inducible component (N.C. Huang et al. 1996). In the complete absence of external nitrate, rather than low external concentrations as described in Table 5, the uptake capacity is very low. In *Hordeum vulgare* (barley) and *Lotus japonicum* (birdsfoot-trefoil) the mRNA for the HATS is almost absent after 72 hours of nitrate deprivation. Upon reexposure of the roots to nitrate, this is first taken up by the constitutive HATS. After 30 minutes, there is a huge rise in mRNA encoding the HATS and after 2 to 4 hours the inducible HATS is reassembled in the plasma membrane (Trueman et al. 1996a) and the rate of nitrate uptake increases (Siddiqi et al. 1990). The general experience, however, is that plants receiving nitrate, but in amounts inadequate for supporting maximum growth, de-

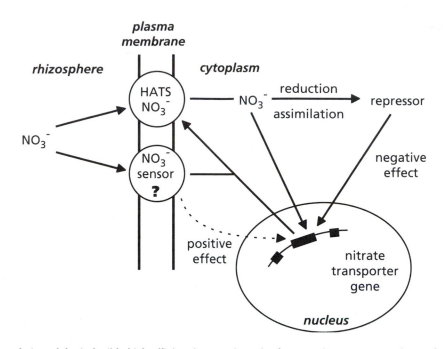

FIGURE 7. Regulation of the inducible high-affinity nitrate uptake system (HATS) by nitrate. The HATS is affected both by external nitrate supply and by internal demand. *Situation 1*: If *no nitrate* is present in the rhizosphere, then there is no positive effector. The gene encoding the nitrate transporter is repressed and the system cannot respond immediately to the addition of nitrate. *Situation 2*: If there is an *inadequate nitrate* concentration in the rhizosphere, then nitrate is sensed (probably by the constitutive HATS) and the gene encoding the nitrate transporter is transcribed and the system responds to the addition of nitrate. Products of the reduction and assimilation of nitrate (amino acids, organic acids) have a negative effect on the transcription of the gene encoding the HATS.

repress their nitrate transport activity (Table 5), so that net nitrate uptake increases in experimental conditions where the plants are given a sudden dose of nitrate (Fig. 7).

The significance of the low-affinity uptake systems that only function at external nitrate concentrations well above that normally found in soil is puzzling. Concentrations in the range of 5 to 20mM, however, do occur in the rhizosphere of crop plants and ruderals. In *Arabidopsis thaliana* (common wall cress), the gene encoding the inducible low-affinity system is expressed in epidermal cells close to the root tip, and in cells beyond the epidermis and even the endodermis further away from the tip, but it is never expressed in the vascular cylinder (N.C. Huang et al. 1996). The low-affinity uptake systems cannot be passive systems because transport occurs against an electrochemical potential gradient even at an external nitrate concentration of 1mM (Siddiqi et al. 1991). The constitutive system may serve as a **nitrate-sensing system**, because it is associated with a plasma membrane-bound nitrate reductase. The concerted action of the constitutive system and its associated

nitrate reductase may lead to the production of intermediates that induce both the inducible high-affinity system and cytosolic nitrate reductase (Martins-Louçao & Cruz 1998). Both the constitutive and the inducible system are carrier-mediated proton-cotransport systems, requiring the entry of two protons for every nitrate taken up (Mistrik & Ullrich 1996, Trueman et al. 1996b).

C_{min} for a given ion decreases in minutes to hours in response to decreases in supply of that ion, due to decreases in its cytoplasmic concentration, which reduce leakage across the plasma membrane and therefore efflux rates (Kronzucker et al. 1997). The increase in I_{max} when plants acclimate to low availability of a given nutrient increases the plant's capacity to absorb nutrients from solutions of low concentration (Fig. 8). This compensation is always less than 100%, however, so that tissue concentrations increase under conditions of high nutrient supply (**luxury consumption**) and decrease under conditions of low nutrient supply (high nutrient-use efficiency; see Sect. 4) (Chapin 1988).

The nature of genetic adaptation to infertile soils differs among ions. Plants adapted to infertile soils

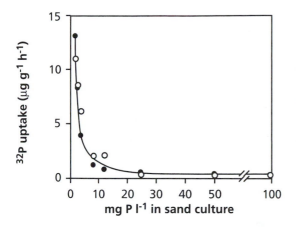

FIGURE 8. Phosphate uptake rate *Betula verrucosa* (birch) seedlings from a phosphate solution (5 µM) in relation to the phosphorus concentration in the sand culture in which plants had previously been grown. Open and solid symbols show results of separate experiments (Harrison & Helliwell 1979). Copyright Blackwell Science Ltd.

2.2.3.3 Response to Other Environmental and Biotic Factors

The responses of nutrient uptake kinetics to changes in water, light, and other factors are readily predicted from changes in plant demand for nutrients. **Water stress** may reduce the capacity of roots to absorb nutrients if it reduces growth and therefore plant demand for nutrients (Table 5, Fig. 9A). Plants adapted to dry environments typically have low relative growth rates (chapter on growth and allocation) and consequently low capacities to absorb nutrients. The effect of **irradiance** on nutrient-uptake kinetics depends on nutrient supply. With adequate nutrition, low light availability reduces nutrient uptake (Table 5, Fig. 9B). By contrast, nutrient-limited plants are not strongly affected by light availability.

Low temperature directly reduces nutrient uptake by plants, as expected for any physiological process that is dependent on respiratory energy (Fig. 10A). Plants compensate through both acclimation and adaptation for this temperature inhibition of uptake by increasing their capacity for nutrient uptake (Fig. 10B,C). In contrast to dry and infertile environments, arctic and alpine plants often grow quite rapidly to exploit the short growing season; therefore, they have a substantial demand for nutrients.

When plants are grown with an adequate nutrient supply, **grazing** of leaves reduces plant nutrient demand and therefore reduces nutrient uptake capacity (Clement et al. 1978). By contrast, grazing of nutrient-stressed plants can deplete plant nutrient

typically have a low capacity to absorb immobile ions like phosphate. A low I_{max} for immobile ions is presumably advantageous in a low-nutrient environment because (1) diffusion so strongly limits uptake in infertile soils that I_{max} has little influence on nutrient uptake (Sect. 2.3), (2) mycorrhizal transfer and other processes may be more important than nutrient influx across the plasma membrane, and (3) a high I_{max} may entail a substantial nitrogen investment in roots (Chapin 1988).

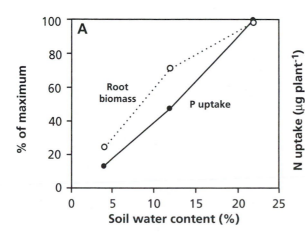

FIGURE 9. Effect of soil water content on root biomass and phosphate uptake per unit root biomass in *Lycopersicon esculentum* (tomato) (A) and of growth irradiance on ammonium uptake per plant in *Oryza sativa* (rice) (B) (Chapin 1991). Copyright Academic Press.

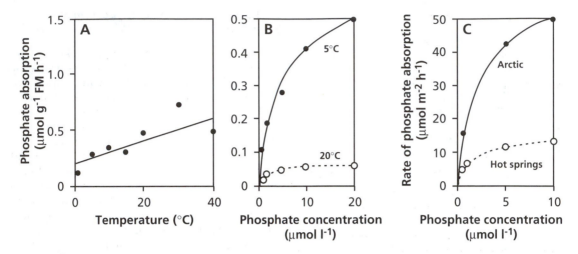

FIGURE 10. Response of phosphate uptake by *Carex aquatilis* (a tundra sedge) to temperature at different time scales: (A) immediate, (B) acclimation, and (C) adaptation measured at 5°C (Chapin 1974; copyright Ecological Society of America, and Chapin & Bloom 1976; copyright Oikos).

stores, so that plants respond by increasing nutrient uptake capacity (Chapin & Slack 1979). Plants that are adapted to frequent grazing, such as grasses from the Serengeti Plains of Africa, similarly increase their capacity to absorb phosphate when clipped to simulate grazing (McNaughton & Chapin 1985).

2.2.4 Acquisition of Nitrogen

Nitrogen can be absorbed by plants in three distinct forms: **nitrate**, **ammonium**, and **amino acids**. There is a substantial carbon cost in nitrogen assimilation (i.e., the conversion of inorganic to organic nitrogen): Nitrate must first be reduced to ammonium, which must then be attached to a carbon skeleton before it can be used in biosynthesis. Thus, the carbon cost of assimilation, which is generally large, is nitrate >> ammonium > amino acids (Clarkson 1985). Depending on the species, nitrate is reduced either in the roots or transported to the leaves, where it is reduced in the light. The first step in the reduction is catalyzed by **nitrate reductase**, which is an inducible enzyme; the gene encoding nitrate reductase is transcribed in response to nitrate application (Campbell 1996). The protein is rather short-lived, being degraded with a half-time of a few hours (Li & Oaks 1993). In addition, the activity of the enzyme is controlled by phosphorylation. In the leaf, the enzyme is turned off at night, which involves a calcium-dependent protein kinase that phosphorylates the enzyme. Phosphorylation allows inactivation of nitrate reductase by an inhibitor protein. A protein phosphatase reactivates the enzyme when the irradiance increases (Huber et al. 1996, Kaiser & Huber 1994). Nitrate assimilation is energetically expensive because of the costs of nitrate reduction. Ammonium is toxic to plant cells (partly because it is in equilibrium with NH_3, which rapidly diffuses across membranes, picks up a H^+ at the other side and so affects the H^+ gradient across membranes) and therefore must be assimilated rapidly to amino acids. Rough calculations suggest that nitrate reduction to ammonium requires approximately 15% of plant-available energy when it occurs in the roots (2% in plants that reduce nitrate in leaves), with an additional 2 to 5% of available energy for ammonium assimilation (Bloom et al. 1992). One might think that the lower costs when ammonium, rather than nitrate, is used as the source of nitrogen by the plant would allow for a higher growth rate. This is not invariably the case, however, either because of adjustments in leaf area ratio (Sect. 2.1.1 in the chapter on growth and allocation) or a lower efficiency of root respiration (Sect. 2.6 in the chapter on respiration).

The distribution of nitrate reductase activity and the presence/absence of nitrate in xylem sap suggest the following ecological patterns (Andrews 1986):

1. All species increase the proportion of nitrate reduced in the shoot as nitrate supply increases, which suggests a limited capacity for nitrate reduction in the root system.

2. Temperate perennials and annual legumes reduce most nitrate in the roots under low nitrate supply.

3. Temperate nonlegume annuals vary considerably among species in the proportion of nitrate reduced in roots under low nitrate supply.

4. Tropical and subtropical species, both annuals and perennials, reduce a substantial proportion of their nitrate in the shoot, even when growing at a low nitrate supply. This is not surprising because nitrogen is frequently nonlimiting in tropical ecosystems (Vitousek & Howarth 1991), and warm temperatures favor nitrification.

Despite these general patterns, some nitrate reduction occurs in leaves of most plants, particularly those of ruderal species (i.e., species that occupy disturbed sites where nitrification rates are generally high). Some shade plants (*Chrysosplenium oppositifolium*, opposite-leaved golden saxifrage) develop nitrate reductase activity only in their leaves when grown in full sun (Smirnoff & Stewart 1985). Leaf nitrate reductase activity is typically highest at midday in association with high light intensities. Some plants, particularly those in the Ericaceae, show low levels of nitrate reductase (Smirnoff et al. 1984), presumably because nitrate availability is generally low in habitats occupied by these species. Most woody species assimilate most of the nitrate in the roots (Martins-Louçao & Cruz 1998). Leaves of most Gymnospermae and Proteaceae reduce nitrate only after induction by feeding leaves with nitrate, which suggests that these species may reduce most nitrate in the roots (Smirnoff et al. 1984).

Plant species differ in their preferred forms of nitrogen absorbed, depending on the forms available in the soil. For example, arctic plants, which experience high amino acid concentrations in soil, preferentially absorb and grow on amino acids, whereas barley preferentially absorbs inorganic nitrogen (Chapin et al. 1993); spruce preferentially absorbs ammonium (Kronzucker et al. 1997). Much of the early work on nitrate and ammonium preference is difficult to interpret because of inadequate pH control (Sect. 2.2.6) or low light intensity. Species from habitats with high nitrate availability (e.g., calcareous grasslands), however, often show preference for nitrate and have higher nitrate reductase activities than do species from low-nitrate habitats. Most plants, however, are capable of absorbing any form of soluble nitrogen, especially if acclimated to its presence (Atkin 1996).

2.2.5 Acquisition of Phosphate

There are numerous traits involved in acquiring sufficient quantities of phosphate from the soil. Some of these traits are specific for phosphate (e.g., root phosphatase); other traits (e.g., root hairs and root mass ratio) promote uptake of all ions but are most critical for phosphate because of the low diffusion coefficient of phosphate in soil and therefore the small volume of soil that each root can exploit. The specialized association with a mycorrhizal fungus will be discussed in Section 2.3 of the chapter on symbiotic associations; mycorrhizae enhance the phosphate-absorbing surface. The same result can be achieved without a microsymbiont, albeit possibly at a greater cost.

2.2.5.1 Plants Can Also Use Some Organic Phosphate Compounds

In agricultural soils, 30 to 70% of all the phosphate is present in an **organic** form; in nutrient-poor grasslands and forest soils this may be as much as 80 to 95% (Häussling & Marschner 1989, Macklon et al. 1994), or 99% in organic tundra soils (Kielland 1994). *Lupinus* (lupin) species can use phytate (inositol phosphate, a major component of the organic P fraction in soil), RNA, and glycerophosphate (also present in the soil), in addition to inorganic phosphate, due to the activity of **phosphatases** in the soil (Adams & Pate 1992). Production of phosphatases by the roots provides an additional source of phosphate; these enzymes hydrolyze organic phosphate-containing compounds, releasing inorganic phosphate that is absorbed by roots (Kroehler & Linkins 1991). Phosphatase production is enhanced by a low phosphate supply to the plants. Roots may alternatively exude organic substances that act as substrates for microorganisms producing enzymes hydrolyzing organic phosphate (Richardson 1994). Whatever the exact mechanism by which organic P is hydrolyzed, the concentration of organic P near the root surface may decrease by as much as 65% in *Trifolium alexandrinum* (clover) and 86% in *Triticum aestivum* (wheat) (Tarafdar & Jungk 1987). This shows that these roots do have access to organic forms of P in the soil (Fig. 11).

The capacity to use organic phosphate varies among species and also depends on soil conditions. It may range from almost none to a capacity similar to that of the rate of inorganic phosphate uptake (Hübel & Beck 1993).

2.2.5.2 Excretion of Phosphate-Solubilizing Compounds

Some plants adapted to low-phosphorus soils excrete **acidifying** and **chelating** compounds (e.g.,

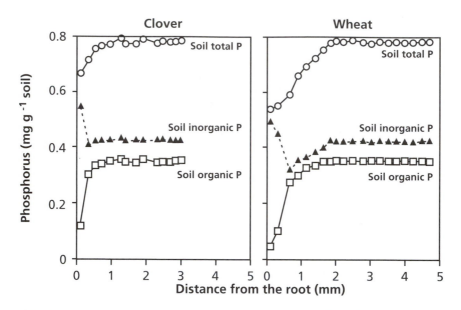

FIGURE 11. Distribution of total, inorganic, and organic phosphorus in the rhizosphere of *Trifolium alexandrinum* (clover, 10 days old) and *Triticum aestivum* (wheat, 15 days old) grown in a silt loam (Tarafdar & Jungk 1987).

citric acid, malic acid, and piscidic acid). Acidification enhances the solubility of phosphate in basic soils (Fig. 1). Chelating compounds bind cations and thus release phosphate from sparingly soluble inorganic substrates. Both processes enhance the diffusion gradient for phosphate between the bulk soil and the root surface. Citric acid releases phosphate from calcium phosphate complexes, whereas piscidic acid (p-hydroxybenzyl tartaric acid), in combination with reducing phenolics, is more effective in releasing phosphate from iron phosphate complexes (Marschner 1995).

The capacity to excrete organic acids is very pronounced in members of the Proteaceae, which do not form a mycorrhizal association, but have **proteoid roots** (Fig. 12). The term proteoid roots was given because the structures were first discovered in the family of the Proteaceae. Similar structures were later found in other species, and now the term **cluster roots** is also used. Cluster roots consist of clusters of longitudinal rows of extremely hairy rootlets, which originate during root development, 1 to 3 cm from the root tip. One lateral branch may contain several clusters, centimeters apart from each other. The cluster roots excrete chelating acids and phenolics. Cluster roots are almost universal in the family Proteaceae; they also occur in species belonging to the Betulaceae, Casuarinaceae, Cyperaceae, Dasypogonaceae, Fabaceae, Mimosaceae, Moraceae, Myricaceae,

Papilonaceae, and Restionaceae (Lamont 1993, Dinkelaker et al. 1995). In Australia, the families with species having cluster roots have their centers of distribution in the oldest, most leached sands and laterites of the continent. Cluster roots are a very effective morphological structure to acquire phosphate from soils in which phosphate is sparingly soluble. In *Lupinus albus* (white lupin) the development of proteoid roots is suppressed by an excess supply of phosphate (Keerthisinghe et al. 1998).

Proteoid roots of *Lupinus albus* release 40, 20, and 5 times more citric, malic, and succinic acid, respectively, than lupin roots in which the development of proteoid roots is suppressed by excess phosphate. The greatest citrate efflux occurs 1 to 3 cm from the root tip (i.e., the zone of the youngest cluster roots). Citrate efflux by more mature root tissue (7 to 9 cm from the root tip) is approximately 80% less (per unit root length). The greater efflux is not related to higher citrate concentrations in the proteoid roots. Hence, the greater export must be due to greater rates of **citrate synthesis** (Keerthisinghe et al. 1997). The capacity to excrete substantial amounts of citric acid and other acids requires an increased activity of phospho*enol*pyruvate (PEP) carboxylase, malate dehydrogenase, and citrate synthase (Johnson et al. 1994). **PEP carboxylase** allows the "dark CO_2 fixation," the rate of which is increased by about fourfold in proteoid roots of *Lupinus albus*, when

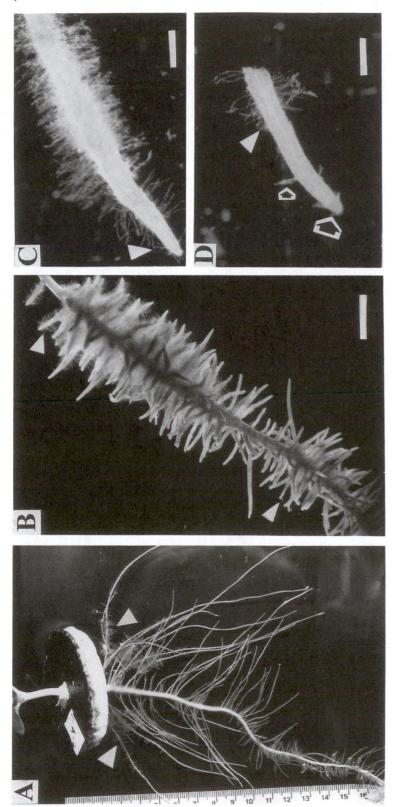

FIGURE 12. Proteoid or cluster roots are bottlebrushlike clusters of rootlets, which arise on some regions of lateral roots (courtesy M. Watt, Research School of Biological Sciences, Australian National University, Canberra, Australia, and G. Keerthisinghe, Division of Plant Industry, CSIRO, Canberra, Australia). (A) Proteoid roots of *Lupinus albus* (white lupin), grown in nutrient solution; the white collar around the root/shoot junction supports the plant. The oldest, first-order lateral roots, which emerge from the base of the primary root, have become proteoid roots. These proteoid roots have regions with short, closely spaced, second-order laterals called *proteoid rootlets* (indicated by white arrow heads). (B) Detail of a region of a proteoid root that is covered in closely spaced rootlets. The youngest rootlets (indicated by arrowhead towards the bottom, left corner) are closest to the tip and are still growing and have developed few root hairs. These young rootlets are very active in citrate production and are completely covered in root hairs. Compared with the younger rootlets, the older rootlets export much less citrate; scale bar = 4mm). (C) Detail of an older proteoid rootlet shown in (B). Many root hairs cover the rootlet, and have even developed over the rootlet tip (arrowhead) (scale bar = 0.345 mm). (D) Detail of a younger proteoid rootlet shown in (B). The root cap and mucilage (indicated by black arrows) still cover the growing tip and the root hairs have only started to develop behind the elongating zone (white arrowhead) (scale bar = 0.345 mm).

compared with roots in which the development of proteoid roots is suppressed by excess phosphate. This is associated with increased levels of mRNA encoding PEP-carboxylase and of the specific activity of PEP-carboxylase (Johnson et al. 1996a, 1996b). As expected, the activity of both PEP-carboxylase and malate dehydrogenase is highest 1 to 3 cm from the root tip, where citrate excretion is greatest. The mechanism that allows the massive and rapid release of organic acids is as yet unknown. Although the excretion of citrate is highest close to the root tip, the capacity to absorb phosphate from the medium is equally high close to the root tip and further away from the tip; however, in situ most of the phosphate in the soil will be depleted by root cells close to the tip, leaving little to be absorbed by the older zones. The decrease in citrate excretion in older zones therefore saves carbon, without negative effects on phosphate acquisition.

The capacity to excrete acidifying and chelating compounds is not restricted to species with morphological structures such as cluster roots. Species that belong to the Brassicaceae also excrete major amounts of citric acid, thus enhancing the capacity to solubilize rock phosphate (Hoffland et al. 1989). Figure 13 provides an overview of some herbaceous species capable of releasing phosphate from a sparingly soluble source, presumably due to the excretion of organic acids. Other species that occupy low-phosphorus chalk substrate (*Brachypodium pinnatum, Briza media, Dactylis glomerata*) lack this capacity (Fig. 13), but have the ability to form mycorrhizal associations.

Neighboring plants may profit from the capacity to release inorganic phosphate from sparingly soluble sources. *Triticum aestivum* (wheat), grown with a sparingly soluble source of phosphate, produces less dry matter than *Lupinus albus* (white lupin), grown in the same pot in such a way that their roots remain separated (Table 6). The higher production of white lupin is associated with its proteoid roots and capacity to excrete citric acid. When grown in

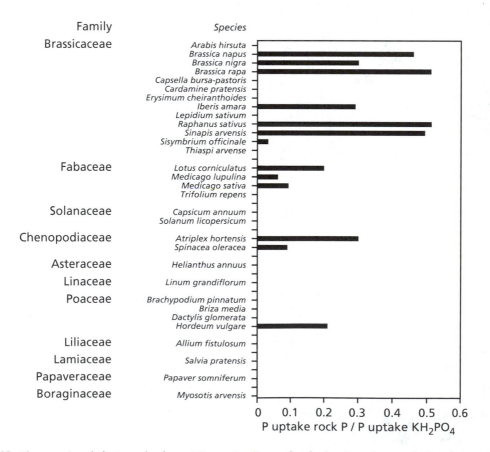

FIGURE 13. The capacity of plant species from different families to absorb phosphate from rock phosphate, relative to their uptake from dissolved phosphate (Hoffland 1991). Reproduced with the author's permission.

TABLE 6. Shoot dry mass of *Triticum aestivum* (wheat) and *Lupinus albus* (white lupin) plants, grown for 58 days in pots in mixed culture in such a way that either their roots remained separated or they could mix.*

P-Supply	Root System	Shoot Dry Mass (g)		
		wheat	lupin	together
No extra phosphate	separated	20	33	54
supplied	mixed	38	29	66
Rock phosphate	separated	24	27	51
supplied	mixed	40	25	64

Source: Horst & Waschkies 1987.

*Additional phosphate was provided as rock phosphate, which is sparingly soluble.

such a way that their roots overlap, the production of wheat is enhanced, at the expense of lupin. The effects can be ascribed entirely to the phosphate-solubilizing activity of the lupin roots.

2.2.6 Changing the Chemistry in the Rhizosphere

The availability of several **micronutrients** in the rhizosphere is greatly affected by physiological processes of the roots (Table 7). For example, **proton extrusion** by roots may reduce rhizosphere pH by more than 2 units from that in the bulk soil; the capacity to affect the pH is strongest at a soil pH of 5 to 6. Roots also have the capacity to **reduce** compounds in the rhizosphere or at the plasma membrane, which is particularly important for the acquisition of iron, when available in its less-mobile oxidized state in the soil. On the other hand roots in flooded soils can **oxidize** compounds in the rhizosphere, largely by the release of oxygen (cf.

Sect. 3.5 of this chapter). This can reduce the solubility of potentially toxic ions like aluminum and sulfide. Roots often excrete exudates that **mobilize** sparingly soluble micronutrients, or stimulate the activity of rhizosphere microorganisms and therefore the mineralization of nitrogen and phosphorus.

2.2.6.1 Changing the Rhizosphere pH

The pH in the rhizosphere is greatly affected by the **source of nitrogen** used by the plant because nitrogen is the nutrient required in largest quantities by plants and can be absorbed as either a cation (ammonium) or an anion (nitrate). Roots must remain electrically neutral, so when plants absorb more cations than anions, as when **ammonium** is the major nitrogen source, more **protons** must be extruded (reducing rhizosphere pH) than when **nitrate** is the major nitrogen source, in which case the pH tends to rise slightly. An additional cause of the

TABLE 7. The availability of a number of micronutrients and toxic elements for plants when the pH decreases and the reason for the change in availability.

Microelement	Effect of Decreased pH on Availability of the Micronutrient for the Plant	Cause of the Effect
Aluminum	increase	increased solubility
Boron	increase	desorption
Cadmium	increase	
Copper	no effect	
Iron	increase	reduction, increased solubility
Manganese	increase	desorption, reduction
Molybdenum	decrease	adsorption
Zinc	increase	desorption

Source: Marschner & Römheld 1996.

decline in rhizosphere pH when ammonium is the source of N is that for each N that is incorporated into amino acids, one H^+ is produced. Because ammonium is exclusively assimilated in the roots, whereas nitrate is assimilated partly in the roots and partly in the leaves, the production of H^+ is greatest with ammonium. A somewhat smaller decrease in pH also occurs when atmospheric nitrogen is supplied as the sole source of N to legumes or other **N$_2$-fixing** systems (cf. sect. 3 in the chapter on symbiotic associations). The drop in pH with ammonium as nitrogen source is due to exchange of NH_4^+ for H^+ (or uptake of NH_3, leaving H^+ behind). The rise in pH with nitrate as the source of N is associated with the generation of hydroxyl ions during its reduction according to the overall equation: $NO_3^- + 8 e^- + 1.5 H_2O \rightarrow NH_3 + 3 OH^-$. Some of the hydroxyl ions are excreted; some of them are neutralized by the formation of organic acids (mainly malic acid) from neutral sugars. As a result, plants grown with nitrate contain more organic acids (mainly malate) than those using NH_4^+ or N_2.

Application of ammonium fertilizers can create major agricultural problems because both the pH in the rhizosphere and that of the bulk soil will decline in the longer term. This may mobilize potentially toxic ions, including aluminum and zinc, and reduce the availability of required nutrients (cf. Fig. 1 and Sect. 3.1 in this chapter).

Rhizosphere pH affects the availability of both soil micronutrients and potentially toxic elements that are not essential for plant growth (Al) (Table 7). With decreasing pH, the availability of zinc, manganese, and boron is enhanced by desorption from soil particles. Manganese and iron also become available due to reduction (to Mn^{2+} and Fe^{2+}, re-

spectively). The availability of ferric iron (Fe^{3+}) is also enhanced because its solubility is higher at a low pH. Fe deficiency symptoms in calcareous soils can be prevented by supplying ammonium, which acidifies the rhizosphere, rather than nitrate, which tends to further increase the pH around the roots. It is only effective, however, in the presence of nitrification inhibitors that prevent the microbial transformation of ammonium to nitrate. Net nitrification is often favored by a high pH, which increases nitrification more than nitrate immobilization by soil microbes. In practice, supplying iron in a chelated form is more effective (Table 8).

The availability of molybdenum decreases with a decreasing pH, and that of copper, which tends to be complexed in the soil, is unaffected by pH. As a result, when grown in soil with ammonium sulfate, the concentrations of Fe, Mn, Zn, and B are higher in plant biomass than those in plants given calcium nitrate (Table 8). In substrates, such as quartz sand, where Mn^{2+} is the only form of manganese, lowering the pH in the rhizosphere may have the opposite effect on manganese uptake and concentrations in the plant (Marschner & Römheld 1996).

Plants can strongly reduce rhizosphere pH by excreting organic acids (cf. Sect. 2.2.6). A decrease in pH can also be brought about by net proton excretion, when the uptake of major cations (e.g., K^+) exceeds that of anions and net H^+ excretion is required to balance the charge. In calcareous soils, this acid excretion occurs to the extent that bulk soil pH is lowered. It is particularly pronounced in, but not restricted to, species with proteoid roots. Leachates from the rhizosphere soil adhering to such proteoid roots of *Lupinus albus* (white lupins) dissolve ten times more MnO_2 than do leachates

TABLE 8. The effect of the form of nitrogen applied to a sandy loam (*Triticum aestivum*, wheat, and *Brassica oleracea* var. *botrytis*, cauliflower) or a calcareous soil (*Arachis hypogaea*, peanut) on concentrations of micronutrients or chlorophyll.*

N-Source	Micronutrient Concentration [mg (kg DM)$^{-1}$]				Chlorophyll Concentration [mg (g FM)$^{-1}$]
	Fe	Mn	Zn	B	
Nitrate	55	23	18	3.5	0.89
Ammonium	68	45	24	12.9	0.85
Ammonium + nitrification inhibitor					1.76
Nitrate + FeEDDHA					2.96

Source: Marschner 1991a.

*Chlorophyll concentration is a good indicator for the availability of Fe in the rhizosphere. To inhibit the transformation of ammonium into nitrate by nitrifying bacteria, a nitrification inhibitor (nitrapyrin) was added. FeEDDHA is a chelated form of iron, which is readily available to the plant. Concentrations were measured in mature leaves (B), young leaves (chlorophyll), or the entire shoot (other micronutrients).

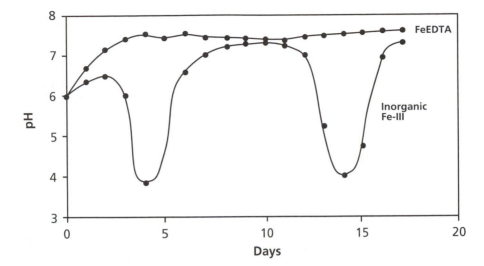

FIGURE 14. Changes in pH in the root environment of *Helianthus annuus* (sunflower). The pH was measured in an unbuffered nutrient solution. Plants were supplied with iron as an Fe-chelate (Fe-EDTA), or as $FeCl_3$ (Marschner et al. 1978, as cited in Marschner 1983).

from bulk soil. They dissolve 100 times more Fe^{3+} than the rhizosphere soil of *Brassica napus* (oil seed rape) (Gardner et al. 1982). The shoots of plants in the immediate vicinity of white lupins also contain more Mn (Gardner & Boundy 1983).

Some nutrient deficiencies cause plants to reduce **rhizosphere pH**. When the Fe supply is insufficient, *Helianthus annuus* (sunflower) plants lower the pH of the root solution from approximately 7 to 4 (Fig. 14). Similar responses have been found for *Zea mays* (maize) and *Glycine max* (soybean) genotypes with a low susceptibility to Fe-deficiency ("lime-induced chlorosis"). Zn deficiency can also cause a lowering of the rhizosphere pH (Römheld 1987). Organic acid-mediated dissolution of iron plays a significant role in elevating the concentration of Fe-complexes in the rhizosphere, especially when iron occurs as $Fe(OH)_3$, but less so when it is present as Fe-oxides (Fe_2O_3 and Fe_3O_4) (Jones et al. 1996a).

Lowering the pH in response to Fe deficiency may coincide with an increased capacity to reduce iron at the root surface, due to the activity of a specific **reductase** in the plasma membrane (Bienfait 1985). Reducing and chelating compounds (phenolics) may be excreted, solubilizing and reducing Fe^{3+} (Deiana et al. 1992). This is the typical response of iron-efficient dicots and monocots other than grasses ("strategy I" in Fig. 15). Excretion of reducing and chelating compounds also enhances the availability and uptake of manganese. In calcareous soils with a low concentration of iron and a high concentration of manganese, this strategy may lead to manganese toxicity. When the buffering capacity of the soil is large and the pH is fairly high, strategy I is not very effective.

2.2.6.2 Excretion of Organic Chelates

Grasses exude very effective chelating compounds, particularly when iron or zinc are in short supply (Fig. 15). These chelators were originally called **phytosiderophores**, because of their role in the ac-

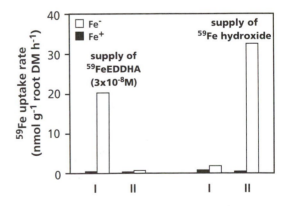

FIGURE 15. The response to Fe-deficiency of species following two contrasting "strategies." Strategy II is restricted to grasses. Strategy I is found in monocots, with the exception of grasses, and in dicots. Plants were grown with and without iron and then were supplied with ^{59}FeEDDHA or ^{59}Fe hydroxide (Römheld 1987). Copyright Physiologia Plantarum.

quisition of iron. Because it was discovered that these chelators are also important for the uptake of metals like zinc, when these are in short supply, the term **phytometallophore** would seem more appropriate (Cakmak et al. 1996). Iron diffuses in the form of an iron-phytosiderophore chelate to the root surface and is absorbed as such by root cells ("strategy II"). The system responsible for uptake of the iron-chelate is induced by iron deficiency. In strategy II iron reduction takes place after uptake into the root cells, rather than prior to uptake as in strategy I. The capacity of a genotype to release phytometallophores is inversely related to its sensitivity to iron or zinc deficiency. For example, barley (*Hordeum vulgare*) is less sensitive to iron deficiency and excretes more phytometallophores than sorghum (*Sorghum bicolor*) and maize (*Zea mays*) (Marschner & Römheld 1996). Genotypes of wheat (*Triticum aestivum*, bread wheat, and *T. durum*, durum wheat) that are more resistant to zinc deficiency excrete more phytometallophores than do more sensitive genotpes (Cakmak et al. 1996).

Phytometallophores are similar to, and sometimes derived from, nicotinamine (Fig. 16). Nicotinamine itself is also an effective chelator and probably plays a role in chelating iron inside the cell, in both strategies I and II (Scholz et al. 1992). Phytometallophores are specific for each species and more effective in chelating iron than many synthetic chelators used in nutrient solutions. They also form stable chelates with copper, zinc, and, to a lesser extent, manganese, and enhance the availability of these nutrients in calcareous soils. Iron-efficient species belonging to strategy I or II show an enhanced capacity to absorb iron upon withdrawal of Fe from the nutrient solution (Fig. 17).

Organic acids are a common component of root exudates. They are excreted in response to a shortage of phosphate, iron, and some other cations. Organic acids transform high-molecular humus compounds into low-molecular ones (molecular mass less than 10,000). Upon transformation of the humus complex, calcium, magnesium, iron, and zinc are released from the humus complex. It is a reversible process: Neutralization with KOH produces the high-molecular humus compounds again. Organic acids are far more effective than their potassium salts or inorganic acids, and their action is likely to be a combination of acidification and chelation (Albuzzio & Ferrari 1989).

2.2.7 Rhizosphere Mineralization

Root exudation of organic acids, carbohydrates, and amino acids and the sloughing of polysaccharides from growing root tips mostly accounts for less than 5% of total carbon assimilation, but it may increase substantially when phosphate has low availability (Table 1 in the chapter on respiration). The controls over the quantity and chemical composition of these exudates, however, are poorly understood. Root exudates have major effects on microbial processes in soils, which are often carbon-limited (Paul & Clark 1990). The densities and activity of microorganisms, especially bacteria, and of microbial predators are much greater in the rhizosphere than they are in bulk soil, and they are enhanced by factors, such as elevated atmospheric CO_2 concentrations, that increase root exudation. The effects of root exudates depend on soil fertility (Hungate et al. 1997). In infertile soils, stimulation of root exudation by elevated CO_2 concentrations increases nitrogen immobilization by rhizosphere microbes and reduces plant uptake (Diaz et al. 1993). By contrast, in more fertile soils, where microbes are more carbon-limited, the stimulation of root exudation by elevated $[CO_2]$ increases nitrogen mineralization and plant uptake (Zak et al. 1994). There are still too few studies to know how broadly these results can be generalized. An additional mechanism by which exudation can enhance nutrient supply is through the stimulation of microbial grazers such as amoebae and nematodes, which feed on bacteria and excrete excess nitrogen, which is then available for absorption by plants (Clarholm 1985). Annual nitrogen uptake by vegetation is often twice the nitrogen mineralization estimated from incubation of soils in the absence of roots (Chapin et al. 1988). Much of this discrepancy could involve the more rapid nutrient cycling that occurs in the rhizosphere, as fueled by root exudation.

2.2.8 Root Proliferation in Nutrient-Rich Patches: Is It Adaptive?

When nitrogen, potassium, or phosphate are limiting for plant growth and only available in localized root zones, roots tend to **proliferate** in these zones more than they do in microsites with low nutrient availability. This response, however, is found only if the elongating tip of the axis from which the laterals emerge has experienced these favorable local conditions while elongating. If it has not, or if the plant as a whole does not experience nutrient deficiency, then no laterals emerge in favorable zones (Drew 1975, Drew et al. 1973).

It would seem that the proliferation of roots in response to a localized nutrient supply is functional, but is it really? When *Triticum aestivum* (wheat) plants are grown with a localized

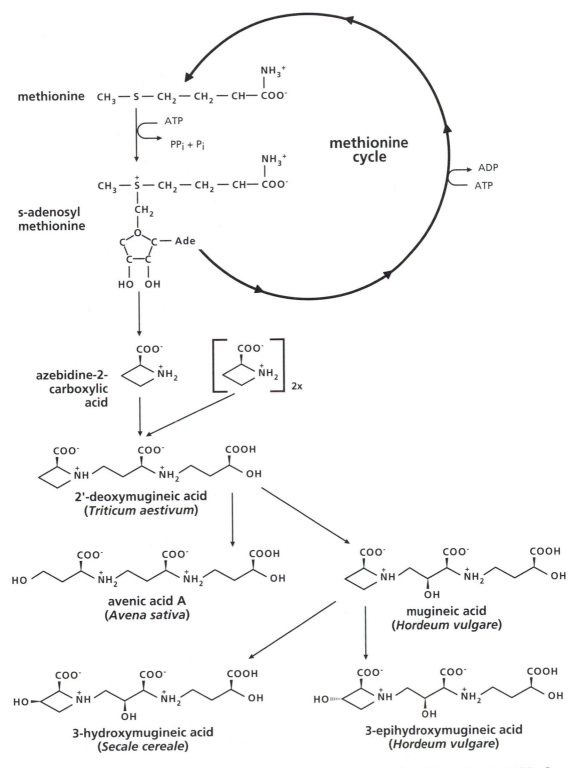

FIGURE 16. Scheme for the biosynthesis of phytometallophores. They are hydroxy- and amino-substituted imino-carboxylic acids exuded by graminaceous monocotyledonous plants (Ma & Nomoto 1996). Copyright Physiologia Plantarum.

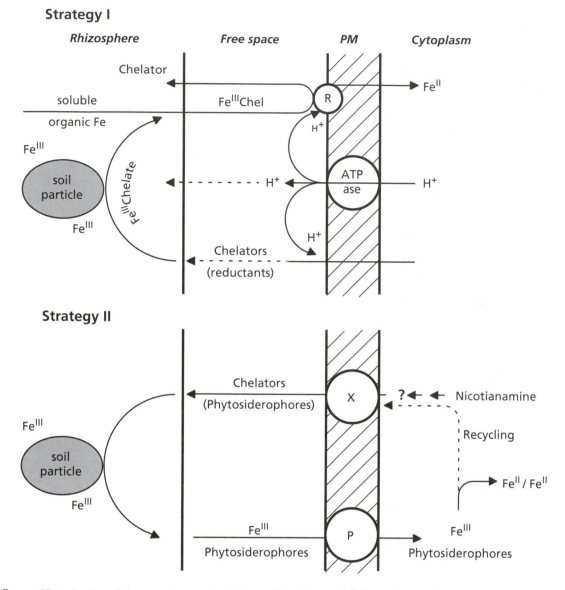

FIGURE 17. Induction of the capacity to absorb Fe as affected by Fe-deficiency in cucumber (strategy I) and barley (strategy II) (Römheld 1987). Copyright Physiologia Plantarum.

[15]N-labeled organic residue in soil, rates of N uptake per unit root length greatly increase during growth through the localized source of nitrogen. During the first 5 days of exploitation of the localized source, 8% of all the N the plants ultimately obtained from the residue is captured. Only then the roots proliferate in the residue and over the next 7 days 63% of the total N obtained from the local source is absorbed. After that time, massive proliferation occurs in the residue, but relatively little

further N is captured (Fig. 18). This would indicate that the local proliferation is of only limited importance for the capture of the nitrogen released from locally decomposing organic matter. Proliferation, triggered by the local source of nitrogen, might be advantageous in the longer term to take up nutrients other than nitrogen.

The extent of the response to a localized supply depends on the overall nutrient status of the plant. Thus, if one half of the roots receives no nutrients at all, then the response is considerably stronger than if that half is supplied with a moderate amount (Table 9) (Robinson 1994). The development of an individual root obviously depends on both the nutrient availability in its own environment and on other roots of the same plant.

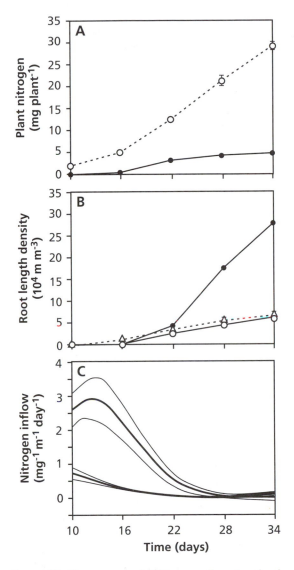

FIGURE 18. The response of *Triticum aestivum* to a localized organic residue, enriched with ^{15}N. (A) Total N in the plant (open symbols) and N in the plant derived from the organic residue; (B) Total root length density in the residue (filled symbols) and in the soil above (squares) and below (triangles) the residue; (C) Nitrogen uptake for the whole root system (lower curve) and for the part of the roots that proliferated in the localized residue (upper curve) (Van Vuuren et al. 1996). Copyright Kluwer Academic Publishers.

2.3 Sensitivity Analysis of Parameters Involved in Phosphate Acquisition

The contribution of different parameters involved in the uptake of phosphate can be appreciated using **simulation models**. Such models are used increasingly to analyze ecophysiological problems.

Nye and co-workers analyzed the significance of root hairs using an experimental and a mathematical approach. They measured the (labeled) phosphate concentration at the root surface of an oil seed root with dense root hairs. In addition they simulated the phosphate concentration under one of the following two assumption: (1) root hairs are not involved in phosphate uptake, and (2) root hairs effectively increase the cylinder intercepted by the root. There was good agreement between the simulated and the experimental data, assuming that root hairs are effective. No such agreement was found under the assumption that root hairs do not play a significant role (Fig. 19). This work has corroborated earlier ideas based on the significance of root hairs for the acquisition of immobile ions, including phosphate (cf. Table 4).

Barber and co-workers analyzed the sensitivity of phosphate uptake by pot-grown soybean plants to various soil and root factors (Fig. 20). The simulated uptake agreed well with their experimental results. Their results demonstrated that phosphate uptake is much more responsive to changes in the rate of root elongation (k in Fig. 20) and root diameter (r_o) than to changes in kinetic properties of the uptake system: K_m, I_{max}, and C_{min}. Soil factors such as the diffusion coefficient (D_e) and the buffer power (b) have greater effects if their values are decreased than if they are increased. Transpiration (v_o) has no effect at all on the rate of phosphate uptake. The spacing between roots (r_1) was such that there was no interroot competition; hence, changes in the value for this parameter had no effect. It is clear, that for a relatively immobile nutrient such as phosphate, kinetic parameters are of considerably less importance than are root traits such as the rate of elongation and root diameter. This is consistent with the generalization that diffusion to the root surface rather than uptake kinetics is the major factor determining nutrient acquisition. For more mobile ions, such as nitrate, kinetic properties play a somewhat more important role (Clarkson 1985).

Hoffland and co-workers (1990a, 1990b) developed a simulation model for phosphate uptake in *Brassica napus* (oilseed rape) plants to evaluate the quantitative significance of the excretion of organic acids in mobilizing rock phosphate. As an example

TABLE 9. Root development of *Pisum sativum* (garden pea) in a split-root design, in which root halves were grown in different pots and supplied with different nutrient concentrations from the time they were 24-mm long.*

Nutrient strength pot 1-pot 2	Root dry mass (mg)			Ratio pot 1/pot 2	Shoot dry mass (mg)
	pot 1	pot 2	Total		
0–50	51	450	501	0.11	806
1–50	60	427	487	0.14	847
10–50	142	370	512	0.38	874
25–50	194	269	463	0.72	935
50–50	300	283	583	1.05	1032
10–0	225	61	286	3.77	463
25–0	343	52	395	6.76	670

Source: Gersani & Sachs 1992.
*Plants were harvested when they were 3 weeks old.

to illustrate the importance of simulation models in modern ecophysiological research, this model is treated in greater detail in this section. Uptake of nutrients by the roots, transport of nutrients to the root surface, and root density are considered as the three main components of this model (cf. Sect. 3 in the chapter on plant water relations).

The rate of nutrient supply to the root surface (F, $mol\,m^{-2}s^{-1}$) by diffusion and mass flow is given by:

$$F = -D_e.dC/dr + v.C \qquad (1)$$

where D_e is the effective diffusion coefficient (m^2s^{-1}), C is the nutrient concentration ($mol\,m^{-3}$), r is the radial distance from the root axis (m), v is the inward water flux ($m^3\,m^{-2}s^{-1}$); the minus sign is added because C at the root surface is less than C in the soil, so that dC is negative. The effective diffusion coefficient depends on the volumetric moisture content (θ, no dimension), the "tortuosity factor" (f, no dimension), which refers to the diffusion path length, and the diffusion coefficient of the nutrient in free solution (D_o, m^2s^{-1}):

$$D_e = \theta.f.D_o \qquad (2)$$

Competition for nutrients between roots of the same plant is accounted for by assigning a finite cylindrical volume with radius r_1 to each root. Water can move from the cylinder assigned to one

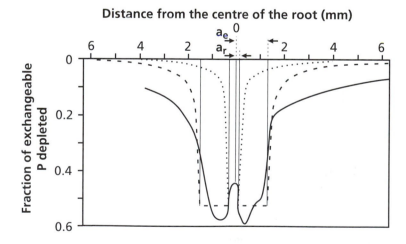

FIGURE 19. Calculated and measured phosphate concentration profiles around an oil seed root. Phosphate profiles were calculated under the assumption that root hairs do (outer broken lines) or do not (inner broken lines) play a role in phosphate uptake. The solid line gives the experimentally determined profile. The radii given are the radius of the root axis only (a_r) and that of the root plus root hairs (a_e) (Bhat & Nye 1973, and Nye & Tinker 1977, as cited in Pitman & Lüttge 1983).

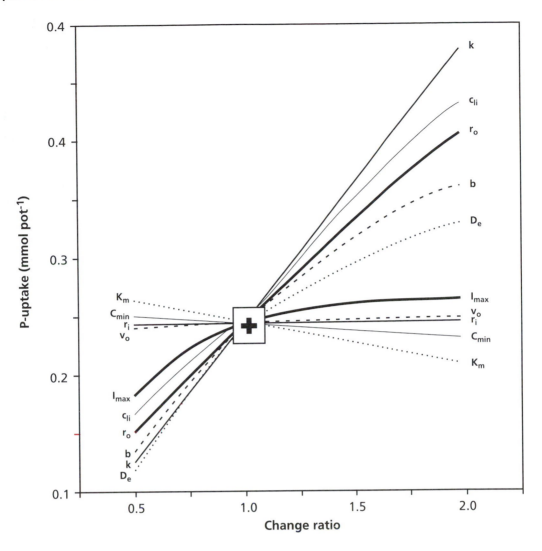

FIGURE 20. Effects of changing parameter values (from 0.5 to 2.0 times the standard value) on simulated phosphate uptake by roots of *Glycine max* (soybean). k is the rate of root elongation, C_{li} is the initial phosphate concentration in solution, r_o is the root diameter, b is the buffer power of the soil, D_e is the diffusion coefficient of phosphate in the soil, I_{max} is the maximum phosphate inflow rate, v_o is the rate of transpiration, r_i is the spacing between individual roots, C_{min} is the lowest concentration at which phosphate uptake is possible, and K_m is the phosphate concentration at which the rate of phosphate uptake is half of that of I_{max} (Silberbush & Barber 1983, as cited in Clarkson 1985). With permission, from the Annual Review of Plant Physiology, Vol. 36, copyright 1985, by Annual Reviews Inc.

root to that assigned to another, but nutrients cannot. The initial radius of each soil cylinder around a root is:

$$r_l = \sqrt{(A/(\pi.n)} \qquad (3)$$

assuming that each plant starts with one root growing in vertical direction; A is the total surface of the roots (m²), and n is the number of plants per pot. Each of these parallel soil cylinders is divided into a number of concentric compartments (shells). The time course of the concentration of nutrients in the soil solution in each of these shells is described using the preceding equations. For each time step, the change in concentration is solved:

$$\delta C/\delta t = -(1/r).\delta/\delta r(r.F) + S \qquad (4)$$

where S is the uptake of nutrients by the roots (which is supposed to decrease the nutrient concen-

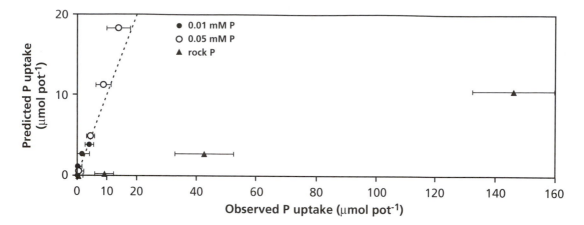

FIGURE 21. Comparison of observed P uptake by *Brassica napus* (oilseed rape) with predicted uptake for the 0.01 and 0.05 mM soluble phosphate (circles) and rock phosphate (triangles) treatment. The line refers to the situation when the predicted and observed rate of P uptake are identical (Hoffland et al. 1990b). Copyright International Society of Soil Science.

tration at the root surface to zero); the initial conditions are described by $t = 0$, $r > r_o$ (r_o being the effective root radius, including root hairs) and $C = C_i$ (the initial nutrient concentration in soil solution).

This model was first used to compare the measured and predicted rates of P uptake when phosphate is supplied in a readily available form. There is a good agreement between theory and experiment (Fig. 21). When the model is used to simulate phosphate uptake from a soil to which sparingly soluble rock phosphate is added, the simulated values are only 6% of those found experimentally. This leads to the conclusion that as much as 94% of the phosphate absorbed is released from rock phosphate due to its mobilization by excreted acids (Sect. 3.2).

This example shows how simulation models can be helpful to explore our intuitive ideas elegantly, if they are used in combination with experiments.

3. Nutrient Acquisition from "Toxic" or "Extreme" Soils

The term *toxic* or *extreme* soil is clearly an anthropomorphic one. For example a soil of a rather high or rather low pH may be toxic for some species, but still be the preferred habitat for others. The presence of fairly large concentrations of "heavy metals" similarly may prevent the establishment of one species and be essential for the completion of the life cycle of another. As pointed out in the introduction to this chapter, the occurrence of species on sites that we tend to call "toxic" does not necessarily mean that these plants grow better on such sites. When we use terms like **halophytes** and **calcifuges** in this section, then these refer to the ecological amplitude of the species. The physiological amplitude of a species is mostly considerably wider than its ecological amplitude. The restriction of a species to its ecological amplitude might indicate that these plants are the only ones that can survive in these soils, due to their specialized mechanisms, and that they are outcompeted on soils that we consider less extreme (see Introduction).

The following sections discuss specialized plant traits associated with phenotypic **acclimation** and genotypic **adaptation** to extreme soils and their consequences for species distribution.

3.1 Acid Soils

Soils tend to become acid naturally, as a result of the following processes (Bolan et al. 1991):

1. **Decomposition of minerals** by weathering, followed by leaching of cations, such as K^+, Ca^{2+}, and Mg^{2+} by rain. This is particularly important in humid regions.

2. **Production of acids** in soils (e.g., due to hydration and dissociation of CO_2, formation of organic acids, oxidation of sulfide to sulfuric acids and nitrification of ammonia).

3. Plant-induced production of acidity, when an **excess of cations** over anions is taken up (e.g., when N_2 or ammonium, rather than nitrate, is used as nitrogen source for plant growth).

In addition soils may acidify due to human activity: input of nitric and sulfuric acids from "acid rain", or the addition of acidic fertilizer, such as ammonium sulfate.

The soil acidity modifies the availability of many mineral nutrients (Fig. 1) as well as the solubility of aluminum. Although a low pH per se may limit the development of plants, aluminum toxicity is considered the most important yield-limiting factor in many acid soils, especially in the tropics and subtropics (Kochian 1995). In acid soils, concentrations of manganese may also increase to toxic levels, whereas phosphate, calcium, magnesium, potassium, and molybdenum may decline to an extent that deficiency symptoms arise (cf. Table 7 in Sect. 2.2.6).

3.1.1 Aluminum Toxicity

Aluminum is the most abundant metal in the earth's crust and the third most abundant element. Like all trivalent cations, it is toxic to plants. Aluminum hydrolyzes in solution, such that the trivalent cation dominates at **low pH** (Fig. 22). In addition, at low pH aluminum is released from chelating compounds.

Many species have a distinct preference for a soil with a particular pH. **Calcifuge** ("chalk-escaping"; also called "acidophilous," acid-loving) species, resist higher levels of soluble Al^{3+} in the root environment. The mechanism of Al toxicity is presently

unclear. Important toxic effects of aluminum occur even outside or at the plasma membrane. These are partly due to the inhibition of the uptake of calcium and magnesium (Table 10), due to blockage of channels in the plasma membrane (J.W. Huang et al. 1996). Some of the symptoms of aluminum toxicity are very similar to those of a deficiency of other ions. This may be due to competition for the same site in the cell walls (some cations), precipitation of aluminium complexes (with phosphate), or inhibition of root elongation, which reduces the absorption capacity (Marschner & Römheld 1996). Aluminum toxicity also resembles boron deficiency, but the reason for this is not clear (Lenoble et al. 1996a).

The **root apex** appears to be the sensitive region for toxicity, and inhibition of root elongation is the primary Al toxicity symptom, but this is not invariably associated with inhibition of calcium uptake (Ryan et al. 1994). Cell division is also inhibited; mitosis appears to be arrested in the S-phase of DNA replication. Inhibition of root elongation in the root tip is due to interference with the formation of cell walls, decreasing cell-wall elasticity by crosslinking with pectin (Rengel 1992b). Whether this is a direct effect of Al or is due to Al entry into the symplast, where it interferes with processes involved in cell-wall formation, is uncertain. Recent findings have suggested that Al might interfere with a signal-transduction pathway (Jones & Kochian 1996). Due to inhibition of cell elongation,

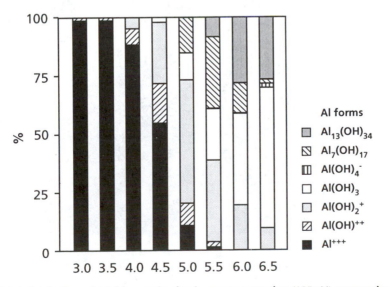

FIGURE 22. Calculated distribution of total inorganic aluminum concentration (185 μM) over various monomeric and polymeric forms as a function of pH. Calculations are based on parameters given by Nair & Prenzel (1978).

TABLE 10. Aluminum, phosphorus, calcium, and magnesium concentration [μmol g^{-1} (dry mass)] in roots and shoot of *Sorghum bicolor*, grown for 35 days at three levels of Al (zero, low: 0.4 mg L^{-1}, high 1.6 mg L^{-1}) and P [low, medium, and high: 285, 570 and 1140 μmol plant^{-1} (35 days)$^{-1}$].*

P-level	Al-level	Shoot				Root			
		Al	P	Ca	Mg	Al	P	Ca	Mg
low	zero	—	26	171	69	—	29	28	22
medium	zero	—	30	151	63	—	34	21	20
high	zero	—	38	139	63	—	39	19	23
low	medium	1	27	127	36	7 (29)	30	20	16
medium	medium	1	29	108	37	5 (40)	34	18	16
high	medium	1	40	85	36	5 (40)	46	20	19
low	high	1	93	61	23	11 (36)	70	16	14
medium	high	1	108	51	21	13 (31)	76	15	15
high	high	1	335	65	25	131 (45)	263	16	16

Source: Tan & Keltjens 1990.
*Values in brackets indicate the percentage removable with 0.05 M H$_2$SO$_4$ (i.e., the fraction in the apoplast).

root cells become shorter and wider. As a consequence, root elongation is impaired and the roots have a "stubby" appearance (Fig. 23) and a low specific root length, when grown in the presence of aluminum (Delhaize & Ryan 1995; Table 11). When most of the roots are exposed to Al, but root tips are in a solution without Al, plant growth is not affected by Al. On the other hand, when only the root tips are exposed to Al, toxicity symptoms are readily visible, confirming that the root tips are the primary site of aluminum toxicity (Kochian 1995).

Some aluminum is rapidly taken up in the symplast as well, possibly, via carriers whose function is to take up magnesium or iron, or via endocytosis (Kochian 1995). In the cytosol, with a neutral pH, it is no longer soluble, but it is bound to proteins and phosphate-containing compounds;

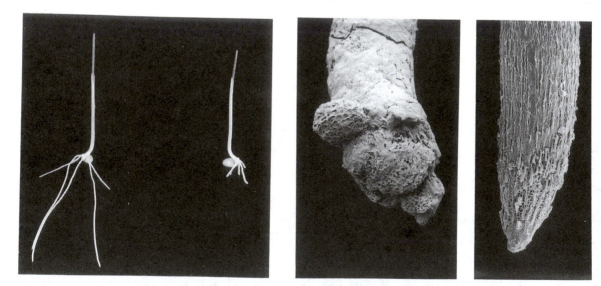

FIGURE 23. Scanning electron micrograph of the root tips of two near-isogenic line of *Triticum aestivum* (wheat). The panels on the left show seedlings of an aluminum-resistant (left) and an aluminum-sensitive (right) genotype; the ones on the right show root tips of the same aluminum-sensitive (left) and aluminium-resistant (right) genotypes. The seedlings were grown for 4 days in a solution containing 5 μM AlCl$_3$ in 200 mM CaCl$_2$ at pH 4.3 (courtesy E. Delhaize, CSIRO, Canberra, Australia; Delhaize & Ryan 1995). Copyright American Society of Plant Physiologists.

TABLE 11. The effects of aluminum concentration on various root parameters of *Mucuna pruriens* (velvet bean).*

$[Al^{3+}]$ (mg L^{-1})	DM (g)	FM (g)	D (mm)	L (m)	SRL (m g^{-1})
0	6.4	126	0.37	1160	175
0.1	6.6	155	0.44	1100	166
0.2	6.6	126	0.46	931	141
0.4	3.3	55	0.51	253	76

Source: Hairiah et al. 1990.
Note: The increase in root dry mass was not statistically significant.
* DM, dry mass; FM, fresh mass; D, diameter; L, root length per plant; SRL, specific root length (root length per gram dry mass of roots).

binding to DNA has also been shown. Aluminum may also replace calcium and/or magnesium from sites where they have a vital function in activation of enzymes; interference with calmodulin (a major component of signal-transduction pathways in plants) and the cytoskeleton may be particularly harmful. Most of these effects occur *after* the very rapid (1 to 2 hours) inhibition of root elongation. They are therefore not the primary cause of inhibition of plant growth (Kochian 1995).

Inhibition by aluminum of the uptake of calcium and magnesium decreases the concentration of these cations in the cell, causing calcium and/or magnesium deficiency symptoms. Calcium is required during cell division for spindle formation and to initiate metaphase/anaphase transition. Hence, the presence of aluminum prevents cell division and root development (Rengel 1992b). Interference with magnesium uptake causes magnesium deficiency symptoms (i.e., chlorotic leaves with brown spots) and stubby discolored roots (Tan et al. 1993).

3.1.2 Alleviation of the Toxicity Symptoms by Soil Amendment

Aluminium toxicity symptoms can be diminished by addition of extra **magnesium** or **calcium**. **Phosphate** addition also has a positive effect, since it precipitates Al, either outside or in the roots. There is some evidence that the toxicity symptoms can be alleviated by magnesium in monocotyledons, calcium having a smaller effect, whereas calcium has a stronger effect with dicotyledons (Fig. 24). It is interesting that the requirement for calcium is also higher for dicots (cf. Sect. 4 in this chapter).

The ability of high-molecular mass organic acids, such as **humic acid** and **fulvic acid**, to bind aluminum is well documented. These substances form much more stable complexes with Al than do citrate and malate, which are excreted by roots of some Al-resistant plants (Sect. 3.1.3). Fulvic acid and humic acid are normal constituents of humus, peat, and leaf litter, which can be added to alleviate toxic effects of Al (Harper et al. 1995).

Some of the symptoms of aluminum toxicity (e.g., inhibition of root elongation) of *Cucurbita pepo* (squash) which is grown in nutrient solution, are relieved by the addition of boron (Lenoble et al. 1996a). Incorporation of boron in an acidic high-Al subsoil promotes the depth of rooting and total root growth in *Medicago sativa* (alfalfa) (Lenoble et al. 1996b).

3.1.3 Aluminum Resistance

Different mechanisms can be discerned to account for a plant's resistance to potentially toxic levels of aluminum:

1. aluminum exclusion from the root apex
2. aluminum tolerance

There is clear evidence that **exclusion** confers Al resistance, but no convincing evidence for Al tolerance (Kochian 1995). Recent work on aluminum-resistant and sensitive genotypes of *Phaseolus vulgaris* (snapbean) and *Triticum aestivum* (wheat) has highlighted the importance of **citrate** and **malate** release by the roots, especially by the root tips (Fig. 25). In resistant genotypes, aluminum appears to activate a channel which allows the exudation of organic anions. Higher rates of exudation are not due to higher concentrations in the root tips; obviously, they are matched by higher rates of synthesis. The excretion of citrate and malate is accompanied by K$^+$ efflux, so that the positive effect of the chelator is not counteracted by a negative effect of lowering the pH. Mucilage exuded by the root cap may allow the malate concentration to remain sufficiently high over extended periods to protect the root tip (Delhaize & Ryan 1995). Wheat genotypes that excrete both malate and **phosphate** at the root tip show a threefold greater resistance to aluminum. Contrary to the inducible release of malate, the release of phosphate is constitutive (Pellet et al. 1996). Micobial degradation of the malate released by the roots of aluminum-resistant plants will limit the effectiveness of these compounds in sequestering aluminum. The half life of the released organic acids is less than 2 hours. For

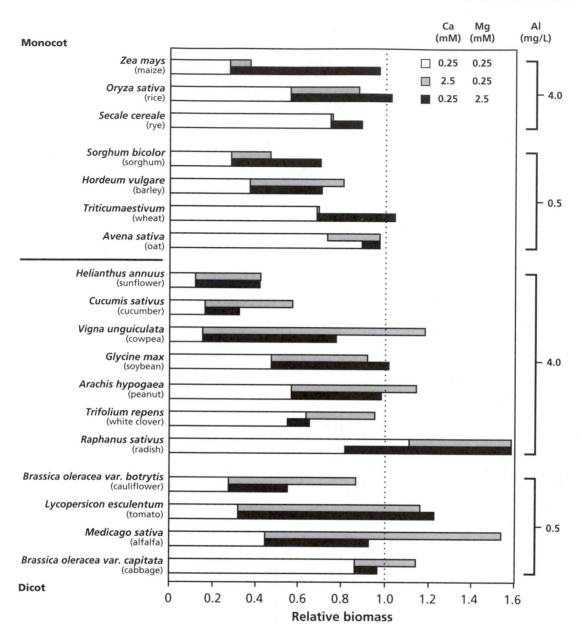

FIGURE 24. Biomass of a range of species in the presence of aluminum, in combination with different concentrations of calcium and magnesium, expressed as the ratio of the measured biomass relative to that in the absence of aluminum. White bars: in the presence of aluminum, with 0.25 mM Ca and Mg; gray bars: in the presence of aluminum, with 2.5 mM Ca and 0.25 mM Mg; black bars: in the presence of aluminum, with 0.25 mM Ca and 2.5 mM Mg (Keltjens & Tan 1993). Copyright Kluwer Academic Publishers.

roots growing at a rate of 15 mm day^{-1}, however, the residence time of the malate-releasing root tips in any zone of soil is around 5 hours. If the rate of microbial growth in soil is considered, then the size of the microbial biomass in the rhizosphere of the root tip will not change much from the time the tip enters a zone. Electron microscope and physiological studies confirm that there is little microbial proliferation at the root apex. It is therefore concluded that malate release protects the root tip from

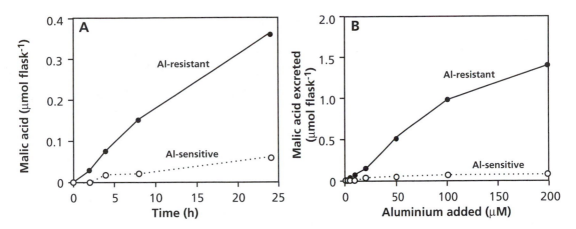

FIGURE 25. (A) Malate release from the roots of seedlings of an Al-resistant and an Al-sensitive genotype of *Triticum aestivum* (wheat) incubated in nutrient solution containing 50 μM Al. (B) Effect of Al concentration in the nutrient solution on malate release of the same genotypes as shown in (A) (Delhaize et al. 1993). Copyright American Society of Plant Physiologists.

the toxic effects of aluminum, despite microbial breakdown of malate in the rhizosphere (Jones et al. 1996b).

At a high pH calcifuge species may show **iron-deficiency** symptoms. This is probably associated with their aluminum-resistance mechanism, which may immobilize other ions as well, including iron. Root growth of calcifuge species may be stimulated by low aluminum concentrations. This growth-enhancing effect of aluminum is most pronounced at low pH. It is likely to be associated with the alleviation of the toxic effects of a low pH, which is a general effect of cations; trivalent cations have the strongest effect, followed by divalent, and then by monovalent ones (Kinraide 1993). The growth of **calcicole** ("chalk-loving"; also called acidifuge, "acid-escaping") species, which naturally occur on soils with a high pH (cf. Sect. 3.2), may also be stimulated by aluminum, but the optimum Al concentration for such species is about 5 μM, as opposed to 20–30 μM for calcifuge species, such as *Nardus stricta* (mat-grass) and *Ulex europaeus* (gorse).

3.2 Calcareous Soils

Calcifuge ("chalk-escaping") species have a distinct preference for a soil with a low pH. They tend to have a very low ability to solubilize the phosphate and/or iron in limestone, but resist higher levels of soluble Al^{3+} in the root environment (cf. Sect. 3.1). The lack of a high capacity to utilize the

forms of iron and other trace elements that prevail in alkaline soils (cf. Sect. 2.2.6) may be the cause of failure of establishment of calcifuge species in such soils. In addition, calcifuges tend to lack the capacity to access the prevalent poorly soluble phosphate sources in alkaline substrates (cf. Sect. 2.2.5) (Tyler 1996). **Calcicole** species are associated with soils of high pH. Their growth may be stimulated by high calcium concentrations, which are saturating for calcifuge species, however, this is not the major factor explaining their distribution. Calcicole species do not resist aluminum in their root environment.

Calcicole species differ from calcifuges in the spectrum of organic acids in their root exudates (Table 12). Calcicole species excrete far more oxalic and citric acid, compared with calcifuges. When determined at the same concentration, oxalic acid is very effective in solubilizing phosphate, whereas citric acid is by far the most effective in solubilizing iron from limestone soil (Ström et al. 1994). It is likely that these differences in iron- and phosphate-solubilizing capacity are of major importance in explaining the distribution of calcifuge species because addition of phosphate or iron in a readily available form enhances the biomass production of many calcifuge plants on calcareous soils (Tyler 1992, 1994).

The solubilization of phosphate and iron inevitably also enhances the concentration of calcium. Indeed, high calcium concentrations may be found in the xylem sap of calcicole species. Because calcium is an important "second messenger" (e.g., in the

TABLE 12. Total exudation of low-molecular mass organic acids from the roots of nine calcifuge and nine acidifuge species from northern Europe, calculated as the molar percentage of each acid exuded by the species.

Organic acid	Exudation (% mol)	
	Calcifuge species	Acidifuge species
Lactic	36.8	24.6
Acetic	35.3	5.3
Formic	1.9	0.3
Pyruvic	1.6	1.3
Malic/succinic	2.9	3.1
Tartaric	1.3	4.5
Oxalic	17.1	42.0
Citric	1.8	13.7
Isocitric	0	3.4
Aconitic	1.3	1.9
Total (μmol g^{-1} DM)	2.2	2.5

Source: Ström et al. 1994.

regulation of stomatal conductance) (cf. Sect. 5.4.2 in the chapter on plant water relations), how does a calcicole plant avoid being poisoned by calcium? It appears that calcicoles have the ability to store excess calcium as crystals, sometimes in leaf trichomes (De Silva et al. 1996).

3.3 Soils with High Levels of Heavy Metals

Heavy metals are characterized by their density, which is greater than 5 g ml^{-1}. Some of these (e.g., cobalt, copper, iron, manganese, molybdenum, nickel, and zinc) are essential micronutrients for plants, but they become toxic at high concentrations. Their role as an essential micronutrient may be as a cofactor or activator of specific enzymes or to stabilize organic molecules. Other heavy metals (e.g., cadmium, lead, chromium, uranium, mercury, silver, and gold), are not essential for plant functioning and are toxic even at low concentrations.

3.3.1 Why Are the Concentrations of Heavy Metals in Soil High?

High levels of heavy metals in soils may have a geological or anthropogenic origin. **Serpentine** soils naturally have high levels of nickel, chromium, cobalt, and magnesium, but low concentrations of calcium, nitrogen, and phosphate. The flora associated with these soils is rich in specially adapted endemic species (Arianoutsou et al. 1993). It has been known in Europe for centuries that rock formations that contain high levels of certain metals (e.g., copper) are characterized by certain plant species associated with these sites (**metallophytes**). This is also true for southern Africa, where only certain herbaceous species establish on the metal-rich sites. Such metal-**hyperaccumulating** plants may contain very high levels of heavy metals (e.g., up to 1% of the leaf dry mass may be copper). These metal-resistant species can be used as indicators to identify potential mining sites (e.g., *Hybanthus floribundus* to find nickel in Australia).

On sites close to mines, where the remains of the mining activity have enriched the soil with heavy metals, metal-resistant genotypes emerge (e.g., of *Agrostis tenuis*, *Deschampsia caespitosa*, *Festuca ovina*, or *Silene cucubalus* (which is synonymous for *Silene vulgaris*). The shoots of such plants may contain as much as 0.25% copper on a dry mass basis, which is a level that is highly toxic to other plants. Along roadsides, which are often enriched in lead from automobile exhaust, Pb-resistant genotypes occur. Some *Agrostis tenuis* (common bent-pass) genotypes grow even better in soils that contain as much as 1% lead than in unpolluted control soil (Newman 1983). Such genotypes are usually resistant to only one metal, unless more than one heavy metal is present in high level at such a site.

3.3.2 Using Plants to Clean Polluted Water and Soil: Phytoremediation

Some metal-accumulating species have been used to remove heavy metals from polluted water (e.g., the water hyacinth, *Eichhornia crassipes*). Terrestrial metallophytic species are also potentially useful to remove heavy metals from polluted sites, a process termed **phytoremediation** (Salt et al. 1995a). It requires plants that show both a high biomass production and metal accumulation to such high levels that extraction is economically viable. This combination of traits is often found in Brassicaceae (cabbage family). For example, *Brassica juncea* (Indian mustard) accumulates high levels of cadmium even when the cadmium level in solution is as low as 0.1 mg l^{-1} (Salt et al. 1995b). Consumption of parts of these crop plants may cause problems for human health. After accumulation of heavy metals from the polluted soil, the plants have to be removed and destroyed, taking care that the toxic metal is removed from the environment.

3.3.3 Why Are Heavy Metals So Toxic to Plants?

The biochemical basis of metal toxicity is not always clear. Cadmium, copper, and mercury affect **sulfydryl groups** in proteins and thus inactivate these. For a redox-active metal such as copper, an excess supply may result in uncontrolled redox reactions, giving rise to the formation of toxic **free radicals** that may lead to lipid peroxidation and membrane leakage (De Vos et al. 1989). Other heavy metals may inactivate major enzymes by **replacing the activating cation**. For example, zinc may replace magnesium in Rubisco, reducing the activity of this enzyme and hence the photosynthetic capacity (Clijsters & Van Assche 1984). Like zinc, cadmium also affects photosynthesis. Fluorescence measurements indicate that the Calvin cycle is the major process affected and that this subsequently leads to a "down-regulation" of photosystem II (Krupa et al. 1993). Cadmium affects the mineral composition even in cadmium-resistant species such as *Brassica juncea* (Brassicaceae). It reduces the concentration of Mn, Cu, and chlorophyll in the leaves, even at a concentration in solution that has no effect on biomass production (Salt et al. 1995b).

Most primary effects of heavy metals occur in the **roots**, which show reduced elongation upon exposure. Metal resistance is often quantitatively assessed by determining the effect of the metal on root elongation (Table 13). The increment in root dry mass tends to be affected less than that in root length, leading to "stubby" roots (Brune et al. 1994). Zinc toxicity is probably due to the binding of this metal to the plasma membrane, leading to reduced water uptake (Rygol et al. 1992). Because some chemical characteristics of zinc are like those of mercury, zinc inhibition of water uptake might be due to binding of the metal to a water-channel protein (cf. Sect. 5.2 in the chapter on plant water relations). Manganese toxicity leads to interveinal chlorosis and reduced photosynthesis (Macfie & Taylor 1992).

3.3.4 Heavy–Metal–Resistant Plants

Resistance in higher plants has been demonstrated for the following heavy metals: Cd, Cu, Fe, Mn, Ni, Pb, and Zn. Resistance to silver and mercury, which are both very toxic, has not been found. **Metal resistance** is sometimes partly based on **tolerance**. For example, damage by lead outside the plasma membrane can be prevented by modification of extracellular enzymes so that they are no longer affected by lead. This has been shown for extracellular phosphatases in lead-resistant genotypes of *Agrostis tenuis* (common bent-grass). **Avoidance** mechanisms generally account for resistance in a range of species. These mechanisms include:

1. **Exclusion** of the metal:

   a. Some microorganisms exude organic compounds that precipitate heavy metals outside the cells. There is no evidence for such a mechanism in ecotypes of higher plants that tolerate high concentrations of heavy metals in the root environment (Verkleij & Schat 1990); higher plants may exclude metals through precipitation, as a result of the excretion of hydroxyl ions into the rhizosphere;

   b. By chemical transformation (e.g., oxidation of reduced Mn in inundated soils);

   c. Through binding to cell walls.

TABLE 13. The effect of zinc on root elongation of a Zn-sensitive and a Zn-resistant ecotype of *Deschampsia caespitosa*.

Zn-sensitive		Zn-resistant	
Zinc concentration (μM)	Root elongation rate (%)	Zinc concentration (μM)	Root elongation rate (%)
1	100	1	100
25	82	250	82
50	78	500	64
100	62	1000	53

Source: Godbold et al. 1983.
Note: The plants were exposed to different Zn concentrations in solution for 10 hours.

2. Uptake followed by storage, typically occurring in plants designated as **hyperaccumulators**:

  a. Storage may occur in the vacuole and in the apoplasmic space (e.g., for zinc, cadmium, and copper);

  b. Epidermal cells may also be used for storage of the metals (Brune et al. 1994);

  c. Cadmium, manganese, and lead are preferentially accumulated in leaf trichomes (hairs) (e.g., in Indian mustard, *Brassica juncea*) (Salt et al. 1995b).

Cadmium resistance is associated with the presence of SH-containing peptides: **phytochelatins** (PCs) (Fig. 26). PCs are poly(γ-glutamyl-cysteinyl)-glycines, which bind metals. Unlike other peptides, with an α-carboxyl peptide bond, they are not made on ribosomes, but via a specific pathway from glutathione. Upon exposure of tobacco (*Nicotiana rustica*) plants to cadmium in the root environment, Cd-binding peptides, (γ-(Glu-Cys)₃-Gly, and γ(Glu-Cys)₄Gly), are produced. Inhibition of PC synthesis leads to loss of the cadmium-detoxification mechanism. Together with cadmium some of the PCs are almost exclusively located in the vacuole (Vögeli-

Lange & Wagner 1990), and an ATP-dependent mechanism transporting the cadmium-PC complex has been identified in tonoplasts of *Avena sativa* (oat) (Salt & Rauser 1995). It is likely that the formation of PCs followed by uptake of the cadmium-PC complex in the vacuole plays a crucial role in cadmium resistance.

Copper resistance in *Silene cucubalus* (= *S. vulgaris*, bladder campion) is based on **exclusion** (Fig. 27). We have no inkling as to whether exclusion is based on the excretion of copper that entered the cell or a mechanism that prevents copper from entering the cell. Upon exposure to copper, both resistant and sensitive *S. cucubalus* plants accumulate phytochelatins (Fig. 27). When compared at a copper concentration in the tissue that gives a similar physiological effect, rather than at the same copper concentration in the root environment, the phytochelatin concentrations in sensitive and resistant genotypes are rather similar (Table 14). Phytochelatin synthesis is likely to be essential to bind the toxic copper, but because phytochelatins are produced in both copper-resistant and sensitive plants, it is apparently not the basis for copper resistance in *S. cucubalus*. In ecotypes of *Arabidopisis*

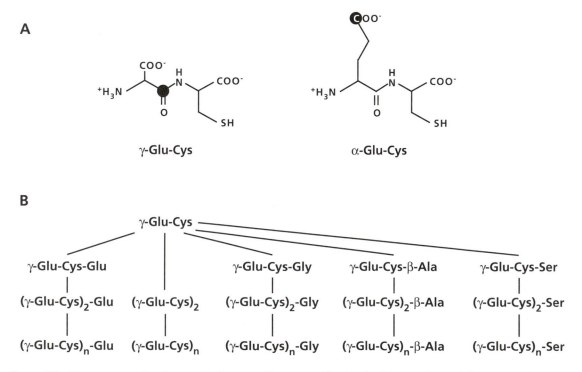

FIGURE 26. The structure of various γ(Glu-Cys) peptides. The γ-carboxyl-C of Glu is highlighted in (A) to indicate the difference between α- and γ-carboxyamide linkages. (B) is a model that summarizes the five families of γ(Glu-Cys) peptides involved in metal immobilization in plants and yeasts; the lines indicate family relationships and do not necessarily specify biosynthetic sequences (Rauser 1995). Copyright American Society of Plant Physiologists.

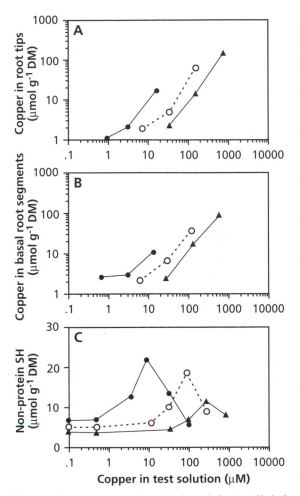

FIGURE 27. Copper (A, B) and phytochelatin sulfydryl concentration (C) in the roots of one copper-sensitive (filled circles) and two copper-resistant (open circles and filled triangles) ecotypes of *Silene cucubalus*. Copper was measured in the apical 10 mm (A) and the adjacent 10 mm (B). Phytochelatin was measured for the entire roots (Reprinted with permission from Schat & Kalff 1992). Copyright Lewis Publishers, an imprint of CRC Press, Boca Raton, Florida © 1992.

thaliana, copper tolerance correlates with the expression of a gene encoding **metallothionein**. Metallothioneins (MTs) are low-molecular mass metal-binding proteins; like other proteins, but unlike phytochelatins, they are synthesized on ribosomes (Murphy & Taiz 1995).

When compared at the same external zinc concentration (100 μM), a **zinc-resistant** ecotype of *Deschampsia caespitosa* (tufted hair-grass) accumulates less zinc in the apical parts of its roots (especially the 0 to 10 mm zone, but also in the 10 to 50 mm zone), but more in the basal parts (further than 50 mm from the apex) (Godbold et al. 1983). At the same external zinc concentration, whole roots of both ecotypes absorb zinc at the same rate. When compared at an external zinc concentration that has a similar effect on root growth (cf. Table 13), the resistant ecotype accumulates more zinc than does the sensitive one (Fig. 28). As found for other Zn-resistant genotypes, it also binds a greater fraction of the zinc to its cell walls than does the sensitive one. Inside the cell, the zinc is probably stored in the vacuole (as a complex with oxalate or citrate). There is very little transport of zinc to the shoot, especially in the resistant ecotypes. Typical zinc-hyperaccumulating species (e.g., *Thlaspi caerulescens*) accumulate and tolerate up to $40\,mg\,Zn\,g^{-1}$ (DM) in their shoots. When exposed to zinc levels that are toxic for most plants, *T. caerulescens* shows both enhanced zinc influx into the roots and increased transport to the shoots, which makes it a promising species to be used for phytoremediation (Lasat et al. 1996).

Nickel-resistance in *Alyssum lesbiacum* (Brassicaceae) is associated with the presence of high concentrations of the amino acid **histidine**. Histidine plays a role in the detoxification of absorbed nickel and transport of a nickel–histidine complex in the xylem to the leaves. In some *Alyssum* species nickel may accumulate to 3% of leaf dry mass (Krämer et al. 1996).

TABLE 14. Phytochelatin sulfydryl concentration [μmol g⁻¹ (dry mass)] and molar ratio of phytochelatin to copper in the roots of a copper-sensitive and a copper-resistant ecotype of *Silene cucubalus*.

Copper exposure level	Phytochelatin concentration		Phytochelatin/copper ratio	
	Sensitive	Resistant	Sensitive	Resistant
Highest concentration without any effect	3.7	2.9	3.7	1.6
Concentration giving 50% inhibition of root growth	7.6	7.5	3.7	1.7
Concentration giving 100% inhibition of root growth	19.0	16.0	1.2	0.3

Source: Schat & Kalff 1992.
Note: The same data were used as given in Figure 25 for the apical 10 mm.

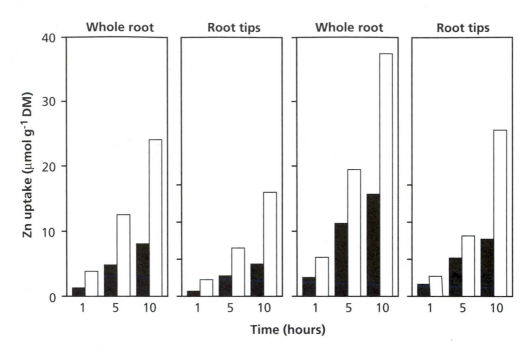

FIGURE 28. Uptake of ^{65}Zn by roots of a sensitive (filled bars) and a resistant (open bars) ecotype of *Deschampsia caespitosa*. The plants were compared at a low and high external concentration, which gives the same effect on root elongation (cf. Table 14). The low [Zn] (panels at left) was 25 mM and 250 mM, and the high [Zn] (panels at right) was 100 mM and 500 mM for the sensitive and resistant ecotype, respectively. At the end of the experiment, desorption into a nonlabeled solution of zinc was allowed for 30 minutes. The data therefore show uptake into the root cells only rather than a combination of uptake and binding of labeled zinc to the cell walls (Godbold et al. 1983). Copyright International Society of Soil Science.

3.3.5 Biomass Production of Sensitive and Resistant Plants

The biomass production of metal-resistant ecotypes tends to be less than that of sensitive ones, even when compared at a concentration of the heavy metal that is optimal for the plants (i.e., a higher concentration for the resistant plants) (Table 15). This might be due to the **costs** associated with the resistance mechanism. Environments with a high concentration of heavy metals, however, are fairly nutrient-poor. Because **nutrient-poor habitats** are generally inhabited by inherently slow-growing species, the low productivity of the resistant ecotypes may be associated with the low nutrient supply in their natural environment, rather than with the costs of their metal resistance per se (cf. the chapter on growth and allocation).

When grown in nontoxic soil, copper-resistant and copper-sensitive ecotypes of *Agrostis tenuis* have a similar yield in monoculture. In mixtures, the yield of the resistant ecotype is reduced (Table 16). This explains why resistant ecotypes are exclusively found in environments containing high levels of heavy metals, although the physiological basis of the effect, as presented in Table 16, is still obscure.

TABLE 15. Dry mass (mg per two plants) of roots and shoot of a copper-sensitive and a copper-resistant ecotype of *Silene cucubalus* (= *S. vulgaris*), after growth in nutrient solution with two Cu concentrations.

Ecotype		0.5 μM	40.5 μM
Sensitive	roots	64	8
	shoot	523	169
	total	587	177
Resistant	roots	22	33
	shoot	146	237
	total	168	270

Source: Lolkema et al. 1986.
Note: The different ecotypes were grown separately.

TABLE 16. Dry mass (mg per plant) of a copper-sensitive and a copper-resistant ecotype of *Agrostis tenuis*, grown in nontoxic soil.*

Ecotype	Separate	Mixture
Sensitive	89	88
Resistant	89	41

* Plants were grown either separately (42 plants per pot) or in mixtures (21 plants of each ecotype).

3.4 Saline Soils: An Ever-Increasing Problem in Agriculture

The presence of high concentrations of Na^+, Cl^-, Mg^{2+}, and SO_4^{2-} ions in saline soils inhibits growth of many plants. On a world scale, there is an area of around 380 million hectares that is potentially usable for agriculture, but where production is severely restricted by salinity. These areas occur predominantly in regions where evaporation exceeds precipitation. The problem of saline soils is ever-increasing, due to poor irrigation and drainage practices and to the expansion of irrigated agriculture into arid zones with high evapotranspiration rates (Flowers 1977).

3.4.1 Glycophytes and Halophytes

Most of our crop species are relatively salt-sensitive (**glycophytes**). A notable exception is sugar beet (*Beta vulgaris*). In saline areas, such as salt marshes, species occur with a high resistance to salt in their root environment (**halophytes**). The problems associated with high salinity are threefold:

1. A high salinity is associated with a low soil **water potential**, giving rise to symptoms similar to those of water stress

2. Specific ions, especially Na^+ and Cl^-, may be toxic

3. High levels of NaCl may give rise to an **ion imbalance** (predominantly calcium), and lead to deficiency symptoms

Plant adaptation and acclimation to salinity involve all these aspects; we discussed acclimation associated with the low water potential in Section 3 of the chapter on plant water relations.

Toxicity effects may include inhibition of nitrate uptake by chloride, probably because both ions are transported across the plasma membrane by the same carrier. High Na^+ may replace Ca^{2+} on root cell membranes, which may give rise to leakage of K^+ from the root cells. It may also reduce the influx and enhance the efflux of calcium. The decreased influx probably results from competition for binding sites in the cell wall, which decreases the concentration at the protein in the plasma membrane responsible for calcium influx. The toxicity of specific ions may subsequently lead to an ion imbalance and ion deficiency, especially calcium deficiency (Rengel 1992a).

At a moderate NaCl concentration in the root environment, Na^+ uptake occurs down an electrochemical potential gradient and higher sodium concentrations are expected inside than outside (Table 17). Roots of glycophytes, however, maintain a low Na^+ concentration in the presence of 1 mM Na^+ in their medium. This indicates that either their plasma membranes are highly impermeable for this ion, or that Na^+ is actively excreted from these roots.

The plasma membrane composition of root cells may indeed affect the entry of Na^+ and Cl^-, at least in glycophytes. Salinity resistance of grape varieties correlates with the solubility of chloride in their membrane lipids. Roots with a relatively high

TABLE 17. Experimentally determined concentrations of sodium and potassium ions in *Avena sativa* (oat) and *Pisum sativum* (pea) roots, compared with values predicted on the basis of the Nernst-equation.*

Ion	Oat Predicted	Oat Experimentally determined	Pea Predicted	Pea Experimentally determined
K^+	27	66	73	75
Na^+	27	3	73	8

Source: Higginbotham et al. 1967.
*The latter values assume that no metabolic energy-dependent mechanism is involved in the transport of these cations. The membrane potential of oat and pea was −84 and −110 mV, respectively.

TABLE 18. Net uptake of labeled sodium in a glycophyte, *Plantago media*, and a halophyte, *P. maritima*, in the presence and absence of DES (diethylstilboestrol, an inhibitor of the plasma membrane ATP-ase).*

	P. media		P. maritima	
NaCl (mM)	−DES	+DES	−DES	+DES
1	0.5	**2.8**	5.9	**2.7**
10	6.6	**27.7**	21.6	25.5
50	37.3	**121.1**	68.1	**82.5**

Source: De Boer 1985.

*The uptake was measured at three levels of NaCl in the nutrient solution and are expressed as [µmol g^{-1} (root dry mass)] hr^{-1}. Values printed bold are significantly different from those to their immediate left.

amount of phosphatidyl choline in their membranes restrict chloride transport to the shoot, as opposed to roots with membranes rich in glycolipids. Addition of phospholipids reduces chloride transport to the shoot. The effects of these lipids might be on passive chloride entry or on proteins in the membranes associated with transport (Kuiper 1968).

3.4.2 Energy-Dependent Salt Exclusion from Roots

The low Na$^+$ concentration inside the cells of glycophytes is mostly due to energy-dependent transport. At an external NaCl concentration of 1 mM, inhibition of the plasma membrane H$^+$-ATPase increases net Na$^+$ uptake in the glycophyte *Plantago media* (hoary plantain), but decreases it in the halopyte *P. maritima* (sea plantain). This illustrates that both Na$^+$ **excretion** and Na$^+$ uptake are **ATP-dependent** processes in these *Plantago* species (Table 18). At higher (10, 50 mM) NaCl concentrations, the roots of the glycophyte continue to excrete sodium, but not to the extent that accumulation in the plant is avoided. At 10 mM there is no evidence for ATPase-mediated uptake in the halophyte, and at 50 mM there is excretion (Table 18).

3.4.3 Energy-Dependent Salt Exclusion from the Xylem

At 10 and 50 mM NaCl, when the inhibitor of the plasma-membrane ATPase has no positive effect on the sodium concentration in the roots, it enhances the sodium concentration in the leaves of *Plantago* (Fig. 29). This indicates ATP-dependent exclusion from the xylem in both the glycophyte (*P. media*) and the halophyte (*P. maritima*). Glycophytes therefore, maintain a lower sodium concentration in their leaves, partly due to excretion by their roots, as well as because of energy-dependent **exclusion from the xylem**. Such an exclusion may be based on reabsorption of sodium from the xylem by surrounding cells. For example, in the glycophyte *Zea mays* (maize), Na$^+$ enters the xylem in the younger parts of the roots, but is resorbed from the xylem in more mature regions. Similar resorption may occur in stems (Shone et al. 1969). For the halophyte *Plantago maritima* exclusion from the xylem is similar to that for the glycophyte.

Using labeled Na$^+$, it has been shown for *Glycine max* (soybean) that salt that leaks into the xylem can be reabsorbed and excreted back into the root environment; however, the extent to which this happens in this glycophyte is rather small (Lacan & Durand 1994).

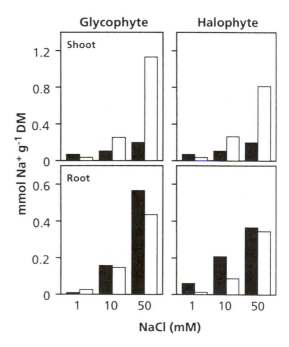

FIGURE 29. The effect of an inhibitor of the plasma membrane ATPase (DES, diethylstilboestrol; open bars) on the accumulation of labelled sodium in roots and shoots of a glycophyte (*Plantago media*) and a halophyte (*P. maritima*). Results for control plants are shown with filled bars (De Boer 1985). Reproduced with the author's permission.

3.4.4 Transport of Na$^+$ from the Leaves to the Roots and Excretion via Salt Glands

Salt transported to the shoot via the transpiration stream may be exported again, via the phloem, to the roots. Using labeled Na$^+$, this was shown for *Phaseolus vulgaris* (common bean). For another glycophyte (*Lupinus albus*, white lupin) this was analyzed by collecting phloem sap, which exudes from the stem upon cutting it. Export of Na$^+$ to the roots may be followed by excretion, as shown for *Plantago media* (Table 18).

True halophytes may have **salt glands**, which excrete salt from their leaves. These may remove a major part of the salt that arrives in the shoot via the transpiration stream, as shown for two mangrove species, especially when growing at a high salinity (Table 19). Salt exclusion in the roots of these mangroves can be derived from the difference in net chloride uptake and the product of the transpiration rate and the chloride concentration in the root environment (Fig. 30). It is substantial in all treatments, increasing from 90% at the lowest salinity level to 97% at 500 mM NaCl; exclusion may have been due to active excretion from the roots, as in *Plantago* (Table 18), or be associated with highly impermeable membranes. Active excretion must incur respiratory costs, as discussed in Section 4.2 of the chapter on plant respiration.

Salt excretion from the leaves involves specialized structures, which are designated as salt glands. In *Atriplex* species, salt that arrives in the transpiration stream is transported via plasmodesmata to the cytosol of epidermal cells and then to bladderlike cells on stalks (special trichomes) on the epidermal surface (Fig. 31). The salt is pumped into the large vacuole of this bladder cell. The trichomes may ultimately collapse, and the salt is deposited on the leaf surface, where it gives the leaves a white appearance until washed away by rain. The salt may also be excreted in such a way that concentrated droplets fall from the leaves, as in some *Tamarix* species. True salt glands, as opposed to the trichomes of *Atriplex*, are found in *Tamarix aphylla* (Fig. 31). The *Tamarix* salt glands consist of eight cells, six of which are involved in pumping the salt to the leaf surface. Two basal cells collect the salt from the leaf and transport it to the secreting cells. The secreting cells are surrounded by a lipophilic layer, except where they are connected to the basal cells via plasmodesmata. The mesophyll cells transport salt via plasmodesmata to the collecting cells, which then allow transport to the secreting cells. In these secreting cells, salt is pumped into microvacuoles. These merge with the plasma membrane, and the salt is then exported to the apoplast. The invaginations in these cells suggest that active membrane transport is involved as well. The salt diffuses via the apoplast to a pore in the cuticle, and it is then deposited on the leaf surface. The waxy layer that surrounds the secreting cells prevents back-diffusion to the mesophyll cells (Popp 1995).

What might be the advantage of salt excretion from the leaves over salt excretion from the roots? If all the salt that arrives via mass flow at the root

TABLE 19. The salt balance of two mangrove species, *Aegiceras corniculatum* and *Avicennia marina*, grown at three salinity levels.*

Parameter	A. corniculatum			A. marina		
NaCl concentration during growth (mM)	50	250	500	50	250	500
Water-use efficiency (mg dry mass mol^{-1} water)	74	45	41	81	79	92
Net chloride uptake (μmol Cl$^-$ mol^{-1} water)						
Accumulation						
Roots	16	25	19	44	35	113
Stems	21	8	2	11	17	28
Leaves	26	26	17	34	35	47
Total	63	59	38	89	87	189
Secretion	29	86	157	13	64	95
Total uptake	91	144	195	102	151	283
Shoot salt balance (% of shoot uptake)						
Cl$^-$ accumulation	62	28	11	77	45	44
Cl$^-$ secretion	38	72	89	23	55	56

Source: Ball 1988.

*The highest level is similar to that of seawater. The net uptake of Cl$^-$ is expressed per amount of water taken up and transpired.

FIGURE 30. Mother mangrove's advice to her offshoot (Ball 1988). Copyright CSIRO Australia.

surface were to be excluded, then the salt concentration in the rhizosphere would rapidly rise to very high levels. In the absence of a substantial removal from the rhizosphere, due to infiltration with less saline water, the local accumulation of salt would rapidly further reduce the water potential and aggravate the problems associated with water uptake (Passioura et al. 1992). A high water-use efficiency in combination with salt exclusion therefore has advantages over active or passive exclusion only. The mangrove species with the highest water-use efficiency are also the most salt-resistant ones (Ball 1988). A lower water-use efficiency at increased salinity may be compensated by a higher excretion by leaves, at least in those mangrove species that have salt glands (Fig. 30).

3.4.5 Compartmentation of Salt Within the Cell and Accumulation of Compatible Solutes

Salt resistance also involves the compartmentation of the potentially toxic ions in the vacuole and the capacity to produce nontoxic, compatible solutes in the cytoplasm (cf. Sect. 3 in the chapter on plant water relations). Compartmentation in the **vacuole**

is achieved by an active mechanism that is induced in halophytes such as *Plantago maritima* (Fig. 32) and *Mesembryanthemum crystallinum* (Barkla et al. 1995), but not in glycophytes, such as *Plantago media*, in the presence of NaCl in the root medium.

Some moderately salt-resistant glycophytes (e.g., *Hordeum vulgare*, barley, cultivars) also accumulate some salt in their leaves. It was shown using X-ray diffraction that chloride predominantly accumulates in the vacuoles of the epidermis cells of leaf blades and sheaths. To a smaller extent chloride is also found in the mesophyll cells of the leaf sheath, whereas the concentration remains low in the mesophyll cells of the leaf blade, even after exposure to 50 mM NaCl in the root environment for 4 days (Huang & Van Steveninck 1989).

3.5 Flooded Soils

The absence of oxygen in the soil causes a drop in redox potential, due to microbial activity. At a low redox potential, nitrate rapidly disappears due to its use as an electron acceptor by **denitrifying bacteria**, and **ammonium** is the predominant source of

A

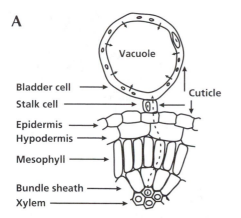

B

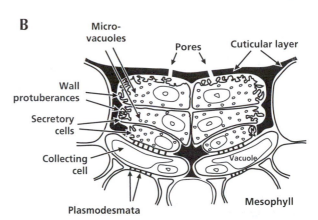

C

D

E

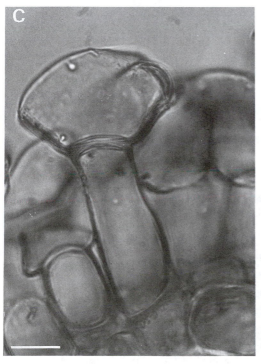

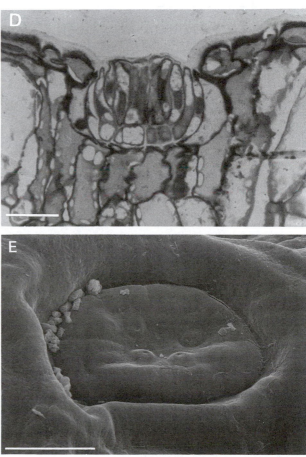

FIGURE 31. Two schematic diagrams of structures involved in the excretion of salt to the leaf surface. A: diagram of a trichome of a leaf of *Atriplex* (saltbush) species. B: diagram of a salt-excreting gland of *Tamarix aphylla* (tamarisk) (Esau 1977). Copyright by John Wiley & Sons. Three electron micrographs of salt glands. C: a bladder hair on the abaxial leaf surface of the mangrove *Avicennia germinans* (scale bar = 15 μm). D: cross section of a salt gland of *Aegialitis annulata*. The gland consists of collecting cells, basal cells, neighboring cells and the central secretion cells (scale bar = 30 μm). E: salt gland of the mangrove *Aegialitis annulata*. Only the central cells secrete the salt through pores (scale bar = 10 μm) (courtesy M. Weipert, Institute of Plant Ecology, University of Münster, Germany).

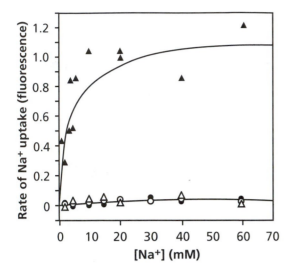

FIGURE 32. Uptake of Na$^+$ in tonoplast vesicles of the glycophyte *Plantago media* (circles) and the halophyte *P. maritima* (triangles). Tonoplast vesicles were isolated from plants grown in the absence (open symbols) or in the presence (filled symbols) of 50 mM NaCl. Uptake was quantified by following the sodium-dependent fluorescence (Staal et al. 1991). Copyright Physiologia Plantarum.

4. Plant Nutrient-Use Efficiency

Plants differ both in their capacity to acquire nutrients from the soil and in the amount of nutrients they need per unit growth (Sect. 4.1 and 4.2), the nutrient concentrations in their tissue (Sect. 4.3), and the time and extent to which they withdraw nutrients during leaf senescence before leaf abscission (Sects. 4.2 and 4.3). We discuss several approaches that have been used to analyze the efficiency with which plants utilize nutrients to produce new biomass. Whole-plant nutrient-use efficiency (NUE) considers processes related to carbon gain and loss, whereas photosynthetic nutrient-use efficiency (see Sect. 6 in the chapter on photosynthesis) considers only the instantaneous use of nutrients for photosynthetic carbon gain.

4.1 Variation in Nutrient Concentration

4.1.1 Tissue Nutrient Concentration

Plants differ in the concentration of mineral nutrients in their tissue, depending on environment, allocation to woody and herbaceous tissues, developmental stage, and species. Figure 33 gives the range of concentrations found in plant tissues, excluding nitrogen, that is lost during ashing. Chemical analysis of the nitrogen concentration of dried plant material gives a range of 5 to 50 mg g^{-1}, with most of it being present as organic N and some as

nitrogen for the plant. Iron and manganese are similarly reduced. These reduced forms are much more soluble and potentially toxic to the plant. Sulfate is also used as an alternative electron acceptor by specialized bacteria, leading to the formation of **sulfide**, which is an inhibitor of cytochrome oxidase (Sect. 3.6 in the chapter on plant respiration). Thus, the availability of many ions is affected by the redox potential, which leads to shortage of some nutrients and potentially toxic levels of others (Armstrong 1982).

Toxicity is largely prevented due to the oxidation, possibly followed by precipitation of these ions in the oxygenated rhizosphere. **Oxygenation of the rhizosphere** of flooding-resistant species is due to the presence of an aerenchyma, which is a continuous system of air spaces that connects the roots with the aerial parts of the plants. Such an aerenchyma both allows the root respiration to continue and leads to detoxification of potentially toxic ions in the rhizosphere (cf. Sect. 4.1.4 in the chapter on plant respiration and Sect. 2.3 in the chapter on growth and allocation).

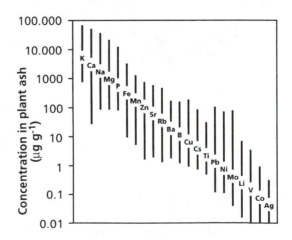

FIGURE 33. The range of concentrations of minerals as determined in plant ash. Because nitrogen disappears as nitrogen oxides during ashing, no figures are included for this mineral. The ash weight is about 10% of the dried mass of plants. Based on numerous references.

nitrate N. Nitrogen, phosphorus, and potassium are the nutrients that most frequently limit plant growth. The presence of a specific mineral does not imply that the plant needs this ion for growth. For example, cadmium is found in plant tissues when plants are growing on cadmium-polluted soil, but it is *not* an essential nutrient for any plant; similarly, high sodium concentrations are not required for growth.

Nutrient concentrations change predictably with plant development. Nutrients associated with metabolism (e.g., nitrogen, phosphorus, and potassium) have highest concentrations when a leaf or other organ is first produced, then decline, first as the concentration becomes diluted by increasing quantities of cell-wall material during leaf expansion, then by resorption of nutrients during **senescence** (Fig. 34). By contrast, calcium, which is largely associated with cell walls and is not resorbed (Sect. 4.3), increases continuously through leaf development.

Tissues differ predictably in nutrient concentrations; leaves have higher concentrations of nutrients associated with metabolism (N, P, K) and lower concentrations of calcium than do woody stems; roots have intermediate concentrations. Whole-plant nutrient concentration, therefore, differs among species and environments, depending on relative allocation to these tissues.

Environment strongly affects plant nutrient concentration by changing both allocation among organs and the composition of individual tissues. The major environmental effect on tissue nutrient com-

position is to alter the concentration of nutrients associated with metabolism. Plants have high concentrations of N, P, and K when conditions are favorable for growth (e.g., with adequate water and nutrients). The balance of available nutrients in the environment then alters the proportions of these nutrients. The *ratio* of nitrogen to phosphorus remains surprisingly constant (8 to 10:1) when plants receive nutrients in a ratio similar to that in their tissues (Ingestad & Ågren 1988), regardless of whether nutrients or light limit plant growth. A 10:1 N:P ratio is also found in plants sampled in the field for nonvascular plants, aquatic macrophytes, and vascular plants (Garten 1976). This is slightly lower than the Redfield ratio of 14:1 for balanced N:P nutrition in algae (Redfield 1958). In aquatic systems, both algae and water have an N:P ratio greater than 14:1 when growth is phosphorus-limited, and a ratio of less than 14:1 when algal growth is nitrogen-limited. When terrestrial plants deviate from the 10:1 N:P ratio, this generally reflects a similar nutritional imbalance caused by reduced uptake of the growth-limiting nutrient (Koerselman & Meuleman 1996, Verhoeven et al. 1996), which is sometimes combined with luxury consumption of nutrients that do not limit growth.

There are no striking differences among species in biochemical allocation of nitrogen and phosphorus among classes of chemical compounds (e.g., protein N, nucleic acid N, lipid N) (Chapin 1988). The major differences among species relate to accumulation of certain compounds in the cytoplasm for

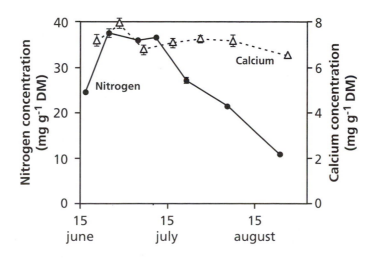

FIGURE 34. Typical seasonal pattern of leaf nitrogen and calcium concentrations of leaves of *Salix pulchra* (willow) from an Alaskan Arctic tundra meadow (Chapin et al. 1980). Copyright Blackwell Science Ltd.

osmotic functions (N-containing compatible solutes) and in vacuoles for storage functions (e.g., nitrate and vegetative storage proteins; Sect. 4.3 in growth and allocation) or chemical defense (e.g., alkaloids and cyanogenic glycosides; Sect. 3 in ecological biochemistry).

When nutrient supply declines relative to plant demand, most plants show the following sequence of events (Chapin 1980): (1) decrease in vacuolar reserves with little effect on growth; (2) continued reduction in tissue nutrient concentrations, especially in older leaves and stems, reduced rates of leaf growth and photosynthesis (in that order), increased nonstructural carbohydrate concentrations, senescence of older leaves, and reallocation of reserves to compensate for reduced nutrient status (increased root mass ratio and increased root absorption capacity); (3) greatly reduced photosynthesis and nutrient absorption, dormancy or death of meristems.

4.1.2 Tissue Nutrient Requirement

Species differ in their nutrient requirement for maximum growth, but the physiological mechanisms for this are largely unknown. For example, the tissue **calcium concentration** at which 90% of the maximum yield is achieved is about twice as high for **dicots** as for **monocots** (Table 20). In addition, when comparing graminoids and forbs at similar sites, the forbs invariably have higher concentrations of both calcium and magnesium (Meerts 1997). The reason for this difference is likely the greater cation exchange capacity of the cell walls of dicotyledonous species (i.e., the amount of free calcium-binding carboxylic acid groups in

pectins) (Woodward et al. 1984). The tissue **phosphate concentration** at which 90% of the maximum yield occurs is greater for many **legumes** than for **nonlegumes** (cf. Fig. 3 in the chapter on symbiotic associations). The physiological basis of this difference is again unclear, but it is likely associated with the high energetic requirement and use of phosphorylated intermediates to fix atmospheric nitrogen in legumes (Israel 1987).

From a biochemical point of view, all species will need similar amounts of N, P, S, and so on, to make a unit of growth simply because they are constructed in a similar manner. Thus, the idea that there are different *metabolic* requirements is erroneous, except that specific enzymes may require a specific ion. For instance, nickel is an essential element for urease, which hydrolyzes urea to CO_2 and H_2O. Urease is required in all plants, but in greater amounts in those legumes that produce ureides when grown symbiotically with rhizobia (Sect. 3.4 in symbiotic assoviations) (Walker et al. 1985). Apart from these exceptional differences, variation in nutrient requirement and nutrient productivity (Sect. 4.2.1) will depend much more on the balance between requirements for protein synthesis for new growth and nitrogen storage (Sect. 4 in growth and allocation).

4.2 Nutrient Productivity and Mean Residence Time

4.2.1 Nutrient Productivity

A useful measure of the efficiency of nutrient use to produce new biomass is **nutrient productivity**

TABLE 20. Effect of the calcium concentration in the nutrient solution on the growth and the calcium concentration in the shoots of a monocotyledonous (*Lolium perenne*, perennial ryegrass) and a dicotyledonous (*Lycopersicon esculentum*, tomato) species.

Species	Calcium supply (µM)				
	0.8	2.5	10	100	1000
Growth rate (% of maximum value)					
Lolium perenne	42	100	94	94	93
Lycopersicon esculentum	3	19	52	100	80
Calcium concentration ($\mu mol\, g^{-1}$ dry mass)					
Lolium perenne	15.0	17.5	37.4	92.3	269.5
Lycopersicon esculentum	49.9	32.4	74.9	321.9	621.3

Source: Loneragan 1968 and Loneragan & Snowball 1969, as cited in Marschner 1983.

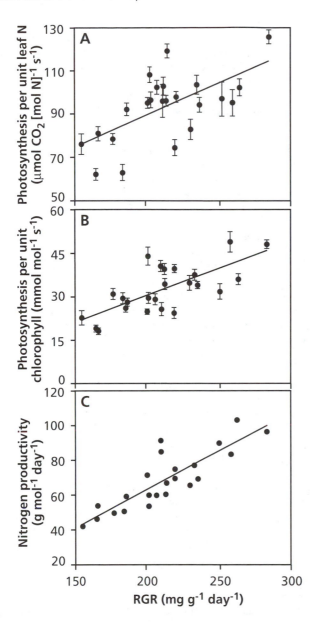

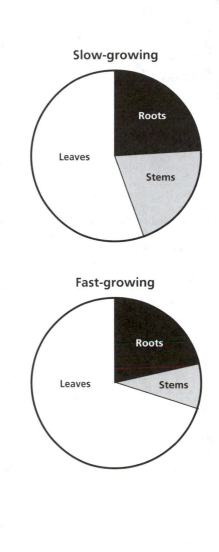

FIGURE 35. The nitrogen productivity of fast- and slow-growing herbaceous plant species, grown with free access to nutrients in a growth room in the same study. The physiological background of the higher nitrogen productivity of fast-growing species is their greater investment of N in leaves, as opposed to roots and stems and their higher rate of photosynthesis per unit nitrogen in the leaves (photosynthetic nitrogen-use efficiency, PNUE). The rate of photosynthesis per unit chlorophyll in the leaves is also higher for the fast-growing herbaceous species, as shown in the middle panel of the figure at the left (Poorter et al. 1990). Copyright American Society of Plant Physiologists.

(Ingestad 1979), the ratio of relative growth rate (RGR, $mg\,g^{-1}\,day^{-1}$) to whole plant nutrient concentration in the plant tissue ($mol\,g^{-1}$). For example, nitrogen productivity (NP, $mg\,mol^{-1}\,N\,day^{-1}$) is:

$$NP = RGR/PNC \qquad (5)$$

where PNC is the plant nitrogen concentration (i.e., total plant N per total plant mass). When grown at an optimum nutrient supply, plants differ widely in their nitrogen productivity (Fig. 35). A higher nitrogen productivity is associated with rapid growth, a relatively large investment of nitrogen in photosynthesizing tissue, an efficient use of the

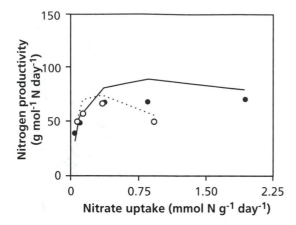

FIGURE 36. **The nitrogen productivity of *Briza media* (open symbols and broken line) and *Dactylis glomerata* (cocksfoot, filled symbols, and continuous line), as a function of the rate of nitrate uptake. The nitrate uptake was varied through different exponential rates of nitrogen addition in order to maintain a constant RGR at each rate of nitrate supply. The symbols give the actual experimental data and the lines refer to results of a simulation model (Van der Werf et al. 1993). Copyright Blackwell Science Ltd.**

nitrogen invested in the leaves for the process of photosynthesis, and a relatively small use of carbon in respiration (Fig. 35). C_4 species also have a high nitrogen productivity under optimal nitrogen supply, which is apparently a result of their lower nitrogen requirement for photosynthesis (high PNUE) (Sect. 6.1 in the chapter on photosynthesis).

The nitrogen productivity shows saturation, and sometimes an optimum curve, when plotted as a function of the nitrogen supply to the plant (Fig. 36). The decrease in NP above the maximum value for NP is due to a decrease in the rate of photosynthesis per unit of nitrogen in the leaf at high leaf N, which reflects increased allocation of nitrogen to storage (cf. Sect. 4 in growth and allocation). The decrease when the nitrogen supply is less than that at the maximum value for NP is largely due to greater investment of N in nonphotosynthetic tissue (Sect. 5.4 in growth and allocation).

4.2.2 The Mean Residence Time of Nutrients in the Plant

Although the nutrient productivity gives a good indication of a plant's **instantaneous nutrient-use efficiency** (NUE), it does not provide insight into a plant's **long-term performance** in a natural habitat. To develop such insight, the concept of **NUE** was

expanded to consider the time during which nutrients remain in the plant to support productivity. Plant NUE ($g\,g^{-1}N$), which is defined in this way, is the product of the NP ($g\,g^{-1}N\,yr^{-1}$) (as defined earlier, but is now determined over much longer periods; say 1 year), and the **mean residence time** (MRT; yr) of that nutrient in the plant (Berendse & Aerts 1987):

$$NUE = NP \cdot MRT$$

The mean residence time is the average time the nitrogen remains in the plant, before it is lost due to leaf shedding, herbivory, root death, and so on.

The nitrogen-use efficiencies of evergreen-heathland shrub species and that of a co-occurring deciduous grass species are remarkably similar, but the underlying components are rather different (Table 21). Evergreen species achieve their NUE with a low nitrogen productivity and a high mean residence time, whereas deciduous species have a considerably higher nitrogen productivity, but a lower mean residence time. In competition experiments with the species from Table 21, the grass wins at a relatively high nitrogen supply because of its higher nitrogen productivity. At a low nitrogen supply, the competitive ability of the evergreen shrub is higher because of its long mean residence time of nitrogen in the plant. A high mean residence time is probably the most important mechanism for nutrient conservation in infertile sites (Aerts 1995).

Both adaptation and acclimation contribute to the greater mean residence time on infertile sites. These sites are typically dominated by evergreen shrubs and trees, and both grasses and evergreen trees and shrubs adapt to low nutrient supply through increases in leaf longevity (Reich et al. 1992).

TABLE 21. **The long-term nitrogen productivity (NP), the mean residence time of nitrogen (MRT), and the nitrogen-use efficiency (NUE) of an evergreen-heathland shrub species (*Erica tetralix*) and a co-occurring deciduous grass species (*Molinia caerulea*).**

	Erica	*Molinia*
Nitrogen productivity ($g\,g^{-1}N\,yr^{-1}$)	77	110
Mean residence time (yr)	1.2	0.8
Nitrogen-use efficiency ($g\,g^{-1}N$)	90	89

Source: Aerts 1990.

TABLE 22. Above-ground nitrogen-use efficiency (NUE), mean residence time of nitrogen (MRT), and long-term nitrogen productivity (NP) of a deciduous grass species (*Molinia caerulea*) at a range of nitrogen supply rates ($gNm^{-2}yr^{-1}$).

N-addition	5	10	20
Nitrogen-use efficiency ($gg^{-1}N$)	345	238	227
Mean residence time (yr)	2.4	1.7	1.9
Nitrogen productivity ($gg^{-1}Nyr^{-1}$)	141	141	123

Source: Aerts 1990.

With increasing nitrogen availability, the nitrogen-use efficiency of individual species tends to decrease because of a decline in both the nitrogen productivity and the mean residence time (Table 22). Due to the replacement of one species by others, however, this does not necessarily lead to a decrease in nitrogen-use efficiency at the community level.

It is interesting that the plant features that favor a low rate of **nutrient loss** (high mean residence time) also decrease the rate of **decomposition** of the leaf litter. This tends to aggravate the low availability of nutrients in the already nutrient-poor environments (Aerts 1995, Hobbie 1992). As will be discussed in Section 2.4 of the chapter on symbiotic associations and in the chapter on decomposition, however, some species can make use of nutrients in leaf litter even if this has not been fully decomposed.

4.3 Nutrient Loss from Plants

Nutrient loss is just as important as nutrient uptake in determining the nutrient budgets of perennial plants; however, much less is known about the controls over nutrient loss.

4.3.1 Leaching Loss

Leaching accounts for about 15% of the nitrogen and phosphorus and half the potassium returned from above-ground plant parts to soil (Table 23), with the remainder coming from senesced leaves and stems. Use of experimental "miniumbrellas" to prevent rain from contacting leaves suggests that leaching losses can be an even larger proportion (25 to 55% of nutrient loss from leaves; Chapin & Moilanen 1991). Leaching occurs most readily

when there are high concentrations of soluble nutrients in the intercellular spaces of leaves (e.g., during rapid leaf production or senescence and when plants grow under conditions of high **nutrient availability**). Leaching rate is highest when rain first hits a leaf, then declines exponentially with continued exposure to rain (Tukey 1970). The frequency of rainfall, therefore, is more important than its intensity in determining leaching loss. **Deciduous** leaves have a higher rate of leaching loss than do **evergreens** because of their higher tissue nutrient concentrations. This is compensated, however, by the greater time of exposure to leaching in evergreen plants (Thomas & Grigal 1976) so that leaching constitutes a similar proportion of above-ground nutrient loss by evergreen and deciduous forests (Table 23).

The magnitude of nutrient loss by leaching decreases in the order potassium > calcium > nitrogen = phosphorus, which reflects the greater mobility of monovalent than divalent cations and the greater susceptibility to loss of inorganic than of organically bound nutrients. It was initially thought that one explanation for the scleromorphic leaves with thick cuticles in nutrient-poor sites was prevention of leaching loss (Loveless 1961). There is no clear relationship, however, between cuticle thickness or sclerophylly and susceptibility of leaves to leaching loss (see Sect. 5.4.5 in the chapter on plant water relations). These leaf traits are more likely selected for their importance in withstanding unfavorable conditions during the nongrowing season and reducing leaf loss to herbivores and pathogens.

Plants with low nutrient status can absorb nutrients from rainfall. This is an important avenue of ammonium and nitrate uptake by nitrogen-limited forests exposed to rain that has high nitrate due to

TABLE 23. Nutrients leached from the canopy (throughfall) as a percentage of the total above-ground nutrient return from plants to the soil for 12 deciduous and 12 evergreen forests.

Nutrient	Throughfall (% of annual return)	
	Evergreen forests	Deciduous forests
N	14	15
P	15	15
K	59	48
Ca	27	24
Mg	33	38

Source: Chapin 1991.

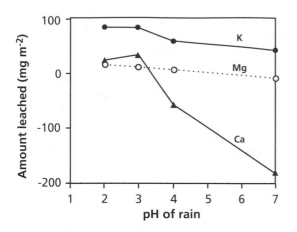

FIGURE 37. Effects of pH of simulated rain on leaching and absorption of calcium, potassium, and magnesium from spruce crowns (Chapin 1991). Copyright Academic Press.

fossil fuel combustion or high ammonium due to volatilization from agricultural lands and stockyards (Clarkson et al. 1986).

Acid rain increases leaching of cations, particularly of calcium (Fig. 37), because hydrogen ions in the rain exchange with cations held on the cuticular exchange surface and because acidity alters the chemical nature of the cuticle so that it is more susceptible to diffusion and mass flow of nutrients

to the leaf surface (Reuss & Johnson 1986, Shriner & Johnston 1985).

4.3.2 Nutrient Loss by Senescence

Approximately half of the nitrogen and phosphorus content of leaves is resorbed during senescence and is used to support further plant growth (Aerts 1996, Killingbeck 1996). By contrast, calcium, which is immobile in the phloem, cannot be resorbed and reutilized. Nitrogen and phosphorus **resorption efficiency** (proportion of maximum nutrient pool resorbed) ranges from 0 to 80% among species and environmental conditions (Killingbeck 1996, Reich et al. 1995; Table 24). Similar variation occurs with respect to the terminal nutrient concentration in senesced leaves (**resorption proficiency**). Concentrations of 0.3% nitrogen and 0.01% phosphorus in senesced leaves represent the ultimate potential resorption of these nutrients in woody perennials. Resorption efficiency of both nitrogen and phosphorus is highest in graminoids; N-resorption efficiency is higher in deciduous shrubs and trees than it is in evergreens, although the differences are small compared with differences in mean residence time (Aerts 1996; Table 25). Evergreen species have a greater ability to reduce the mass-based P concentration in senescing leaves than do deciduous species (greater phosphorus-resorption profi-

TABLE 24. Nutrient withdrawal from senescing leaves of trees growing on nutrient-poor sandy sites.*

Species	Location	Leaf longevity	Resorption (%)	
			P	N
Goupia glabra	Guyana	Lower	53	0
Cecropia obtusa	Guyana	Lower	63	0
Dicymbe altsonii	Guyana	Higher	60	33
Chlorocardium rodiei	Guyana	Higher	73	0
Banksia menziesii	Australia	Lower	82	73
Eucalyptus gomphocephala	Australia	Higher	55	61
Larix laricina	Minnesota	Lower	0	48
Populus tremuloides	Minnesota	Lower	42	65

*For the tree species from a rainforest in Guyana, nitrogen was not a factor limiting growth. Phosphorus was available in critically low amounts, which may have been limiting for growth; however, productivity may also have been limited by other nutrients or the low pH of the soil. For the Australian tree species from an open sclerophyll nutrient-poor woodland, where bushfires regularly remove large amounts of nitrogen from the ecosystem, growth of the investigated trees was limited by nitrogen (Raaimakers 1995). In Minnesota nitrogen was the growth-limiting nutrient (Tilton 1977, Verry & Timmons 1976). In addition, most other nutrients were also scarcely available. Nutrient resorption was calculated from the amount of P or N present in senesced leaves and the peak amount found in the leaves of each species.

TABLE 25. Nitrogen- and phosphorus-resorption efficiency of different growth forms.

Growth form	Resorption efficiency (% of maximum pool)	
	N	P
All data	50 (287)	52 (226)
Evergreen trees and shrubs	47 (108)[a]	51 (88)[a]
Deciduous trees and shrubs	54 (115)[b]	50 (98)[a]
Forbs	41 (33)[a]	42 (18)[a]
Graminoids	59 (31)[b]	72 (22)[b]

Source: Aerts 1996.
Note: Results are mean values, with the number of species in parentheses. Different letters within a column indicate statistical difference between growth forms at $P < 0.05$.

ciency) (Fig. 38). In spite of the large range observed in nutrient resorption and the importance of resorption to plant nutrient budgets, no clear patterns of physiological and ecological controls over nutrient resorption have emerged. About 60% of studies show no relationship of resorption efficiency to nutrient availability, with most of the remaining studies showing small decreases in resorption efficiency in fertile sites (Aerts 1996, Demars & Boerner 1997). On nutrient-rich sites larger *quantities* of nutrients are generally withdrawn from the leaves and larger *quantities* remain in senesced leaves, compared with leaves of plants growing on infertile sites (Killingbeck 1996), but the proportion of nitrogen and phosphorus resorbed is similar across sites (Aerts 1996). Thus, the nutrient concentration of litter is higher on more fertile sites, which has important consequences for decomposition (see the chapter on decomposition).

Resorption is the net result of several processes: enzymatic breakdown of nitrogen- and phosphorus-containing compounds in the leaves, phloem loading and transport, and the formation of an abscission layer that cuts off the transport path and causes the leaf to fall. Resorption is positively correlated with leaf mass loss during senescence, which suggests a link with export via the phloem (Chapin & Kedrowski 1983). Leaves that are darkened during senescence to reduce source strength have low resorption, whereas leaves with strong sinks (e.g., nearby developing fruits or new leaf growth) have high resorption, which again suggests a role for source–sink interactions and phloem transport in explaining proportional resorption (Chapin & Moilanen 1991, Nambiar & Fife 1987). Both graminoids (Aerts 1996) and evergreens (Nambiar & Fife 1987) that have active growth of new leaves (a strong sink) at the time of leaf senescence have

high resorption efficiency. Comparing different species, all major N and P chemical fractions are broken down to the same extent during autumn senescence (Chapin & Kedrowski 1983). It is therefore unlikely that there is some recalcitrant nutrient fraction that limits resorption efficiency in some species more than in others. Strong winds and early frosts can reduce resorption efficiency (Oland 1963), but leaves typically abscise only after resorption has ceased (Chapin & Moilanen 1991). Species with gradual leaf fall may have low resorption efficiencies (del Arco et al. 1991). Drought can also reduce resorption efficiency (Boerner 1985, Pugnaire & Chapin 1993).

There is very little information available on nutrient resorption from senescing stems and roots. Berendse & Aerts (1987) assume 25% resorption from woody stems of shrubs, but in the few studies of roots, no resorption has been reported (Aerts 1990, Nambiar 1987).

4.4 Ecosystem Nutrient-Use Efficiency

Our definitions of **nutrient-use efficiency** have so far been based on individual plants. The same concept has been applied to ecosystems that are approximately in steady state [i.e., where aboveground production is approximately equal to litterfall (leaves, twigs, small branches and reproductive parts)]. Ecosystem NUE is the ratio of litterfall mass to litterfall nutrient content (i.e., the inverse of the nutrient concentration of litterfall) (Vitousek 1982). This is equivalent to the biomass produced per unit of nutrient gained or lost. Ecosystem NUE is rather similar among sites, or greater on sites with low availability of nutrients (e.g., for nitrogen, which is the element that most strongly limits productivity in most terrestrial ecosystems). The data for ecosystem NUE, however, have to be interpreted with care: NUE and nutrient concentration in the litter are inversely and negatively correlated and are not independent. If all things are equal, then a high nutrient concentration of the litter (i.e., a low dry mass to N ratio) is inevitably associated with a high N loss in litterfall.

The three processes that might cause differences in ecosystem NUE are:

1. photosynthesis per unit nutrient
2. mean residence time (MRT) during which the nutrient contributed to production
3. proportion of nutrients resorbed prior to senescence

We have seen that PNUE is low in slow-growing plants from low-nitrogen environments (Fig. 34,

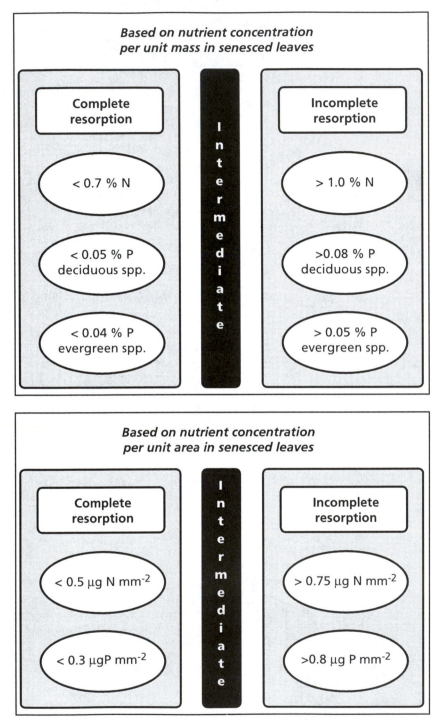

FIGURE 38. Ranges of N and P concentrations, expressed per unit mass (top) or per unit leaf area (bottom). Complete and incomplete resorption are synonymous with high and low resorption proficiency, respectively. Mass-based P concentrations are segregated between deciduous and evergreen species because of the large difference between these life-forms in ability to reduce P in senescing leaves (Killingbeck 1996). Copyright Ecological Society of America.

Sect. 6 of the chapter on photosynthesis), and that resorption is similar across sites or slightly higher in infertile sites (Sect. 4.3), so the only cause of any variation across sites in ecosystem NUE is differences in tissue longevity. The greater leaf longevity observed in infertile sites (Aerts 1995, Berendse & Aerts 1987), however, is almost entirely compensated by a lower nitrogen productivity (Reich et al. 1992, 1995). As a result, upon closer inspection, ecosystem NUE is rather similar in infertile and fertile sites. Given that ecosystem NUE is the inverse of litter nutrient concentration, an explanation for the pattern is that plants in infertile sites produce leaves with lower nitrogen concentrations and similar resorption efficiencies to fertile sites and therefore have low litter nitrogen concentrations (Aerts 1996). An ecosystem consequence of the long leaf lifespan and low nutrient productivity in infertile sites is that the differences in NUE across soil fertility gradients are smaller than what would be predicted from patterns of nitrogen availability.

In summary, plants differ in their capacity for nutrient uptake and efficiency of nutrient use. Genetic differentiation and acclimation, however, vary in their relative importance to different processes. Acclimation is probably the major factor that accounts for the high root mass ratio in infertile sites. Due to low availability, rates of nutrient acquisition are low for plants in infertile sites. These plants generally have low leaf nitrogen concentrations and a low photosynthetic nitrogen use efficiency (Table 25), due primarily to effects of environment on tissue concentration and to both genetic and phenotypic differences in PNUE. Plants on infertile sites generally keep their nutrients for a longer period; for example, the mean residence time of nutrients is higher for evergreens than it is for deciduous leaves, and any given species retains its leaves longer on infertile sites. Plants also differ in the extent to which they withdraw nutrients from senescing leaves, but the variation in the extent to which nutrients are withdrawn shows a less consistent difference between fertile and infertile sites. The high ecosystem NUE in infertile sites reflects low tissue nitrogen concentrations and high mean residence time.

5. Mineral Nutrition: A Vast Array of Adaptations and Acclimations

Nutrients move in the soil to root surfaces, by mass flow and diffusion, but selective systems (channels, carriers) are then needed to transport the nutrients into the symplast. Because anion transport mostly occurs up an electrochemical potential gradient, metabolic energy is required to import these nutrients from the rhizosphere. Although cation transport may occur down an electrochemical gradient, metabolic energy is also required to import these nutrients from the rhizosphere because the maintenance of the electrochemical gradient requires ATP. When essential nutrients move too slowly to the roots' surface, adaptive mechanisms are required, especially for the acquisition of phosphate, iron, and zinc.

Species have adapted to adverse or favorable soil conditions, and individual plants have some capacity to acclimate to a range in soil conditions. Some of these acclimations are physiological (e.g., an induction of ion-uptake systems when nutrients are in short supply and an excretion of phosphate-hydrolyzing enzymes). Others are anatomical (e.g., the formation of more or longer root hairs when phosphate is in short supply) or morphological (e.g., the increase in root mass ratio when nitrogen is limiting for growth). These anatomical and morphological acclimations, however, also have a physiological basis and often require induction of specific genes, after a shortage of nutrients has been sensed.

Plants need many macronutrients and micronutrients, but the concentration of the various elements in plant tissues does not necessarily give us a correct estimate of a plant's requirements. Rather, elements may accumulate because the plant lacks mechanisms to keep these out and store these elements in compartments where they are least harmful. In this chapter we have encountered numerous plants that grow on sites that are practically inaccessible to others. These adapted plants include halophytes, metallophytes, calcifuges, and calcicoles. Halophytes and metallophytes do not need high concentrations of NaCl and heavy metals, respectively, for maximum growth, but they are among the few species that can cope with such adverse soil conditions; that is, their ecological amplitude is much narrower than their physiological amplitude. Calcifuges and calcicoles, conversely, are largely restricted to acid and alkaline soils, respectively, because they lack the capacity to acquire some nutrients from alkaline soils and are adversely affected by toxic compounds in acid soil, respectively. Understanding plant distribution as dependent on soil type clearly requires a full appreciation of physiological mechanisms.

Plants differ not only in the mechanisms employed to acquire nutrients from various soils, but also in the requirement for these nutrients and in their long-term nutrient-use efficiency. Plants from nutrient-rich sites tend to produce more biomass per unit nutrient in the plant, whereas plants from nutrient-poor sites tend to keep the nutrients they have acquired for a longer time. There is remarkably little variation among species in the extent to which they resorb nutrients from senescing leaves, as opposed to variation that depends on nutrient availability (i.e., a smaller proportion of the nitrogen invested in leaf mass is remobilized on nutrient-rich sites than it is on nitrogen-poor sites.

Knowledge of a plant's mineral nutrition is pivotal to understanding the distribution of plant species and it is essential for modern agriculture and forestry (e.g., to avoid nutrient deficiency disorders). It is also important to resolve environmental problems (e.g., through phytoremediation). Mixed cultures can be highly beneficial in cropping situations. Intercrop species (i.e., plants that are used because of their favorable effect on the actual crop that is of agronomic interest) can be selected on the basis of ecophysiological information presented in this chapter. For example, if the intercrop plant solubilizes rock phosphate that becomes available to the crop, then it might reduce the need for phosphate fertilization. This chapter should inspire us to think of traits that might be exploited in future agriculture.

References and Further Reading

Adams, M.A. & Pate, J.S. (1992) Availability of organic and inorganic forms of phosphorus to lupins (*Lupinus* spp.). Plant Soil 145:107–113.

Aerts, R. (1989) Nitrogen use efficiency in relation to nitrogen availability and plant community composition. In: Causes and consequences of variation in growth rate and productivity of higher plants, H. Lambers, M.L. Cambridge, H. Konings, & T.L. Pons (eds). SPB Academic Publishing, The Hague, pp. 285–297.

Aerts, R. (1990) Nutrient use efficiency in evergreen and deciduous species from heathlands. Oecologia 84:391–397.

Aerts, R. (1995) The advantages of being evergreen. Trends Ecol. Evol. 10:402–407.

Aerts, R. (1996) Nutrient resorption from senescing leaves of perennials: Are there general patterns? J. Ecol. 84:597–608.

Andrews, M. (1986) The partitioning of nitrate assimilation between root and shoot of higher plants. Plant Cell Environ. 9:511–519.

Albuzzio, A. & Ferrari, G. (1989) Modulation of the molecular size of humic substances by organic acids of the root exudates. Plant Soil 113:237–241.

Arianoutsou, M., Rundel, P.W., & Berry, W.L. (1993) Serpentine endemics as biological indicators of soil elemental concentrations. In: Plants as biomonitors,

B. Markert (ed). VCH Weinheim, New York, pp. 179–189.

Armstrong, W. (1982) Waterlogged soils. In: Environment and plant ecology, J.R. Etherington (ed). John Wiley & Sons, New York, pp. 290–330.

Aslam, M., Travis, R.L., & Rains, D.W. (1996) Evidence for substrate induction of a nitrate efflux system in barley roots. Plant Physiol. 112:1167–1175.

Atkin, O.K. (1996) Reassessing the nitrogen relations of arctic plants: A mini-review. Plant Cell Environ. 19:695–704.

Ball, M.C. (1988) Ecophysiology of mangroves. Trees 2:129–142.

Barber, S.A. (1984) Soil nutrient bioavailability. John Wiley & Sons, New York.

Barkla, B.J., Zingarelli, L., Blumwald, E., & Smith, A.C. (1995) Tonoplast Na^+/H^+ antiport activity and its energization by the vacuolar H^+-ATPase in the halophytic plant *Mesembryanthemum crystallinum*. Plant Physiol. 109:549–556.

Bates, T.R. & Lynch, J.P. (1996) Stimulation of root hair elongation in *Arabidopsis thaliana* by low phosphorus availability. Plant Cell Environ. 19:529–538.

Berendse, F. & Aerts, R. (1987) Nitrogen-use efficiency: A biologically meaningful definition? Funct. Ecol. 1:293–296.

Berendse, F. & Elberse, W.T. (1989) Competition and nutrient losses from the plant. In: Causes and consequences of variation in growth rate and productivity of higher plants, H. Lambers, M.L. Cambridge, H. Konings, & T.L. Pons (eds). SPB Academic Publishing, The Hague, pp. 269–284.

Bienfait, H.F. (1985) Regulated redox processes at the plasmalemma of plant root cells and their function in iron uptake. J. Bioenerget. Biomembr. 17:73–83.

Bloom, A.J., Sukrapanna, S.S., & Warner, R.L. (1992) Root respiration associated with ammonium and nitrate absorption and assimilation by barley. Plant Physiol. 99:1294–1301.

Boerner, R.E.J. (1985) Foliar nutrient dynamics, growth, and nutrient use efficiency of *Hamamelis virginiana* in three forest microsites. Can. J. Bot. 63:1476–1481.

Bolan, N.S., Hedley, M.J., & White, R.E. (1991) Processes of soil acidification during nitrogen cycling with emphasis on legume based pastures. Plant Soil 134:53–63.

Borstlap, A.C. (1983) The use of model-fitting in the interpretation of "dual" uptake isotherms. Plant Cell Environ. 6:407–416.

Brouwer, R. (1962) Nutritive influences on the distribution of dry matter in the plant. Neth. J. Agric. Sci. 10:399–408.

Brune, A., Urbach, W., Dietz, K.-J. (1994) Compartmentation and transport of zinc in barley primary leaves as basic mechanisms involved in zinc tolerance. Plant Cell Environ. 17:153–162.

Cakmak, I., Sari, N., Marschner, H., Ekiz, H., Kalayci, M., Yilmaz, A., & Braun, H.J. (1996) Phytosiderophore release in bread wheat genotypes differing in zinc efficiency. Plant Soil 180:183–189.

Campbell, W.H. (1996) Nitrate reductase biochemistry comes of age. Plant Physiol. 111:355–361.

Chapin III, F.S. (1974) Morphological and physiological mechanisms of temperature compensation in phosphate absorption along a latitudinal gradient. Ecology 55:1180–1198.

Chapin III, F.S. (1980) The mineral nutrition of wild plants. Annu. Rev. Ecol. Syst. 11:233–260.

Chapin III, F.S. (1988) Ecological aspects of plant mineral nutrition. Adv. Min. Nutr. 3:161–191.

Chapin III, F.S. (1991) Effects of multiple environmental stresses on nutrient availability and use. In: Response of plants to multiple stresses, H.A. Mooney, W.E. Winner, & E.J. Pell (eds). Academic Press, San Diego, pp. 67–88.

Chapin III, F.S. & Bloom, A. (1976) Phosphate absorption: Adaptation of tundra graminoids to a low temperature, low phosphorus environment. Oikos 27:111–121.

Chapin III, F.S., Fetcher, N., Kielland, K., Everett, K.R., & Linkins, A.E. (1988) Productivity and nutrient cycling of Alaskan tundra: Enchancement by flowing soil water. Ecology 69:693–702.

Chapin III, F.S., Moilanen, L., & Kielland, K. (1993) Preferential use of organic nitrogen for growth by non-mycorrhizal arctic sedge. Nature 361:150–153.

Chapin III, F.S. & Slack, M. (1979) Effect of defoliation upon root growth, phosphate absorption, and respiration in nutrient-limited tundra graminoids. Oecologia 42:67–79.

Chapin III, F.S. & Kedrowski, R.A. (1983) Seasonal changes in nitrogen and phosphorus fractions and autumn retranslocation in evergreen and deciduous taiga trees. Ecology 64:376–391.

Chapin III, F.S. & Moilanen, L. (1991) Nutritional controls over nitrogen and phosphorus resorption from Alaskan birch leaves. Ecology 72:709–715.

Chapin III, F.S., Johnson, D.A., & McKendrick, J.D. (1980) Seasonal movement of nutrients in plants of differing growth form in an Alaskan tundra ecosystem: Implications for herbivory. J. Ecol. 68:189–209.

Chapin III, F.S., Moilanen, L., & Kielland, K. (1993) Preferential use of organic nitrogen for growth by non-mycorrhizal arctic sedge. Nature 361:150–153.

Clarholm, M. (1985) Interactions of bacteria, protozoa and plants leading to mineralization of soil nitrogen. Soil Biol. Biochem. 17:181–187.

Clarkson, D.T. (1981) Nutrient interception and transport by root systems. In: Physiological factors limiting plant productivity, C.B. Johnson (ed). Butterworths, London, pp. 307–314.

Clarkson, D.T. (1985) Factors affecting mineral nutrient acquisition by plants. Annu. Rev. Plant Physiol. 36:77–115.

Clarkson, D.T. (1996) Root structure and sites of ion uptake. In: Plant roots: The hidden half, Y. Waisel, A. Eshel, & U. Kafkaki, (eds). Marcel Dekker, Inc., New York, pp. 483–510.

Clarkson, D.T., Lüttge, U., & Kuiper, P.J.C. (1986) Mineral nutrition: Sources of nutrients for land plants from outside the pedosphere. Prog. Bot. 48:80–96.

Clement, C.R., Hopper, M.J., Jones, L.H.P., & Leafe, E.L. (1978) The uptake of nitrate by Lolium perenne from flowing nutrient solution. II. Effect of light, defoliation, and relationship to CO_2 flux. J. Exp. Bot. 29:1173–1183.

De Boer, A.H. (1985) Xylem/symplast ion exchange: Mechanism and function in salt-tolerance and growth. PhD Thesis, University of Groningen, Groningen, the Netherlands.

De Boer, A.H. & Wegner, L.H. (1997) Regulatory mechanisms of ion channels in xylem parenchyma cells. J. Exp. Bot. 48:441–449.

Deina, S., Gessa, C., Manunza, B., Marchetti, M., & Usai, M. (1992) Mechanism and stoichiometry of the redox reaction between iron(III) and caffeic acid. Plant Soil 145:287–294.

del Arco, J.M., Escudero, A., & Garrido, M.V. (1991) Effects of site characteristics on nitrogen retranslocation from senescing leaves. Ecology 72:701–708.

Delhaize, E. & Ryan, P.R. (1995) Aluminium toxicity and tolerance in plants. Plant Physiol. 107:315–321.

Delhaize, E., Ryan, P.R., & Randall (1993) Aluminium tolerance in wheat (Triticum aestivum L.). II. Aluminium-stimulated excretion of malic acid from root apices. Plant Physiol. 103:695–702.

Demars, B.G. & Boerner, R.E.J. (1997) Foliar nutrient dynamics and resorption in naturalized Lonicera maackii (Caprifoliaceae) populations in Hhio, USA. Am. J. Bot. 84:112–117.

De Silva, D.L.R., Hetherington, A.M., & Mansfield, T.A. (1996) Where does all the calcium go? Evidence of an important regulatory role for trichomes in two calcicoles. Plant Cell. Environ. 19:880–886.

De Vos, C.H.R., Vooijs, R., Schat, H., & Ernst, W.H.O. (1989) Copper-induced damage to the permeability barrier in roots of Silene cucubalus. J. Plant Physiol. 135:165–169.

Diaz, S.A., Grime, J.P., Harris, J., & McPherson, E. (1993) Evidence of a feedback mechanism limiting plant response to elevated carbon dioxide. Nature 364:616–617.

Dinkelaker, B., Römheld, V., & Marschner, H. (1989) Citric acid excretion and precipitation of calcium citrate in the rhizosphere of white lupin. Plant Cell Environ. 12:285–292.

Dinkelaker, B., Hengeler, C., & Marschner, H. (1995) Distribution and function of proteoid roots and other root clusters. Bot. Acta 108:183–200.

Drew, M.C. (1975) Comparison of the effects of a localized supply of phosphate, nitrate, ammonium and potassium on the growth of the seminal root system, and the shoot, in barley. New Phytol. 75:479–490.

Drew, M.C. & Saker, L.R. (1978) Nutrient supply and the growth of the seminal root system in barley. III. Compensatory increase in growth of lateral roots, and in rates of phosphate uptake, in response to a localized supply of phosphate. J. Exp. Bot. 29:435–451.

Drew, M.C., Saker, L.R., & Ashley, T.W. (1973) Nutrient supply and the growth of the seminal root system in barley. I. The effect of nitrate concentration on the growth of axes and laterals. J. Exp. Bot. 24:1189–1202.

Erskine, P.D., Stewart, G.R., Schmidt, S., Turnbull, M.H., Unkovich, M.H., & Pate, J.S. (1996) Water availability—

a physiological constraint on nitrate utilization in plants of Australian semi-arid mulga woodlands. Plant Cell Environ. 19:1149–1159.

Esau, K. (1977) Anatomy of seed plants. 2nd edition. John Wiley & Sons, New York.

Eviner, V.T. & Chapin III, F.S. (1997) Plant-microbial interactions. Nature 385:26–27.

Flowers, T.J., Troke, P.F., & Yeo, A.R. (1977) The mechanism of salt tolerance in halophytes. Annu. Rev. Plant Physiol. 28:89–121.

Gardner, W.K. & Boundy, K.A. (1983) The acquisition of phosphorus by *Lupinus albus* L.: 4 The effect of interplanting wheat and white lupin on the growth and mineral composition of the two species. Plant Soil 70:391–402.

Gardner, W.K., Parbery, D.G., & Barber, D.A. (1981) Proteoid root morphology and function in *Lupinus albus*. Plant Soil 60:143–147.

Gardner, W.K., Parbery, D.G., & Barber, D.A. (1982) The acquisition of phosphorus by *Lupinus albus* L. I. Some characteristics of the soil/root interface. Plant Soil 67:19–32.

Garten, C.T. Jr. (1976) Correlations between concentrations of elements in plants. Nature 261:686–688.

Gersani, M. & Sachs, T. (1992) Development correlations between roots in heterogeneous environments. Plant Cell Environ. 15:463–469.

Godbold, D.L., Horst, W.J., Marschner, H., & Collins, J.C. (1983) Effect of high zinc concentrations on root growth and zinc uptake in two ecotypes of *Deschampsia caespitosa* differing in zinc tolerance. In: Root ecology and its practical application, W. Böhm, L. Kutschera, & E. Lichtentegger (eds). Bundesanstalt für alpenländische Landwirtscaft, Gumpenstein, pp. 165–172.

Guerinot, M.L. & Yi, Y. (1994) Iron: Nutritious, noxious, and not readily available. Plant Physiol. 104:815–820.

Gutierrez, F.R. & Whitford, W.G. (1987) Chihuahuan desert annuals: Importance of water and nitrogen. Ecology 68:2032–2045.

Hairiah, K., Stulen, I., & Kuiper, P.J.C. (1990) Aluminium tolerance of the velvet beans *Mucuna pruriens* var. *utilis* and *M. deeringiana*. I. Effects of aluminium on growth and mineral composition. In: Plant nutrition-Physiology and applications, M.L. van Beusichem (ed). Kluwer Academic Publishers, Dordrecht, pp. 365–374.

Harper, S.M., Edwards, D.G., Kerven, G.L., & Asher, C.J. (1995) Effects of organic acid fractions extracted from *Eucalyptus camaldulensis* leaves on root elongation of maize (*Zea mays*) in the presence and absence of aluminium. Plant Soil 171:189–192.

Harrison, A.F. & Helliwell, D.R. (1979) A bioassay for comparing phosphorus availability in soils. J. Appl. Ecol. 16:497–505.

Häussling, M. & Marschner, H. (1989) Organic and inorganic soil phosphates and acid phosphatase activity in the rhizosphere of 80-year-old Norway spruce (*Picea abies* (L.) Karst.) trees. Biol. Fertil. Soils 8:128–133.

Hedin, L.O., Granat, L., Likens, G.E., Buishand, A., Galloway, J.N., Butler, T.J., & Rodhe, H. (1994) Steep declines in atmospheric base cations in regions of Europe and North America. Nature 367:351–354.

Higginbotham, N., Etherton, B., & Foster, R.J. (1967) Mineral ion contents and cell transmembrane electropotentials of pea and oat seedling tissue. Plant Physiol. 43:37–46.

Hobbie, S.E. (1992) Effects of plant species on nutrient cycling. Trends Ecol. Evolu. 7:336–339.

Horst, W.J. & Waschkies, C. (1987) Phosphatversorgerung von Sommerweizen (*Triticum aestivum* L.) in Mischkultur mit Weiszer Lupine (*Lupinus albus* L.). Z. PflanzenernŠhr. Bodenk. 150:1–8.

Hoffland, E. (1991) Mobilization of rock phosphate by rape (*Brassica napus*). PhD Thesis, Wageningen Agricultural University, Wageningen, the Netherlands.

Hoffland, E., Findenegg, G.R., & Nelemans, J.A. (1989) Solubilization of rock phosphate by rape. II. Local root exudation of organic acids as a response to P-starvation. Plant Soil 113:161–165.

Hoffland, E., Bloemhof, H.S., Leffelaar, P.A., Findenegg, G.R., & Nelemans, J.A. (1990a) Simulation of nutrient uptake by a growing root system considering increasing root density and inter-root competition. Plant Soil 124:149–155.

Hoffland, E., Findenegg, G.R., Leffelaar, P.A., & Nelemans, J.A. (1990b) Use of a simulation model to quantify the amount of phosphate released from rock phosphate by rape. Transactions of the 14th International Congress of Soil Science (Kyoto) II, pp. 170–175.

Huang, C.X. & Van Steveninck, R.F.M. (1989) Maintenance of low Cl-concentrations in mesophyll cells of leaf blades of barely seedlings exposed to salt stress. Plant Physiol. 90:1440–1443.

Huang, J.W., Pellet, D.M., Papernik, L.A., & Kochian, L.V. (1996) Aluminium interactions with voltage-dependent calcium transport in plasma membrane vesicles isolated from roots of aluminium-sensitive and -resistant wheat cultivars. Plant Physiol. 110:561–569.

Huang, N.-C., Chiang, C.-S., Crawford, N.M., & Tsay, Y.F. (1996) *Chl1* encodes a component of the low-affinity nitrate uptake system in *Arabidopsis* and shows cell type-specific expression in roots. Plant Cell 8:2183–2191.

Hübel, F. & Beck, F. (1993) In-situ determination of the P-relations around the primary root of maize with respect to inorganic and phytate-P. Plant Soil 157:1–9.

Huber, S.C., Bachman, M., & Huber, J.L. (1996) Post-translational regulation of nitrate reductase activity: A role for Ca^{2+} and 14-3-3 proteins. Trend Plant Sci. 1:432–438.

Hungate, B.A. (1998) Ecosystem responses to rising atmospheric CO_2: Feedbacks through the nitrogen cycle. In: Interactions of elevated CO_2 and environmental stress, J. Seeman & Y. Luo (eds). Academic Press, San Diego, in press.

Ingestad, T. (1979) Nitrogen stress in birch seedlings II. N, P, Ca and Mg nutrition. Physiol. Plant. 52:454–466.

Ingestad, T. & Ågren, G.I. (1988) Nutrient uptake and allocation at steady-state nutrition. Physiol. Plant. 72:450–459.

Israel, D.W. (1987) Investigation of the role of phosphorus in symbiotic dinitrogen fixation. Plant Physiol. 84:835–840.

Jackson, P.J., Delhaize, E., & Kuske, C.R. (1992) Biosynthesis and metabolic roles of cadystins (γ-EC)$_n$G and their precursors in *Datura innoxia*. Plant Soil 146:281–289.

Jenny, H. (1980) The soil resources. Origin and behavior. Springer-Verlag, New York.

Johnson, J.F., Allan, D.L., & Vance, C.P. (1994) Phosphorus stress-induced proteoid roots show altered metabolism in *Lupinus albus*. Plant Physiol. 104:657–665.

Johnson, J.F., Allan, D.L., Vance, C.P., & Weiblen, G. (1996a) Root carbon dioxide fixation by phosphorus-deficient *Lupinus albus*. Contribution to organic acid exudation by proteoid roots. Plant Physiol. 112:19–30.

Johnson, J.F., Vance, C.P., & Allan, D.L. (1996b) Phosphorus deficiency in *Lupinus albus*. Altered lateral root development and enhanced expression of phospho*enol*pyruvate carboxylase. Plant Physiol. 112:31–41.

Johnson, M.N., Reynolds, R.C., & Likens, G.E. (1972) Atmospheric sulfur: Its effect on the chemical weathering of New England. Science 177:514–515.

Jones, D.L. & Kochian, L.V. (1996) Aluminium inhibition of the inositol, 1,4,5-triphosphate signal transduction pathway in wheat roots: A role in aluminium toxicity. Plant Cell 7:1913–1922.

Jones, D.L., Darrah, P.R., & Kochian, L.V. (1996a) Critical evaluation of organic acid mediated iron dissolution in the rhizosphere and its potential role in iron uptake. Plant Soil 180:57–66.

Jones, D.L., Prabowo, A.M., & Kochian, L.V. (1996b) Kinetics of malate transport and decomposition in acid soils and isolated bacterial populations: The effects of microorganisms on root exudation of malate under Al stress. Plant Soil 182:239–247.

Jungk, A.O. (1996) Dynamics of nutrient movement at the soil-root interface. In: Plant roots: The hidden half, Y. Waisel, A. Eshel, & U. Kafkaki, (eds). Marcel Dekker, Inc., New York, pp. 529–556.

Kaiser, W.M. & Huber, S.C. (1994) Posttranslational regulation of nitrate reductase in higher plants. Plant Physiol. 106:817–821.

Keerthisinghe, G., Hocking, P., Ryan, P.R., & Delhaize, E. (1998). Proteoid roots of lupin (*Lupinus albus* L.): Effect of phosphorus supply on formation and spatial variation in citrate efflux and enzyme activity. Plant Cell Environ., in press.

Keltjens, W.G. & Tan, K. (1993) Interactions between aluminium, magnesium and calcium with different monocotyledonous and dicotyledonous plant species. Plant Soil 155/156:485–488.

Kielland, K. (1994) Amino acid absorption by Arctic plants: Implications for plant nutrition and nitrogen cycling. Ecology 75:2373–2383.

Killingbeck, K.T. (1996) Nutrients in senesced leaves: Keys to the search for potential resorption and resorption proficiency. Ecology 77:1716–1727.

King, B.J., Siddiqui, N.Y., Ruth, T.J., Warner, R.L., & Glass, A.D.M. (1993) Feedback regulation of nitrate influx in barley roots by nitrate, nitrite, and ammonium. Plant Physiol. 102:1279–1286.

Kinraide, T.B. (1993) Aluminium enhancement of plant growth in acid rooting media. A case of reciprocal alleviation of toxicity by two toxic cations. Physiol. Plant. 88:619–625.

Kochian, L. (1995) Cellular mechanisms of aluminium toxicity and resistance in plants. Annu. Rev. Plant Physiol. Plant Mol. Biol. 46:237–260.

Koerselman, W. & Meuleman, A.F.M. (1996) The vegetation N:P ratio: A new tool to detect the nature of nutrient limitation. J. Appl. Ecol. 33:1441–1450.

Krämer, U., Cotter-Howels, J.D., Charnock, J.M., Baker, A.J.M., & Smith, J.A. (1996) Free histidine as a metal chelator in plants that accumulate nickel. Nature 379:635–638.

Kroehler, C.J. & Linkins, A.E. (1991) The absorption of inorganic phosphate from ^{32}P-labelled inositol hexaphosphate by *Eriophorum vaginatum*. Oecologia 85:424–428.

Kronzucker, H.J., Siddiqui, M.Y., & Glass, A.D.M. (1997) Conifer root discrimination against soil nitrate and the ecology of forest succession. Nature 385:59–61.

Krupa, Z., Oquist, G., & Huner, N.P.A. (1993) The effect of cadmium on photosynthesis of *Phaseolus vulgaris*—a fluorescence analysis. Physiol. Plant. 88:626–630.

Kuiper, P.J.C. (1968) Ion transport characteristics of grape root lipids in relation to chloride transport. Physiol. Plant. 65:245–250.

Lacan, D. & Durand, N. (1994) Na$^+$ and K$^+$ transport in excised soybean roots. Physiol. Plant. 93:132–138.

Lamont, B. (1982) Mechanisms for enhancing nutrient uptake in plants, with particular reference to mediterranean South Africa and Western Australia. Bot. Rev. 48:597–689.

Lamont, B. (1993) Why are hairy root clusters so abundant in the most nutrient-impoverished soils of Australia. Plant Soil 155/156:269–272.

Lasat, M.M., Baker, A.J.M., & Kochian, L.V. (1996) Physiological characterization of root Zn^{2+} absorption and translocation to shoots in Zn hyperaccumulator and nonaccumulator species of *Thlaspi*. Plant Physiol. 112:1715–1722.

LeNoble, M.E., Blevins, D.G., Sharp, R.E., & Cumbie, B.G. (1996a) Prevention of aluminium toxicity with supplemental boron. I. Maintenance of root elongation and cellular structure. Plant Cell Environ. 19:1132–1142.

LeNoble, M.E., Blevins, D.G., & Miles, R.J. (1996b) Prevention of aluminium toxicity with supplemental boron. II. Stimulation of root growth in an acidic, high-aluminium subsoil. Plant Cell Environ. 19:1143–1148.

Li, X.Z. & Oaks, A. (1993) Induction and turnover of nitrate reductase in *Zea mays*. Influence of NO$_3^-$. Plant Physiol. 102:1251–1257.

Lipton, D.S., Blanchar, R.W., & Blevins, D.G. (1987) Citrate, malate, and succinate concentration in exudates from P-sufficient and P-stressed *Medicago sativa* L. seedlings. Plant Physiol. 85:315–317.

Lolkema, P.C., Doornhof, M., & Ernst, W.H.O. (1986) Interaction between a copper-tolerant and a copper-sensitive population of *Silene cucubalus*. Physiol. Plant. 67:654–658.

Loneragan, J.F. (1968) Nutrient requirements of plants. Nature 220:1307–1308.

Loveless, A.R. (1961) A nutritional interpretation of sclerophylly based on differences in chemical composition of sclerophyllous and mesophytic leaves. Ann. Bot. 25:168–184.

Ma, J.F. & Nomoto, K. (1996) Effective regulation of iron acquisition in graminaceous plants. The role of mucigeneic acids as phytosiderophores. Physiol. Plant. 97:609–617.

Macfie, S.M. & Taylor, G.J. (1992) The effect of excess manganese on photosynthetic rate and concentration of chlorophyll in *Triticum aestivum* grown in solution culture. Physiol. Plant. 85:467–475.

Macklon, A.E.S., Mackie-Dawson, L.A., Sim, A., Shand, C.A., & Lilly, A. (1994) Soil P resources, plant growth and rooting characteristics in nutrient poor upland grasslands. Plant Soil 163:257–266.

Marschner, H. (1983) General introduction to the mineral nutrition of plants. In: Encyclopedia of plant physiology, N.S., Vol 15A, A. Läuchli & R.L. Bieleski (eds). Springer-Verlag, Berlin, pp. 5–60.

Marschner, H. (1991a) Root-induced changes in the availability of micronutrients in the rhizosphere. In: Plant roots: The hidden half, Y. Waisel, A. Eshel, & U. Kafkaki, (eds). Marcel Decker, Inc., New York, pp. 503–528.

Marschner, H. (1991b) Mechanisms of adaptation of plants to acid soils. Plant Soil 134:1–20.

Marschner, H. (1995) Mineral nutrition of higher plants. 2nd edition. Academic Press, London.

Marschner, H. & Römheld, V. (1996) Root-induced changes in the availability of micronutrients in the rhizosphere. In: Plant roots: The hidden half, Y. Waisel, A. Eshel, & U. Kafkaki (eds). Marcel Decker, Inc., New York, pp. 557–580.

Martins-Louçao, M. & Cruz, C. (1998) The role of nitrogen source in carbon balance. In: Modes of nitrogen nutrition in higher plants, H.S. Srivastava (ed). Associated Publishing Company, in press.

McNaughton, S.J. & Chapin III, F.S. (1985) Effects of phosphorus nutrition and defoliation on C_4 graminoids from the Serengeti Plains. Ecology 66:1617–1629.

Meerts, P. (1997) Foliar macronutrient concentrations of forest understorey species in relation to Ellenberg's indices and potential relative growth rate. Plant Soil 189:257–265.

Mistrik, I. & Ullrich, C.I. (1996) Mechanism of anion uptake in plant roots: Quantitative evaluation of H^+/ NO_3^- and H^+/$H_2PO_4^-$ stoichiometries. Plant Physiol. Biochem. 34:621–627.

Murphy, A. & Taiz, L. (1995) Comparison of metallothionein gene expression and nonprotein thiols in ten *Arabidopsis* ecotypes. Plant Physiol. 109:945–954.

Nair, V.D. & Prenzel, J. (1978) Calculations of equilibrium concentration of mono- and polynuclear hydroxyaluminium species at different pH and total aluminium concentrations. Z. PflanzenernŠhr. Bodenk. 141:741–751.

Nambiar, I.K.S. (1987) Do nutrients retranslocate from fine roots? Can. J. For. Res. 17:913–918.

Nambiar, I.K.S. & Fife, D.N. (1987) Growth and nutrient retranslocation in needles of radiata pine in relation to nitrogen supply. Ann. Bot. 60:147–156.

Oland, K. (1963) Changes in the content of dry matter and major nutrient elements of apple foliage during senescence and abscission. Physiol. Plant. 16:682–694.

Oscarson, P., Ingemarsson, B., af Ugglas, M., & Larsson, C.-M. (1987) Short-term studies of NO_3^- uptake in *Pisum* using $^{13}NO_3^-$. Planta 170:550–555.

Paul, E.A. & Clark,, F.E. (1989) Soil microbiology and biochemistry. Academic Press, San Diego.

Pitman, M.G. & Lüttge, U (1983) The ionic environment and plant ionic relations. In: Encyclopedia of plant physiology, N.S., Vol 12C, O.L. Lange, P.S. Nobel, C.B. Osmond, & H. Ziegler (eds). Springer-Verlag, Berlin, pp. 5–34.

Pons, T.L., Van der Werf, A., & Lambers, H. (1994) Photosynthetic nitrogen use efficiency of inherently slow- and fast-growing species: Possible explanations for observed differences. In: A whole-plant perspective of carbon-nitrogen interactions, J. Roy & E. Garnier (eds). SPB Academic Publishing, pp. 61–77.

Poorter, H., Remkes, C., & Lambers, H. (1990) Carbon and nitrogen economy of 24 wild species differing in relative growth rate. Plant Physiol. 94:621–627.

Popp, M. (1995) Salt resistance in herbaceous halophytes and mangroves. Prog. Bot. 56:416–429.

Powell, C.L. (1974) Effect of P-fertilizer on root morphology and P-uptake of *Carex coriacea*. Plant Soil 41:661–667.

Pugnaire, F.I. & Chapin III, F.S. (1993) Controls over nutrient resorption from leaves of evergreen Mediterranean species. Ecology 74:124–129.

Raaimakers, T.H.M.J. (1995) Growth of tropical rainforest trees as dependent on P-availability. Tree saplings differing in regeneration strategy and their adaptations to a low phosphorus environment in Guyana. PhD Thesis, Utrecht University, Utrecht, the Netherlands.

Rauser, W.E. (1995) Phytochelatins and related peptides. Structure, biosynthesis, and function. Plant Physiol. 109:1141–1149.

Redfield, A.C. (1958) The biological control of chemical factors in the environment. Am. Scient. 46:205–221.

Reich, P.B., Walters, M.B., & Ellsworth, D.S. (1992) Leaf life-span in relation to leaf, plant and stand characteristics among diverse ecosystems. Ecol. Monogr. 62:365–392.

Reich, P.B., Ellsworth, D.S., & Uhl, C. (1995) Leaf carbon and nutrient assimilation and conservation in species of differing succesional status in an oligotrophic Amazonian forest. Funct. Ecol. 9:65–76.

Rengel, Z. (1992a) The role of calcium in salt toxicity. Plant Cell Environ. 15:625–632.

Rengel, Z. (1992b) Disturbance of cell Ca^{2+} homeostasis as a primary trigger of Al toxicity syndrome. Plant Cell Environ. 15:931–938.

Reuss, J.O. & Johnson, D.W. (1986) Acid Deposition and the Acidification of Soils and Waters. Springer-Verlag, New York.

Reynolds, H.L. & D'Antonio, C. (1996) The ecological significance of plasticity in root weight ratio in response to nitrogen. Opinion. Plant Soil 185:75–97.

Richardson, A.E. (1994) Soil microorganisms and phosphorus availability. In: Soil Biota. Management in sustainable farming systems, C.E. Pankhurst, B.M. Doube, V.V.S.R. Gupta, & P.R. Grace (eds). CSIRO, East Melbourne, pp. 50–62.

Robinson, D. (1994) The responses of plants to non-uniform supplies of nutrients. New Phyol. 127:635–674.

Robinson, D. (1996) Variation, co-ordination and compensation in root systems in relation to soil variability. Plant Soil 187:57–66.

Römheld, V. (1987) Different strategies for iron acquisition in higher plants. Physiol. Plant. 70:231–234.

Ryan, P.R. & Kochian, L.V. (1993) Aluminium differentially inhibits calcium uptake into the root apex of near-isogenic lines of wheat. A possible mechanism of toxicity. Plant Physiol. 102:975–982.

Ryan, P.R., Kinraide, T.B., & Kochian, L.V. (1994) Al^{3+}-Ca^{2+} interactions in aluminium rhizotoxicity. I. Inhibition of root growth is not caused by reduction of calcium uptake. Planta 192:98–103.

Ryan, P.R., Delhaize, E., & Randall, P.J. (1995) Malate efflux from root apices and tolerance to aluminium are highly correlated in wheat. Aust. J. Plant Physiol. 22:531–536.

Salt, D.E. & Rauser, W.E. (1995) MGATP-dependent transport of phytochelatins across the tonoplast of oat roots. Plant Physiol. 107:1293–1301.

Salt, D.E., Blaylock, M., Kumar, P.B.A.N., Dushenkov, V., Ensley, B.D., Chet, I., & Raskin, I. (1995a) Phytoremediation: A novel strategy for the removal of toxic metals from the environment using plants. Biotechnology 13:468–474.

Salt, D.E., Prince, R.C., Pickering, I.J., & Raskin, I. (1995b) Mechanisms of cadmium mobility and accumulation in Indian mustard. Plant Physiol. 109:1427–1433.

Scholz, G., Becker, R., Pich, A., & Stephan, U.W. (1992) Nicotinamine—a common constituent of strategies I and II of iron acquisition by plants: A review. J. Plant Nutr. 15:1647–1665.

Shaver, G.R. & Chapin III, F.S. (1991) Production: Biomass relationships and element cycling in contrasting arctic vegetation types. Ecol. Monogr. 61:1–31.

Shone, M.G.T., Clarkson, D.T., & Sanderson, J. (1969) The absorption and translocation of sodium by maize seedlings. Planta 86:301–314.

Shriner, D.S. & Johnston Jr., J.W. (1985) Acid rain interactions with leaf surfaces: A review. In: Acid deposition: Environmental, economic, and policy Issues, D.D. Adams & W.P Page (eds). Plenum Publishing Corporation, New York, pp. 241–253.

Siddiqi, M.Y., Glass, A.D.M., & Ruth, T.J. & Rufty, T.W. (1990) Studies of the nitrate uptake system in barley. I. Kinetics of $^{13}NO_3^-$ influx. Plant Physiol. 93:1426–1432.

Siddiqi, M.Y., Glass, A.D.M., & Ruth, T.J. (1991) Studies of the uptake of nitrate in barley. III. Compartmentation of NO_3^-. J. Exp. Bot. 42:1455–1463.

Smart, C.J., Garvin, D.F., Prince, J.P., Lucas, W.J., & Kochian, L.V. (1996) The molecular basis of potassium nutrition. Plant Soil 187:81–89.

Smirnoff, N. & Stewart, G.R. (1985) Nitrate assimilation and translocation by higher plants: Comparative physiology and ecological consequences. Physiol. Plant. 64:133–140.

Smirnoff, N., Todd, P., & Stewart, G.R. (1984) The occurrence of nitrate reduction in the leaves of woody plants. Ann. Bot. 54:363–374.

Smith, F.W., Ealing, P.M., Hawkesford, M.J., & Clarkson, D.T. (1995) Plant members of a family of sulfate transporters reveal functional subtypes. Proc. Natl. Acad. Sci. USA 92:9373–9377.

Stark, J.M. & Hart, S.C. (1997) High rates of nitrification and nitrate turnover in undisturbed coniferous ecosystems. Nature 385:61–64.

Staal, M., Maathuis, F.J.M., Elzenga, T.J.M., Overbeek, J.H.M., & Prins, H.B.A. (1991) Na^+/H^+ antiport activity in tonoplast vesicles from roots of the salt-tolerant Plantago maritima and the salt-sensitive Plantago media. Physiol. Plant. 82:179–184.

Stark, J.M. & Hart, S.C. (1997) High rates of nitrification and nitrate turnover in undisturbed coniferous ecosystems. Nature 385:61–64.

Ström, L., Olsson, T., & Tyler, G. (1994) Differences between calcifuge and acidifuge plants in root exudation of low-molecular organic acids. Plant Soil 167:239–245.

Tarafdar, J.C. & Jungk, A. (1987) Phosphatase activity in the rhizosphere and its relation to the depletion of soil organic phosphorus. Biol. Fert. Soils 3:199–204.

Ter Steege, M. (1996) Regulation of nitrate uptake in a whole plant perspective. PhD thesis, University of Groningen, the Netherlands.

Thomas, W.A. & Grigal, D.F. (1976) Phosphorus conservation by evergreenness of mountain laurel. Oikos 27:19–26.

Tilton, D.L. (1977) Seasonal growth and foliar nutrients of Larix laricina in three wetland ecosystems. Can. J. Bot. 55:1291–1298.

Touraine, B., Clarkson, D.T., & Muller, B. (1994) Regulation of nitrate uptake at the whole plant level. In: A whole-plant perspective on carbon-nitrogen interactions, J. Roy & E. Garnier (eds). SPB Academic Publishing, pp. 11–30.

Trueman, L.J., Richardson, A., & Forde, B.G. (1996a) Molecular cloning of higher plant homologues of the high-affinity nitrate transporters of Chlamydomonas reinhardtii and Aspergillus nidulans. Gene 175:223–231.

Trueman, L.J., Onyeocha, I., & Forde, B.G. (1996b) Recent advances in the molecular biology of a family of eukaryotic high affinity nitrate transporters. Plant Physiol. Biochem. 34:621–627.

Tukey Jr., H.B. (1970) The leaching of substances from plants. Annu. Rev. Plant Physiol. 21:305–324.

Tyler, G. (1992) Inability to solubilized phosphate in limestone soil—key factors controlling calcifuge habit of plants. Plant Soil 145:65–70.

Tyler, G. (1994) A new approach to understanding the calcifuge habit of plants. Ann. Bot. 73:327–330.

Tyler, G. (1996) Soil chemical limitations to growth and development of *Veronica officinalis* L. and *Carex pilulifera* L. Plant Soil 184:281–289.

Ullrich, W.R. (1992) Transport of nitrate and ammonium through plant membranes. In: Nitrogen metabolism of plants, K. Mengel & D.J. Pilbeam (eds). Clarendon Press, Oxford, U.K., pp. 121–137.

Van Assche, F. & Clijsters, H. (1984) Substitution in vivo of Ma^{2+} by Zn^{2+} in Rubisco-CO_2-Me^{2+} complexes as a result of toxic zinc nutrition to *Phaseolus vulgaris* L. Arch. Internat. Physiol. Biochim. 92:V18–V19.

Van der Werf, A.K., Visser, A.J., Schieving, F., & Lambers, H. (1993) Evidence for optimal partitioning of biomass and nitrogen at a range of nitrogen availabilities for a fast- and slow-growing species. Funct. Ecol. 7:63–74.

Van Vuuren, M.M.I., Robinson, D., & Griffiths, B.S. (1996) Nutrient inflow and root proliferation during the exploitation of a temporally and spatially discrete source of nitrogen in the soil. Plant Soil 178:185–192.

Verhoeven, J.T.A., Koerselman, W., & Meuleman, A.F.M. (1996) Nitrogen- or phosphorus-limited growth in herbaceous, wet vegetation: relations with atmospheric inputs and management regimes. Trend Ecol. Evol. 11:495–497.

Verkleij, J.A.C. & Schat, H. (1990) Mechanisms of metal tolerance in higher plants. In: Heavy metal tolerance in plants, A.J. Shaw (ed). CRC Press Inc., Boca Raton, pp. 179–193.

Verry, E.S. & Timmons, D.R. (1976) Elements in leaves of a trembling aspen clone by crown position and season. Can. J. For. Res. 6:436–440.

Vitousek, P. (1982) Nutrient cycling and nutrient use efficiency. Am. Nat. 119:553–572.

Vitousek, P.M. & Howarth R.W. (1991) Nitrogen limitation on land and in the sea: How can it occur? Biogeochemistry 13:87–115.

Vögeli-Lange, R. & Wagner, G.J. (1990) Subcellular localization of cadmium and cadmium-binding peptides in tobacco leaves. Plant Physiol. 92:1086–1093.

Walker, C.D., Graham, R.D., Madison, J.T., Cary, E.E., & Welch, R.M. (1995) Effects of Ni deficiency on some nitrogen metabolites in cowpea (*Vigna unguiculata* L. Walp). Plant Physiol. 79:474–479.

Woodward, R.A. Harper, K.T., & Tiedemann, A.R. (1984) An ecological consideration of the significance of cation-exchange capacity of roots of some Utah range plants. Plant Soil 79:169–180.

Zak, D.R., Pregitzer, K.S., Curtis, P.S., Teeri, J.A., Fogel, R., & Randlett, D.A. (1993) Elevated CO_2 and feedback between carbon and nitrogen cycles. Plant Soil 151:105–117.

7
Growth and Allocation

1. Introduction: What Is Growth?

Growth of a plant is a consequence of the inter-action of all the processes discussed in previous chapters: photosynthesis, long-distance transport, respiration, water relations, and mineral nutrition. By the same token, these physiological processes may be controlled themselves by the growth rate of the plants, as discussed in the preceding chapters; however, what exactly do we mean by plant growth? **Growth** is the increment in dry mass, volume, length, or area, and it mostly involves the **division**, **expansion**, and **differentiation** of cells. Increment in dry mass, however, may not occur at the same time as increment in one of the other parameters. For example, leaves often expand and roots elongate at night, when the entire plant is decreasing in dry mass because of carbon use in respiration. On the other hand, a tuber may gain dry mass without concomitant change in volume. When we discuss "growth" in the context of this chapter, therefore, we must define the term for each context. It is also important to stress that, whereas cell divisions occur when an organ is growing, this process cannot lead to growth by itself. In addition, cell elongation and the deposition of mass in the cytoplasm and cell walls determine the increment in volume or mass. To appreciate ecophysiological aspects of plant growth, we must understand the cellular basis of it. Although this is a fascinating and rapidly moving field, many questions remain unanswered, as will be revealed in this chapter.

This chapter will also deal with the question of why some plants grow more rapidly than others. A plant's growth rate is the result of its genetic background as well as the environment in which it grows. Plants are the product of natural selection, which has led to genotypes with different **suites of traits** that allow them to perform in specific habitats. Such a suite of traits is often referred to as a "strategy." The term is used here, as well as elsewhere in this text, to indicate the functionality of traits in an ecological context and to help to explain a plant's performance in an ecological and evolutionary context (Box 10 in the chapter on interactions among plants). In this chapter we will discuss how genetic and environmental factors affect the growth of plants. A major factor appears to be the pattern of biomass allocation: Are the plant's acquired resources invested in leaves, roots, or stems? Are they allocated toward machinery involved in resource acquisition or toward storage in order to be available for future growth?

2. Growth of Whole Plants and Individual Organs

Plant growth can be analyzed in terms of an increase in total plant dry mass and its distribution ("allocation") among organs involved in acquisition of above-ground or below-ground resources. In such an approach, the pattern of resource allocation plays a pivotal role in determining a plant's growth rate. Plant growth can also be studied at the level of individual organs or cells. Using this approach we can ask why the leaves of one plant grow faster or bigger than those of another. The two approaches are complementary, although most authors tend to follow only one of them. For a full appreciation of a plant's functioning, they should be integrated to highlight traits that determine a plant's growth potential.

2.1 Growth of Whole Plants

Growth analysis provides considerable insight into the functioning of a plant as dependent on genotype or environment. Different growth analyses can be carried out, depending on what is considered a key factor for growth (Lambers et al. 1989). Leaf area and net assimilation rate are most commonly treated as the driving variables. As discussed in Section 4.2 of the chapter on mineral nutrition, however, we can also consider the plant's nutrient concentration and nutrient productivity as driving variables.

2.1.1 A High Leaf Area Ratio Enables Plants to Grow Fast

We will first concentrate on the plant's leaf area as the driving variable for the relative growth rate (RGR, the rate of increase in plant mass per unit of plant mass already present) (Evans 1972). According to this approach, RGR is factored into two components: the **leaf area ratio** (**LAR**), which is the amount of leaf area per unit total plant mass, and the **net assimilation rate** (**NAR**), which is the rate of increase in plant mass per unit leaf area (see Table 1 for a list of abbreviations and the units in which they are expressed):

$$RGR = LAR \cdot NAR \qquad (1)$$

LAR and NAR, in turn, can each be subdivided into additional components. The LAR is the product of the **specific leaf area** (**SLA**), which is the amount of leaf area per unit leaf mass, and the **leaf mass ratio** (**LMR**), which is the fraction of the total plant biomass allocated to leaves:

$$LAR = SLA \cdot LMR \qquad (2)$$

The NAR, which is the rate of dry mass gain per unit leaf area, is largely the net result of the rate of carbon gain in **photosynthesis** per unit leaf area (A) and that of carbon use in **respiration** of leaves, stems, and roots (LR, SR, and RR), which in this

case is also expressed per unit leaf area. If these physiological processes are expressed in moles of carbon, then the net balance of photosynthesis and respiration has to be divided by the carbon concentration of the newly formed material, [C], to obtain the increase in dry mass. The balance can be completed by subtracting losses due to volatilization and exudation per unit time, again expressed on a leaf area basis. For simplicity's sake, volatilization and exudation will be ignored in this Section, although these processes can be ecologically important to the plant's carbon budget under some circumstances. We will discuss volatile losses in Section 5.2, whereas the process of exudation has been treated in Sections 2.2.5, 2.2.6, 3.1.3, and 3.2 of the chapter on mineral nutrition. The simplified equation for the net assimilation rate is:

$$NAR$$
$$= \frac{\left\{ A_a - LR_a - (SR \cdot SMR/LAR) - (RR \cdot RMR/LAR) \right\}}{[C]}$$
$$\qquad (3)$$

The subscript a indicates that the rates are expressed on a leaf area basis. This is a common way to express rates of CO_2 assimilation (chapter on photosynthesis). Of course, stem and root respiration are not directly related to leaf area, but rather to the biomass of the different organs. This has been resolved by multiplying the rate of stem respiration (SR) and root respiration (RR) by SMR/LAR and RMR/LAR, respectively; SMR and RMR are the stem mass ratio and the root mass ratio, which is the fraction of plant biomass allocated to stem and roots, respectively (Table 1). Although the net assimilation rate is relatively easy to estimate from harvest data, it is not really an appropriate parameter to gain insight into the relation between physiology and growth. Rather, we should concentrate on the underlying processes: **photosynthesis, respiration**, and **allocation**.

For the relative growth rate, we can now derive the following equation:

$$RGR$$
$$= \frac{A_a \cdot SLA \cdot LMR - LR_m \cdot LMR - SR \cdot SMR - RR \cdot RMR}{[C]}$$
$$\qquad (4)$$

This equation has been widely used to identify traits that are associated with genetic variation in a plant's RGR at an optimum nutrient supply as well as variation caused by environmental factors, such as light, temperature, or nutrient supply.

TABLE 1. Abbreviations related to plant growth analysis and the units in which they are expressed.

Abbreviation	Meaning	Preferred units
A	Rate of CO_2 assimilation	$\mu mol\, CO_2\, m^{-2}\, s^{-1}$
[C]	Carbon concentration	$mmol\, C\, g^{-1}$
LAR	Leaf area ratio	$m^2\, kg^{-1}$
LMA	Leaf mass per unit leaf area	$kg\, m^{-2}$
LMR	Leaf mass ratio	$g\, g^{-1}$
LR_a (LR_m)	Rate of leaf respiration	$\mu mol\, CO_2\, m^{-2}$ (leaf area) s^{-1} $[nmol\, CO_2\, g^{-1}$ (leaf mass) $s^{-1}]$
NAR	Net assimilation rate	$g\, m^{-2}\, day^{-1}$
NP	Nutrient productivity	g (plant mass) mol^{-1} (plant nutrient) day^{-1}
PNC	Plant nutrient concentration	mol (nutrient) g^{-1} (plant mass)
RGR	Relative growth rate	$mg\, g^{-1}\, day^{-1}$
RMR	Root mass ratio	$g\, g^{-1}$
RR	Rate of root respiration	$nmol\, CO_2\, g^{-1}$ (root mass) s^{-1}
SLA	Specific leaf area	$m^2\, kg^{-1}$
SR	Rate of stem respiration	$nmol\, CO_2\, g^{-1}$ (stem mass) s^{-1}
SRL	Specific root length	$m\, g^{-1}$
SMR	Stem mass ratio	$g\, g^{-1}$

2.1.2 Plants with High Nutrient Concentrations Can Grow Faster

In an alternative approach, the plant's nutrient concentration (mostly **plant nitrogen concentration**, **PNC**) is assumed to be a driving variable, as discussed in Section 4 of the chapter on mineral nutrition. PNC, in combination with the nutrient productivity (mostly **nitrogen productivity**, **NP**) determines plant growth. Thus, we arrive at:

$$RGR = NP \cdot PNC \qquad (5)$$

As pointed out in the chapter on mineral nutrition, plants differ widely in their nitrogen productivity, when grown with free access to nutrients. A higher nitrogen productivity is associated with a relatively large investment of nitrogen in photosynthesizing tissue, an efficient use of the nitrogen invested in the leaves for the process of photosynthesis, and a relatively low carbon use in respiration (Garnier et al. 1995, Poorter et al. 1990).

2.2 Growth of Cells

Insights into the cellular basis of growth analysis come from studying the actual processes of growth (cell division, cell expansion, mass deposition) in greater detail, as described in the following section.

2.2.1 Cell Division and Cell Expansion: The Lockhart Equation

Growth of leaves and roots, like that of other organs, is determined by **cell division**, **cell expan-**

sion, and **deposition** of cell material. Cell division, however, cannot result in an increase in volume and does not drive growth by itself. Rather, it provides the raw materials for subsequent cell expansion (Green 1976).

The processes of cell division and cell expansion are not mutually independent. Cells probably divide when they reach a certain size (i.e., they elongate after division and then divide again, before they have elongated substantially). This limits the developmental phase at which cell division can occur. This implies that any process that slows down cell expansion inevitably leads to fewer cells per leaf or root and hence smaller leaves or roots. For example, consider a newly formed meristematic leaf cell that differentiates to produce epidermal leaf cells. Suppose this cell divides only after it doubles in cell volume, and the cell has 240h left to undergo repeated mitoses at the point of determination. If the cell doubled in volume every 10h, then cell divisions will occur 24 times, producing 2^{24} cells. If an environmental factor slows the rate of cell expansion such that the cells now take 12h to double in volume, however, then only 20 division cycles will occur, which gives rise to 2^{20} cells. Such a reduction in cell number could substantially reduce leaf surface area (Van Volkenburgh 1994).

Once a cell has divided, it can elongate and expand, provided the turgor pressure (Ψ_p, MPa) exceeds a certain **yield threshold** (Y, MPa). In cells capable of expansion, this threshold value is around 15 to 50% of the turgor pressure under normal conditions (no stress) (Pritchard 1994). The proportional growth rate (r, s^{-1}) is measured as the rate of increase in volume (dV, m^3) per unit volume (V, m^3); r is proportional to the difference between **turgor** and **yield threshold**. The proportional rate of expansion (dV/V·dt, s^{-1}) is described by the simplified **Lockhart equation** (Cosgrove 1986):

$$r = dV/(V·dt) = \phi\left(\Psi_p - Y\right) \qquad (6)$$

where ϕ is the cell-wall **yield coefficient** ($MPa^{-1}s^{-1}$), which is a proportionality constant that depends on physical properties of the cell wall. Plant cell expansion, therefore, is a turgor-driven process believed to be controlled, both in extent and in direction, by the physical properties of the primary cell wall. If cells expand more in one direction than in another, then the cell walls are more **extensible** (looser) in the direction in which they expand most. This simple analysis using the Lockhart equation assumes that neither water flow nor solute influx is limiting. This assumption appears to be met when plants are growing under favorable conditions. In later sections of this chapter we will discuss whether this assumption still applies under conditions of environmental stress.

Because cell expansion and cell division are closely linked, the increase in length or volume of entire leaves and other organs can be analyzed with a similar equation (Passioura & Fry 1992). Both the cell-wall yield coefficient, ϕ, and the yield threshold, Y, reflect the **extensibility** of the cell walls, as determined by their biochemical and biophysical properties. The turgor pressure, P, or, more precisely, the difference between P and Y, allows cell expansion. Uptake of ions into the cell maintains the turgor pressure, which tends to drop slightly as the cell volume increases.

Turgor tends to be **tightly regulated**, particularly in growing cells (Pritchard 1994). This tight regulation of cell turgor is most likely due to **stretch-activated channels** in the plasma membrane (Ramahaleo et al. 1996) that close when cell turgor drops and open when it increases. There are several examples, however, where a step-change in turgor does *not* lead to (full) readjustment to the original turgor pressure (Passioura 1994, Zhu & Boyer 1992). This might reflect differences between species and/or tissue-specific behavior. These results also point out that growth is not really controlled by turgor in the simple manner suggested by the Lockhart equation. Above the turgor threshold, the rate of cell enlargement is controlled by metabolic reactions, which causes synthesis and/or extension of wall polymers. Inside the cell, sufficient solutes need to be generated to maintain turgor above the threshold.

2.2.2 Cell-Wall Acidification and Removal of Calcium Reduce Cell-Wall Rigidity

Cell walls are strongly cross-linked gels that contain numerous charged and uncharged polymers (Carpita & Gibeaut 1993). The charged molecules include the negatively charged cation-binding **polygalacturonic acids**. **Cellulose microfibrils**, which consist of bundles of around 50 cellulose molecules, provide the tensile strength of the cell wall. In expanding cells, the microfibrils tend to be arranged transversely, which favors expansion in a longitudinal, rather than in a radial, direction. **Glycoproteins** add further strength to the cell walls. **Hemicelluloses** (i.e., polysaccharides with a glucan or similar backbone) probably bind to cellulose microfibrils and to each other by means of hydrogen bonds. Because there are many hemicellulose molecules per cellulose microfibril, the microfibrils are completely coated, making a three-dimensional net. Finally, there are several enzymes that can cleave covalent bonds that link the sugar residues of the noncellulosic polymers in the walls, and other enzymes that can join loose ends of similar polymers (Carpita & Gibeaut 1993).

Individual cellulose macromolecules of the cell walls are cross-linked via noncellulosic polysaccharides, and expansion requires that load-bearing cross-links be enzymatically altered. That is, a bond is broken first, which allows the cell to expand, after which a new cross-link is made. The enzyme that catalyzes this reaction is **xyloglucan endotransglycosylase** (XET). **Calcium-pectate complexes** also contribute to the rigidity of the cell wall. Both protons and calcium play an important role in the breaking of cross-linkages. For example, the light-induced growth of leaves (phototropism) is preceded by extrusion of protons from the cytosol into the cell wall. Enhancement of stem elongation by shade results from removal of calcium from the cell walls. A low pH in the cell wall, through activation of specific proteins (**expansins**), induces disruption of hydrogen bonding between cell-wall macromolecules (Cosgrove 1993, McQueen-Mason 1995). Hydrolytic enzymes, especially xyloglucan endotransglycosylase, catalyze breakage of some of the hemicellulose cross-links between cellulose molecules. Calcium enhances cell-wall rigidity by binding to pectin components (Pritchard 1994).

The **light-induced enhancement of leaf growth**, which is preceded by the perception of light by both a red-light receptor (phytochrome) and a blue-light receptor, is due to **cell-wall acidification**, which increases the extensibility of the cell walls (Sect. 5.1.1). Cells of stems may also respond to light,

which is perceived by phytochrome, in that red light suppresses stem elongation and far-red light enhances it. Like far-red light, gibberellins also enhance cell elongation. In lettuce hypocotyls this effect of gibberellin is associated with the removal of **calcium** from the cell walls rather than with cell-wall acidification. In gibberellin-sensitive dwarfs of *Hordeum vulgare* (barley) gibberellins induce a specific **xyloglucan endotransglycosylase**, which presumably catalyzes the breakage of load-bearing bonds in expanding leaf cells (Smith et al. 1996). **Cytokinins** promote and **abscisic acid** reduces the rate of **leaf expansion**, but, as with gibberellins, this is unlikely to be due to cell-wall acidification. Cytokinins and abscisic acid have either no effect or the opposite effect on **root elongation** (i.e., cytokinins tend to inhibit and ABA tends to promote root growth).

Phototropic reactions, which allow coleoptiles to grow toward the light, are based on greater **acidification** of the walls of cells furthest away from the light source as compared with the more proximal cells. Such a difference in acidification is based on a difference in **auxin activity** in the distal and proximal cells (see Box 8). These examples show that cells can respond to light and hormones, sometimes in interaction, by changes in cell-wall properties that in turn affect growth of leaf, stem, or root cells. Genetic or environmental factors that affect the cell-wall cross-linkages, and hence ϕ or Y, affect the rate of cell expansion and the extent to which an organ will grow. Environmental factors such as hypoxia, water stress, and light affect leaf or stem growth exactly in this manner (Sect. 5).

Cell-wall extensibility declines with age of the cells and the walls of older cells no longer respond to cell-wall acidification. This is associated with changes in chemical composition (e.g., incorporation of more galactose). Formation of **phenolic cross-links** between wall components might also play a role, as do rigid cell-wall proteins (**extensins**). A wide range of environmental factors, including water-stress, flooding, and soil compaction, affect leaf growth through their effect on cell-wall extensibility, as will be discussed elsewhere in this chapter (Pritchard 1994).

2.2.3 Cell Expansion in Meristems Is Controlled by Cell-Wall Extensibility and Not by Turgor

The growth rate of individual cells along a growing root tip varies considerably (Fig. 1). A pressure probe that measures the **turgor pressure** in individual growing root cells shows that the turgor

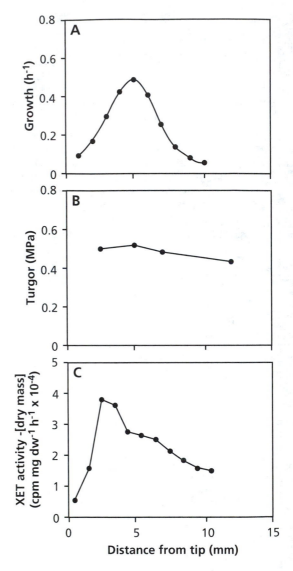

FIGURE 1. Longitudinal profiles of growth, turgor and xyloglucan endotransglycosylase (XET) activity in unstressed roots of *Zea mays* (maize). (A) local elongation rate; (B) turgor-pressure profile; (C) XET activity (Pritchard et al. 1993). Copyright Journal of Experimental Botany.

varies little along the growing root (Fig. 1). Changes in **cell-wall mechanical properties**, rather than in turgor, must therefore be responsible for the immediate control of the expansion rate of roots (Pritchard 1994).

Removal of minute quantities of sap from these cells has shown that the **osmotic component of the water potential** becomes less negative by approximately 15% during cell expansion. This change is small, compared with that in cell volume during expansion, and it results from the drop

Box 8
Phytohormones

Many aspects of plant growth and development are controlled by internal messengers: phytohormones. In the animal literature, the term **hormone** refers to a molecule that is produced in cells of a specific organ (gland) and which has specific effects on other cells (target cells). Phytohormones are not produced in specific glands, but in organs and tissues that serve other functions as well. The effect of phytohormones is also less specific than that of their animal counterparts. They may mediate between several environmental factors and lead to several plant responses.

Phytohormones are characterized as:

1. organic molecules produced by the plant itself
2. compounds that affect growth and development (either positively or negatively) at very low concentrations

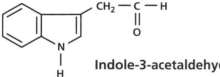

Indole-3-acetaldehyde

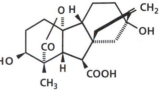

Gibberellin A$_1$ or GA$_1$

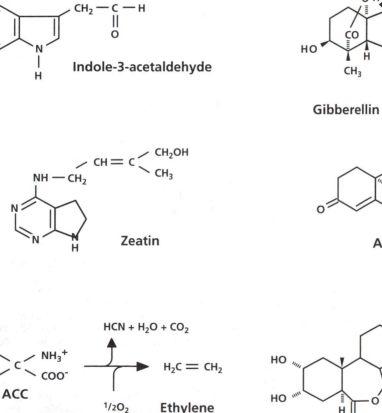

Zeatin

Abscisic acid

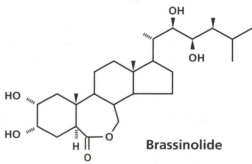

ACC

$\frac{1}{2}O_2$ **Ethylene**

Brassinolide

FIGURE 1. The chemical structure of a representative of the six groups of phytohormones: indole-3-acetaldehyde (IAA, an auxin), gibberellin A$_1$ (GA$_1$, one of many gibberellins, of which only a small number is physiologically active; GA$_1$ is the gibberellin that is ordinarily responsible for stem elongation), zeatin (a cytokinin occurring in *Zea mays*), abscisic acid (ABA), ethylene (the only gaseous phytohormone) and its water-soluble precursor: 1-amino-cyclopropane-1-carboxylic acid (ACC), and brassinolide, which is the most biologically active brassinosteroid.

continued

Box 8. *Continued*

3. compounds that act primarily in a part of the plant that differs from the site they are produced
4. compounds whose action depends on their chemical structure, rather than the elements they contain

There are six groups of phytohormones (Fig. 1). The first phytohormone was discovered in the 1930s by F.W. Went, who was doing a PhD with his father, F.A.F.C. Went, at Utrecht University. It is indoleacetic acid (IAA), termed **auxin**, because of its involvement in the growth of *Avena* coleoptiles toward the light (auxin comes from the greek verb to grow). In many plants, this phototropic effect ("auxin activity") is due to a combination of IAA itself and of antagonists of IAA.

The **gibberellins** derived their name from the fungus *Gibberella fujikori*, which turns dwarf rice cultivars into tall ones. Gibberellic acid (GA) has a profound effect on stem elongation, stimulating both cell division and cell elongation.

The action of **cytokinins** is illustrated by making a leaf cutting. The leaf tends to senesce, but this is often reversed as soon as roots emerge. Root tips are a major site of cytokinin production, and this group of phytohormones delays senescence. They are also involved in the stimulation of cell division and elongation.

Abscisic acid (ABA) derives its name from its effect on leaf abscision, but it has a wide range of effects (e.g., stomatal closing, inhibition of leaf growth, and induction of senescence).

Ethylene is the only gaseous hormone. It is produced from the water-soluble precursor 1-amino-cyclopropane-1-carboxylic acid (ACC) in an oxygen-requiring step, catalyzed by ACC oxidase. It induces senescence and influences cell growth.

The hormonal status of **brassinosteroids** has been established most recently. It was first isolated 20 years ago from *Brassica napus* (oilseed rape) pollen, and has since been found in a broad range of species. Mutants with defects in brassinosteroid biosynthesis or sensitivity are all dwarfs. Application of exogenous brassinosteroids, which promote cell elongation, changes the dwarfs into normal phenotypes.

Phytohormones are important both to internally coordinate the growth and development of different organs and as chemical messengers whose synthesis may be affected when plants are exposed to certain environmental factors. Such factors need to be **sensed** first, which is the first step in a **signal-transduction pathway**, ultimately leading to the plant's **response**. The plant's response is not necessarily due to an effect on the rate of production of the phytohormone, but it may involve its rate of **breakdown** or the **sensitivity** of the target cells for the hormone. At a molecular level, a plant's response may involve upregulation or down-regulation of genes coding for enzymes involved in synthesis or breakdown of the phytohormone, or genes coding for a **receptor** of the phytohormone.

References and Further Reading

Davies, P.J. (1995) (ed). Plant hormones and their role in plant growth and development. Kluwer Academic Publishers, Dordrecht.

Jackson, M.B. (1993) Are plant hormones involved in root to shoot communication. Adv. Bot. Res. 19:103–187.

Jackson, M.B. (1996) Hormones from roots as signals for the shoots of stressed plants. Trends Plant Sci. 1:22–28.

Voesenek, L.A.C.J. & Blom, C.W.P.M. (1996) Plants and hormones: An ecophysiological view on timing and plasticity. J. Ecol. 84:111–119.

Yokota, T. (1997) The structure, biosynthesis and function of bassinosteroids. Trends Plant Sci. 2:137–143.

in concentration of K^+ by about 50%. The concentration of other solutes is constant, showing that solute uptake into the expanding cells occurs at just about the same rate as that of water. There is very little information to indicate which processes affect the cell-wall properties of roots. It may be similar to the situation in leaves, where cell-wall acidification plays a major role. On the other hand, calcium might play a role as it does in hypocotyls.

As the cells expand, more cell-wall material is deposited, so that the cell-wall thickness remains approximately the same during the expansion phase. Further **deposition of cell-wall material** may occur after the cells have reached their final size, which causes the cell wall to become thicker.

2.2.4 The Physical and Biochemical Basis of Yield Threshold and Cell-Wall Yield Coefficient

From a physical point of view, the parameters ϕ, the cell-wall yield-coefficient, and Y, the yield threshold, in the Lockhart equation intuitively make sense. They can also be demonstrated experimentally by using a pressure probe to determine turgor in the growing zone. The Lockhart "parameters," however, often behave as "variables" (i.e., the relationship between r and P is often nonlinear) (Cosgrove 1986, Passioura 1994). What exactly do these "parameters" mean?

In hypocotyl segments of *Vigna unguiculata* (cowpea) the cell-wall mechanical properties are affected by the phytohormones auxin and gibberellin (Box 8). In segments that are deficient in endogenous gibberellin, **auxin** only affects the **yield threshold**, but not the yield coefficient. As a result, the effect of auxin is only half that in segments with normal gibberellin levels. After pretreatment with **gibberellin**, auxin does affect the **yield coefficient**. These results suggest that auxin decreases the yield threshold independently of gibberellin, but that it increases the yield coefficient only in the presence of gibberellin (Okamoto et al. 1995). In the same tissue, both the yield coefficient and the yield threshold are affected by the **pH** in the cell wall. Both parameters are also affected by exposure to high temperature and proteinase, but not in the same manner. That is, a brief exposure to 80°C affects the yield threshold, but not the yield coefficient. Exposure to proteinase affects the yield coefficient, but not the yield threshold. These results suggest that the two cell-wall mechanical properties are differentially controlled by two proteins, both of which are activated by low pH (Okamoto & Okamoto 1995).

So far we can merely speculate what the wall-loosening enzymes really catalyze. The basis of the yield threshold might be that a bond between cellulose molecules in the cell wall must be under tension before its conformation is suitable for enzymatic attack. Increased wall extensibility could be the result of an enzyme breaking bonds within the wall. An increase in its activity would loosen the wall, allowing faster elongation for a given increase in turgor (Pritchard 1994).

Enzymes involved in breaking bonds between cell-wall components are loosely termed **expansins**. It has been suggested that xyloglucan chains, which can form strong hydrogen bonds with cellulose and thus tether the cellulose molecules together, need to be broken down to allow wall loosening.

XET is thought to cleave xyloglucan molecules and so loosen the wall. XET could also re-form the xyloglucan tethers, thus restoring the strength of the wall (Fig. 1). Wall tightening, which is the opposite of loosening, may be due to decreased activity of XET (Pritchard 1994). Other processes, however, appear to play a role as well (Cosgrove 1993).

2.2.5 The Importance of Meristem Size

As discussed in previous sections, cell elongation depends on an increase in cell-wall extensibility. A more rapid rate of cell elongation may lead to a higher rate of leaf expansion or root elongation. A higher rate of leaf expansion or root elongation, however, is not invariably due to greater cell-wall extensibility. If more cells in the meristem divide and elongate at the same rate, then this also results in higher rates of expansion. Indeed, variation in growth can be associated with variation in **meristem size** (i.e., the number of cells that divide and elongate at the same time). In a comparison of the growth of *Festuca arundinacea* (tall fescue) at high and low supply of nitrogen, the major factor contributing to variation in leaf elongation is the size of the meristem (i.e., the number of cells elongating at the same time) (Fig. 2A). Along these lines, two genotypes of *Festuca arundinacea* that differ in their rate of leaf elongation by 50% when grown at high nutrient supply differ in the number of cells that elongate at the same time, whereas the rate of elongation of the expanding cells is fairly similar (Fig. 2B).

3. The Physiological Basis of Variation in RGR—Plants Grown with Free Access to Nutrients

Plant species characteristic of **favorable environments** often have inherently higher maximum relative growth rates (**RGR$_{max}$**) than do species from less favorable environments (Chapin 1980, Grime & Hunt 1975, Lambers & Poorter 1992, Parsons 1968). For example, inherently slow growth has been observed in species characteristic of nutrient-poor (Grime & Hunt 1975), saline (Ball 1988, Ball & Pidsley 1995), and alpine (Atkin et al. 1996) environments. It is clear from Equations 1 and 2 that a high RGR could be associated with a high NAR (reflecting high photosynthesis and/or low whole-plant respiration), a high SLA (i.e., high leaf area per unit leaf mass), and/or a high LMR (high allo-

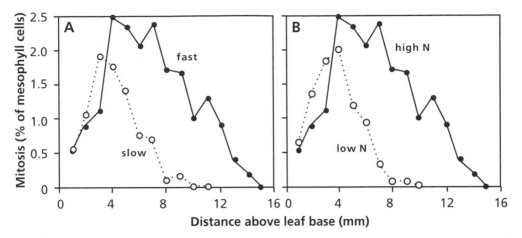

FIGURE 2. Percentage of mesophyll cells that are in mitosis as observed in longitudinal sections from the basal 40 mm of elongating leaf blades of *Festuca arundinacea* (tall fescue). A greater area under the curves indicates a larger meristem. (A) A comparison of meristem size of a fast-elongating and slow-elongating genotype. (B) Effects of nitrogen supply on leaf meristem size in the fast-elongating genotype (MacAdam et al. 1989). Copyright American Society of Plant Physiologists.

cation to leaf mass). Which of these traits is most strongly associated with a high RGR?

3.1 SLA Is a Major Factor Associated with Variation in RGR

Several extensive surveys have shown that the main trait associated with inherently **slow growth** in temperate lowland species from nutrient-poor habitats is their **low SLA**, both in monocotyledonous and in dicotyledonous species (Garnier 1992, Marañón & Grub 1993, Poorter & Remkes 1990). The same conclusion holds for a wide range of woody species. Low SLA values decrease the amount of leaf area available for light interception and hence photosynthetic carbon gain, therefore reducing RGR. Although this conclusion follows logically from Equation 4, it may not provide insight into the exact **mechanisms** that account for slow growth. A further understanding of these mechanisms requires a thorough analysis of the processes discussed in Section 2.2.

When comparing a large number of herbaceous C$_3$ species, there are significant positive correlations of RGR with LAR, LMR, and SLA, but not with NAR (Fig. 3). In a wide comparison using 80 woody species from the British Isles and northern Spain, ranging widely in leaf habit and life-form, RGR is also tightly correlated with LAR (Cornelissen et al. 1996). When comparing more productive cultivars of tree species with less productive ones, SLA,

rather than photosynthesis, is the main factor that accounts for variation in RGR (Ceulemans 1989). In addition, leaf and twig architecture of the more productive trees is such that more of the light is harvested throughout the entire day (Leverenz 1992). In some cases, RGR may also correlate positively with NAR (Garnier 1992), but this correlation is not as strong, nor does it occur as consistently, as does the correlation of RGR with SLA.

LMR does not correlate with RGR in monocotyledons, but it may account for some of the variation in RGR among dicotyledonous species. This reflects the phylogenetic constraints on a plant: A change in LMR appears to require a greater genetic change than that allowed by the genetic variation within a species, genus, or perhaps even family (Marañón & Grub 1993).

Fast-growing species allocate relatively less to their stems, both in terms of biomass and nitrogen, when compared with slower-growing ones. A high allocation to stem growth reflects a diversion of resources from growth to storage in slower-growing species (Sect. 4).

In broad comparisons, **NAR** is often not correlated with RGR in dicots, whereas it is in monocots. The effect of variation in SLA on the RGR of monocots is invariably stronger than that of variation in NAR. When pairs of annual and perennial grass species that belong to the same genus are compared, the highest RGR is invariably associated with the **annual life form**. Because annuals are thought to have descended from perennial

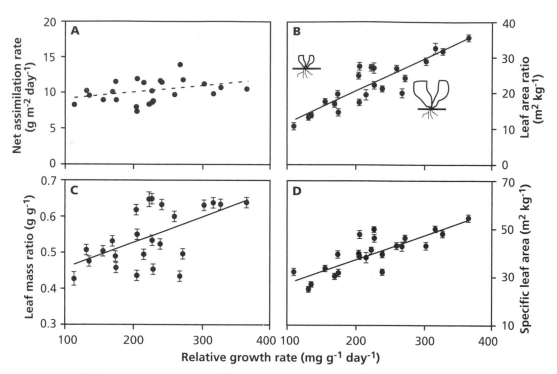

FIGURE 3. A comparison of (A) NAR, (B) LAR, (C) LMR, and (D) SLA of 24 herbaceous C_3 species that differ in their RGR as determined on plants grown with free access to nutrients. The broken line indicates a nonsignificant regression; solid lines indicate significant regressions (Poorter & Remkes 1990).

ancestors, it has been suggested that the same morphological changes that enhance a genotype's RGR have occurred repeatedly in different genera and that a high RGR is the more recent development (Garnier 1992, Garnier & Vancaeyzeele 1994).

3.2 Leaf Thickness and Leaf-Mass Density

Variation in **SLA**, or its inverse [the leaf mass per unit leaf area (**LMA**, $kg\,m^{-2}$)], must be due to variation in **leaf thickness** (m) or in **leaf-mass density** ($kg\,m^{-3}$) (Witkowski & Lamont 1991).

$$LMA = (leaf\ thickness) \cdot (leaf\ mass\ density) \tag{7}$$

When **shade leaves** and **sun leaves** are compared, leaf thickness is a major parameter in determining variation in LMA, and it reflects increased **thickness of palisade parenchyma** in sun leaves (Sect. 3.2.2 in photosynthesis). In addition, comparing alpine and congeneric lowland species, variation in LMA is associated with that in leaf thickness. In comparisons of closely related species from nutrient-poor and nutrient-rich sites, however, variation in LMA is due to differences in leaf mass

density (Garnier & Laurent 1994, Van Arendonk & Poorter 1994). In addition, leaf mass density also accounts for a part of the variation in LMA between shade leaves and sun leaves, between widely contrasting woody species (Cornelissen et al. 1996), and especially when comparing congeneric lowland and alpine species (Atkin et al. 1996).

3.3 Anatomical and Chemical Differences Associated with Leaf-Mass Density

The inherent variation in LMA and **leaf mass density** is associated with differences in both leaf **anatomy** and **chemical composition**. Fast-growing species with a low LMA have relatively **large epidermal cells**. Because these cells lack chloroplasts, which are a major component of the mass in the cytoplasm of mesophyll cells, they have a low mass density, which contributes to the low leaf mass density of the fast-growing species. Slow-growing plants with a high LMA have **thicker cell walls** and contain more **sclerenchymatic cells**. These cells are small and are characterized by very thick cell walls; therefore, they have a high mass density. Associ-

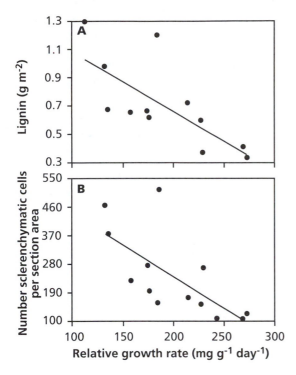

FIGURE 4. The lignin concentration per unit leaf area (A) and the number of sclerenchymatic cells, associated with the epidermis (B), as related to the growth rate of a number of grass species (drawn from data in Van Arendonk & Poorter 1994). Copyright Blackwell Science Ltd.

ated with these and other anatomical differences, the leaves of slow-growing species have more **lignin** and **cell-wall components** per unit leaf mass or area (Fig. 4).

3.4 Net Assimilation Rate, Photosynthesis, and Respiration

As explained in Section 2.1.1, the **net assimilation rate** (NAR) is related to the balance of carbon gain in **photosynthesis** and carbon use in whole-plant **respiration**. Variation in NAR, therefore, may be due to variation in photosynthesis, respiration, or a combination of the two. In a wide comparison of herbaceous species, which is shown in Figure 3A, there is no clear trend of NAR with RGR. Rate of photosynthesis per unit leaf area also shows no correlation with RGR (Fig. 5A). **Slow-growing species**, however, use relatively more of their carbon for **respiration**, especially in their roots (Fig. 5B), whereas **fast-growing** species invest a relatively greater proportion of assimilated carbon in **new growth**, especially **leaf growth**. Next to the variation in LAR (SLA and LMR), this difference in the amount of carbon required for respiration is the second-most important factor that is associated with inherent variation in RGR.

If widely different **tree species** are compared, then rates of **photosynthesis** per unit leaf area are higher in **fast-growing pioneer species** than they are in **slower-growing climax species** (Evans 1989). SLA and allocation, however, also differ strikingly among these taxa. The lack of a correlation between photosynthesis and RGR among closely related taxa or among morphologically similar taxa (Fig. 3) indicates that these broad differences in photosynthetic rate are not the *major* cause of differences in RGR (Lambers et al. 1998).

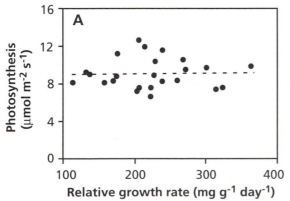

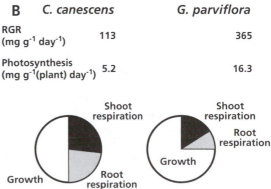

FIGURE 5. (A) The rate of photosynthesis per unit leaf area in fast and slow-growing herbaceous species (Poorter et al. 1990). Copyright American Society of Plant Physiologists. (B) The carbon budget of a slow-growing species (*Corynephorus canescens*) and a fast-growing species (*Galinsoga parviflora*). In the upper line the RGR of these species is given. The second line gives the daily gross CO_2 fixation (Lambers & Poorter 1992). Copyright Academic Press Ltd.

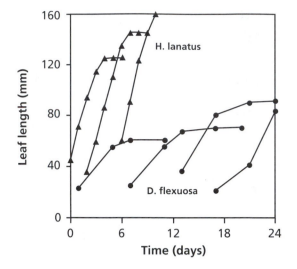

FIGURE 6. The rate of leaf elongation of a slow-growing grass species (*Deschampsia flexuosa*, circles) and a fast-growing grass (*Holcus lanatus*, triangles) (after Groeneveld & Bergkotte 1996). Copyright Blackwell Science Ltd.

3.5 RGR and the Rate of Leaf Elongation and Leaf Appearance

The higher RGR and SLA of fast-growing grass species is associated with a more **rapid leaf elongation** (Fig. 6). The extent to which this difference in leaf expansion is associated with variation in cell-wall properties of the elongating cells is not known. Does cell-wall acidification or the removal of calcium from the cell walls play a role? Are the cells of rapidly elongating leaves more responsive to changes in pH or calcium? Does it reflect a difference in meristem size, as shown in Figure 2? It is also apparent from Figure 6 that in the fast-growing grass (*Holcus lanatus*) the next leaf starts to grow just before the previous one has reached its final size. This typically contrasts with the pattern in slow-growing grasses (e.g., *Deschampsia flexuosa*), where the next leaf does not start elongating until well after the previous one has stopped. Is this akin to apical dominance? It could be, but very little is explained by making this comparison because the phenomenon of apical dominance is still poorly understood.

3.6 RGR and Activities Per Unit Mass

The growth analysis discussed earlier clearly demonstrates that SLA "explains" much more of the variation in RGR than do area-based measures of

NAR and photosynthesis. This area-based measure is the most logical way to describe the environmental controls over capture of light and CO_2. **Economic analyses of plant growth** (the return on a given biomass investment in leaves or roots), however, more logically express resource capture (photosynthesis or nutrient uptake) per unit plant mass. This is achieved by multiplying the area-based measures of carbon gain by SLA, for example:

$$NAR_m = NAR_a \cdot SLA \qquad (8)$$

Because of the strong correlation between SLA and RGR, RGR also has a strong positive correlation with NAR_m (Fig. 7A). The low NAR_m of slow-growing species in part reflects their high carbon requirement for root respiration (Fig. 5; Sect. 5.2.3 of the chapter on respiration). Both the V_{max} for nitrate uptake and the net rate of nitrate inflow show a strong correlation with RGR_{max} (Fig. 7B,C). It is likely that this is more a result than the cause of variation in growth rates (Sect. 2.2.3.2 of the chapter on mineral nutrition; Touraine et al. 1994). The positive correlations between RGR_{max} and mass-based activity of both roots and leaves hold for monocots and dicots (Fig. 7A–C). By contrast, there is no correlation of RGR_{max} with biomass allocation to roots and leaves for monocotyledonous species (Fig. 7D), whereas RGR_{max} decreases with increasing biomass allocation to roots in dicotyledonous species (Fig. 7E).

These correlations result from **rapidly growing plants** producing leaves and roots with relatively large allocation to metabolically active components, rather than to cell walls and storage (Fig. 4). As a result, they have leaves with a high mass-based photosynthetic capacity and roots with a high mass-based capacity for nitrogen inflow. It is the balance of net mass-based carbon gain (leaf photosynthesis minus total plant respiration, NAR_m) and mass-based root activity (NIR_m) in combination with the pattern of root:leaf allocation that accounts for differences in RGR_{max}. It is interesting that the limited data available suggest that NIR_m has a stronger correlation with RGR_{max} than does NAR_m. This is evident from the positive correlation between RGR_{max} and the ratio of mass-based specific ion uptake rate and mass-based net assimilation rate (Fig. 7F).

3.7 RGR and Suites of Plant Traits

Our analysis of the correlation of RGR_{max} with plant traits suggests that SLA is the key trait because it enables the plant to expose a large leaf area to light

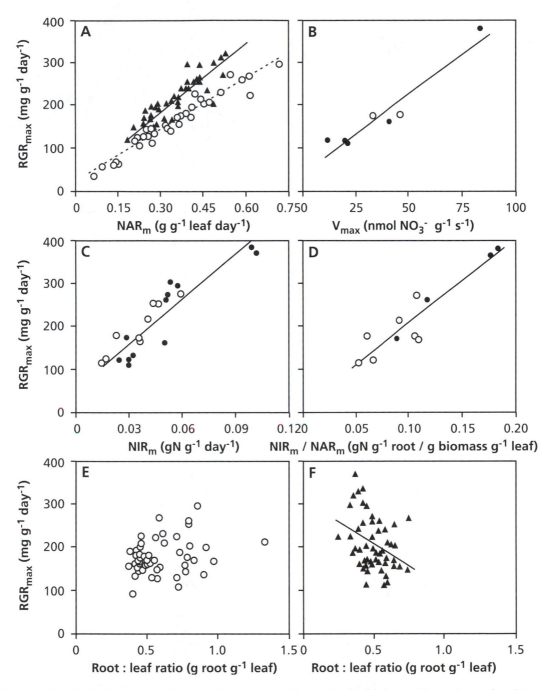

FIGURE 7. Correlation between maximum relative growth rate (RGR$_{max}$) and (A) mass-based net assimilation rate (NAR$_m$), (B) mass-based maximum rate of nitrate uptake (V$_{max}$), (C) mass-based specific nitrate inflow rate (NIR$_m$), (D) the ratio of NIR$_m$/NAR$_m$ and (E, F) the ratio of biomass allocation to roots and leaves for 51 monocotyledonous (E), and 53 dicotyledonous (F) species. Each point represents a separate species of monocot (open symbols) or dicot (closed symbols) grown with free access to nutrients (data synthesized by Garnier 1991). Copyright Elsevier Science Ltd.

and CO_2 per given biomass invested in leaves. Certain other traits, however, also correlate positively with RGR_{max} (e.g., mass-based measures of photosynthesis and nutrient uptake), whereas other traits are negatively associated with RGR_{max} (e.g., leaf mass density due to support tissues and root respiration). These observations suggest that there is a **suite of plant traits** associated with rapid growth (high SLA, high mass-based rates of photosynthesis, and nutrient uptake), whereas other traits are associated with slow growth (greater investment in cell walls and fiber) (Lambers & Poorter 1992). These dichotomies suggest a **trade-off** between traits that promote rapid growth and those that promote persistence.

Due to their greater investment in carbon-rich compounds, such as lignin, and less accumulation of minerals, the carbon concentration of the slow-growing species is higher than that of fast-growing ones. This is an additional, albeit minor, factor that contributes to their low growth potential. There may well be differences in exudation and volatilization, but their quantitative significance in explaining variation in RGR is probably small.

4. Allocation to Storage

Up to now in this chapter, we have only dealt with allocation of resources to structural components of the plant, during vegetative growth. Plants, however, also channel some of their resources to storage compartments, where the stored resources are available for future growth. Plants store both carbon and nutrients, but there is a wide variation in the amount and kind of resources that are stored, and in the organ where the storage predominantly takes place: leaves, stems, roots, or specialized storage organs. In this section we will first discuss the concept of storage and its chemical nature, then describe differences in the role of storage in annuals, biennials, and perennials.

4.1 The Concept of Storage

We define **storage** as resources that build up in the plant and can be mobilized in the future to support biosynthesis (Chapin et al. 1990). There are three general categories of storage:

1. **Accumulation** is the increase in compounds that do not directly promote growth. Accumulation occurs when resource acquisition exceeds demands for growth and maintenance (Millard 1988).

2. **Reserve formation** involves the metabolically regulated synthesis of storage compounds that might otherwise directly promote growth. Reserve formation directly competes for resources with growth and defense (Rappoport & Loomis 1985).

3. **Recycling** is the reutilization of compounds whose immediate physiological function contributes to growth or defense, but which can subsequently be broken down to support future growth (Chapin et al. 1990).

Accumulation, also termed **interim deposition** (Heilmeier & Monson 1994), accounts for much of the short-term fluctuations in chemical composition of plants (e.g., the daily fluctuation of starch in chloroplasts) (Sect. 2.1.4 of the chapter on photosynthesis) or of nitrate in vacuoles (Martinoia & Wiemken 1981). Accumulation allows a relatively constant export rate of carbohydrates from source leaves throughout the 24-hour cycle, despite the obvious diurnal pattern of photosynthetic carbon gain (Fondy & Geiger 1985). Carbohydrate accumulation also occurs when conditions favor photosynthesis more than nutrient acquisition (Heilmeier & Monson 1994). This accounts for accumulation of starch during sunny weather and its depletion under cloudy conditions. On the other hand, nitrogen accumulation, also termed **luxury consumption**, occurs after pulses of nitrogen availability or when nitrogen supply exceeds the capacity of the plant to utilize nitrogen in growth. Although accumulation may explain many of the short-term changes in storage, it is less important over time scales of weeks to years. Over these longer time scales, capacities for photosynthesis and nutrient uptake adjust to plant demand, thus minimizing large long-term imbalance between carbon and nutrient stores.

Reserve formation diverts newly acquired carbon and nutrients from growth or respiration into storage. This can occur when rates of acquisition are high and vegetative growth is slow, and during periods of rapid vegetative growth, often in competition with it. Grafting experiments clearly demonstrate this competition between storage and growth. For example, roots of sugar beet (*Beta vulgaris*), which allocate strongly to storage in a taproot, decrease shoot growth when grafted to shoots of a leafy variety of the same species (chard). On the other hand, chard roots, which have a small capacity for storage, cause grafted sugar beet shoots to grow larger than normal (Rappoport & Loomis 1985). Stored reserves make a plant less dependent on current photosynthesis or nutrient uptake from the soil and provide resources at times when growth demands are large but there are few leaves or roots present to acquire these resources, such as early spring. Stored reserves also enable plants to recover following catastrophic loss of leaves or roots to fire, herbivores, or other disturbances. Finally, stored reserves enable plants to shift rapidly from a vegetative to a reproductive mode, even at times of year when conditions are not favorable for resource acquisition.

Recycling of nutrients following **leaf senescence** allows reutilization of about half of the nitrogen and phosphorus originally contained in the leaf (Sect. 4 of the chapter on mineral nutrition), but it is a relatively unimportant source of carbon for growth (Chapin et al. 1990). These stored nutrients are then a nutrient source for developing leaves. For example, in arctic and alpine plants 30 to 60% of the nitrogen and phosphorus requirement for new growth comes from retranslocated nutrients. Reserve formation and recycling allow plants to achieve rapid growth following snow release, despite low soil temperatures that may limit nutrient uptake from the soil (Atkin 1996, Chapin et al. 1986).

4.2 Chemical Forms of Stores

In the previous section we demonstrated that there are several types of controls over carbon and nutrient stores (accumulation, reserve formation, and recycling). The chemistry and location of stored reserves, however, may be similar for each of these processes.

Carbohydrates are stored as **soluble sugars** (predominantly sucrose), **starch**, or **fructans**, depending on the species; fructans (polyfructosylsucrose) are only found in some taxa: Asterales, Poales, and Liliales. The capacity for storage depends on the presence of a specific organ, such as a rhizome, tuber, bulb, or taproot. Thus, an important cost of storage is production of the storage structure, in addition to the stores themselves. In a comparison of 92 species (15 genera) that belong to the Epacridaceae in a fire-dominated Australian habitat, species that regenerate from seeds ("seeder species") have low starch levels in their roots ($2\,mg\,g^{-1}$ DM) when compared with "resprouter species" ($14\,mg\,g^{-1}$ DM), whereas no differences occur in their shoots (Bell et al. 1996). It is interesting that when the capacity to store carbohydrates in the taproot increases with increasing plant age, the rate of root respiration decreases greatly. Young roots

of *Daucus carota* (carrot) probably respire large amounts of carbohydrates via the alternative respiratory path (see Sect. 2.5 of the chapter on plant respiration), whereas older roots, with a larger capacity to store carbohydrates, no longer use this pathway (Steingröver 1981). This indicates that storage of carbohydrates does not invariably occur at the expense of vegetative growth but may involve a decline in carbon expenditure in respiration by the alternative path.

Nitrogen is stored as **nitrate** (especially in petioles and shoot axes of fast-growing species), when plants are supplied with rather high levels of nitrate from the soil. At a moderate or low nitrogen availability, nitrogen is stored as **amino acids** (often of a kind not found in proteins), **amides** (asparagine, e.g., in *Asparagus*, and glutamine), or **protein** (normal enzymes such as **Rubisco**, sometimes special **storage proteins**) (Chapin et al. 1986, Heilmeier & Monson 1994). Storage as protein involves the additional costs of protein synthesis, but has no effects on the cell's osmotic potential. In addition, proteins may serve a catalytic or structural function, as well as being a store of nitrogen. Leaves contain vast amounts of Rubisco, of which some may be inactivated and not contribute to photosynthesis (Section 4.2 of the chapter on mineral nutrition). Rubisco is not a storage protein in a strict sense, but it is nonetheless available as a source of amino acids that are exported to other parts of the plant (Chapin et al. 1990). Storage of nitrogenous compounds is sometimes considered a reflection of "luxury consumption." This is misleading, however, because N-deficient plants also store some nitrogen, which they later use to support reproductive growth (Millard 1988).

Phosphorus is stored as **inorganic phosphate** (orthophosphate or polyphosphate) as well as in **organic phosphate-containing compounds** [e.g., phytate (inositol hexaphosphate)] (Chapin et al. 1982, Hübel & Beck 1996).

4.3 Storage and Remobilization in Annuals

Annuals allocate relatively little of their acquired resources (carbon and nutrients) to storage, which contributes to their high growth rate (Schulze & Chapin 1987). Annuals are generally short-lived, and the rapid formation of a large seed biomass ensures survival of the population and avoids periods of low resource supply.

During seed filling, carbohydrate reserves in stems are depleted, and the nitrogen invested in the photosynthetic apparatus is exported, after hydrolysis of the proteins to amino acids, which are exported via the phloem. The gradual breakdown and export of resources invested in leaves occurs during leaf **senescence**. This is a **controlled process** in plants, and it is rather different from the uncontrolled collapse with increase in age of animal cells. It ensures remobilization of resources previously invested in vegetative structures to developing reproductive structures. Roots and some parts of the reproductive structures also show a net loss of nitrogen and a decrease in nutrient uptake during some stages of seed filling (Table 2).

In addition to the use of proteins that first functioned in the plant's primary metabolism during vegetative growth, *Glycine max* (soybean) also has specific **vegetative storage proteins**. These vacuolar glycoproteins accumulate abundantly in bundle sheath and associated mesophyll cells and in the upper epidermis of leaves (Staswick 1990). In hypocotyls, the storage proteins accumulate in epidermal and vascular tissues. As these organs mature, the storage proteins are hydrolyzed, and the amino acids are exported (Staswick 1988, 1990). In soybean, the amount of vegetative storage proteins and the level of mRNA encoding this protein depend on the supply of nitrogen to the plants. Wounding, water deficit, blockage of export via the phloem, and exposure to jasmonic acid (a molecule signaling stress in plants; see Sect. 4.3 of the chapter on ecological biochemistry) all enhance the accumulation of the proteins in leaves of soybean (Staswick et al. 1991) and *Arabidopsis thaliana* (Berger et al. 1995).

4.4 The Storage Strategy of Biennials

Biennials represent a specialized life history that enables them to exploit habitats where resources are available intermittently and where a small change in these environmental conditions may tip the balance toward either annuals or perennials (Hart 1977). In their first year, biennials develop a storage organ, as do perennials. In their second year, they invest all available resources into reproduction, in a manner similar to annuals.

The storage organ contains both **carbohydrates** and **nitrogen**. Do the stored reserves of carbon or nitrogen add significantly to seed yield? In the biannual thistle, *Arctium tomentosum*, the carbohydrates stored in the taproot are important to sustaining root respiration, but they contribute less than 0.5% to the formation of new leaves. Carbohydrate storage only primes the growth of the first leaves, after which the next leaves grow independently of stored carbon. Of all the nitrogen invested into growth of new leaves, however, about half originates from the nitrogen remobilized from the storage root. The nitrogen stored in roots contributes 20% to the total nitrogen requirement during the second season. Under shaded conditions, this fraction is as high as 30%. Seed yield is most significantly correlated with total plant nitrogen content early in the second year. In shaded plants, the amount of N in the seeds is very similar to the amount stored after the first year, whereas in plants grown at normal levels of irradiance the amount of N in the seeds is about twice that which was stored. These data indicate that storage of nitrogen is of far greater importance than that of carbohydrates (Fig. 8).

4.5 Storage in Perennials

Perennials have a large capacity for storage of both **nutrients** and **carbohydrates**, which reduces their growth potential in the early vegetative stage (Rosnitschek-Schimmel 1983). Once storage of resources has been achieved, however, it enables these plants to start growth early in a seasonal climate and to survive unfavorable conditions for CO_2 assimilation or nutrient absorption. The stored products allow rapid leaf development when annuals depend on recently acquired carbon and nutrients.

In the tundra sedge, *Eriophorum vaginatum*, amino-acid **nitrogen** and organic **phosphorus** reserves vary nearly fourfold during the growing season and provide all the nutrients required

TABLE 2. Net export of nitrogen (mainly as amino acids and amides after protein hydrolysis) from senescing glumes, leaves, stem, and roots, and accumulation of the same amount in the grains of *Triticum aestivum* (wheat), between 9 and 15 days after flowering.

Plant Part µg (plant part)$^{-1}$ day^{-1}	Change in nitrogen content
Glumes	−192
Leaves	−335
Stem	−193
Roots	−132
Total	−852
Grains	+850

Source: Simpson et al. 1983.

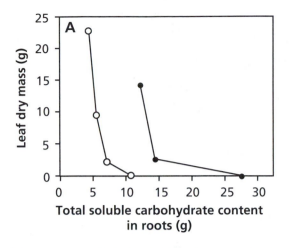

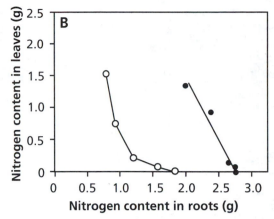

FIGURE 8. The relation between (A) the decrease with time of the content of total soluble carbohydrates in the taproot and the increase with time of leaf dry mass and (B) the decrease with time of the nitrogen content of the taproot and the increase with time of the nitrogen content of the leaves, at the beginning of the second season in the biennial herbaceous thistle, *Arctium tomentosum*. Filled circles refer to control plants, grown under natural light in the field, and open circles to plants growing in shade, 20% of the irradiance of control plants (Heilmeier et al. 1986).

to support leaf growth in early summer, when the arctic soil is largely frozen (Chapin et al. 1986). Plants whose roots are experimentally isolated from the soil are able to grow just as rapidly as plants rooted in soil for an entire growing season, based entirely on stored nutrient reserves (Jonasson & Chapin 1985).

As in the annual *Glycine max* (Sect. 4.1), some perennial herbaceous species may also accumulate specific **storage proteins** [e.g., in the taproots of *Taraxacum officinale* (dandelion) and *Cichorium intybus* (cichory)] (Cyr & Bewley 1990). Storage proteins have the advantage over amino acids and amides as storage products in that they allow storage at a lower cellular water content and thereby reduce the danger of freezing damage. Upon defoliation, the storage proteins are remobilized during regrowth of the foliage (e.g., in the taproot of *Medicago sativa*, alfalfa), where they constitute approximately 28% of the soluble protein pool. Several weeks after defoliation, the storage proteins may again comprise more than 30% of the soluble protein pool (Avice et al. 1996).

Special storage proteins occur in woody plants, especially in structural roots, bark, and wood tissue of trees, where they may constitute 25 to 30% of the total extractable proteins. In *Populus canadensis* (poplar), **storage glycoproteins** accumulate in protein bodies in ray parenchyma cells of the wood in autumn and disappear again in spring (Sauter & Cleve 1990). In *Populus trichocarpa* the synthesis of

storage proteins is induced by exposure to short-day conditions (Fig. 9), most likely under the control of the phytochrome system (Coleman et al. 1992).

4.6 Costs of Growth and Storage: Optimization

Costs of storage include **direct costs** for **translocation** of storage compounds to and from storage sites, **chemical conversions** to specific storage compounds, and **construction of special cells**, **tissues**, or **organs** for storage as well as their protection. There are also "**opportunity costs**" (i.e., diminished growth as a result of diverting metabolites from resources that might have been used for structural growth) (Bloom et al. 1985). The construction of storage cells and tissue does not necessarily occur at the same time as the accumulation of the stored products, which makes it difficult to assess whether vegetative growth and storage are competing processes. If the storage compounds are derived from recycling of leaf proteins (e.g., Rubisco), which functioned in metabolism during the growing season, then storage does not compete with vegetative growth. Use of accumulated stores similarly does not compete with growth and has negligible opportunity cost. If carbohydrates accumulate during the period of most vigorous vegetative growth, particularly when plants are limited by the level of irradiance, then there is no competition

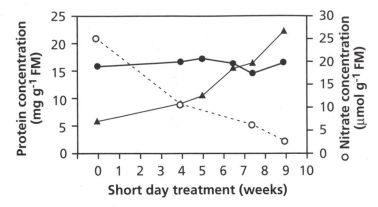

FIGURE 9. The effect of exposure to short days (8h light) preceded by growth under long-day conditions (16h light) on the protein concentration in bark (triangles) and leaves (filled circles) and on the nitrate concentration (open circles) in leaves of *Populus trichocarpa* (poplar). The plants were grown in a full nutrient solution in a growth chamber under a temperature regime of 22°C during the day and 18°C at night, both before and after exposure to short days (Langheinrich & Tischner 1991). Copyright American Society of Plant Physiologists.

between storage and vegetative growth (Heilmeier & Monson 1994).

5. Environmental Influences

In earlier sections we discussed the causes of inherent differences among species in growth rate under favorable conditions. Natural conditions, however, are seldom optimal for plant growth, so it is critical to understand the patterns and mechanisms by which growth responds to variation in environmental factors, including water supply, irradiance, oxygen supply, and temperature. Plants may acclimate to different environments, or they may differ genetically in their programmed response to the environment. Aspects of both acclimation and adaptation will be discussed in this section.

Plants generally respond to suboptimal conditions through reductions in growth rate and changes in allocation to minimize the limitation of growth by any single factor. Arguments based on economic analogies suggest that plants can minimize the cost of growth (and therefore maximize growth rate) if allocation is adjusted such that all resources are equally limiting to growth (Bloom et al. 1985). Thus, we might expect greater allocation to leaves when light strongly limits growth and greater allocation to roots in response to water or nutrient limitation (Brouwer 1963). The net result of these adjustments, through both adaptation and acclimation, should be a functional balance between the activity of roots and shoots in which below-ground resources are acquired in approxi-

mate balance with above-ground resources (Garnier 1991):

$$\text{root mass} \cdot \text{NIR}_m = k \cdot \text{leaf mass} \cdot \text{NAR}_m \quad (9)$$

where NIR_m is the net inflow of ions per unit root mass, NAR_m is the net assimilation rate, which is now expressed per unit leaf mass rather than leaf area, and k is the concentration of the nutrient for which the intake rate NIR_m appears in the formula. The accumulation of nutrients under conditions of carbon limitation and of carbohydrates under conditions of nutrient or water limitation (Sect. 4) shows that plants never achieve perfect functional balance.

Growth is arguably the most important process to understand in predicting plant **responses to environment**, and we therefore need to understand the **basic mechanisms** by which growth responds to environment. Does growth decline in direct response to reductions in resource supply and acquisition or does the plant anticipate and respond to specific signals before any single resource becomes overwhelmingly limiting to all physiological processes? In other words, is growth **source-controlled** or do specific signals modulate sink activity (growth) that then governs rates of resource acquisition (**feedforward control**)? For example, if growth responds directly to reduced source strength, low availability of light or CO_2 would act primarily on photosynthesis, which would reduce the carbon supply for growth; similarly, water or nitrogen shortage would restrict acquisition of these resources such that water potential or nitrogen supply would directly determine growth rate. On the other hand, if unfavorable environmental conditions are sensed and trigger signals that reduce growth rate directly, then this would lead to a feedforward response that would reduce rates of acquisition of nonlimiting resources before the plant experienced severe resource imbalance.

Unfavorable environmental conditions tend to reduce growth. For example, unfavorable conditions below ground often trigger changes in the balance among abscisic acid, cytokinins, and gibberellins, which lead to changes in growth rate that precede any direct detrimental effects of these changes in environment. This **feedforward response** minimizes the physiological impact of the unfavorable environment on plant growth. In the following sections, we will describe the evidence for the relative importance of direct environmental effects on resource acquisition (source control) versus those mediated by feedforward responses. All current simulation models of plant growth in agriculture and ecology assume that source control is the major mechanism of plant response to environment. If this is incorrect, then it is important to know whether the feedforward responses of plants lead to qualitatively different predictions of how plants respond to their environment.

5.1 Growth as Affected by Irradiance

The **level of irradiance** is a major ecological factor that influences plant growth. Plants respond to different levels of irradiance through both genetic adaptation and phenotypic acclimation. Both the total level of irradiance and the **photoperiod**, **spectral composition**, and **direction** of the light (phototropism) affect plant development. These effects of light are the topics of this section, whereas effects of UV radiation are discussed in Section 2.2 of the chapter on effects of radiation and temperature. Nitrogen allocation to different leaves, as dependent on incident irradiance will be discussed in Section 5.4.6, after discussing the involvement of cytokinins in nitrogen allocation (Sect. 5.4.4) (see also Box 7 in the chapter Scaling up).

5.1.1 Growth in Shade

Shade caused by a leaf canopy reduces the irradiance predominantly in the photosynthetically active region of the spectrum (400 to 700 nm). Apart from a reduction of irradiance in this range, therefore, there is a shift in the spectral composition of the light (Box 9).

5.1.1.1 Effects on Growth Rate, Net Assimilation Rate, and Specific Leaf Area

Plants that grow in a shady environment invest relatively more of the products of photosynthesis and other resources in leaf area: They have a **high LAR**. Their leaves are relatively thin: They have a **high SLA** and **low leaf mass density**. This is associated with relatively few, and small **palisade mesophyll** cells per unit area. The leaves have a high **chlorophyll concentration** per unit fresh mass, which results in a rather similar chlorophyll concentration per unit leaf area, as in sun leaves, but relatively **less protein** per unit chlorophyll (cf. Sect. 3.2.3 of the chapter on photosynthesis).

Table 3 summarizes the results of morphological acclimation and adaptation to a low irradiance. The RGR of the shade-tolerant *Impatiens parviflora* (touch-me-not) is reduced less by growth in shade as compared with full sun, when compared with *Helianthus annuus* (sunflower), which is a shade-avoiding species. This is due to a stronger increase of the LAR in the shade in *Impatiens*, which is due to a large increase in SLA and a small increase in LMR (the various abbreviations used in growth analysis have been further explained in Table 1 and Sect. 2.1). The regulation of the increase in LAR is not fully understood; however, it serves to capture more of the growth-limiting resource in the shade. Table 3 also shows that at a moderately low irradiance, the RGR of the shade-tolerant *Impatiens* is higher than that of the shade-avoiding *Helianthus*. This is a somewhat extreme example, however, of widely contrasting species. The RGR of moderately shade-tolerant species is not that much higher than that of shade-avoiding species, when plants are compared at a low irradiance level (Fig. 10). These patterns indicate that changes in allocation and leaf morphology in response to shade maximize capture of the growth-limiting resource (light), and that this shade acclimation is more extreme in the shade-adapted species.

At a very low irradiance, such as under a dense canopy, many shade-avoiding plants do not survive, even though they may exhibit a positive RGR in short-term growth experiments (Table 3). Thus, there must be additional factors that account for the distribution of sun-adapted and shade-adapted species. What other plant traits play a role in the distribution of shade-avoiding and shade-tolerant species? First, **leaf longevity** appears to be important. Shade-tolerant species tend to keep their leaves for a longer time and so increase the potential photosynthetic return (Reich et al. 1991, 1992a,b). When grown in shade, fast-growing tropical trees show a higher leaf area ratio and lower root mass ratio, as well as a greater mortality than do slower-growing ones (Kitajima 1994). Shade-tolerant plants also minimize leaf loss through their greater allocation to chemical defenses against pathogens and herbivores than in shade-avoiding species (chapter on ecological biochemistry). In addition, the enhanced rate of stem elongation (Sect. 5.1.1.3) may weaken the shade-avoiding plants.

The sun or shade character of leaves is determined during leaf development. In some species, such as many trees, this is determined at a very

---

Box 9
Phytochrome

Plants can monitor various aspects of the light climate, and they use this information to adjust their growth and reproduction to environmental conditions. Phytochrome is one of the systems in plants that allows it to gain this information about its light environment. It was discovered by Butler et al. (1959) as the photoreceptor involved in red to far-red reversible reactions, and it has been characterized better than other photoreceptors in plants (i.e., cryptochrome and the UV-light receptors). In vivo the phytochrome chromophore exists in two different photoconvertible forms (Fig. 1): the red-light-absorbing form (P_r) and the far-red-light-absorbing form (P_{fr}). Phytochrome is involved in the perception of the presence of light per se, the spectral composition of light, its irradiance level, its direction, and its duration. It plays a key role throughout the life cycle of plants, from seed germination and seedling development, during vegetative growth, and on to the control of flowering.

In *Arabidopsis thaliana* five genes encoding phytochrome have been identified: *PHYA-PHYE*. Phytochromes A and B have partly antagonistic effects, whereas the functions of phytochromes C, D, and E have yet to be discovered (Smith 1995). The phytochrome species have identical chromophores but different apoproteins.

In continuous light, phytochrome B, unlike phytotochrome A, is necessary for red light perception. On the other hand, phytochrome A, unlike phytochrome B, is necessary for far-red light perception (the so-called red and far-red high-irradiance responses; HIR) (Quail et al. 1995). Thus, although phytochrome A and phytochrome B each absorb red and far-red, the two phytochromes monitor distinct facets of the light environment (Fig. 1).

Phytochromes are also involved in the detection of first exposure to light after a long period of darkness in soil, such as in the case of seeds and etiolated seedlings. During these "induction" reactions, phytochrome A is activated by a broad range of wavelengths of environmental light (e.g., ultraviolet, visible, and far-red; all at very low fluence) and it irreversibly triggers physiological and molecular responses (very low fluence response; VLFR). In contrast, phytochrome B photoreversibly switches induction responses to "on" or "off" upon alternate irradiation with red and far-red light, respectively; both of these responses occur at a fluence that is four orders of magnitude higher than that required by phytochrome A (low fluence response; LFR) (Furuya & Schäfer 1996).

References and Further Reading

Butler, W.L., Norris, H.W., & Hendricks, S.B. (1959) Detection, assay, and preliminary purification of the pigment controlling photoresponsive developments of plants. Proc. Natl. Acad. Sci., USA 45:1703–1708.

Furuya, M. (1993) Phytochromes: Their molecular species, gene families, and functions. Annu. Rev. Plant Physiol. Plant Mol. Biol. 44:617–645.

Furuya, M. & Schäfer, E. (1996) Photoperception and signalling of induction reactions by different phytochromes. Trends Plant Sci. 1:301–307.

Quail, P.H., Boylan, M.T., Parks, B.M., Short, T.W., Xu, Y., & Wagner, D. (1995) Phytochromes: Photosensory perception and signal transduction. Science 268:675–680.

Smith, H. (1995) Physiological and ecological function within the phytochrome family. Annu. Rev. Plant Physiol. Plant Mol. Biol. 46:289–315.

continued

Box 9. *Continued*

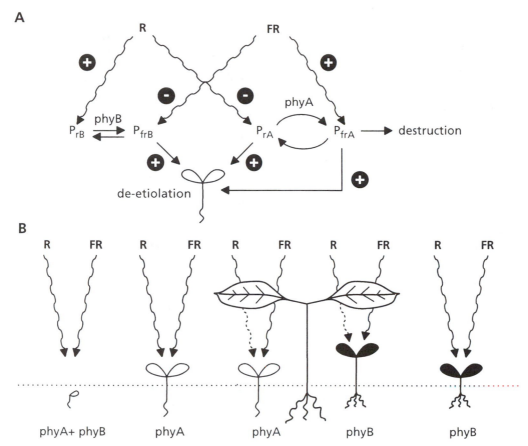

A

B

phyA+ phyB   phyA   phyA   phyB   phyB

FIGURE 1. Transduction of partly antagonistic light signals by phytochrome A (phyA) and phytochrome B (phyB) with respect to seedling development (modified after Quail et al. 1995, with information from Smith 1995 and Furuyama & Schäfer 1996). (A) Simplified scheme showing the action of red (R) and far-red (FR) radiation (wavy lines) absorbed separately by phyA and phyB. R absorbed by phyB induces de-etiolation (+) through the maintenance of high amounts of P_{frB}. This response is reversed by FR (–) (low fluence response; LFR). Very low fluences of R absorbed by phyA induce de-etiolation, probably through tiny amounts of P_{frA} (+) (very low fluence response; VLFR). High fluences of continuous FR also induce de-etiolation through phyA (+) (high irradiance response; HIR). The mechanisms of HIR is not fully understood. Transduction of high irradiance FR may either proceed through the maintenance of low levels of the labile P_{frA} or high levels of P_{rA} depending on the active species of phyA. High fluences of R then counteract the FR effect (–) by creating high levels of P_{frA} that degrade rapidly, thus depleting the phyA pool, or by creating a low level of P_{rA}. (B) Simpli-

fied scheme of apparent photosensory functions of phyA and phyB in seedling development in the natural environment. Under the soil surface in total darkness, the phytochromes are present in the P_r form; phyA in large amounts. The upward growing etiolated seedling, still under the soil surface, meets the first daylight. The R component of daylight creates small amounts of P_{frA} which triggers the VLFR. At somewhat higher irradiances closer to the soil surface, sufficient P_{frB} is formed for action of the LFR; both induce de-etiolation responses (left). Once above the surface, high irradiances and still large amounts of phyA, cause a predominant action of phyA through the HIR in the further de-etiolation process. This is independent of whether the seedling emerges in exposed or shaded places (nr 2 and 3 from left). The photolabile phyA is rapidly depleted and the FR-HIR response is reduced. PhyB dominates the regulation of stem extension in the de-etiolated seedling. The FR-enriched shade light causes increased stem extension through low levels of P_{frB} and/or high levels of P_{rB} (2 seedlings, at the right).

TABLE 3. Effects of the irradiance level on growth parameters of a sun-adapted species, *Helianthus annuus* (sunflower), and a shade-adapted species *Impatiens parviflora* (touch-me-not).*

Growth parameter	Relative irradiance level				
	100	50	22	10	5
Net assimilation rate ($g\,m^{-2}\,day^{-1}$)					
Impatiens	8.7	7.4	4.4	2.9	1.7
Helianthus	9.7	7.9	4.1	2.9	0.7
Specific leaf area ($m^2\,kg^{-1}$)					
Impatiens	32	42	53	71	80
Helianthus	26	32	43	41	36
Leaf mass ratio[1] ($g\,g^{-1}$)					
Impatiens	0.41	0.43	0.44	0.45	0.45
Helianthus	0.61	0.57	0.54	0.47	0.46
Leaf area ratio ($m^2\,kg^{-1}$)					
Impatiens	13.2	18.0	23.5	32.0	36.0
Helianthus	16.4	19.0	22.0	19.0	17.0
Relative growth rate ($mg\,g^{-1}\,day^{-1}$)					
Impatients	114	133	104	91	61
Helianthus	157	144	90	53	13

Source: Evans & Hughes 1961, and Hiroi & Monsi 1963, as cited in Björkman 1981.
*Daily irradiances (100% values) were similar for both species: 40–45 $mol\,m^{-2}\,day^{-1}$.
   Note that there is a small increase in LMR with decreasing irradiance for one of the species and a decline in the other; in wider comparisons, the general trend is for LMR to increase with decreasing irradiance.

early stage. For example, in deciduous trees, this may already be determined the year before new leaves appear by the light climate to which the leaves on similar positions in the canopy were exposed. Some parameters (e.g., SLA) may still change in a mature leaf, but others, such as cell size and number, are more rigid. Even these parameters, however, are still rather plastic for the long-lived leaves of *Hedera helix* (ivy). Shade leaves of this species can make an additional layer of palisade parenchyma upon exposure to increased irradiance (Bauer & Thöni 1988).

5.1.1.2 Adaptations to Shade

Apart from **acclimation** to a specific light environment, there are also specific **adaptations**. That is, there are species with a genetic constitution that restricts their distribution to an environment with a specific light climate. To put it simply, three "plant strategies" are discerned (Smith 1981):

1. Plants avoiding shade, or obligate sun plants
2. Plants tolerating shade, or facultative sun or shade plants
3. Plants requiring shade, or obligate shade plants

Many weedy species and most crop species are **obligate sun species. Obligate shade plants** include some mosses, ferns, club mosses, and a few higher plant species in tropical rainforests (e.g., young individuals of *Monstera* and *Philodendron*). Among higher plants, obligate shade species are rare in temperate regions and will not be discussed here. Most understory species are **facultative** rather than obligate shade plants.

5.1.1.3 Stem and Petiole Elongation: The Search for Light

Stem and petiole elongation of **shade-avoiding** plants growing in the shade are greatly enhanced, branching is reduced (increased apical dominance), total leaf area and **leaf thickness** are less, and **SLA** is increased. The effects of leaf canopy shade can be separated into those due to **reduced irradiance** and those affected by the **red/far-red ratio**.

   Plants that tolerate shade do not respond with increased stem elongation; instead, they increase their leaf area. Their leaf thickness is reduced to a smaller extent than it is in shade-avoiding species, and their chlorophyll concentration per unit leaf area often increases. The increased chlorophyll concentration gives these plants (e.g., *Hedera* spp.,

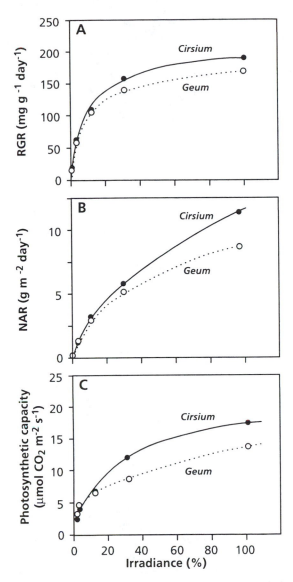

FIGURE 10. The relative growth rate (RGR) (A), net assimilation rate (NAR) (B), and photosynthetic capacity (C) of the shade-avoiding species *Cirsium palustre* and the shade-tolerant species *Geum urbanum*, grown at a range of light intensities. Full daylight is 100% (Pons 1977). Copyright Royal Botanical Society of the Netherlands.

centration per unit area is not enhanced (sometimes even less), and they do not appear dark-green.

The **red/far-red ratio** (R/FR) is the ratio of the irradiance at 655 to 665 nm and that at 725 to 735 nm. Comparison of a number of species from open habitats [e.g., *Chamaenerion angustifolium* (fireweed), *Sinapis alba* (white mustard), *Senecio vulgaris* (groundsel)], from intermediate habitats [*Urtica dioica* (stinging nettle)], and from closed habitats (shade in forest understory) [*Geum urbanum* (avens), *Oxalis acetosella* (soursop), *Silene dioica*] shows that the stem elongation of sun-adapted species responds much more strongly to R/FR than does that of shade species. The effect of a change in R/FR on stem elongation can be recorded within 10 to 15 minutes, as shown for *Sinapis alba* (white mustard) in Figure 11.

5.1.1.4 The Role of Phytochrome

Perception of R/FR involves the **phytochrome** system (Box 9 and Sect. 2.2.2). In *Vigna sinensis* (cowpea) the response of **stem elongation** to R/FR is similar to that of **gibberellins**. It is very likely that the phytochrome effect is mediated through increased levels of or sensitivity to gibberellins and that it is entirely due to cell elongation rather than to cell division (Martinez-Garcia & Garcia-Martinez 1992). These phytochrome responses clearly demonstrate that many of the light responses of shade plants are hormonally mediated (sink-controlled) rather than direct responses to irradiance level.

5.1.1.5 Phytochrome and Cryptochrome: Effects on Cell-Wall Elasticity Parameters

Both red light and **blue light** inhibit stem elongation. A blue photoreceptor (**cryptochrome**) is involved in the perception of blue light. Both red light and blue light affect cell-wall properties rather than the osmotic or turgor potential of the cells (Table 4). As explained in Section 2.2, stem elongation is the result of cell expansion ($dV/V.t$), which is related to the cell-wall yield coefficient, the turgor pressure, and the yield threshold. Red light inhibits elongation mainly by lowering the **cell-wall yield coefficient** (ϕ), whereas blue light predominantly acts by enhancing the **yield threshold** (Y) (Table 4). This indicates that shade affects growth through **feedforward responses** rather than through direct supply of photosynthate.

ivy; and species from the understory of tropical rainforests) their dark-green color. Less extreme shade-tolerant species (e.g., *Geum urbanum*, avens) also enhance their chlorophyll concentration per unit fresh mass. Because their SLA is increased at the same time, however, the chlorophyll con-

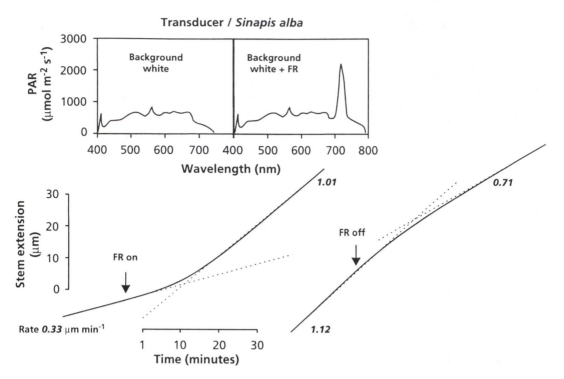

FIGURE 11. Continuous measurements of stem extension rate by a position-sensitive transducer. A seedling attached to the transducer and exposed to background white fluorescent light was given far-red (FR) light via a fiberoptic probe. The FR source was switched on and off as indicated. The insets show the spectral composition of the irradiance of the background white light with and without FR (data of D.C. Morgan, as presented in Smith 1981).

5.1.1.6 Effects of Total Level of Irradiance

The total level of irradiance is the major factor that determines the LAR and SLA of shade-avoiding species, although the spectral composition of the irradiance also has an effect in some species. Shade-avoiding species respond to the spectral composition in the shade primarily with enhanced stem elongation, at the expense of their leaf mass ratio. Shade-tolerant species tend to invest relatively more resources in their leaves when exposed to shade, primarily as a response to the level of irradiance (Smith 1981).

There is no evidence that these responses to the level of irradiance are a direct consequence of photosynthate supply enhancing the raw materials for growth. They are most likely mediated through **sugar-sensing systems** (Sects. 4.3 and 12.1 of the chapter on photosynthesis and Sect. 4.4 of the chapter on respiration).

TABLE 4. Effects of darkness, red light, and blue light on in vivo cell-wall properties of stems of etiolated pea (*Pisum sativum*) seedlings.*

	Dark	Red light	Blue light
Elongation rate, $\mu m\,s^{-1}$	9.2	3.3	3.0
Turgor potential, MPa	0.53	0.59	0.58
Osmotic potential, MPa	0.84	0.82	0.83
Yield threshold (Y), MPa	0.05	0.16	0.33
Yield coefficient (ϕ), $Pa^{-1}s^{-1}$	19.1	7.8	15.6

Source: Kigel & Cosgrove 1991.

* In darkness the P_{fr} configuration of phytochrome reverts to the P_r configuration.

5.1.2 Effects of the Photoperiod

The length of the photoperiod affects the flowering response of long-day and short-day plants (Sect. 3.3.1 of the chapter on life cycles) as well as aspects of vegetative plant development that are not directly related to reproduction. These effects are mediated by the **phytochrome** system and differ from those that result from changes in the total level of irradiance received by the plants.

For temperate-zone species, acclimation to low temperatures (**cold hardening**) is important, especially in woody species. The length of the photoperiod is an important signal in this acclimation process. In a Norwegian ecotype of *Dactylis glomerata* the dry matter production is enhanced under long days at low temperature, compared with short days at the same low temperature (Fig. 12). In a Portuguese ecotype at higher temperatures, photoperiod has little effect. The greater production at a low temperature and long days in the Norwegian ecotype reflects a higher RGR because of a higher LAR, which itself reflects a higher SLA. The net assimilation rate is reduced in long days, at all temperatures and in both ecotypes (Fig. 12). Leaves tend to be thinner in long days and their cells are longer. It is common for populations of a species from different latitudes to differ in their photoperiodic cues, indicating that changes in photoperiodic requirement is a relatively easy evolutionary adjustment that is differentially selected at different latitudes.

There is some evidence that increased levels of **gibberellins** and/or an enhanced sensitivity to these hormones are involved in the growth response of grasses to long days (Heide et al. 1985). The photoperiod also affects the plant's chemical composition, again independent of the total level of irradiance received by the plant. The percentage of total nitrogen in dry matter declines with increasing photoperiod, which is the likely cause for a decrease in NAR at long days (Fig. 12). The nitrate concentration also declines (Bakken 1992).

5.2 Growth as Affected by Temperature

Temperature affects a range of enzymatically catalyzed and membrane-associated processes in the plant and is a major factor affecting plant distribution. The **activation energy** of different reactions may differ widely. Growth, development, and allocation are affected in different ways in different species. The temperature optimum of root growth tends to be lower than that of the shoot. In spring, therefore, roots start growing before the leaves do. Temperature also affects the uptake of nutrients and water by the roots. The optimum temperature for root growth of plants from temperate regions is between 10 and 30°C, but growth may continue around 0°C. Subtropical species have a higher optimum temperature for root growth, and growth may cease below 10 to 15°C (Bowen 1991). In tropical species damage may occur at temperatures of 12°C or less. How exactly does a low temperature affect root and leaf growth and the pattern of allocation to roots and leaves? This is a highly relevant question, in view of the current rise in global temperature.

5.2.1 Effects of Low Temperature on Root Functioning

Exposure to a low temperature reduces **root extension**, without an effect on turgor in the elongation zone. In *Zea mays* (maize) the reduction in elongation rate is associated with a decrease in cell-wall extensibility, more specifically in the **cell-wall yield coefficient**. Reduced elongation may lead to an increased number of rather small cells, immediately behind the root tip. These resume expansion upon exposure of the roots to a more favorable temperature (Pritchard 1994).

For a proper functioning of roots at low temperature, their membranes must remain fluid and semipermeable. The **lipid composition** of the membranes in the roots affects membrane fluidity and interactions with membrane-bound proteins, and, therefore, the transport of both ions and water. Cold-acclimated plants tend to have a higher degree of unsaturation of phospholipids, which causes their membranes to remain fluid at lower temperatures.

The major resistances for **water flow** in the roots are probably in the **exodermis**, if present, and the **endodermis**. At the exodermis or endodermis, water must enter the **symplast** before it can arrive in the xylem vessels. Water passes the membranes in a single file through specific **water-channel proteins** (Sect. 5.2 of the chapter on plant water relations). The effect of temperature on the rate of water uptake by roots, therefore, possibly reflects direct effects on these water-channel proteins and indirect effects on membrane fluidity.

The effects of temperature on the roots' capacity to absorb water largely account for temperature ef-

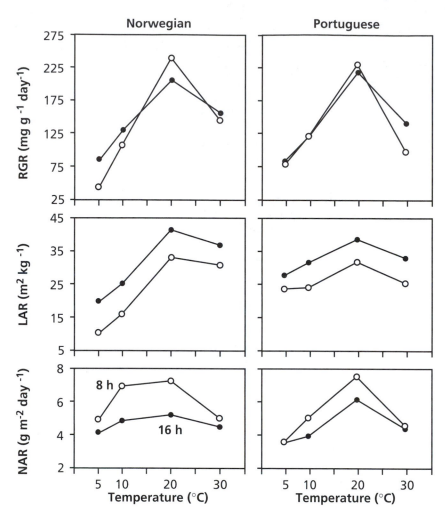

FIGURE 12. Results of a growth analysis of seedlings of *Dactylis glomerata* (cocksfoot) from two different origins at four temperatures and two daylengths (Eagles 1971, as cited in Hay 1990). Copyright Trustees of The New Phytologist.

fects on plant growth. Increasing the root temperature of *Glycine max* (soybean) in the range that is suboptimal for growth, while maintaining a constant shoot temperature, increases the water potential of the whole plant (Kuo & Boersma 1971). It is likely that the effects of temperature on the relative investment of biomass in roots and leaves reflect the roots' ability to take up water, at least in the range of temperatures around the optimum (Li et al. 1994).

Does this imply that effects of temperature on the allocation pattern are accounted for by an effect of root temperature on the roots' capacity to transport water and that temperature effects on nutrient uptake are not a cause for changes in the allocation pattern? Current evidence does indeed support this contention. Whereas growth at a low root temperature does affect the rate of absorption of both nitrate and ammonium, this appears to be a response to the decline in growth rate (Clarkson et al. 1992). That is, the decline in the rate of nutrient absorption at low root temperatures is, in part, a response to the decreased nutrient demand of the plant (Sect. 2.2.3.2 of the chapter on mineral nutrition).

5.2.2 Changes in the Allocation Pattern

Variation in growth rate with temperature is associated with changes in plant carbon balance. A positive carbon balance can be maintained at adverse temperatures by changes in the pattern of resource allocation to leaves and nonphotosynthetic plant parts. Acclimation to different temperatures, therefore, may affect the rate of photosynthesis per unit leaf area (Fig. 25 in the chapter on photosynthesis) or the plant's allocation pattern. In very general terms, the effect of temperature on biomass allocation in the vegetative stage is that the relative investment of biomass in roots is lowest at a certain optimum temperature and that it increases at both higher and lower temperatures (Fig. 13). This is found both when the temperature of the entire plant is varied and when only root temperature is changed (constant shoot temperature) (Bowen 1991).

It has been suggested that an increase in root temperature in the suboptimal range increases the demand for respiratory substrate in roots, which results in lower carbohydrate concentrations in the whole plant or in the shoots. These effects of root temperature on root respiration, however, are often only transient, with values returning to control rates within 1 day.

Temperature strongly affects the uptake of both nutrients and water by the roots. Although **nutrient uptake** does depend on root temperature, at least in short-term experiments, it is unlikely that long-term temperature effects on biomass partitioning are due to effects on nutrient uptake. Upon prolonged exposure to low root temperature, the uptake system acclimates (Chapin 1974, Clarkson et al. 1988); there is compelling evidence that, at a low root temperature, **growth controls the rate of nutrient uptake**, rather than being controlled by it (e.g., Clarkson 1986, Clarkson et al. 1990). Effects of root temperature, through the plant's water relations, are probably mediated by **ABA** (Sect. 5.3), but further evidence is needed to support this contention.

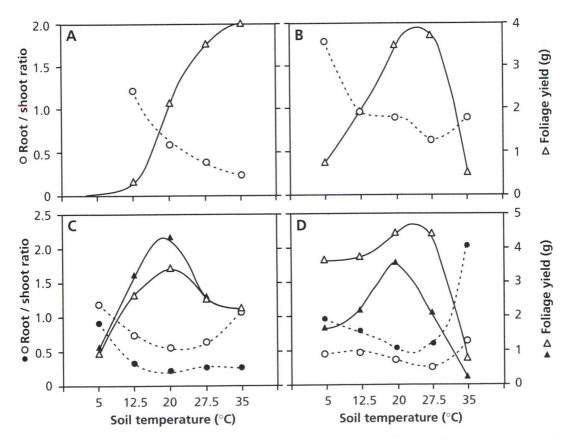

FIGURE 13. Biomass in roots relative to that in the shoots (circles, broken line) and foliage yield (triangles, solid line), as a function of root temperature. (A) Subtropical species; (B) warm-temperate species; (C) cool-temperate species; (D) winter annual species (various authors, as cited in Bowen 1991). By courtesy of Marcel Dekker, Inc.

There are also indirect effects of temperature on nutrient availability, in that rates of mineralization decline at low temperatures (e.g., in arctic and alpine environments).

5.3 Growth as Affected by Soil Water Potential and Salinity

Many processes in the plant are far more sensitive to a low water potential than are stomatal conductance and photosynthesis. The growth reduction at a low soil water potential is largely due to inhibition of more sensitive processes, such as **leaf cell elongation** and **protein synthesis**. At a low soil water potential, the rate of leaf expansion decreases, whereas the rate of **root elongation** is much less affected (Fig. 14). In glycophytes (e.g., *Zea mays*) root elongation is inhibited by exposure to high concentrations of NaCl. It is interesting that this inhibition is not associated with a loss of turgor of the growing tip, but with an **increased yield threshold pressure** (Neumann et al. 1994). Current evidence indicates that salinity-induced inhibition of root elongation in *Zea mays* is not associated with a decreased capacity to acidify the cell walls (Zidan et al. 1990).

Maintenance of root elongation at a low soil water potential may occur despite a (transient) decline in turgor of the root cells, suggesting that the yielding ability of the elongating cells has increased. Two processes account for this increased yielding ability: The **amount** and **activity** of **expansins** in the root tip of plants grown at low soil water potential is increased, and the **susceptibility** of the cell wall to expansin is increased (Wu et al. 1996).

It may be tempting to believe that the reduction in leaf expansion as shown in Figure 14 is due to a loss in turgor of the leaf cells. Such a turgor loss, however, is often not found, and the reduction in leaf growth is due to leaf cell-wall stiffening (Van Volkenburgh & Boyer 1985) in response to (**chemical**) **signals** arriving from the roots in contact with the drying soil (Davies & Zhang 1991). How do we know that chemical signals play a role?

5.3.1 Do Roots Sense Dry Soil and Then Send Signals to the Leaves?

To answer this question, Passioura (1988) used a pressure vessel placed around the roots of a *Triticum aestivum* (wheat) seedling growing in drying soil. As the soil dried out, the hydrostatic pressure in the vessel was increased to maintain shoot water relations similar to those of well-watered

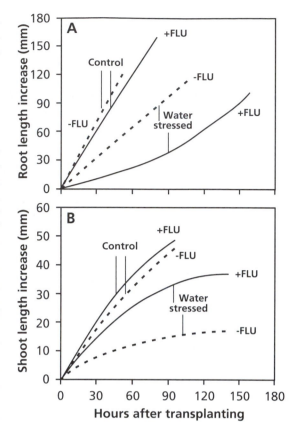

FIGURE 14. Elongation of the primary root and shoot of *Zea mays* (maize) seedlings that were well-watered or grown at a low water potential. Also shown is the effect of fluridone, which is an inhibitor of the synthesis of ABA. (A) root growth of seedlings soaked in water for 36 h and then transplanted to −1.6 MPa; (B) shoot growth of seedlings soaked in water for 60 h and then transplanted to −0.3 MPa (Saab et al. 1990). Copyright American Society of Plant Physiologists.

plants. The treated wheat plants showed reductions in leaf growth similar to those of plants in drying soil outside a pressure chamber. Additional evidence has come from experiments with small apple trees (*Malus x domestica*) with their roots growing in two separate containers, one with moist and one with dry soil. Soil drying in one container restricted leaf expansion and initiation, although the roots in the moist soil continued to maintain shoot water relations similar to those of control plants. Leaf growth recovered upon severing the roots in contact with the drying soil (Gowing et al. 1990). These effects on leaves of wheat seedlings and apple trees must therefore be attributed to effects of soil drying that do not require a change in shoot water status (Davies et al. 1994).

As with effects of soil drying on stomatal conductance (Sect. 5.4.1 in the chapter on plant water relations), **hydraulic signals**, in addition to **chemical messengers** from the roots, probably play a role in effects of drying soils on leaf growth (Dodd & Davies 1996). Thus, there are multiple signal-transduction pathways by which water shortage reduces plant growth.

5.3.2 ABA and Leaf Cell-Wall Stiffening

The effect of water stress on leaf growth is mediated by the phytohormone abscisic acid (**ABA**) (Munns & Cramer 1996). Soil drying and salinity enhance the concentration of this hormone in the leaves (Tardieu et al. 1992, He & Cramer 1996).

Above-ground plant parts respond more strongly to a decreased soil water potential than do roots. This is due to a greater **inhibition by ABA of leaf growth**, as compared with that of the roots (Saab et al. 1990). If, and to what extent, ABA is responsible for the decline in cell-wall acidification upon water stress (Van Volkenburgh & Boyer 1985) and acid-induced wall loosening (Cleland 1967) remains to be investigated (Munns & Sharp 1993). We do know that leaves tend to have stiffer walls when the plants are exposed to water stress (Chimenti & Hall 1994). The leaves also show higher endogenous ABA concentrations and reduced leaf growth. ABA probably affects the growth of roots and leaves through its effect on **ethylene** biosynthesis.

Salt-sensitive species respond more strongly, both in terms of ABA level and in leaf expansion, than do resistant species (Fig. 15). ABA seems to harden the cell wall of leaf cells by increasing the yield threshold, Y, and decreasing wall extensibility, ϕ. Both the carbohydrate and the protein component of cell walls are affected (Munns & Cramer 1996).

5.3.3 Effects on Root Elongation

Roots that experience a moderate water stress may loosen their walls and increase their extension growth rate. **Wall loosening** is possibly due to an ABA-induced increase in activity of **XET**, which is the enzyme putatively involved in wall loosening (Sect. 2.2.4) (Pritchard 1994). An increased wall-loosening capacity in response to water stress is widespread and is presumably an adaptation to growth in drying soils that allows exploitation of a falling water table. As in leaves, osmotic stress has no effect on the **turgor** of *Zea mays* (maize) root cells; however, it increases the **concentration of osmotic solutes** to the extent that the difference in cell water potential and that of the root environment is restored (Pritchard et al. 1996).

Lowering the water potential around the roots also enhances sugar transport to the roots, probably due to the growth reduction of the leaves. Because photosynthesis is less affected than leaf growth, sugar transport as well as root growth may be enhanced in both a relative and an absolute sense, at

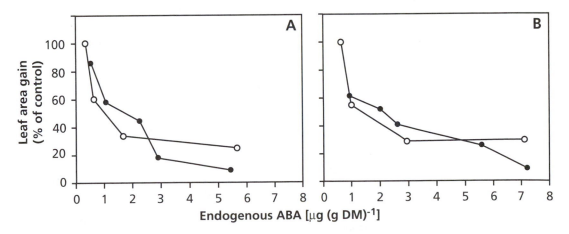

FIGURE 15. Leaf expansion of the more salt-resistant *Brassica napus* (A) and the less resistant *Brassica carinata* (B) as affected by endogenous ABA. Plants were either grown on nutrient solution (open symbols) or on a nutrient solution to which 5.2 g l^{-1} "Instant Ocean" (largely NaCl) was added (filled symbols). The variation in endogenous [ABA] was due to the salt treatment and to the addition of ABA to the nutrient solution: 0, 1, 10 and 80 μM for the control, and 0, 1, 10, 20, and 80 μM for the salt-treated plants. Note that the endogenous ABA level was higher for the salt-treated plants when no ABA was added to the nutrient solution and that the two curves are similar for control and salt-treated plants (He & Cramer 1996). Copyright Kluwer Academic Publishers.

least in the early stages of the stress. The unresolved question, however, remains: How does an increased concentration of sugars affect the growth of roots? This probably requires a sugar-sensing mechanism similar to the one discussed for leaves where a specific hexokinase senses hexose levels and affects the repression of genes that encode photosynthetic enzymes (Sect. 4.3 of the chapter on photosynthesis). Gene transcription in roots *is* affected by sugar levels, as discussed for respiratory enzymes in Section 4.4 of the chapter on plant respiration, but we are still searching for genes that affect root elongation. We are still far from understanding the entire signal–transduction pathway from elevated sugar levels in roots cells on the one hand to stimulation of root elongation on the other. This is clearly a major challenge for molecular ecophysiologists!

5.3.4 A Hypothetical Model That Accounts for Effects of Water Stress on Biomass Allocation

The effects of water stress on phytohormone production in the roots, leaf expansion, and root growth are tentatively summarized in Figure 16.

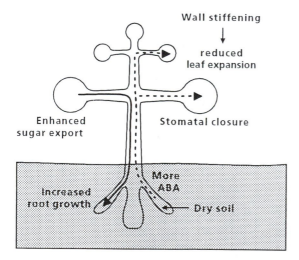

FIGURE 16. Hypothetical model to account for the effects of water stress on plant growth and biomass allocation. Roots sensing dry soil enhance the export of ABA, which is transported in the xylem and moves to the leaves. Here, ABA reduces stomatal conductance and wall extensibility of growing cells. The effects are a reduction in the rate of transpiration and photosynthesis as well as in leaf expansion. As long as photosynthesis is affected less than leaf expansion, the export of assimilates to the roots is enhanced. The increased import of assimilates, in combination with ABA-enhanced wall loosening of growing root cells may enhance the rate of root growth.

Whatever the exact signal–transduction pathway, the overall effect of inhibition of leaf area expansion while root elongation is inhibited less, or even stimulated, is that the LAR and/or the LMR decrease and that the root mass ratio (RMR) increases in response to a decrease in soil water potential. The increased respiratory costs of such an increase in RMR may contribute to reduced growth of droughted plants; they also reduce the dry mass gain per unit of water lost in transpiration (Van den Boogaard et al. 1996).

5.4 Growth at a Limiting Nutrient Supply

It is well-documented that plants allocate relatively less biomass to leaves and more to their roots when nitrogen or phosphate are in short supply (e.g., Brouwer 1963, 1983). Like the response to water stress (Sect. 5.3), the response to nutrient shortage is also functional. In both situations the investment in plant parts that acquire the limiting resource is favored at the expense of allocation to plant parts that have a requirement for the limiting resource. As we have seen in Section 5.1, the opposite and equally functional response is found when plants are growing at a low irradiance.

In this section we will concentrate on the response to **nitrogen shortage** because the effect of nitrogen shortage on biomass allocation is stronger than that of other nutrients. Phosphate may have similar effects, possibly acting through an effect on nitrogen acquisition (Kuiper et al. 1989). This may also be the case for sulfate, whereas the pattern is less clear for other nutrients. Leaf expansion rates are decreased at a low nitrogen supply (Gastal et al. 1992). Leaves of plants grown with a limiting nitrogen supply are smaller, compared with those of plants grown with an optimum nutrient supply, predominantly due to an effect on **meristem size** and **cell number** (Fig. 2b) (Terry 1970). How are the changes in biomass allocation pattern brought about?

5.4.1 Cycling of Nitrogen Between Roots and Leaves

In vegetative plants, whether grown with an optimum or a limiting supply of nitrogen, much of the nitrogen transported from the roots via the xylem to the leaves is exported back to the roots via the phloem (Fig. 17). Such a process of continuous **cycling** of nitrogen between roots and leaves makes it highly unlikely that the transport of nitrogen to the leaves itself can be a controlling factor. Rather, we

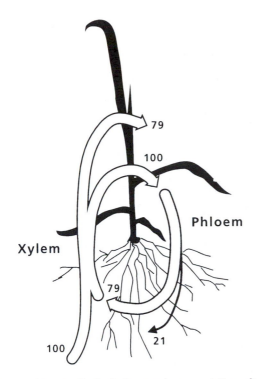

79

100

Phloem

Xylem

79

100   21

FIGURE 17. "Cycling" of nitrogen in a vegetative wheat plant (*Triticum aestivum*, wheat). Much of the nitrogen (nitrate, amino acids, and amides) that arrives in the leaves via the xylem is exported again in the phloem (amino acids and amides). Upon arrival in the roots, some of the nitrogen may be used for root growth, whereas the remainder cycles back to the shoot (Simpson et al. 1982a). Copyright Physiologia Plantarum.

should search for signals that change concomitantly with the nitrogen supply.

5.4.2 Hormonal Signals That Travel via the Xylem to the Leaves

The response of plants to a low nutrient supply is akin to that to a limiting supply of water: reduced leaf growth while root growth is maintained or enhanced. This response is generally described in terms of a **functional equilibrium** between leaves and roots (Brouwer 1963, 1983). That is, when resources that have to be acquired by the roots are in short supply, the growth of the roots is favored over that of the leaves. It is interesting that nitrogen deficiency reduces the roots' **hydraulic conductivity** (Chapin et al. 1988, Radin & Boyer 1982), but this is not associated with a decline in the turgor of leaf cells (Palmer et al. 1996). The rapid decline (within hours) in leaf growth of *Zea mays* (maize) upon transfer to a low-nutrient solution is associated

with a decreased extensibility of the cell walls of expanding leaf cells. Transfer to high-nutrient conditions enhances this extensibility. The transfer has no effect on the osmotic potential of the leaf cells or on cell production (Snir & Neumann 1997).

Contrary to what has been found for plants exposed to water stress, there is no evidence that ABA plays a role as a signal between roots and leaves of plants exposed to a nutrient supply that is limiting to plant growth (Munns & Cramer 1996). Rather, a reduced nutrient supply to the roots reduces the synthesis of **cytokinins** in the root tips and their subsequent export to the leaves (Fetene & Beck 1993, Van der Werf & Nagel 1996). Nitrogen appears to be the predominant nutrient that leads to this response (Kuiper et al. 1989). Due to the lower cytokinin import into leaves of plants grown with a limiting nitrogen supply, growth of the leaves is reduced (Simpson et al. 1982b). Because cytokinins affect the growth of leaves and roots in an opposite manner (Sect. 2.2.2), root growth is either stimulated or unaffected by a low nitrogen supply.

In plants grown with a limiting supply of nitrogen, the level of cytokinin can be maintained, by the addition of benzyladenine, a synthetic cytokinin, to the roots (Table 5). It is interesting that this maintains the RGR of the leaves of plants transferred to a low nutrient supply to a rate close to that in plants grown with a full nutrient supply; of course, this effect can only last for a few days, after which the plants start to collapse. Addition of cytokinin reduces the root growth to the level of plants well-supplied with nutrients.

What kind of effects do cytokinins have on leaf metabolism? First, cytokinins promote the synthe-

TABLE 5. Cytokinin (zeatin) concentrations (pmol g^{-1} FM) and the relative growth rate (RGR, mg g^{-1} day^{-1}) of *Plantago major* plants, exposed to a full-nutrient solution, or transferred to a diluted solution, plus or minus 10^{-8}M benzyladenine (BA), a synthetic cytokinin.

Treatment	Cytokinin concentration		Growth (RGR)	
	shoot	roots	shoot	roots
Full nutrients	110	160	220	160
Diluted solution				
Without BA	25	23	150	180
with BA	100	140	190	160

Source: Kuiper & Staal 1987 and Kuiper et al. 1989.

sis of several proteins that are involved in photosynthesis. They also have a specific effect on a gene encoding a protein involved in the cell cycle and hence promote cell division. **Cytokinins** also promote cell expansion (Sect. 2.2.2). To put it simply, cytokinins promote **leaf cell division** and **leaf cell expansion**, increase the **photosynthetic capacity**, **delay leaf senescence**, and enhance **leaf expansion**. Thus, as with water and temperature, nutrient supply governs growth through hormonal signals (**feedforward control**) rather than through a direct effect on the availability of substrates for protein synthesis (source control). The hormonal signals that regulate growth in response to nutrient shortage (cytokinins), however, differ from those associated with water and salinity stress (ABA) and light shortage (phytochrome-induced changes in gibberellins).

5.4.3 Signals That Travel from the Leaves to the Roots

Leaves that experience a low import of nutrients probably send signals back to the roots, which accounts for their enhanced growth. What is the nature of these signals? The signal might well be the amount of **carbohydrates** exported via the phloem (Lambers & Atkin 1995). When the low nutrient

supply reduces leaf growth, products of photosynthesis accumulate. These probably affect the **sugar-sensing mechanism** (Sect. 4.3 of the chapter on photosynthesis). Genes encoding photosynthetic enzymes are subsequently suppressed, leading to down-regulation of photosynthesis. The increased level of carbohydrates in the leaves, however, implies that more photosynthate is available for translocation to the roots. There it may act as a signal and affect sugar-sensing mechanisms. Rather than suppressing genes, it is likely to derepress genes coding for respiratory enzymes (Sect. 4.4 of the chapter on plant respiration) and possibly others (Farrar 1996).

5.4.4 Integrating Signals from the Leaves and the Roots

The results presented in Section 5.4.2 lead to the hypothetical model depicted in Figure 18 (Van der Werf & Nagel 1996). An early response of a plant to a decline in the nitrogen supply is the decrease in synthesis and export of **cytokinins**. This reduces the rate of protein synthesis, cell division, and expansion in the growing leaves. Carbohydrates accumulate, leading to suppression of photosynthetic genes and down-regulation of photosynthesis. Plenty of carbohydrates are available for export to

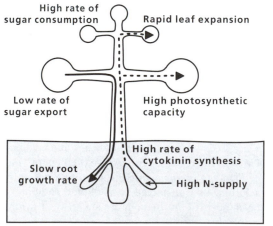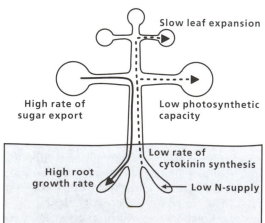

FIGURE 18. Hypothetical model to account for the effects of nitrogen supply on plant growth and biomass allocation. (Left) Roots sensing a high availability of nitrogen produce large amounts of cytokinins, which are exported via the xylem to the leaves. Here, the cytokinins enhance the photosynthetic capacity and leaf expansion. Hence, a large fraction of the photosynthates are consumed in the leaves, and a relatively small fraction is available for export to the roots. (Right) Roots sensing a low availability of nitrogen produce only small amounts of cytokinins. The import of

cytokinins into leaves is small, so that their photosynthetic capacity and rate of leaf expansion is reduced. Only a small fraction of the photosynthates are consumed in the leaves, so that the concentration of sugars in the leaves is high and a relatively large fraction is available for export to the roots. The high level of sugars in leaves suppresses genes encoding photosynthetic enzymes. In roots, high sugar levels induce genes coding for respiratory and possibly other enzymes.

the roots. In the roots they derepress genes that encode respiratory enzymes and possibly other enzymes. The roots may either grow at the same rate as those of control plants, or their growth may be increased (Van der Werf 1996).

It appears that the relative increase in biomass allocation to roots with nitrogen shortage is largely accounted for by the decrease in production of cytokinins in the roots. This phytohormone then sets in motion the change in biomass partitioning, which leads to a new **functional equilibrium** between roots and leaves. It would seem that roots have very little *direct* control over the rate of carbon import from the leaves. They do exert *indirect* control, however, via their effect on leaf growth, which depends on the supply of cytokinins from the roots.

The lack of direct control over carbon import into the roots might require a respiratory system in the roots that both functions to generate ATP and allows the oxidation of sucrose to proceed with little ATP production (see Sect. 2.3.2 of the chapter on plant respiration). Convincing evidence to support this contention, however, will have to come from experiments with genotypes that lack alternative path activity.

5.4.5 Effects of Nitrogen Supply on Leaf Anatomy and Chemistry

In a comparison of four congeneric grass species (*Poa annua*, *Poa trivialis*, *Poa compressa*, and *Poa pratensis*) grown at both an optimum and a limiting nitrogen supply, RGR and nitrogen concentrations decrease with low nitrogen supply (Van Arendonk et al. 1997). The decrease in RGR is accounted for by the decrease in **LAR**; both **SLA and LMR**. The changes are largest in the fastest-growing *Poa annua*. The anatomical basis of the decrease in SLA and of changes in chemical composition has also been analyzed, using transverse leaf sections of all four *Poa* species. Nitrogen shortage invariably enhances the proportion of leaf tissue that is occupied by **sclerenchymatic cells**, from about 0.5 to 6%, predominantly due to an increase in number of these sclerenchymatic cells. The area occupied by **veinal tissue** doubles, from approximately 4.5 to 9%, whereas that occupied by epidermal cells is more or less constant (25%), despite a substantial decrease in **size of the epidermal cells**, especially in *P. annua*. Mesophyll + intercellular spaces occupy a variable area of about 60% in all species and treatments. Nitrogen stress decreases the concentration of protein and enhances that of (hemi) cellulose and lignin.

It is not known whether cytokinins are involved in the control of these anatomical and chemical features by nutrient supply. These anatomical changes, however, are probably ecologically important in that the increase in sclerenchymatic and veinal tissue is likely to give better protection of the leaves (Lambers & Poorter 1992).

Nitrogen shortage also has a major effect on allocation to nonstructural secondary metabolites such as **lignin and tannins** (Sect. 4.1 of the chapter on ecological biochemistry). Because these compounds slow down the rate of litter decomposition, this response aggravates the nitrogen shortage in the environment (Sects. 2 and 3 of the chapter on decomposition).

5.4.6 Nitrogen Allocation to Different Leaves, as Dependent on Incident Irradiance

Different leaves of a plant may differ widely with respect to their nitrogen concentration. This may be due to the **nitrogen withdrawal** from older, senescing leaves (Sect. 4). It is more interesting in the present context that leaves adjust their nitrogen concentration to the **level of incident irradiance**. That is, leaves at the top of the canopy that are exposed to full daylight have higher concentrations per leaf area than leaves near the ground surface, where they are shaded by higher leaves (Hirose & Werger 1987a).

To appreciate the functional significance of the correlation between leaf nitrogen concentration and the level of irradiance the leaf is exposed to, we have to recall that the lion's share of leaf nitrogen is associated with the photosynthetic apparatus (Sect. 3.2.3 of the chapter on photosynthesis). Investment of greater amounts of nitrogen in the leaf will lead to a greater photosynthetic capacity, but this capacity can only be fully used when levels of irradiance are sufficiently high. Because levels of irradiance are higher for the top leaves than they are for the bottom ones, it would appear to be advantageous for a plant to have a **gradient in leaf nitrogen concentrations** rather than a uniform distribution (Hirose & Werger 1987b, Pons et al. 1989, Field 1991). Mathematical models have been developed to assess the significance of a gradient in leaf nitrogen concentration, as opposed to a uniform distribution (Box 7 in the chapter on scaling-up).

What might be the plant's mechanism to achieve a nitrogen gradient that tends to follow the gradient of irradiance in the canopy? Leaves exposed to higher levels of irradiance, high in the canopy, will have higher rates of transpiration than the shaded ones lower in the canopy. This occurs partly because stomata respond to the level of irradiance (Sect. 5.4.4 of the chapter on plant water relations), partly because of the greater vapor pressure difference between leaf and air higher in the canopy, and possibly also because the temperature of the top leaves is higher, which increases the partial pressure of water vapor inside the leaf. The higher rate of transpiration will cause a greater influx of solutes imported via the xylem, including amino acids and root-produced phytohormones. The greater import of nitrogen is probably not the immediate cause of enhanced incorporation of nitrogen into the photosynthetic apparatus because far more nitrogen is imported via the xylem in leaves than is required for biosynthesis (Fig. 17). It is more likely that other xylem-transported compounds control the differential incorporation of nitrogen in the leaves. As depicted in Figure 18, it has been hypothesized that **cytokinins** are transported in greater amounts to rapidly transpiring leaves that are exposed to high levels of irradiance, compared with slowly transpiring leaves that are lower in the canopy. In the top leaves, the greater inflow of cytokinins enhances the net incorporation of nitrogen into the photosynthetic apparatus (Sect. 5.4.4; Fig. 19).

The mechanism depicted in Figure 19 leads us to the following question: To what extent does the plant achieve its nitrogen allocation to different leaves so as to maximize its rate of photosynthesis? To answer this question, ecophysiological experiments have to be combined with a modeling approach.

To assess whether plants optimize the allocation of nitrogen to the different leaves, we need to know (1) the gradient of light within the canopy, (2) the relationship between photosynthesis and the level of irradiance, and (3) the relationship between photosynthesis and leaf nitrogen concentration. The optimal pattern of nitrogen distribution is the one that maximizes the rate of photosynthesis of the entire plant. The problem is analyzed in detail in Box 7 (attached to the chapter Scaling up). The outcome can be summarized as follows. Although plants do not quite achieve the pattern of nitrogen allocation to their leaves that would yield the highest possible rate of canopy photosynthesis, both monocotyledons and dicotyledons, as well as both C_3 and C_4 plants, have a nitrogen allocation pattern that approaches the optimal pattern. In this way the plants have a higher rate of canopy photosynthesis than could have been achieved with a

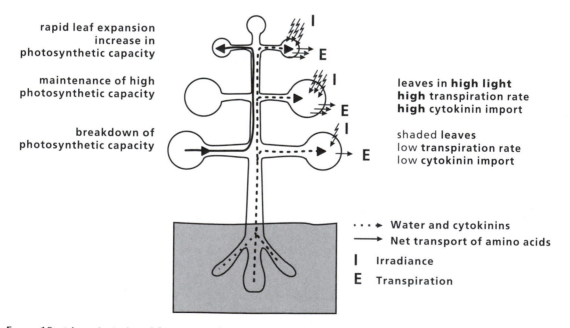

FIGURE 19. A hypothetical model to account for the differential allocation of nitrogen to leaves exposed to high or low levels of irradiance. Cytokinins are imported in greater amounts by rapidly transpiring leaves high in the canopy than by leaves lower in the canopy, which have lower rates of transpiration. Cytokinins then promote the incorpora- tion of nitrogen into the photosynthetic apparatus. In the absence of a large inflow of cytokinins, much of the nitrogenous compounds imported via the xylem are exported again via the phloem. Based on information in Pons & Bergkotte (1996).

uniform distribution of the same amount of nitrogen (Anten et al. 1994, Hirose & Werger 1987a,b, Pons et al. 1989).

5.5 Plant Growth as Affected by Soil Compaction

Soil structure affects plant performance in many ways, both reducing leaf growth and changing root morphology. Roots are smooth and cylindrical in friable soil, but they become **stubby** and **gnarled** upon soil compaction and explore less soil, with potentially deleterious effects on the supply of water and nutrients (Bengough & Mullins 1990a,b).

5.5.1 Effects on Biomass Allocation: Is ABA Involved?

Plants that grow in compacted soil have a **reduced LMR**, even in the presence of adequate nutrients and water. Soil compaction sometimes enhances the concentration of ABA in the xylem sap (Munns & Cramer 1996). ABA might then be responsible for a reduced stomatal conductance, but this remains to be established. Is ABA also the cause of the reduction in leaf growth, as it is under water stress? This is doubtful because ABA-deficient mutants of both *Lycopersicon esculentum* (tomato) and *Zea mays* (maize) show exactly the same response as wildtype plants (Munns & Cramer 1996).

The effects of plants that grow in compacted soil are similar to those of plants that are **pot-bound** (i.e., grown in pots that are too small for their roots). The roots somehow sense the walls of the pots to be "impenetrable soil." Leaf area expansion is reduced, even when sufficient water and nutrients are provided. The xylem sap of pot-bound sunflower plants contains far more ABA than does that of control plants (Table 6), but in bean no such effect

is observed (Munns & Cramer 1996). The cause of the increased concentration of ABA in the xylem is therefore still unclear. It might be synthesized in the leaves, exported via the phloem to the roots, and subsequently find its way to the xylem.

5.5.2 Changes in Root Length and Diameter: A Modification of the Lockhart Equation

Mechanical resistance (impedance) of the soil can be an important factor that limits root growth (Boone 1986). The resulting increase in the rate of **ethylene** production has been suggested to trigger the observed reduction in root elongation and an increase in root diameter and (sometimes) number of cortical cells (Harpham et al. 1991). There is also a change in the branching pattern. When ethylene production is inhibited, however, soil compaction still induces the same root morphology. The effects of soil compaction on root morphology may therefore be accounted for by physical effects.

For the roots to be able to elongate, the mechanical impedance of the soil matrix acting against the cross-section of the root tip must be less than the pressure exerted by the root itself. To expand on Equation 6 (Sect. 2.2), the proportional root elongation (r) is the result of cell expansion, which is related to the cell-wall yield coefficient (ϕ, $MPa^{-1}s^{-1}$), the turgor pressure (Ψ_P, Pa), the **yield threshold of the root** (Y_r, MPa), and the **yield threshold of the soil** (Y_s, MPa) (Pritchard 1994):

$$r = \phi(\Psi_P - Y_r - Y_s) \qquad (10)$$

Maximum axial and radial root growth pressures range from 0.24 to 1.45 and from 0.51 to 0.90 MPa, respectively, and vary with plant species. Because it is impractical to measure the mechanical impedance of the soil directly by using actively growing roots, a penetrometer has been developed that

TABLE 6. **Effects of root confinement on yield and physiology of 14-day-old *Helianthus annuus* (sunflower) plants.**

Treatment	Fresh mass (mg) Shoot	Fresh mass (mg) Root	RMR	Transpiration (mm day^{-1})	K$^+$ transport (pmol g^{-1}s^{-1})	Plant water potential (MPa)	[ABA] in xylem (nM)
Control	163	9.5	0.055	0.054	97	−0.51	10
Confined	112	7.3	0.061	0.053	136	−0.51	70

Source: Ternesi et al. 1994.

*The root mass ratio (RMR) is the root fresh mass as a fraction of total plant mass; K$^+$ transport (expressed per unit root fresh mass) was calculated from the concentration of K$^+$ in the xylem exudate and the rate of exudation). Plants were grown in such a way as to ensure that water and nutrients were supplied at an optimum level.

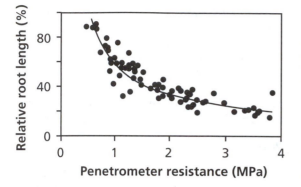

FIGURE 20. The relative root length of 70-day-old plants of *Zea mays* (maize), *Gossypium hirsutum* (cotton), *Triticum aestivum* (wheat), and *Arachis hypogaea* (groundnut) as dependent on mechanical impedance of the soil, as determined with a penetrometer (Bennie 1996). By courtesy of Marcel Dekker, Inc.

measures the pressure required to force a steel probe, with a 60 or 30° conical tip (i.e., 30 or 15° semiangle), into the soil.

Root elongation is primarily determined by the rate at which files of cells are produced and by the cell elongation rate in the apex. Root elongation and total root length are reduced by mechanical impedance (Fig. 20) due to inhibition of cell elongation. The root diameter commonly increases because of radial cell expansion of cortical cells (Fig. 21), and the solute concentration of the root cells is enhanced (Atwell 1989). Thicker and more rigid roots,

which result from radial root expansion, are thought to exert higher pressure on the surrounding soil and deform the soil ahead of the root, which facilitates subsequent penetration (Pritchard 1994). Turgor measurements showed **turgor pressures** of 0.78 MPa in impeded root tips of *Pisum sativum* (pea), as opposed to 0.55 MPa in unimpeded root tip cells (Clark et al. 1996).

The smaller root system under conditions of soil compaction may be detrimental for the uptake of nutrients and water, and hence reduce the plant's growth rate and productivity. There are also effects on leaf expansion, however, that are not accounted for by the plant's water or nutrient status. The roots perceive soil compaction as such, and they send inhibitory signals to the leaves, which causes a **feedforward response** (Stirzaker et al. 1996). There is no conclusive evidence that species differ in their ability to grow in compacted soil. Rather, they differ in their capacity to find less compacted sites in the same soil (Bennie 1996). They may also differ in the size of their root system and hence in the extent to which they explore the soil, including the compacted part (Materechera et al. 1993).

5.6 Growth as Affected by Soil Flooding

Flooding or inundation of the soil leads to filling with water of the soil pores that are normally filled with air. This reduces the supply of soil oxygen, which may reduce aerobic respiration (Sect. 4.1 of

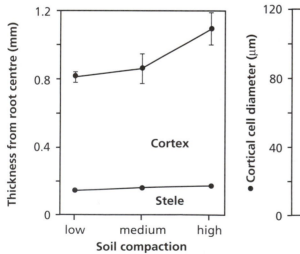

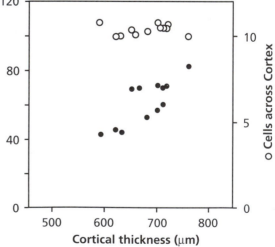

FIGURE 21. The radius of the stele and cortex in roots of *Lupinus angustifolius*, grown at three levels of soil compaction (left) and the diameter and number of cortical cells and mean cortical cell diameter of the same plants (right). Increasing cortical thickness on the abscissa in the right-hand figure is the result of increased soil compaction, as illustrated in the left-hand figure (Atwell 1989). Copyright Kluwer Academic Publishers.

the chapter on respiration). Flooding also affects the roots' hormone metabolism. Concentrations of **ethylene** in the roots increase, largely because this gas diffuses more slowly in a flooded than soil in it does a well-aerated soil so that it gets trapped in the roots, and partly because of an enhanced production of this hormone (Brailsford et al. 1993).

5.6.1 The Pivotal Role of Ethylene

Ethylene inhibits root elongation and induces the formation of **aerenchyma** in roots (Fig. 22).

Aerenchyma formation is preceded by enhanced transcription of a gene that encodes a **xyloglucan endotransglycosylase**, which is a cell-wall loosening enzyme involved in the hydrolysis of cell walls (Sect. 2.2) and ultimately in the **lysis** of some **cortical cells** (Saab & Sachs 1996). The ethylene-induced aerenchyma facilitates **gas diffusion** between roots and aerial parts (Sect. 4.1.4 of the chapter on plant respiration) and enhances the storage capacity for oxygen. Many **hydrophytes** such as *Oryza sativa* (rice) and *Senecio congestus* possess extensive aerenchyma even when growing in well-drained conditions. In mesophytes such as *Zea mays* (maize)

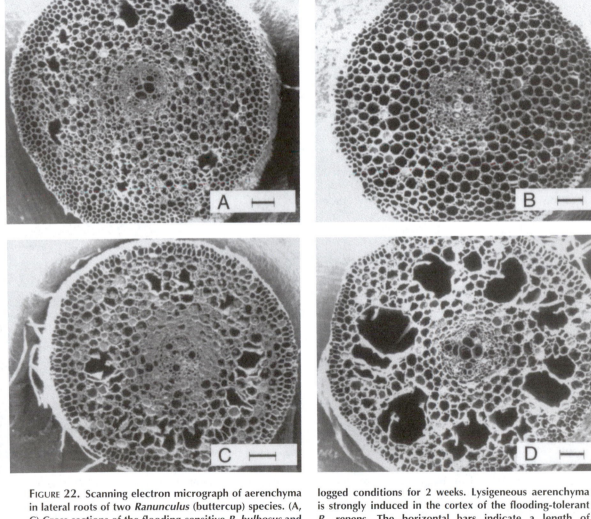

FIGURE 22. Scanning electron micrograph of aerenchyma in lateral roots of two *Ranunculus* (buttercup) species. (A, C) Cross-sections of the flooding-sensitive *R. bulbosus* and the flooding-tolerant *R. repens*, respectively, grown under well-drained conditions. Both species have only a few air channels in their root cortex. (B, D) Cross-sections of *R. bulbosus* and *R. repens*, respectively, grown under water-logged conditions for 2 weeks. Lysigeneous aerenchyma is strongly induced in the cortex of the flooding-tolerant *R. repens*. The horizontal bars indicate a length of 100 μm (courtesy G.M. Bögemann, L.A.C.J. Voesenek and C.W.P.M. Blom, Department of Ecology, University of Nijmegen, the Netherlands).

and *Helianthus annuus* (sunflower), however, cortical aerenchyma formation by cell breakdown is minimal in well-aerated conditions and is promoted by poor aeration (Jackson 1985, Konings & Lambers 1991).

Ethylene also increases the **elongation of the coleoptile and mesocotyl** in *Oryza sativa* (rice), which enables it to reach the surface of the water more rapidly. Such a "snorkeling" response is characteristic of most flood-tolerant species. A similar response has been found for petioles and lamina in the flood-tolerant *Rumex palustris* during submergence of entire plants. The flood-sensitive *R. acetosa*, on the other hand, responds to flooding with enhanced ethylene concentrations in the shoot, but not with enhanced elongation rates (Voesenek et al. 1993). This indicates that it is the greater **responsiveness to ethylene**, and not the enhanced ethylene production, that increases petiole elongation in the flood-tolerant *Rumex* species (Banga et al. 1996). The increased responsiveness of the flood-tolerant *Rumex* species is associated with an increased transcription of the gene coding for an **ethylene receptor** upon submergence. High concentrations of ethylene and exposure to high concentrations of carbon dioxide and low concentrations of ethylene increased the levels of transcripts coding for the ethylene receptor. This suggests that flood-tolerant *Rumex* species respond to flooding stress by increasing the number of their ethylene receptors, which subsequently enhances their responsiveness to ethylene, leading to leaf elongation (Vriezen et al. 1997).

In water chestnut it is the root that shows a "snorkeling" response to flooding; it reaches the surface of the water by growing upward, rather than showing the normal positive gravitropism.

5.6.2 Effects on Water Uptake and Leaf Growth

The responses of leaf growth and metabolism to soil inundation are similar to those of droughted plants. Flooding delays the normal daily increase in root **hydraulic conductance** (L_p, $mm^3 s^{-1} MPa^{-1}$) in flooding-sensitive *Lycopersicon esculentum* (tomato) plants (Else et al. 1995). This may be due to fewer water-channel proteins in the roots' plasma membranes, due to lack of metabolic energy to sustain their synthesis. **Stomatal conductance** declines and the rate of **leaf elongation** is reduced (Fig. 23). If the lower L_p is compensated by pressurizing the roots (see Sect. 5.3), however, then both the stomatal conductance and the rate of leaf expansion remain low. As in plants exposed to water shortage, **chemical**

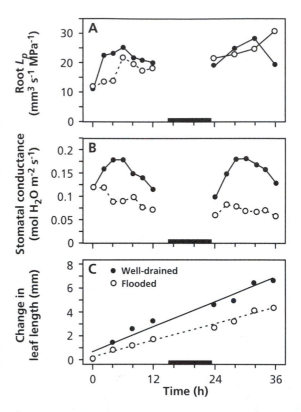

FIGURE 23. Effects of soil flooding for 24 to 36h on (A) root hydraulic conductance, (B) stomatal conductance, and (C) leaf elongation of *Lycopersicon esculentum* (tomato) (Else et al. 1995). Copyright American Society of Plant Physiologists.

signals are probably responsible for the early responses to flooding in sensitive plants. **ABA** is just one of the chemical signals arriving from the roots that cause stomatal closure (Else et al. 1996). Exposure of roots to hypoxia also reduces leaf **cell-wall extensibility**, and it is parallelled by a decreased capacity to **acidify leaf cell walls** (Van Volkenburgh 1994). It remains to be demonstrated, however, whether the two phenomena, chemical signals and cell-wall acidification, are linked.

5.6.3 Effects on Adventitious Root Formation

Whenever the effects of soil flooding become too severe, plants with some degree of flooding tolerance make new, aerenchymatous adventitious roots with air channels that are connected to the shoot that permit oxygen diffusion to the new roots (Armstrong et al. 1994). Endogenous auxin is the phytohormone that is generally responsible for adventitious root formation, even in flooding-sensitive plants. Auxin accumulates at the base of

TABLE 7. The effect of exposure to hypoxia and treatment with auxin, ethylene, or a combination of ethylene and an inhibitor of auxin transport on the formation of adventitious roots in the flooding-tolerant *Rumex palustris*.

Treatment	Number of adventitious roots
Aerobic control	4
Anaerobic control	43
Auxin	45
Ethylene	44
Ethylene + inhibitor	8

Source: Visser et al. 1996.

the shoot, possibly due to inhibition of the energy-dependent transport of auxin to the roots. In the flood-tolerant *Rumex palustris* both **ethylene** and **auxin** enhance the formation of new adventitious roots (Table 7). Because ethylene has no effect in the presence of an inhibitor of auxin transport, however, it must exert its effect through auxin. Because the concentration of auxin is not increased, it has been concluded that ethylene, which accumulates upon flooding the plants, enhances the tissue's sensitivity for endogenous auxin, allowing root primordia to develop where they would otherwise remain dormant (Visser et al. 1996).

5.7 Growth as Affected by Touch and Wind

Some plants can "move" when touched. Unless *Mimosa pudica* (touch-me-not) has just been assaulted by a classroom of school children, its petioles and pinnate leaves will respond to touch, due to the movement of ions in the pulvinus (Sect. 5.4.6 of the chapter on plant water relations). These movements in response to touch are *not* related to growth. The growth of some plant organs, however, does respond to touch (e.g., the tendrils of climbing plants like *Clematis* or *Lathyrus*). Upon contact, these tendrils enhance their growth at the side away from the place of contact, sometimes in combination with a growth reduction at the side where contact occurred. Another response of the tendril to contact may be a strong reduction in the rate of elongation, as in the tendrils of *Cucumis sativus* (cucumber) (Ballaré et al. 1995). The phytohormone **ethylene** may be involved in this growth response, but no clear answers are available. Susceptibility of plants to contact was already recognized by Theophrastus, around 300 BC, and by Darwin (1880), who described this phenomenon for

the apex of the radicle of *Vicia faba* (broad bean). Since then, its has been shown that wind, vibrations, rain, and turbulent water flow affect a plant's physiology and morphology, which is a phenomenon generally termed **thigmomorphogenesis** (Jaffe 1973). Wind exposure may make plants less susceptible to other forms of stress. An extreme form of thigmomorphogenesis is that found in trees at high altitude, which show the typical "Krumholz" sculpture (i.e., a wind-induced deformation). Trees at the edge of a plantation or forest tend to be hardened by wind and have thicker and shorter trunks. Whenever these trees are removed, the weaker, slender trees are easily knocked over (Jaffe & Forbes 1993).

Plant growth may decline in response to careful touching or stroking of leaves, much to the disappointment of some students who have tried to carry out a **nondestructive growth analysis**. Although not all species or genotypes of a species show thigmomorphogenesis, it is a common and often underestimated phenomenon, generally associated with a reduction in plant growth (Table 8). Touching the leaves may also affect leaf respiration, in some species by as much as 56% (Todd et al. 1972), transpiration, and chemical composition, sometimes even in plants whose growth is not reduced by such a treatment (Kraus et al. 1994).

Exposure of the grasses *Lolium perenne* (perennial ryegrass) and *Festuca arundinacea* (tall fescue) to a **high wind** of $8.4\,\mathrm{m\,s^{-1}}$, as compared with $1.0\,\mathrm{m\,s^{-1}}$ for control plants, reduces their rate of leaf elongation by about 25%, which is partially reversible. The wind-exposed plants are shorter and less leafy. Although wind speed reduces leaf temperature of these grasses, this effect is small and cannot account

TABLE 8. The effect of "handling" on growth of a touch-sensitive and a touch-insensitive population of *Lolium perenne*.

	Sensitive population	Insensitive* population
Dry mass per organ or plant		
leaf blades	60	96
leaf sheath	54	112
roots	68	109
Leaf area per plant	55	94

Source: Kraus et al. 1993.

*Treated plants were handled once a week and the data are expressed as a percentage of the values found for control plants. None of the effects were significant for the insensitive population, whereas all effects were significant for the sensitive one.

for the large effects on leaf elongation. Wind speed reduces the LAR, mainly due to a decrease in SLA. The RGR of the grasses is also reduced, although not to the same extent due to a 15% increase in NAR by this wind treatment (Russel & Grace 1978, 1979). In other species, where leaf respiration is increased (Todd et al. 1972), a decrease in NAR may also be associated with the growth reduction in windy conditions.

Thigmomorphogenetic effects may vary between genotypes of the same species (Table 8). An alpine ecotype of *Stellaria longipes*, which is characterized by a short erect habit, produces substantial amounts of **ethylene** in response to wind, and stem growth is inhibited by ethylene (Emery et al. 1994). By contrast, the prairie ecotype of *S. longipes* produces substantial amounts of ethylene even in the absence of wind stress, but stem growth is not inhibited by ethylene. This demonstrates that ethylene dwarfs stems in alpine *S. longipes* in response to wind stress, whereas ethylene may not affect or enhance stem growth in the prairie ecotype, independent of wind stress. To an alpine plant, wind is an important selective force, whereas in the prairie habitat it is far less important than competition for light. For prairie plants it is important that their stems elongate rapidly in order to avoid being overtopped by competitors. On the other hand, suppression of stem growth by wind may prevent future damage. It appears that both the production of ethylene and a genotype's sensitivity to this phytohormone play an important role in adaptation to the contrasting environment of the ecotypes. Such genetic differentiation is likely to affect a genotype's success in contrasting environment, as will be further discussed in the chapter on interactions among plants.

Exposure of *Arabidopsis thaliana* to wind, rain, or touch led to the serendipitous discovery that these stimuli rapidly (within 10 min) induce several **touch-specific** (*TCH*) **genes**, three of which encode calmodulin, which is a calcium-binding protein that turns on several cellular processes, or calmodulin-related proteins (Braam & Davis 1990). Another *TCH* gene encodes XET (Braam et al. 1996). Using in planta expression of the jellyfish apoaequorin gene, which encodes a calcium-dependent luminescent protein, Knight et al. (1991, 1992) showed that touch immediately increases cytosolic free-calcium levels. Calcium has therefore been implicated as the **second messenger** that induces the expression of the *TCH* genes. It has been postulated that stretch-activated ion channels open upon exposure to touch and wind, thus acting as the sensing mechanism (Jaffe & Forbes 1993). The increased produc-

tion of calmodulin and calmodulin-related proteins probably starts many calcium-regulated events. For example, wind-induced production of calmodulin reduces the rate of elongation of petioles and of bolting in *Arabidopsis thaliana*, modifies callose deposition, and induces auxin-enhanced growth and mitosis. Up-regulation of the XET-encoding gene may play a critical role in determining properties of the cell wall, including extensibility (Sect. 2.2).

5.8 Growth as Affected by Elevated Concentrations of CO_2 in the Atmosphere

On average, the final mass of C_3 plants, grown at high nutrient supply without shading by neighboring plants, increases by 47% when the atmospheric CO_2 concentration is doubled to 700 µmol mol^{-1} (70 Pa) (Poorter 1993, Poorter et al. 1996). When plants have **numerous sinks**, such as tillers or side shoots, the stimulation can be considerably higher (several hundred percent). The average enhancement is rather small, considering the extent of the stimulation of the rate of **photosynthesis** in short-term experiments (see Fig. 6 and Sect. 2.2.1 of the chapter on photosynthesis). To analyze this discrepancy, it is helpful to examine the impact of elevated [CO_2] on each growth parameter in the equation describing plant growth (Sect. 2.1.1):

$$RGR = \frac{(A_a.SLA.LMR - LR_m.LMR - SR_m.SMR - RR_m.RMR)}{[C]}$$

(11)

where A_a is the rate of photosynthesis per unit leaf area; RGR is the plant's relative growth rate; SLA is the specific leaf area; LMR, SMR, and RMR are the leaf mass ratio, stem mass ratio, and root mass ratio, respectively; LR_m, SR_m, and RR_m are the rate of respiration per unit mass of the leaves, stems, and roots, respectively; [C] is the carbon concentration of the plant biomass. If the RGR and final mass of the plants are enhanced less than expected from the increase in rate of photosynthesis, one or more of the parameters in the equation must have been affected by elevated atmospheric CO_2 concentrations. In other words, growth at 70 Pa CO_2 leads to a number of changes in the plant that may compensate for the higher rate of photosynthesis as found in A versus p_i curves. Photosynthetic **acclimation** to high CO_2 concentrations was addressed in Section 12.1 of the chapter on photosynthesis. Here we will discuss some additional changes that counteract the initial stimulation of photosynthesis.

There are numerous examples where exposure of plants to a high atmospheric CO_2 concentration transiently enhances the plant's RGR, whereupon the RGR returns to the level found for control plants (e.g., Fonseca et al. 1996, Wong 1993; Fig. 24). The **transient increase in RGR** often accounts for the increase in final mass of the plants grown at elevated [CO_2] (Fig. 24). Other species show a continuously enhanced RGR, but some degree of acclimation is common. Which component(s) of the growth equation accounts for such acclimation?

A decrease in SLA is the major adjustment found upon prolonged exposure to 70 Pa CO_2. This is partly due to the accumulation of nonstructural carbohydrates (see Sect. 3.4 of the chapter on long-distance transport). LMR, SMR, and RMR are not, or are only marginally, affected (Stulen & Den Hertog 1993). If they are affected, then it is due to the more rapid depletion of nutrients in the soil of the faster-growing plants exposed to elevated [CO_2].

Leaf respiration is inhibited by high [CO_2], at least in short-term experiments (see Sects. 2.3 and 3.6 of the chapter on plant respiration). Upon prolonged exposure, respiration may also be affected, but it increases in one plant and decreases in another. The carbon concentration varies with CO_2 concentration, but again without a distinct trend (Poorter et al. 1992). Results from short-term measurements on single leaves clearly cannot simply be extrapolated to the growth of whole plants over a long period.

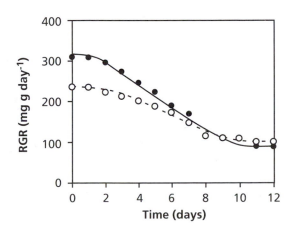

FIGURE 24. The relative growth rate (expressed on a fresh mass basis) of *Plantago major* grown at 35 Pa CO_2 (open symbols) or at 70 Pa from day zero onwards, when the plants were 4 weeks old (Fonseca et al. 1996). Copyright Trustees of The New Phytologist.

Because the rate of net CO_2 assimilation in C_3 plants is not CO_2-saturated at 35 Pa CO_2, the rise in CO_2 concentration may enhance the rate of photosynthesis. In addition, as discussed in Section 5.4.2 of the chapter on plant water relations, stomata respond to CO_2: when p_a increases, g_c declines. As a result, the rate of **transpiration is reduced** at a high atmospheric CO_2 concentration. The increased rate of photosynthesis, in combination with the reduced rate of transpiration, enhances the water-use efficiency. For C_4 plants, the rate of CO_2 assimilation is virtually saturated at a CO_2 partial pressure of 35 Pa. Their stomatal conductance, however, also decreases; hence, their water-use efficiency should also increase.

When plants are studied in a vegetation, rather than as single plants, additional factors may play a role. The reduced stomatal conductance tends to reduce the plant's transpiration. As a result, leaf temperature may rise and the vapor pressure of the air surrounding the plants may be less when growth occurs at elevated [CO_2]. Both effects counteract the decreased stomatal conductance, so that the effect of elevated [CO_2] on the transpiration of a vegetation is less than expected from single-leaf measurements (Kimball et al. 1993).

Different types of plants may respond to varying degrees to elevated CO_2. For example, **C_4 plants**, whose rate of photosynthesis is virtually saturated at 35 Pa, respond to a smaller extent (Poorter et al. 1996). It is surprising that elevated [CO_2] does not consistently affect the **competitive balance between C_3 and C_4 plants**, as discussed in Section 5.4 of the chapter on interactions among plants.

CO_2-enrichment of the atmosphere also has indirect effect on vegetation. Heteroatomic gasses absorb **infrared radiation**. In the absence of water vapor and carbon dioxide in the atmosphere all infrared radiation emitted from the earth would be lost to space and the earth's temperature would be $-20°C$ rather than the more current value of $+15°C$. CO_2 enrichment of the atmosphere has almost certainly contributed to the recent **increase in global temperature** that has been observed over the past century, especially in the temperate and polar regions (Santer et al. 1996). This, in turn, will affect the global weather pattern (precipitation, storms, etc.) and plant growth. The increase in global temperature has not been as great as would be expected from increases in atmospheric concentration of CO_2 and other radiatively active trace gases, probably because of the counteracting cooling effects of **sulfate aerosols** from fossil fuel combustion and of dust from expansion of agriculture. These anthropogenic effects on climate are superimposed on natural variations in climate.

6. Adaptations Associated with Inherent Variation in Growth Rate

6.1 Fast-Growing and Slow-Growing Species

In **unpredictable but productive environments**, where "catastrophes" like fire, inundation, or other forms of disturbance occur, **fast-growing short-lived species** are common. In more **predictable environments** with a low incidence of disturbance, **longer-lived slow-growing species** predominate. Apart from their life span, these short- and long-lived species differ in many other traits and, broadly generalizing, have been termed **r-species** and **K-species**, where r and K are constants in a logistic growth curve (McArthur & Wilson 1967, Pianka 1970). Such a classification, once proposed for both plants and animals, has been questioned, but it provides a useful context in which to understand the ecological performance of vastly different species (Table 9).

Grime (1979) extended this concept by suggesting that there are two major categories of selective factors: **stress**, which is an environmental factor that reduces the growth rate of plants, and **disturbance**, which is a factor that destroys plant biomass. High-stress environments include those with low availability of water, nutrients, and light, or where other conditions are unfavorable for growth (low temperature, high salinity, low oxygen, heavy metal contamination, etc.). Disturbance can result from herbivory or from environmental

factors like fire or wind. Grime describes three extreme types of plant strategies: **competitors**, which exist under conditions of low stress and low disturbance, **stress-tolerant** species, which occupy habitats with high stress and low disturbance, and **ruderals** (= weeds), which occur in highly disturbed nonstressful environments. There is no viable plant strategy that can deal with the combination of high stress and high disturbance. Most plants actually fall at intermediate points along these continua of stress and disturbance, so it is most useful to use the scheme in a comparative sense, with some species being more stress-tolerant than others, and where some species are more tolerant of disturbance than others. Although this classification has also been seriously questioned, it has led to the recognition that plants characteristic of low-resource and stressful environments consistently have a lower RGR than do plants from more favorable environments (Box 10 in the chapter on interactions among plants).

The close association between a species' growth potential and the quality of its natural habitat (Fig. 25) raises two questions. First, how are the differences in growth rate between species brought about? Second, what ecological advantage is conferred by a plant's growth potential? These two questions are in fact closely related. Before evaluating the **ecological significance** of the inherent RGR of a species, it is important to analyze the **physiological basis** of the genetic variation in RGR (Lambers & Poorter 1992). Numerous plant characteristics contribute to a plant's absolute growth rate in its natural habitat (e.g., seed size, germination time, or plant size after overwintering). In view of the close correlation between a plant's inherent RGR and environmental parameters (Fig. 25), we will restrict the present discussion to traits that con-

TABLE 9. Some of the characteristics of r- and K-species and the habitats in which they occur.

	r selection	K selection
Climate	Variable and/or unpredictable; uncertain	Fairly constant and/or predictable; more certain
Mortality	Often catastrophic; density-independent	Density-dependent
Population size	Variable; usually well below carrying capacity; frequent recolonization	Fairly constant; at or near carrying capacity; norecolonization required
Intra- and interspecific competition	Variable; often minor	Usually severe
Traits favored by selection	Rapid development	Slower development
	High growth rate	Competitive ability
	Early reproduction	Delayed reproduction
	Single reproduction	Repeated reproductions
Life span	Relatively short	Longer

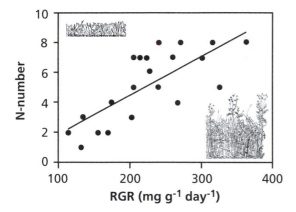

FIGURE 25. The relationship between the relative growth rate (RGR) of 24 herbaceous C_3 species and the nitrogen index of the species' habitat (high values correspond to habitats of high nitrogen availability). The RGR was determined under identical conditions for all species: free access to nutrients and an irradiance of $320\,\mu mol\,m^{-2}s^{-1}$ (Poorter & Remkes 1990).

tribute to variation in RGR. Finally, we will discuss the ecological implications of inherent differences in the various traits and in the growth rate itself.

6.2 Growth of Inherently Fast- and Slow-Growing Species Under Resource-Limited Conditions

In Section 2.1 we compared plants under conditions favorable for growth. How do fast- and slow-growing species perform at a low nutrient concentration?

6.2.1 Growth at a Limiting Nutrient Supply

Although the RGR of potentially fast-growing species is reduced more than that of slow-growing ones, when nutrients are in short supply, the inherently fast-growing species still tend to grow fastest (Fig. 26). Similar results are obtained in a situation where a fast-growing species competes with a slow-growing one under **nutrient stress**, at least when the duration of the experiment is short relative to the plant's life span.

The higher RGR of inherently fast-growing species at a low nutrient supply, in comparison with slow-growing ones, is largely "explained" by differences in LAR (SLA), which is similar to the situation with a free access to nutrients (Table 10). (Note that "explained" is used here in a statistical sense and that it does not refer to physiological mechanisms.)

6.2.2 Growth in the Shade

In a comparison of tropical tree species, fast-growing species with a high LAR and low RMR maintain a higher RGR when grown in the shade; however, they also show greater mortality (Kitajima 1994). This trend can be accounted for by greater investment in defense against herbivores and pathogens (dense and tough leaves) in the slower-growing trees, which have a large root system and a high wood density (Kitajima 1996).

6.3 Are There Ecological Advantages Associated with a High or Low RGR?

The ecological advantage of a high RGR seems straightforward: Fast growth results in the rapid occupation of space, which is advantageous in a situation of competition for limiting resources. A high RGR may also maximize the reproductive output in plants with a short life span, which is particularly important for ruderals. What, then, is the possible survival value of slow growth? Grime and Hunt (1975) and Chapin (1980, 1988) offered several explanations, which we will evaluate in this section.

6.3.1 Various Hypotheses

It has been suggested that slow-growing species make modest demands and are therefore less likely to exhaust the available nutrients (Parsons 1968).

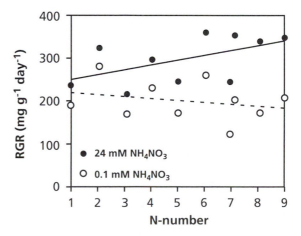

FIGURE 26. The RGR of 10 annual herbaceous C_3 species grown at a high and a low nitrogen supply. The 10 species were from habitats differing in "N-number" (higher values indicating a higher nitrogen availability as well as an inherently higher RGR_{max}) (Fichtner & Schulze 1992).

TABLE 10. The effect of a nutrient solution with a high or a low nitrate concentration on a number of growth parameters of an inherently slow-growing species (*Deschampsia flexuosa*) and a fast-growing one (*Holcus lanatus*).

Parameter	High [NO$_3$]		Low [NO$_3$]	
	Deschampsia	*Holcus*	*Deschampsia*	*Holcus*
RGR (mg g^{-1} day^{-1})	97	172	47	66
NAR (g m^{-2} day^{-1})	6.9	8.5	5.2	4.6
LAR (m^2 kg^{-1})	13	20	9	14
SLA (m^2 kg^{-1})	28	51	24	44

Source: Poorter et al. 1995.

This is not, however, a stable evolutionary strategy because a neighboring individual with a faster nutrient uptake could absorb most nutrients (cf. Schulze & Chapin 1987). In addition, these modest demands cannot explain slow growth as an adaptation to saline environments or other situations where conditions are stressful for reasons other than low resource supply.

Slow-growing species have also been suggested to function closer to their optimum than fast-growing ones in an adverse environment (Chapin 1980). This explanation suggests that allocation or some other aspects of the plant's physiology at a low nutrient supply is closer to the optimal pattern for inherently slow-growing species than it is for fast-growing ones. Information on the pattern of allocation, however, indicates that both fast- and slow-growing species allocate their carbon and nitrogen in a manner that will maximize their RGR (Van der Werf et al. 1993).

Slow-growing species were thought to incorporate less photosynthates and nutrients into structural biomass. This might allow them to form reserves for later growth, thereby enabling them to maintain physiological integrity during periods of low nutrient availability. As we discussed in Sections 5.3.3 and 5.4.3, however, under such adverse conditions growth is restricted before photosynthesis is, and sugars tend to accumulate. Hence, it is unlikely that survival during periods of nutrient shortage depends on storage of photosynthates. There is also no convincing evidence that slow-growing species have a greater capacity to accumulate nutrients, perhaps with the exception of phosphate.

Finally, it has been suggested that a high growth rate cannot be realized in a low-resource environment; therefore, a high potential RGR is a selectively neutral trait. As discussed in Section 6.2, however, potentially fast-growing species still grow faster than slow-growing ones, even in low-resource environments. This indicates that the potential RGR is not a selectively neutral trait. Even in low-resource environments, fast-growing species attain a larger size more rapidly, which has advantages in terms of their competitive ability and fitness. Although a very high RGR is not attainable, a slightly higher RGR might therefore still be advantageous.

6.3.2 Selection on RGR$_{max}$ Itself, or on Traits That Are Associated with RGR$_{max}$?

Having scrutinized the various hypotheses accounting for variation in growth potential, we conclude that a low potential growth rate per se does *not* confer ecological advantage. Why, then, do slow-growing species occur more frequently in unfavorable habitats than fast-growing ones? An alternative explanation for the observed differences in potential growth rate is that one of the **components linked with RGR**, and not RGR itself, has been the target of natural selection (Lambers & Poorter 1992).

The most likely traits selected for are those that protect the tissue (**quantitative defense**; Sect. 3.2 of the chapter on ecological biochemistry). In leaves this is associated with a **low SLA,** which is accounted for by variation in **leaf mass density** (i.e., the amount of dry mass per unit fresh mass). As pointed out in Sections 3.2 and 3.3, variation in leaf biomass density is largely accounted for by variation in **cell-wall thickness**, number of **sclerenchymatic cells**, and the concentration of **quantitatively important secondary plant compounds**. Variation in these traits is closely correlated with that in RGR (Figs. 3 and 4 in Sect. 3). In a situation where nutrients are limiting, conservation of the scarce resource is at least as important as its capture (Sect. 4 of the chapter on mineral nutrition).

Hence, plants growing under severe nutrient limitation are expected to **conserve their nutrients**. Indeed, low-productivity species are more successful due to less leaf turnover; therefore, nutrient losses are restricted (cf. Sects. 4.3 and 4.4 of the chapter on mineral nutrition) (Berendse & Aerts 1987). Comparing tree seedlings, a close negative correlation exists between relative growth rate and leaf life-span (Chapin 1980, Reich et al. 1992a,b).

How can leaf longevity be increased? This depends on the environmental factor that affects leaf longevity. Herbivory can be reduced by increasing **leaf toughness** and accumulating **palatability-reducing compounds** (Sect. 3 of the chapter on ecological biochemistry). The abrasive effects of high wind speeds can be reduced by investment in **fiber** and **sclerenchyma** (Sect. 3.3 of this chapter). Trampling resistance may be the result of a large amount of **cell-wall material** per cell. Transpiration can be decreased and water-use efficiency can be increased by the construction of **leaf hairs** or **epicuticular waxes** (Sect. 2 of the chapter on the plant's energy balance). Epicuticular waxes may also confer disease resistance and diminish deleterious effects of salt spray. Each of these additional investments increases the leaf's longevity, but each also decreases SLA and therefore diminishes the plant's growth potential, but positively influences its fitness under adverse conditions.

There is considerably less information on root turnover than on leaf turnover, and not enough to generalize about inherent differences associated with a plant's growth potential. We do know, however, that the **biomass density** tends to be higher in roots of slow-growing grass species, when compared with that in fast-growing ones, which is similar to what has been found for leaves (Ryser & Lambers 1995); this higher root mass density is associated with thicker cell walls. The high mass density might be associated with slow root turnover, but this still needs further study (Ryser 1998).

Is there any indication that plants without the types of leaf and root adjustment discussed in this section could not survive in unfavorable habitats? This would require introduction of plants that only differ in one specific trait in different environments. Such isogenic genotypes are rarely available, however, and variation in one trait could be expected to affect related traits. The best ecological information available does support the contention that a decrease in SLA enhances the capacity to survive in more stressful environments (Lambers & Poorter 1992).

6.3.3 An Appraisal of Plant Distribution Requires Information on Ecophysiology

It was pointed out in Section 3 of this chapter that a plant's growth potential is part of a strategy that explains the distribution of a species. Various hypotheses have been proposed to account for the ecological advantage of a high or low RGR_{max}. As we have learned before, however, when discussing the ecology and physiology of C_4 and CAM plants (Sects. 9 and 10 of the chapter on photosynthesis), detailed information of the biochemistry and physiology is essential to fully appreciate a plant's functioning in different environments as well as a species' distribution.

In the present context, we conclude that a thorough **ecophysiological analysis** of inherent variation in RGR has led to greater insight in the **ecological significance** of this trait. Rather than RGR per se, one or more associated components has been the target of natural selection. This natural selection has inevitably led to variation in maximum RGR and an associated **suite of traits** (Table 11). This analysis also serves to illustrate that a thorough ecophysiological analysis is essential for a full appreciation of a species' strategy.

7. Growth and Allocation: The Messages about Plant Messages

The numerous examples in this chapter, as well as in some before, provide a wealth of information about how plants cope with their environment. Plant responses to mild stress are not merely the direct effect of resource deprivation on growth rate. Intricate physiological adjustments that minimize major disturbances in plant metabolism take place. Upon sensing water or nutrient shortage in the root environment, signals are sent to the leaves, which respond in such a way as to minimize deleterious effects. This is a **feedforward response**: An anticipating response in which the rate of a process is affected before large deleterious effects of that process have occurred. Low levels of irradiance are, similarly, detected, and the signal leads to a feedforward response that minimizes the effect of growth in the shade.

What do all these examples have in common? They suggest that a plant is continuously **sensing** its changing **environment** and using this information to control its physiology and allocation pattern. They indicate that, in general, environment affects growth via chemical or hydraulic messages (**sink control**). We assume that all plants have this capac-

TABLE 11. Typical characteristics of inherently fast-growing and slow-growing herbaceous C_3 species, summarizing information presented in the text.

Characteristic	Fast-growing species	Slow-growing species
Habitat:		
• nutrient supply	high	low
• potential productivity	high	low
Morphology and allocation:		
• leaf area ratio	high	low
• specific leaf area	high	low
• leaf mass ratio	higher	lower
• root mass ratio	lower	higher
Physiology:		
• photosynthesis		
(per unit leaf area)	equal	equal
(per unit leaf mass)	high	low
• carbon use in respiration		
(% of total C fixed)	low	high
• ion uptake rate		
(per unit root mass)	high	low
Chemical composition:		
• concentration of quantitative secondary compounds	low	high
• concentration of qualitative secondary compounds	variable	variable
Other aspects:		
• leaf mass density	low	high
• root mass density	low	high
• leaf turnover	high	low
• root turnover	high?	low?
• leaf longevity	low	high
• root longevity	low?	high?

Note: Unless stated otherwise, the differences refer to plants grown with free access to nutrients. A ? indicates that further study is needed.

ity to sense their environment. What makes species different from one another is perhaps the manner in which they are able to respond, and not so much the variation in their capacity to sense the environment. The typical response of a ruderal species upon sensing nutrient shortage is to slow down leaf expansion and allocate more resources to root growth; it will promote leaf senescence and so withdraw nutrients from older leaves and use these for its newly developing tissues. A species naturally occurring on nutrient-poor sand will use the same signal to slow down the production of new tissues, with less dramatic effects on leaf senescence and allocation pattern. Upon sensing water shortage some plants may similarly respond by severely reducing leaf expansion, and others by shedding some leaves, whereas facultative CAM plants switch from the C_3 or C_4 pathway to the CAM mode. Shade is perceived by shade-avoiding and shade-tolerant plants, but the response to promote stem elongation is typical only for shade-avoiding species.

It is this **variation in responses**, rather than the actual sensing mechanism itself, that must be of paramount importance accounting for a species' ecological amplitude as well as in such ecological processes as succession and competition (Aphalo & Ballaré 1995). Ignoring the capacity of plants to process and respond to environmental information (and assuming that plants grow until they run out of resources) leads to a distorted view of the process of competition. As neighbors interact, how do the continuous changes in plant form and function, elicited by information-sensing systems, contribute to competitive success? To what extent does the capacity of an individual to adjust its allocation and development contribute to the outcome of competition?

It is not our aim to promote the "Panglossian" view, which is referred to in the Introduction chapter, that just because a species exhibits certain traits in a particular environment, these traits must be beneficial and have resulted from natural selection in that environment (Gould & Lewontin 1979). We do wish to stress, however, that plants are information-acquiring systems rather than passively responding organisms, and that this trait must not be ignored, as we will discuss in the chapter on interactions among plants.

If we aim to understand plant functioning in different environments, information at the cellular and molecular level is of vital importance. Perception of the environment by specific molecules (e.g., phytochrome), followed by transduction of the information and effects on cell growth (e.g., through cell-wall acidification), allow the plant to acclimate to its environment (e.g., shade). In the past decade our understanding of numerous intricate processes has increased enormously. It is to be expected that fascinating progress will be made in the next decade that will allow us both to deepen our understanding of plant performance in an ecological context and to apply this information in breeding new varieties for adverse environments.

References and Further Reading

Ahn, J.H., Choi, Y., Kwon, Y.M., Kim, S.-G., Choi, Y.D., & Lee, J.S. (1996) A novel extensin gene encoding for a hydroxyprolin-rich glycoprotein requires sucrose for its wound-inducible expression in transgenic plants. Plant Cell 8:1477–1490.

Anten, N.P.R., Schieving, F., & Werger, M.J.A. (1995) Patterns of light and nitrogen distribution in relation to whole canopy carbon gain in C_3 and C_4 mono- and dicotyledonous species. Oecologia 101:504–513.

Aphalo, P.J. & Ballaré, C.L. (1995) On the importance of information-acquiring systems in plant-plant interactions. Funct. Ecol. 9:5–14.

Armstrong, W., Jackson, M.B., & Brändle, R. (1994) Mechanisms of flood tolerance in plants. Acta Bot. Neerl. 43:307–358.

Atkin, O.K. (1996) Reassessing the nitrogen relations of arctic plants: A mini-review. Plant Cell Environ. 19:695–704.

Atkin, O.K., Botman, B., & Lambers, H. (1996) The causes of inherently slow growth in alpine plants: An analysis based on the underlying carbon economies of alpine and lowland Poa species. Funct. Ecol. 10:698–700.

Atwell, B.J. (1989) Physiological responses of lupin roots to soil compaction. In: Structural and functional aspects of transport in roots, B.C. Loughman, O. Gasparikova, & J. Kolek (eds). Kluwer Academic Publishers, Dordrecht, pp. 251–255.

Atwell, B.J., Drew, M.C., & Jackson, M.B. (1988) The influence of oxygen deficiency on ethylene synthesis, 1- amino-cyclopropane 1-carboxylic acid levels and aerenchyma formation in roots of Zea mays. Physiol. Plant. 72:15–22.

Avice, J.-C., Ourry, A., Volenec, J.J., Lemaire, G., & Boucaud, J. (1996) Defoliation-induced changes in abundance and immuno-localiztion of vegetative storage proteins in taproots of Medicago sativa. Plant Physiol. Biochem. 34:561–570.

Bakken, A.K. (1992) Effect of daylength on the nitrogen status of timothy (Phleum pratense L.). Acta Agric. Scand. 42B:62–68.

Ball, M.C. (1988) Salinity tolerance in the mangroves Aegiceras corniculatum and Avicennia marina. I. Water use in relation to growth, carbon partitioning and salt balance. Aust. J. Plant Physiol. 15:447–464.

Ball, M.C. & Pidsley, S.M. (1995) Growth responses to salinity in relation to distribution of two mangrove species, Sonneratia alba and S. lanceolata, in northern Australia. Funct. Ecol. 9:77–85.

Ballaré, C.L., Scopel, A.L., Roush, M.L., & Radosevich, S.R. (1995) How plants find light in patchy canopies. A comparison between wild-type and phytochrome-B-deficient mutant plants of cucumber. Funct. Ecol. 9:859–868.

Banga, M., Blom, C.W.P.M., & Voesenek, L.A.C.J. (1996) Sensitivity to ethylene: The key factor in ethylene production by primary roots of Zea mays L. in submergence-induced shoot elongation of Rumex. Plant Cell Environ. 19:1423–1430.

Bauer, H. & Thöni, W. (1988) Photosynthetic light acclimation in fully developed leaves of the juvenile and adult life phases of Hedera helix. Physiol. Plant. 73:31–37.

Beemster, G.T.S., Masle, J., Williamson, R.E., & Farquhar, G.D. (1996) Effects of soil resistance to root penetration on leaf expansion in wheat (Triticum aestivum L.): Kinematic analysis of leaf elongation. J. Exp. Bot. 47:1663–1678.

Belanger, G., Gastal., F., & Warembourg, F.R. (1994) Carbon balance of tall fescue (Festuca arundinacea Schreb.): Effects of nitrogen fertilization and the growing season. Ann. Bot. 74:653–659.

Bell, T.L., Pate, J.S., & Dixon. K.W. (1996) Relationship between fire response, morphology, root anatomy and starch distribution in south-west Australian Epacridaceae. Ann. Bot. 77:357–364.

Bennie, A.T.P. (1996) Growth and mechanical impedance. In: Plant roots: The hidden half, Y. Waisel, A. Eshel, & U. Kafkaki (eds). Marcel Dekker, New York, pp. 453–470.

Bengough, A.C. & Mullins, C.E. (1990a) The resistance experienced by roots growing in a pressurized cell. Plant Soil 123:73–82.

Bengough, A.C. & Mullins, C.E. (1990b) Mechanical impedance to root growth: A review of experimental techniques and root growth responses. J. Soil Sci. 41: 341–358.

Berger, S., Bell, E., Sadka, A., & Mullet, J.E. (1995) Arabidopsis thalina Atvsp is homologous to soybean vspA and vspB, genes encoding vegetative storage protein acid phosphatases, and is regulated similarly by methyl jasmonate, wounding, sugars, light and phosphate. Plant Mol. Biol. 27:933–942.

Berry, J.A. & Raison, J.K. (1981) Responses of macrophytes to temperatue. In: Encyclopedia of plant physiology, N.S., Vol. 12A, O.L. Lange, P.S. Nobel, C.B. Osmond, & H. Ziegler (eds). Springer-Verlag, Berlin, pp. 277–338.

Björkman, O. (1981) Responses to different quantum flux densities. In: Encyclopedia of plant physiology, N.S., Vol 12A, O.L. Lange, P.S. Nobel, C.B. Osmond, & H. Ziegler (eds). Springer-Verlag, Berlin, pp. 57–107.

Bloom, A.J., Chapin III, F.S., & Mooney, H.A. (1985) Resource limitation in plants—An economic analogy. Annu. Rev. Ecol. Syst. 16:363–392.

Boese, S.R. & Huner, N.P. (1990) Effect of growth temperature and temperature shifts on spinach leag morphology and photosynthesis. Plant Physiol. 94:1830–1835.

Boone, F.R. (1986) Towards soil compaction limits for crop growth. Neth. J. Agric. Sci. 34:349–360.

Bowen, G.D. (1991) Soil temperature, root growth, and plant function. In: Plant roots: The hidden half. Y. Waisel, A. Eshel, & U. Kafkaki (eds). Marcel Dekker, New York, pp. 309–330.

Braam, J. & Davis, R.W. (1990) Rain- wind-, and touch-induced expression of calmodulin and calmodulin-related genes in Arabidopsis. Cell 60:357–364.

Braam, J., Sistrunk, M.L., Polisensky, D.H., Xu, W., Purugganan, M.M., Antosiewicz, D.M., Campbell, P., & Johnson, K.A. (1996) Life in a changing world: TCH gene regulation of expression and responses to environmental signals. Physiol. Plant. 98:909–917.

Brailsford, R.W., Voesenek, L.A.C.J., Blom. C.W.P.M., Smith, A.R., Hall, M.A., & Jackson, M.M. (1993) Enhanced ethylene production by primary roots of Zea mays L. in response to sub-ambient partial pressures of oxygen. Plant Cell Environ. 16:1071–1080.

Brouwer, R. (1963) Some aspects of the equilibrium between overground and underground palant parts. Meded. Inst. Biol. Scheikd. Onderzoek Landbouwgewassen 213:31–39.

Brouwer, R. (1983) Functional equilibrium: Sense or nonsense? Neth. J. Agric. Sci. 31:335–348.

Carpit, N.C. & Gibeaut, D.M. (1993) Structural modelks of primry cell walls in flowering plants: Consistency of molecular structure with the physical properties of the walls during growth. Plant J. 3:1–30.

Ceulemans, R. (1989) Genetic variation in functional and structural productivity components in Populus. In: Causes and consequences of variation in growth rate and productivity of higher plants, H. Lambers, M.L. Cambridge, H. Konings, & T.L. Pons (eds). SPB Academic Publishing, The Hague, pp. 69–85.

Chapin III, F.S. (1974) Morphological and physiological mechanisms of temperature compensation in phosphate absorption along a latitudinal gradient. Ecology 55:1180–1198.

Chapin III, F.S. (1980) The mineral nutrition of wild plants. Annu. Rev. Ecol. Syst. 11:233–260.

Chapin III, F.S. (1988) Ecological aspects of plant nutrition. Adv. Min. Nutr. 3:161–191.

Chapin III, F.S., Follet J.M., & O'Connor, K.F. (1982) Growth, phosphate absorption, and phosphorus chemical fractions in two Chionochloa species. J. Ecol. 70:305–321.

Chapin III, F.S., Shaver, G.R., & Kedrowski, R.A. (1986) Environmental controls over carbon, nitrogen and phosphorus fractions in Eriophorum in Alaskan tussock tundra. J. Ecol. 74:167–195.

Chapin III, F.S., Schulze, E.-D., & Mooney, H.A. (1990) The ecology and economics of storage in plants. Annu. Rev. Ecol. Syst. 21:423–447.

Chapin III, F.S., Walter, C.H.S., & Clarkson, D.T. (1988) Growth response of barley and tomato to nitrogen stress and its control by abscisisc acid, water relations and photosynthesis. Planta 173:352–366.

Chimenti, C.A. & Hall, A.J. (1994) Responses to water stress of apoplastic water fraction and bulk elastic modulus of elasticity in sunflower (Helianthus annuus L.) genotypes of contrasting capacity for osmotic adjustment. Plant Soil 166:101–107.

Clark, L.J., Whalley, W.R., Dexter, A.R., Barraclough, P.B., & Leight, R.A. (1996) Complete mechanical impedance increases the turgor of cells in the apex of pea roots. Plant Cell Environ. 19:1099–1102.

Clarkson, D.T. (1986) Regulation of the absorption and release of nitrate by plant cells: A review of current ideas and methodology. In: Fundamental, ecological and agricultural aspects of nitrogen metabolism in higher plants, H. Lambers, J.J. Neeteson, & I. Stulen (eds). Martinus Nijhof/Dr W. Junk, The Hague, pp. 3–27.

Clarkson, D.T., Earnshaw, M.J., White, P.J., & Cooper, H.D. (1988) Temperature dependent factors influencing nutrient nutrient uptake: An analysis of responses at different levels of organization. In: Plants and temperature, S.P. Long & F.I. Woodward (eds). Company of Biologists, Cambridge, pp 281–309.

Clarkson, D.T., Jones, L.H.P., & Purves, J.V. (1992) Absorption of nitrate and ammonium ions by Lolum perenne from flowing solution cultures at low root temperatures. Plant Cell Environ. 15:99–106.

Cleland, R.E. (1967) Extensibility of isolated cell walls: Measurrements and changes during cell elongation. Planta 74:197–209.

Coleman, G.D., Chen, T.H.H., & Fuchigami, L.H. (1992) Complementary DNA cloning of poplar bark storage protein and control of its expression by photoperiod. Plant Physiol. 98:687–693.

Cornelissen, J.H.C., Castro Diez, P., & Hunt, R. (1996) Seedling growth, allocation and leaf attributes in a wide range of woody plant species and types. J. Ecol. 84:755–765.

Cosgrove, D. (1986) Biophysical control of plant cell growth. Annu. Rev. Plant Physiol. 37:377–405.

Cosgrove, D.J. (1993) How do plant cell walls extend? Plant Physiol. 24:1–6.

Creelman, R.A., Mason, H.S., Bensen, R.J., Boyer, J.S., & Mullet, J.E. (1990) Water deficit and abscisic acid cause differential inhibition of shoot versus root growth in soybean seedlings. Analysis of growth, sugar accumulation, and gene expression. Plant Physiol. 92:205–214.

Cyr, D.R. & Bewley, J.D. (1990) Proteins in the roots of perennial weeds chicory (*Cichorium intybus* L.) and dandelion (*Taraxacum officinale* Weber) are associated with overwintering. Planta 182:370–374.

Cyr, D.R., Bewley, J.D., & Dumbroff, E.B. (1990) Seasonal dynamics of carbohydrate and nitrogenous components in the roots of perennial weeds. Plant Cell Environ. 13:359–365.

Dale, J.E. (1988) The control of leaf expansion. Annu. Rev. Plant Physiol. Plant Mol. Biol. 39:267–295.

Darwin, C. (1880) The power of movement in plants. John Murray, London.

Davies, W.J. & Zhang, J. (1991) Root signals and the regulation of growth and development of plants in drying soil. Annu. Rev. Plant Physiol. Mol. Biol. 42:55–76.

Davies, W.J., Tardieu, F., & Trejo, C.L. (1994) How do chemical signals work in plants that grow in drying soil? Plant Physiol. 104:309–314.

Dodd, I.C. & Davies, W.J. (1996) The relationship between leaf growth and ABA accumulation in the grass leaf elongation zone. Plant Cell Environ. 19:1047–1056.

Else, M.A., Davies, W.J., Malone, M., & Jackson, M.B. (1995) A negative hydraulic message from oxygen-deficient roots of tomato plants? Influence of soil flooding on leaf water potential. leaf expansion, and synchrony between stomatal conductance and root hydraulic conductivity. Plant Physiol. 109:1017–1024.

Else, M.A., Tiekstra, A.E., Croker, S.J., Davies, W.J., & Jackson, M.B. (1996) Stomatal closure in flooded tomato plants involves abscisic acid and a chemically unidentified anti-transirant in xylem sap. Plant Physiol. 1012:239–247.

Emery, R.J.N., Reid, D.M., & Chinnappa (1994) Phenotypic plasticity of stem elongation in two ecotypes of *Stellara longipes*: The role of ethylene and reponse to wind. Plant Cell Environ. 17:691–700.

Evans, G.C. (1972) The quantitative analysis of plant growth. Blackwell Scientific Publications. Oxford.

Evans, J.R. (1989) Photosynthesis and nitrogen relationships in leaves of C_3 plants. Oecologia 78:9–19.

Farrar, J.F. (1996) Regulation of root weight ratio is mediated by sucrose. Plant Soil 185:13–19.

Fetene, M. & Beck, E. (1993) Reversal of direction of photosynthate allocation in *Urtica dioica* L. plants by increasing cytokinin import into the shoot. Bot. Acta. 106:235–240.

Fichtner, K. & Schulze, E.D. (1992) The effct of nitrogen nutrition on growth and biomass partitioning of annual plants originating from habitats of different nitrogen availability. Oecologia 92:236–241.

Field, C.B. (1991) Ecological scaling of carbon gain to stress and resourse availability In: Integrated responses of plants to stress, H.A. Mooney, W.E. Winner, & E.J. Pell (eds). Academic Press, San Diego, pp. 35–65.

Fondy, B.R. & Geiger, D.R. (1985) Diurnal changes in allocation of newly fixed carbon in exporting sugar beet leaves. Plant Physiol. 78:753–757.

Fonseca, F., Den Hertog, J., & Stulen, I. (1996) The response of *Plantago major* ssp. *pleiosperma* to elevated CO_2 is modulated by the formation of secondary shoots. New Phytol. 133:627–635.

Garnier, E. (1991) Resource capture, biomass allocation and growth in herbaceous plants. Trends Ecol. Evol. 6:126–131.

Garnier, E. (1992) Growth analysis of congeneric annual and perennial grass species. J. Ecol. 80:665–675.

Garnier, E. & Laurent, G. (1994) Leaf anatomy, specific leaf mass and water content in congeneric annual and perennial grass species. New Phytol. 128:725–736.

Garnier, E. & Vancaeyzeele, S. (1994) Carbon and nitrogen content of congeneric annual and perennial grass species: Relationships with growth. Plant Cell Environ. 17:399–407.

Garnier, E., Gobin, O., & Poorter, H. (1995) Interspecific variation in nitrogen productivity depends on photosynthetic nitrogen use efficiency and nitrogen allocation within the plant. Ann. Bot. 76:667–672.

Gastal, F. & Belanger, G. (1993) The effects of nitrogen fertilization and the growing season on photosynthesis of field-grown tall fescue (*Festuca arundinacea* Schreb.) canopies. Ann. Bot. 72:401–408.

Gould, S.J. & Lewontin, R.C. (1979) The spandrels of San Marco and the Panglossian paradigm: A critique of the adaptationists programme. Proc. R. Soc. Lond. B. 205:581–598.

Gowing, D.J.G., Davies, W.J., & Jones, H.G. (1990) A positive root-sourced signal as an indicator of soil drying in apple, *Malus x domestica* Borkh. J. Exp. Bot. 41:1535–1540.

Green, P.B. (1976) Growth and cell pattern formation on an axis: Critique of concepts, terminology, and mode of study. Bot. Gaz. 137:187–202.

Grime, J.P. (1979) Plant strategies and vegetation processes. John Wiley & Sons, Chichester.

Grime, J.P. & Hunt, R. (1975) Relative growth-rate: Its range an adaptive significance in a local flora. J. Ecol. 63:393–422.

Groeneveld, H.W. & Bergkotte, M. (1996) Cell wall composition of leaves of an inherently fast- and a slow-growing grass species. Plant Cell Environ. 19:1389–1398.

Harpham, N.V.J., Berry, A.W., Knee, E.M., Roveda-Hoyos, G., Raskin, I., Sanders, I.O., Smith, A.R., Wood, C.K., & Hall, M.A. (1991) The effect of ethylene on the growth and development of wild-type and mutant *Arabidopsis thaliana* (L.) Heynh. Ann. Bot 68:55–61.

Hart, R. (1977) Why are biennials so few? Am. Nat. 111:792–799.

Hay, R.K.M. (1990) The influence of photoperiod on the dry-matter production of grasses and cereals. New Phytol. 116:233–254.

He, T. & Cramer, G.R. (1996) Abscisic acid concentrations are correlated with leaf area reductions in two salt-stressed rapid-cycling *Brassica* species. Plant Soil 179:25–33.

Heide, O.M., Bush, M.G., & Evans, L.T. (1985) Interaction of photoperiod and gibberellin on growth and photosynthesis of high-latitude *Poa pratensis*. Physiol. Plant. 65:135–145.

Heilmeier, H. & Monson, R.K. (1994) Carbon and nitrogen storage in herbaceous plants. In: A whole-plant perspective on carbon-nitrogen interactions, J. Roy & E. Garnier (eds). SPB Academic publishing, The Hague pp. 149–171.

Heilmeier, H., Schulze, E.-D., & Whale, D.M. (1986) Carbon and nitrogen partitioning in the biennial monocarp *Arctium tomentosum* Mill. Oecologia 70:466–474.

Hirose, T. & Werger, M.J.A. (1987a) Maximizing daily canopy photosynthesis with respect to leaf nitrogen allocation pattern in the canopy. Oecologia 72:520–526.

Hirose, T. & Werger, M.J.A. (1987b) Nitrogen use efficiency in instantaneous and daily photosynthesis of leaves in the canopy of a *Solidago altissima* stand. Physiol. Plant. 70:215–222.

Hübel, F. & Beck, F. (1996) Maize root phytase. Purification, characterization, and localization of enzyme activity and its putative substrate. Plant Physiol. 112:1429–1436.

Hunt, R. (1982) Plant growth curves. The functional approach to growth analysis. Edward Arnold, London.

Jackson, M.B. (1985) Ethylene and responses of plants to soil waterlogging and submergence. Annu. Rev. Plant Physiol. 36:145–174.

Jaffe, M.J. (1973) Thigmomorphogenesis: The response of plant growth and development to mechanical stimulation. Planta 114:143–157.

Jaffe, M.J. & Forbes, S. (1993) Thigmomorphogenesis: The effects of mechanical perturbation on plants. Plant Growth Regul. 12:313–324.

Jonasson, S. & Chapin III, F.S. (1985) Significance of sequential leaf development for nutrient balance of the cotton sedge, *Eriophorum vaginatum* L. Oecologia 67:511–518.

Kendrick, R.E. & Kronenberg, H.H.M. (eds) (1994) Photomorphogenesis in plants. Kluwer Academic Publishers. Dordrecht.

Keyes, G., Sorrells, M.E., & Setter, T.L. (1990) Gibberellic acid regulates cell wall extensibility in wheat (*Triticum aestivum* L.). Plant Physiol. 92:242–245.

Kigel, J. & Cosgrove, D.J. (1991) Photoinhibition of stem elongation by blue and red light. Effects on hydraulic and cell wall properties. Plant Physiol. 95:1049–1056.

Kimball, B.A., Mauney, J.R., Nakayama, F.S., & Idso, S.B. (1993) Effects of increasing atmospheric CO_2 on vegetation. Vegetation 104/105:65–75.

Kitajima, K. (1994) Relative importance of photosynthetic traits and allocation patterns as correlates of seedling shade tolerance of 13 tropical trees. Oecologia 98:419–428.

Kitajima, K. (1996) Ecophysiology of tropical tree seedling. In: Tropical forest plant ecophysiology, S. Mulkey, R. Chazdon, & A. Smith (eds). Chapman & Hall, New York, pp. 559–596.

Knight, M.R., Campbell, A.K., Smith, S.M., & Trewawas, A.J. (1991) Transgenic plant aequorin reports the effects of touch and cold-shock and elicitors on cytoplasmic calcium. Nature 352:524–526.

Knight, M.R., Smith, S.M., & Trewawas, A.J. (1992) Wind-induced plant motion immediately increases cytosolic calcium. Proc. Natl. Acad. Sci. USA 89:4967–4971.

Konings, H. & Lambers, H. (1991) Respiratory metabolism, oxygen transport and the induction of aerenchyma in roots. In: Plant life under low oxygen: Ecology, physiology and biochemistry, M.B. Jackson, D.D. Davies, & H. Lambers (eds). SPB Academic Publishing, The Hague, pp. 247–265.

Kraus, E., Lambers, H., & Kollöffel, C. (1993) The effect of handling on the yield of two populations of *Lolium perenne* selected for differences in mature leaf respiration rate. Physiol. Plant. 89:341–346.

Kraus, E., Kollöffel, C., & Lambers, H. (1994) The effect of handling on photosynthesis, transpiration, respiration, and nitrogen and carbohydrate content of populations of *Lolium perenne*. Physiol. Plant. 91:631–638.

Kuiper, D. & Staal, M. (1987) The effect of exogenously supplied plant growth substances on the physiological plasticity in *Plantago major* ssp *major*: Responses of growth, shoot to root ratio and respiration. Physiol. Plant. 69:651–658.

Kuiper, D., Kuiper, P.J.C., Lambers, H., Schuit, J.T., & Staal, M. (1989) Cytokinin contents in relation to mineral nutrition and benzyladenine addition in *Plantago major* ssp. *pleiosperma*. Physiol. Plant. 75:511–517.

Kuo, T. & Boersma, L.L. (1971) Soil water suction and root temperature effects on nitrogen fixation in soybeans. Agron. J. 63:901–904.

Lambers, H. & Atkin, O.K. (1995) Regulation of carbon metabolism in roots. In: Carbon partitioning and source-sink interactions in plants, M.A. Madore & W.J. Lucas (eds). American Society of Plant Physiologists, Rockville, MD, pp. 226–238.

Lambers, H. & Poorter, H. (1992) Inherent variation in growth rate between higher plant: A search for physiological causes and ecological consequences. Adv. Ecol. Res. 22:187–261.

Lambers, H., Cambridge, M.L., Konings, H., & Pons, T.L. (eds). (1989) Causes and consequences of variation in growth rate and productivity of higher plants. SPB Academic Publishing, The Hague.

Lambers, H., Poorter, H., & Van Vuuren, M.M.I. (eds). (1998) Inherent variation in plant growth. Physiological mechanisms and ecological consequences. Backhuys, Leiden.

Langheinrich, U. & Tischner, R. (1991) Vegetative storage proteins in poplar. Induction and characterization of a 32- and a 36-kilodalton polypeptide. Plant Physiol. 97:1017–1025.

Leverenz, J.W. (1992) Shade shoot structure and productivity of evergreen conifer stands. Scand. J. For. Res. 7:345–353.

Li, X., Feng, Y., & Boersma, L. (1994) Partitioning of photosynthates between shoot and root in spring wheat (*Triticum aestivum* L.) as a function of soil water potential and root temperature. Plant Soil 164:43–50.

Li, Z.-C. Durachko, D.M., & Cosgrove, D.J. (1993) An oat coleoptile wall protein that induces wall extension *in vitro* and that is antigenetically related to a simular protein from cucumber hypocotyls. Planta 191:349–356.

MacAdam, J.W., Volenec, J.J., & Nelson, C.J. (1989) Effects of nitrogen supply on mesophyll cell division and epidermal cell elongation in tall fescue leaf blades. Plant Physiol. 89:549–556.

Maranon, T. & Grub, P.J. (1993) Physiological basis and ecological significance of the seed size and relative growth rate relationship in Mediterranean annuals. Funct. Ecol. 7:591–599.

Martinez-Garcia, J.F. & Garcia-Martinez, J.L. (1992) Interaction of gibberellins and phytochrome in the control of cowpea elongation. Physiol. Plant. 86:236–244.

Martinoia, E. & Wiemken, A. (1981) Vacuoles as storage compartments for nitrate in barley leaves. Nature 289:292–293.

Masle, J. & Passioura, J.B. (1987) The effect of soil strength on the growth of young wheat plants. Aust. J. Plant Physiol. 14:643–656.

Materechera, S.A., Alston, A.M., Kirby, J.M., & Dexter, A.R. (1993) Field evaluation of laboratory techniques for predicting the ability of roots to penetrate strong soil and of the influence of roots on water absorptivity. Plant Soil 149:149–158.

McArthur, R.H. & Wilson E.O. (1967) The theory of island biogeography. Princeton Univ. Press, Princeton, New Jersey.

McDonald, A.J.S. & Davies, W.J. (1996) Keeping in touch: Responses of the whole plant to deficits in water and nitrogen supply. Adv. Bot. Res. 22:229–300.

McQueen-Mason, S.J. (1995) Expansions and cell wall expansion. J. Exp. Bot. 46:1639–1650.

McQueen-Mason, S.J., Durachko, D.M., & Cosgrove, D.J. (1992) Two endogenous proteins that induce cell wall extension. Plant Cell 4:1425–1433.

Millard, P. (1988) The accumulation and storage of nitrogen by herbaceous plants. Plant Cell Environ. 11:1–8.

Millard, P. & Neilson, G.H. (1989) The influence of nitrogen supply on the uptake and remobilization of stored N for the seasonal growth of apple trees. Ann. Bot. 63:301–309.

Munns, R. & Cramer, G.R. (1996) Is coordination of leaf and root growth mediated by abscisic acid? Plant Soil 185:33–49.

Munns, R. & Sharp, R.E. (1993) Involvement of abscisic acid in controlling plant growth in soil of low water potential. Aust. J. Plant Physiol. 20:425–437.

Neumann, P.M., Azaizeh, H., & Leon, D. (1994) Hardening of root cell walls: A growth inhibitory response to salinity stress. Plant Cell Environ. 17:303–309.

Okamoto, A. & Okamoto, H. (1995) Two proteins regulate the cell-wall extensibility and the yield threshold in glycerinated hollow cylinders of cowpea hypocotyl. Plant Cell Environ. 18:827–830.

Okamoto, A., Katsumi, M., & Okamoto, H. (1995) The effects of auxin on the mechanical properties *in vivo* of cell wall in hypocotyl segments from gibberellin-deficient cowpea seedlings. Plant Cell Physiol. 36:645–651.

Palmer, S.J., Berridge, D.M., McDonald, A.J.S., & Davies, W.J. (1996) Control of leaf expansion is sunflower (*Helianthus annuus* L.) by nitrogen nutrition. J. Exp. Bot. 47:359–368.

Parsons, R.F. (1968) The significance of growth-rate comparisons for plant ecology. Am. Nat. 102:595–597.

Passioura, J.B. (1988) Root signals control leaf expansion in wheat seedlings growing in drying soil. Aust. J. Plant Physiol. 15:687–693.

Passioura, J.B. (1994) The physical chemistry of the primary cell wall: Implications for the control of expansion rate. J. Exp. Bot. 45:1675–1682.

Passioura, J.B. & Fry, S.C. (1992) Turgor and cell expansion: beyond the Lockhart equation. Aust. J. Plant Physiol. 19:565–576.

Peters, W.S. & Tomos, D. (1996) The epidermis still in control? Bot. Acta 109:264–267.

Pianka, E.R. (1970) On r and K selection. Am. Nat. 104:592–597.

Pons, T.L. (1977) An ecophysiological study in the field layer of ash coppice. II. Experiments with *Geum urbanum* and *Cirsium palustre* in different light intensities. Acta Bot. Neerl. 26:29–42.

Pons, T.L. & Bergkotte, M. (1996) Nitrogen allocation in response to partial shading of a plant: Possible mechanisms. Physiol. Plant. 98:571–577.

Pons, T.L., Schieving, F., Hirose, T., & Werger, M.J.A. (1989) Optimization of leaf nitrogen allocation for canopy photosynthesis in *Lysimachia vulgaris*. In: Causes and consequences of variation in growth rate and productivity of higher plants, H. Lambers, M.L. Cambridge, H. Konings, & T.L. Pons (eds). SPB Academic Publishing, The Hague, pp. 175–186.

Poorter, H. (1993) Interspcific variation in the growth response of plants to an elevated ambient CO_2 concentration. Vegetatio 104/105:77–97.

Poorter, H. & Remkes, C. (1990) Leaf area ratio and net assimilation rate of 24 wild species differing in relative growth rate. Oečologia 83:553–559.

Poorter, H., Remkes, C., & Lambers, H. (1990) Carbon and nitrogen economy of 24 wild species differing in relative growth rate. Plant Physiol. 94:621–727.

Poorter, H., Gifford, R.M., Kriedemann, P.E., & Wong, S.C. (1992) A quantitative analysis of dark respiration and carbon content as factors in the growth response of plants to elevated CO_2. Aust. J. Bot. 40:501–513.

Poorter, H., Van de Vijver, C.A.D.M., Boot, R.G.A., & Lambers, H. (1995) Growth and carbon economy of a fast-growing and a slow-growing grass species as dependent on nitrate supply. Plant Soil 171:217–227.

Poorter, H., Roumet, C., & Campbell, B.D. (1996) Interspecific variation in the growth response of plants to elevated CO_2: A search for functional types. In: Biological diversity in a CO_2-rich world, C. Körner & F.A. Bazzaz (eds). Physiological Ecology Series, Academic Press, San Diego, pp. 375–412.

Pritchard, J. (1994) The control of cell expansion in roots. New Phytol. 127:3–27.

Pritchard, J., Hethrington, P.R., Fry, S.C., & Tomos, A.D. (1993) Xyloglucan endotransglycosylase activity, microfibril orientation and the profiles of cell wall properties along growing regions of maize roots. J. Exp. Bot. 44:1281–1289.

Pritchard, J., Fricke, W., & Tomos, D. (1996) Turgor-regulation during extension growth and osmotic stress of maize roots. An example of single-cell mapping. Plant Soil 187:11–21.

Radin, J.W. & Boyer, J.S. (1990) Control of leaf expansion by nitrogen nutrition in sunflower plants: Role of hydraulic conductivity and turgor. Plant Physiol. 69:771–775.

Ramahaleo, T., Alexandre, J., & Lasalles, J.-P. (1996) Stretch activated channels in plant cells. A new model for osmoelastic coupling. Plant Physiol. Biochem. 34:327–334.

Rappoport, H.F. & Loomis, R.S. (1985) Interaction of storage root and shoot in grafted sugarbeet and chard. Crop Sci. 25:1079–1084.

Reich, P.B. (1993) Reconciling apparent discrepancies among studies relating life span, structure and function of leaves in contrasting plant life forms and climates: "The blind men and the elephant retold." Funct. Ecol. 7:721–725.

Reich, P.B., Uhl, C., Walters, M.B., & Ellsworth, D.S. (1991) Leaf life-span as a determinant of leaf structure and function among 23 amazonian tree species. Oecologia 86:16–24.

Reich, P.B., Walters, M.B., & Ellsworth, D.S. (1992a) Leaf life-span in relation to leaf, plant and stand characteristics among diverse ecosystems. Ecol. Monogr. 62:365–392.

Reich, P.B., Walters, M.B., & Ellsworth, D.S. (1992b) Leaf life-span in relation to leaf, plant and stand characteristics among diverse ecosystems. Ecol. Monogr. 62:365–392.

Rosnitschek-Schimmel, I. (1983) Biomass and nitrogen partitioning in a perennial and an annual nitrophilic species of Urtica. Z. Pflanzenphysiol. 109:215–225.

Rozema, J., Lambers, H., Van de Geijn, S.C., & Cambridge, M.L. (1992) CO_2 and Biosphere. Kluwer, Dordrecht.

Russel, G. & Grace, J. (1978) The effects of wind on grasses. V. Leaf extension, diffusive conductance, and photosynthesis in the wind tunnel. J. Exp. Bot. 29:1249–1258.

Russel, G. & Grace, J. (1979) The effects of windspeed on the growth of grasses. J. Appl. Ecol. 16:507–514.

Ryser, P. (1998) Intra- and interspecific variation in root length, root turnover and the underlying parameters. In: Inherent variation in plant growth. Physiological mechanisms and ecological consequences, H. Lambers, H. Poorter, & M.M.I. Van Vuuren (eds). Backhuys, Leiden, pp. 441–502.

Saab, I.N. & Sachs, M.N. (1996) A flooding-induced xyloglucan endo-transglycosylase homolog in maize is responsive to ethylene and associated with aerenchyma. Plant Physiol. 112:385–391.

Saab, I.N., Sharp, R.R., Pritchard, J., & Voetberg, G.S. (1990) Increased endogenous abscisic acid maintains primary root growth and inhibits shoot growth of maize seedlings at low water potentials. Plant Physiol. 93:1329–1336.

Santer, B.D., Wigley, T.M.L., Barnett, T.P., & Anyamba, E. (1996) Detection of climate change and attribution of causes. In: Climate change 1995: The science of climate change, J.T. Houghton, L.G. Meira Filho, B.A.Callander, N. Harris, A. Kattenberg, & K. Maskell (eds). Cambridge University Press, Cambridge, UK, pp. 407–443.

Sauter, J.J. & Van Cleve, B. (1990) Biochemical, immunochemical, and ultrastructural studies of protein storage in poplar (Populus x canadensis "robusta") wood. Planta 183:92–100.

Schulze, E.-D. & Chapin III, F.S. (1987) Plant specialization to environments of different resource availability. In: Potentials and limitations of ecosystem analysis, E.-D. Schulze & H. Zwölfer (eds). Springer-Verlag, Berlin, pp. 120–148.

Simpson, R.J., Lambers, H., Beilharz, V.C., & Dalling, M.J. (1982a) Translocation of nitrogen in a vegetative wheat plant (Triticum aestivum). Physiol. Plant. 56:11–17.

Simpson, R.J., Lambers, H., & Dalling, M.J. (1982b) Kinetin application to roots and its effects on uptake, translocation and distribution of nitrogen in wheat (Triticum aestivum) grown with a split root system. Physiol. Plant. 56:430–435.

Simpson, R.J., Lambers, H., & Dalling, M.J. (1983) Nitrogen redistribution during grain growth in wheat (Triticum aestivum L.). IV. development of a quantitative model of the translocation of nitrogen to the grain. Plant Physiol. 71:7–14.

Smith, H. (1981) Adaptation to shade. In: Physiological processes limiting plant productivity, C.B. Johnson, (ed). Butterworths, London, pp. 159–173.

Smith, R.C., Matthews, P.R., Schunmann, & Chandler, P.M. (1996) The regulation of leaf elongation and xyloglucan endotransglycosylase by gibberellin in "Himalaya" barley (Hordeum vulgare L.). J. Exp. Bot. 47:1395–1404.

Snir, N. & Neumann, P.M. (1997) Mineral nutrient supply, cell wall adjustment and the control of leaf growth. Plant Cell Environ. 20:239–246.

Staswick, P.E. (1988) Soybean vegetative storage protein structure and gene expression. Plant Physiol. 87:250–254.

Staswick, P.E. (1990) Novel regulation of vegetative storage protein genes. Plant Cell 2:1–6.

Staswick, P.E., Huang, J.-F., & Rhee, Y. (1991) Nitrogen and methyl jasmonate induction of soybean vegetative storage protein genes. Plant Physiol. 96:130–136.

Steingröver, E. (1981) The relationship between cyanide-resistant root respiration and the storage of sugars in the taproot in Daucus carota L. J. Exp. Bot. 32:911–919.

Steponkus, P.L. (1981) Responses to extreme temperatures. Cellular and sub-cellular bases. In: Encyclopedia of plant physiology, N.S., Vol. 12A, O.L. Lange, P.S. Nobel, C.B. Osmond, & H. Ziegler (eds). Springer-Verlag, Berlin, pp. 371–402.

Stirzaker, R.J., Passioura, J.B., & Wilms, Y. (1996) Soil structure and plant growth: Impact of bulk density and biopores. Plant Soil 185:151–162.

Stulen, I. & Den Hertog, J. (1993) Root growth and functioning under atmospheric CO_2 enrichment. Vegetatio 104/105:99–115.

Tardieu, F., Zhang, J., Katerji, N., Bethenod, O., Palmer, S., & Davies, W.J. (1992) Xylem ABA controls the stomatal conductance of field-grown maize subjected to soil compaction or soil drying. Plant Cell Environ. 15:193–197.

Ternesi, M., Andrade, A.P., Jorrin, J., & Benloloch, M. (1994) Root-shoot signalling in sunflower plants with confined root systems. Plant Soil 166:31–36.

Terry, N. (1970) Developmental physiology of sugar-beet. II. Effect of temperature and nitrogen supply on the growth, soluble carbohydrate content and nitrogen content of leaves and roots. J. Exp. Bot. 21:477–496.

Todd, G.W., Chadwick, D.L., & Tsai, S.-D. (1972) Effect of wind on plant respiration. Physiol. Plant. 27:342–346.

Touraine, B., Clarkson, D.T., & Muller, B. (1994) Regulation of nitrate uptake at the whole plant level. In: A whole-plant perspective on carbon-nitrogen interactions, J. Roy & E. Garnier (eds). SPB Academic Publishing, The Hague, pp. 11–30.

Van Arendonk, J.J.C.M., Niemann, G.J., Boon, J.J., & Lambers, H. (1997) Effects of N-supply on anatomy and chemical composition of leaves of four grass species, belonging to the genus *Poa*, as determined by image-processing analysis and pyrolysis-mass spectrometry. Plant Cell Environ. 20:881–897.

Van Arendonk, J.J.C.M. & Poorter, H. (1994) The chemical composition and anatomical structure of leaves of grass species differing in relative growth rate. Plant Cell Environ. 17:963–970.

Van den Boogaard, R., Goubitz, S., Veneklaas, E.J., & Lambers, H. (1996) Carbon and nitrogen economy of four *Triticum aestivum* cultivars differing in relative growth rate and water use efficiency. Plant Cell Environ. 19:998–1004.

Van der Werf, A. (1996) Growth analysis and photoassimilate partitioning. In: Photoassimilate distribution in plants and crops: Source-sink relationships, E. Zamski & A.A. Schaffer (eds). Marcel Dekker, New York, pp. 1–20.

Van der Werf, A., Visser, A.J., Schieving, F., & Lambers, H. (1993) Evidence for optimal partitioning of biomass and nitrogen at a range of nitrogen availabilities for a fast-growing and slow-growing species. Funct. Ecol. 7:63–74.

Van der Werf, A. & Nagel, O.W. (1996) Carbon allocation to shoots and roots in relation to nitrogen supply is mediated by cytokinins and sucrose. Plant Soil 185:21–32.

Van Volkenburgh, E. (1994) Leaf and shoot growth. In: Physiology and determination of crop yield, K.J. Boote, J.M. Bennet, T.R. Sinclair, & G.M. Paulsen (eds). American Society of Agronomy, Crop Science Society of America, Soil Science Society of America, Madison, pp. 101–120.

Van Volkenburgh, E. & Boyer, J.S. (1985) Inhibitory effects of water deficit on maize leaf elongation. Plant Physiol. 77:190–194.

Visser, E.J.W., Cohen, J.D., Barendse, G.W.M., Blom, C.W.P.M., & Voesenek, L.A.C.J. (1996) An ethylene-mediated increase in sensitivity to auxin induces adventitious root formation in flooded *Rumex palustris* Sm. Plant Physiol. 112:1687–1692.

Voesenek, L.A.C.J., Banga, M., Thier, R.H., Mudde, C.M., Harren, F.J.M., Barendse, G.W.M., & Blom, C.W.P.M. (1993) Submergence-induced ethylene synthesis, entrapment, and growth in two plant species with contrasting flooding resistance. Plant Physiol. 103:783–791.

Vriezen, W.H., Van Rijn, C.P.E., Voesenek, L.A.C.J., & Mariani, C. (1997) A homologue of the *Arabidopsis thaliana* ERS gene is actively regulated in *Rumex palustris* upon flooding. Plant J. 11:1265–1271.

Wong, S.C. (1993) Interaction between elevated atmpspheric concentration of CO_2 and humidity on plant growth: comparison between cotton and radish. Vegetatio 104/5:211–221.

Witkowski, E.T.F. & Lamont, B.B. (1991) Leaf specific mass confounds leaf density and thickness. Oecologia 88:486–493.

Wu, Y., Sharp, R.E., Durachko, D.M., & Cosgrove, D.J. (1996) Growth maintenance of the maize primary root at low water potentials involves increases in cell-wall extension properties, expansin activity, and wall susceptibility to expansins? Plant Physiol. 111:765–772.

Zhu, G.L. & Boyer, J.S. (1992) Enlargement in *Chara* studied with a turgor clamp. Growth rate is not determined by turgor. Plant Physiol. 100:2071–2080.

Zidan, I., Azaizeh, H., & Neumann, P.M. (1990) Does salinity reduce growth in maize root epidermal cells by inhibiting their capacity for cell wall acidification? Plant Physiol. 93:7–11.

8
Life Cycles: Environmental Influences and Adaptations

1. Introduction

Previous chapters have emphasized the physiological responses of mature plants to their environment. The environmental stresses encountered by plants and optimal physiological solutions, however, can change dramatically from the seedling to mature to reproductive phases of plants. Following germination most species pass through several distinctive life phases: **seedling** (loosely defined as the stage during which cotyledons are still present), **vegetative** (sometimes with a juvenile phase preceding the adult phase), and **adult reproductive**. This chapter addresses the major ecophysiological changes that occur in the life cycles of plants. These involve changes in **development** (i.e., the initiation and occurrence of organs), in **phenology** (i.e., the progress of plants through identifiable stages of development), and in allocation to different plant parts. The pattern and duration of developmental phases depend on environmental conditions and acclimation to specific environments. The developmental pattern also varies genetically, which may reflect special adaptations to specific abiotic or biotic environments. This chapter discusses plant development and processes associated with transition between developmental stages.

2. Seed Dormancy and Germination

Germination is the event that marks the transition between two developmental stages of a plant: seed and seedling. The seed has a package of food reserves that makes it largely independent of environmental resources for its survival. This changes dramatically in the photoautotrophic seedling, which depends on a supply of light, CO_2, water, and inorganic nutrients from its surroundings for autotrophic growth. In this section we will discuss the mechanisms by which some seeds sense the suitability of the future seedling's environment. For example, how does a seed acquire information about the expected light, nutrient, and water availabilities.

Most seeds require water, oxygen, and a suitable temperature for germination, which is a process that occurs when part of the embryo, usually the radicle, penetrates the seed coat. **Dormancy** is defined as a state of the seed that does not permit germination, although conditions for germination may be favorable with respect to temperature, water, and oxygen. Conditions required to break dormancy and allow subsequent germination are often quite different from those that are favorable for growth of the autotrophic life stage of a plant.

Timing of seed germination can be critical for the survival of natural plant populations, and dormancy mechanisms play a major role in such timing. These mechanisms are pronounced in some species (e.g., many ruderals and other species from disturbed habitats), but absent in others (e.g., many tropical tree species). In a dormant seed, the chain of events that leads to germination of the seed is blocked. This block, and hence dormancy itself, can be relieved by a specific factor or combination of

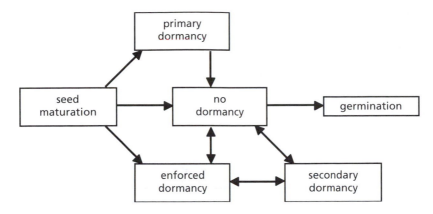

FIGURE 1. Schematic representation of changes in dormancy after seed maturation.

factors (e.g., light, temperature regime, or specific compounds).

In some cases environmental factors, such as the absence of light, nitrate, and/or a diurnally fluctuating temperature, may keep seeds in a dormant state (**enforced dormancy**). Enforced dormancy is sometimes not considered a true dormancy mechanism; rather, it is seen as a mechanism that prevents germination (Vleeshouwers et al. 1995). As soon as conditions become conducive for breaking enforced dormancy, germination can take place. Seeds are in a true dormant state when they do not germinate even if given the stimuli for breaking enforced dormancy and favorable conditions for germination. Breaking of this type of dormancy occurs gradually over weeks or months. Seeds may be dormant upon release from the mother plant (**primary or innate dormancy**), but dormancy can also be induced in a nondormant seed (**secondary or induced dormancy**), if conditions become unfavorable for germination. Transitions among the various forms of dormancy are illustrated in Figure 1.

2.1 Hard Seed Coats

The hard testa (seed coat) of many species (e.g., many legumes) prevents germination because it is largely impermeable to water. Water uptake occurs only when the seed coat is sufficiently deteriorated (Fig. 2).

Deterioration of the seed coat may be due to the activity of microorganisms when seeds are buried in the soil. It may also be due to physical processes, such as exposure to strong temperature fluctuations at the soil surface, as occurs in a desert. In both conditions the breakdown of the seed coat is gradual and, consequently, germination is spread in time. Exposure to short periods of extremely high

temperatures, such as during a fire (approximately 100°C) may lead to synchronous breaking of dormancy and thus account for the massive germination of some species after a fire. Chemical signals, however, may be involved in the response of seeds to fire as well (see Sect. 2.4).

In the testa there is a preformed "weak site," the **strophiole**, where damage first occurs and through which water uptake starts. Dormancy associated with a hard seed coat often complicates germination for plant cultivation purposes. Dormancy in hard-coated seeds can be relieved artificially by mechanical (sanding or breaking the seed coat) or chemical (concentrated sulfuric acid) treatments.

2.2 Germination Inhibitors in the Seed

Arid climates are characterized by little precipitation, often concentrated in just a few showers in an unpredictable manner. After such a shower, massive seed germination of short-lived species may take place. How can the seeds perceive that the environment has become more favorable for germination and subsequent seedling development?

A common trait of many species germinating under such conditions is the presence of **water-soluble inhibitors** in the pericarp (matured ovulary wall) and/or testa. A small shower does not remove these inhibitors, so germination cannot take place (Fig. 3). Germination occurs only after a major shower or prolonged rain that washes away the inhibitor; in this case the emerged seedling has access to sufficient water to enhance its chances to survive and complete its life cycle. The substance that inhibits germination may be either a specific organic compound or accumulated salts. Following moderate rains, those seeds that fail to germinate synthesize additional inhibitors; therefore, sub-

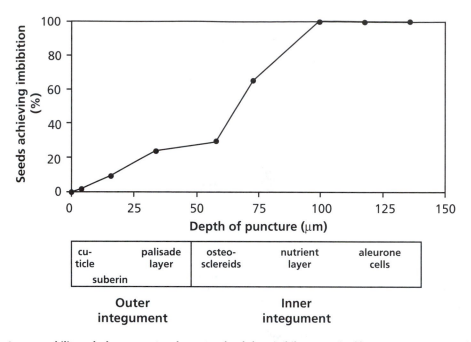

FIGURE 2. Impermeability of the testa (seed coat) of *Coronilla varia*. The testa was pierced to varying depths by a 0.4-mm diameter indentor, after which the seeds were left to imbibe on moist filter paper (McKee et al. 1977, as presented in Bewley & Black 1982).

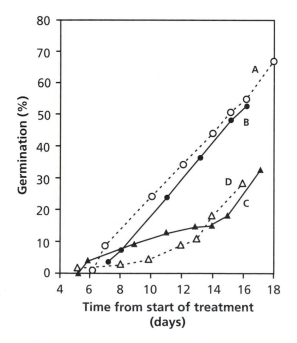

FIGURE 3. Time course of germination of *Oryzopsis miliacea* as affected by duration of a drip treatment. The origin of the x-axis represents the start of the drip treatment. Curves A, B, C, and D refer to a duration of the treatment of 93, 72, 48, and 24 hours, respectively. Control seeds did not germinate (Koller & Negbi 1959). Copyright Ecological Society of America.

sequent rains must still be substantial to trigger germination.

Germination inhibitors also play an important role in preventing germination in seeds in a fleshy fruit. The high solute concentration of many fruits plays an important role, but ABA can also have an essential function here, as illustrated by the germination of the seeds of *Lycopersicon esculentum* (tomato) inside the fruit of ABA-deficient mutants (Karssen & Hilhorst 1992).

2.3 Effects of Nitrate

Germination of many seeds is stimulated by nitrate (Fig. 4). This role of nitrate as an environmental trigger is not associated with a need for nitrate for protein synthesis because no nitrate reductase activity is detected in seeds. Rather, nitrate functions as a factor breaking dormancy, especially in many weedy and other ruderal species. When the mother plant has grown at a nitrogen-rich site, seeds may accumulate nitrate and then lose the requirement for nitrate to trigger germination. Why would weedy and ruderal species use nitrate as an environmental cue?

Nitrate requirement may function as a **gap-detection** mechanism, just like other factors involved in enforced dormancy: light and diurnal

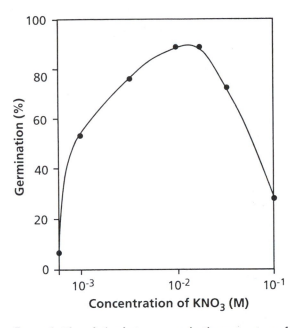

FIGURE 4. The relation between germination percentage of seeds of *Epilobium montanum* and KNO_3 concentration. Germination took place in the dark for 14 days at 16 to 20°C (Hesse 1924, as cited in Mayer & Polyakoff-Mayber 1982). Copyright Butterworth-Heinemann.

temperature fluctuation. Seeds in soil where a large plant biomass depletes soil nitrate experience a low-nitrate environment, which enforces dormancy. When the vegetation is destroyed, mineralization and nitrification continue, but absorption by plants is reduced. This increases soil nitrate concentrations to levels that break dormancy (Fenner 1985, Pons 1989).

2.4 Other Chemical Signals

Various compounds in the natural environment of seeds may have stimulating or inhibiting effects on seed germination (Karssen & Hilhorst 1992). The inhibition of buried seeds often cannot be explained by the absence of light or alternating temperatures alone. The gaseous environment may play a role (low O_2 and high CO_2), but specific organic compounds, such as allelochemicals, may also be involved.

A stimulating effect on germination of specific compounds, for instance, was found in the case of smoke. De Lange and Boucher (1990) discovered that smoke derived from the combustion of plant material stimulates seed germination of *Audouinia capitata*, a fire-dependent South African fynbos

species. Exposure of dormant seed to cold smoke derived from burnt native vegetation also promotes seed germination of many species from Western Australia (Dixon et al. 1995), the Californian chaparral (Keeley 1991), and the English moorlands (Legg et al. 1992). Chemicals in cold smoke promote germination of seeds that are normally hard to germinate (Fig. 5). Several components play a role (e.g., ethylene, ammonia and octanoic acid) (Sutcliffe & Whitehead 1995). It is remarkable that triggering of germination by chemicals in smoke is not restricted to plants in fire-dominated ecosystems (Strydom et al. 1996, Thomas & Van Staden 1995), and commercial "smoke" is now available to enhance the germination rate of seeds that are difficult to germinate and to promote seed germination for mine rehabilitation in Western Australia (Roche et al. 1997).

Any ecological advantage of the capacity to respond to compounds present in smoke for species that do not occur in a fire-dominated system remains to be demonstrated. The fact that some of the chemicals in smoke (e.g., ammonia and octanoic acid) also occur in soil where no burning has taken place indicates that seeds might use the chemical cues as a mechanism to detect major changes in the vegetation, as suggested for a seed's capacity to respond to nitrate (Sect. 2.3).

2.5 Effects of Light

Light is an important factor that determines seed dormancy (Pons 1992). A wide variety of light responses of seeds have been described. These can be subdivided in two main types: responses to light doses of short duration and effects of long-duration exposures. The light response of seeds depends strongly on other simultaneously occurring environmental factors, such as temperture, water potential, and nitrate. The responses also depend on earlier conditions that affect dormancy status, such as temperature regime.

The light climate under natural conditions has many aspects, and some of these are used by seeds for regulation of dormancy. Three major roles for light responses can be distinguished.

1. A **light-requirement** prevents germination of seeds that are buried too deep **in soil**. Such seeds, which only germinate when exposed to light, do not germinate below the soil depth where no light penetrates. This prevents "fatal germination" of the predominantly small seeds in which this mechanism is frequent.

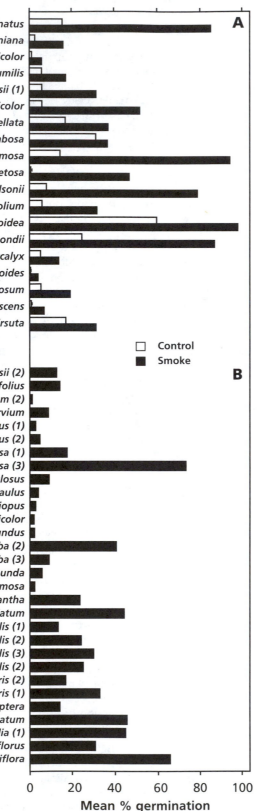

FIGURE 5. Glasshouse germination studies with Western Australian species. (A) Species for which there is a significant difference in germination between control (open bars) and smoked (filled bars). (B) Species that did not germinate in the absence of smoke (Dixon et al. 1995).

2. A **light-requirement** further breaks dormancy after **soil disturbance**. Germination occurs only when the soil is turned over and the seeds reach the soil surface where they are exposed to light. A short exposure is sufficient to trigger the response. This often coincides with damage or the complete disappearance of the established vegetation. The emergent seedlings thus have a more favorable position among established plants than they would have otherwise.

3. The **spectral composition** of daylight as modified by a leaf canopy is also important for the **timing of germination after disturbance** of vegetation. Light under a leaf canopy is depleted in red compared with that above the canopy (Fig. 6), resulting in a low red : far-red ratio. This enforces dormancy in many species (Fig. 7). Germination of such seeds is prevented, as long as the seeds remain under the leaf canopy on the soil surface. This is particularly important shortly after seed shedding. The seeds may subsequently end up in the soil, where a light-requirement further enforces dormancy and where the chances to be predated are smaller than at the soil surface. Litter, especially dry litter, also decreases the red/far-red ratio, which further reduces the probability of germination (Vazquez-Yanes et al. 1990).

Perception of light *per se* as well as the response to the spectral composition of the light involves the

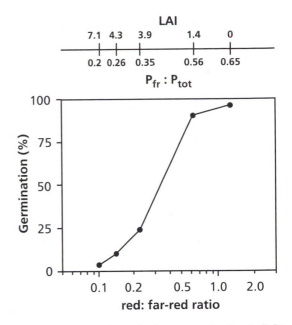

FIGURE 7. Germination of *Plantago major* in daylight under stands of *Sinapis alba* of different densities resulting in different red/far-red photon ratios of the transmitted light. Corresponding leaf area index (LAI) and phytochrome photoequilibria (P_{fr} : P_{total} ratios) are shown (Frankland and Poo 1980, Pons 1992). Copyright C.A.B. International.

phytochrome system (Box 9 in the chapter on growth and allocation). Seeds with a dormancy mechanism involving phytochrome require a minimum amount of the far-red-absorbing form of phytochrome (P_{fr}) to break dormancy. Light with a high red : far-red ratio enhances the formation of P_{fr}. When the seeds are exposed to light with a low red : far-red ratio, less P_{fr} is formed. The amount of P_{fr} is also determined by dosage of light in the nonsaturating region. The amount of P_{fr} required for germination depends on environmental conditions and dormancy status, and it also strongly varies among species. Hence, a low red : far-red ratio inhibits seed germination only in some species under certain conditions.

If, after exposure to light of appropriate spectral composition, germination is subsequently impaired by some other environmental factor, then a new exposure to light is required to break dormancy. This is due to the decay of P_{fr} in the dark. This mechanism also explains why seeds that are initially not light-requiring upon ripening do become so after burial in the soil (Pons 1991b). A requirement for light for breaking dormancy is clearly not a fixed characteristic of a species.

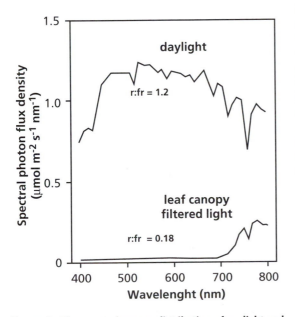

FIGURE 6. The spectral energy distribution of sunlight and light filtered through a leaf canopy. Red/far-red ratios (660/730) are also shown (Pons 1992). Copyright C.A.B. International.

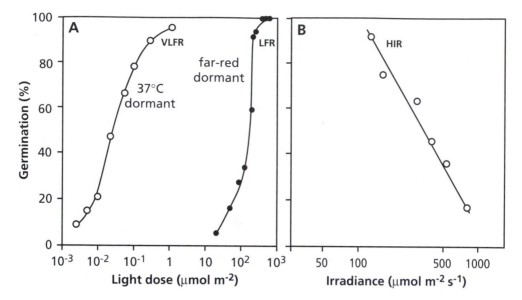

FIGURE 8. The three light responses of seed germination demonstrated in one species *Lactuca sativa* (lettuce). (A) Fluence response to red light of seeds pretreated at 37°C and with far-red showing the very low fluence response (VLFR) and the low fluence response (LFR), respectively (Blaauw-Jansen & Blaauw 1975). Copyright Royal Botanical Society of the Netherlands. (B) Irradiance response to daylight showing the high-irradiance response (HIR) (Gorski & Gorska 1979).

Seeds that are not obviously light-requiring may still have a dormancy mechanism that is regulated by phytochrome. In such seeds there may be sufficient P_{fr} in the ripe seeds to allow germination in the dark.

The light responses described earlier are typically referred to as the **low-fluence response** (LFR). That is, a rather low light dosage is required to give the response. The total light dosage integrated over a period of time is relevant here. Some seeds under certain conditions respond to much lower light doses (three to four orders of magnitude) with the breaking of dormancy (Fig. 8). Such a response is called the **very low fluence response** (VLFR). The two responses can be found in the same seeds, depending on pretreatment as illustrated for lettuce seeds in Figure 8. Transition between LFR and VLFR is also found to vary seasonally during burial of seeds in soil (Derkx & Karssen 1993). The significance of the VLFR under natural conditions is probably related to the response to the short exposures to light that occur during soil disturbance (Scopel et al. 1994).

Recently, different forms of phytochrome have been identified. Their roles in developmental processes in plants have been studied using mutants deficient in one form of phytochrome. Phytochrome A is required for the VLFR and phytochrome B for the LFR (Botto et al. 1996).

Most seeds can also be inhibited by exposure to light. This is evident only when exposure times are long. The inhibiting effect increases with increasing irradiance (Fig. 8), and the maximum effective wavelength region is 710 to 720 nm. This response is called the **high-irradiance response** (HIR). The cycling between P_r and P_{fr} is somehow involved in the HIR, but the mechanism is not fully understood. Seeds that are negatively photoblastic, which means that their germination is prevented by light, have a strongly developed HIR. Short exposures and low irradiances are not inhibitory, and they sometimes even stimulate germination in such seeds. Hence, their germination is not always inhibited by light, as the term suggests.

Light responses of seeds have been extensively studied with short exposures to light (LFR and VLFR). Seeds, however, mostly experience long exposure times under natural conditions. For seeds under a leaf canopy, both the photoequilibrium of phytochrome and the HIR are important because seeds experience daylength exposure times to wavelengths that are effective. Hence, the inhibiting effect of a leaf canopy can be stronger than expected from the spectral composition alone (Frankland and Poo 1980).

Seeds on the surface of bare soil may be inhibited by the HIR due to the prevailing high irradiances. In light-requiring seeds, this may restrict germination to the upper few millimeters of the soil profile where light penetrates, but does not reach a high intensity. The combination of multiple light responses causes germination to occur in the uppermost millimeters of the soil, where moisture is available and the seedling can reach the soil surface.

2.6 Effects of Temperature

Temperature influences seed dormancy in several ways:

1. The **diurnal fluctuation** in temperature controls the dormancy of many seeds. This form of enforced dormancy is relieved by increasing temperature fluctuation. The response is independent of the absolute temperature, which illustrates that it is the amplitude that causes the response (Fig. 9). This mechanism prevents germination of seeds **buried deep** in the soil, where temperature fluctuations are damped. Seeds in unvegetated soil are similarly exposed to larger temperature fluctuations than are seeds under a canopy. Hence, the ability to perceive temperature fluctuations allows the detection of soil depth and of gaps in the vegetation.

Most small-seeded marsh plants also respond to diurnally fluctuating temperature, which indicates the absence of deep water over the seed. Hence, in these plants temperature fluctuation functions as a mechanism to detect **water depth**.

2. The **temperature range** over which germination occurs is an indication of the degree of true dormancy of the seed. If this range is narrow, then the seed is strongly dormant. If it is wider, then the seed is less dormant or nondormant. Variation in this temperature range may occur as a result of a shift in the upper and/or lower critical temperature limits for germination.

3. The **temperature** to which the seed is exposed when no germination takes place is a major factor in determining release and induction of true dormancy. Two main types of responses are discerned in climates with seasonally changing temperatures:

a. **Summer annuals** and other species that produce seeds in autumn that germinate in spring. A long exposure (1 to 4 months) of imbibed seeds to low temperature (approximately 4°C) breaks dormancy; this process is termed **stratification**, or chilling. In many species with a persistent seed bank, secondary dormancy is subsequently induced by exposure to higher summer temperatures (e.g., 20°C), which causes large seasonal changes in the degree of dormancy (Fig. 10). This seasonal change in dormancy restricts germination to spring, which is the beginning of the most suitable season for growth in temperate climates (Fig. 11).

b. **Winter annuals.** Exposure to relatively high temperatures relieves the dormancy, even without imbibition having taken place. In this case,

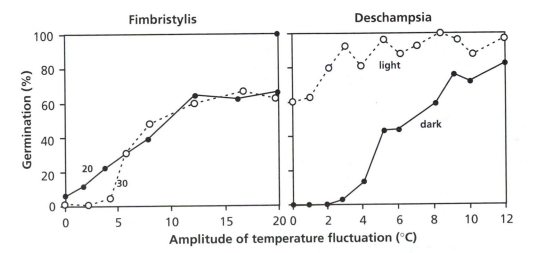

FIGURE 9. Germination responses to various amplitudes of diurnal temperature fluctuations. (Left) The light-requiring rice-field weed *Fimbristylis littoralis* at mean temperatures of 20° and 30°C (Pons & Schröder 1986). (Right) The grass species *Deschampsia flexuosa* in light and darkness (Thompson et al. 1977). Reprinted with permission from Nature © copyright 1997 Macmillan Magazines Ltd.

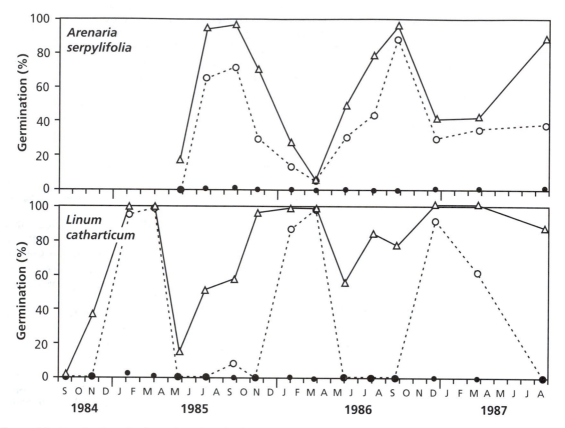

FIGURE 10. Germination of exhumed seeds under laboratory conditions after different burial times in a chalk grassland in South Limburg, the Netherlands. *Arenaria serpylifolia*, which is a winter annual, and *Linum catharticum*, which is a biennial that emerges in spring. Germination in darkness (closed symbols) and in light (open symbols), solid line final germination percentage, dashed line, germination after 1 week at 22/12°C (Pons 1991a). Copyright Blackwell Science Ltd.

low temperatures induce dormancy (Fig. 10). This seasonal dormancy pattern causes the seeds to germinate in autumn (Fig. 11), which is the beginning of the most suitable season for many species from Mediterranean climates.

Seeds may go through several cycles of induction and release of dormancy if enforced dormancy prevents germination (e.g., by the light-requirement of seeds buried in the soil) (Fig. 1).

It is interesting that water supply is the factor that makes winter the most favorable season for growth of winter annuals and, thus, autumn the best period for germination; however, the dormancy is controlled by temperature. In many seasonal climates, such as the Mediterranean climate, temperature and water supply are closely correlated, but temperature is a better predictor of the beginning of the wet season than is moisture itself. In summer annuals, it is the low temperature in winter that releases dormancy in the seeds and, hence, it is used as a signal; however, the subsequently oc-

curring high temperatures in summer form the suitable conditions for growth of the autotrophic plant.

2.7 Physiological Aspects of Dormancy

Many studies have examined the mechanisms, particularly the role of **phytohormones** (Box 8), that regulate dormancy and germination of seeds. Little progress was made, however, until mutants became available that are deficient in the synthesis of a phytohormone or that have a reduced sensitivity to a phytohormone. On the basis of this type of research, a fascinating view has emerged that probably applies to many seeds (Hillhorst & Karssen 1992).

During seed development on the mother plant, there is an increase in concentration of **abscisic acid** (ABA). This phytohormone is involved in the prevention of precocious germination, synthesis of reserve proteins, the development of

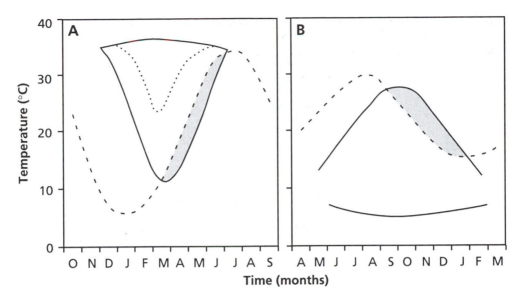

FIGURE 11. Widening and narrowing of the temperature range of germination in relation to the temperature in the natural habitat during the season. The broken line gives the mean daily maximum temperature in the field; the continuous line gives the temperature range for germination in light; the dotted line represents the minimum temperature for germination in darkness. In the hatched area, the actual and the required temperatures in light overlap. The arrow indicates the time that the threshold temperature for germination is reached in spring (A, summer annual) and autumn (B, winter annual) (Karssen 1982).

desiccation tolerance, and the induction of primary dormancy.

Induction of and release from primary dormancy involves changes in both the concentration of ABA and the sensitivity to this phytohormone. The

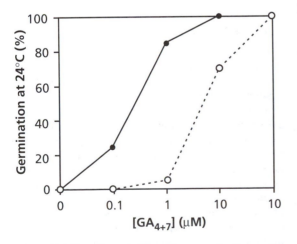

FIGURE 12. The effect of gibberellin concentration on the germination of a GA-deficient mutant of *Arabidopsis thaliana* in darkness at 24°C. Seeds directly sown (open symbols), or preincubated at 2°C for 7 days (filled symbols) (Hillhorst & Karssen 1992). Copyright Kluwer Academic Publishing.

phytohormone **gibberellic acid** (GA) is also involved in the release of dormancy (Steinbach et al. 1997). Release from true dormancy is typically accompanied by an increase in sensitivity to GA (Fig. 12), whereas, with release from enforced dormancy, GA is synthesized de novo. GA is involved in the induction of enzymatic hydrolysis of reserve carbohydrates, especially of galactomannan-rich endosperm cell walls. Cell-wall hydrolysis weakens the endosperm layer, so that the radicle of the embryo can penetrate the seed coat leading to the germination event.

Induction of secondary dormancy, as they occur in buried seeds, is accompanied by a decrease in the sensitivity to GA. Phytohormone receptors in the plasma membrane could be affected by the temperature-dependent state of membranes, thus at least partly explaining the effect of temperature on dormancy. The change in sensitivity to GA is reflected in sensitivity for environmental stimuli that break enforced dormancy by stimulating GA synthesis.

2.8 Summary of Ecological Aspects of Seed Germination and Dormancy

In this chapter we have described the possible ecological significance of the environmental factors

TABLE 1. A summary of the possible ecological significance of a range of mechanisms that control dormancy.

Type of dormancy	Environmental factor involved	Ecological role
Enforced dormancy	• light	• gap detection • sensing depth in soil
	• diurnal temperature fluctuation	• gap detection • sensing depth in soil and water
	• nitrate	• gap detection
	• inhibitor in seed coat	• detection of water availability
True dormancy	• smoke signals • seasonal temperature regime • hard seed coat	• response to fire • detection of suitable season • response to fire • spreading risks in time

that control dormancy. These environmental cues lead to a **timing** of germination that maximizes the chances of seedling survival and subsequent reproductive success. Table 1 provides a summary of these germination cues. Those germination cues that indicate presence of disturbance (light, diurnal temperature fluctuation, nitrate, and other compounds) are typically best developed in early-successional species. In the absence of these cues, these species enter long-lasting seed reserves ("seed banks") in the soil, where they can remain for tens to hundreds of years until the next disturbance occurs. By contrast, late-successional species have short-lived seeds that are produced regularly and have poorly developed seed dormancy mechanisms. As a result, these species are poorly represented in the seed bank. The viability of seeds in the seed bank declines with time, but it is quite common for the seed bank to be a major source of germinants, even when disturbance occurs more than a century after the previous disturbance that gave rise to the seed bank.

3. Developmental Phases

Most species pass through several distinctive life phases after germination. Plants grow most rapidly, but they are most vulnerable to environmental stress and to the effects of competition during the seedling phase. There is then a gradual transition from the seedling to the juvenile phase, where many species allocate significant resources to defense and storage. Finally, there is an abrupt hormonally triggered shift to the reproductive phase, where some shoot meristems produce reproductive rather than vegetative organs. The response of plants to the environment often differs among these developmental phases, and species differ substantially in the timing and triggers for phase shifts. For example, annuals rapidly switch to their reproductive phase, whereas perennials remain vegetative for a longer time, sometimes many years. Biennials are programmed to complete their life cycle within 2 years, but this may last longer if environmental conditions are less favorable. What are the physiological differences between plants with these contrasting strategies, and how is the program in biennials modified by the environment?

3.1 Seedling Phase

Seedlings are susceptible to many abiotic and biotic stresses after germination. Due to their small root systems seedlings are vulnerable to desiccation from minor soil drying events, so there is strong selection for rapid root extension. Where seedling densities are high, there is also strong competition for light, and even 1 or 2 days advantage in time of germination is a strong determinant of competitive success (Harper 1977). Most plant mortality occurs in the seedling phase through the interactive effects of environmental stress, competition, pathogens, and herbivory, so there is strong selection for rapid growth at this vulnerable phase to acquire resources (leaves and roots) and to grow above neighbors (stem) (Cook 1979). In most species, this can be achieved only through minimal allocation to storage or defense.

Seed size is a major determinant of initial size and absolute growth rate (g day^{-1}) of seedlings (Leishman et al. 1995) (Fig. 13). Species that colonize disturbed open sites with minimal competition typically produce abundant small seeds, which maximizes the probability of a seed encountering a disturbed patch but minimizes the reserves available to support initial growth and survivorship (Leishman & Westoby 1994) (Fig. 14). Trees, shrubs, and woodland herbs, however, which confront stronger competition at the seedling stage, often produce a few large seeds (Fenner 1985, Shipley & Dion 1992). Thus, for a given reproductive allocation, there is a clear **trade-off** between **seed size** and **seed number**, with seed size generally favored in species that establish in closed vegetation. It is interesting that a small seed size is one of the few traits that differentiates rare from common species of grass (Rabinowitz 1978), perhaps because of the longer dispersal distance associated with rare species.

Many tropical trees and some temperate trees produce extremely large nondormant seeds that germinate, grow to a small size, and then cease growth until a branch or treefall opens a gap in the canopy. This **seedling bank** is analogous to the seed bank of ruderal species in that it allows new recruits to persist in the environment until disturbance creates an environment favorable for seedling establishment. Large **seed reserves** to support maintenance respiration are essential to species that

form a seedling bank. There is a strong negative relationship between seed size and survivorship in shade (Fig. 15). In contrast to the situation in rapidly growing seedlings, the leaves of seedlings in the seedling bank are extremely well defended against herbivores and pathogens. These seedlings quickly resume growth following disturbance and have a strong initial competitive advantage over species that persist as a seedbank in the soil.

3.2 Juvenile Phase

There is a gradual transition from a seedling phase with minimal reserves to a juvenile phase with accumulation of some reserves to buffer the plant against unfavorable environmental conditions. There are striking differences among plants, however, in the length of the juvenile phase and the extent of reserve accumulation. At one extreme, *Chenopodium album* (pigweed) can be induced to flower at the cotyledon stage immediately after germination, whereas some trees may grow for decades before switching to reproduction (e.g., 40 years in *Fagus sylvatica*, beech). The switch to reproduction is typically hormonally mediated.

Annuals allocate relatively little of their acquired resources (carbon and nutrients) to storage, whereas perennials are characterized by storage of both nutrients and carbohydrates. The greater resource allocation to storage, rather than to leaf area,

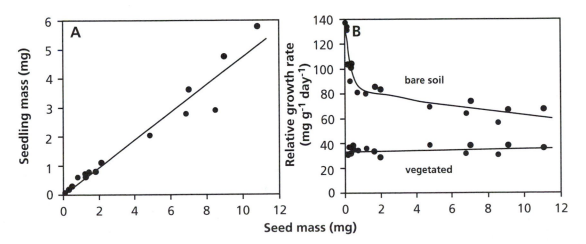

FIGURE 13. Relationship between seed mass of prairie perennials and (A) mass of newly emerged seedlings (<12 h) or (B) relative growth rate of seedlings on bare soil and in a mat of *Poa pratensis* in the glasshouse. Absolute plant size increases with increasing seed mass. Relative growth rate decreases with increasing seed size in absence of competition, but is independent of seed size in presence of competition. Species are *Verbascus thapsus, Oenothera biennis, Daucus carota, Dipsacus sylvestris, Tragopogon dubius,* and *Arctium minus* (Gross 1984). Copyright Blackwell Science Ltd.

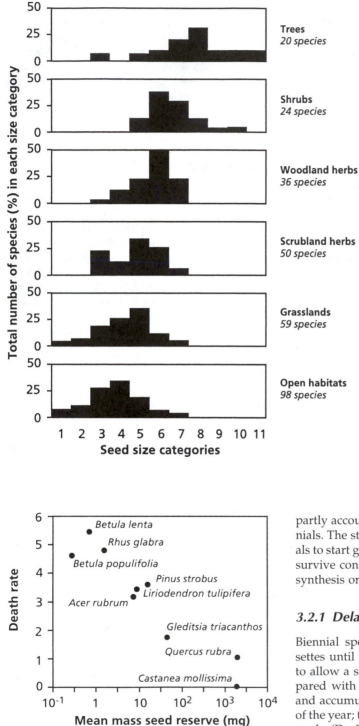

FIGURE 14. Frequency distribution of seed size in different ecological groups of plants (Fenner 1985, based on data of Salisbury 1942). Species that establish in closed habitats tend to have larger seeds than open-habitat plants. Copyright Chapman and Hall.

FIGURE 15. Relationship between death rate (mean number of fatalities per container in 12 weeks in shade) and mean mass of seed reserve in nine North American tree species (Grime & Jeffrey 1965). Copyright Blackwell Ltd.

partly accounts for the lower growth rate of perennials. The stored reserves, however, allow perennials to start growth early in a seasonal climate and to survive conditions that are unfavorable for photosynthesis or nutrient acquisition.

3.2.1 Delayed Flowering in Biennials

Biennial species typically grow as vegetative rosettes until the storage pools are sufficiently filled to allow a switch to the reproductive phase. Compared with an annual, biennials are able to grow and accumulate nutrients throughout a larger part of the year; therefore, they are able to produce more seeds (De Jong et al. 1987). Biennials may grow longer than 2 years at a low irradiance (Pons & During 1987) or low nutrient supply if their stores are not filled sufficiently to induce a switch to flowering (Table 2). In general, shifts from one developmental phase to another correlate more closely with

TABLE 2. Probability of flowering in *Cirsium vulgare* (spear thistle) of small rosettes after transfer from the field to a long-day regime in a growth room in February.

Treatment	Probability of flowering (%)	Average time before bolting (days)
Without nutrients	25	45
With nutrients	80	40

Source: Klinkhamer et al. 1986.
Note: A control group in the field showed 13% flowering.

plant size than with plant age. Hence, the term **biennial** is not appropriate; **monocarpic perennial** is used instead to indicate that the plant terminates its life cycle once the transition to the reproductive stage has been made. Vegetative growth in monocarpic perennials can also be very long [e.g., in palm (sago) and agave species].

3.2.2 Juvenile and Adult Traits

In woody plants there is a distinctive suite of morphological and chemical traits that disappear when the plant becomes reproductively mature. Juvenile plants are typically more strongly defended against herbivores, either by producing spines (e.g., apple or orange trees) or a variety of chemical defenses (Bryant & Kuropat 1980). Many woody species show a difference in morphology between their **juvenile** and **adult foliage**. For example, the young foliage of many *Acacia* (wattle) species in Australia and elsewhere is characterized by bipinnate leaves, whereas older individuals produce "phyllodes" (compressed petioles) (New 1984). Phyllodinous species in which the juvenile foliage persists longest are generally native to moist regions, whereas phyllodes that are reduced to small whorled spines are common in *Acacia* species from many (semi)arid zones. *Acacia* species commonly show a mosaic of bipinnate leaves and phyllodes, with the highest frequency of bipinnate leaves under more favorable conditions. In *A. pycnantha*, a shade-tolerant forest species, seedlings produce predominantly juvenile foliage for more than 9 months if growing in the shade, and they show a high survival rate and high leaf area ratio. When grown in full sun, they become entirely phyllodinous after a few months. Treatment with the phytohormone gibberellic acid favors production of the bipinnate leaves.

Acacia melanoxylon (blackwood) is another Australian forest species with a mosaic of leaves, like the Hawaiian shade-intolerant *A. koa* (koa) that grows at sites characterized by unpredictable drought periods. It has been suggested that the bipinnate *Acacia* leaves function as shade leaves, whereas the phyllodes may be sun leaves. To test this hypothesis, gas-exchange characteristics of the contrasting leaves have been determined (Table 3). The juvenile *Acacia* leaves have higher rates of photosynthesis (on a leaf mass and leaf N basis) and transpiration (leaf area basis), but a lower water-use efficiency and leaf water potential when compared with the adult phyllodes. The traits of the juvenile leaves promote establishment (rapid growth),

TABLE 3. Gas-exchange characteristics and aspects of the leaf chemical composition and anatomy of juvenile bipinnate leaves and adult phyllodes of *Acacia koa* (koa), a shade-intolerant endemic tree from Hawaii.

Parameter	Juvenile bipinnate leaves	Adult phyllodes
Light-saturated rate of CO_2 assimilation ($\mu mol\,m^{-2}\,s^{-1}$)	11.1	12.1
Light-saturated rate of CO_2 assimilation ($\mu mol\,g^{-1}\,s^{-1}$)	0.8	0.5
Stomatal conductance (daily mean) ($mol\,m^{-2}\,s^{-1}$)	0.4	0.3
Transpiration (daily mean) ($mmol\,m^{-2}\,s^{-1}$)	7.5	6.9
Water-use efficiency (daily mean) [$mmol\,CO_2\,(mol\,H_2O)^{-1}$]	1.3	1.5
Internal CO_2 pressure at light saturation (Pa)	28.2	27.4
Carbon isotope discrimination (‰)	19.7	18.0
Leaf water potential (MPa)	−1.2	−0.9
Leaf N concentration ($mmol\,g^{-1}$)	2.1	1.7
Photosynthetic nitrogen-use efficiency [$mmol\,CO_2\,(mol\,N)^{-1}\,s^{-1}$]	0.24	0.20
C/N ($mol\,mol^{-1}$)	19.3	24.6
Leaf mass per unit area (LMA) ($kg\,m^{-2}$)	0.14/0.10*	0.24/0.51*

*The values are for open and understory habitats, respectively.
Source: Hansen 1986, 1996.

whereas the phyllodes are more like the leaves of slow-growing stress-tolerant species.

3.2.3 Vegetative Reproduction

Many plants such as grasses or root-sprouting trees have a modular structure composed of units, each of which has a shoot and root system. This "vegetative reproduction" can be viewed simply as a form of growth, as described in the chapter on growth and allocation, or as a mechanism of producing physiologically independent individuals without going through the bottleneck of reproduction and establishment (Jonsdottir et al. 1996).

Vegetative reproduction is best developed in environments where flowering is infrequent and seedling establishment is a rare event. For example, clones of an arctic sedge, *Carex aquatilis*, are estimated to be thousands of years old as a result of continual production of new tillers by vegetative reproduction (Shaver et al. 1979). In this situation, the carbon cost of producing a new tiller by sexual reproduction is estimated to be 10,000-fold greater than the cost of a new tiller by vegetative reproduction, because of very low rates of seedling establishment (Chapin et al. 1980). Aspen (*Populus tremuloides*) clones in the Rocky Mountains of the central United States are similarly estimated to be of Pleistocene age as a result of root sprouting. This is an effective mechanism of maintaining a given genotype under conditions where sexual reproduction is a rare event. The **trade-off** is that vegetative clones often lack the genetic diversity for long-term evolutionary change.

Clonal growth is one mechanism by which plants can explore patchy habitats. For example, daughter ramets (i.e., a unit composed of a shoot and root) of *Fragaria chiloensis* (strawberry) draw on reserves of the parental ramet to grow vegetatively. If the daughter ramet encounters a resource-rich patch, it produces additional ramets, whereas ramets that move into resource-poor patches fail to reproduce vegetatively. Resource translocation can also occur between established ramets of clonal plants, supporting damaged or stressed ramets growing under relatively unfavorable conditions (Chapman et al. 1992, Jondottir et al. 1996). When the roots of one ramet of *Trifolium repens* (white clover) are in a dry patch, whereas those of another are well supplied with water, relatively more roots are produced in the wet patch. On the other hand, when leaves of one ramet are exposed to high irradiance, whereas those of another are in the shade, the ramet exposed to high irradiance produces relatively more leaf mass (Fig. 16).

The data on *Trifolium repens* (Fig. 16) suggest that ramets can exchange captured resources. To test this in another clonal plant, *Potentilla anserina* (silverweed), phloem transport was interrupted by "steam girdling"; this leaves the xylem intact.

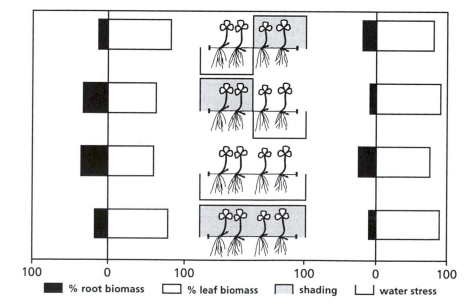

FIGURE 16. Percentage biomass allocation to leaves and roots of two interconnected ramets of *Trifolium repens* (white clover) (Stuefer et al. 1996). Copyright Blackwell Science Ltd.

The shaded ramet produces less shoot and root biomass than does the control, with its phloem connection still intact, whereas the growth of the ramet in dry soil is not affected by the treatment. In fact, the same amount of water is removed from the wet soil in treated and control plants. This experiment confirms that carbohydrates can be exported from the sun-exposed ramet to the shaded one (Stuefer 1995).

The developmental process by which vegetative reproduction occurs differs among taxonomic groups. These mechanisms include production of new tillers (a new shoot and associated roots) in grasses and sedges, initiation of new shoots from the root system (root suckering) in some shrubs and trees, production of new shoots at the base of the parental shoot (stump sprouting) in other shrubs and trees, initiation of new shoots from below-ground stems or burls, as in many Mediterranean shrubs, and rooting of lower limbs of trees that become covered by soil organic matter (layering) in many conifers.

3.2.4 Delayed Greening During Leaf Development in Tropical Trees

Many tropical, shade-tolerant rainforest species initiate leaves that are white, red, blue, or light-green during the stage of leaf expansion, which reflects their low chlorophyll concentration. This pattern of **delayed greening** is typical of shade-tolerant species and is less common in gap specialist (Table 4). The pattern of delayed greening is distinctly different from the shift from juvenile to adult foliage because it is typical of all young leaves, even those on mature plants. Leaves that show delayed greening function below the light-compensation point for photosynthesis at saturating light until fully expanded. After full expansion,

TABLE 4. **The color of young leaves of 175 species, common in a tropical rain forest in Panama.***

Leaf color	Gap-specialist (%)	Shade-tolerant (%)
White	0	8
Red	3	33
Light-green	3	41
Delayed greening	7	82
Green	93	18

Source: Kursar & Coley 1991b.
*Values are the number of species in each category. Percentages are calculated for gap-specialist and shade-tolerant species separately.

their rate of dark respiration is very high, presumably due to the high rates of metabolism associated with the development of chloroplasts. The completion of this development may take as long as 30 days after the leaves have fully expanded. In contrast, normally greening leaves achieve maximum photosynthetic capacity at the end of leaf expansion (Kursar & Coley 1992b).

There is obviously a cost involved in delayed greening: During leaf expansion species showing this pattern of chloroplast development absorb only 18 to 25% of the maximum possible absorption, compared with 80% for leaves that show a normal developmental pattern. At the irradiance level that is typical of the forest understory, the quantum yield of photosynthesis is also less than half that of green leaves, largely due to their low absorption (Kursar & Coley 1992a). What might be the advantages associated with delayed greening?

It has been hypothesized that delayed greening is a mechanism to reduce herbivory of young leaves. All young leaves lack toughness, which is provided by cell-wall thickening and lignification, which are processes that tend to be incompatible with cell expansion and leaf growth. Because toughness provides protection against both biotic and abiotic factors, young leaves are poorly protected (Table 5). The accumulation of nitrogen and other nutrients associated with chloroplast development in species without delayed greening presumably makes young unprotected leaves even more attractive to herbivores.

Hence, although delayed greening may represent a loss of potential carbon gain, it also reduces carbon losses associated with **herbivory**. In the high-irradiance environment losses incurred by delayed greening could be substantial. In the low-light environment of shade-adapted species, where the irradiance is only about 1% of full sunlight, losses by herbivory could be relatively more important (Table 6).

We have so far discussed the delayed greening in terms of lack of chlorophyll; however, the red or blue appearance also reflects the presence of specific pigments: **anthocyanins**. Early hypotheses suggest that these anthocyanins raise leaf temperature have been rejected. The suggestion that these anthocyanins protect against damage by ultraviolet light (cf. Sect. 2.2.2 in the chapter on effects of radiation and temperature) does not appear valid, considering the very low irradiance level in understory habitats. Bioassays using leaf-cutter ants suggest that these anthocyanins may protect the leaves because of their **antifungal** properties. These leaf-cutter ants collect leaves,

TABLE 5. Rates of herbivory of young leaves, measured during the 3 days prior to full expansion (when they lack toughness) and 4 to 6 days after full expansion (when their toughness has increased substantially).*

Species	Number of leaves	During expansion	After expansion
Ouratea lucens	274	3.08	1.63
Connarus panamensis	179	0.22	0.03
Xylopia micrantha	90	0.57	0.01
Desmopsis panamensis	262	0.75	0.27
Annona spragueii	204	0.37	0.08

Source: Kursar & Coley 1991b.
* The values are expressed as the percent of the leaves that was eaten per day; they were all significant at $p < 0.01$.

store them underground as substrate for fungi, which are fed on by ants. Leaves that contain anthocyanins, either naturally or experimentally added, are collected to a lesser extent than are leaves with lower anthocyanin concentrations (Coley & Aide 1989).

3.3 Reproductive Phase

We know that some plants flower in spring, when days are getting longer, whereas others flower in autumn, when days are shortening. How do plants sense that it is spring or autumn? Depending on the species, plants may use either the **daylength** or the **temperature** as environmental cues. Many plants from temperate regions use a combination of both cues and are thus able to distinguish between spring and autumn. Our understanding of the timing mechanisms of plants has not only led to greater insight in how plants time their switch from the vegetative to the reproductive phase, but also to important applications in the glasshouse industry.

3.3.1 Timing by Sensing Daylength: Long-Day and Short-Day Plants

In the chapter on growth and allocation (Sect. 5.1.2) we discuss how vegetative growth can be affected by daylength. This environmental cue is pivotal in triggering flowering in many species. Daylength does not play a role in so-called **day-neutral plants**, like *Cucumis sativus* (cucumber), *Ilex aquifolium* (holly), *Lycopersicon esculentum* (tomato), *Impatiens balsamina* (touch-me-not), and *Poa annua*. It is most important, however, in plants whose flowering is triggered by the short days in autumn (**short-day plants**, which require a photoperiod less than about 10 to 12 h) or the long days in spring (**long-day plants**, require a photoperiod longer than about 12 to 14 h). Examples of short-day plants include *Chrysanthemum* species, *Euphorbia pulcherrima*, *Fragaria* species (strawberry), *Glycine max* (soybean), *Nicotiana tabacum* (tobacco), and *Xanthium strumarium* (cocklebur), which is one of the best-studied short-day species. Long-day plants include *Avena sativa* (oat), *Hordeum vulgare* (barley), *Fuchsia* species, *Lolium perenne* (perennial ryegrass),

TABLE 6. Hypothetical carbon budgets for white and green young developing leaves in sun and shade environments.

Habitat	Leaf color	CO_2 assimilation (carbon gain)	Herbivory (carbon loss)	Net carbon* gain/loss
Sun	Green	High	High	+++
	White	Low	Low	−
Shade	Green	Low	High	−−
	White	Low	Low	−

Source: Kursar & Coley 1991b.
*Positive signs indicate a net carbon gain during this stage; negative signs indicate net carbon loss.

Trifolium pratense (strawberry clover), *Triticum aestivum* (wheat), and *Hyoscyamus niger* (black henbane), which is a much-researched long-day species. Some species [e.g., *Bouteloua curtipendula* (side-oats grama)] have short-day ecotypes at the southern end of their distribution and long-day ecotypes at the northern end (Olmsted 1944). The requirement for a certain daylength may be **qualitative**, meaning that plants will not flower at all without exposure to at least 1 day of the appropriate photoperiod. It may also be **quantitative**, which means that flowering will occur more quickly when exposed to the appropriate photoperiod (Fig. 17). Do plants really sense the daylength, or is it the duration of the night period that is perceived?

The answer to this question has come from experiments in which the night was interrupted with either white or red light. A short interruption of the dark period prevents or delays flowering in a short-day plant, whereas the same treatment promotes flowering in long-day plants. Interrupting the light period has no effect on either short-day or long-day plants. The period between two light periods, normally the *night*, clearly, must be the **critical time** that is perceived by the plant. But *how* do plants perceive the duration of the night?

The answer again, has come from experiments in which the night was interrupted, now using light of a specific wavelength: red (660 nm) or far-red

(730 nm). A short flash is generally sufficient to obtain the effect: red light has the same effect as white light and this effect is reversed by exposure to far-red light. This points to **phytochrome** as the photoreceptor involved in perception of the photoperiod (cf. Box 9 attached to the chapter on growth and allocation). In fact, phytochrome was discovered in the first place through these sorts of experiments (Bernier et al. 1981a).

The photoperiod is perceived by leaves that have just matured. Exposure of just one leaf to the inducing photoperiod may be enough. The presence of older leaves inhibits flower induction. If a plant subjected to the inducing photoperiod is grafted onto one that has not received the photoinductive light, both plants will flower, which indicates that a signal is transmitted from the induced plant to the graft. The signal may be a chemical compound or compounds, but the exact nature remains unclear. Gibberellins and ethylene can induce flowering in some long-days plants, whereas ABA inhibits the process. In the short-day plant *Pharbitis nil*, ABA both promotes and inhibits flowering, depending on addition before or after the 14-hour inductive dark period (Takeno & Maeda 1996). Cytokinin levels in the short-day plant *Chenopodium rubrum* (lambsquarters) are also affected by exposure to a photoperiod inductive for flowering (Machackova et al. 1996). The signals may therefore involve the classical phytohormones, although it is not yet possible to account for all the observed effects (Bernier et al. 1981b).

Because interruption of the photoperiod at different times of the night has different effects on induction or prevention of flowering, a biological clock with a rhythm of about 24 hours (a **circadian clock**) has been postulated in plants. Such a circadian clock also plays a role in plants that fold their leaves at night and in many other processes. The biological clock presumably controls the sensitivity for P_{fr}. If the ability of plants from temperate climates to sense the length of the night is impressive, then that of some tropical species is truly astounding. Here the variation in daylength may be very short, and a change of 20 to 30 minutes may suffice to trigger flowering.

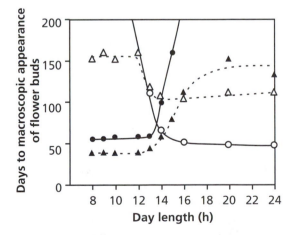

FIGURE 17. Days to macroscopic appearance of flower buds as a function of daylength in four species grown at 20°C in fluorescent light. The indicated daylength was applied since sowing. *Perilla nankinensis* (●), a qualitative short-day plant; *Cosmos variegata* (▲), a quantitative short-day plant; annual *Hyoscyamus niger* (black henbane) (○), a qualitative long-day plant; *Coleus blamei* (△), a quantiatative long-day plant (Bouillenne 1963). Copyright Académie Royale de Belgique.

3.3.2 Do Plants Sense the Difference Between a Certain Daylength in Spring and Autumn?

Daylength is a tricky environmental cue because days of the same length occur in both spring and autumn. How do plants sense the difference between the two seasons? It was once thought that plants could sense the lengthening or the shortening of days; however, experiments have not confirmed the existence of such a mechanism. How, then do they do it?

In addition to daylength, plants need a second environmental cue (e.g., temperature) (Sect. 3.3.3). Such a combination is required to induce flowering in *Fragaria ananassa* (strawberry) and *Beta vulgaris* (sugar beet). Flower primordia are induced in autumn, when daylength is reduced to a critical level. Further development of the primordia is stopped by low temperature in winter and only continues when the temperature increases in spring (Bernier et al. 1981a).

3.3.3 Timing by Sensing Temperature: Vernalization

In seasonal climates, changes in daylength may coincide with changes in temperature. Many species that flower in spring are not long-day plants; rather, they use temperature as an environmental cue. Exposure of the entire plant or of the moist seed induces flowering. We owe much of the information on effects of temperature on flower induction to the Russian botanist **Lysenko**. He showed that exposure of moist seeds of winter wheat (*Triticum aestivum*) to low temperatures allowed the plants to flower, without exposure of the seedlings to the harsh Russian winter. The physiological changes triggered by exposure to low temperature are called **vernalization** (from the Latin word for spring, ver) and, to honor Lysenko, also as jarowization (from the Russian word for spring) (Atkinson & Porter 1996).

Lysenko unfortunately did not place his important findings in the right scientific perspective. Rather than concluding that phenotypic changes in the seeds exposed to low temperature accounted for the flowering of the mature wheat plants, he insisted that the changes were genetic. Inspired and supported by the political flavor of the 1930s in his country, he stuck to his genetic explanation, much to the detriment of genetics and geneticists in the Soviet Union.

Vernalization is essential, both for crop species such as *Triticum aestivum* (winter wheat) and for winter annuals in general, which survive during winter as seedlings. Vernalization also triggers flowering in biennials that overwinter as a rosette, such as *Digitalis purpurea* (fox glove), *Lunaria annua* (honesty), *Daucus carota* (carrot), Beta vulgaris (beet), and in perennials such as *Primula* (primrose) and Aster species, and plants that overwinter as a bulb, tuber, or rhizome.

Vernalization is believed to require perception of low temperature in the vegetative apex. Cold treatment supposedly induces the breakdown of a compound that accumulated during exposure to short days in autumn and which inhibits flower induction; this might be ABA. At the same time, a chemical compound is produced that promotes flower induction, most likely gibberellic acid (Bernier 1988).

The practical applications of our ecophysiological knowledge on environmental cues that trigger flowering are enormous. Many flowers that used to be available during specific seasons only, can now be produced all year round. Building on fundamental ecophysiological experiments, in the Netherlands the flower industry has become a flourishing branch of horticulture.

3.3.4 Effects of Temperature on Plant Development

In the previous section, we discussed the effects of a low temperature as a **trigger** for flower induction of biennials. Temperature, however, also affects plant development (Atkinson & Porter 1996). Reaumur (1735) introduced the concept of a **thermal unit** to predict plant development. This concept assumes that plants need a fixed number of temperature sum to fulfill a developmental phase. This assumption implies that the rate of crop development, expressed as the inverse of the duration in days for a given phase, is a linear function of temperature. Although the concept of thermal unit is widely applied, it has no physiological basis (Horie 1994).

3.3.5 Attracting Pollinators

Pollination of flowers by insects, birds, or bats requires attraction of pollinators. Attraction may occur through secondary phenolic compounds (flavonoids) in the petals (Shirley 1996). These **UV-absorbing compounds** are invisible to the human eye, but they are perceived by pollinating bees. The flowers of many species change color with pollination, thus guiding potential pollinators to those flowers that are still unpollinated and provide a **nectar** reward (Weiss 1991). The change in color may be due to a change of the pH in the vacuole, in which the phenolics compounds are located [e.g., in *Ipomoea caerulea* (morning glory)]. Following pollination, most flowers cease nectar production. Pollinators quickly learn which colors provide a nectar reward.

The quantity of nectar provided by a flower depends on the number of flowers in an inflorescence and the type of pollinator that a flower is "designed" to attract. For example, long-tubed red flowers pollinated by hummingbirds typically produce more nectar than short-tubed flowers pollinated by small insects; this makes sense in view of the 140-fold greater energy requirement of hummingbirds (Heinrich & Raven 1972). Those species that produce many flowers in an inflorescence typically produce less nectar per flower than do species that produce a single flower. In general, plants produce enough nectar to attract pollinators, but not to satiate them, thus forcing pollinators to visit additional flowers to meet their energetic requirements and increasing the probability of effective pollen transfer (Heinrich 1975).

Secondary compounds play a role as **visual cues** for insects. Others, with specific **scents**, are often released only at a specific time of the day or night. These scents may be an olfactory delight for humans, or they may represent the foul smell of rotting meat, such as in the flowers of many species of the Araceae (Meeuse 1975). In these flowers odoriferous compounds are volatilized due to an increase in temperature of the spadix. As pointed out in Section 3.1 of the chapter on plant respiration, the cyanide-resistant alternative path increases in activity prior to **heat production** and is partly responsible for it. During its "respiratory crisis" the respiration rate of this reproductive organ may approach the incredibly high rate found in the flying muscles of hummingbirds. As a result the temperature rises to approximately 10°C above ambient, and odoriferous amines are volatilized, and pollinators are attracted (Skubatz et al. 1991). During heat production the respiration of the spadix is largely cyanide-resistant. Although this is not the only reason for thermogenesis (high respiration rates per se are also important), it definitely contributes to the heat production because the lack of proton extrusion coupled to electron flow allows a large fraction of the energy in the substrate to be released as heat.

The temperature of the flower, compared with that of the ambient air, can also be enhanced by **solar tracking**, which is a common phenomenon in alpine and arctic species that belong to the Asteraceae, Papaveraceae, Ranunculaceae, and Rosaceae, and involves the perception of blue light (Stanton & Galen 1993). This may raise flower temperature by several degrees above the ambient temperature, as long as the wind speed is not too high (see Sect. 2.2 in the chapter on the plant's energy balance). Solar tracking might therefore affect fitness in many ways. When solar tracking is prevented in *Dryas octopetala*, by tethering the plants, lighter seeds are produced, but the seed set is not affected (Kjellberg et al. 1982). A similar treatment decreased both seed set and seed mass in *Ranunculus adoneus* (snow buttercup) (Stanton & Galen 1989). The flowers of the solar-tracking Norwegian alpine buttercup, *Ranunculus acris*, traverse an arc of about 50 degrees, with speed of movement and solar tracking accuracy being highest at midday (between 11 A.M. and 5 P.M.). This solar tracking enhances flower temperature by about 3.5°C. Solar tracking decreases with flower aging and stops completely as the petals wither, so that it cannot have effects on postanthesis events. Tethering the flowers does not affect the attractiveness to pollinating insects, seed:ovule ratio, seed mass, or seed abortion rate (Totland 1996). If solar tracking has any selective advantage in this species, then it is probably only under special weather conditions (e.g., when pollinator activity is limited by low temperatures).

3.4 Fruiting

Allocation to reproduction varies substantially among plant species and with environmental conditions, ranging from 1 to 30% of net primary production, with median values of perhaps 10%. This modest allocation to reproduction—the process that most directly governs plant fitness—is less than typical allocation to root exudation under nutrient stress or nutrient uptake under favorable conditions (Table 1 in the chapter on plant respiration), which suggests that the processes of resource acquisition under conditions of environmental stress and competition with neighboring plants often leave relatively few resources for reproduction.

Wild plants generally produce fewer fruits than flowers. Low allocation to reproduction sometimes reflects poor pollination, when weather conditions are bad for pollinators or for appropriate pollen-producing plants. Even when the flowers are artificially pollinated, however, the ratio between fruits and flowers, commonly referred to as **fruit set**, may still be substantially below 1. In addition, increased pollination may have more seeds setting, but at the expense of seed size, which indicates that seed production may be both "pollen-limited" and "resource-limited" (Stanton et al. 1987).

Allocation to reproduction differs substantially among species. In general, annuals and other short-lived species allocate a larger proportion of annual production to reproduction than do long-lived per-

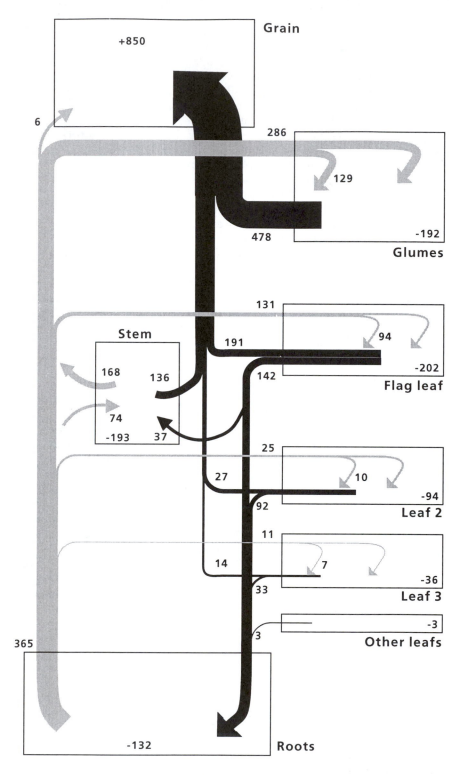

FIGURE 18. Transport of nitrogen in *Triticum aestivum* (wheat) 15 days after anthesis. The grains accumulated 850 µg of nitrogen per day, all of which is derived from senescing plant parts; the nitrogen in the soil was completely depleted at the time of anthesis. All vegetative organs show a net loss of nitrogen, but they continue to import nitrogenous compounds via the xylem (shown in gray; above-ground organs) or phloem (shown in black; roots). Leaves export some of their nitrogen directly to the developing grains, but some of it "cycles" through the roots before arriving in the grains (Simpson et al. 1983). Copyright American Society of Plant Physiologists.

ennials, which suggests a trade-off between reproduction and traits that promote survival or growth (Bazzaz et al. 1987). For example, many conifers and other tree species reproduce prolifically once in several years. These "mast years" are correlated with years of low wood production and are often synchronized among individuals in a population. Mast reproduction may be possible only after several years of reserve accumulation. This pattern of reproduction serves to "swamp" seed predators in years of abundant seed production and to limit the population growth of seed predators in intervening years (Eis et al. 1965).

Allocation to female function is generally considered the most costly component of reproduction, because of the large investments of carbon and nutrient required to produce seeds. This may explain why female individuals of dioecious species are generally underrepresented in sites of low water availability (Bazzaz et al. 1987). Male function, however, also entails substantial costs. For example, *Phacelia linearis* has both female and hermaphroditic individuals. Those individuals that have both male and female function (hermaphrodites) grow more slowly than do females, particularly at low nutrient supply, which suggests that it is the nitrogen investment in male function that accounts for the slower growth of hermaphrodites (Eckhart 1992a,b). During the vegetative phase, hermaphroditic genotypes of *Plantago lanceolata* (plantain) have exactly the same growth rate and photosynthetic characteristics as the ones with only female function. When grown at a nutrient supply that resembles that in their natural environment, however, the female plants have a three- to fivefold higher reproductive output. Female genotypes invest three times more biomass in each flower, with an even greater difference in terms of nitrogen investment, because the stamens contain relatively more nitrogen than do the female components of flowers (sepals and petals). The female plants use the nitrogen saved by not producing pollen for additional vegetative as well as reproductive growth, showing that resource compensation is a primary mechanism that accounts for the persistence of genotypes that are exclusively female (Poot 1996, 1997).

Allocation to reproduction is difficult to quantify because the inflorescence can often meet much of its own carbon requirement and because some structures serve both reproductive and nonreproductive roles. A substantial proportion of the energetic costs of reproduction are met by photosynthesis in the inflorescence and associated leaves. For example, photosynthesis by the inflorescence accounts for 2 to 65% (median 22%) of the carbon required for reproduction of temperate trees (Bazzaz et al. 1979). In cereals, the ear accounts for up to 75% of the photosynthate required for grain production, and the inflorescence plus the closest leaf (the flag leaf) provide all of the photosynthate required for reproduction (Evans & Rawson 1970). The high photosynthetic rate of the flag leaf is a result of the strong sink strength of adjacent reproductive parts (Chapin & Wardlaw 1988). When vegetative leaves are removed by herbivores, an increased proportion of flag-leaf photosynthate goes to vegetative organs, whereas damage to the flag leaf increases carbon transport from other leaves to the inflorescence. Thus, the role of each leaf in supporting reproduction depends on the integrated carbon supply and demand of the entire plant. Stem growth often increases during reproduction of herbaceous plants, which increases the probability of pollen exchange and the dispersal distance of wind-dispersed fruits. The greatest gains in yield of crops (e.g., cereals, peanuts, sugar beet) have come from breeding for a higher **harvest index** [i.e., the ratio between harvestable biomass and total (aboveground) biomass]. In cereals this has been achieved by selection for varieties with reduced stem allocation which was due to a low production of or sensitivity to GA. There has been no increase in photosynthetic capacity during crop breeding (Evans 1980, Gifford 1984).

After flowering, phloem-mobile nutrients are exported from the senescing leaves and roots to the developing fruits, as discussed in Section 4.3.2 of mineral nutrition (Fig. 18). Unlike "getting old and wearing out," senescence in plants is a carefully programmed, hormonally controlled developmental process: **programmed cell death** (Jones & Dangl 1996). It is an integral part of plant development that is affected by environmental factors (e.g., irradiance level, photoperiod, and nitrogen supply). It is promoted by **ethylene** and **ABA**, and it is slowed down or reversed by **cytokinins** and or **GA**. A number of specific genes are induced or upregulated during leaf senescence (Smart 1994). An early visible symptom of leaf senescence is leaf yellowing, due to loss of chlorophyll. Rubisco and other chloroplast proteins are hydrolyzed by proteolytic enzymes, and free amino acids are exported via the phloem. Mitochondrial proteins tend to be hydrolyzed in a later phase and tissues around the vascular system, which are required for nutrient export, are the last to senesce. The breakdown of the nucleus, whose activity is essential for senescence to proceed, is a relatively late event in the developmental process (Gan & Amasino 1997).

Nitrogenous compounds are remobilized, as are most other compounds that can move in the phloem. Unlike phloem-mobile elements, calcium concentration in phloem sap is very low.

Considering the driving force for phloem transport (i.e., a gradient in hydrostatic pressure between source and sink) it is not surprising that some of the compounds remobilized from senescing leaves are transported to roots, even though these may show a net export of nutrients. Figure 18 shows the pattern of nitrogen loss from vegetative organs of wheat (*Triticum aestivum*) and the pathways along which nitrogen is transported from these organs to developing grains. The diagram is somewhat similar to that in vegetative wheat plants, which show a continuous cycling of nitrogen between leaves and roots, via both phloem and xylem (Sect. 5.4.1 in the chapter on growth and allocation). The rather indirect manner in which nitrogen moves from senescing leaves to developing kernels probably reflects the way the systems for long-distance transport (i.e., xylem and phloem) operate. That is, phloem sap will move in the sieve tubes from a site where the phloem is loaded, thus creating a high turgor, to a site where phloem unloading takes place, thus decreasing the turgor pressure. Xylem sap will move in the xylem conduits, down a gradient in hydrostatic pressure. There is some exchange between the transport pathways, especially in the stem (Fig. 18), but this is obviously not sufficient to stop the need for a continuous cycling process in plants (Simpson 1986).

4. Seed Dispersal

Seeds are often well-protected, either physically, by a hard seed coat (Sect. 2.1), or chemically, due to poisonous compounds like cyanogenic glycosides or specific inhibitors of digestive enzymes (Sects. 3.1 and 3.2 of the chapter on ecological biochemistry).

Numerous plant attributes are associated with seed dispersal (e.g., floating designs in aquatics, sticky seed parts in mistletoes that ensure deposition on a host branch, "ballistic" structures, plumes, and wings that allow transfer through air, etc.) (Murray 1986). Some of these mechanisms involve aspects of the plant's physiology, of which a few examples will be presented in this section.

4.1 Dispersal Mechanisms

Explosive or **ballistic** seed dispersal occurs in many plant species. Such dispersal mechanisms are highly undesirable in crop plants because they cause "shattering" and loss of seed during harvest (e.g., in *Brassica* species). In the tropical rainforest legume tree, *Tetrabelinia moreliana*, such a mechanism allows seeds to be launched and transferred over as much as 50 meters (Van der Burgt 1997). It is a consequence of drying of the pod walls, which creates tension that builds up between the two valves of the pod. Once the tension exceeds a threshold value, the pod explodes and the seed is launched.

Tension in the tissue may also occur without drying of the reproductive structure (as, e.g., in *Impatiens*). In this case the tissue tension reflects an aspect of tissue water relations, which we alluded to in Section 4 of the chapter on plant water relations. That is, within the reproductive tissue, the water relations of individual cells must differ widely, creating **tissue tension**. Touch or wind may cause a threshold be exceeded, which causes rupture in the reproductive structure and launching of the seeds.

4.2 Life-History Correlates

Plants have an ancient and uneasy relationship with vertebrate animals that eat their fruits, and either digest or disperse their seeds. As early as 300 million years ago, Carboniferous progenitors of modern cycads bore fleshy fruits that were apparently adapted for consumption by primitive reptiles, which then dispersed the seeds (Howe 1986). Many species (e.g., *Acacia*) produce a lipid-rich morphological structure, termed **aril** or **elaiosome**, that allows dispersal via ants (Hughes et al. 1994). These transport the seeds to their nest, thus burying the *Acacia* seeds, safe from fire (O'Dowd & Gill 1985).

5. The Message to Disperse: Perception, Transduction, and Response

Plants continuously sense their environment, both as adults and as seeds, before germination has started. Seeds acquire information about the suitability of their environment for seedling growth, and they use this information to germinate or to remain dormant. There are numerous environmental cues, with plants from different environments using different cues. At a later stage, plants similarly sense their environment to change from the vegetative to the reproductive stage and to time their flowering. Daylength and low temperature are major cues, with irradiance level and nutrient supply occasionally playing an additional role in the switch to the reproductive phase in biennials.

There are also changes during development that are programmed, with environmental factors playing at most a moderating role. For example, leaf senescence is part of a scenario of programmed cell death that can be hastened by low irradiance and limiting nitrogen supply. The switch from juvenile to adult foliage is also programmed, but it can be affected by irradiance and plant water status.

Once flowering has started, the plant may require pollinating animals to produce seeds. Olfactory and visual cues are produced to attract these pollinators. The seeds that are subsequently produced may end up close to the mother plant, but there are also numerous mechanisms that ensure dispersal of the seeds over relatively great distances. One of the mechanisms that is of ecophysiological interest is that of plants that "launch" their seeds. Other dispersal mechanisms require allocation of reserves to elaiosomes (i.e., producing food for dispersing ants). Ants both disperse and bury the seeds; therefore, the seeds are safe during a bushfire. Surviving seeds remain dormant until the right environmental cues have been perceived, and the life cycle continues.

Plants sense their environment during their entire life, and the acquired information determines what is going to happen in several steps of the plant's life cycle. We now have a reasonable understanding of important environmental cues and plant responses. This contrasts strongly with our lack of knowledge of signal-transduction pathways that connect the environmental cue and the plant's response.

References and Further Reading

Atkinson, D. & Porter, J.R. (1996) Temperature, plant development and crop yields. Trends Plant Sci. 1:119–124.

Bazzaz, F.A., Carlson, R.W., & Harper, J.L. (1979) Contribution to reproductive effort by photosynthesis of flowers and fruits. Nature 279:554–555.

Bazzaz, F.A., Chiariello, N.R., Coley, P.D., & Pitelka, L.F. (1987) Allocating resources to reproduction and defense. BioScience 37:58–67.

Bernier, G. (1988) The control of floral evocation and morphogenesis. Annu. Rev. Plant Physiol. 39:175–219.

Bernier, G., Kinet, J.-M., & Sachs, R.M. (1981a) The physiology of flowering. Vol. I. CRC Press, Boca Raton.

Bernier, G., Kinet, J.-M., & Sachs, R.M. (1981b) The Physiology of Flowering. Vol. II. CRC Press, Boca Raton.

Bewley, J.D. & Black, M. (1982) Physiology and biochemistry of seeds. Vol. 2. Springer-Verlag, Berlin.

Blaauw-Jansen, G. & Blaauw, O.H. (1975) A shift in the response threshold to red irradiation in dormant lettuce seeds. Acta Bot. Neerl. 24:199–202.

Bouillenne, R. (1963) Recherche de la photopériode critique chez diverses espèces de jours longs et de jours courts cultivées en milieu conditionné. Bulletin de la Classe des Sciences de l'Académie Royale de Belgique, 5e série, 49:337–345.

Bryant, J.P. & Kuropat, P.J. (1980) Selection of winter forage by subarctic browsing vertebrates: The role of plant chemistry. Annu. Rev. Plant Physiol. 11:261–285.

Chapin III, F.S. & Wardlaw, I.F. (1988) Effect of phosphorus deficiency on source-sink interactions between the flag leaf and developing grain in barley. J. Exp. Bot. 39:165–177.

Chapin III, F.S., Tieszen, L.L., Lewis, M., Miller, P.C., & McCown, B.H. (1980) Control of tundra plant allocation patterns and growth. In: An arctic ecosystem: The coastal tundra at Barrow, Alaska, J. Brown, P. Miller, L. Tieszen, & F. Bunnell (eds). Dowden, Hutchinson and Ross, Stroudsburg, pp. 140–185.

Chapman, D.F., Robson, M.J., & Snaydon, R.W. (1992) Physiological integration in the perennial herb Trifolium repens L. Oecologia 89:338–347.

Coley, P.D. & Aide, T.M. (1989) Red coloration of tropical young leaves: A possible antifungal defence? J. Trop. Ecol. 5:293–300.

Cook, R.E. (1979) Patterns of juvenile mortality and recruitment in plants. In: Topics in plant population Biology, O.T. Solbrig, S. Jain, G.B. Johnson, & P.H. Raven (eds). Columbia University Press, New York, pp. 207–231.

De Jong, T.J., Klinkhamer, P.G.L., Nell, H.W., & Troelstra, S.J. (1987) Growth and nutrient accumulation of the biennials Cirsium vulgare and Cynoglossum officinale under nutrient-rich conditions. Oikos 48:62–72.

De Lange, J.H. & Boucher, C. (1990) Autecological studies on Audouinia capitata (Bruniaceae). I. Plant-derived smoke as a seed germination cue. S. Afr. J. Bot. 56:700–703.

Derkx, M.P.M. & Karssen, C.M. (1993) Changing sensitivity to light and nitrate but not to gibberellins regulates seasonal dormancy patterns in Sisymbrium officinale seeds. Plant Cell Environ. 16:469–479.

Eckhart, V.M. (1992a) The genetics of gender and the effects of gender on floral characteristics in gynodioecius Phacelia linearis (Hydrophyllaceae). Am. J. Bot. 79:792–800.

Eckhart, V.M. (1992b) Resource compensation and the evolution of gynodioecy in Phacelia linearis (Hydrophyllaceae). Evolution 46:1313–1328.

Eis, S., Garman, E.H., & Ebell, L.F. (1965) Relation between cone production and diameter increment of Douglas-fir (Pseudotsuga menziesii (Mirb.) Franco), grand fir (Abies grandis (Dougl.) Lindl.), and western white pine (Pinus monticola Dougl.). Can. J. Bot. 43:1553–1559.

Evans, L.T. (1980) The evolution of crop yield. Am. Sci. 68:388–397.

Evans, L.T. & Rawson, H.M. (1970) Photosynthesis and respiration by the flag leaf and components of the ear during grain development in wheat. Aust. J. Biol. Sci. 23:245–254.

Fenner, M. (1985) Seed ecology. Chapman and Hall, London.

Fenner, M. (1992) Seeds, The ecology of regeneration in plant communities. CAB International, Wallingford.

Frankland, B. & Poo, W.K. (1980) Phytochrome control of seed germination in relation to natural shading. In: Photo receptors in plant development, J. de Greef (ed). University Press, Antwerpen, pp. 357–366.

Gan, S. & Amasino, R.M. (1997) Making sense of senescence. Plant Physiol. 113:313–319.

Gifford, R.M., Thorne, J.H., Hitz, W.D., & Giaquinta, R.T. (1984) Crop productivity and photoassimilate partitioning. Science 225:801–808.

Gorski, T & Gorska, K. (1979) Inhibitory effects of full daylight on the germination of Lactuca sativa. Planta 144:121–124.

Grime, J.P. (1979) Plant strategies and vegetation processes. Wiley, Chicester.

Grime, J.P. & Jeffrey, D.W. (1965) Seedling establishment in vertical gradients of sunlight J. Ecol. 53:621–642.

Gross, K.L. (1984) Effects of seed size and growth form on seedling establishment of six monocarpic perennial plants. J. Ecol. 72:369–387.

Hansen, D.H. (1986) Water relations of compound leaves and phyllodes in Acaeia koa var. latifolia. Plant Cell Environ. 9:439–445.

Hansen, D.H. (1996) Establishment and persistence characteristics in juvenile leaves and phyllodes of Acacia koa (Leguminosae) in Hawaii. Int. J. Plant Sci. 157:123–128.

Harper, J.L. (1977) Population biology of plants. Academic Press, London.

Heinrich, B. (1975). Energetics of pollination. Annu. Rev. Ecol. Syst. 6:139–170.

Heinrich B. & Raven, P.H. (1972) Energetics and pollination ecology. Science 176:597–602.

Hesse, O. (1924) Untersuchungen über die Einwirkung chemischer Stoffe auf die Keimung lichtempfindlicher Samen. Botanisches Archiv 5:133–171.

Hillhorst, H.W.M. & Karssen, C.M. (1992) Seed dormancy and germination: The role of abscisic acid and gibberellins and the importance of hormone mutants. Plant Growth Regul. 11:225–238.

Horie, T. (1994) Crop ontogeny and development. In: Physiology and determination of crop yield, K.J. Boote, J.M. Bennet, T.R. Sinclair, & G.M. Paulsen (eds). American Society of Agronomy, Crop Science Society of America, Soil Science Society of America, Madison, pp. 153–180.

Howe, H.F. (1985) Seed dispersal by fruit-eating birds and mammals. In: Seed dispersal, D.R. Murray (ed). Academic Press, Sydney, pp. 123–189.

Hughes, L., Westoby, M., & Jurado, E. (1994) Convergence of elaiosomes and insect prey: Evidence from ant foraging behaviour and fatty acid composition. Funct. Ecol. 8:358–365.

Jones, A.M. & Dangl, J.L. (1996) Logjam at the Styx: Programmed cell death in plants. Trends Plant Sci. 1:114–119.

Jonsdottir, I.S., Callaghan, T.V., & Headly, A.D. (1996) Resource dynamics within arctic clonal plants. Ecol. Bull. 45:53–64.

Kahn, A.A. (1982) The physiology and biochemistry of seed development, dormancy and germination. Elsevier, Amsterdam.

Karssen, C.M. (1982) Seasonal patterns of dormancy in weed seeds. In: The physiology and biochemistry of seed development, dormancy and germination, A.A. Kahn (ed). Elsevier, Amsterdam, pp. 243–270.

Karssen, C.M. & Hilhorst, H.W.M. (1992) Effect of chemical environment on seed germination. In: Seeds, the Ecology of Regeneration in Plant Communities, M. Fenner (ed.). C.A.B. International, Wallingford, pp. 327–348.

Keeley, J.E. (1991) Seed germination and life history syndromes in the California chaparral. Bot. Rev. 67:81–116.

Kjellberg, B., Karlsson, S., & Kerstensson, I. (1982) Effects of heliotropic movements of flowers of Dryas octopetala on gynoecium temperature and seed development. Oecologia 70:155–160.

Klinkhamer, P.G.L., De Jong, T.J., & Meelis, E. (1986) Delay of flowering in spear thistle (Cirsium vulgare (Savi) Ten.): Size-effects and devernalization. In: Population ecology of the biennials Cirsium vulgare and Cynoglossum officinale: An experimental approach. PhD Thesis, University of Leiden, Leiden, pp. 121–131.

Koller, D. & Negbi, M (1959) The regulation of germination in Oryzopsis miliacea. Ecology 40:20–36.

Kursar, T.A. & Coley, P.D. (1991a) Nitrogen content and expansion rate of young leaves of rain forest species: Implications for herbivory. Biotropica 23:141–150.

Kursar, T.A. & Coley, P.D. (1991b) Delayed greening in tropical trees: An antiherbivore defense? Biotropica 24:256–262.

Kursar, T.A. & Coley, P.D. (1992a) The consequences of delayed greening during leaf development for light absorption and light use efficiency. Plant Cell Environ. 15:901–909.

Kursar, T.A. & Coley, P.D. (1992b) Delayed development of the photosynthetic apparatus in tropical rain forest species. Funct. Ecol. 6:411–422.

Leishman, M.R. & Westoby, M. (1994) The role of large seed size in shaded conditions: Experimental evidence. Funct. Ecol. 8:205–214.

Leishman, M.R., Westoby, M., & Jurado, E. (1995) Correlates of seed size variation: A comparison among five temperate floras. J. Ecol. 83:517–530.

Machackova, I., Eder, J., Motyka, V., Hanus, J., & Krekule, J. (1996) Photoperiodic control of cytokinin transport and metabolism in Chenopodium rubrum. Physiol. Plant. 98:564–570.

Mayer, A.M. & Polyakoff-Mayber, A. (1982) The Germination of Seeds. 3d ed. Pergamon Press, Oxford.

McKee, G.W., Pfeiffer, R.A., & Mohsenin, N.N. (1977) Seedcoat structure in Coronilla varia L. and its relations to hard seed. Agronomy J. 69:58.

Meeuse, B.J.D. (1975) Thermogenic respiration in aroids. Annu. Rev. Plant Physiol. 26:117–126.

Murray, D.R. (ed) (1986) Seed dispersal. Academic Press, Sydney.

New, T.R. (1984) A biology of acacias. Oxford University Press, Melbourne.

O'Dowd, D.J. & Gill, A.M. (1985) Seed dispersal syndromes in Australian *Acacia*. In: Seed dispersal, D.R. Murray (ed). Academic Press, Sydney, pp. 87–121.

Olsen, J.E., Jensen, E., Junttila, O., & Moritz, T. (1995) Photoperiodic control of endogenous gibberellins in seedlings of *Salix pentandra*. Physiol. Plant. 93:639–644.

Olmsted, C.E. (1944) Growth and development in range grasses. IV. Photoperiodic responses in twelve geopgraphic strains of side-oats grama. Bot. Gaz. 106: 46–74.

Pons, T.L. (1989) Breaking of seed dormancy by nitrate as a gap detection mechanism. Ann. Bot. 63:139–143.

Pons, T.L. (1991a) Dormancy, germination and mortality of seeds in a chalk-grassland flora. J. Ecol. 79:765–780.

Pons, T.L. (1991b) Induction of dark dormancy in seeds: Its importance for the seed bank in the soil. Funct. Ecol. 5:669–675.

Pons, T.L. (1992) Seed responses to light. In: Seeds, the ecology of regeneration in plant communities, M. Fenner (ed). C.A.B. International, Wallingford, pp. 259–284.

Pons, T.L. & During, H.J. (1987) Biennal behaviour of *Cirsium palustre* in ash coppice. Holarctic Ecol. 10:40–44.

Pons, T.L. & Schröder, H.F.J.M. (1986) Significance of temperature fluctuation and oxygen concentration for germination of the rice field weeds *Fimbristylis littoralis* and *Scirpus juncoides*. Oecologia 68:315–319.

Poot, P. (1996) Ecophysiological aspects of maintenance of male sterility in *Plantago lanceolata*. PhD thesis, Utrecht University, Utrecht, the Netherlands

Poot, P. (1997) Reproductive allocation and resource compensation in male-sterile, partially male sterile and hermaphroditic plants of *Plantago lanceolata*. Am. J. Bot. 84:1256–1265.

Rabinowitz, D. (1978) Abundance and diaspore weight in rare and common prairie grasses. Oecologia 37:213–219.

Reaumur, R.A.F. (1735) Observations du thermomètre faites à Paris pendant l'anneé 1735, comparées avec celles qui ont été faites sous la Ligne, à l'Isle de France, a Algeres, & en quelquesunes de nos Isles de l'Amerique. Histoire de l'Academie Royale des Sciences, avec les Mémoires de Mathematique & de Physique pour la même année (Paris) 545–580.

Salisbury, E.J. (1942) The reproductive capacity of plants. Bell, London.

Scopel, A.L., Ballaré, C.L., & Radosevich, S.R. (1994) Photostimulation of seed germination during soil tillage. New Phytol. 126:145–152.

Shaver, G.A., Chapin III, F.S., & Billings, W.D. (1979) Ecotypic differentiation in *Carex aquatilis* on ice-wedge polygons in the Alaskan coastal tundra. J. Ecol. 67:1025–1046.

Shipley, B. & Dion, J. (1992) The allometry of seed production in herbaceous angiosperms. Am. Nat. 139:467–483.

Shirley, B.W. (1996) Flavonoid biosynthesis: "New" functions for an "old" pathway. Trends Plant Sci. 1:377–382

Simpson, R.J. (1986) Translocation and metabolism of nitrogen: Whole plant aspects. In: Biochemical, ecological and agricultural aspects of nitrogen metabolism in higher plants, H. Lambers, J.J. Neeteson, & I. Stulen (eds). Martinus Nijhoff Publishers, The Hague, pp. 71–96.

Simpson, R.J., Lambers, H., & Dalling, M.J. (1983) Nitrogen redistribution during grain growth in wheat (*Triticum aestivum* L.). IV. Development of a quantitative model of the translocation of nitrogen to the grain. Plant Physiol. 71:7–14.

Skubatz, H., Nelson, T.A., Meeuse, B.J.D., & Bendich, A.J. (1991) Heat production in the voodoo lily (*Sauromatum guttatum*) as monitored by infrared thermography. Plant Physiol. 95:1084–1088.

Smart, C. (1994) Gene expression during leaf senescence. New Phytol. 126:419–448.

Stanton, M. & Galen, C. (1989) Consequences of flower heliotropism for reproduction in an alpine buttercup (*Ranunculus adoneus*). Oecologia 78:477–485.

Stanton, M. & Galen, C. (1993) Blue light controls solar tracking by flowers of an alpine plant. Plant Cell Environ. 16:983–989.

Stanton, M.L., Bereczky, J.K., & Hasbrouck, H.D. (1987) Pollination thoroughness and maternal yield regulation in wild radish, *Raphanus raphanistrum* (Brassicaceae). Oecologia 74:68–76.

Steinbach, H.S., Benech-Arnold, R.L., & Sanchez, R.A. (1997) Hormonal regulation of dormancy in developing sorghum seeds. Plant Physiol. 113:149–154.

Strydom, A., Jäger, A.K., & Van Staden, J. (1996) Effect of a plant-derived smoke extract, N^6-benzyladenine and gibberellic acid on the thermodormancy of lettuce seeds. Plant Growth Regul. 19:97–100.

Stuefer, J.F. (1995) Separating the effects of assimilate and water integration in clonal fragments by the use of steam-girdling. Abstr. Bot. 19:75–81.

Stuefer, J.F., De Kroon, H., & During, H.J. (1996) Exploitation of environmental heterogeneity by spatial division of labour in a clonal plant. Funct. Ecol. 10:328–334.

Takeno, K. & Maeda, T. (1996) Abscisic acid both promotes and inhibits photoperiodic flowering of *Pharbitis nil*. Physiol. Plant. 98:467–470.

Thomas, T.H. & Van Staden, J. (1995) Dormancy break of celery (*Apium graveolens* L.) seeds by plant derived smoke extract. Plant Growth Regul. 17:195–198.

Thompson, K., Grime, J.P., & Mason, G. (1977) Seed germination response to diurnal fluctuations of temperature. Nature 267:147–149.

Totland, O. (1996) Flower heliotropism in an alpine population of *Ranunculus acris* (Ranunculaceae): Effects on flower temperature, insect visitation, and seed production. Am. J. Bot. 83:452–458.

Van der Burgt, X.M. (1997) Explosive seed dispersal of the rainforest tree *Tetrabelinia moreliana* (Leguminosae —Caesalpiniodeae) in Gabon. J. Trop. Ecol. 13:145–151.

Vazquez-Yanes, C., Orozco-Segovia, A., Rincón, E., Sánchez-Coronado, M.E., Huante, P., Toledo, J.R., & Barradas, V.L. (1990) Light beneath the litter in a tropical forest: Effect on seed germination. Ecology 71:1952–1958.

Vleeshouwers, L.M., Bouwmeester, H.J., & Karssen C.M. (1995) Redefining seed dormancy: An attempt to integrate physiology and ecology. J. Ecol. 83:1031–1037.

Weiss, M.R. (1991) Floral colour changes as cues for pollinators. Nature 354:227–229.

9
Biotic Influences

9A. Symbiotic Associations

1. Introduction

Symbiosis is the "living together" of two or more organisms. In its widest sense, symbiotic associations include parasitic and commensal as well as mutually beneficial partnerships. As is common in the ecophysiological literature, however, we use the term **symbiosis** in a narrow sense and only refer to **mutually beneficial associations** between higher plants and microorganisms. Mutual benefits may not always be easy to determine, and certainly not for the microsymbiont. In this chapter benefits for the macrosymbiont ("host") are often expressed in terms of biomass. In an ecological context, benefits in terms of "fitness" may be more relevant, but this is rarely done. In the mutually beneficial associations discussed in this chapter nutrients or specific products of the partners are shared between two or three partners: the macrosymbiont and the microsymbiont(s). Parasitic associations between different higher plants are dealt with in a separate chapter, but parasitic associations between microorganisms and higher plants will be discussed briefly in this chapter and more elaborately in effects of microbial pathogens.

In the chapter on mineral nutrition, we discussed numerous special mechanisms that allow some higher plants to acquire sparingly soluble nutrients from soils (e.g., excretion of organic acids and phytometallophores). We also pointed out (Sects. 2.2.5 and 2.2.6 of the chapter on mineral nutrition) that some species are quite capable of growing on soils where phosphate is poorly available, without

having a large capacity to excrete protons or organic acids. How do these plants manage to grow? It is also obvious that special mechanisms to acquire nutrients (e.g., nitrogen) are of little use, if the nitrogen is simply not there. Such plants must have ways to acquire nitrogen in an alternative manner.

This chapter discusses associations between higher plants and microorganisms that are of vital importance for the acquisition of nutrients. Such symbiotic associations play a major role in environments where the supply of phosphate, nitrogen, or immobile cations limits plant growth. In the rhizosphere (or elsewhere in the plant's immediate surroundings), mycorrhiza-forming fungi and dinitrogen-fixing bacteria or cyanobacteria may form symbiotic associations. For those species that are capable of such symbioses, it tends to be profitable for both the higher plant (**macrosymbiont**) and the microorganism (**microsymbiont**). Indeed, it is so profitable for the macrosymbiont that some plants are associated with two symbioses at the same time.

2. Mycorrhizas

A vast majority of higher plant species can form symbiotic associations with **mycorrhizal fungi**. Mycorrhizas are the structures arising from the association of roots and fungi. Like root hairs (see Sects. 2.2.1 and 2.2.5 of the chapter on mineral nutrition), the mycorrhizal association enhances the symbiotic plant's below-ground absorbing surface.

For endomycorrhizas (see Sect. 2.1), this is the primary mechanism by which mycorrhizal plants are able to acquire scarcely available, poorly mobile nutrients, especially phosphate (Jakobsen et al. 1994). For ectomycorrhiza (Sect. 2.1) additional mechanisms, such as excretion of organic acids and hydrolytic enzymes, may also play a role.

The mycorrhizal association enhances plant growth, especially when phosphate or other immobile nutrients are in short supply; they may also be beneficial when water is in short supply. As such, mycorrhizae are of great agronomic and ecological significance. When the nutrient supply is high, however, they are a potential carbon drain on the plant, with little benefits in return. Plants, however, have mechanisms to suppress the symbiotic association at a high supply of phosphate (Sect. 2.3.1).

It is interesting that some species never form a mycorrhizal association, even when phosphate is in short supply. Some of these (e.g., Proteaceae and Brassicaceae) (Sect. 2.2.5.2 of the chapter on mineral nutrition), perform well when phosphate is severely limiting. Such nonmycorrhizal plants are often harmed by mycorrhizal fungi. On the other hand, some nonmycorrhizal plants severely inhibit the growth of mycorrhizal hyphae. Before dealing with these nonmycorrhizal species in Section 2.2, we will first discuss some general aspects of the mycorrhizal associations.

2.1 Endo- and Ectomycorrhizas: Are They Beneficial for Plant Growth?

Mycorrhizal associations consist of three vital parts:

1. the root
2. the fungal structures in close association with the root
3. the external mycelium

Mycorrhizas are predominantly classified as **ectomycorrhizas** and **endomycorrhizas**. This classification is entirely based on the site of the fungal mycelium in close association with the root and does not imply a functional difference. Mycorrhizas occur in 83% of dicotyledonous and 79% of monocotyledonous species so far investigated; all gymnosperms are mycorrhizal (Trappe 1987, Wilcox 1991).

In the **ectomycorrhizas**, the fungal tissue is largely outside the root. This symbiotic association is frequently found between trees and basidiomycetes (Dipterocarpaceae: 98%; Pinaceae:

95%; Fagaceae: 94%; Myrtaceae: 90%; Salicaceae: 83%; Betulaceae: 70%; Fabaceae: 16%). Although ectomycorrhizas predominantly occur in woody angiosperms and gymnosperms, they have also been found in some monocotyledons and ferns. In the **endomycorrhizas** a large fraction of the fungal tissue is within the root cortical cells. An important group is indicated as the vesicular-arbuscular mycorrhizas. They frequently occur on herbaceous plants, but they are also found on trees, especially in tropical forests. *Eucalyptus*, *Cupressus*, *Salix*, and *Populus* are genera that have both endo- and ectomycorrhizas. Somewhat different structures are found in the Orchidaceae, Ericaceae, and Epacridaceae, but we will only refer to these in passing.

Much attention has been given to the **vesicular arbuscular mycorrhiza** (**VAM**). The fungi belong to the Glomales, with *Glomus* being the largest genus. The VAM has been named after the **vesicles** and **arbuscules**, (which are treelike structures that are found intracellularly in the cortex of the roots (Fig. 1); in some cases, the vesicular structures are absent, which is the reason the associations are also referred to as AM. VA mycorrhiza are considered the most ancient type of mycorrhizal symbiosis and occur in the most phylogenetically advanced groups. Few species have evolved mechanisms to completely prevent infection by VAM fungi (Sect. 2.2). In nature the roots of more than 80% of all plant species are infected with VAM-forming fungi (Smith & Read 1997). Even species that are typically ectomycorrhizal form VAM associations in the absence of ectomycorrhizal inoculum (e.g., after fire).

2.1.1 The Infection Process

During the establishment of VAMs, fungal hyphae that grow from spores in the soil or from adjacent plant roots contact the root surface, where they differentiate to form an **appressorium** and initiate the internal colonization phase (Peterson & Bonfante 1994). This is the first indication of recognition between the fungus and the plant, and appressoria are not formed on the roots of nonhost plants. Penetration of the root occurs via the appressoria, and the fungus frequently enters by forcing its way between two epidermal cells. The point of entry often is the **passage cells** in the exodermis (Fig. 12 in the chapter on plant water relations). It has been speculated that endomycorrhizal fungi receive some signal from the exodermal passage cells, but the nature of the signal is unknown (Peterson & Enstone 1996). Once inside the root, the fungus may

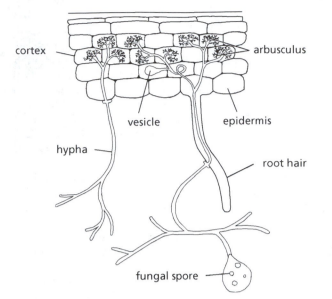

cortex

arbusculus

vesicle

epidermis

hypha

root hair

fungal spore

Figure 1. (Top) Schematic structure of a vesicular-arbuscular mycorrhiza (VAM). (Middle) Intraradical hyphae of *Glomus etunicatum* excised after enzymatic digestion of a root of *Tagetes patula* (marigold). The hyphae were stained with trypan blue. C, collapsed arbuscule; M, mature arbuscule; V, vesicle; scale bar = 20 μm. (Bottom) Detail of the intraradical hyphae of *Glomus mosseae* in a root of *Tagetes patula* (marigold), after enzymatic digestion of the root, showing fine branches and trunks of an arbuscule; scale bar = 40 μm (Ezawa et al. 1995) (courtesy T. Ezawa, Faculty of Horticulture, Nagoya University, Japan). Copyright Kluwer Academic Publishers.

produce intracellular coils in the subepidermal cell layers, followed by intercellular growth into the inner cortex of the root. On reaching the inner cortical cells, hyphal branches penetrate the cortical cell walls, without disrupting the plasma membrane. The hyphae differentiate within the cells to form the previously mentioned arbuscular structures, surrounded by the host plasma membrane. The function of the **arbuscules** is most likely to increase the surface of membranes over which exchange of metabolites occurs and so to enhance active transport between the plasma membrane of the host and the hyphae of the fungus (Smith et al. 1994). The hyphae proliferate both in the cortex and in the soil. **Vesicles**, in which lipids are stored, are formed in many VAM in a later stage. Vesicles can form between or within cells. The VAM fungus does not penetrate into the endodermis, stele, or meristems. It has so far proved impossible to grow the VAM fungus in the absence of its host (Harrison 1997).

The mycorrhizas of the Orchidaceae have also been studied intensively. Their structure is fairly similar to that of the VAM in that there is extensive intracellular growth of the fungus; however, in Orchidaceae the intracellular fungal tissue appears as **coils**, rather than arbuscules (Fig. 2). The fungi forming the mycorrhizas are basidiomycetes, and many belong to the genus *Rhizoctonia*. As soon as they have germinated, the orchid seedlings, which have very few reserves, depend on organic matter in the soil or from other host plants via the mycorrhizal fungus. For example, many *Rhizoctonia* species form associations with both orchids and conifers. The orchids are therefore not saprophytic, but "mycoheterotrophic," (i.e., parasitic on the fungus) (Leake 1994); the association between host and fungus does not appear to be mutually beneficial. In those orchids that remain nonphotosynthetic during their entire life cycle (e.g., the Western Australian fully subterranean *Rhizanthella gardneri*) (Dixon et al. 1987), the fungus continues to play this role. In all orchids, including those that are green, (photosynthetic) as adults, the fungi also absorb mineral nutrients from soil (like VAM, see later).

In the mycorrhizas of the Ericaceae and Epacridaceae, a large number of infection points are found: up to 200 per mm root in *Calluna*, as opposed to approximately 2 to 10 per mm in the grass *Festuca ovina*, infected by a VAM fungus. Up to 80% of the volume of these mycorrhizas may be fungal tissue (this value does not include the external mycelium). This percentage is considerably greater than that for the arbuscular mycorrhiza, possibly because members of the Ericaceae and Epacridaceae have very fine roots ("hair roots"), with very few cell layers (often only one or two) outside the stele. In the ectomycorrhiza the fraction of fungal origin is approximately 40%. The infecting fungi of the Ericaceae are ascomycetes [e.g., *Hymenoscyphus ericae* (= *Pezizella ericae*)]. Several fungal species, possibly ascomycetes, are involved in the mycorrhizal symbiosis with Epacridaceae, but none of them appear to be the same as those that infect Ericaceae, despite their similarity in morphology (Hutton et al. 1994).

Spores of ectomycorrhizal fungi in the rhizosphere may germinate to form a **monokaryotic mycelium**. This needs to fuse with another hypha in order to form a **dikaryotic mycelium**, which can then colonize the root, forming a mantle of fungal hyphae that encloses the root. The hyphae usually penetrate intercellularly into the cortex, where they form the **Hartig net** (Fig. 3). As hyphae contact the root surface, roots may respond by increasing their diameter and switching from apical growth to precocious branching (Peterson & Bonfante 1994). Numerous fungal species have the capacity to form ectomycorrhizas. Most of these belong to the basidiomycetes and ascomycetes, and they are often species with which we are familiar as toadstools.

In contrast to infection by pathogenic fungi, colonization with mycorrhizal fungi never causes disease symptoms. In the presence of mycorrhizal fungi in the rhizosphere, flavonoids accumulate in the roots of *Medicago sativa* (alfalfa) host roots, which is similar to, but much weaker than, the response upon fungal attack (see Sect. 3 of the chapter on effects of microbial pathogens). VAM fungi initiate a host defense response in the early stages of colonization; the defense response is subsequently suppressed (Volpin et al. 1994). The production of defense-related gene products is restricted to arbusculated cells, whereas intercellular hyphae and vesicles elicit no such defense response (Blee & Anderson 1996). The rate and location of fungal growth within the root could be controlled through activation of plant defense mechanisms.

Some mutants of *Pisum sativum* (pea) and *Vicia faba* (faba bean) and other legumes are characterized by aborted mycorrhizal infections, after formation of appressoria (Duc et al. 1989, Harrison 1997). The mutations are recessive and the genes appear to be closely linked with genes that control nodulation by rhizobia (Sect. 3.3) (Smith et al. 1992). This association may point to a genetic linkage and presumably tight control of two

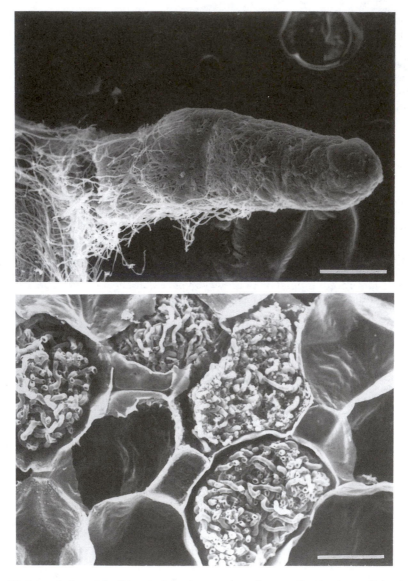

FIGURE 2. Mycorrhizal association of the Western Australian fully subterranean orchid *Rhizanthella gardneri*. (Top) rootlet covered in fungal tissue. (Bottom) transverse section of a root showing intracellular fungal coils (courtesy J. Kuo, The University of Western Australia, Australia).

FIGURE 3. (Top) Schematic representation of an ectomycorrhiza, showing the fungal sheath around the root and the hyphae in the cortex, which form the Hartig net. (Bottom) Ectomycorrhizal association between *Pinus resinosa* and *Pisolithus tinctorius*. The smaller magnification (grid squares = 1.0 mm) shows thickened branched rootlets, covered in a fungal sheath, and external hyphae. The larger magnification (scale bar is 10 μm) is a longitudinal section through the mycorrhizal root, showing the Hartig net in between epidermal and outer cortical cells in the middle, with fungal tissue at the far left and inner cortical root cells at the far right (courtesy R.L. Peterson, University of Guelph, Canada).

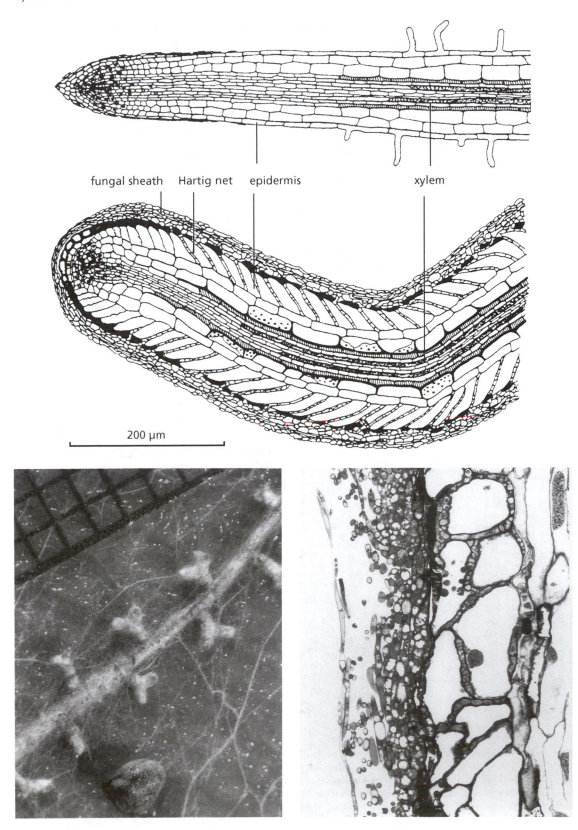

fungal sheath   Hartig net   epidermis   xylem

200 µm

carbon-consuming and potentially competing symbioses (Sect. 2.6).

2.1.2 Mycorrhizal Dependency

In soils with a low phosphate availability, plants vary widely in the extent to which their growth responds to root colonization by mycorrhizal fungi (Hetrick et al. 1989). **Mycorrhizal dependency** is defined as the ratio of the dry mass of the mycorrhizal plants to that of nonmycorrhizal plants. Phylogenetically advanced groups that are less dependent on VA mycorrhiza for their nutrient acquisition (e.g., Poaceae) are generally colonized to a lesser extent in the field than are more primitive, mycorrhizal-dependent orders (e.g., Magnoliales).

A high mycorrhizal dependency is commonly associated with lack of well-developed root hairs and coarse fibrous roots (cf. Sect. 2.2.1 of the chapter on mineral nutrition). Among *Citrus* rootstocks, mycorrhizal dependency in phosphate-deficient soil is positively correlated with the extent to which the rootstocks are colonized by the VA fungus in the field at high phosphate supply (Fig. 4). There is no correlation with root diameter or rate of root extension. This suggests that species that have evolved root systems with low dependency on mycorrhiza have also evolved mechanisms to suppress mycorrhizal colonization (Sect. 2.3.1) (Graham & Eissenstat 1994).

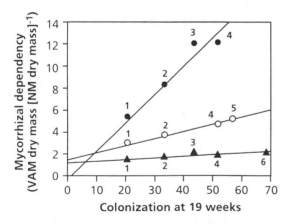

FIGURE 4. The relationship between mycorrhizal dependency of six citrus rootstocks in a soil with low phosphate availability in three glasshouse trials and root colonization, assessed by determining intraradical VAM colonization (% of root length colonized) (Graham et al. 1991). Copyright Blackwell Science Ltd.

2.2 Nonmycorrhizal Species and Their Interactions with Mycorrhizal Species

Although mycorrhizal associations are very common, some species cannot be colonized, or only marginally so (Ocampo et al. 1980). These **nonmycorrhizal species** include many that belong to the Brassicaceae, Caryophyllaceae, Chenopodiaceae, Lecythideceae, Proteaceae, Restionaceae, Sapotaceae, Urticaceae, and Zygophyllaceae, as well as *Lupinus* spp. (Fabaceae). It is interesting, as discussed in Section 2.2.5.2 of the chapter on mineral nutrition, that many of these species have "cluster roots" (Proteaceae) and almost all of them have the capacity to excrete protons or organic acids when the phosphate supply is low. This is probably not the case for *Urtica dioica* (Urticaceae), but this species only occurs on phosphate-rich soils and would benefit very little from mechanisms that enhance the availability of phosphate from insoluble sources.

The mechanisms that prevent colonization in nonmycorrhizal species are unclear (Tester et al. 1987). In some species, the exudation of fungi-toxic compounds, such as glucosinolates in Brassicaceae (Koide & Schreiner 1994) or a chitin-binding agglutinin in *Urtica dioica* (Vierheilig et al. 1996), may prevent infection. In others (e.g., Chenopodiaceae) the correct chemical cues necessary for spore germination and the subsequent development of arbuscules may be lacking (Koide & Schreiner 1992).

Whereas mycorrhizal fungi enhance growth of mycorrhizal plants, at least at a low phosphate supply, the opposite is found for nonmycorrhizal species (Sanders & Koide 1994). In these species, the mycorrhizal fungus may reduce stomatal conductance (Allen & Allen 1984) and inhibit root hair elongation, probably via its exudates. This mechanism may well explain why nonmycorrhizal species show poor growth in a community dominated by mycorrhizal species, unless the phosphate level is increased (Francis & Read 1994).

2.3 Phosphate Relations

Like root hairs, mycorrhizas increase the roots' absorptive surface. In fact, the effective root length of the mycorrhizal associations may increase one hundred-fold or more per unit root length (Table 1).

2.3.1 Mechanisms That Account for Enhanced Phosphate Absorption by Mycorrhizal Plants

When the availability of phosphate in the soil is low, the mycorrhizal association mostly enhances

TABLE 1. The length of mycorrhizal hyphae per unit colonized root length as measured for a number of plant species, infected with different mycorrhiza-forming fungal species.

Fungus	Host	Hyphal length (m cm^{-1} root)
Glomus mosseae	onion	0.79–2.5
Glomus mosseae	onion	0.71
Glomus macrocarpum	onion	0.71
Glomus microcarpum	onion	0.71
Glomus sp.	clover	1.29
Glomus sp.	rye grass	1.36
Glomus fasciculatum	clover	2.50
Glomus tenue	clover	14.20
Gigaspora calospora	onion	0.71
Gigaspora calospora	clover	12.30
Acaulospora laevis	clover	10.55

Various authors, as cited in Smith & Gianinazzi-Pearson 1988.

phosphate uptake and growth (Fig. 5). There is little evidence that VA mycorrhizal plants have access to different chemical pools of phosphate in the soil (Read 1991). They are capable, however, of acquiring phosphate outside the depletion zone that surrounds the root because of the **widely ramified hyphae**. These hyphae allow phosphate transport over as much as 10 cm from the root surface (Fig. 6). In addition, they may also get into smaller soil pores and compete effectively with other microorganisms (Joner & Jakobsen 1995). Ectomycorrhizal hyphae may extend for much greater distances, possibly several meters. Ectomycorrhizal and ericoid mycorrhizal roots also have access to additional phosphate sources; they may release **phosphatases**, which enhance the availability of organic phosphate, exude **organic acids**, which increase the availability of sparingly soluble phosphate, and use organic N (Fig. 6). It is interesting

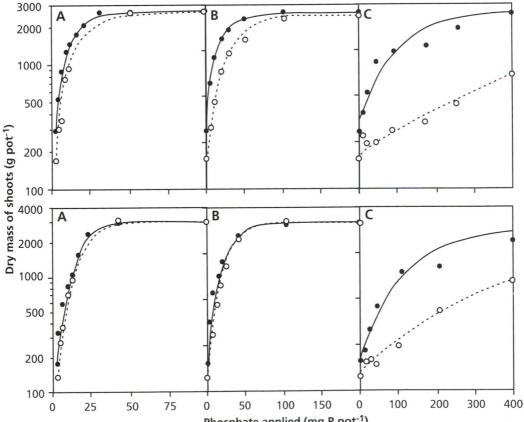

FIGURE 5. Effect of phosphate application on the dry mass of shoots of plants that were inoculated with *Glomus fasciculatum* (solid lines) or not inoculated (broken lines). *Trifolium subterraneum* (subclover) (top); (bottom) *Lolium rigidum* (stiff darnel). The phosphate sources were: A: KH$_2$PO$_4$, B: colloidal iron phosphate, C: stengite (they are listed in order of decreasing solubility) (Bolan et al. 1987). Copyright Kluwer Academic Publishers.

that these ectomycorrhizas and ericoid mycorrhizas frequently occur in organic soils, whereas VAM are more typical of mineral soils.

As in roots, phosphate uptake by the mycorrhizal hyphae occurs via active transport, against an electrochemical potential gradient, presumably with a proton-cotransport mechanism (cf. Sect. 2.2.2 in the chapter on mineral nutrition). Once absorbed by the external hyphae, phosphate is polymerized into inorganic polyphosphate (poly P). This reaction is catalyzed by polyphosphate kinase, which is induced when excess inorganic phosphate is absorbed. Some of the poly P, together with K^+, is stored in vacuoles. Translocation of phosphate and other immobile ions through the hyphae to the plant is relatively rapid, bypassing the very slow diffusion of these ions in soil. The mechanism of translocation is unknown, but it is the subject of active research. Once the polyphosphate has

arrived near the plant cells, polyphosphate needs to be degraded, catalyzed by a phosphatase. The mechanisms that account for polyphosphate production at one end of the hypha and breakdown at the other are unknown. It might involve sensing a gradient in plant-derived carbohydrates. Transfer of phosphate and other ions from fungus to plant must be a two-step process across the membranes of the two symbionts that probably involves passive efflux from the fungus and active uptake by the plant (Smith et al. 1994).

2.3.2 Suppression of Colonization at High Phosphate Availability

Colonization by the mycorrhizal fungus occurs with greater frequency in phosphate-poor soils than in soils that contain more phosphate (Fig. 7). To some extent this greater frequency may be asso-

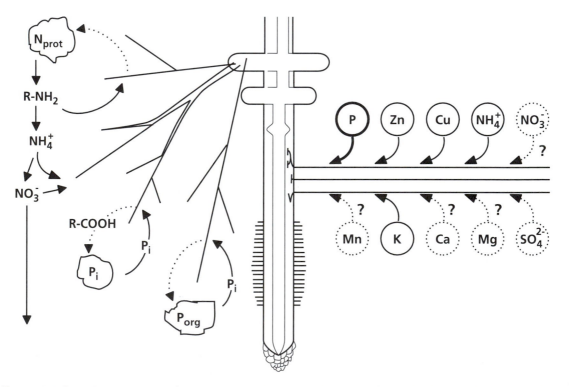

FIGURE 6. Schematic presentation of components of the nutrient acquisition from the soil by vesicular-arbuscular mycorrhizal roots (right). The thickness of the circle indicates the importance of VAM in acquiring this nutrient (? indicates lack of definitive information). Additional components in ectomycorrhizal roots are also shown (left). Note that all mycorrhizas enhance the availability of soil nutrients by enlarging the soil volume that is exploited and

that this is most relevant for those nutrients that are least mobile (e.g., phosphate). Ectomycorrhizas excrete hydrolytic enzymes, which allows them to use organic forms of both P and N, and chelating organic acids, which allows the use of poorly soluble forms of inorganic P (after Marschner & Dell 1994). Copyright Kluwer Academic Publishers.

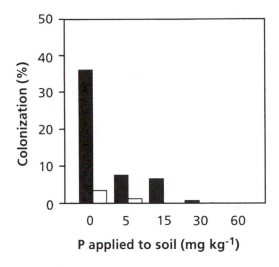

FIGURE 7. The effect of applied phosphate on colonization (% of root length colonized) by mycorrhizal fungi in *Triticum aestivum* (wheat) grown in two soils with different propagule densities (26 and 0.9 propagules g⁻¹; filled bars and open bars, respectively) (Baon et al. 1992, as cited in Smith et al. 1992). Copyright CSIRO, Australia.

ciated with a decreased rate of root elongation so that the colonization by the fungus keeps up with the growth of the root. Phosphate, however, may also have a direct effect on the fungus. To establish whether it is the phosphate concentration in the plant or that in the soil that determines the degree of colonization, plants have been grown with a divided root system. That is, part of the root system grows in a pot without phosphate, and the remaining part in a separate pot supplied with phosphate. In such a plant some of the phosphate acquired by the roots supplied with phosphate ends up in the roots deprived of phosphate, as a result of transport via the xylem to the shoot, followed by export via the phloem from the shoot. Hence, the phosphate concentration in these roots is considerably higher than that in roots of plants that did not receive any phosphate. The infection percentage of such roots is relatively low; much lower than that of plants not receiving any phosphate. This result suggests that the **phosphate concentration** in the **plant**, rather than that in the soil, determines the extent of colonization (Smith & Read 1997).

What is the explanation for enhanced colonization by VAM fungi when the phosphate concentration in the roots is lowest? When grown in soil low in phosphate, roots contain very little phosphate, especially very few **phospholipids**. These phospholipids are an important component of the

root cells' plasma membranes, and they determine their permeability for a number of compounds, including potassium, sugars, and amino acids. If the phospholipid concentration in the cells is low, then the roots exude relatively large amounts of organic compounds (Fig. 8). It has been speculated that these organic compounds may promote germination of fungal spores, followed by colonization by the VAM fungi, which subsequently enhances the phosphate status of the macrosymbiont. It is likely that more specific **signaling** compounds, possibly of a phenolic nature, also play a role in the regulation of the infection process as dependent on the plant's phosphate status (Koide & Schreiner 1992). In ectomycorrhizal species **palmitic acid** may play a role similar to that of specific phenolic compounds in arbuscular mycorrhizas (Marschner 1995).

Arbuscular mycorrhizal fungi may infect many plants at the same time, even plants of different species. In this way they may transport carbon or nutrients from one plant to the other (Heap & Newman 1980); for example, nitrogenous compounds from an N_2-fixing soybean plant (*Glycine max*) to a maize plant (*Zea mays*) (Bethlenfalvay et al. 1991). Interplant nutrient transfer might be an important ecological process in grasslands and significantly influence plant neighborhood interactions. It is likely that some of such transfer is mediated through mycorrhizal connections (Chiariello et al. 1982). In tallgrass prairie, Fischer Walter et al. (1996) found that transfer of (labeled) phosphate occurs over distances of up to 0.5 m. Although tracer experiments have shown that interplant transfer does occur, the rate of transfer of ^{32}P between mycorrhizal plants of *Lolium perenne* and *Plantago lanceolata* appears to be too slow to be ecologically significant. Nitrogen, however, is transferred more freely between mycorrhizal plants (Fig. 9), and this could be important in transfer from legumes to nonlegumes. Treatments that induce a net loss of nutrients from the roots enhance the transfer of nutrients, but the fraction transferred via fungal hyphae remains the same (Fig. 9). Net nutrient transfer is possibly greater between ectomycorrhizal plants.

2.4 Effects on Nitrogen Nutrition

Unlike VAM, some **ectomycorrhiza** have the capacity to utilize **organic nitrogen**, including proteins. This is not an artefact that only occurs in laboratory experiments, which show utilization of complex nitrogen-containing molecules by ectomycorrhizal fungi (e.g., Turnbull et al. 1995), but actually occurs in situ, as revealed by a comparison of ^{15}N discrimination in plants with and without ectomycorrhiza. Ectomycorrhizal plants may have a 1.0 to 2.5‰ more positive $\delta^{15}N$ value than do plants infected with arbuscular mycorrhiza (Table 2). Such a comparison proves that different N-sources are used. The different N-source, of course, might reflect the use of the same inorganic compound, which is absorbed from different regions in the soil. In a study on a Tanzanian woodland, however, which is illustrated in Table 2, there was no evidence for a difference in isotope composition in different soil layers. Discrimination against the heavy nitrogen isotope (^{15}N) occurs during mineralization and nitrification, therefore organic N becomes enriched with [15]N. The data in Table 2 provide evidence that the ectomycorrhizal plants use a significant amount of nitrogen from a pool that was not decomposed

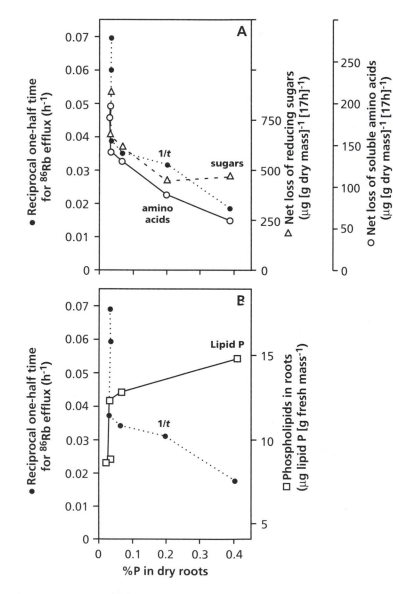

FIGURE 8. Efflux of potassium (assayed by using radioactive rubidium as a substitute), sugars, and amino acids and the concentration of phospholipids in the roots as a function of the phosphate concentration in the roots. Efflux was quantified as the reciprocal half-time (Ratnayake et al. 1978). Copyright Trustees of The New Phytologist.

TABLE 2. ^{15}N abundance of leaf samples collected in different years in Tanzania.*

Species	Symbiotic status	δ^{15}N		
		1980	1981	1984
Brachystegia boehmii	EC	1.64	1.32	1.23
B. microphylla	EC	1.53	1.51	1.73
Julbernardia globiflora	EC	2.81	1.63	1.60
Pterocarpus angolensis	VAM + NO	−0.81	−0.87	−0.93
Diplorynchus condylocarpon	VAM	—	−0.36	−0.60
Xeroderris stuhlmannii	VAM + NO	—	0.01	0.62
Dichrostachys cinerea	VAM + NO	—	0.45	−0.38

Source: Högberg 1990.
*EC = ectomycorrhizal; VAM = arbuscular mycorrhizal; NO = nodulated. The experiments summarized here were actually carried out with the aim of determining the extent of symbiotic N$_2$ fixation of the nodulated plants. Because nodulated plants have access to dinitrogen from the atmosphere, they are expected to have δ^{15}N values closer to atmospheric N$_2$ than do plants that do not fix dinitrogen. The data presented here stress that control plants need to be sampled to allow a proper comparison. This table shows that the choice of the control plants is highly critical (see also Sect. 3 of this chapter).

and nitrified (i.e., organic nitrogen). In boreal forest and arctic tundra, ectomycorrhizal plant species also have distinctive ^{15}N signatures, with ^{15}N concentrations that are higher than those of species with ericoid mycorrhizas, but lower than those of nonmycorrhizal or VAM species (Nadelhoffer et al. 1996, Schulze et al. 1995). In these studies either rooting depth or form of nitrogen utilized could have contributed to the different ^{15}N signatures. It still remains to be established to what extent the conclusions derived from these studies can be generalized (Högberg & Alexander 1995).

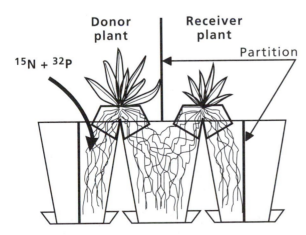

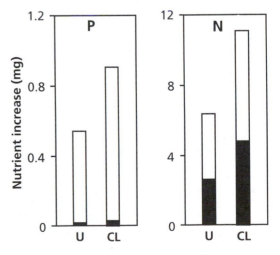

FIGURE 9. (Left) Planting arrangement in a split-root design. Using measurements of the ^{15}N and ^{32}P content of the shoots of the "receiver" plants, it was possible to calculate how much of the increase in N and P was by transfer from the "donor" plants and how much was by uptake from the soil by the roots of the "receiver." (Right) Increase in phosphorus (P) and nitrogen (N) in shoot of "receiver" *Plantago lanceolata* during 19 days after shoot of neighboring "donor" plants had been clipped near the base soon after the isotopes had been fed (CL) or not (U). The black portion of the bar is the amount transferred via fungal hyphae; the white portion refers to transfer of nutrients first released to the soil and then reabsorbed (Eissenstat 1990).

Ericoid mycorrhizas, like the ectomycorrhizas, can use quite **complex organic sources of nitrogen and phosphate** (Fig. 6; Marschner & Dell 1994). It has been suggested that this ability allows members of the Ericaceae to dominate in cold and wet soils, where rates of decomposition and mineralization are low. The arbuscular mycorrhizas are at the other extreme of the continuum of mycorrhizal associations. Their predominant significance lies in the acquisition of sparingly available inorganic nutrients, especially phosphate. VAMs are relatively unimportant for acquisition of nitrogen, if this is available as nitrate, but they enhance N-acquisition when mineral N is present as the less mobile ammonium (Johansen et al. 1994). VAM may enhance the uptake of nitrate from dry soils, when mass flow and diffusion are limited, but not in wet soils (Tobar et al. 1994). Ectomycorrhizas are thought to function somewhere in between the extremes of the mycorrhizas associated with the Ericaceae and arbuscular mycorrhiza (Wilcox 1991).

2.5 Effects on the Acquisition of Water

Mycorrhizal plants may have an enhanced capacity to acquire water from the root environment (Allen & Allen 1986). Several hypotheses have been put forward to explain this increased capacity (Krikun 1991). A likely one includes an indirect effect via the **improved phosphate status** of the plant, which increases the hydraulic conductance of the roots or affects the plant's hormone metabolism. Stomatal conductance of mycorrhizal plants, however, may also be higher when their leaf phosphate concentration is lower than that of nonmycorrhizal control plants. This higher stomatal conductance is associated with a lower level of ABA in mycorrhizal plants (Sanchez-Diaz & Honrubia 1994). There is also evidence (Ruiz-Lozana & Azcón 1995) that the fungal hyphae themselves absorb significant amounts of water, which subsequently moves to the plant. The multihyphal strands of ectomycorrhiza are thought to have a particularly high capacity to transport water.

2.6 Carbon Costs of the Mycorrhizal Symbiosis

Since the higher plant supplies carbon to the microsymbiont, there are **costs** associated with this symbiosis for the macrosymbiont. These have been estimated in various ways (e.g., by comparing plants with and without the mycorrhizal symbiont at the same growth rate). This can be achieved by providing more phosphate to the nonmycorrhizal plant, compared to the supply to the mycorrhizal plant. The carbon use for growth and respiration by the roots of both types of plants can then be used to quantify costs of the mycorrhizal symbiosis (Baas et al. 1989, Snellgrove et al. 1982; Fig. 10). The problem with this method is that it assumes steady-state rates of phosphate acquisition and carbon consumption, whereas in fact these may vary following active root colonization. A variation of the approach shown in Figure 10 is to grow nonmycorrhizal plants at a range of phosphate supplies so that a P-response curve can be constructed with which to compare the mycorrhizal plants (Eissenstat et al. 1993, Rousseau & Reid 1991).

An alternative approach to quantifying the costs of the mycorrhizal symbiosis has been to grow

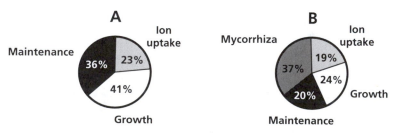

FIGURE **10.** Respiratory energy requirement in (A) nonmycorrhizal and (B) mycorrhizal roots of *Plantago major* (plantain). The mycorrhizal plants were inoculated with *Glomus fasciculatum*. The total amount of ATP produced per unit of time and mass in the roots of the two groups of plants was 76 and 137 nmol O_2 g^{-1} (dry mass of roots) s^{-1} for nonmycorrhizal and mycorrhizal plants, respectively (Baas et al. 1989, as cited in Lambers et al. 1996). Copyright Marcel Dekker, Inc.

TABLE 3. Comparison of accumulated ^{14}C and fresh mass in mycorrhizal and nonmycorrhizal halves of root system.*

Species	^{14}C recovered from below-ground tissue $(dpm\,g^{-1})$		Fresh mass $(mg\,plant^{-1})$	
	+	−	+	−
Sour orange	66.4	33.6	1580	1240NS
Carrizo citrange	67.7	32.3	1990	1520NS

Source: Koch & Johnson 1984.

* + and − denote mycorrhizal and nonmycorrhizal plants, respectively; NS indicates that there was no significant difference.

plants with a divided root system. That is, part of the root system is grown in one pot, and the remaining part in a separate pot. One part of the divided root is then inoculated with a mycorrhizal fungus, while the other is not and remains nonmycorrhizal. The shoot is then given $^{14}CO_2$ to assimilate in photosynthesis and the partitioning of the label over the two root parts is measured (Douds et al. 1988, Koch & Johnson 1984; Table 3). It is also possible to calculate carbon costs of the mycorrhizal symbiosis by measuring the flow of ^{14}C-labeled assimilates into soil and external hyphae (Jakobsen & Rosendahl 1990).

The estimates of the carbon costs of the mycorrhizal symbiosis vary between 4 and 20% of the carbon fixed in photosynthesis (Lambers et al. 1996). Only a minor part (15%) of the increased rate of root respiration is associated with an increased rate of ion uptake by the mycorrhizal roots. The major part (83%) is explained by the respiratory metabolism of the fungus and/or other effects of the fungus on the roots' metabolism (Fig. 10). Construction costs of fibrous roots are also higher for mycorrhizal than they are for nonmycorrhizal roots because of their higher fatty acid concentration (Peng et al. 1993) (cf. Sect. 5.2.1 of the chapter on plant respiration).

In addition to a higher carbon expenditure, mycorrhizal plants also tend to have a **higher rate of photosynthesis** per plant, partly due to higher rates of photosynthesis per unit leaf area and partly to their greater leaf area. The higher rate of CO_2 assimilation is most pronounced when the soil water potential is low (Sanchez-Diaz et al. 1990). When phosphate and water are limiting for growth, therefore, benefits outweigh the costs and the mycorrhizal plants usually grow faster, despite the large carbon sink of the symbiotic system. The

relatively high costs of the mycorrhizal association, however, may help to explain why mycorrhizal plants sometimes grow less than their mycorrhizal counterparts (Fredeen & Terry 1988, Thompson et al. 1986), especially when a second microsymbiont (*Rhizobium*) plays a role (Fig. 11). Under drought, however, mycorrhizal plants may still show more benefit from an association with *Rhizobium* than do nonmycorrhizal control plants (Pena et al. 1988).

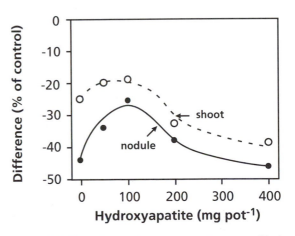

FIGURE 11. The relative host response to mycorrhizal infection as dependent on the supply of hydroxyapatite $[Ca_{10}(PO_4)_6(OH)_2]$. *Phaseolus vulgaris* (common bean) plants were infected with *Rhizobium phaseoli*, a nitrogen-fixing bacteria. Half of the plants were also infected with the mycorrhizal fungus *Glomus fasciculatum*. The difference in mass of the parts of the mycorrhizal plants relative to the nonmycorrhizal control plants was calculated as percentage of the difference. Negative values indicate that the shoot or nodule mass was less in the mycorrhizal plants (Bethlenfalvay et al. 1982). Copyright American Society of Plant Physiologists.

2.7 Agricultural and Ecological Perspectives

From an ecological point of view, information on the mycorrhizal status of plants in a community is most important. In a mixed community nonmycorrhizal species quite possibly profit most from **fertilization** with phosphate because the mycorrhizal association is often suppressed at a higher phosphate supply and not because their growth is more severely phosphate-limited (see Sect. 2.3). Suppression of the mycorrhiza might then reduce the harmful effect of the mycorrhizal fungus on nonmycorrhizal species (Sect. 2.2). We should therefore be warned against a too hasty interpretation of the effects of phosphate fertilization on the growth of certain plants in a community.

Close proximity between the roots of a seedling and those of an established, infected plant may speed up VAM infection, but, for unknown reasons, this is not invariably the case (Newman et al. 1992). Mycorrhizas may have profound effects on interactions between plants in a community, as discussed in Section 7 of the chapter on interactions among plants.

Mycorrhizas can, obviously, never enhance growth and productivity of crop plants in the absence of any phosphate. In addition, if wheat breeding occurs under conditons of a high rate of phosphate application, then this will virtually exclude any positive effects that mycorrhizal colonization might have on phosphate acquisition or growth (Smith et al. 1992). Mycorrhizal associations, however, do have great potential in improving crop production when phosphate or other immobile nutrients are in short supply. Introduction of spores of the best microsymbiont and breeding for genotypes with a more efficient mycorrhizal symbiosis are tools that can be used to enhance food production in countries where immobile nutrients restrict crop production. As such, mycorrhizas allow good crop growth and may reduce nutrient losses to the surrounding environment (Tisdall 1994).

3. Associations with Nitrogen-Fixing Organisms

Nitrogen is the major limiting nutrient for the growth of many plants in many environments (Bohlool et al. 1992). Terrestrial nitrogen is subject to rapid turnover, and, because it is eventually lost as nitrogen gas into the atmosphere, its maintenance requires a continuous reduction of atmospheric dinitrogen. Biological reduction of dinitrogen gas to ammonia can be performed only by some prokaryotes and is a highly **oxygen-sensitive process**. The most efficient dinitrogen-fixing microorganisms establish a symbiosis with higher plants, in which the energy for dinitrogen fixation and the oxygen-protection system are provided by the plant partner (Mylona et al. 1995).

Symbiotic associations with microorganisms that fix atmospheric dinitrogen may be of major importance for a symbiotic plant's nitrogen acquisition, especially in environments where nitrogen is severely limiting for plant growth. As such, the symbiosis is also of agronomic importance because it reduces the need for costly fertilizers. A nonsymbiotic association, [e.g., with *Azospirillum* in the rhizosphere of tropical grasses or *Acetobacter diazotrophicus* in the apoplast of the stems of *Saccharum officinarum* (sugarcane)] is sometimes found. Contrary to the strictly symbiotic systems, no special morphological structures are induced.

Symbiotic N_2-fixing systems require a carbon input from the host, which is far greater than the carbon requirements for the acquisition of N in the combined form (e.g., NO_3^-, NH_4^+, or amino acids). Are there mechanisms to suppress the symbiosis when there is plenty of combined nitrogen around? How does a plant discriminate between a symbiotic guest and a pathogenic microorganism? What is the significance of nonsymbiotic N_2 fixation for plants? To answer these ecological questions we will first provide a basic understanding of some physiological aspects of this symbiotic association.

3.1 Symbiotic N_2 Fixation Is Restricted to a Fairly Limited Number of Plant Species

Because of its overwhelming economic importance, the most widely studied associations between N_2-fixing microorganisms and vascular plants are those that involve a symbiosis between bacteria of the genera *Rhizobium*, *Bradyrhizobium*, *Sinorhizobium*, *Mesorhizobium*, or *Azorhizobium* (collectively know as **rhizobia**) and more than 3000 species of **Fabaceae**. *Parasponia* is the only nonlegume species known to have a symbiotic association with *Rhizobium*. Invariably **root nodules** are formed (Fig. 12), with the exception of *Azorhizobium*, which induces nodules on both stems and roots (of *Sesbania rostrata*). The Fabaceae family is comprised of three subfamilies—Caesalpinioideae, Mimosoideae, and Papilionoideae—each of which contains genera able to form nodules. The less-specialized

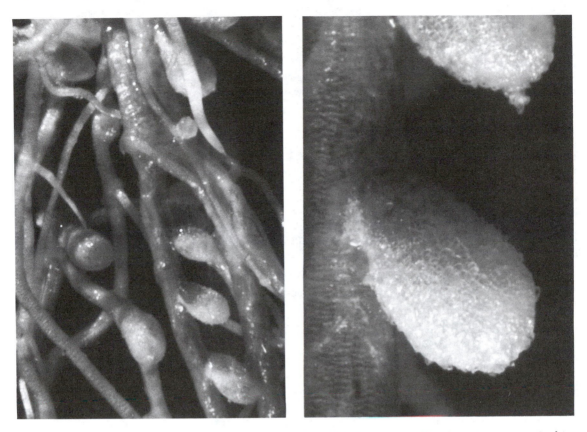

FIGURE 12. A nodulated root system and a close-up of a nodule of *Trifolium repens*, white cloves (courtesy A. Cookson and F.R. Minchin, Institute of Grassland and Environmental Research, Aberystwyth, UK).

subfamily Caesalpinioideae includes far more nonnodulating species than do the other two subfamilies (Van Rhijn & Vanderleyden 1995). The symbiosis between rhizobia and legume crops is of enormous agronomic importance, especially where fertilizer inputs are low.

There are also **nonlegume species** that are capable of forming a symbiotic association with N$_2$-fixing organisms, other than rhizobia. First, there is the **actinorhizal symbiosis** between soil bacteria (*Frankia*) and more than 200 species from eight nonlegume families of angiosperms [e.g., *Alnus* (alder), *Hippophae*, *Myrica*, *Elaeagnus*, *Dryas*, and *Casuarina* (sheoak)]. In all these symbioses **root nodules** are formed (Berry 1994). Second, there are symbioses between **cyanobacteria** (*Nostoc*, *Anabaena*) and species of the genus *Macrozamia* and *Gunnera*. Special morphological structures are sometimes formed on the roots (e.g., the "**coralloid roots**" in *Macrozamia* species) (Pate et al. 1988). The

endosymbiont only fixes N$_2$, not CO$_2$, although cyanobacteria are photosynthetically active when free-living (Lindblad et al. 1991). In addition in the symbiosis between fungi of the genus *Collema* and cyanobacteria (*Nostoc*), the cyanobacteria are photosynthetically active; this symbiosis occurs in **lichens**.

Table 4 gives an overview of major symbiotic associations between plants and microorganisms capable of fixing atmospheric dinitrogen. It shows that N$_2$-fixing organisms can be very significant for the input of nitrogen into natural and agricultural systems.

3.2 Host–Guest Specificity in the Legume–Rhizobium Symbiosis

The associations between legumes and rhizobia have been studied most elaborately. Many are

TABLE 4. Symbiotic associations between plants and microorganisms capable of fixing atmospheric dinitrogen.*

Plant type	Genus	Microorganism	Location	Amount of N_2 fixed (kg N ha^{-1} season^{-1})
Fabaceae	*Pisum, Glycine, Trifolium*	*Rhizobium, Bradyrhizobium*	Root nodules	10–350
	Medicago	*Bradyrhizobium*	Root nodules	440–790
Fabaceae	*Sesbania*	*Azorhizobium*	Stem nodules	nd
Ulmaceae	*Parasponia*	*Bradyrhizobium*	Root nodules	20–70
Betulaceae	*Alnus*	*Frankia*	Root nodules	15–300
Casuarinaceae	*Casuarina*	(Actinomycete)	Root nodules	10–50
Eleagnaceae	*Eleagnus*	(Actinomycete)		nd
Rosaceae	*Rubus*	(Actinomycete)		nd
Pteridophytes	*Azolla*	*Anabaena*	Heterocysts in cavities of dorsal leaf lobes	40–120
Cycads	*Ceratozamia*	*Nostoc*	Modified coralloid shaped roots	19–60
Lichens	*Collema*	*Nostoc*	Interspersed between fungal hyphae	nd

Source: Gault et al. 1995, Kwon & Beevers 1992, Vance 1996.
*Only a limited number of species are listed, just to provide an example; nd is not determined.

highly specific. For example, *Rhizobium meliloti* will infect *Medicago*, *Melilotus*, and *Trigonella*, but not *Trifolium* or *Glycine*. *Bradyrhizobium japonicum* will nodulate *Glycine max*, but not *Pisum* and *Medicago*. Other rhizobia, for example *Rhizobium* strain NGR234 may infect up to 100 host species, from different genera, including *Parasponia andersonii*, a nonlegume. What determines the specificity and why does this specificity vary among different rhizobia? To answer these questions, we first need to discuss the infection process in more detail.

3.3 The Infection Process in the Legume–Rhizobium Association

Nodule formation is preceded by the release of **specific phenolic compounds** (**flavonoids**: flavones, flavanones or isoflavones) and betaines from the legume roots (Phillips et al. 1994; Fig. 13). It is interesting that the same or similar flavonoids are induced as antibiotics (phytoalexins) upon infection by pathogenic microorganisms (Sect. 3 of the chapter on effects of microbial pathogens). Subtle

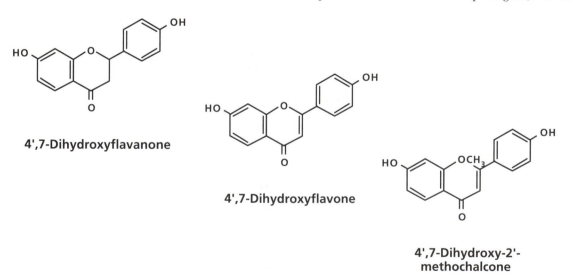

4',7-Dihydroxyflavanone

4',7-Dihydroxyflavone

4',7-Dihydroxy-2'-methochalcone

FIGURE 13. The structures of a number of flavonoids (phenolic compounds) isolated from the root exudate of *Medicago sativa* (alfalfa). These flavonoids activate transcription of the *nod* genes in *Rhizobium meliloti*; from top to bottom, they are a weak, moderate, and strong *nod*-gene inducer (Maxwell et al. 1989). Copyright American Society of Plant Physiologists.

differences between host plant species, of which we are only just beginning to understand some details, may therefore determine if an interaction between a bacterium and a plant results in symbiosis or pathogenesis. The flavonoids bind with a bacterial gene product, and then interact with a specific promoter in the genome of *Rhizobium*. This promoter is associated with the genes that are responsible for inducing nodulation (the nodulation, or *nod* genes). As detailed in following sections, the products of these genes, the **nod-factors**, induce root hair curling on the plant and cortical cell divisions, which are among the earliest, microscopically observable events in the nodulation.

3.3.1 The Role of Flavonoids

To some extent **specificity** between the host and rhizobium is determined by the type of **flavonoids** released by the host and by the sensitivity of the rhizobium promoter for a given type of flavonoid (Table 5). Rhizobium species with a broad specificity respond to a wider range of flavonoids than those species that are more specific, but if the flavonoid concentration is increased a response may still be found in those more specific rhizobia. In addition, nonlegumes may also exude flavonoids, and several legumes exude flavonoids that also activate the promoter of rhizobium species that are unable to establish a symbiosis. Therefore, apart from flavonoids other factors must also contribute to specificity (Spain 1995). What ultimate effects do the flavonoids have in the rhizobium?

To study the effect of flavonoids on rhizobium an appropriate assay is required, which is less elaborate than measuring root hair curling. This is briefly explained in the legend of Figure 14, which shows the relative effect of different flavonoids. In the following section we examine the kind of products produced by the bacterial *nod* genes.

3.3.2 Rhizobial nod Genes

There are three types of *nod* genes. First, all rhizobia have a *nod* gene that appears to be transcribed constitutively; it probably confers some host specificity. The product of this gene is a **lipo–polysaccharide** with chitin-like substituents. Then there are the common *nod* genes, which are found in all *Rhizobium* species. Finally, there are the host-specific *nod* genes, conferring the specificity of a certain *Rhizobium* for a certain plant species. The common *nod* genes encode enzymes involved in the synthesis of chitinlike lipo-oligosaccharides, whereas the specific *nod* genes encode enzymes that "decorate" this bacterial lipo-oligosaccharide. The "decorated" lipo-oligosaccharides are known as the **nod-factors**. The lipid component of the nod-factor allows penetration through membranes. Different side groups are added to the backbone of this molecule, and this confers the specificity of a certain *Rhizobium* (Fig. 15). Species with a broad specificity produce many different nod-factors, as opposed to ones with a narrow host range. That is, the structure of the lipo-oligosaccharide determines if it will be recognized as a symbiont or a pathogen by a potential host plant. Because the nod-factor is effective at concentrations as low as 10^{-12} M, it is likely that a plant **receptor** is involved, but such a receptor has not yet been identified (Downie 1994).

Much less is known about plant factors determining if a *Rhizobium* will recognize a plant as an appropriate host. As discussed before, flavonoids

TABLE 5. Comparisons of indeterminate and determinate nodules.

Parameter	Indeterminate	Determinate
Nodule initiation	Inner cortex	Outer cortex
Cell infection through	Infection threads	Infection threads and cell division
Meristem	Persistent (months)	Nonpersistent (days)
Bacteroid size	Larger than bacteria	Variable, although usually not too much larger than bacteria
Peribacteroid membrane	One bacteroid per symbiosome	Several bacteroids per symbiosome
N$_2$-fixation products transported	Amides usually	Ureides usually
Infected cells	Vacuolate	Nonvacuolate
Geographical origin	Temperate	Tropical to subtropical
Genera	*Medicago, Trifolium, Pisum, Lupinus*	*Glycine, Phaseolus, Vigna*
nod Gene inducers	Flavones, isoflavones	Isoflavones

Source: Vance 1996.

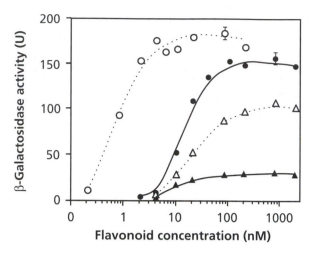

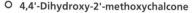

O 4,4'-Dihydroxy-2'-methoxychalcone

● Luteolin

△ 4,7'-Dihydroxyflavone

▲ 4,7'-Dihydroxyflavanone

FIGURE 14. The effect of various flavonoids from the root exudate of *Medicago sativa* (alfalfa, lucerne) on nodulation (*nod*) genes of *Rhizobium meliloti*. Luteolin has been identified as an inducing compound, but it was not a constituent of the isolated root exudate. It was included as a control in these experiments (Maxwell et al. 1989). Copyright American Society of Plant Physiologists. Although the effect of compounds released from legume roots on the *nod* promoter of rhizobium can be quantified by determi-nation of root hair curling, this rather tedious assay was not used in this experiment. Here, first a construct was made, coupling the gene from *Escherichia coli*, which codes for β-galactosidase, to the promoter of the *nod* genes. The activity of the enzyme β-galactosidase can be measured in a simple and quick spectrophotometric assay. In this way, the relative effect of various flavonoids can be assessed by determining the activity of β-galactosidase rather than the extent of root curling.

offer only a partial explanation. When rhizobial genes conferring host specificity are transferred to a rhizobium strain with a different specificity, these genes alter the bacterial acidic polysaccharide structure and in situ binding to the host's root hairs. Furthermore, introducing the genes encoding specific root-hair proteins (**lectins**) into *Trifolium repens* (white clover) allows nodulation of clover roots by a *Rhizobium* strain, which is usually specific for *Pisum sativum* (pea). It has therefore been thought that the host-*Rhizobium* specificity involves the interaction of the root hair lectins with specific carbohydrates on the bacterial surface (Diaz et al. 1989).

Searches have been made of *Frankia* DNA for sequence homology to the common rhizobial *nod* genes, but no definitive demonstration of such genes exists (Berry 1994).

3.3.3 Entry of the Bacteria Through Root Hairs

Following the release of flavonoids by the host and the release of the nod-factors by *Rhizobium*, the bacteria multiply rapidly in the rhizosphere. The bacteria adhere to root hairs and affect those that have just stopped growing. Younger and older root hairs are not affected. The cell wall of the affected root hair is then partly hydrolyzed at the tip. In this process the root hairs curl, attaching the bacteria to the root hair. On those locations on the root hairs that have become deformed due to the presence of rhizobia, the cell wall is degraded, allowing the bacteria to enter. An **infection thread** is formed by invagination of the cell wall. This thread consists of cell-wall components similar to those that form the normal root-hair cell wall. The infection thread grows down the root hair at a rate of 7 to 10 μm h^{-1} and provides a conduit for bacteria to reach the root cortex. The tip of the thread appears to be open; sealing of the thread tip results in abortion of the infection thread. The formation of the infection thread may well be analogous to the enlargement of epidermal cell walls, in response to a pathogen's attempted penetration (Vance 1996).

Only 1 to 5% of all root hairs become infected, and only 20% of these infections result in nodules. Why are most of the infections unsuccessful? This is likely to be due to the production of **chitinases** by the plant. These enzymes hydrolyze the chitinlike nod-factor (Mellor & Collinge 1995). Legumes contain different chitinases. In an early stage of infection the host produces a chitinase that breaks down nod-factors of *Rhizobium* species that are not suitable guests. In this way the plant prevents the entry of bacteria that cannot form a symbiosis. Chitinases are therefore another factor that confers host–guest specificity. At a later stage, different chitinases are produced that are effective against the nod-factor of "homologous" *Rhizobium* bacteria, that is suitable guests. It is likely that this is a mechanism to control the entry of *Rhizobium* and to prevent more nodules being formed than can be supported by the host. In addition the breakdown of nod-factors prevents entering bacteria from being erroneously recog-

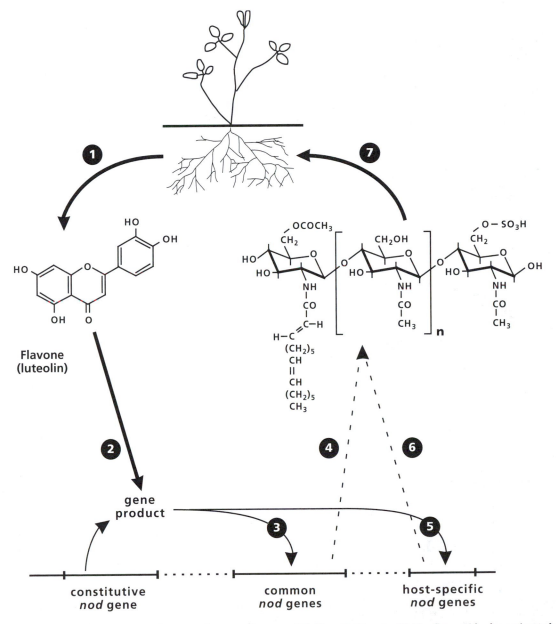

FIGURE 15. Symbiotic signaling between legume plants and rhizobia. (1) Flavonoids are exuded by the legume roots. (2) The flavonoids bind to the gene product of a constitutively expressed nodulation (*nod*) gene. (3) After this binding, common *nod* genes are activated. (4) This leads to the production of a lipo-polysaccharide with chitinlike substituents. (5) The flavonoids also activate the transcription of specific *nod* genes. (6) The products of the specific *nod* genes lead to modification of the lipo-oligosaccharides and the formation of nodulation (nod) factors, which confer specificity. (7) The nod-factors are recognized by the legume-host's roots.

nized as pathogens. *Rhizobium* strains that overproduce the nod-factor do indeed lead to a defense response. Not only is the nod-factor broken down by plant chitinase activity, but expression of the bacterial *nod* genes is also suppressed at a later stage of infection. Plant phenolics may play a role in this suppression. If the *Rhizobium* fails to recognize the plant-derived suppressor molecule, then the bacteria may be recognized as a pathogen and further development stops. This offers another possibility for host–guest specificity. The train of events that likely plays a role in the early stages of infection and the role of chitinases is outlined in Figure 16.

If the infection is successful, then specific genes are activated in the inner cortex and pericycle,

which allows the formation of an infection thread through which the bacteria enter. Cell divisions start in the inner cortex, opposite protoxylem poles, so that a new **meristem** is formed due to the presence of the rhizobia. This meristem gives rise to the **root nodule**. The infection thread grows inward, and, finally, the bacteria are taken up into the cytoplasm of the parenchyma cells of the center of the developing nodule. Inside the infected host plant cell, the bacteria continue to divide for some time, now differentiating into **bacteroids**, which have a diminished ability to grow on laboratory culture medium and may be greatly enlarged with various shapes. In most legumes, bacteroids are enclosed within a **peribacteroid membrane**, to form a **symbiosome** (Whitehead & Day 1997). Most symbiosomes are of a similar volume, but in some nodules each symbiosome contains a single, enlarged pleomorphic bacteroid, whereas there may be several (up to 20) smaller, rod-shaped bacteroids in others. The former type of symbiosome is more typical of elongate, cylindrical nodules of the so-called **indeterminate** (meristematic) class, as found on *Trifolium* (clover) or *Pisum* (pea) (Table 5). The latter type of symbiosome is usually found in spherical, **determinate** nodules (with no meristem), such as those of *Vigna unguiculata* (cowpea) or *Glycine max* (soybean). There are many variations in symbiosome structure, however, and a few legumes have nodules in which there are no symbiosomes, the bacteria being retained within multiply branched infection threads. Mature nodules of the determinate and indeterminate type are strikingly different, but their initiation is rather similar (Vance 1996).

3.3.4 Final Stages of the Establishment of the Symbiosis

Each **infected cell** may contain many hundreds of symbiosomes. The symbiosome membrane (**peribacteroid membrane**) originates from an invagination and endocytosis of the plasma membrane of the infected cortical cells. This membrane acts as a selective permeability barrier to metabolite exchange between the bacteroids and the cytosol of the infected cells. Interspersed between the infected cells of many nodules are smaller, **uninfected cells**, which occupy about 20% of the total volume of the central zone of soybean nodules. **Plasmodesmata** connect uninfected with infected cells and with other uninfected cells in the central zone of the nodule. These plasmodesmata allow for the massive transport of a carbon source from the uninfected

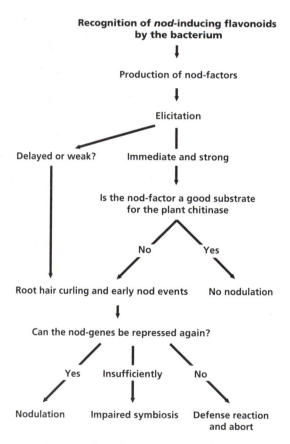

FIGURE 16. Tentative scheme to account for events that determine the establishment of a functional symbiosis (effective nodulation) between *Rhizobium* and a host legume. "Elicitation" is a combination of the elicitor activity of a specific bacterial nod-factor and the sensitivity of the plant to that elicitor (Mellor & Collinge 1995). Copyright Journal of Experimental Botany.

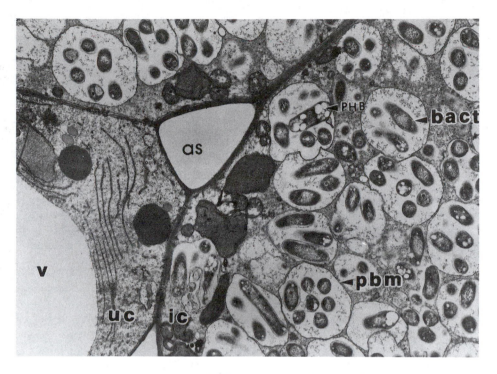

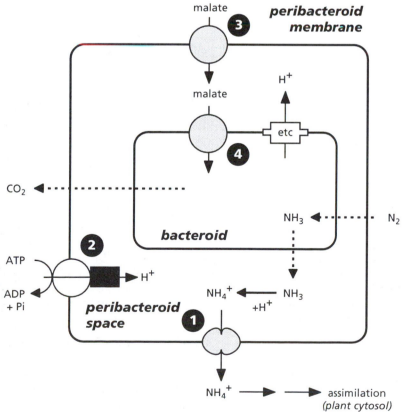

cells to the infected ones and of nitrogenous compounds in the reverse direction (Brown et al. 1995). Both infected and uninfected cells contain numerous plastids and mitochondria. Uninfected and infected cells in nodules have different metabolic roles in symbiotic dinitrogen fixation (Day & Copeland 1991). The central tissue of some nodules, however, contains no uninfected cells.

In soybean nodules, an outer layer of cortical cells surrounds an endodermal cell layer, which in turn encloses several layers of subcortical cells. The central zone of the nodules contains several thousand infected cells. The nodule is connected with the vascular tissue in the stele, due to the proliferation of cells from the pericycle.

The pattern of gene expression in the host cells that are part of the nodules is altered by the presence of the bacteria, resulting in the synthesis of as many as 30 different proteins, known as **nodulins**. Only a few of these nodulins have been characterized biochemically, including the O_2 carrier **leghemoglobin** and nodule-specific forms of the enzymes uricase, glutamine synthetase, and sucrose synthase.

3.4 Nitrogenase Activity and Synthesis of Organic Nitrogen

Biological reduction of dinitrogen gas to ammonia is catalyzed by **nitrogenase**, in a highly **oxygen-sensitive process**. This oxygen sensitivity accounts for the pink color of the nodule tissue, which is due to the presence of **leghemoglobin** in the cytoplasm of the infected cells. This heme-protein comprises as much as 35% of the total nodule soluble protein. It is related to the myoglobin of mammalian muscle, and its protein component is encoded by the DNA of the plant. The enzymes responsible for the synthesis of the O_2-binding heme-group of the protein are encoded in rhizobium. Leghemoglobin plays a role in the **oxygen supply** to the bacteroid. It ensures sufficiently rapid oxygen supply for the highly active respiratory processes in the plant and bacteroid compartment while maintaining a low free oxygen concentration (between 3 and 30 nM). The latter is very important because **nitrogenase**, which is the enzyme responsible for the fixation of atmospheric dinitrogen to NH_3, is rapidly damaged by free oxygen.

The bacteroids contain nitrogenase, which is an enzyme complex that consists of two proteins. One, nitrogenase-reductase, is an Fe-S-protein which accepts electrons, via an intermediate electron carrier,

from NADPH, and then binds ATP. At the same time, the other subunit (an Fe-Mo-protein) binds N_2. Reduction of this N_2 occurs if the two subunits have formed an active complex. A minimum of 12, and possibly as many as 16 ATP are required per N_2; therefore the overall equation is:

$$N_2 + 8\,H^+ + 8\,e^- + 16\,ATP$$
$$\rightarrow 2\,NH_3 + H_2 + 16\,ADP + 16\,P_i \qquad (1)$$

Almost all the dinitrogen fixed by the bacteroids is released as NH_3 to the peribacteroid space and then as NH_4^+ to the cytosol of the nodule cells. This may be due to the relatively low activity or absence of the ammonium-assimilating enzyme, glutamine synthetase, in the bacteroids. A nodule-specific glutamine synthetase is expressed in the cytoplasm of infected cells. Glutamine 2-oxoglutarate aminotransferase (GOGAT) then catalyzes the formation of two molecules of glutamate from one molecule of glutamine (Fig. 17). The major N-containing products exported via the xylem are the **amides** asparagine and glutamine in such species as *Pisum* (pea), *Medicago* (alfalfa), and *Trifolium* (clover). These products are typical for nodules that are elongate-cylindrical with **indeterminate** apical meristematic activity (Fig. 18, Table 5). In common bean and soybean, the products are predominantly **ureides**: allantoin and allantoic acid (Fig. 18, Table 5). Nodules exporting these compounds are spherical with **determinate** internal meristematic activity (Rolfe & Gresshoff 1988).

The ureides released to the xylem of plants with determinate nodules are products that are only found when the symbiotic plants are fixing N_2 (Pate 1980). Hence, the concentration of these compounds in the xylem sap, relative to the total amount of N transported in the xylem, has been used to estimate the proportion of N derived from N_2-fixation, as opposed to the assimilation of combined nitrogen (Bergersen 1989, Peoples et al. 1996). The amides released to the xylem of plants with indeterminate nodules are also found when these plants grow nonsymbiotically. In fact, they are not even typical for legumes. Hence, they cannot be used as "markers" for symbiotic N_2-fixation.

3.5 Carbon and Energy Metabolism of the Nodules

Carbohydrates are supplied via the phloem to the nodules, where they are rapidly converted in the plant compartment to dicarboxylic acids (malate, succinate), predominantly in the uninfected cells in

Ureides

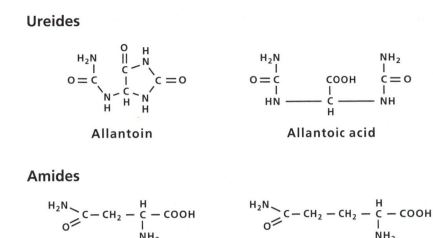

Allantoin   Allantoic acid

Amides

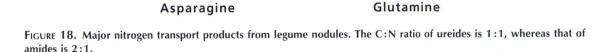

Asparagine   Glutamine

FIGURE 18. Major nitrogen transport products from legume nodules. The C:N ratio of ureides is 1:1, whereas that of amides is 2:1.

the nodules (Udvardi & Day 1997). **Malate** and **succinate** are the major substrates for the bacteroids (Fig. 17). Accumulation of dicarboxylic acids is often associated with increased engagement of the nonphosphorylating, alternative respiratory pathway (cf. Sect. 2.3 of the chapter on plant respiration). How do the infected cells prevent a large part of the organic acids from being oxidized via the nonphosphorylating alternative path in a situation where the demand for organic acids the bacteroids is very large? It is interesting that mitochondria from nodules have very little **alternative path capacity**; the little capacity they have is restricted to the uninfected cortical cells, rather than to the infected ones (Table 6). There is no risk, therefore of oxidizing the organic acids destined for the bacteroids.

Apart from N_2, H^+ is also reduced by nitrogenase (see Eq. 1 in Sect. 3.4), leading to the production of H_2. Most rhizobia contain **hydrogenase**, however, which is an enzyme that consumes H_2, using it as an electron donor. The characteristic of nitrogenase to reduce acetylene (ethyne) to ethylene (ethene) is frequently used to assay nitrogenase activity in vivo. Because the assay itself interferes with the process of N_2 fixation, however, it should only be considered as a *qualitative* indicator for the occurrence of nitrogenase activity, rather than as a good *quantitative* measure of its actual activity (Hunt & Layzell 1993).

3.6 Quantification of N_2 Fixation In Situ

The contribution of symbiotic dinitrogen fixation to the total accumulation of nitrogen in the aboveground biomass can be determined as follows. [15]N-

FIGURE 17. (Top) Electron micrograph of a nodule of *Glycine max* (soybean), infected with *Bradyrhizobium japonicum*. as, air space; bact, bacteroid; ic, infected cell; pbm, peribacteroid membrane; phb, poly-β-hydroxy butyrate (storage compound in bacteroids); uc, uninfected cell; v, vacuole. Note that the uninfected cell is vacuolated and much smaller than the infected ones, and that the bacteroids are grouped as a "symbiosome", surrounded by a peribacteroid membrane (courtesy D.A. Day, Australian National University, Canberra, Australia; Day & Copeland 1991). Plant Physiology and Biochemistry—copyright Gauthier-Villars. (Bottom) Scheme of N_2 fixation and NH_3 production in bacteroids. NH_3, being an uncharged mol-ecule, can cross the bacteroid membrane and arrives in the peribacteroid space, where it picks up a proton and becomes NH_4^+. Subsequently, NH_4^+ leaves the symbiosome via a monovalent cation channel (1). An ATPase in the peribacteroid membrane (2) creates a membrane potential (positive inside of the symbiosome), which drives the uptake of organic acids (predominantly malate and succinate) into the symbiosome (3). An electrontransport chain in the bacteroid membrane (etc) similarly creates a proton-motive force, which drives the uptake of malate (and succinate) in the bacteroids (4) [Whitehead et al. (1995)]. Copyright Balaban Publishers.

TABLE 6. Cyanide-resistant, SHAM-sensitive respiration (V_{alt} = alternative pathway capacity) in *Glycine max* (soybean).*

Mitochondria from	Oxygen consumption (nmol min^{-1} mg^{-1} protein)		KCN-resistant respiration (%)
	Control	V_{alt}	
Green cotyledons	99	43	43
Leaves	177	83	47
Roots	153	66	43
Nodules	86	6	6

Cells from root nodules	Oxygen consumption (nmol min^{-1})		Resistance (%)
	Control	V_{alt}	
Infected	60	0	0
Uninfected	45	22	49

Source: Kearns et al. 1992.

*Measurements were made on isolated mitochondria from different tissues, as well as on infected and uninfected cells from root nodules.

labeled inorganic nitrogen ($^{15}NO_3^-$ or $^{15}NH_4^+$) is given to a plant community that consists of both N_2-fixing species (e.g., clover) and other species (e.g., grasses) (^{15}N is a nonradioactive isotope of nitrogen). The grasses have a $^{15}N/^{14}N$ ratio, which is used as a reference. N_2 fixation in the N_2-fixing clovers will "dilute" their ^{15}N concentration. The extent of the dilution is used to calculate the contribution of fixation to the total amount of nitrogen that accumulates in the clover plants. The contribution of N_2-fixation to the total amount of N that accumulates in the plant may amount to 75 and 86% in *Trifolium repens* (white clover) and *T. pratense* (strawberry clover). The contribution depends on the amount of inorganic combined nitrogen that the plants receive from soil and fertilizer and also varies with the developmental stage of the plant and the time of the year (Fig. 19). The overwhelming importance of symbiotic N_2 fixation in some agricultural systems is illustrated in Table 7.

Sometimes the **natural abundance** of ^{15}N in the soil is used to quantify N_2 fixation (cf. Table 2 in Sect. 2.4). Instead of adding ^{15}N-labelled inorganic combined nitrogen, the natural abundance of nitrogen in the soil can be used. The natural abundance of soil-N is likely to differ from that of N_2 in the atmosphere, due to discrimination against the heavy isotope in various biological processes (e.g., mineralization, nitrification, and denitrification). Atmospheric dinitrogen, therefore, is depleted with

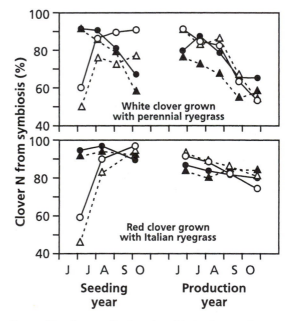

FIGURE 19. The contribution of symbiotic nitrogen fixation to the total N accumulation in the above-ground biomass of *Trifolium repens* (white clover) (top) and *Trifolium pratense* (strawberry clover) (bottom). The white clover plants grew in a mixed culture with *Lolium perenne* (perennial ryegrass) and the strawberry clover with *Lolium multiflorum* (Italian ryegrass). Circles: no N fertilizer; triangles: 30 kg N ha^{-1} per cut; open and closed symbols refer to different years (Boller et al. 1987). Copyright Kluwer Academic Publishers.

TABLE 7. Symbiotic dinitrogen fixation by some legume crops[1] and native species in their natural environment in Brazil.[2]

Species	N_2 fixed (kg ha^{-1} per season from atmosphere)	Plant N absorbed (%)
Medicago sativa	440–780	65–96
Glycine max	120	53
Lotus corniculatus	92	55
Lupinus angustifolius	170	65
Medicago sativa	180	70
Phaseolus vulgaris	65	40
Pisum sativum	72	35
Trifolium pratense	170	59
Vicia faba	151	nd
Vigna angularis	80	70
Chamaecrista species	nd	66–79
Mimosoid legumes	nd	42–63
Papilionoid legumes	nd	68–79

Sources: [1] Gault et al. 1995, Vance 1996. [2] Sprent et al. 1996.

the heavy isotope, relative to the nitrogen in the soil. To apply this technique in situ, reliable control plants have to be used. As discussed in Section 2 (Table 2), this is not always easy, if it is at all possible (Handley et al. 1993).

The ^{15}N technique referred to earlier has also been used to demonstrate a significant transfer of nitrogen from the symbiotic plants to neighboring grasses (up to 52 kg N ha^{-1} yr^{-1}; on average a value of 17 kg ha^{-1} yr^{-1} was found). Conditions favoring dinitrogen fixation by the legume, such as a high irradiance, a favorable temperature, long days, and a relatively high phosphate supply, enhance the transfer of nitrogen from the legume to the

nonfixing neighbors. This transfer is to some extent the result of the uptake of nitrogenous compounds released after decomposition of parts of the legumes. Some of it is also due, however, to the exudation of nitrogenous compounds by the legumes, followed by absorption by the nonfixing plants. As discussed in Section 2.3.1 of this chapter, some transfer of nitrogen may also occur through mycorrhizal hyphae.

As shown in Figure 5 in Section 2, the maximum yield of *Lolium rigidum* (stiff darnel) is reached at a much lower supply of phosphate than that of *Trifolium subterraneum* (subclover), at least when both species are grown in monoculture). This demand for a higher phosphate supply is fairly common for legumes, although it is by no means universal (Bobbink 1991, Koide et al. 1988). The high demand for phosphate of many legumes may reflect their adaptation to soils with a high availability of phosphate, and it points out that a benefit from such legumes can only be expected when these have access to sufficient phosphate. Apart from P, Mo and S have to be available to the legume to allow symbiotic N_2 fixation.

3.7 Ecological Aspects of the Nonsymbiotic Association with N_2-Fixing Microorganisms

Next to the truly symbiotic association that leads to N_2 fixation as discussed earlier, a somewhat looser association between *Azospirillum* and higher plants (especially grasses) has been investigated. Inoculation of the soil in which *Zea mays* (maize) plants are grown with *Azospirillum* bacteria significantly enhances the yield of the maize plants, especially when the nitrogen supply is relatively low (Table 8). It is by no means certain, however, that this is a

TABLE 8. The effect of inoculation with *Azospirillum brasiliense* on the production of *Zea mays* (maize) plants as dependent on the N supply.*

N supply g L^{-1}	Shoot dry mass (g)		Root dry mass (g)		Relative increment of total plant mass (%)
	Inoculated	Control	Inoculated	Control	
0	0.49	0.32	0.36	0.27	44
0.04	0.97	0.66	0.76	0.53	45
0.08	1.84	1.23	0.97	0.86	34
0.16	2.93	2.52	1.96	1.70	16

Source: Cohen et al. 1980.
*N was supplied as NH_4NO_3.

direct result of the fixation of N_2 by the *Azospirillum* bacteria. These organisms are more likely to enhance the growth of higher plants in a different manner, such as the production of phytohormones. In a comparison of three cultivars of *Triticum aestivum* (wheat), grown with *Azospirillum brasiliense*, most N_2 is fixed in the rhizosphere of aluminum-resistant cultivars. Because these cultivars also exude more dicarboxylic acids (cf. Sect. 3.1 of the chapter on mineral nutrition), it has been suggested that fixation is enhanced by the excretion of these organic molecules (Christiansen-Weniger et al. 1992).

N_2-fixing bacteria in the rhizosphere might affect plants in many ways, other than through an increased supply of nitrogen. To discriminate between the bacterial effects associated with their capacity to fix atmospheric dinitrogen and possible other effects, *Poa pratensis* (Kentucky bluegrass) and *Triticum aestivum* (wheat) have been inoculated with a number of different bacteria: *Klebsiella pneumonia*, *K. terrigena*, *Enterobacter agglomerans*, and *Azospirillum lipoferum* (Haahtela et al. 1988). All these species have the ability to colonize the roots and to fix atmospheric dinitrogen. Transfer of the fixed nitrogen from the bacteria to the higher plants, however, did not occur in all combinations. The positive effects on the growth of the plant correlated very poorly with the dinitrogen-fixing activity of the bacteria. These results show that the positive effects of nonsymbiotic dinitrogen-fixing microorganisms in the rhizosphere are by no means all due to the dinitrogen-fixing capacity of these organisms. The microorganisms might promote plant growth in many other ways, such as **suppression of pathogenic organisms** and the **production of vitamins**.

In some areas in Brazil, sugarcane has been grown continuously for more than a century without any nitrogenous fertilizer. Although it had long been suspected that substantial N_2 fixation occurs in such systems, none of the N_2-fixing bacteria isolated from the rhizosphere of *Saccharum officinarum* (sugarcane) occur in large enough numbers to account for the high rates of N_2 fixation found in these crops. An acid-tolerant N_2-fixing bacterium has been found to be closely associated with sugarcane (Cavalcante & Döbereiner 1988). N_2-fixing bacteria (*Acetobacter diazotrophicus*) have only recently been identified in the **intercellular spaces** of sugarcane stem parenchyma (Dong et al. 1994, James et al. 1994). These spaces are filled with a solution that contains 12% sucrose (pH 5.5). *A. diazotrophicus* has most unusual growth requirements; it shows optimal growth with 10% sucrose and pH 5.5. It will grow in a medium with 10% sucrose and rapidly acidifies its surroundings by the formation of acetic acid. It has been isolated from sugarcane tissues, but was not found in the soil between rows of sugarcane or in grasses from the same location. The apoplasmic fluid occupies approximately 3% of the stem volume, which is equivalent to 3 tons of fluid per hectare of the sugarcane crop. It has been suggested that this amount suffices to make the sugarcane independent of N fertilizers. Other bacteria, including *Enterobacter agglomerans*, *Herbaspirillum seropedicae*, *H. rubrisubalbicans*, and *Klebsiella terrigena*, are also believed to be able to fix atmospheric N_2 in the apoplast of plants that have high apoplasmic sugar concentrations. Some of these (e.g., *Herbaspirillum seropedicae*), are pathogens on certain grass species, which restricts their potential use as inoculants in agriculture (Palus et al. 1996, Triplett 1996).

3.8 Carbon Costs of the Legume–Rhizobium Symbiosis

Because all the organic acids required for symbiotic dinitrogen fixation by rhizobium and for maintenance of the root nodules come from the plant (Fig. 17), there are **costs** involved for this symbiotic system for the higher plant. These costs exceed those required for assimilation of nitrate or ammonia. These costs have been estimated for a clover–*Rhizobium* association, in which the clover plants were totally dependent on the microsymbiont for their supply of nitrogen (Fig. 20). In this symbiotic system, the dinitrogen fixation was briefly interrupted by decreasing the oxygen concentration that surrounds the plants. The oxygen concentration was kept sufficiently high to maintain aerobic metabolism of the plant fully. It was sufficiently low, however, to block completely the respiration and dinitrogen fixation by the bacteroids. By relating the decrease in respiration upon blocking the dinitrogen fixation to the activity of dinitrogen fixation as determined from the accumulation in the clover plants, **carbon costs** per unit fixed dinitrogen are calculated. The costs for N_2 fixation amount to approximately 25% of all the carbon fixed in photosynthesis per day. This proportion is rather high when compared with the figures for N acquisition by nonsymbiotic plants given in Table 1 of the chapter on respiration: 4 to 13% when plants are grown at an optimum nutrient supply. At a limiting nutrient supply, the percentage is likely to be more similar to that of the costs of N_2 fixation.

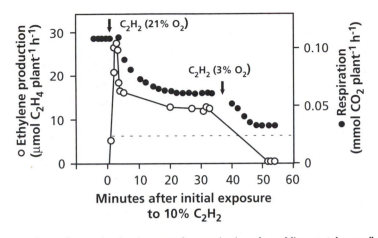

FIGURE 20. Respiration and acetylene reduction in roots of *Trifolium repens* (white clover) at either high or low oxygen concentration. The low oxygen concentration was sufficiently high to maintain aerobic metabolism of the plant cells, but virtually completely abolished the activity of the dinitrogen-fixing activity of the bacteroids. The decline in respiration after adding acetylene reflects the sensitivity of the nodules for "manipulations". It also shows that this technique cannot be use to give reliable quantitative estimates of the rate of dinitrogen fixation (Ryle et al. 1985). Copyright Journal of Experimental Botany.

3.9 Suppression of the Legume–Rhizobium Symbiosis at Low pH and in the Presence of a Large Supply of Combined Nitrogen

At **low pH**, nodule formation on legumes tends to be inhibited. Because fixation of dinitrogen tends to lower the soil pH (see Sect. 3.1 of the chapter on mineral nutrition), continued use of legumes in agriculture requires **regular liming**. Why is nodulation impaired at a low soil pH? The assay system depicted in Figure 14 has been used to establish that a relatively acid or alkaline, as opposed to a neutral pH of the soil, leads to a less effective root exudate. This offers an explanation for the common observation of a poor infection of legumes by rhizobium in acid soils. Survival of rhizobia is also lower in soils with a low pH, but some degree of adaptation of rhizobia strains has been observed.

N_2 fixation is an energetically expensive process as compared with the assimilation of nitrate or ammonia. Reminiscent of the effect of phosphate on the formation of the mycorrhizal symbiosis, **nitrate** inhibits the **infection** of legumes by rhizobia. When *Medicago sativa* (alfalfa) plants are grown under nitrogen-limiting conditions, the expression of the genes involved in flavonoid biosynthesis and the production of root flavonoids are enhanced. This may account for greater infection by *Rhizobium meliloti* under conditions when nitrogen is in short supply, as opposed to suppression of nodulation in the presence of high nitrate concentrations (Coronado et al. 1995).

Nitrate also inhibits the process of **fixation** itself (Table 9). Several mechanisms have been

TABLE 9. Apparent nitrogenase activity and the oxygen limitation coefficient, 2 days after addition of nitrate to the root environment of nodulated 21-day-old plants of *Pisum sativum* (pea).*

$[NO_3]$ (mM)	Apparent nitrogenase activity [nmol $H_2 g^{-1}$ (nodule dry mass) s^{-1}]	Oxygen limitation coefficient
0	45	0.89
5	38	0.64
10	22	0.45
15	24	0.49

Source: Kaiser et al. 1997.

*The apparent nitrogenase activity was measured as the rate of H_2 evolution. As explained in Section 3.4, nitrogenase activity leads to the production of H_2. There is, normally no net evolution of H_2 because rhizobia have an hydrogenase (i.e., an enzyme that uses H_2 as an electron donor). In the present experiment, a rhizobium strain was used that lacks hydrogenase so that the evolution of H_2 could be measured. The oxygen limitation coefficient is calculated as the ratio between total nitrogenase activity (H_2 evolution in the absence of N_2) and potential nitrogenase activity (H_2 evolution in the absence of N_2 at an optimum concentration of O_2).

proposed to account for this inhibition (Hunt & Layzell 1993):

1. Competition for **carbohydrates** between nitrogenase and nitrate reductase, located in leaves, roots, or nodules.

2. Inhibitory effects of **nitrite**, the product of nitrate reductase. NO_2^- may inhibit nitrogenase directly, by irreversibly binding to the enzyme, or indirectly, by forming a bond with leghemoglobin, so that it can no longer function in O_2 transport.

3. A decrease of the partial pressure of **O_2** in the nodule, due to a decrease in the conductance for gas transport in the pathway between the outside air and the infected cells.

4. **Feedback-inhibition** of nodule metabolism by nitrogenous compounds that arrive via the phloem.

There is evidence for a decrease in the conductance for oxygen transport to the infected cells, which leads to a more severe limitation of nitrogenase activity by oxygen (Table 9). It is unlikely, however, that this is the only mechanism that accounts for nitrate inhibition of nodule activity. Rather, all four mechanisms are likely to occur at one stage or another in some species.

In the last decade, several mutants of a number of legumes have been produced, of which neither infection nor dinitrogen fixation itself is inhibited by nitrate. These mutants are expected to enhance input of nitrogen through the legume–rhizobium symbiosis in agricultural systems.

4. Endosymbionts

Many plants are infected by **fungal endophytes** (family Clavicipitaceae, ascomycetes) that live their entire life cycle within the plant (Bacon & De Battista 1991). The fungi form nonpathogenic and usually intercellular associations in living plant tissue. The endophytes are often transmitted through the plant seed, particularly in grasses and sedges, but seeds may lose their endophytes upon prolonged storage. Infection through germinating spores is an alternative way to enter the macrosymbiont. The association between higher plants and endosymbiotic fungi has been well studied in grasses in which the fungi may produce **alkaloids** in the tissue of their hosts, many of which have a neurotoxic effect, and hence make the infected plants toxic to domestic mammals and increase their resistance to insect herbivores (Table 10).

In some species, plant growth and seed production can be increased by infection with the endo-

TABLE 10. Antiherbivore effects of fungal endophytes that infect grasses.

Animal	Host grass genus	Fungal endophyte genus	Comments
Mammals			
Cattle, horses	*Festuca*	*Acremonium*	Reduced mass gain, gangrene, spontaneous abortion
Cattle, sheep, deer	*Lolium*	*Acremonium*	Reduced mass gain, tremors, staggers, death
Cattle, goats	*Andropogon*	*Balansia*	Reduced milk production, death
Cattle	*Paspalum*	*Myriogenospora*	Reduced mass gain, tremors, gangrene
Insects			
Fall armyworm	*Cenchrus*	*Balansia*	Avoidance, reduced survival, reduced growth,
	Cyperus	*Balansia*	increased development time
	Festuca	*Acremonium*	
	Lolium	*Acremonium*	
	Paspalum	*Myriogenospora*	
	Stipa	*Atkinsonella*	
Aphids	*Festuca*	*Acremonium*	Avoidance
Billbugs	*Lolium*	*Acremonium*	Reduced feeding and oviposition
Crickets	*Lolium*	*Acremonium*	Complete mortality
Cutworms	*Dactylis*	*Epichloe*	Reduced survival and mass gain
Flour beetles	*Lolium*	*Acremonium*	Reduced population growth
Sod webworms	*Lolium*	*Acremonium*	Reduced feeding and oviposition
Stem weevils	*Lolium*	*Acremonium*	Reduced feeding and oviposition

Source: Clay 1988.
Note: The examples are representative, but not exhaustive.

TABLE 11. Survival and development of the larvae of *Spodoptera frugiperda* (fall armyworm) on grasses infected with the endosymbiont *Acremonium* spp.

Food plant	Infection	Larval mass after 10 days (mg)	Survival to pupation (%)	Pupal mass (mg)	Days to pupation
Lolium perenne	+	26.3	75	167	18.4
Lolium perenne	−	35.6	65	155	18.7
Festuca arundinacea	+	18.0	8	181	23.7
Festuca arundinacea	−	37.0	63	178	21.0
Festuca rubra	+	—	0	—	—
Festuca rubra	−	33.4	43	163	20.9

Source: Clay et al. 1993.

phyte. The symbiotic associations between grasses and fungal endophytes may be an association in which the fungi derive carbohydrates from their host and defend their host against herbivory, thereby defending their own resources. Similar to the effect that mycorrhizal fungi have on interactions among mycorrhizal and nonmycorrhizal plants (Sect. 2.2), fungal endophytes may also influence competitive interactions between plants. For example, grass plants infected with fungal endosymbionts are less nutritious (Table 11). They are also less preferred than the uninfected plants of the same species. The presence of endophytes may also affect competition among grasses in interaction with herbivory (Clay et al. 1993) and suppress the fungal take-all disease in *Triticum aestivum* (wheat) (Dewan & Sivasithamparam 1988).

The presence or absence of fungal endophytes is not a specific trait of a plant species; rather, it also depends on environmental conditions in an as-yet-unclear manner. For example, in Western Australian heaths (Epacridaceae), the number of fungal associates is considerably smaller on a mesic wetland site when compared with a dryland habitat, even when comparing the endophytes associated with the same plant species (*Lysinema ciliatum*). This appears to reflect the response of different fungal endophytes to water stress (Hutton et al. 1996).

Bacteria as well as **fungi** may act as endosymbionts. Some plant-growth-promoting endosymbiotic bacteria have already been discussed in Section 3.7: *Acetobacter diazotrophicus*, which fixes N_2 in the tissues of *Saccharum officinarum* (sugarcane). Common endophytic bacteria from healthy tubers of *Solanum tuberosum* (potato) belong to six genera (*Pseudomonas, Bacillus, Xanthomonas, Agrobacterium, Actinomyces,* and *Acinetobacter*). As discussed in Section 3 of the chapter on effects of microbial patho-

gens, many bacterial endophytes make the host plant more resistant to pathogen attack (induced resistance) or they enhance growth. There are also endophytic bacteria, however, that are plant-growth-neutral or plant-growth-retarding (Sturz 1995).

5. Plant Life among Microsymbionts

At one stage in the history of plant ecophysiology it may have seemed most appropriate to discuss the mineral nutrition and performance of plants devoid of their microsymbionts. If we wish to unravel basic principles of plant mineral nutrition (e.g., the nature of a nitrate transporter or the function of a micronutrient), then this is still a valid approach. If we really wish to understand plant functioning in its real environment, whether a natural ecosystem or an agricultural field, however, then we simply cannot ignore the existence and overwhelming importance of the microsymbionts that interact with higher plants in such an intricate manner. This is most certainly true for mycorrhizal fungi, which affect both mycorrhizal and nonmycorrhizal species, although in a very different manner.

In many basic textbooks on plant physiology, these interactions do not receive the attention that they deserve. Interactions with dinitrogen-fixing symbionts have been the target of plant physiological research for much longer. In recent years, there has been an enormous development in the process of signaling between rhizobia and legumes. Similar signaling processes are likely to exist between other dinitrogen-fixing microsymbionts and their nonlegume hosts, and between mycorrhizal fungi and their macrosymbiotic partners, but we still have very little understanding of those interactions.

In recent years, endophytes other than the "classical" mycorrhizal fungi and dinitrogen-fixing microorganisms have gradually been discovered. These include the fascinating dinitrogen-fixing microorganisms in the apoplast of sugarcane and the toxin-producing endophytes in grasses. We are only just beginning to understand the agronomic and ecological significance of these endophytes. Another question that remains to be answered is how symbiotic microorganisms are allowed entry into the plant when we know that plants have a wide array of defense mechanisms to keep microorganisms at bay.

In this chapter we are faced with one of those many areas in plant physiological ecology where "established" terms like *ecology* and *molecular plant physiology* have become obsolete. We can only expect to further our basic understanding of interactions between plants and their microsymbionts if we abolish barriers that hinder the developments in this field. Many applications of a basic understanding of symbiotic associations between plants and microorganisms are to be expected.

References and Further Reading

Allen, E.B. & Allen, M.F. (1984) Competition between plants of different successional stages: Mycorrhizae as regulators. Can J. Bot. 62:2625–2629.

Allen, E.B. & Allen, M.F. (1986) Water relations of xeric grasses in the field: Interactions of mycorrhizas and competition. New Phytol. 104:559–571.

Baas, R., Van der Werf, A., & Lambers, H. (1989) Root respiration and growth in *Plantago major* as affected by vesicular-arbuscular mycorrhizal infection. Plant Physiol. 91:227–232.

Bacon, C.W. & De Battista, J. (1991) Endophytic fungi of grasses. In: Handbook of applied mycology. Vol. 1: Soil and plants, D.K. Arora, B. Rai, K.G. Mukerji, & G.R. Knudsen (eds). Marcel Dekker, New York, pp. 231–256.

Bergersen, F.J., Brockwell, J., Gault, R.R., Morthorpe, L., Peoples, M.B., & Turner, G.L. (1989) Effects of available soil nitrogen and rates of inoculation on nitrogen fixation by irrigated soybeans and evaluation of the $\delta^{15}N$ methods for measurements. Aust. J. Agric. Res. 40:763–780.

Berry, A.M. (1994) Recent developments in the actinorhizal symbioses. Plant Soil 161:135–145.

Bethlenfalvay, G.J., Pacovsky, R.S., Bayne, H.G., & Stafford, A.E. (1982) Interactions between nitrogen fixation, mycorrhizal colonization, and host-plant growth in the *Phaseolus-Rhizobium-Glomus* symbiosis. Plant Physiol. 70:446–450.

Bethlenfalvay, G.J., Reyes-Solis, M.G., Camel, S.B., & Ferrera-Cerrato, R. (1991) Nutrient transfer between the root zones of soybean and maize plants connected by common mycorrhizal mycelium. Physiol. Plant. 82:423–432.

Blee, K.A. & Anderson, A.J. (1996) Defense-related transcript accumulation in *Phaseolus vulgaris* L. colonized by the arbuscular mycorrhizal fungus *Glomus intraradices* Schenck & Smith. Plant Physiol. 110:675–688.

Bobbink, R. (1991) Effects of nutrient enrichment in dutch grassland. J. Appl. Ecol. 28:28–41.

Bohlool, B.B., Ladha, J.K., Garrity, D.P., & George, T. (1992) Biological nitrogen fixation for sustainable agriculture. Plant Soil 141:1–11.

Bolan, N.S. (1992) A critical review on the role of mycorrhizal fungi in the uptake of phosphorus by plants. Plant Soil 134:189–207.

Bolan, N.S., Robson, A.D., & Barrow, N.J. (1987) Effect of vesicular-arbuscular mycorrhiza on the availability of iron phosphates to plants. Plant Soil 99:401–410.

Boller, B.C. & Nösberger, J. (1987) Symbiotically fixed nitrogen from field-grown white and red clover mixed with ryegrass at low levels of ^{15}N-fertilization. Plant Soil 104:219–226.

Brown, S.M., Oparka, K.J., Sprent, J.I., & Walsh, K.NB. (1995) Symplasmic transport in soybean root nodules. Soil Biol. Biochem. 27:387–399.

Cavalcante, V.A. & Döbereiner, J. (1988) A new acid-tolerant nitrogen fixing bacterium associated with sugarcane. Plant Soil 108:23–31.

Chiarello, N., Huckman, J.C., & Mooney, H.A. (1982) Endomycorrhizal role for interspecific transport of phosphorus in a community of annual plants. Science 217:941–943.

Christiansen-Weniger, C., Groneman, A.F., & Van Veen, J.A. (1992) Associative N_2 fixation and root exudation of organic acids from wheat cultivars of different aluminium tolerance. Plant Soil 139:167–174.

Clay, K. (1988) Fungal endophytes of grasses: A defensive mutualism between plants and fungi. Ecology 69:10–16.

Clay, K., Marks, S., & Cheplick, G.P. (1993) Effects of insect herbivory and fungal endophyte infection on competitive interactions among grasses. Ecology 74:1767–1777.

Cohen, E., Okon, Y., Kigel, J., Nur, I., & Henis, Y. (1980) Increase in dry weight and total nitrogen content in *Zea mays* and *Setaria italica* associated with nitrogen-fixing *Azospirillum*. Plant Physiol. 66:746–749.

Coronado, C., Zuanazzi, J.A.S., Sallaud, C., Quirion, J.-C., Esnault, R., Husson, H.-P., Kondorosi, A., & Ratet, P. (1995) Alfalfa root flavonoid production is nitrogen regulated. Plant Physiol. 108:533–542.

Day, D.A. & Copeland, L. (1991) Carbon metabolism and compartmentation in nitrogen-fixing legume nodules. Plant Physiol. Biochem. 29:185–201.

Dewan, M.M. & Sivasithamparam, K. (1988) A plant-growth-promoting sterile fungus from wheat and ryegrass roots with potential for suppressing take-all. New Phytol. 91:687–692.

Diaz, C.L., Melchers, L.S., Hooykaas, P.J.J., Lugtenberg, B.J.J., & Kijne, J.W. (1989) Root lectin as a determinant of host-plant specificity in the *Rhizobium*-legume symbiosis. Nature 338:579–581.

Dixon, K.W., Pate, J.S., & Kuo, J. (1987) The Western Australian fully subterranean orchid *Rhizanthella gardneri*. In: Orchid biology. Reviews and perspectives, J. Arditti (ed). Timber Press, Portland, pp. 37–62.

Dong, Z., Canny, M.J., McCully, M.E., Roboredo, M.R., Cabadilla, C.F., Ortega, E., & Rodes, R. (1994) A nitrogen-fixing endophyte of sugarcane stems. A new role for the apoplast. Plant Physiol. 105:1139–1147.

Douds, D.D., Johnson, C.R., & Koch, K.E. (1988) Carbon cost of the fungal symbiont relative to net leaf P accumulation in a split-root VA mycorrhizal symbiosis. Plant Physiol. 86:491–496.

Downie, J.A. (1994) Signalling strategies for nodulation of legumes by rhizobia. Trends Microbiol. 9:318–324.

Duc, G., Trouvelot, A., Gianinazzi-Pearson, V., & Gianinazzi, S. (1989) First report of non-mycorrhizal plant mutants (Myc⁻) obtained in pea (*Pisum sativum*) and fababean (*Vicia faba* L.). Plant Sci. 60:215–222.

Eissenstat, D.M. (1990) A comparison of phosphorus and nitrogen transfer between plants of different phosphorus status. Oecologia 82:342–347.

Eissenstat, D.M., Graham, J.H., Syvertsen, J.P., & Drouillard, D.L. (1993) Carbon economy of sour orange in relation to mycorrhizal colonization and phosphorus status. Ann. Bot. 71:1–10.

Ezawa, T., Saito, M., & Yoshida, T. (1995) Comparison of phosphatase localization in the intraradical hyphae of arbuscular mycorrhizal fungi, *Glomus* spp. and *Gigaspora* spp. Plant Soil 176:57–63.

Fischer Walter, L.E., Hartnett, D.C., Hetrick, B.A.D., & Schwab, A.P. (1996) Interspecific nutrient transfer in a tallgrass prairie plant community. Am. J. Bot. 83:180–184.

Francis, R. & Read, D.J. (1994) The contribution of mycorrhizal fungi to the determination of plant community structure. Plant Soil 159:11–25.

Fredeen, A.L. & Terry, N. (1988) Influence of vesicular-arbuscular mycorrhizal infection and soil phosphorus level on growth and carbon metabolism of soybean. Can. J. Bot. 66:2311–2316.

Gault, R.R., Peoples, M.B., Turner, G.L., Lilley, D.M., Brockwell, J., & Bergersen, F.J. (1995) Nitrogen fixation by irrigated lucerne during the first three years after establishment. Aust. J. Agric. Res. 56:1401–1425.

Graham, J.H. & Eissenstat, D.M. (1994) Host genotype and the formation of VA mycorrhizae. Plant Soil 159:179–185.

Graham, J.H., Eissenstat, D.M., & Drouillard, D.L. (1991) On the relationship between a plant's mycorrhizal dependency and rate of vesicular-arbuscular mycorrhizal colonization. Funct. Ecol. 5:773–779.

Haahtela, K., Laakso, T., Nurmiaho-Lassila, E.-L., & Korhonen, T.K. (1988) Effects of inoculation of *Poa pratensis* and *Triticum aestivum* with root-associated, N₂-fixing *Klebsiella*, *Enterobacter* and *Azospirillum*. Plant Soil 106:239–248.

Handley, L.L. & Raven, J.A. (1992) The use of natural abundance of nitrogen isotopes in plant physiology and ecology. Plant Cell Environ. 15:965–985.

Handley, L.L., Daft, M.J., Wilson, J., Scrimgeour, C.M., Ingelby, K., & Sattar, M.A. (1993) Effects of the ecto- and VA-mycorrhizal fungi *Hydnagium carneum* and *Glomus clarum* on the δ¹⁵N and δ¹³C values of *Eucalyptus globulus* and *Ricinus communis*. Plant Cell Environ. 16:375–382.

Harrison, M.J. (1997) The arbuscular mycorrhizal symbiosis: An underground association. Trends Plant Sci. 2:54–60

Hartwig, U.A., Maxwell, C.A., Joseph, C.M., & Phillips, D.A. (1990) Chrysoeriol and luteolin released from alfalfa seeds induce *nod* genes in *Rhizobium meliloti*. Plant Physiol. 92:116–122.

Heap, A.J. & Newman, E.I. (1980) Links between roots by hyphae of vesicular-arbuscular mycorrhizas. New Phytol. 85:169–171.

Hetrick, B.A.D., Wilson, G.W., & Hartnett, D.C. (1989) Relationship between mycorrhizal dependence and competitive ability of two tallgrass prairie species. Can J. Bot. 67:2608–2615.

Högberg, P. (1990) ¹⁵N natural abundance as a possible marker of the ectomycorrhizal habit of trees in mixed African woodlands. New Phytol. 115:483–486.

Högberg, P. & Alexander, I.J. (1995) Roles of root symbioses in African woodland and forest: Evidence from ¹⁵N abundance and foliar analysis. J. Ecol. 83:217–224.

Hunt, S. & Layzell, D.B. (1993) Gas exchange of legume nodules and the regulation of nitrogenase activity. Annu. Rev. Plant Physiol. Plant Mol. Biol. 44:483–511.

Hutton, B.J., Dixon, K.W., & Sivasithamparam, K. (1994) Ericoid endophytes of Western Australian heaths (Epacridaceae). New Phytol. 127:557–655.

Hutton, B.J., Sivasithamparam, K., Dixon, K.W., & Pate, J.S. (1996) Pectic zymograms and water stress tolerance of endophytic fungi isolated from Western Australian heaths (Epacridaceae). Ann. Bot. 77:399–404.

Israel. D.W. (1987) Investigation of the role of phosphorus in symbiotic dinitrogen fixation. Plant Physiol. 84:835–840.

Jakobsen, I. & Rosendahl, L. (1990) Carbon flow into soil and external hyphae from roots of mycorrhizal cucumber plants. New Phytol. 115:77–83.

Jakobsen, I., Joner, E.J., & Larsen, J. (1994) Hyphal phosphorus transport, a keystone to mycorrhizal enhancement of plant growth. In: Impact of arbuscular mycorrhizas on sustainable agriculture and natural ecosystems, S. Gianinazzi & H. Schuepp (eds). Birkhäuser Verlag, Basel, pp. 133–146.

James, E.K., Reis, V.M., Olivars, F.L., Baldani, J.I., & Döbereiner, J. (1994). Infection of sugar cane by the nitrogen-fixing bacterium *Acetobacter diazotrophicum*. J. Exp. Bot. 45:757–766.

Johansen, A., Jakobsen, I., & Jensen, E.S. (1994) Hyphal N transport by a vesicular-arbuscular fungus associated with cucumber grown at three nitrogen levels. Plant Soil 160:1–9.

Joner, E.J. & Jakobsen, I. (1995) Uptake of ^{32}P from labelled organic matter by mycorrhizal and nonmycorrhizal subterranean clover (*Trifolium subterraneum* L.). Plant Soil 172:221–227.

Kaiser, B.N., Layzell, D.B., & Shelp, B.J. (1997) Role of oxygen limitation and nitrate metabolism in the nitrate inhibition of nitrogen fixation by pea. Physiol. Plant. 101:45–50.

Kapulnik, Y. (1991) Non-symbiotic nitrogen fixing micro-organisms. In: Plant roots: The hidden half, Y. Waisel, A. Eshel & U. Kafkaki (eds). Marcel Dekker, New York, pp. 703–716.

Kearns, A., Whelan, J., Young, S., Elthon, T.E., & Day, D.A. 1992. Tissue-specific expression of the alternative oxidase in soybean and siratro. Plant Physiol. 99:712–717.

Kennedy, I.R. & Tchan, Y.-T. (1992) Biological nitrogen fixation in non-leguminous field crops: Recent advances. Plant Soil 141:93–118.

Koch, K.E. & Johnson, C.R. (1984) Photosynthetic partitioning in split-root citrus seedlings with mycorrhizal and nonmycorrhizal root systems. Plant Physiol. 75:26–30.

Koide, R.T. & Schreiner, R.P. (1992) Regulation of the vesicular-arbuscular mycorrhizal symbiosis. Annu. Rev. Plant Physiol. Plant Mol. Biol. 43:557–581.

Koide, R.T., Huenneke, L.F., Hamburg, S.P., & Mooney, H.A. (1988) Effects of applications of fungicide, phosphorus and nitrogen on the structure and productivity of an annual serpentine plant community. Funct. Ecol. 2:335–344.

Krikun, J. (1991) Mycorrhizae in agricultural crops. In: Plant roots: The hidden half, Y. Waisel, A. Eshel & U. Kafkaki (eds). Marcel Dekker, New York, pp. 767–786.

Kwon, D.-K. & Beevers, H. (1992) Growth of *Sesbania rostrata* (Brem) with stem nodules under controlled conditions. Plant Cell Environ. 15:939–945.

Lambers, H., Atkin, O.K., & Scheurwater, I. (1996) Respiratory patterns in roots in relation to their functioning. In: Plant roots: The hidden half, Y. Waisel, A. Eshel, & U. Kafkaki (eds). Marcel Dekker, New York, pp. 323–362.

Leake, J.R. (1994) The biology of myco-heterotrophic "saprophytic" plants. New Phytol. 127:171–216.

Lindblad, P., Atkins, C.A., & Pate, J.S. (1991) N_2-fixation by freshly isolated *Nostoc* from coralloid roots of the cycad *Macrozamia riedlei* (Fisch. ex Gaud.) Gardn. Plant Physiol. 95:753–759.

Marschner, H. (1995) Mineral nutrition of higher plants. Second edition. Academic Press, London.

Marschner, H. & Dell, B. (1994) Nutrient uptake in mycorrhizal symbiosis. Plant Soil 159:89–102.

Maxwell, C.A., Hartwig, U.A., Joseph, C.M., & Phillips, D.A. (1989) A chalcone and two related flavonoids released from alfalfa roots induce *nod* genes of *Rhizobium meliloti*. Plant Physiol. 91:842–847.

Mellor, R.B. & Collinge, D.B. (1995) A simple model based on known plant defence reactions is sufficient to explain most aspects of nodulation. J. Exp. Bot. 46:1–18.

Mylona, P., Pawlowski, K., & Bisseling, T. (1995) Symbiotic nitrogen fixation. Plant Cell 7:869–885.

Nadelhoffer, K., Shaver, G., Fry, B., Giblin, A., Johnson, L., & McKane, R. (1996) ^{15}N natural abundances and N use by tundra plants. Oecologia 107:386–394.

Newman, E.I., Eason, W.R., Eissenstat, D.M., & Ramos, M.I.F.R. (1992) Interactions between plants: The role of mycorrhizae. Mycorrhiza 1:47–53.

Ocampo, J.A., Martin, J., & Hayman, D.S. (1980) Influence of plant interactions on vesicular-arbuscular mycorrhizal infections. I. Host and non-host plants grown together. New Phytol. 84:27–35.

Osborne, B.A. (1989) Comparison of photosynthesis and productivity of *Gunnera tinctora* Molina (Mirbel) with and without the phycobiont *Nostoc punctiforme* L. Plant Cell Environ. 12:941–946.

Palus, J.A., Borneman, J., Ludden, P.W., & Triplett, E.W. (1996) A diazotrophic bacterial endophyte isolated from stems of *Zea mays* L. and *Zea luxurians* Iltis and Doebley. Plant Soil 186:135–142.

Pate, J.S. (1980) Transport and partitioning of nitrogenous solute. Annu. Rev. Plant Physiol. 31:313–340.

Pate, J.S., Lindblad, P., & Atkins, C.A. (1988) Pathway of assimilation and transfer of fixed nitrogen in coralloid roots of cycad-*Nostoc* symbioses. Planta 176:461–471.

Penas, J.I., Sanchez-Diaz, M., Aguirreola, J., & Becana, M. (1988) Increased stress tolerance of nodule activity in *Medicago-Rhizobium-Glomus* symbiosis under drought. J. Plant Physiol. 79:79–83.

Peng, S., Eissenstat, D.M., Graham, J.H., Williams, K., & Hodge, N.C. (1993) Growth depression in mycorrhizal citrus at high-phosphorus supply. Analysis of carbon costs. Plant Physiol. 101:1063–1071.

Peoples, M.B., Palmer, B., Lilley, D.M., Duc, L.M., & Herridge, D.F. (1996) Application of ^{15}N and xylem ureide methods for assessing N_2 fixation of three shrub legumes periodically pruned for forage. Plant Soil 182:125–137.

Peterson, C.A. & Enstone, D.E. (1996) Functions of passage cells in the endodermis and exodermis of roots. Physiol. Plant. 97:592–598.

Peterson, R.L. & Bonfante, P. (1994) Comparative structure of vesicular-arbuscular mycorrhizas and ectomycorrhizas. Plant Soil 159:79–88.

Phillips, D.A., Dakora, F.D., Sande, E., Joseph, C.M., & Zon, J. (1994) Synthesis, release, and transmission of alfalfa signal to rhizobial symbionts. Plant Soil. 161:69–80.

Ratnayake, M., Leonard, R.T., & Menge, J.A. (1978) Root exudation in relation to supply of phosphorus and its possible relevance to mycorrhizal formation. New Physiol. 67:543–552.

Read, D.J. (1991) Mycorrhizas in ecosystems. Experientia 47:376–391.

Richardson, A.E., Djordjevic, M.A., Rolfe, B.G., & Simpson, R.J. (1988) Effects of pH, Ca and Al on the exudation from clover seedlings of compounds that induce the expression of nodulation genes in *Rhizobium trifolii*. Plant Soil 109:37–47.

Rolfe, B.G. & Gresshoff, P.M. (1988) Genetic analysis of legume nodule initiation. Annu. Rev. Plant Physiol. Plant Mol. Biol. 39:297–319.

Rousseau, J.V.D. & Reid, C.P.P. (1991) Effects of phosphorus fertilization and mycorrhizal development on phosphorus nutrition and carbon balance of loblolly pine. New Phytol. 92:75–87.

Ruiz-Lozana, J.M. & Azcón, R. (1995) Hyphal contribution to water uptake in mycorrhizal plants as affected by the fungal species and water status. Physiol. Plant. 95:472–478.

Ryle, G.J.A., Powell, C.E., & Gordon, A.J. (1985) Short-term changes in CO_2-evolution associated with nitrogenase activity in white clover in response to defoliation and photosynthesis. J. Exp. Bot. 36:634–643.

Sanchez-Diaz, M. & Honrubia, M. (1994) Water relations and alleviation of drought stress in mycorrhizal plants. In: Impact of arbuscular mycorrhizas on sustainable agriculture and natural ecosystems, S. Gianinazzi & H. Schupp (eds). Birkhäuser Verlag, Basel, pp. 167–178.

Sanchez-Diaz, M., Pardo, M., Antolin, M., Pena, J., & Aguirreola, J. (1990) Effect of water stress on photosynthetic activity in the *Medicago-Rhizobium-Glomus* symbiosis. Plant Sci. 71:215–221.

Sanders, I.R. & Koide, R.T. (1994) Nutrient acquisition and community structure in co-occurring mycotrophic and non-mycotrophic oldfield annuals. Funct. Ecol. 8:77–84.

Schulze, E.-D., Chapin III, F.S., & Gebauer, G. (1995) Nitrogen nutrition and isotope differences among life forms at the northern treeline of Alaska. Oecologia 100:406–412.

Smith, S.E. & Gianinazzi-Pearson, V. (1988) Physiological interactions between symbionts in vesicular-arbuscular mycorrhizal plants. Annu. Rev. Plant Physiol. Mol. Biol. 39:221–244.

Smith, S.E. & Read, D.J. (1997) Mycorrhizal symbiosis. Academic Press, London.

Smith, S.E., Gianinazzi-Pearson, V., Koide, R., & Cairney, J.W.G. (1994) Nutrient transport in mycorrhizas: Structure, physiology and consequences for efficiency of the symbiosis. Plant Soil 159:103–113.

Smith, S.E., Robson, A.D., & Abbott, L.K. (1992) The involvement of mycorrhizas in assessment of genetically dependent efficiency of nutrient uptake and use. Plant Soil 146:169–179.

Snellgrove, R.C., Splittstoesser, W.E., Stribley, D.P., & Tinker, P.B. (1982) The distribution of carbon and the demand of the fungal symbiont in leek plants with vesicular-arbuscular mycorrhizas. New Phytol. 92:75–87.

Spaink, H.P. (1995) The molecular basis of infection and nodulation by rhizobia: The ins and outs of sympathogenesis. Annu. Rev. Phytopathol. 33:345–368.

Sprent, J.I., Geoghegan, I.E., Whitty, P.W., & James, E.K. (1996) Natural abundance of ^{15}N and ^{13}C in nodulated

legumes and other plants in the cerrado and neighbouring regions of Brazil. Oecologia 105:440–446.

Sturz, A.V. (1995) The role of endophytic bacteria during seed piece decay and potato tuberization. Plant Soil 172:257–263.

Ta, T.C. & Faris, M.A. (1988) Effects of environmental conditions on the fixation and transfer of nitrogen from alfalfa to associated timothy. Plant Soil 107:25–30.

Tester, M., Smith, S.E., & Smith, F.A. (1987) The phenomenon of "nonmycorrhizal" plants. Can. J. Bot. 65:419–431.

Thompson, B.D., Robson, A.D., & Abbott, L.K. (1986) Effects of phosphorus on the formation of mycorrhizas by *Gigaspora calospora* and *Glomus fasciculatum* in relation to root carbohydrates. New Phytol. 103:751–765.

Tisdall, J.M. (1994) Possible role of soil microorganisms in aggregation in soils. Plant Soil 159:115–121.

Tobar, R., Azcón, R., & Barea, J.-M. (1994) Improved nitrogen uptake and transport from ^{15}N-labelled nitrate by external hyphae of arbuscular mycorrhiza under water-stressed conditions. New Phytol. 126:119–122.

Trappe, J.M. (1987) Phyllogenetic and ecological aspects of mycotrophy in angiosperms from an evolutionary standpoint. In: Ecophysiology of VA mycorrhizas, G.R. Safir (ed). CRC Press, Boca Raton, pp. 5–25.

Triplett, E.W. (1996) Diazotrophic endophytes: progress and prospects for nitrogen fixation in monocots. Plant Soil 186:29–38.

Turnbull, M.H., Goodall, R., & Stewart, G.R. (1995) The impact of mycorrhizal colonization upon nitrogen source utilization and metabolism in seedlings of *Eucalyptus grandis* Hill ex Maiden and *Eucalyptus maculata* Hook. Plant Cell Environ. 18:1386–1394.

Udvardi, M.K. & Day, D.A. (1997) Metabolite transport across symbiotic membranes of legume nodules. Annu. Rev. Plant Physiol. Plant Mol. Biol. 48:493–523.

Vance, C.P. (1996) Root-bacteria interactions. Symbiotic nitrogen fixation. In: Plant roots: The hidden half, Y. Waisel, A. Eshel, & U. Kafkaki (eds). Marcel Dekker, New York, pp. 723–755.

Vance, C.P., Egli, M.A., Griffith, S.M., & Miller, S.S. (1988) Plant regulated aspects of nodulation and N_2 fixation. Plant Cell Environ. 11:413–427.

Van Rhijn, P. & Vanderleyden, J. (1995) The *Rhizobium*-plant symbiosis. Microbiol. Rev. 59:124–142.

Vierheilig, H., Iseli, B., Alt, M., Raikhel, N., Wiemken, A., & Boller, T. (1996) Resistance of *Urtica dioica* to mycorrhizal colonization: A possible involvement of *Urtica dioica* agglutinin. Plant Soil 183:131–136.

Volpin, H., Elkind, Y., Okon, Y., & Kapulnik, Y. (1994) A vesicular arbuscular mycorrhizal fungus (*Glomus intraradices*) induces a defense response in alfalfa roots. Plant Physiol. 104:683–689.

Whitehead, L.F. & Day, D.A. (1997) The peribacteroid membrane. Physiol. Plant. 100:30–44.

Whitehead, L.F., Tyerman, S.D., Salom, C.L., & Day, D.A. (1995) Transport of fixed nitrogen across symbiotic membranes of legume nodules. Symbiosis 19:141–154.

Wilcox, H.E. (1991) Mycorrhizae. In: Plant roots: The hidden half, Y. Waisel, A. Eshel, & U. Kafkaki (eds). Marcel Dekker, New York, pp. 731–765.

Wilson, D. (1993) Fungal endophytes: Out of sight but should not be out of mind. Oikos 68:279–384.

Yoneyama, T., Muraoka, T., Kim, T.H., Decanay, E.V., & Nakanishi, Y. (1997) The natural 15N abundance of sugarcane and neigbouring plants in Brazil, the Philippines and Miyako (Japan). Plant Soil 189:239–244.

Zaat, S.A.J., Wijfelman, C.A., Mulders, I.H.M., Van Brussel, A.A.N., & Lugtenberg, B.J.J. (1988) Root exudates of various host plants of *Rhizobium leguminosarum* contain different sets of inducers of *Rhizobium* nodulation genes. Plant Physiol. 88:1298–1303.

9B. Ecological Biochemistry: Allelopathy and Defense Against Herbivores

1. Introduction

Plants contain a vast array of compounds that do not play a role in primary catabolic or biosynthetic pathways. These are commonly referred to as **secondary metabolites**. Many of these metabolites play a role in numerous ecological interactions (e.g., deterring herbivores, protection against pathogens, allelopathy, symbiotic associations, seed germination of parasites, or interactions with pollinators). Others provide protection against ultraviolet radiation or high temperatures. Some of these roles have already been discussed in the pertinent chapters. This chapter will discuss the role of secondary compounds in allelopathic and plant–herbivore interactions. Plant–pathogen interactions are discussed in a separate chapter.

2. Allelopathy

Some plants harm the growth or development of surrounding plants by the excretion of chemical compounds: **allelopathic compounds**. These effects (**allelopathy**) are invariably negative, and the compounds may come from living roots or leaves, or from decomposing plant remains. There are also positive effects of compounds released from living or dead plant material, such as the example of fulvic acid and humic acid released from *Eucalyptus camaldulensis* (river gum), which alleviates alu-

minum toxicity in *Zea mays* (maize) (Sect. 3.1.2 of the chapter on mineral nutrition). Such compounds may also have a positive effect because of their role in complexing heavy metals. Another example of a positive effect is that of plants that excrete phosphate-solubilizing acids, which make poorly available phosphate sources available, even for neighboring plants (cf. Sect. 2.2.5.2 of the chapter on mineral nutrition). Note that these positive effects are not referred to as allelopathy.

Numerous publications have appeared on allelopathy, many of which need to be taken with a grain of salt; upon further investigations many of the alleged allelopathic interactions could be explained in an alternative way. For example, the absence of seedlings near a bush that produces volatile terpenes suggested that allelopathy might be involved, but closer investigation showed that seed-eating animals prefer to graze in the shelter of the bush, where they are in less danger from predatory birds. There is general agreement in the literature, however, that allelopathic interactions do exist. Water-soluble compounds (mainly of a phenolic nature) as well as volatiles (predominantly terpenoids) may have an allelopathic effect (Fig. 1). They may originally have arisen in the course of plant evolution as compounds deterring pathogens or herbivores, and subsequently become involved in interactions between higher plants. Secretory glands were well developed in the early gymnosperms and angiosperms of the Paleozoic, before there were terrestrial herbivores, but after the evolution of terrestrial fungi, which

Terpenes concerned in plant allelopathy

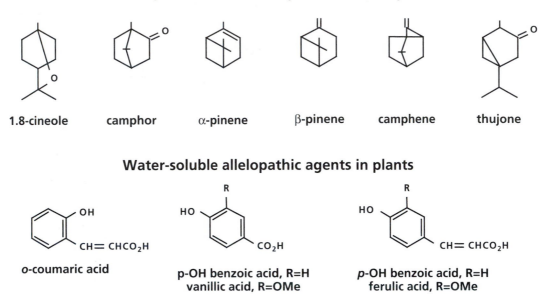

1.8-cineole   **camphor**   **α-pinene**   **β-pinene**   **camphene**   **thujone**

Water-soluble allelopathic agents in plants

o-coumaric acid

p-OH benzoic acid, R=H
vanillic acid, R=OMe

p-OH benzoic acid, R=H
ferulic acid, R=OMe

FIGURE 1. Some examples of volatile and water-soluble allelopathic compounds (Harborne 1988).

suggests that early defense systems may have been directed at pathogens.

The mode of action of the various allelopathic compounds is largely unknown. Many phenolic compounds inhibit seed germination of grasses and herbs, and they may inhibit ion uptake. Volatile terpenoids may inhibit cell division. Very little is known of the mechanisms by which potentially allelopathic compounds are detoxified by some species.

A classic example of allelopathy concerns *Juglans nigra* (black walnut). In a zone up to 27 m from the tree trunk, many plants [e.g., *Lycopersicon esculentum* (tomato) *Medicago sativa* (alfalfa)] die. The toxic effects are due to the leaching from the leaves, stems, branches, and roots of a bound phenolic compound, which undergoes hydrolysis and oxidation in the soil. The bound compound, which is nontoxic itself, is the 4-glucoside of 1,4,5-trihydroxy-naphthalene. It is converted to the toxic compound juglone (5-hydroxynaphtoquinone) (Fig. 2). Some species are tolerant of juglone [e.g., *Poa pratensis* (Kentucky blue grass)], but the nature of this tolerance is unknown. *Sorghum* species have a reputation for suppressing weed growth, due to the exudation of allelochemicals. One of these is a dihydroxyquinone (sorgoleone) that inhibits mitochondrial respiration (Rasmussen et al. 1992). Similarly, under the trees of *Leucaena leucocephala* plantations in Taiwan, very few weeds are found.

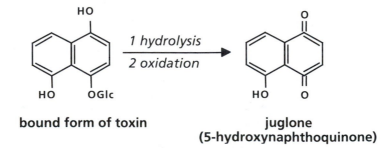

bound form of toxin

juglone
(5-hydroxynaphthoquinone)

FIGURE 2. The chemical structure of a precursor of juglone, which is released from *Juglans nigra* (black walnut), and its conversion into the toxic juglone (Harborne 1988).

Table 1. The effects of *Leucaena leucocephala* leaves mixed with 150 g of soil or mulched and spread on the soil surface on growth and survival of a number of plant species. The data are expressed as percent survival relative to that in the soil alone.

Species	Survival (% of the control)		
	Leaves mixed with soil		Leaf mulch added
	1 g	2 g	5 g
Leucaena leucocephala	100	100	87
Alnus formosana	72	44	37
Acacia confusa	30	19	14
Liquidamber formosana	5	9	31
Casuarina glauca	0	0	0
Mimosa pudica	0	0	0

Source: Chou & Kuo 1986.
Note: The data are expressed as percentage.

This has been ascribed to the presence of high concentrations of mimosine (a toxic nonprotein amino acid) as well as a range of phenolic compounds, which originate from the tree leaves and which inhibit germination and growth of many forest species (Table 1). Allelopathic interactions also appear to play a major role in desert plants (e.g., between *Encelia farinosa* and its surrounding plants in the Mojave desert in California). In many of these plants, a simple benzene derivative is produced, primarily in the leaves (Fig. 3). It is released when the leaves fall to the ground and decompose.

One of the few examples of growth inhibition by a toxin produced in roots, rather than leaves, is that of the rubber plant guayule (*Parthenium argentatum*). The aromatic compound (Fig. 3), causes inhibition of plants of the same species (**autotoxicity**). A similar example of autotoxicity has been found for *Grevillea robusta* (silky oak), which is a tree from subtropical rainforests in Australia.

Allelopathic and antotoxic effects probably play a role in many environments; however, it is hard to estimate their ecological significance. Some of the released compounds are probably decomposed rather rapidly by microorganisms, thus diminishing their potential effects. On the other hand, some allelopathic compounds decompose rather slowly, including a group of phenolic compounds mostly referred to as tannins (cf. Sect. 3.1). The consequence of this slow decomposition are discussed in the chapter on decomposition.

3. Chemical Defense Mechanisms

Many secondary plant compounds play a role in deterring herbivores. Some herbivores, however, have found ways "to get around the problem" or even prefer the plants that contain specific secondary compounds: **food selection**. Both topics will be discussed in this section.

3.1 Defense Against Herbivores

Chemical defense is quite obvious in the stinging nettle (*Urtica dioica*) and closely related members of the Urticaceae. Touching the plant breaks off the tip of the hairs on leaves or stem. The walls of these hairs are thin and contain silica, which gives the cut hair a sharp end to penetrate the skin. The contents of the hair are then released, locally giving pain and swelling of the skin. The exact nature of the content of the stinging hairs of *U. dioica* is unknown; the older literature suggests biogenic amines, including serotonin, whereas a tropical member of the Urticaceae, *Laporta moroides*, accumulates peptides, including a tricyclic octapeptide (moroidin) (Leung et al. 1986). The number of stinging hairs varies widely in *U. dioica*; some plants have none at all. Grazing by large herbivores is negatively correlated with the number of hairs (Pollard & Briggs 1984).

Some secondary compounds inhibit specific steps in mitochondrial respiration, including, for example, HCN, released from **cyanogenic com-**

3-acetyl-6-methoxybenzaldehyde
Encelia farinosa

***trans*-cinnamic acid**
Parthenium argentatum

Figure 3. Two examples of toxins produced in desert shrubs (Harborne 1988).

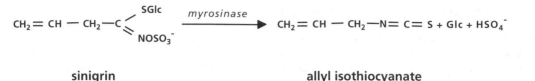

sinigrin   **allyl isothiocyanate**

FIGURE 4. The chemical structure of sinigrin, which is a glucosinolate in *Brassica* species, and allyl isothiocyanate, into which it can be converted enzymatically (Harborne 1988).

pounds present in a wide range of species, which blocks cytochrome oxidase. **Fluoroacetate** blocks aconitase, which is an enzyme in the TCA cycle. Platanetin, which is a flavonoid from the bud scales of *Platanus acerifolia* (plane tree), inhibits the external NADH dehydrogenase (Roberts et al. 1996; cf. Sect. 2.3.1 of the chapter on plant respiration). Seeds of a wide range of *Phaseolus* (bean) species contain a specific inhibitor of α-amylase, which is a digestive enzyme to hydrolyze starch (Pueyo & Delgado-Salinas 1997). Other secondary plant compounds are much less specific; for example, **tannins**, which precipitate proteins and thus interfere with food digestion. Toxic **phenolic glycosides** in *Salix* (willow) species deter herbivores. Others [e.g., glucosinolates in Brassicaceae (cabbage family)], probably evolved as secondary metabolites in plants because they are toxic for most herbivores. Some herbivores have evolved, however, mechanisms that defy this chemical defense and use glucosinolates as **attractants**. The glucosinolate sinigrin is converted enzymatically to allyl isothiocyanate, which gives mustard its distinct sharpness (Fig. 4). In *Brassica* (cabbage) species and other Cruciferae, sinigrin attracts butterflies of *Pieris brassicae* (cabbage moth) as well as certain aphids (e.g., *Brevicoryne brassicae*) and cabbage-root flies (*Delia radicum*). Cabbage moths normally deposit their eggs only on plants that contain sinigrin, but accept filter paper that contains this compound as a substitute. Their larvae exclusively eat food that contains sinigrin, either naturally or experimentally added.

Both *Populus* (poplar) and *Salix* (willow) plants contain a wide range of toxic phenolic glycosides, including salicin (Clausen et al. 1989). After ingestion, salicin is hydrolyzed and oxidized, producing salicylic acid (Fig. 5) which uncouples oxidative phosphorylation in mitochondrial preparations. The structure of phenolic glycosides resembles that of many allelopathic compounds, which suggests that the driving force in evolution for the formation of allelopathic compounds may well have been their role in deterring herbivores or pathogens. Both the total phenolic glycoside concentration in

the leaves and the spectrum of these compounds varies among *Salix* species (Table 2).

The role of phenolic glycosides in the food selection pattern of beetles feeding on willow leaves has been investigated extensively. Leaves of the eight willow species shown in Table 2 were used for laboratory feeding experiments with four beetle species. In all cases, the leaves of the willow species that was chemically most related to the preferred species were fed on to the highest degree (Fig. 6). Both the total amount and the quality of the phenolic glycosides determine the food selection pattern of the investigated beetles.

Mammals have been important selective influences for the patterns of defense in woody plants, which are vulnerable to mammalian herbivory throughout the winter. Mammalian herbivory is a major cause of mortality in woody plants (Bryant & Kuropat 1980), in part because mammals remain active and often have highest energy demand in winter, when plants cannot grow to compensate for tissues lost to herbivores. Woody plant defenses are better developed in regions with a long history of vertebrate browsing than they are in regions that were glaciated during the Pleistocene (Bryant et al. 1989). There is strong developmental control over defenses in woody plants, with these being most strongly expressed in juvenile woody plants that grow in a height range where they are vulnerable to mammalian herbivores. Following browsing, juvenile shoots are produced that have higher levels of secondary metabolites that deter further browsing.

salicylic acid

FIGURE 5. The chemical structure of salicylic acid, which is produced after ingestion from some of the phenolic glycosides that regularly occur in *Populus* and *Salix* species. Salicylic acid is closely related to acetyl salicylic acid, which is the active ingredient of aspirin (Harborne 1988).

Table 2. Phenolic glycoside concentration [mg g^{-1} (dry mass)] in the leaves of eight *Salix* species native to Finland or introduced to this area.*

	Sal1	Sal2	Frag	Tria	Sal3	Pic	Total
Wild willows							
S. nigricans	48	3	0.2				51
S. phylicifolia	0.5	0.1	0.1	0.3	0.1		1.8
S. caprea	0.3	0.2	0.1	0.1			1.2
S. pendandra		0.7	0.7				7.6
Introduced willows							
S. cv. *aquatica*	6.4	1.3	0.1				7.8
S. dascylados	9.9	2.0	0.2				12.1
S. viminalis	0.1		0.1	0.1		0.2	1.5
S. triandra		0.3			7.4		7.8

Source: Tahvanoli 1985.

*Apart from these identified compounds, some others were present, so that the total amount differs from the sum of the identified ones. Sal1, Salicortin; Sal2, Salicin, Frag, Fragilin; Tria, Triandrin; Sal3, Salidroside; Pic, Picein.

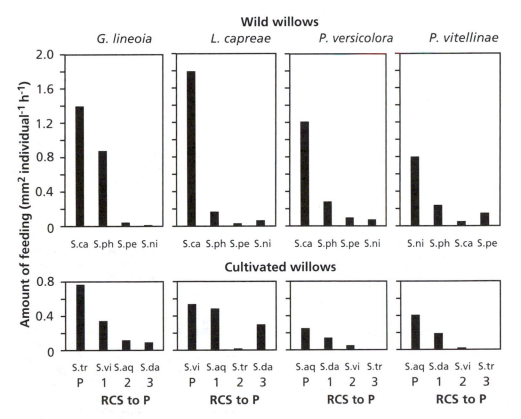

Figure 6. Food selection pattern by four beetle species (*Galerucell lineola Lochmaea capreae, Plagiodera versicolora,* and *Pratora vitellinae*) when leaves of four native and four introduced *Salix* species were offered in two separate food choice experiments. The preferred species is placed at the left; the others are ranked according to their chemical similarity (RCS) relative to the preferred (P) species. The species are the same as those presented in Table 2 (Tahvanainen et al. 1985).

These defenses include ether-soluble terpenes, such as papyriferic acid in *Betula resinifera* (birch) and pinosylvin in *Alnus crispa* (alder), that deter feeding below levels required for weight maintenance and, if consumed, result in negative nitrogen and sodium balance (Bryant et al. 1992).

3.2 Qualitative and Quantitative Defense Compounds

Secondary metabolites involved in deterring herbivores are generally divided into two categories:

1. **Qualitatively** important secondary plant compounds. These are **toxins**, which are usually present in low concentrations, but may constitute up to 10% of the fresh weight of some seeds. Numerous compounds belong to this category, including alkaloids (Fig. 7), cyanogenic glycosides, nonprotein amino acids, cardiac glycosides, glucosinolates, and proteins. Their mode of action varies widely.

2. **Quantitatively** important secondary plant compounds. These reduce the **digestibility** and/or **palatability** of the food source and invariably make up a major fraction of the biomass. They are mostly phenolic compounds (phenolic acids, tannins, lignin; Fig. 8). Tannins and some other phenolics reduce the digestibility of plant tissues by blocking the action of digestive enzymes, binding to proteins being digested, or interfering with protein activity in the gut wall. Tannins, as well as lignin, also increase the leaf's toughness.

If you consider that relatively few resources are required to acquire protection against herbivores by toxic compounds, then one may wonder why the alternative strategy of the digestibility reducing compounds, which requires far greater investment of carbon resources, has evolved at all. The answer to this question is that there are numerous examples of herbivores that cope with the toxic compounds that are effective against the majority of herbivores. They may metabolize the toxin to an extent that it is used as a food source, or they may store the toxin, sometimes after slight modification, and thus gain protection themselves. Such combinations of toxic plants and animals that cope with the toxin provide examples of **coevolution** of plants and animals in an ever-continuing "arms race."

Although the distinction between qualitative and quantitative defenses is a useful starting point, it is not a clearcut dichotomy. Many phenolic compounds also have toxic effects on herbivores and

may be more toxic against some herbivores than others (Ayres et al. 1997).

3.3 The Arms Race of Plants and Herbivores

The expression *arms race* graphically describes the continuous evolution of ever more toxic defense compounds in plants and of more mechanisms to cope with these compounds in herbivores. Numerous examples of such a **coevolution** exist, but they appear to be restricted to mechanisms in herbivores that store or detoxify qualitative defense compounds, with very little evidence for mechanisms that break through the quantitative defense. We will first present a number of striking examples of coevolution of predators coping with the qualitative defense.

Whereas the stinging hairs on members of the Urticaceae protect the plants against large herbivores, some caterpillars (e.g., those of *Inachis io*, *Vanessa atalanta* and *Aglais urticae*) are not affected by them. Some of these caterpillars simply bite the hairs off. Snails (e.g., *Arion ater* and *Agriolimax columbianus*) are also little affected by the leaf hairs on *Urtica dioica* (Cates & Orians 1975). Plants and herbivores, particularly insects, are in a continuous battle. From a plant's perspective, success in this interaction is determined by its ability to defend itself from devastation by insect feeding. From an insect's perspective, success is measured by its ability to protect itself from a variety of toxic plant defense compounds, thereby allowing it to use plants as its sole food source.

------------------------------------>

FIGURE 7. Alkaloid subclasses. (A) Isoquinoline alkaloids. These are synthesized [e.g., carnegine and gigantine in a species-specific manner in Saguaro (*Carnegia gigantea*) and cardon (*Pachycereus pringlei*) cacti]. Sentia cactus (*Lophocereus schottii*) contains as much as 30–150 mg g⁻¹ (DM) lophocerine and its trimers, pilocereine and piloceredine. (B) bisbenzylisoquinoline alkaloids. Examples include berbamunine from barberry (*Berberis stolonifera*) and tubocurarine, an arrow poison, from *Chondrodendron tomentosum*. (C) Monoterpene indole alkaloids, including quinine (from *Cinchona officinalis*), strychnine (from *Strychnos nux-vomica*) and nepetalactone (from *Nepeta racemosa*, catnip). (D) Nicotine and tropane alkaloids. These are naturally occurring insecticides and feeding deterrents in Solanaceae [e.g., nicotine in tobacco (*Nicotiana tabacum*), scopolamine in *Hyoscyamus niger* (henbane) and atropine in *Atropa belladonna* (deadly nightshade)] (Harborne 1988, Schuler 1996).

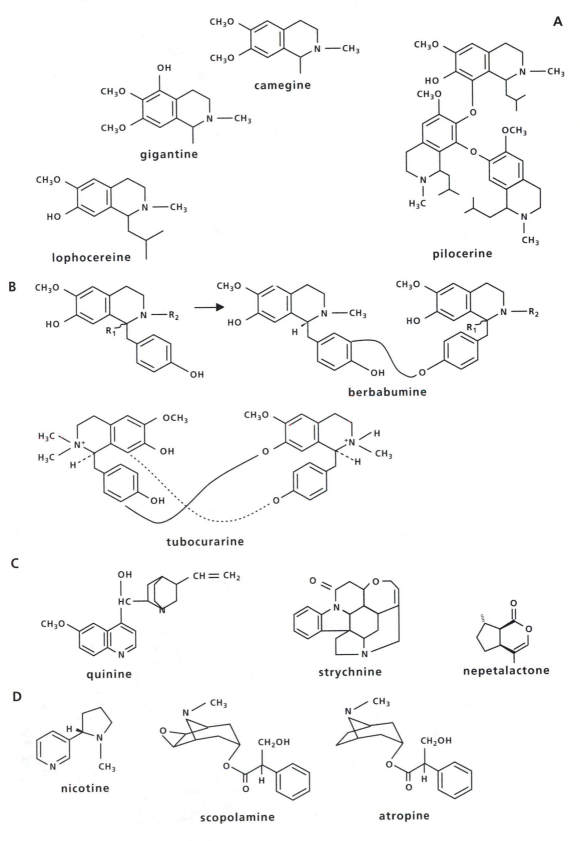

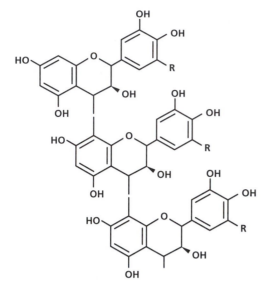
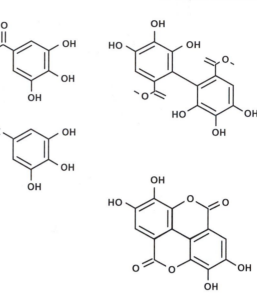

FIGURE 8. The chemical structure of proanthocyanidin (condensed tannin) (left). Gallotannin (top, middle) and ellagitannin (top, right) are hydrolyzable tannins, releasing gallic acid (bottom, middle) and ellagic acid (bottom, right), respectively, and the esterified sugar(s), mostly glucose, upon hydrolysis.

One example of coevolution that involes defensive secondary plant compounds is that of small marsupials in Australia that are resistant to the very poisonous **fluoroacetate**, which is a potent inhibitor of an enzyme of the TCA cycle (aconitase). Fluoroacetate occurs in some leguminous shrubs (*Gastrolobium* and *Oxylobium*) of the native Australian flora and is poisonous to the introduced cows and sheep. Another well-studied example of coevolution is the combination of *Senecio jacobaea* (Tansy ragwort) and *Tyria jacobaea* (cinnabar moth). The *Senecio jacobaea* plants contain at least six pyrrolizidine **alkaloids** (Fig. 9). Alkaloids are characterized by a nitrogen-containing heterocyclic ring and their alkaline reaction. They represent the largest (>10,000 structures) and one of the most structurally diverse groups of substances that serve as plant defense agents (Schuler 1996). The highly toxic alkaloids from *Senecio* may cause damage to the liver. The larvae of *Tyria jacobaea* are not harmed by these alkaloids and use *Senecio* as a much preferred food source; they sometimes consume the leaves of *Senecio vulgaris* (groundsel) or *Petasites hybridus* (coltsfoot) as alternative food sources. They accumulate the toxins, which even end up in the mature butterfly. Both the larvae and the butterfly are poisonous for birds. The toxic nature of these animals coincides with black and bright yellow warning coloration (visual advertisement). In addition to the larvae of *Tyria*, there are some other animals that cope with the toxic alkaloids in *Senecio* [e.g., the tiger moth (*Arctia caja*) and the flea beetle *Longitarsus jacobaea*].

The interaction of *Asclepias curassavica* (milkweed) and *Danaus plexippus* (the monarch butterfly) is similar to that of *Senecio* and *Tyria*; the *A. curassavica–D. plexippus* interaction has an interesting additional dimension in that it is exploited by *Limenitis archippus* (the viceroy butterfly). The milk sap of *A. curassavica* plants contains **cardiac glycosides** (calotropine and calactine). Cardiac glycosides (= cardenolides) are bitter compounds that stimulate the heart when applied in small doses, but are lethal in slightly higher doses; the structure of some cardiac glycosides is given in Figure 10. The presence of these toxic compounds in the larvae of *D. plexippus* is again advertized; moreover, caterpillars of the viceroy butterfly have similar colors, but without containing any cardiac glycosides (mimicry). While most predators are deterred by the cardiac glycoside in milkweed, there are two bird species from Mexico that are able to consume the caterpillars of the monarch butterfly, possibly because they have a detoxification mechanism (Harborne 1988).

Being able to cope with toxic plants does not invariably lead to accumulation of the toxin. Larvae of the beetle *Caryedes brasiliensis* from Costa Rica largely feed on the seeds of *Dioclea megacarpa*. These seeds contain canavanine, a toxic **nonprotein**

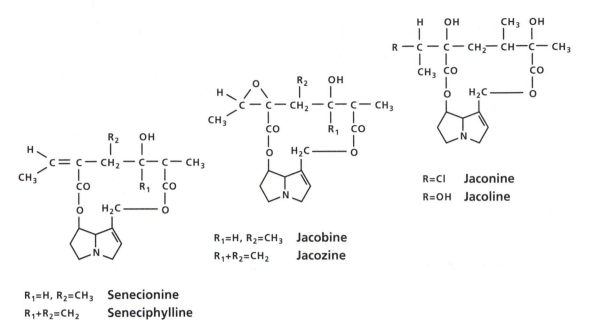

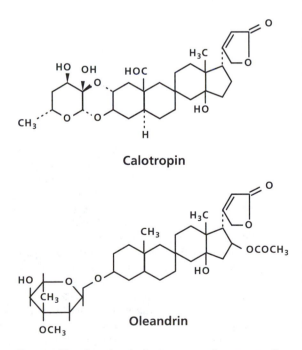

R₁=H, R₂=CH₃ **Senecionine**
R₁+R₂=CH₂ **Seneciphylline**

Inside figure (center): R₁=H, R₂=CH₃ **Jacobine** / R₁+R₂=CH₂ **Jacozine**

Inside figure (right): R=Cl **Jaconine** / R=OH **Jacoline**

FIGURE 9. The chemical structure of some pyrrolizidine alkaloids from *Senecio jacobaea* (Vrieling 1991).

amino acid that resembles arginine (Fig. 11) and may constitute as much as 7 to 10% of the seed fresh mass. Nonprotein amino acids are toxic because they act as "antimetabolites." That is, their structure

Calotropin

Oleandrin

FIGURE 10. The chemical structure of some cardiac glycosides, including calotropin from *Asclepias curassavica* (milkweed) (Harborne 1988).

is recognized as the same as that of the amino acid they resemble, which leads to proteins without the same tertiary structure and function as the protein that contains the normal amino acid. Resistance of the larvae of *Caryedes brasiliensis* is based on two principles. First, the larvae have a slightly different *t*-RNA synthetase, which recognizes arginine as being different from canavanine. Second, they have high levels of the enzyme urease, which breaks down canavanine. Thus, the toxin is a major source of nitrogen for the larvae.

These few examples selected from many show that one or more animal species have invariably coevolved with a plant species producing a toxin. Thus, while **qualitative defense** against herbivores requires relatively little investment of resources, it is also a vulnerable strategy. Although there are also some examples of animals coping with large quantities of digestibility-reducing and unpalatable compounds (**quantitative defense**), these examples appear to be rare. Hence, the strategy that requires a major investment of carbon is most certainly the safest. A large investment of carbon in protective compounds and structures inevitably goes at the expense of the possibility of investment of carbon in growth. It is therefore most predominant in slow-growing species, especially those with evergreen leaves (Bryant et al. 1983, Lambers & Poorter 1992). On the other hand, toxins are found in both fast-growing and slow-growing species. In the ever-green *Ilex opaca* (holly) the toxic saponins are only

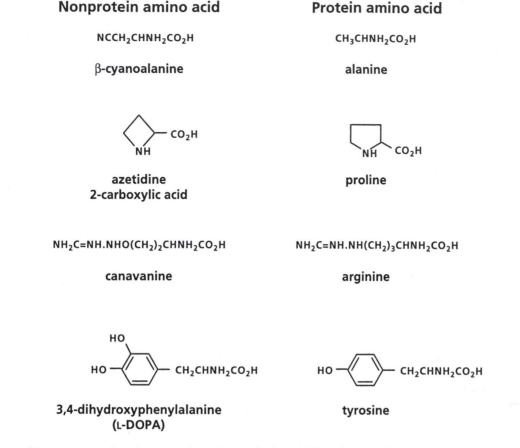

FIGURE 11. Some examples of nonprotein amino acids from higher plants, including canavanine from *Dioclea megacarpa*. The structure of the corresponding ordinary amino acids is also given for comparison (Harborne 1988).

found in young leaves and in the mesophyll cells of older leaves. Nonmesophyll cells of older leaves contain digestibility-reducing compounds like lignin, crystals, and tannin (Kimmerer & Potter 1987).

3.4 How Do Plants Avoid Being Killed by Their Own Poisons?

Most secondary plant compounds that deter herbivores are also toxic to the plants themselves. Prussic acid (HCN) is produced upon ingestion of plant material of approximately 2000 species from some 110 families, including genotypes of *Trifolium* spp. (clover), *Linum usitatissimum* (flax), *Sorghum bicolor* (millet), *Pteridium aquilinum* (bracken fern), and *Mannihot esculenta* (cassava). HCN inhibits a number of enzymes from both animals and plants (e.g., cytochrome oxidase and catalase), and this

also holds for plants that contain the cyanogenic compounds. How, then, do cyanogenic plants protect themselves from this toxic HCN?

Cyanogenic plants do not actually store HCN; rather, they store **cyanogenic glycosides** (i.e., cyanide attached to a sugar moiety or cyanogenic lipids in Sapindaceae), and these only produce HCN upon hydrolysis (Poulton 1990). The reaction is catalyzed by specific enzymes (e.g., linamarase catalyzing hydrolysis of linamarin in some legumes) (Fig. 12). Synthesis of many cyanogenic compounds requires amino acids as precursors, as illustrated in Figure 12 for the synthesis of linamarin from valine. The enzymes responsible for the breakdown of the cyanogenic compound and the cyanogenic compounds themselves occur in different cell compartments. Upon damage of the cells, such as after ingestion, the enzyme and its substrate come into contact. For example, dhurrin, which is a cyanogenic glycoside in *Sorghum* species, occurs exclusively in the vacuole of leaf epidermal cells, whereas the enzyme responsible for its hydrolysis is located in mesophyll cells. Linamarase, hydrolyzing linamarin, occurs in the walls of mesophyll cells, whereas its substrate is stored inside the cell. As long as this strict **compartmentation** is maintained, no problem arises for the plant itself; however, the linamarin present in the roots of *Hevea brasiliensis* (rubber tree) and *Mannihot esculenta* (cassava) is synthesized in the shoot and imported via the phloem. How is HCN production avoided during transport of a cyanogenic compound from leaf cells, via the phloem, to the roots? It is most likely that linustatin, which is a non-hydrolyzable form of linamarin, rather than the hydrolyzable linamarin itself, is transported (McMahon et al. 1995).

Although avoidance of damage by compartmentation is the best strategy, some **detoxification mechanisms** may be needed. Detoxification of HCN in plants is possible; it is catalyzed by β-cyano-alanine synthase, transforming L-cysteine + HCN into β-cyano-alanine. The nitrogen in cyanogenic compounds that are stored in seeds can be remobilized and incorporated into primary nitrogenous metabolites (Selmar et al. 1988, 1990). In addition, in vegetative plant organs, cyanogenic compounds may be subject to some turnover.

Resistance against cyanogenic glucosides in animals is based on the presence of the enzyme rhodanese (e.g., in sheep and cattle). It catalyzes the transformation of cyanide to thiocyanate. The sulfur required for this reaction comes from mercaptopyruvate. Treatment of patients suffering from HCN poisoning is based on the same principle when thiosulphate is administered to the victim.

In *Trifolium* (clover) species, as well as in others, polymorphism for cyanogenesis has been found. Genotypes are cyanogenic only when they are homozygous for both the recessive gene responsible for the production of linamarase and for the recessive gene responsible for linamarin production. In southern Europe cyanogenic genotypes are predominant, except at higher locations. In northern and western Europe, most genotypes are acyanogenic. This correlation (with temperature), however, has not yet been explained in a satisfactory manner. There may be other factors involved, such as in the case of *Hevea brasiliensis* (rubber tree), which releases HCN when it is infected by a pathogenic fungus (*Microcyclus ulei*), apparently hampering its ability to ward off the

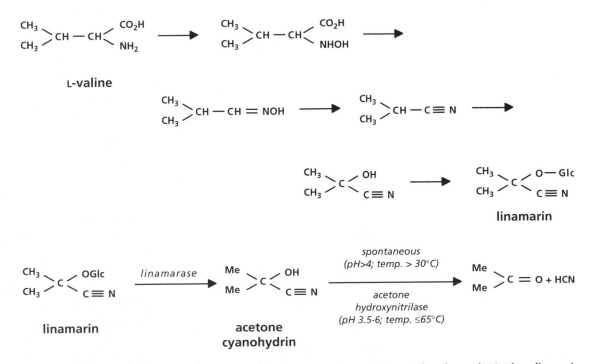

FIGURE 12. In the synthesis of linamarin the amino acid valine is used as a precursor. The release of HCN from linamarin is catalyzed by a specific enzyme, linamarase (McMahon et al. 1995).

fungus (Lieberei et al. 1989). Being cyanogenic would then have a disadvantage. It is therefore possible that the correlation of genotype with temperature reflects the temperature dependence of a pathogenic organism.

Like cyanogenic compounds, many alkaloids are also stored in specific compartments (i.e., either the vacuole or smaller vesicles in which they are produced). In *Papaver somnifera* (opium poppy), laticifers contain abundant vesicles that both contain morphine and the enzymes to synthesize and metabolize it. In *Berberis wilsoniae* (barberry), *Thalictrum glaucum* (rue), and many other species cells have similar "alkaloid vesicles," which contain berberin or other alkaloids and some of the enzymes of the pathway that produces them. The "alkaloid vesicles" may fuse with the central vacuole and thus deposit the alkaloids there (Hashimoto & Yamada 1994).

How does a plant protect itself against a compound such as ricin, which is an abundant protein in seeds of *Ricinus communis* (castor bean) that is highly toxic? Ricin is a ribosome-inactivating protein; similar proteins occur in taxonomically and ecologically diverse species, including domesticated crop plants (Hartley et al. 1996). Ricin is a heterodimeric protein that consists of an enzymatic polypeptide that destroys ribosomal RNA that is covalently bound to a galactose-binding lectin [lectins are proteins that recognize and bind carbohydrates; the first ones were discovered in castor bean more than a century ago; numerous other plants were found to contain lectins since then (Etzler 1985); see also Sect. 3.3]. This bipartite structure and functional properties allow ricin to bind to galactosides on the cell surface. Upon binding, ricin enters the cell via endocytotic uptake and traverses an intracellular membrane to deliver the enzymatic component to the cytosol. Once it is there, it irreversibly inhibits protein synthesis, followed by death of the cell. Ricin is one of the most potently toxic compounds known, and entry of a single toxin molecule into the cytosol may be sufficient to kill the cell. *Ricinus* ribosomes which synthesize ricin are also susceptible to the catalytic action of this protein. How, then, does *Ricinus* avoid suicide? The subunits of which the heterodimer is composed are originally synthesized together in the form of a single precursor protein: proricin. Proricin is an active lectin, but it does not bind to ribosomal RNA. It is transported to the vacuole, where acidic endoproteases remove amino acid residues to generate the heterodimer: ricin. None of the ricin appears to escape from the vacuole (Lord & Roberts 1996).

3.5 Secondary Metabolites for Medicines and Crop Protection

Secondary metabolites that deter herbivores or inhibit pathogens have been used by humans for centuries. The bark of willow (*Salix*) contains salicylic acid (Fig. 5), which is closely related to acetyl salicylic acid (**aspirin**) and has been used as medicine. Quinine, which is an alkaloid from the bark of *Cinchona officinalis*, has been used for centuries to combat **malaria**. Other examples of secondary com-

TABLE 3. Examples of secondary metabolites for which man has found some use.

Chemical compound	Species	Applications
Acetyl salicylic acid (aspirin)	*Salix* sp., *Populus* sp.	Pain killer
Aconitine	*Aconitum napellus*	Pain killer
Atropine	*Atropa bella-donna*	Ophthalmology
Cytisine	*Cytisus laburnum*	Migraine
Germerine, protoveratrine	*Veratrum album*	Muscle diseases, pain killer
Cardiac glycosides	*Digitalis* sp., *Asclepias* sp.	Heart diseases
Linarine, linine	*Linaria vulgaris*	Haemorrhoids
Quinine	*Cinchona officinalis*	Malaria
Atropine	*Atropa bella-donna*	Poisoning
Taxine	*Taxus baccata*	Poisoning (arrowheads of Celts)
Cicutoxin	*Cicuta virosa*	Poisoning (of Socrates)
Hyoscyamine, scopolamine	*Hyoscyamus niger*	Poisoning (in Shakespeare's "Hamlet")
Pyrethrin	*Chrysanthemum cinearifolium*	Insecticide
Rotenone	*Derris* sp.	Rat poisoning
Camphor	*Cinnamonum camphora*	Moth balls

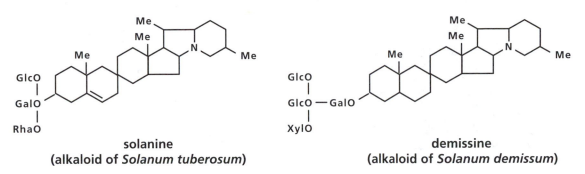

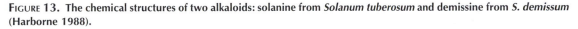

solanine
(alkaloid of *Solanum tuberosum*)

demissine
(alkaloid of *Solanum demissum*)

FIGURE 13. The chemical structures of two alkaloids: solanine from *Solanum tuberosum* and demissine from *S. demissum* (Harborne 1988).

pounds used as medicine are included in Table 3; some of these are still used [e.g., atropine from *Atropa belladonna* (deadly nightshade)]. Others are used because of their antitumor activity [e.g., the diterpene taxol from *Taxus brevifolia* (western yew) and other *Taxus* species] (Heinstein & Chang 1994). Many more compounds, as yet undiscovered, may well be found to have similar effects, as long as the species that contain these do not become extinct.

Humans, however, have also found other uses for secondary metabolites, some of these in ancient history, such as taxine (from *Taxus baccata*, yew) to make arrowheads poisonous, and cicutoxin (from *Cicuta virosa*, hemlock) to poison Socrates. One of the more recent applications includes the now widespread use of pyrethrin from *Chrysanthemum cinearifolium* as an "environmentally friendly" insecticide.

The ancestors of our food plants also contain many toxic compounds, including alkaloids in *Lycopersicon* (tomato) and *Solanum* (potato) (Fig. 13). Breeding has fortunately greatly reduced the alkaloid levels in tomato and potato, so that food poisoning by potatoes, which was known until the beginning of this century, no longer occurs. Whenever wild species are used to make new crosses, however, new cultivars emerge that may produce poisonous solanine. In recent years, one cultivar was banned from the United States because it contained too much alkaloid.

Cyanogenic glycosides in *Mannihot esculenta* (cassava), *Sorghum bicolor* (millet), and *Vicia faba* (broad bean) are made harmless during food preparation. This also holds for many inhibitors of digestive enzymes (proteases, amylases), if the food is properly prepared. Eating raw or insufficiently cooked beans is an unhealthy affair, because they will still contain large amounts of secondary compounds. Some

compounds in herbs which are commonly used to flavour our food, are on the black list. These include safrole (in nutmeg, cacao, black pepper) and capsaicin (in red pepper, hot pepper), but taken in small doses they do not cause problems.

There are certainly compounds, however, that should be avoided at all costs (e.g., **aflatoxin**). This is a fungal compound produced by *Aspergillus flavus* growing on peanuts, maize, or other food plants. This compound may cause severe liver damage or cancer. Other secondary compounds have a distinctly positive effect on our health in that they reduce the risks for certain forms of cancer. These include the **flavonoids** in a so-called fiber-rich diet. These phenolics likely inhibit the production of sex hormones; hence, they appear to reduce the incidence of cancers in which the sex hormones play a role, including breast cancer and prostate cancer.

Breeding or genetically modifying genotypes of crop species that contain antiherbivore compounds is of increasing economic importance and it may lead to more environmentally friendly methods in agriculture and horticulture. The recent tendency to breed for oilseed varieties with low glucosinolate levels is an example how *not* to go about it. Such a breeding approach makes the crop more vulnerable to herbivores and makes agriculture more dependent on pesticides. It would be better instead, to aim for oilseed varieties that have their leaves well-protected against herbivores, while having a reduced level of glucosinolates only in their seeds.

There are also positive developments in breeding resistant cultivars. For example, *Leptinotarsa decemlineata* (Colorado beetle) is a well-known predator of *Solanum tuberosum* (potato) and may cause severe damage to the potato crop in North America and western Europe. A closely related species of our cultivated potato, *S. demissum*, is not

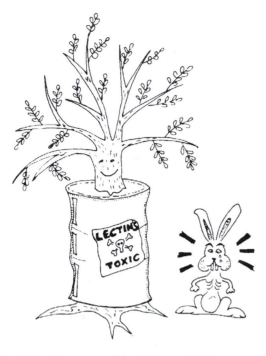

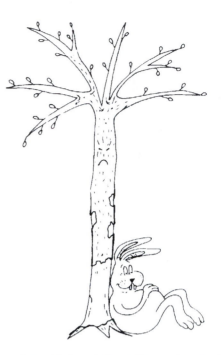

FIGURE 14. Lectins are carbohydrate-precipitating proteins. Some of these give plants protection against insects as well as vertebrates. When present in bark (e.g., in *Sambucus nigra*, elderberry) they offer good protection against rodents and deer (Peumans & Van Damme 1995). Copyright American Society of Plant Physiologists.

affected by the beetle. It contains an alkaloid (demissine) which is slightly different from solanine in *S. tuberosum* (Fig. 13). Further analysis of different alkaloids and their effect on the Colorado beetle has shown that the toxicity of alkaloids for the Colorado beetle involves three parts of the alkaloid molecule: (1) The presence of a tetrasaccharide on the 3-position, (2) the presence of a xylose molecule as one of the four sugar molecules, and (3) the absence of the double bond in solanine. Tomatine, which occurs in *Lycopersicon esculentum* (tomato) as well as in a number of *Solanum* species, contains a tetrasaccharide with xylose and lacks the double bond. It is equally effective as demissine. Such information can be used to target the breeding of resistant crop cultivars. One striking example of the application of ecophysiological information on plant–herbivore interactions is the incorporation of a gene from *Phaseolus vulgaris* (common bean), encoding an amylase inhibitor, into *Pisum sativum* (garden pea). The transgenic plants suffered considerably less from attack by pea weevils (*Bruchus pisorum*) than did the wildtype (Schroeder et al. 1995). Similarly, genes encoding a proteinase inhibitor or for lectins have been inserted.

Herbivores, however, may acclimate and possibly even adapt to an increased level of a specific protease. They do so by producing other enzymes, whose activity is not inhibited by the plant-produced inhibitor. For example, one type of α-amylase inhibitor protects seeds of the common bean (*Phaseolus vulgaris*) against predation by the cowpea weevil (*Callosobruchus maculatus*) and the azuki bean weevil (*C. chinensis*), but not against predation by the bean weevil (*Acanthoscelides obtectus*) or the Mexican bean weevil (*Zabrotus subfasciatus*). A serine protease in midgut extracts of the larvae of the Mexican bean weevil rapidly digests and inactivates α-amylase from *P. vulgaris* as well as from *P. coccineus* (scarlet runner bean), but not the α-amylase from wild common bean accessions or from *P. acutifolius* (tepary bean) (Ishimoto & Chrispeels 1996).

Lectins bind carbohydrates (by definition). As such they play a role as defense compounds (Peumans & Van Damme 1995; Fig. 14). Lectins occur in many plants, including *Sambucus nigra* (elderberry), *Hevea brasiliensis* (rubber tree), *Galanthus nivalis* (snowdrop), and *Datura stramonium* (thorn apple) (Raikhel et al. 1993). In *Sambucus nigra* lectin is located in protein bodies in the phloem parenchyma of the bark (Green et al. 1986). Some lectins are highly toxic to many animals and also offer good protection against viruses and some fungi (see chapter on effects of microbial pathogens). Although some insects appear to tolerate lectins, sucking insects like aphids are highly sensitive.

The gene encoding the lectin from *Galanthus nivalis* (snowdrop) has been linked to a promoter that ensures expression of the gene in the phloem and then inserted in rice. In this way rice contains its own insecticide to enhance its resistance to aphids and brown planthoppers. An ever-increasing number of transgenic plants with resistance genes inserted is now being produced.

4. Environmental Effects on the Production of Secondary Plant Metabolites

Although specific secondary metabolites tend to be specific for certain species, the concentration of these compounds may vary greatly, depending on environmental conditions.

4.1 Abiotic Factors

The concentration of secondary plant compounds depends on plant age as well as on abiotic environmental factors (e.g., light intensity, water stress, waterlogging, frost, pollution, and nutrient supply). For example, in *Leucaena retusa* the production of organic sulfur compounds (COS and CS_2) from crushed roots increases with increasing supply of sulfate, especially in young seedlings (Feng & Hartel 1996). The concentration of caffeine (an alkaloid) in the shoot of *Camellia sinensis* (tea) is higher when the plants are grown at high irradiance, rather than in the shade. Pine trees exposed to water stress produce less resins and are affected more by herbivorous beetles. Defoliation of spruce trees reduces the production of terpenoids, which is associated with an increased attack on their bark. In other plants, stress enhances the production of secondary metabolites. For example, in *Salix aquatica* (willow), as well as in many other species, the concentration of tannin and lignin is enhanced when plants are grown under nitrogen limitation as compared with an optimum supply (Northup et al. 1995, Waring et al. 1985). In a cross between *Festuca* and *Lolium*, the alkaloid concentration declines when plants are exposed to water stress, whereas

that in *Nicotiana tabacum* (tobacco) increases. Possibly, these effects are mediated via carbohydrate-modulated gene expression (Sect. 12.1 of the chapter on photosynthesis). Whereas genes that encode photosynthetic enzymes are down-regulated by carbohydrates, evidence is accumulating that a number of defense genes are positively modulated by carbohydrates (Koch 1996).

Two hypotheses have been advanced to explain patterns of environmental effects on plant secondary metabolites. The **carbon/nutrient balance** (CNB) hypothesis explains the level of investment in **carbon-based secondary metabolites** (i.e., those that contain only C, H, and O) as a balance between photosynthesis and growth, which, in turn, is sensitive to the carbon/nutrient balance of the plant (Bryant et al. 1983, Gershenzon 1984, Tuomi et al. 1984). According to the CNB hypothesis, plants allocate carbon preferentially to growth when nutrients are available. Low nutrient availability, however, constrains growth more than is affects photosynthesis (Sect. 5 of the chapter on growth and allocation), leading to a build-up of carbohydrates that are funneled into production of carbon-based secondary metabolites (broadly synonymous with quantitative defenses). This hypothesis explains the high levels of plant defenses typically found in plants that grow on infertile soils, and the reductions in defense that occur in response to both nutrient addition or shading. For example, tropical trees that grow on infertile soils have higher concentrations of phenolic compounds and less herbivory than do trees that grow on more fertile sites (McKey et al. 1978). The hypothesis also predicts that plants that grow more rapidly should invest less carbon in defense, as observed among seedling of the tropical tree *Cecropia peltata* (Coley 1986).

The **growth-differentiation balance** (GDB) hypothesis was advanced to explain seasonal and interannual variations in rates of production of carbon-based secondary metabolites (Loomis 1932, Lorio 1986). According to this hypothesis, growth is the primary path of carbon investment as long as conditions permit cell division and expansion; however, once water stress, photoperiod, or any other environmental factor constrains growth, cells differentiate, resin ducts form, and plants switch allocation of carbon to production of resins and other secondary metabolites. This hypothesis accounts for the greater vulnerability of pine to attack by southern pine beetle early in the growing season, and it explains why resin production increases late in the year, particularly in years when water stress constrains growth (Lorio 1986).

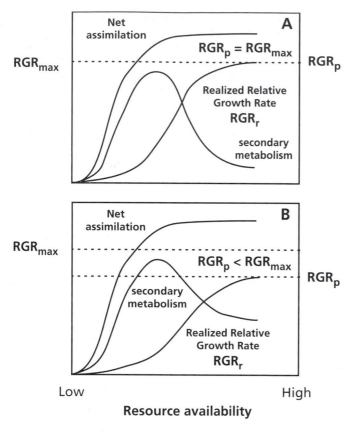

FIGURE 15. A hypothetical model that shows the realized relative growth rate (RGR_r), the net assimilation rates, and the investment of carbon in secondary plant compounds as a function of the availability of resources. Two populations (a and b) are depicted that differ with respect to the RGR that they can achieve at optimal resource availability (RGR_p). RGR_{max} in these figures denotes the maximum possible RGR of population a at the most favorable resource supply in the environment given its investment in secondary plant metabolites. Population b does not reach this RGR_{max}, due to a greater allocation to secondary metabolites (Herms & Mattson 1992). Copyright by The University of Chigaco.

Herms & Mattson (1992) have integrated these two hypotheses into an expanded version of the GDB hypothesis, which suggests that scarcity of any resource that restricts growth more than photosynthesis should enhance secondary metabolite production (Fig. 15). At extremely low resource availability, assimilation rate may be so low that maintenance respiration consumes most carbon, so that both growth and secondary metabolite production are limited (Waring & Pitman 1985). In the expanded GDB model, which is supported by recent evidence (Lambers & Poorter 1992), fast-growing species invest less carbon in secondary plant compounds than do slow-growing ones, when compared at a high resource availability. Herms and Mattson emphasized that further testing of their model is necessary. It may well be valid for one class of secondary compounds only (e.g., the quantitatively important defense compounds of a phenolic nature).

The CNB and GDB hypotheses provide a plausible mechanism for a pattern that should be strongly selected for: long-lived leaves of slow-growing plants should be well protected against pathogens and herbivores to minimize tissue loss

(Sect. 4.1 of the chapter on interactions among plants). The actual biochemical allocation to specific pathways of synthesis of individual secondary metabolites is undoubtedly regulated much more specifically than is implied by the CNB and GDB hypotheses.

4.2 Induced Defense and Communication Between Neighboring Plants

The production of secondary metabolites depends on abiotic environmental factors as well as on the presence of herbivores: **induced defense**. Physical damage of leaves (e.g., due to insect attack) often enhances the formation of tannins and reduces the quality of the leaves as a food source. This response sometimes occurs within minutes to hours (**short-term induction**), as a result of reactions among precursors already present in the leaf. For example, chewing of quaking aspen (*Populus tremuloides*) leaves causes enzymatic hydrolysis of two phenolic glycosides (salicortin to salicin and tremulacin to tremuloidin) with the release of 6-HCH

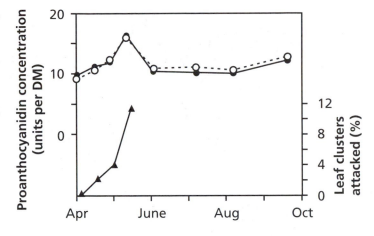

FIGURE 16. Attack of *Salix sitchensis* (Sitka willow) by western tent caterpillars (*Malacosoma californicum pluviale*) (bottom, triangles) and the accumulation of proanthocyanidins (tannins) by the attacked (solid line) and control neighboring (broken line) trees (Rhoades 1985b). Copyright Plenum Publishing Corporation.

(6-hydroxycyclohex-2-ene-1-one), which then becomes converted to phenol or catechol (potent toxins) in the gut of the insect (Clausen et al. 1989). As a result, insects cannot feed continuously on a few leaves; rather, they must constantly move among leaves, which makes them more vulnerable to predators. Short-term induced defenses are effective against those herbivores that cause the initial damage.

There are also **long-term induced defenses** produced by the next cohort of leaves after severe insect outbreaks. These serve to protect plants against catastrophic herbivory by insects with large population outbreaks. Long-term induction is typically associated with increases in phenolics or fiber, less leaf nitrogen, and often smaller leaves. Long-term induced defenses are best developed in tree populations with an evolutionary history of outbreaking insects. In some cases they are induced more strongly by insect feeding than they are by comparable amounts of physical damage, which suggests a tight evolutionary linkage with insect herbivores (Haukioja 1980, Haukioja & Neuvonen 1985). Both long- and short-term induced defenses are best developed in rapidly growing woody plants, whereas slowly growing species have higher levels of background (**constitutive**) defenses that are always present to deter herbivores (Bryant et al. 1991, Coley et al. 1985, Lambers & Poorter 1992).

There is increasing evidence that neighboring, unattacked plants respond by increasing the concentration of defensive compounds (Fig. 16), which indicates that plants communicate with each other after herbivore attack. For example, "thrashing" of the leaves of African trees to simulate browsing by kudus (*Tragelaphus strepciceros*) causes an increase in tannin and a reduction in palatibility of leaves

of untouched **neigboring trees** within 20 minutes (Fig. 17). This is probably important for nature conservation. For example, the size of nature reserves for African browsers needs to be considerably larger than one might expect on the basis of the available biomass (Hughes 1990, Van Hoven 1991).

Effects of leaf damage on neighboring trees of *Acer saccharum* (sugar maple) are not due to transfer via the roots because these effects are also found when plants are grown in separate pots. Volatile compounds supposedly play a role in this type of "**communication** between trees." Likely candidates are volatile terpenoids and phenols (Bruin et al. 1995). Jasmonic acid is probably also involved. It is still largely unknown what kind of transduction pathway enables such a molecule to alter protection against herbivores.

There is a wide variation in the extent to which plants respond to browsing with an increased concentration of phenolics. Of three South African Karoo shrubs, the deciduous species (*Osteospermum sinuatum*) is the most palatable. It contains very few polyphenols, does not enhance this level upon browsing, but has a high regrowth capacity. On the other hand, the evergreen succulent species (*Ruschia spinosa*) shows almost no regrowth after browsing, but contains the highest level of constitutive and browser-induced levels of polyphenols, condensed tannins, and protein-precipitating tannins. The evergreen sclerophyllous species (*Pteronia pallens*) shows an intermediate response in terms of regrowth capacity and browser-induced phenols. It also contains intermediate levels of phenols before browsing (Stock et al. 1993). This suggests a trade-off between allocation to (induced) defense (**avoidance**) and regrowth capacity (**tolerance**) upon attack by herbivores.

FIGURE 17. "The Kudus are coming . . . Quickly, make your leaves bitter or we'll soon be crying like the weeping wattles over there!"After damage of their leaves, due to herbivory or to thrashing, the concentration of tannins increases in several South African trees, which makes their leaves less palatable and digestible [courtesy Chris Ebersohn, reproduced with permission from Custos, vol. 13 (5), p. 16 (August 1984), p. 16].

In *Leucaena* species, damaging the roots or shoots greatly enhances the production of organic sulfur compounds (COS and CS_2), which are foul-smelling compounds that are toxic to fungi and animals (Feng & Hartel 1996). The idea to use some of these species as potential animal fodder should therefore be viewed with some scepticism.

4.3 Communication Between Plants and Their Bodyguards

Volatile compounds play a role in communication between plants and predatory mites or parasitic wasps as well as between neigboring plants. These tritrophic systems offer another fascinating example of **coevolution** in the arms race between plants and herbivores, except now there is an ally involved. The volatiles are released by leaves upon attack by herbivorous mites or caterpillars, and they attract predatory mites or parasitic wasps, respectively. These predatory mites and parasitic wasps then act as **bodyguards**. The attractants produced by plants upon attack are specific in that they are not produced upon artificially damaging the leaves or much less so. Upon attack of *Brassica oleracea* (cabbage) plants by caterpillars of *Pieris brassicae* (cabbage moth) the plant responds to a specific caterpillar enzyme (β-galactosidase) with the synthesis of a mixture of volatiles, which are highly specific for the parasitic wasp, *Cotesia glomerata*. Leaves treated with β-galactosidase from

almonds respond in a similar manner, which shows that this compound acts as an "elicitor" (Mattiacci et al. 1995). *Zea mays* (maize) plants attacked by larvae of *Spodoptera frugiperda* and *S. exigua* (Noctuidae) emit terpenoids and indole that attract a parasitic wasp, *Cotesia marginiventris*. When infested by the larvae of *Pseudaletia separata*, the maize plants emit terpenoids, indole, oximes, and nitriles that attract *Cotesia kariyai*. The production of the attractants is **systemic**. In other words, it is not restricted to the damaged parts of the plant, but also occurs in undamaged leaves; for example, this occurs in *Gossypium hirsutum* (cotton) attacked by larvae of the beet armyworm (*Spodoptera exigua*) (Röse et al. 1996).

Several crop species infested by the same herbivorous two-spotted mite, *Tetranychus urticae*, or the Noctuid larvae of *Spodoptera exigua*, become attractive to a predatory mite, *Phytoseiulus persimilis* and *Cotesia marginiventris*, respectively (Fig. 18). Each species produces its own blend of chemicals, however, that attract the bodyguards. The bodyguards can learn to distinguish between herbivore-induced volatiles emitted by different species. The attractants produced by *Phaseolus lunatus* (Lima bean) are presented in Figure 19. There is a wide genetic variation in the amount of attractants produced upon attack on which natural selection can act. In addition, there is scope for breeding efforts to exploit this aspect of ecological biochemistry.

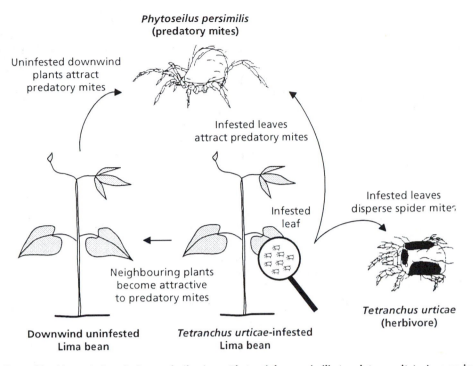

FIGURE 18. Effect of herbivore-induced plant volatiles in the tritrophic system of *Phaseolus lunatus* (Lima bean), *Tetranychus urticae* (two-spotted spider mite), and *Phytoseiulus persimilis* (predatory mite). Arrows indicate flow of chemical information (Takabayashi & Dicke 1996). Copyright Elsevier Science Ltd.

5. The Costs of Chemical Defense

The production of secondary plant compounds requires **investment** of carbon, as well as some other elements. Does this mean that a gram of biomass is more costly to produce if it contains large quantities of secondary plant compounds? This is certainly not so when costs are expressed in terms of grams of glucose required for carbon skeletons and for production of energy to produce the biomass. Approximately equal amounts of glucose are needed to produce 1 g of dry mass in slow-growing herbaceous species (which contain relatively little phenolic compounds) and fast-growing ones (Fig. 20; Sect. 5 of the chapter on plant respiration). Per gram of fresh mass or leaf area, the situation is different, but this is due to the lower water content or thinner leaves of the slow-growing species.

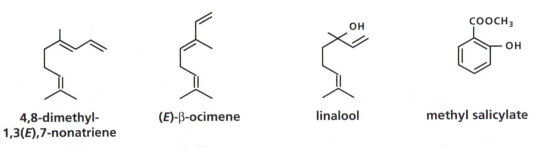

4,8-dimethyl-1,3(*E*),7-nonatriene   **(*E*)-β-ocimene**   **linalool**   **methyl salicylate**

FIGURE 19. Chemical structures of volatile components that are released by the leaves of *Phaseolus lunatus* (Lima bean) upon attack by *Tetranychus urticae* (the herbivorous two-spotted spider mite) and attract *Phytoseiulus persimilis* (a predatory mite). Methyl salicylate is a phenolic compound, the other three are monoterpenes (Dicke & Sabelis 1989). Copyright SPB Academic Publishing.

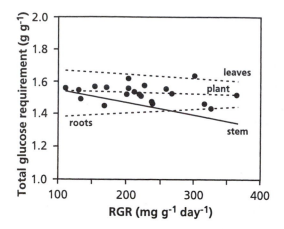

FIGURE 20. The amount of glucose required to produce biomass in slow-growing and fast-growing herbaceous species, all grown with free access to nutrients. Glucose costs include the costs for the carbon in the biomass as well as those associated with the formation of biomass, for which glucose has to be catabolized to generate ATP and NAD(P)H (Poorter & Bergkotte 1992). Copyright Blackwell Science Ltd.

There are **costs**, however, that are certainly associated with the strategy of accumulating vast quantities of secondary plant compounds. This can best be illustrated by imagining a leaf with a certain amount of protein. If half of this protein were to be replaced by lignin or tannin, then its physiological performance would probably be less. It is quite likely that its photosynthetic capacity would decline by approximately half. The higher costs of well-protected leaves, therefore, do not reflect high costs of the production of new leaves. Rather, defense is costly because it diverts resources from primary growth (an **opportunity cost**, i.e., the cost of resources that would otherwise be gained by an alternative allocation; Bloom et al. 1985, Herms & Mattson 1992), which reduces the potential growth rate of the plant.

Investment of large quantities of carbon in secondary plant compounds that reduce herbivory will lead to a greater plant fitness only when the costs of repairing the damage incurred by herbivory exceed those needed for protection. This explains why quantitatively important secondary plant compounds are more pronounced in inherently slow-growing species from unproductive environments than they are in fast-growing ones from more productive habitats. On one hand, costs select against defensive adaptations, whereas on

the other hand herbivore pressure leads to investment in defense. Defensive adaptations may then lead to offensive adaptations in animals (e.g., the coevolution of *Tyria* and *Senecio*) (Fig. 21). When costs of defense have been evaluated by comparing fitness of resistant and susceptible genotypes in the absence of herbivores or pathogens, the costs of resistance have not been large (Bergelson & Purrington 1996, Vrieling 1991), although most of these tests have been done on rapidly growing crop species where we would not expect a large cost of defense.

Two strategies may be discerned among the offensive adaptations of animals (Fig. 22). The evolutionary response upon communication between plants, which leads to the accumulation of protective compounds in neighboring plants, may be to suppress the communication or to emit countersignals. The response to the accumulation of protective compounds in plants upon recognition of a predator may be either to suppress recognition of the predator or to consume the plant quickly and so prevent protection (surprise). Inducible defenses may be counteracted by suppression of the induced defense or by decreasing the defense. Constitutive defense may be counteracted by detoxification or avoidance of the most toxic plant parts. Rhoades (1985a) termed these offensive adaptations of herbivores "stealthy" or "opportunistic."

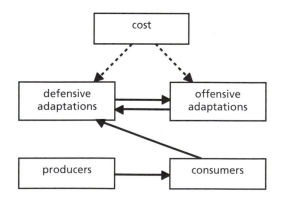

FIGURE 21. Interactions between higher plants and animals involving secondary plant compounds. Attack by herbivores leads to the evolution of protection with defense compounds (defensive adaptations in producers). At the same time, there is a selection against production of defense compounds, since it incurs a cost. Defensive adaptations in plants lead to the evolution of offensive adaptations in consumers. These offensive adaptations are selected against because they incur some costs (Rhoades 1985a). Copyright by The University of Chigaco.

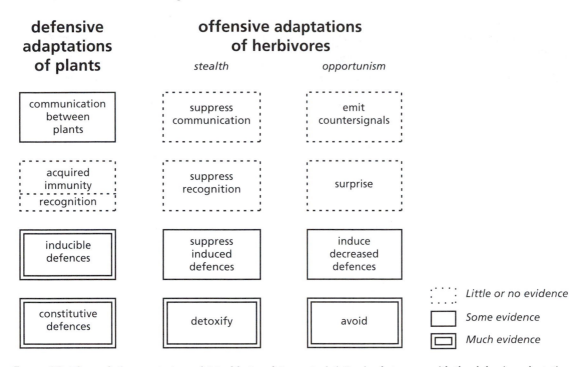

FIGURE 22. The evolutionary strategy of "stealthy" and "opportunistic" animals to cope with the defensive adaptations of plants (Rhoades 1985a). Copyright by The University of Chigaco.

6. Secondary Chemicals and Messages That Emerge from This Chapter

Plants produce a wealth of secondary plant compounds that play a pivotal role in **defense** and **communication**. We are only just beginning to understand how plants communicate with other plants, symbionts, pathogens, herbivores, and with their personal "bodyguards" via chemical signals, which are often very specific. This new area is fascinating from an ecological point of view, and it should also lead to major applications in agriculture and forestry. For example, **intercrops** can be selected that protect a crop in an environmentally friendly manner. For the intercrop to be of maximum benefit, however, intercrops should not compete to any great extent with the crop plant. It is up to ecophysiologists to help define desirable traits of an intercrop, both with respect to its secondary chemistry, and also in terms of root traits that minimize competitive ability of the intercrop or, even better, that are beneficial to the crop. Numerous pertinent traits can be found in this and preceding chapters to help to identify a desirable intercrop.

Knowledge of the chemical compounds that protect plants, preferably with full identification of the genes encoding the traits, will allow us to design crop plants that are better protected against pathogens and herbivores. Such plants will reduce the need for pesticides. We should be aware, however, that the arms race between plants and herbivores will continue, and that for every newly designed crop genotype resistant herbivores will coevolve. A thorough understanding of the intricate chemical interactions between plants and their herbivores is required to optimize the production of new crops.

References and Further Reading

Ayres, M.P. Clausen T.P., MacLean, S.F., Redman, A.M., & Reichardt, P.B. (1997) Diversity of structure and antiherbivore activity in condensed tannins. Ecology 78:1696–1712.

Bergelson, J. & Purrington, C.B. (1996) Surveying patterns in the cost of resistance in plants. Am. Nat. 148:536–558.

Bloom, A.J., Chapin III, F.S., & Mooney, H.A. (1985) Resource limitation in plants—an economic analogy. Annu. Rev. Ecol. Syst. 16:363–392.

Bruin, J., Sabelis, M.W., & Dicke, M. (1995) Do plants tap SOS signals from their infested neighbours. Trends Ecol. Evol. 10:167–170.

Bryant, J.P. & Kuropat, P.J. (1980) Selection of winter forage by subarctic browsing vertebrates: The role of plant chemistry. Annu. Rev. Ecol. Syst. 11:261–285.

Bryant, J.P., Chapin III, F.S., & Klein, D.R. (1983) Carbon/nutrient balance of boreal plants in relation to vertebrate herbivory. Oikos 40:357–368.

Bryant, J.P., Chapin III, F.S., Reichardt, P., & Clausen, T. (1985) Adaptation to resource availability as a determinant of chemical defense strategies in woody plants. In: Chemically mediated interactions plants and other organisms, G.A. Cooper-Driver, T. Swain, & E.E. Conn (eds). Plenum Press, New York, pp. 219–237.

Bryant, J.P., Tahvanainen, J., Sulkinoja, M., Julkunen-Titto, R., Reichardt, P., & Green, T. (1989) Biogeographic evidence for the evolution of chemical defense by boreal birch and willow against mammalian browsing. Am. Nat. 134:20–34.

Bryant, J.P., Heitkonig, I., Kuropat, P., & Owen-Smith, N. (1991) Effects of severe defoliation on the long-term resistance to insect attack and on leaf chemistry in six woody species of the southern African savanna. Am. Nat. 137:50–63.

Bryant, J.P., Reichardt, P.B., Clausen, T.P., Provenza, F.D., & Kuropat, P.J. (1992) woody plant-mammal interactions. In: Herbivores: their interactions with secondary plant metabolites. Vol II, Ecological and evolutionary processes, 2nd edition, G.A. Rosenthal (ed). Academic Press, San Diego, pp. 343–370.

Cates, R.G. & Orians, G.H. (1975) Successional status and the palatability of plants to generalized herbivores. Ecology 56:410–418.

Chou, C.-H. & Kuo, Y.-L. (1986) Allelopathic research of subtropical vegetation in Taiwan. III. Allelopathic exclusion of understory by Leucaena leucophylla (Lam.) de Wit. J. Chem. Ecology 12:1431–1448.

Chrispeels, M.J. & Raikhel, N.V. (1991) Lectins, lectin genes, and their role in plant defense. Plant Cell 3:1–9.

Clausen, T.P., Reichardt, P.B., Bryant, J.P., Werner, R.A., Post, K., & Frisby, K. (1989) A chemical model for short-term induction in quaking aspen (Populus tremuloides) foliage against herbivores. J. Chem. Ecol. 15:2335–2346.

Coley, P.D. (1986) Costs and benefits of defense by tannins in a neotropical tree. Oecologia 70:238–241.

Coley, P.D., Bryant, J.P., & Chapin III, F.S. (1985) Resource availability and plant anti-herbivore defense. Science 230:895–899.

De Jong, T. (1995) Why fast-growing plants do not bother about defence. Oikos 74:545–548.

Dicke, M. & Sabelis, M.W. (1989) Does it pay to advertize for body guards? In: Causes and consequences of variation in growth rate and productivity of higher plants,

H. Lambers, M.L. Cambridge, H. Konings, & T.L. Pons (eds). SPB Academic Publishing, The Hague, pp. 341–358.

Ernst, W.H.O. (1990) Ecological aspects of sulfur metabolism. In: Sulfur nutrition and sulfur assimilation in higher plants, H. Rennenberg, C. Brunold, L.J. De Kok, & I. Stulen (eds). SPB Academic Publishing, The Hague, pp. 131–144.

Ersek, T. & Kiiraly, Z. (1986) Phytoalexins: Warding-off compounds in plants? Physiol. Plant. 68:343–346.

Etzler, M.E. (1985) Plant lectins: Molecular and biological aspects. Annu. Rev. Plant Physiol. 36:209–234.

Feng, Z. & Hartel, P.G. (1996) Factors affecting production of COS and CS_2 in Leucaena and Mimosa species. Plant Soil 178:215–222.

Gershenzon, J. (1984) Changes in the levels of plant secondary metabolites under water and nutrient stress. In: Phytochemical adaptations to stress, B.N. Timmermann, C. Steelink, & F.A. Leowus (eds). Plenum Press, New York, pp. 273–320.

Greenwood, J.S., Stinissen, H.M., Peumans, W.J., & Chrispeels, M.J. (1986) Sambucus nigra agglutinin is located in protein bodies in the phloem parenchyma of the bark. Planta 167:275–278.

Hahn, M.G., Bonhoff, A., & Griesenbach, H. (1985) Quantitative localization of the phytoalexin glyceollin I in relation to fungal hyphae in soybean roots infected with Phytophtora megasperma f. sp. glycinea. Plant Physiol. 77:591–601.

Harborne, J.B. (1988) Introduction to Ecological Biochemistry. Academic Press, New York.

Hartley, M.R., Chaddock, J.A., & Bonness, M.S. (1996) The structure and function of ribosome-inactivating proteins. Trends Plant Sci. 1:254–260.

Hartmann, T., Ehmke, A., Eilert, U., von Bortsel, K., & Theurig, C. (1989) Sites of synthesis, translocation and accumulation of pyrrolizidine alkaloid N-oxides in Senecio vulgaris L. Planta 177:98–107.

Hashimoto, T. & Yamada, Y. (1994) Alkaloid biogenesis: molecular aspects. Annu. Rev. Plant Physiol. Plant Mol. Biol. 45:257–285.

Haukioja, E. (1980) On the role of plant defenses in the fluctuations of herbivore populations. Oikos 35:202–213.

Haukioja, E. & Neuvonen, S. (1985) Induced long-term resistance of birch foliage against defoliators: Defensive or incidental. Ecology 66:1303–1308.

Heinstein, P.F. & Chang, C.-J. (1994) Taxol. Annu. Rev. Plant Physiol. Plant Mol. Biol. 45:663–674.

Herms, D.A. & Mattson, W.J. (1992) The dilemma of plants: To grow or defend. Quart. Rev. Biol. 67:283–325.

Howe, H.F. & Westley, L.C. (1988) Ecological relationships of plants and animals. Oxford University Press, New York.

Hughes, S. (1990) Antelope activate the acacia's alarm system. New Scientist 127:19.

Ishimoto, M. & Chrispeels, M.J. (1996) Protective mechanism of the Mexican bean weevil against high levels of α-amylase inhibitor in the common bean. Plant Physiol. 111:393–401.

Keller, H., Blein, J.-P., Bonnet, P., & Ricci, P. (1996) Physiological and molecular characteristics of elicitin-induced systemic acquired resistance in tobacco. Plant Physiol. 110:365–376.

Kimmerer, T.W. & Potter, D.A. (1987) Nutritional quality of specific leaf tissues and selective feeding by a specialist leafminer. Oecologia 71:548–551.

Koch, K.E. (1996) Carbohydrate-modulated gene expression in plants. Annu. Rev. Plant Physiol. Plant Mol. Biol. 47:509–540.

Lambers, H. & Poorter, H. (1992) Inherent variation in growth rate between higher plants: A search for physiological causes and ecological consequences. Adv. Ecol. Res. 21:187–261.

Leung, T.-W. C., Williams, D.H., Barna, J.C.J., Foti, S., & Oelrichs, P.B. (1986) Structural studies on the peptide moroidin from Laporta moroides. Tetrahedron 42:3333–3348.

Lieberei, R., Biehl, B., Giesemann, A., & Junqueira, N.T.V. (1989) Cyanogenesis inhibits active defense reactions in plants. Plant Physiol. 90:33–36.

Loomis, W.E. (1932) Growth-differentiation balance vs. carbohydrate-nitrogen ratio. Proc. Am. Soc. Hortic. Sci. 29:240–245.

Lord, J.M. & Roberts, L.M. (1996) The intracellular transport of ricin: Why mammalian cells are killed and how Ricinus cells survive. Plant Physiol. Biochem. 34:253–261.

Lorio, P.L., Jr. (1986) Growth-differentiation balance: A basis for understanding southern pine beetle-tree interactions. For. Ecol. Manage. 14:259–273.

Mattiacci, L., Dicke, M., & Posthumus, M.A. (1995) β-galactosidase: An elicitor of herbivore-induced plant odor that attracts host-searching parasitic wasps. Proc. Natl. Acad. Sci. USA 92:2036–2040.

McKey, D., Waterman, P.G., Mbi, C.N., Gartlan, J.S., & Struhsaker, T.T. (1978) Phenolic content of vegetation in two African rain forests: Ecological implications. Science 202:61–63.

McMahon, J.M., White, W.L.B., & Sayre, R.T. (1995) Cyanogenesis in cassava (Mannihot esculenta Crantz. J. Exp. Bot. 46:731–741.

Northup, R.R., Yu, Z., Dahlgren, R.A., & Vogt, K.A. (1995) Polyphenol control of nitrogen release from pine litter. Nature 377:227–229.

Peumans, W.J. & Van Damme, E.J.M. (1995) Lectins as plant defense proteins. Plant Physiol. 109:347–352.

Pollard, A.J. & Briggs, D. (1984) Genecological studies of Urtica dioica L. III Stinging hairs and plant-herbivore interactions. New Phytol. 97:507–522

Poorter, H. & Bergkotte, M. (1992) Chemical composition of 24 wild species differing in relative growth rate. Plant Cell Environ. 15:221–229.

Poulton, J.E. (1990) Cyanogenesis in plants. Plant Physiol. 94:401–405.

Pueyo, J.J. & Delgado-Salinas, A. (1997) Presence of α-amylase inhibitor in some members of the subtribe

Phaselinae (Phaseoleae: Fabaceae). Am. J. Bot. 84:79–84.

Putnam, A. & Tang, C.-S. (eds) The science of allelopathy. John Wiley & Sons, New York.

Raikhel, N.V., Lee, H.-I., & Broekaert, W.G. (1993) Structure and function of chitin-binding proteins. Annu. Rev. Plant Physiol. Plant Mol. Biol. 44:591–615.

Rasmussen, J.A., Hejl, A.M., Einhellig, F.A., & Thomas, J.A. (1992) Sorgoleone from root exudate inhibits mitochondrial functions. J. Chem. Ecol. 18:197–207.

Rhoades, D.F. (1985a) Offensive-defensive interactions between herbivores and plants: Their relevance in herbivore population dynamics and ecological theory. Am. Nat. 125:205–238.

Rhoades, D.F. (1985b) Pheromonal communication between plants. In: Chemically mediated interactions between plants and other organisms, G.A. Cooper-Driver, T. Swain, & E.E. Conn (eds). Plenum Publishing Corporation, New York, pp. 195–218.

Rice, E.L. (1974) Allelopathy. Academic Press, New York.

Roberts, T.H., Rasmusson, A.G., & Moller, I.M. (1996) Platanetin and 7-iodo-acridone-4-carboxylic acid are not specific inhibitors of respiratory NAD(P)H dehydrogenases in potato tuber mitochondria. Physiol. Plant. 96:263–267.

Röse, U.S.R., Manukian, A., Heath, R.R., & Tumlinson, J.H. (1996) Volatile semiochemicals released from undamaged cotton leaves. A systemic response of living plants to caterpillar damage. Plant Physiol. 111:487–495.

Schroeder, H.E., Gollasch, S., Moore, A., Tabe, L.M., Craig, S., Hardie, D.C., Chrispeels, M.J., Spences, D., & Higgins, T.J.V. (1995) Bean α-amylase inhibitor confers resistance to the pea weevil (Bruchus pisorum) in transgenic peas (Pisum sativum L.). Plant Physiol. 107:1233–1239.

Schuler, M.A. (1996) The role of cytochrome P450 monooxygenase in plant-insect interactions. Plant Physiol. 112:1411–1419.

Selmar, D., Liebererei, R., & Biehl, B. (1988) Mobilization and utilization of cyanogenic glycosides. Plant Physiol. 86:711–716.

Selmar, D., Grocholewski, S., & Seigler, D.S. (1990) Cyanogenic lipids. Utilization during seedling development of Ungnadia speciosa. Plant Physiol. 93:631–636.

Stock, W.D., Le Roux, D., & Van der Heyden, F. (1993) Regrowth and tannin production in woody and succulent karoo shrubs in response to simulated browsing. Oecologia 96:562–568.

Tahvanainen, J., Julkumen,-Tiitto, R., & Kettunen, J. (1985) Phenolic glycosides govern the food selection pattern of willow feeding beetles. Oecologia 67:52–56.

Takabayashi, J. & Dicke, M. (1996) Plant-carnivore mutualism through herbivore-induced carnivore attractants. Trends Plant Sci. 1:109–113.

Tuomi, J., Niemela, P., Haukioja, E., & Neuvonen, S. (1984) Nutrient stress: an explanation for plant anti-herbivore responses to defoliation. Oecologia 61:208–210.

Van Hoven, W. (1991) Mortalities in kudu (*Tragelaphus strepciceros*) populations related to chemical defence in trees. J. Afr. Zool. 105:141–145.

Vrieling, K. (1991) Costs and benefits of alkaloids of *Senecio jacobaea* L. PhD Thesis, University of Leiden.

Waring, R.H. & Pitman, G.B. (1985) Modifying lodgepole pine stands to change susceptibility to mountain pine beetle attack. Ecology 66:889–897.

Waring, R.H., McDonald, A.J.S., Larsson, S., Ericsson, T., Wiren, A., Arwidsson, E., Ericsson, A., & Lohammar, T. (1985) Differences in chemical composition of plants grown at constant relative growth rates with stable mineral nutrition. Oecologia 66:157–160.

Yip, W.-K. & Yang, S.F. (1988) Cyanide metabolism in relation to ethylene production in plant tissues. Plant Physiol. 88:473–476.

9C. Effects of Microbial Pathogens

1. Introduction

Plants frequently encounter potentially pathogenic fungi, bacteria, and viruses, yet disease results from relatively few of these exposures. In many cases there is no obvious trace of its occurrence, and the microorganism fails to establish itself due to a low pathogenicity or highly effective plant defense mechanisms. Other encounters leave evidence of an intense plant–microbe interaction that results in the arrest of pathogen development after attempted colonization. In these cases plant tissues often display activated defense functions that produce antimicrobial compounds (phytoalexins), enzymes, and structural reinforcement that may limit pathogen growth (Delaney 1997).

2. Constitutive Antimicrobial Defense Compounds

Plants produce a wide range of compounds with an **antimicrobial** effect (**phytoanticipins**) (VanEtten et al. 1994). Some of these have already been discussed in Section 2.1 of the chapter on ecological biochemistry (e.g., alkaloids, flavonoids, and lignin). **Saponins** are plant glycosides that derive their name from their soaplike properties. A common species that contains saponins is *Saponaria officinalis* (soapwort), which used to be grown near wool mills; the soapy extracts from its leaves and

roots were used for washing wool. Saponins consist of triterpenoid, steroid or steroidal glyco-alkaloid molecules that bear one or more sugar chains (Fig. 1). Saponins have been implicated as preformed determinants of resistance to fungal attack. For example, wounding of *Avena strigosa* (oat) plant tissue, which results from pathogen attack, causes a breakdown of compartmentalization, which allows an enzyme to contact the saponin avenacoside B. The C-26 glucose is then removed by hydrolysis, yielding a fungitoxic compound. The fungitoxic compound causes loss of membrane integrity (Osbourn 1996).

Lipid-transfer proteins, which were discussed in Section 3.5 of the chapter on radiation and temperature, may also be active as plant-defense proteins. Lipid-transfer proteins from *Raphanus sativus* (radish), *Hordeum vulgare* (barley), *Spinacia oleracea* (spinach), *Zea mays* (maize), and several other species are active against several pathogens, with varying degrees of specificity (Kader 1996). In addition, in *Allium cepa* (onion) some of the genes encoding antimicrobial proteins with lipid-transfer activity are up-regulated in response to infection by fungal pathogens (Kader 1997). The name lipid-transfer protein is rather unfortunate in that the transfer of lipids is unlikely to be the (sole) role of these proteins in vivo; a protein from the seeds of *Allium cepa* with homology to lipid-transfer proteins has a strong antimicrobial activity, without being able to transfer lipids. The antimicrobial activity of lipid-transfer proteins is unlikely to be linked with their ability to transfer lipids in vitro

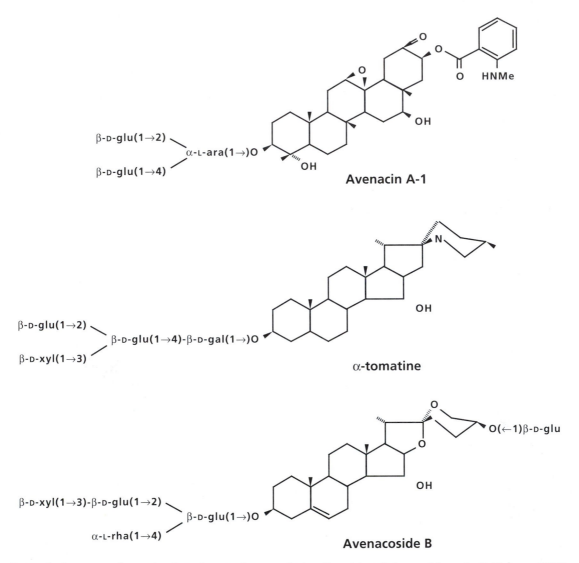

β-D-glu(1→2)
                α-L-ara(1→)O
β-D-glu(1→4)

Avenacin A-1

β-D-glu(1→2)
                β-D-glu(1→4)-β-D-gal(1→)O
β-D-xyl(1→3)

α-tomatine

O(←1)β-D-glu

β-D-xyl(1→3)-β-D-glu(1→2)
                            β-D-glu(1→)O
α-L-rha(1→4)

Avenacoside B

FIGURE 1. Structures of saponins from *Lycopersicon esculentum* (tomato) and *Avena strigosa* (oat) (Osbourn 1996). Copyright Elsevier Science Ltd.

(Cammue et al. 1995). It is as yet unknown how lipid-transfer proteins inhibit the growth of pathogens.

Lectins [i.e., defense compounds against herbivores (Sects. 3.4 and 3.5 of the chapter on ecological biochemistry)] are also effective against pathogens. For example, the lectin in rhizomes of *Urtica dioica* (stinging nettle) hydrolyzes fungal cell walls (Raikhel et al. 1993).

The constitutive defense against microorganisms obviously incurs a **cost**. When a range of cultivars of *Raphanus sativus* (radish), which differ widely in their sensitivity to *Fusarium oxysporum* (fungal wilt disease) are compared, the most resistant ones have

the lowest relative growth rate and vice versa (Fig. 2). The exact nature of the constitutive defense is unknown, but it is probably not based solely on the presence of glucosinolates because these defense compounds tend to be present only in low amounts (cf. Sect. 3.1 of the chapter on ecological biochemistry). Leaves of slow-growing, resistant radish cultivars contain more cell-wall material. The roots of slow-growing, resistant cultivars, however, have significantly less cell-wall material and their higher biomass density is due to more cytoplasmic elements (proteins). It has been speculated that this higher protein concentration accounts for the rapid and adequate resistance reaction, be it at the ex-

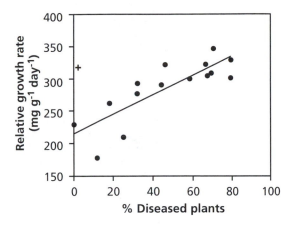

FIGURE 2. **Correlation between the relative growth rate (RGR) of *Raphanus sativus* (radish) grown in absence of pathogens and resistance level to *Fusarium oxysporum* in 15 cultivars. Each symbol refers to a different cultivar; RGR and resistance level correlate significantly (Hoffland et al. 1996a). Copyright Blackwell Science Ltd.**

pense of greater construction and turnover costs (Hoffland et al. 1996a).

In some phytopathogenic fungi, **detoxifying enzymes** have evolved that break through the plant's constitutive or induced defense against fungal attack (VanEtten et al. 1995). A number of fungi avoid the toxicity of plant saponins. Some do so by growing only in extracellular plant compartments. Some fungi that infect *Lycopersicon esculentum* (tomato) **lower the pH** at the infection sites to levels at which the saponin in tomato (α-tomatin; Fig. 1) has no effect on membrane integrity. More important mechanisms involve a change in **membrane composition** of the fungus and the production of **saponin-detoxifying enzymes** (Osbourn 1996).

3. The Plant's Response to Attack by Microorganisms

Plants that are resistant to microbial pathogen attack may show a **hypersensitive response** (Stakman 1915): membrane damage, necrosis, and collapse of cells. This "suicidal" response is often confined to individual penetrated cells. It is considered a sacrifice of locally infected tissue (sometimes only one or a few cells) to protect against the spread of the pathogen into healthy tissue. Mutants that spontaneously form patches of dead tissue (necrosis) occur in many plant species. Further

analysis of these mutants shows that the hypersensitive response is both caused by the production of toxic compounds by the plant or pathogen and results, partly, from genetically programmed cell death (Jones & Dangl 1996). The hypersensitive response differs from cell death that spreads beyond the point of infection, which follows from the interaction of a susceptible plant and a virulent pathogen. In this interaction, cell death does not effectively prevent pathogen multiplication or spread (He 1996).

The hypersensitive response often enhances the production of **active oxygen species** (O_2^-, H_2O_2), which are generated by a signaling pathway similar to that employed by mammalian neutrophils during immune responses (Mehdy et al. 1996). Active oxygen species are involved in cross-linking of cell-wall proteins, rendering these more resistant to attack by enzymes from the pathogen. The active oxygen species are also thought to be toxic for pathogens. In addition they may act as "second messengers" in the induction of defense genes (Boller 1995). These active oxygen species might be the cause of upregulation of the gene encoding the alternative oxidase (Fig. 3). Up-regulation of this gene greatly enhances the rate of cyanide-resistant respiration of the infected plant tissue (Sect. 4.8 of the chapter on plant respiration). Increased activity of the alternative path presumably allows a rapid flux through the oxidative pentose phosphate pathway and NADP-malic enzyme, thus producing carbon skeletons and NADPH that are required in the defense reaction (Simons & Lambers 1998; Fig. 4).

In the hypersensitive response of a resistant plant to an avirulent pathogen, the cells surrounding the site of entry modify their cell walls so that they are less digestible by microbial enzymes. In this way they form a physical barrier to the microbial pathogens. They also produce **phytoalexins** (i.e., low-molecular-mass antibiotics that are not found in uninfected plants). The chemical nature of phytoalexins is extremely variable (Fig. 5); closely related species often have phytoalexins with a similar structure. Microorganisms, or components thereof (**elicitors**), induce the formation of phytoalexins (Boller 1995). Numerous other compounds in phytopathogenic microorganisms (e.g., carbohydrates and lipids) may also cause nonspecific production of phytoalexins. In addition, cell-wall components (e.g., glucans or glucomannans) may elicit rapid synthesis of phytoalexins in resistant cultivars.

Now that phytoalexins have been introduced, we will stress two points. First, accumulation of a

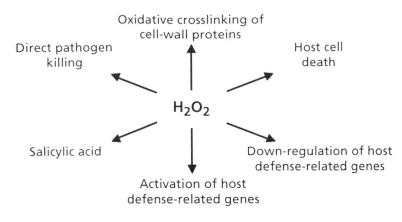

FIGURE 3. A central role for hydrogen peroxide in defense responses of plants to microbial pathogen infection. The responses shown occur in different plant species and may not all occur within a given species (Mehdy et al. 1996). Copyright Physiologia Plantarum.

specific compound upon attack does not prove that this compound is involved in resistance. Rather, accumulation may be a side reaction that has nothing to do with the actual resistance mechanism. To prove that a compound is involved in a resistance

mechanism may require mutants that are unable to make the putative defense compound. Some 30 years ago, Chamberlain and Paxton (1968) demonstrated that the stems of a cultivar of *Glycine max* (soybean), which is susceptible for the fungus

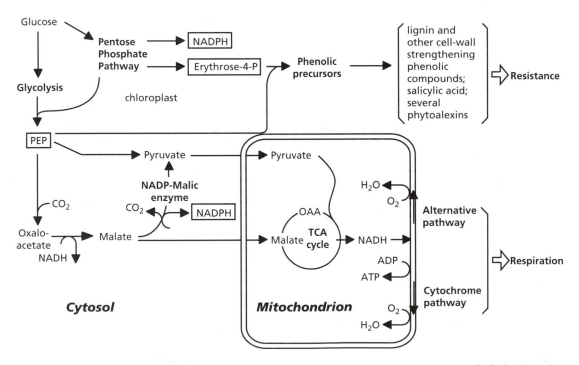

FIGURE 4. Major metabolic pathways involved in the plant's resistance response to pathogens and its association with respiration. Plant defense requires an increased production of erythrose-4-phosphate for numerous phenolic precursors. Erythrose-4-phosphate is produced in the oxidative pentose pathway, which also generates the NADPH that is required for the biosynthesis of, for example, lignin and some phytoalexins. Additional NADPH is produced by NADP-malic enzyme, which decarboxylates malate to pyruvate. Increased activity of these reactions enhances the production of pyruvate, which is postulated to require an increased activity of the alternative path. Up-regulation of the gene encoding the alternative oxidase may be triggered by accumulation of reactive oxygen species (after Simons & Lambers 1998).

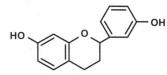

4'7-dihydroxyflavan

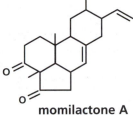

orchinol

momilactone A

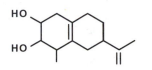

rishitin

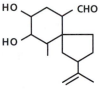

4-hydroxylubimin

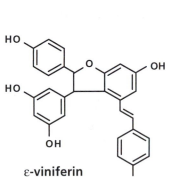

capsidiol

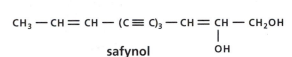

safynol

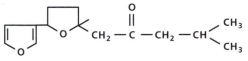

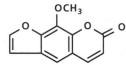

ipomeamarone

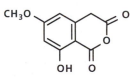

ε-viniferin

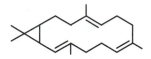

casbene

xanthotoxin

6-methoxymellein

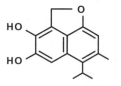

desoxyhemigossypol

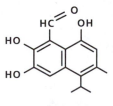

hemigossypol

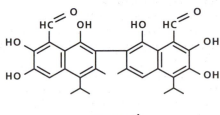

gossypol

FIGURE 5. Some examples of phytoalexins in different plant species (Bell 1981). With permission, from the Annual Review of Plant Physiology, Vol. 32, copyright 1981, by Annual Reviews Inc.

TABLE 1. The effect of a number of chromatogram fractions, isolated from a resistant cultivar of *Glycine max* (soybean), as protectant against a fungus, *Phytophthora megasperma*.

R_F number of chromatogram fraction	Plants killed by the fungus (% of control)
0.0	75
0.0–0.1	48
0.1–0.2	44
0.2–0.3	24
0.3–0.4	31
0.4–0.5	37
0.5–0.6	58
0.6–0.7	65
0.7–0.8	34
0.8–0.9	24
0.9–1.0	34

Source: Chamberlain & Paxton 1968.

Phytophthora megasperma, can become resistant upon addition of a phytoalexin isolated from a resistant cultivar (Table 1). Inhibition of phenylalanine ammonia lyase, which is a key enzyme in the synthesis of isoflavanoids, decreases the concentration of phytoalexins and increases the growth of the infecting fungus. As in the arms race between plants and herbivores, pathogens have also evolved that detoxify phytoalexins. For example, isolates of the phytopathogenic fungus *Nectria haematococca* produce pisatine demethylase that detoxifies the toxic phytoalexin pisatine in *Pisum sativum* (pea) (VanEtten et al. 1995).

The second point we stress is that synthesis of phytoalexins is only one of a range of mechanisms involved in combating the pathogen. Cell-wall reinforcement and production of hydrolytic enzymes (e.g., chitinases) are also part of the defense response (Vierheilig et al. 1993). The proteins induced upon pathogen attack are generally described as **pathogenesis-related (PR)** proteins (Van Loon 1985, Van Loon et al. 1994). Some of these PR proteins confer disease resistance (Broglie et al. 1991) and inhibit fungal growth in vitro as well as in vivo (Boller 1995).

After attack of a resistant host by an avirulent pathogen ("incompatible interaction"), the enzymes required for the synthesis of phytoalexins are first produced de novo and then the phytoalexins accumulate. The chemicals are produced in living plant cells and may well lead to the death of these cells, due to their toxicity for the host plant as well as for the pathogenic microorganism.

Some phytoalexins (e.g., glyceollin from *Glycine max*, soybean) are specific inhibitors of complex I of the mitochondrial electron-transport chain (cf. Sect. 2.2.1 of the chapter on plant respiration). With a specific radioimmunoassay for glyceollin, phytoalexins can be found in roots of *Glycine max* (soybean) as soon as 2 hours after inoculation with the fungus *Phytophthora megasperma*. The phytoalexins accumulate first at the site of inoculation. Glyceollin predominantly accumulates in epidermal cells, but it is also found in the cells of the cortex. Glyceollin accumulation precedes the occurrence of the fungal hyphae at the site. The glyceollin concentration declines toward the edge of the infection zone.

In several pathosystems, the hypersensitive response to an avirulent pathogen as well as the response to a virulent pathogen are accompanied by the induction of **systemic acquired resistance** to infection by other pathogens (Ryals et al. 1994). It involves the accumulation of **salicylic acid** and activation of genes encoding PR proteins (Linthorst 1991, Van Loon 1985). Methyl salicylate, which is a volatile liquid known as oil of wintergreen, is produced from salicylic acid by a number of plant species. It is a major volatile compound released by *Nicotiana tabacum* (tobacco) inoculated with tobacco mosaic virus. Methyl salicylate may act as an airborne signal that activates disease resistance and the expression of defense-related genes in neigboring plants and in the healthy tissues of the infected plants (Shulaev et al. 1997).

Resistance genes are also activated by exposure to ethylene or the vapor of methyl jasmonic acid [e.g., in *Lycopersicon esculentum* (tomato), *Medicago sativa* (alfalfa), or *Nicotiana tabacum* (tobacco)]. Methyl jasmonic acid from either a synthetic solution or from undamaged twigs of *Artemisia tridentata* (sagebush) is equally effective. This compound is a common secondary metabolite that often occurs in higher levels in damaged plants (Bruin et al. 1995). Plants like *Artemisia* are promising for use as an "intercrop" (i.e., a plant used in combination with a crop plant to protect the crop against pests in an environmentally friendly way). "Intercropping" has been proposed as a method to contribute to pest control.

Plants can also become resistant by exposure to nonpathogenic root-colonizing bacteria (e.g., fluorescent *Pseudomonas* sp. in *Dianthus caryophyllus*) (Van Peer et al. 1991) and *Raphanus sativus* (radish) (Hoffland et al. 1996b) and *Serratia plymuthica* in *Cucumis sativus* (cucumber) (Wei et al. 1991). In this **induced systemic resistance**, however, salicylic acid does not play a role (Pieterse et al. 1996).

4. Messages from One Organism to Another

Plants continually receive messages from their environment, including chemical messages (elicitors) released by pathogenic and nonpathogenic microorganisms. Resistant plants respond to these messages by defending themselves. This involves sacrificing a small number of cells in a programmed manner and both a physical and chemical defense of the surviving cells. Upon attack by pathogens, both resistant and surviving sensitive plants acquire greater resistance to subsequent attack, be it by the same or by a different pathogen. In recent years remarkable surveillance mechanisms have been discovered that have evolved in plants to recognize microbial factors and combat pathogenic microbes. The recent discovery of resistance that is induced by nonpathogenic rhizobacteria, which is a process of plant immunization to diseases, is receiving increasing attention. It may help us protect our crops against pathogens in an environmentally friendly manner.

References and Further Reading

Bell, A.A. (1981) Biochemical mechanisms of disease resistance. Annu. Rev. Plant Physiol. 32:21–81.

Benhamou, N., Kloepper, J.W., Quadt-Hallman, A., & Tuzun, S. (1996) Induction of defense-related ultrastructural modifications in pea root tissues inoculated with endophytic bacteria. Plant Physiol. 112: 919–929.

Boller, T. (1995) Chemoperception of microbial signals in plant cells. Annu. Rev. Plant Physiol. Plant Mol. Biol. 46:189–214.

Broglie, K., Holliday, M., Cressman, R., Riddle, P., Knowtown, S., Mauvais, C.J., & Broglie, R. (1991) Transgenic plants with enhanced resistance to the fungal pathogen Rhizoctonia solani. Science 254:1195–1197.

Bruin, J., Sabelis, M.W., & Dicke, M. (1995) Do plants tap SOS signals from their infested neighbours. Trends Ecol. Evol. 10:167–170.

Cammue, B.P.A., Thevissen, K., Hendriks, M., Eggermont, K., Goderis, I.J., Proots, P., Van Damme, J., Osborn, R.P., Guerbette, F., Kader, J.-K., & Broekaert, W.F. (1995) A potent antimicrobial protein from onion seeds showing sequence homology to plant lipid transfer proteins. Plant Physiol. 109:445–455.

Chamberlain, D.W. & Paxton, J.D. (1968) Protection of soybean plants by phytoalexins. Phytopathology 58:1349–1350.

Delaney, T.P. (1997) Genetic dissection of acquired resistance to disease. Plant Physiol. 113:5–12.

He, S.Y. (1996) Elicitation of plant hypersensitive response by bacteria. Plant Physiol. 112:865–869.

Hoffland, E., Niemann, G.J., Van Pelt, J.A., Pureveen, J.B.M., Eijkel, G.B., Boon, J.J., & Lambers, H. (1996a) Relative growth rate correlates negatively with pathogen resistance in radish. The role of plant chemistry. Plant Cell Environ. 19:1281–1290.

Hoffland, E., Hakulinen, I., & Van Pelt, J.A. (1996b) Comparison of systemic resistance induced by avirulent and nonpathogenic Pseudomonas species. Phytopathology 86:757–762.

Jones, A.M. & Dangl, J.L. (1996) Logjam at the Styx: Programmed cell death in plants. Trend Plant Sci. 1:114–119.

Kader, J.-C. (1996) Lipid-transfer proteins in plants. Annu. Rev. Plant Physiol. Plant Mol. Biol. 47:627–654.

Kader, J.-C. (1997) Lipid-transfer proteins: A puzzling family of plant proteins. Trends Plant Sci. 2:66–70.

Linthorst, H. (1991) Pathogenesis-related proteins of plants. Crit. Rev. Plant Sci. 10:123–150.

Mehdy, M.C., Sharma, Y.K., Sathasivan, K., & Bays, N.W. (1996) The role of activated oxygen species in plant disease resistance. Physiol. Plant. 98:365–374.

Osbourn, A. (1996) Saponins and plant defence—a soap story. Trend Plant Sci. 1:4–9.

Pieterse, C.M.J., Van Wees, S.C.M., Hoffland, E., Van Pelt, J.A., & Van Loon, L.C. (1996) Systemic resistance in Arabidopsis induced by biocontrol bacteria is independent of salicylic acid accumulation and pahogenesis-related gene expression. Plant Cell 8:1225–1237.

Raikhel, N.V., Lee, H.-I., & Broekaert, W.F. (1993) Structure and function of chitin-binding proteins. Annu. Rev. Plant Physiol. Plant Mol. Biol. 44:591–615.

Ryals, J., Uknes, S., & Ward, E. (1994) Systemic acquired resistance. Plant Physiol. 104:1109–1112.

Shulaev, V., Silverman, P., & Raskin, I. (1997) Airborne signalling by methyl salicylate in plant pathogen resistance. Nature 385:718–721.

Simons, B.H. & Lambers, H. (1998) The alternative oxidase: is it a respiratory pathway allowing a plant to cope with stress? In: Plant responses to environmental stresses: From phytohormones to genome reorganization, H.R. Lerner (ed). Plenum Publishing Corporation, New York, in press.

Stakman, E.C. (1915) Relation between Puccinia graminis f.sp. tritici and plants highly resistant to its attack. J. Agric. Res. 4:195–199.

VanEtten, H.D., Sandrock, R.W., Wasmnan, C.C., Soby, S.D., McCluskey, K., & Wang, P. (1994) Detoxification of phytoanticipins and phytoalexins by phytopathogenic fungi. Can. J. Bot. 73 (Suppl. 1):S518–S525.

Van Loon, L.C. (1985) Pathogenesis-related proteins. Plant Mol. Biol. 4:111–116

Van Loon, L.C., Pierpoint, W.S., Boller, T., & Conejero, V. (1994) Recommendations for naming plant pathogenesis-related proteins. Plant Mol. Biol. Rep. 12:245–264.

Van Peer, R., Niemann, G.J., & Schippers, B. (1991) Induced resistance and phytoalexin accumulation in biological control of Fusarium wilt of carnation by Pseudomonas sp. strain WCS417r. Phytopathology 81:728–734.

Vierheilig, H., Alt, M., Neuhaus, J.-M., Boller, T., & Wiemken, A. (1993) Colonization of transgenic *Nicotiana sylvestris* plants, expressing different forms of *Nicotiana tabacum* chitinase, by the root pathogen *Rhizoctonia solani* and by the mycorrhizal symbiont *Glomus mosseae*. Mol. Plant-Microbe Int. 6:261–264.

Wei, G., Kloepper, J.W., & Tuzun, S. (1991) Induction of systemic resistance of cucumber to *Colletrotrichium orbiculate* by selected strains of plant growth-promiting rhizobacteria. Phytopathology 81:1508–1512.

9D. Parasitic Associations

1. Introduction

We have so far dealt with **autotrophic** plants that assimilate CO_2 from the atmosphere into complex organic molecules and acquire nutrients and water from the rhizosphere. Mutants of such plants that are unable to photosynthesize or absorb nutrients or that are seriously disturbed in their water relations (e.g., because they do not close their stomata when water-stressed) may survive under protected laboratory conditions, but not in nature. There are also fascinating higher plant species, however, that lack the capacity to assimilate sufficient CO_2 to sustain their growth and which cannot absorb nutrients and water from the rhizosphere in sufficient quantities to allow them to reproduce. These plants are parasitic and rely on a host plant to provide them with the materials they cannot acquire from their abiotic environment.

More than 3000 plant species distributed over 18 families (predominantly Angiospermae) rely on a parasitic association with a host plant for their mineral nutrition, water uptake, and/or carbon supply (Table 1). Some of these species (e.g., *Striga* spp., *Orobanche ramosa*, and *Cuscuta* spp.) are economically important pests, which result in large yield losses of crop plants, especially in African and Mediterranean countries. Ecologically, parasitic plants fill a fascinating niche in their exploitation of other plants to acquire scarcely available resources.

Parasitic angiosperms are generally divided into **holoparasites** and **hemiparasites** (Table 1).

Holoparasites are invariably **obligate** parasites. That is, they totally depend on their host for the completion of their life cycle. They do not contain appreciable amounts of chlorophyll and lack the capacity to photosynthesize; their CO_2-compensation point may be as high as 200 Pa (Dawson et al. 1994), which is extremely high compared with autotrophic plants (cf. Sect. 2.2.1 of the chapter on photosynthesis). Holoparasites also lack the capacity to assimilate inorganic nitrogen. Hemiparasites may be either **facultative** or **obligate** parasites. They contain chlorophyll and have some photosynthetic capacity, but they also depend on their host for the supply of water and nutrients. The distinction between holoparasites and hemiparasites is not sharp (e.g., *Striga* species are considered hemiparasites, but they have very little chlorophyll and show only a limited photosynthetic capacity) (Table 2).

Parasitic angiosperms are further subdivided into **stem parasites**, such as the holoparasitic *Cuscuta* (dodder) and the hemiparasitic *Viscum* (mistletoe), and **root parasites**, such as the holoparasitic *Orobanche* (broomrape) and the hemiparasitic *Striga* (witchweed) (Stewart & Press 1990).

Parasites may be small herbaceous species [e.g., *Rhinantus sclerotinus* (yellow rattle) and *Melampyrum pratense* (cow-wheat)] or large trees [e.g., *Nuytsia floribunda* (Christmas tree) and *Exocarpus cupressiformis* (cherry ballart) in Australia]. Some parasites are highly host-specific, whereas others parasitize a wide range of species.

TABLE 1. Taxonomic survey of families of parasitic vascular plants.

Subclass • Family	Type of parasitism	Representative genus
Angiospermae		
Magnoliidae		
• Lauraceae	hemiparasitism	*Cassytha*
Rosidae		
• Balanophoraceae	holoparasitism	*Balanophora*
		Cynomorium
• Eremolepidaceae	hemiparasitism	*Eremolepis*
• Hydnoraceae	holoparasitism	*Hydnora*
• Krameriaceae	hemiparasitism	*Krameria*
• Loranthaceae	hemiparasitism	*Loranthus*
		Nuytsia
		Tapinanthus
• Misodendraceae	hemiparasitism	*Misodendrum*
• Olacaceae	hemiparasitism	*Olax*
• Opiliaceae	hemiparasitism	*Cansjera*
• Rafflesiaceae	holoparasitism	*Rafflesia*
• Santalaceae	both	*Dendrotrophe*
		Exocarpus
		Santalum
• Viscaceae	both	*Amyema*
		Phoradendron
		Viscum
Asteridae		
• Cuscutaceae	holoparasitism	*Cuscuta*
• Lennoaceae	holoparasitism	*Lennoa*
• Orobanchaceae	holoparasitism	*Conopholis*
		Orobanche
• Scrophulariaceae	both	*Alectra*
		Melampyrum
		Odontites
		Rhinanthus
		Striga
Gymnospermae		
• Podocarpaceae	hemiparasitism	*Podocarpus*
		Parasitaxis

Source: After Atsatt et al. 1983, Kuijt 1969.

2. Growth and Development

2.1 Seed Germination

Many parasitic angiosperms have small seeds with a hard seed coat and remain viable for many years. The seeds have very small reserves so that the seedlings run the risk of dying if they do not quickly find a host to attach to. Germination of *Cuscuta* seeds is completely independent of its host (Dawson et al. 1994), but many other species require a **chemical signal** from their host to trigger germination, which increases their chances to sur-

vive. The first naturally occurring stimulant was identified from *Gossypium hirsutum* (cotton), which stimulates germination of *Striga*, although cotton is not a host for *Striga* itself. The stimulant is a sesquiterpene, which is given the name strigol (Fig. 1); it is active in concentrations as low as 10^{-12} M in the soil solution. A second compound has been isolated from the root exudate of *Vigna unguiculata* (cowpea), which is a host for both *Striga* and *Alectra*. A third stimulant has been isolated from the roots of *Sorghum bicolor* (broom-corn), which is a host for *Striga*. It is a hydroquinone: 2-hydroxy-5-methoxy-3-[(8'Z,11'Z)-8',11',14'-pentadecatriene]-p-hydroquinone (Fate et al. 1990) (Fig. 1). These three stimulants have widely differing structures. When seeds of *Striga asiatica* are placed in agar at a distance of about 5 mm from the root surface of *Sorghum bicolor*, germination takes place. No germination occurs at a distance of 10 mm or more. Germination occurs only after a minimum of 5 hours exposure to 1 mM hydroquinone.

The stimulant from *Sorghum bicolor* enhances the synthesis of the phytohormone **ethylene**, which is an absolute requirement for the germination of the *Striga* seeds. Inhibition of the action or synthesis of ethylene prevents the effect of the germination stimulant, whereas its action can be substituted by ethylene (Babiker et al. 1993, Logan & Stewart 1991).

The release of strigol by the roots of cotton, which is not a host itself, has encouraged the use of this species as a "trap crop" for *Orobanche* or *Striga*. [A "trap crop" is used to stimulate the germination of as many seeds as possible, so that the problems for the next crop, which can act as a host, are mini-

TABLE 2. Some characteristics of *Striga hermonthica*, an obligate root hemiparasite, in comparison with *Antirrhinum majus*, a related nonparasitic species.

Trait	*Striga*	*Antirrhinum*
Stomatal frequency mm^{-2}		
adaxial leaf surface	114	36
abaxial leaf surface	192	132
stem	24	28
Transpiration (mmol m^{-2} s^{-1})	8.5	5.7
Chlorophyll a + b content (g m^{-2})	2.6	7.2
Soluble protein content (g m^{-2})	12	23
Photosynthesis		
per m^2 leaf area (µmol s^{-1})	2.5	15.0
per g chlorophyll (µmol s^{-1})	1.0	2.6
Water-use efficiency	0.3	2.9
[mmol CO$_2$ mol^{-1} (H$_2$O)]		

Source: Shal et al. 1997.

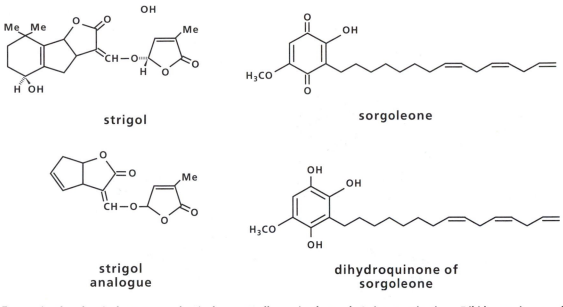

strigol

sorgoleone

strigol
analogue

dihydroquinone of
sorgoleone

FIGURE 1. The chemical structure of strigol, a naturally occurring sesquiterpene from *Gossypium hirsutum* (cotton) that acts as a germination stimulant for *Striga*, and a synthetic analogue (Harborne 1982). Sorgoleone, a natural stimulant of *Striga* germination. Dihidroxyquinone of sergoleone is an allelopathic compound released from several *Sorghum* species. It is rapidly oxidized to the more stable sorgoleone (Einhellig & Souza 1992).

mized.] If strigol is abundant in the soil during seed ripening, then it does not stimulate germination in the normal concentration range. Rather, a much higher concentration of strigol is then required to allow germination. This may be a mechanism of avoiding germination at the end of the season, when the concentration of root exudates may be high.

Analogues of strigol and numerous other, unrelated compounds have been synthesized and tested for their capacity to stimulate germination. Such compounds are potentially useful to reduce the economic problems that parasites cause to crops. Many of the chemicals that act as triggers for germination or haustorium formation are either allelochemicals or related to phytoalexins. It is interesting that the stimulant from *Sorghum bicolor* readily oxidizes to a more stable quinone (sorgoleone), which strongly inhibits the growth of neighboring weeds (Einhellig & Souza 1992).

2.2 Haustoria Formation

All parasitic species, with the exception of members of the Rafflesiaceae, have a **haustorium**, which is an organ that functions in attachment, penetration, and transfer of water and solutes. Most parasitic plants will only develop a functional haustorium in

the presence of a chemical signal derived from the host, which is quite different from the signal that triggers germination. For example, in *Striga asiatica* haustorium formation proceeds only when the signal 2,6-dimethoxy-p-benzoquinone is produced by the host roots, in response to an enzyme from the parasite (Smith et al. 1990). The chemical signals are probably not released into the root environment, but they are tightly bound to cells walls. The compounds, again with very different structures, are biologically active in the concentration range of 10^{-5}–10^{-7} M.

Elaborate work has been done on the ultrastructure of haustoria formation in a range of root parasites (e.g., in the Australian root hemiparasite *Olax phyllanthi*) (Kuo et al. 1989). Walls of parasitic cells that contact host xylem are thickened with polysaccharides rather than with lignin. Host xylem pits are a major pathway for water and solute transport from the host to the haustorium, whereas direct connections between xylem conducting elements of host and parasite are extremely rare. Symplasmic connections between the two partners are absent. Cells of the parasite that are adjacent to host cells often have a similar appearance as **transfer cells**. Electron micrographs suggest that the developing haustorium may act as "scissors," which effectively cut-off the distal part of the host from the rest of the plant (Fig. 2).

The completely encircling haustorium of the root hemiparasite *Nuytsia floribunda* (Christmas tree) is unique in cutting the host root transversely by means of a sclerenchymatic cutting device, which is shaped remarkably like garden shears. Parenchymatous tissue of the parasite then develops tubelike apical extensions into the cut host xylem vessels, thereby possibly facilitating absorp-

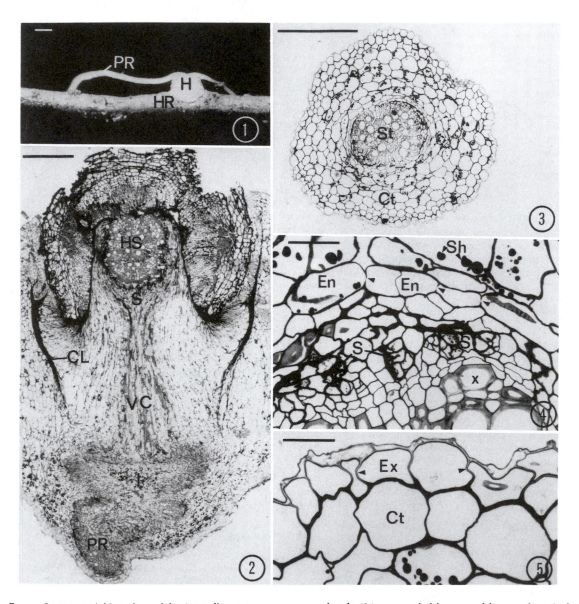

FIGURE 2. Haustorial interface of the Australian root parasite *Olax phyllanthi* (Kuo et al. 1989). Courtesy J. Kuo, The University of Western Australia, Australia. (1) A parasitic root (PR) forms a haustorium (H) on a host root (HR). (2) Transverse section of the root of *Scaveola nitida* at midpoint of attachment of a haustorium of *O. phyllanthi*. HS, host stele; S, sucker of haustorium; CL, collapsed layer; VC, core of tracheary elements in center of sucker; I, interrupted zone of vascular tissue connecting VC to the parent root (PR) of the parasite. Scale bar = 5 μm. (3) Transverse section of fine root of *O. phyllanthi* (PR in 1), showing central stele (St) surrounded by several layers of cortical cells (Ct) with prominent starch reserves. Scale bar = 5 μm. (4) Detailed anatomy of outer part of stele of root shown in (3). Note the endodermis (En) with Casparian band (pointed at by arrows), starch (Sh) in cortex, sieve elements (S), and xylem elements (X) in vascular tissue. Scale bar = 0.5 μm. (5) Outer region of root shown in (3) to show suberin-impregnated exodermis (Ex and arrows) external to cortex (Ct) and immediately inside a collapsed epidermis (Ep). Scale bar = 0.5 μm.

tion of xylem solutes from host xylem sap. Conducting xylem tissue in the haustorium terminates some distance from the interface, so absorbed substances must traverse several layers of parenchyma before gaining access to the xylem stream of the parasite. Water intake may therefore be minimal, but this parasitic tree has its own deep sinker roots

to enable it to tap underground reserves of water (Pate 1995).

After germination in the soil, the seedlings of the obligate stem parasite *Cuscuta* (dodder) start to grow up and circumnutate. Under favorable conditions many stems may grow from a twined seedling after attachment to the host (Fig. 3). Enzymes from

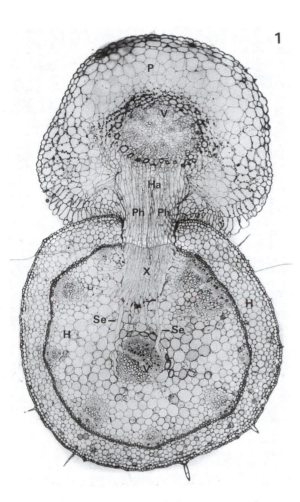
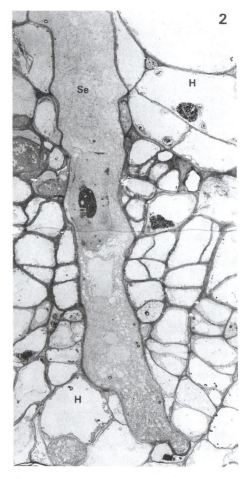

FIGURE 3. Haustorial interface of the stem parasite *Cuscuta odorata* parasitizing on *Pelargonium zonale*. Courtesy I. Dörr, Christian-Albrechts-University, Kiel, Germany. (1) Longitudinal section through a mature haustorium (Ha). Stems of the parasite (P) and petiole of the host (H) are cut transversely. The vascular bundles (V) of the stem of *C. odorata* are surrounded by thick-walled cells that contain latex. Phloem (Ph) of the haustorium accompanies the xylem (X). Searching hyphae (Se) are growing from the haustorial surface into the host tissue. (2) Electron micrograph of a searching hypha (Se) that develops in the tissue of the host. A searching hypha develops by expansion of a single cell from the tip of the haustorium and grows between the cells of the host tissue (H) or penetrates these host cells. Scale bar = 0.1 mm. (3) Light micrograph of a longitudinally cut xylem strand (X) within the haustorium. In addition, note the phloem (Ph), with sieve elements (S) and companion cells (Co). Scale bar = 50 μm. (4) Electron micrograph of one differentiated xylem element (XP) of *C. odorata* contacting several host xylem elements (XH). The large openings between xylem elements of host and parasite show a faint pit membrane. Scale bar = 10 μm. (5) Electron micrograph of a transversely cut sieve element of the host (SH) surrounded by a longitudinally cut protrusion on an "absorbing" hypha (Ab). Note the host companion cell (CoH). Wall labyrinth enlarges the inner surface of the parasitic cell. Smooth-surfaced endoplasmic reticulum is characteristic for absorbing hyphae. Scale bar = 5 μm.

the parasite soften the surface tissue of the host, and the haustorium penetrates the host tissue. Vascular cells of the parasite contact vascular cells of the host, and the contents of the host's sieve tubes and xylem conduits are diverted into the parasite. As the parasite continues to grow, it maintains its support by continually reattaching to host plants (Dawson et al. 1994).

Transfer of solutes via the haustorium may be partly passive, via the apoplast. The presence of parenchyma cells with many mitochondria, dictyosomes, ribosomes, and a well-developed ER,

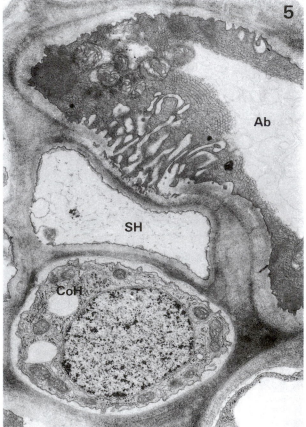

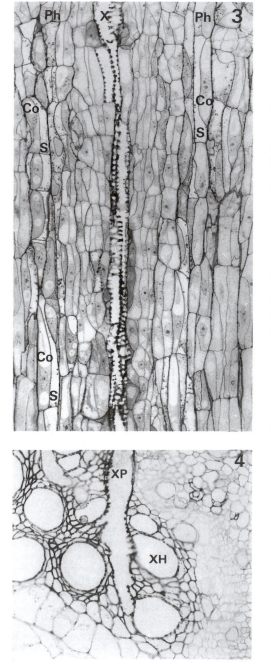

FIGURE 3. *Continued*

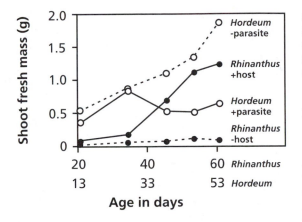

FIGURE 4. The increment of shoot fresh mass of *Hordeum vulgare* (barley), either grown alone or with a hemiparasite attached to its roots, and of *Rhinanthus serotinus* (yellow rattle), a hemiparasite, either grown alone or attached to its host (Klaren 1975). Reproduced with the author's permission.

however, suggests that active processes play a role as well. Indeed, compounds absorbed by the haustoria are processed before entering the shoot. As a result, the carbohydrates, amino acids, and organic acids in the xylem sap of *Striga hermontica* (witchweed) and *Olax phyllanthi* differ from those in their hosts. The major compound in *S. hermonthica* is mannitol, which does not occur in the host *Sorghum bicolor* (broom-corn). Similarly, in xylem sap of *Sorghum* asparagine predominates as a nitrogenous compound, and malate and citrate as organic

acids, whereas the major nitrogenous compound of *Striga* is citrulline, and shikimic acid is the main organic acid. The carbohydrate concentrations in the parasite xylem sap may be five times higher than those in the host (Pate et al. 1994).

2.3 Effects of the Parasite on Host Development

Although some hemiparasitic plants can grow in the absence of a host, their productivity is greatly enhanced when they are attached to a host (Fig. 4). At the same time, the growth of the host is reduced when a parasite is attached to it. The reduction in growth and grain yield of *Sorghum bicolor* infected by the parasitic *Striga hermonthica* is strongest at low nitrogen supply and may disappear completely at optimum nitrogen supply. The parasite is also affected by the low nitrogen supply, with considerably reduced seed germination, reduced attachment, and poor growth of *Striga* plants (Fig. 5).

Even though the root growth of *Ricinus communis* (castor bean) is inhibited when parasitized by *Cuscuta reflexa* (dodder), which is an obligate stem holoparasite, the rate of nitrate uptake per unit root mass is stimulated by 40 and 80% at high and low nitrate supply, respectively (Jeschke & Hilpert 1997). The rate of nitrate uptake in the host plant obviously increases with increasing demand for nitrate of the parasite–host association (cf. Sect. 2.2.3 of the chapter on mineral nutrition). When

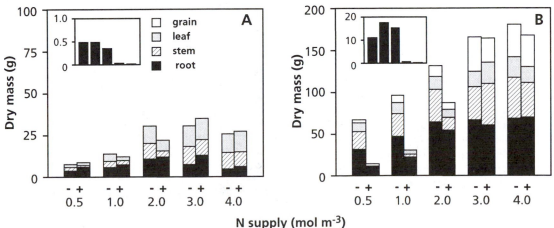

FIGURE 5. Partitioning of dry mass in *Sorghum bicolor* grown at a range of N-supply rates in the absence (−) and presence (+) of *Striga hermonthica*. Dry masses of the parasite are shown in the insets; (A) and (B) refer to 50 and 140 days after planting. Different shades in the columns, from bottom to top, refer to roots, stems, leaves, and seeds (Cechin & Press 1993). Copyright Blackwell Science Ltd.

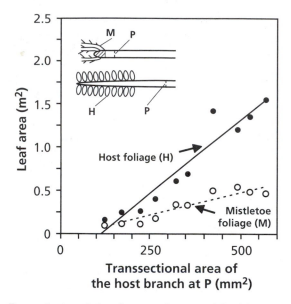

FIGURE 6. Correlations between the area of the foliage of a nonparasitized branch of *Acacia acuminata* or the foliage of the xylem-tapping stem hemiparasite (mistletoe) *Amyema preissii*, parasitizing on *Acacia acuminata* and the transectional area of the branch of the host. Note that a similar transectional branch area supports substantially more foliage of the host than of the parasite (Tennakoon & Pate 1996a). Copyright Blackwell Science Ltd.

parasitized, host plants may also show a higher rate of photosynthesis, greater stomatal conductance, and higher rates of transpiration, despite their smaller root system. Enhanced photosynthesis may be due to higher concentration of nitrogen in the leaves (cf. Sect. 6.1 of the chapter on photosynthesis), a higher sink demand (cf. Sect. 4.2 in the chapter on photosynthesis) or delayed leaf senescence (Jeschke & Hilpert 1997).

Xylem-tapping stem hemiparasites (mistletoes), such as *Phoradendron juniperinum* and *Amyema preissii*, have no phloem connection with their host, and they tend to kill the host shoot beyond the point of infection. In this way, the mistletoe is the only green tissue to be supplied via the xylem by a particular branch. Despite the absence of phloem connections, the growth of the mistletoe and that of xylem of the host are closely correlated. Just like the correlation between leaf area and sapwood area in trees (see Sect. 5.3.5 of the chapter on plant water relations), there is also a close correlation between the leaf area of the mistletoe and the sapwood area of the host branch proximal to the point of attachment (Fig. 6). This indicates that enlargement of the host stem must proceed, despite the impossibility of transport of any signals from the parasites' leaves

via the phloem. For a similar area of foliage, the mistletoe appears to require a substantially greater sapwood area than does the host plant itself. This is likely to be related to a relatively high rate of transpiration of the hemiparasite (Sect. 4).

3. Water Relations and Mineral Nutrition

Most herbaceous root and stem hemiparasites have high stomatal frequencies, high rates of transpiration, and their water-use efficiency is considerably lower than that of their host (Davidson & Pate 1992, Press et al. 1987, Schulze & Ehleringer 1984). The stomata of the herbaceous hemiparasites do respond to water stress, but stomatal closure is induced at much lower relative water contents (Fig. 7). Thus, the gradient in water potential between leaves and roots is steeper for the parasite than it is for its host, facilitating the flux of solutes imported via the xylem (Klaren & Jansen 1976, Klaren & Van de Dijk 1976). This appears to be due to the sensitivity of the stomata to ABA, the hormone associated with stomatal closure during water stress (Sect. 5.4.2 in the chapter on plant water relations), which is much less for parasitic species (*Striga hermonthica*) than it is for related nonparasitic species (*Antirrhinum majus*) (Shah et al. 1987). Leaves of *Zea mays* (maize) plants that are parasitized by *Striga*

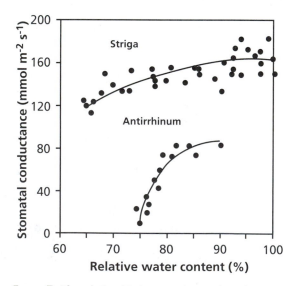

FIGURE 7. The relationship between stomatal conductance and the relative water content of the leaves for the hemiparasite *Striga hermonthica* and the closely related nonparasitic plant *Antirrhinum majus* (Shah et al. 1987). Copyright Physiologia Plantarum.

TABLE 3. Carbon isotope discrimination values (in ‰) for mistletoe-host pairs (number of pairs in brackets) from different continents; mean values and standard errors in brackets.

Region	Host	Mistletoe	Difference between host and mistletoe
Nitrogen-fixing hosts			
United States (7)	−26.3 (0.5)	−26.5 (0.2)	0.2
Australia (28)	−26.9 (0.2)	−28.3 (0.3)	1.4
South Africa (4)	−24.7 (0.3)	−25.7 (1.0)	1.1
Nonfixing hosts			
United States (8)	−23.4 (0.1)	−26.6 (0.1)	3.2
Australia (19)	−26.5 (0.3)	−28.8 (0.2)	2.3
South Africa (11)	−24.7 (0.4)	−26.9 (0.6)	2.2

Source: Ehleringer et al. 1985.

hermonthica have higher levels of ABA than do leaves of control plants, and the concentration of this phytohormone is an order of magnitude higher again in the leaves of the parasite (Taylor et al. 1996). The stomates of *Striga* do not close, however, even when the relative water content of its leaves declines to 70% or less.

The high rates of transpiration have major consequences for the leaf temperature of the parasitic plants. The leaf temperature of *Striga hermonthica* may be as much as 7°C below air temperature. The use of antitranspirants, which reduce transpirational water loss, may enhance the leaf temperature of parasites to an extent that the leaves blacken and die. These compounds have been suggested as tools to control parasitic pests (Stewart & Press 1990).

The high stomatal conductance and high rate of transpiration of parasites allows rapid import of solutes via the xylem. As expected, their p_i is relatively high and the discrimination against $^{13}CO_2$ is stronger in mistletoes than it is in their host because of the high stomatal conductance of the parasite (cf. Sect. 6 of the chapter on plant water relations). It is interesting that the difference in discrimination against $^{13}CO_2$ between host and parasite is less when the host is an N_2-fixing tree than when it is a nonfixing one. It has been suggested that more nitrogenous compounds are imported when the host is fixing nitrogen, which then reduces the transpiration and increases the parasite's water-use efficiency (Ehleringer et al. 1985, Schulze & Ehleringer 1984). The smaller difference in the case of the N_2-fixing hosts, however, is not merely due to the difference between the mistletoes; the hosts also contribute to the difference because they discriminate more strongly (Table 3). Further evidence indicates that the decline in carbon-isotope dis-

crimination with increasing nitrogen concentration is due to enhanced carbon import from the host; a substantial part of the carbon in mistletoes originates from the host via the xylem as organic acids and amino acids (Sect. 4).

To look at this in broad terms, hemiparasitic plants have high rates of transpiration and a low water-use efficiency. The Australian woody root hemiparasites *Nuytsia floribunda* (Christmas tree) and *Olax phyllanthi*, however, do not follow this pattern and have normal transpiration rates and water-use efficiency. *Nuytsia* has sinker roots, which tap deep reserves of water, but *Olax* can presumably survive by contact with deep-rooted species only (Pate 1995). These hemiparasites presumably extract relatively smaller quantities of resources from their hosts.

Holoparasites, which predominantly import compounds from the sieve tubes of the host, have distinctly lower calcium to potassium ratios than do parasites that only tap the xylem (Jeschke et al. 1995, Ziegler 1975). This is due to the fact that calcium is only present in very low concentrations in phloem sap, whereas most other minerals occur in higher concentrations in phloem sap than in xylem fluid. To acquire sufficient calcium for their growth, some additional xylem connections are required. Whereas *Cuscuta reflexa* (dodder) acquires 94% of its nitrogen and 74% of its potassium from the phloem of the host *Lupinus albus* (white lupin), virtually none of its calcium arrives via the phloem (Jeschke et al. 1995).

Because most xylem-tapping mistletoes, possibly with the exception of *Olax phyllanthi* (Tennakoom & Pate 1996b), have no mechanism to selectively import specific ions that arrive via the xylem or to export ions that have arrived in excess of their requirement, mistletoes often accumulate vast amounts of inorganic ions. Increased succulence with increasing leaf age and sequestration of sodium in older leaves appear to be mechanisms to maintain inorganic solute concentrations at a tolerable level (Popp et al. 1995). A consequence of the accumulation of vast amounts of inorganic ions is the need for **compatible solutes** in the cytoplasm (cf. Sect. 4.1 in the chapter on plant water relations). This may well account for the high concentrations of polyols in xylem-tapping mistletoes (Popp et al. 1995, Richter & Popp 1992). Some of the accumulated ions may be excreted via leaf glands (e.g., in *Odontites verna* and *Rhinanthus sclerotinus*) (Govier et al. 1968, Klaren & Van de Dijk 1976).

Rapid import of nitrogen may lead to higher concentrations of organic nitrogen in the leaves of the parasite than in those of the host. This often coincides with a similarity in leaf shape and appearance: **cryptic mimicry**. The nitrogen concentration of the parasite's leaves, however, is sometimes lower than that of the host, which may coincide with differences in leaf shape and appearance between host and parasite: **visual advertisement**. Because many herbivores prefer leaves with a high organic nitrogen concentration, it has been suggested that both "cryptic mimicry" and "visual ad-

vertisement" **reduce herbivory** (Ehleringer et al. 1986).

4. Carbon Relations

Hemiparasites are assumed to rely on their hosts only for water and mineral nutrients, but to fix their own carbon dioxide. Their photosynthetic capacity, however, is often very low (0.5 to 5.0 $\mu mol\,m^{-2}\,s^{-1}$), and in some species there may be substantial carbon import from the host. *Striga gesneroides* (witchweed), which is an obligate root hemiparasite, has a very low photosynthetic capacity coupled with a very high rate of respiration. There is no net CO_2 fixation even at light saturation (Graves et al. 1992), so it imports carbohydrates from its host. In *Striga hermonthica* approximately 27% of the carbon is derived from its host (*Sorghum bicolor*) at a low nitrogen supply; this value declines to approximately 6% at a high nitrogen supply and higher rates of host photosynthesis (Cechin & Press 1993). Xylem-tapping mistletoes also import a large fraction of all their carbon from the host (Schulze et al. 1991) (e.g., 23 to 43% in *Viscum album*) (Richter &

TABLE 4. Heterotrophic carbon gain of the xylem-tapping mistletoe *Tapinanthus oleifolius* on *Euphorbia virosa* and *Acacia nebrownii*.*

	Papinanthus oleifolius on *Euphorbia virosa*		*Papinanthus oleifolius* on *Acacia nebrownii*
	Young leaves	Old leaves	
Carbon-budget method			
Carbon concentration of xylem sap ($mmol\,Cl^{-1}$ xylem sap)	120.8	120.8	116.3
Transpiration [$l\,H_2O\,m^{-2}\,(10h)^{-1}$]	1.3	3.9	1.6
Carbon import via the xylem (C_x) [$mmol\,C\,m^{-2}\,(10h)^{-1}$]	156.6	469.7	188.4
CO_2 assimilation in photosynthesis [$mmol\,CO_2\,m^{-2}\,(10h)^{-1}$]	126.0	108.0	144.0
Total carbon gain [$mmol\,C\,m^{-2}\,(10h)^{-1}$]	282.6	577.7	332.4
Heterotrophic carbon gain (%)	55	81	57
$\delta^{13}C$-difference method			
$\delta^{13}C$ xylem sap (‰)	−16.92	−16.92	−21.05
$\delta^{13}C$ parasite leaves (‰, measured)	−23.73	−18.99	−26.81
$\delta^{13}C$ parasite leaves (‰, predicted from measured c_i/c_a)	−29.60	−33.20	−32.88
Heterotrophic carbon gain (%)	46	87	51

Source: Richter et al. 1995.
* The host-derived part of the mistletoe's carbon was calculated from the carbon flux from the host xylem sap (i.e., carbon concentration in the xylem sap multiplied by the transpiration rate; "carbon-budget method") or from the difference between the predicted and the actual carbon isotope ratios of the parasite ("$\delta^{13}C$-difference method").

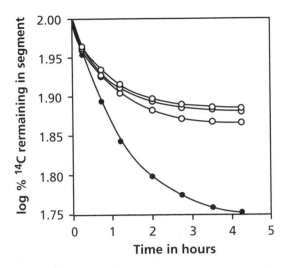

FIGURE 8. The effect of *Cuscuta europaea*, a stem parasite, on the release of ^{14}C-labeled valine from the sieve tubes in the stem of its host, *Vicia faba* (broad bean). The values represent the fraction of the labeled amino acid originally present in the stem segment that was not released from the sieve tubes to the apoplast. Open symbols refer to nonparasitized segments of the stem; filled symbols refer to the release in the apoplast of the segment where the parasite had formed a haustorium (Wolswinkel et al. 1984). Copyright American Society of Plant Physiologists.

Popp 1992). Two methods have been used to assess heterotrophic carbon gain in the African xylem-tapping mistletoe, *Tapinanthus oleifolius*. One method is based on an analysis of xylem sap and transpiration rate (Pate et al. 1991; cf. Sect. 3.4 of the chapter on life cycles); the other is based on an analysis of carbon isotope composition and gas exchange (Marshall & Ehleringer 1990; cf. Sect. 5.3 in the chapter on photosynthesis and Box 2). Both methods agree and yield values in the range of 55 to 80%, with the higher values pertaining to older leaves that have high transpiration rates (Table 4).

The presence of a parasite like *Cuscuta europaea* (dodder) on the stem of a host plant greatly enhances the release of amino acids and other solutes from the phloem of the host (Fig. 8). It is not known how the parasite manages to locally affect phloem unloading in the stem of the host (Wolswinkel 1978, Wolswinkel et al. 1984). It does show, however, that the haustorium is more than an organ that simply channels solutes from the host to the parasite. The compounds that are released from the sieve tubes of the host are rapidly absorbed by the transfer cells of the parasite in the haustorium.

A parasite like *Striga gesneroides* may use up to 70% of all the imported carbohydrates for its respi-

ration; the use of carbon from the host may be even more important for the yield reduction of its host, *Vigna unguiculata* (cowpea), than the reduction in host photosynthesis (Graves et al. 1992). It is not clear why such a large fraction of imported carbon is used in respiration; in the holoparasite *Cuscuta reflexa*, when it grows on the stem of *Lupinus albus*, only 29% of all the incorporated carbon is respired (Jeschke et al. 1994). This value is in the same range as that of heterotrophic plant parts of nonparasitic plants (cf. Sect. 5 in the chapter on plant respiration).

The reduction in photosynthesis of the host *Sorghum bicolor* by *Striga hermontica* is strongest at a low nitrogen supply and high infection rate (approximately 40%). This may be associated with reduced nitrogen concentrations in the host leaves or reduced stomatal conductance, which is due to the high demand for nitrogen and water of the parasite. At a very low infection rate and high nitrogen supply, there may be some enhancement of photosynthesis in the presence of the parasite, which is due to the stimulation of photosynthesis by enhanced sink strength (Cechin & Press 1993). It is quite likely that under more natural conditions, where infection percentages are lower than they are in agricultural monocultures, *Striga* does not have the same detrimental effect on host photosynthesis and performance.

5. What Can We Extract from This Chapter?

The 3000 or so species of parasitic angiosperms of the world flora collectively represent an extraordinarily broad assemblage of taxa mostly from distantly related families of dicotyledons and an equally profuse range of woody forms, morphologies, and life strategies. With the notable exception of some Australian trees, **hemiparasites** tend to have high rates of transpiration and a low water-use efficiency, which ensures rapid intake of xylem solutes. Hemiparasites also import carbon (amino acids, organic acids) via the transpiration stream, which supports their carbon requirement to a varying extent: from almost none to virtually completely.

Holoparasites tap the host's phloem and depend entirely on their host for their carbon requirements. Because the phloem contains very little calcium, holoparasites have distinctly lower Ca/K ratios than do hemiparasites.

Some parasitic plants are notorious **pests**, reducing crop yield in many areas of the world. A thorough understanding of host factors that affect seed germination of some parasitic plants may help to control these pests, either by employing trap crops or by using analogues that stimulate seed germination of the parasite.

References and Further Reading

Atsatt, P.R. (1983) Host-parasite interactions in higher plants. In: Encyclopedia of plant physiology, N.S. Vol. 12C, O.L. Lange, P.S. Nobel, C.B. Osmond, & H. Ziegler (eds). Springer-Verlag, Berlin, pp. 519–535.

Babiker, A.G.T., Ejeta, G., Butler, L.G., & Woodson, W.R. (1993) Ethylene biosynthesis and strigol-induced germination of Striga asiatica. Physiol. Plant. 88:359–365.

Cechin, I. & Press, M.C. (1993) Nitrogen relations of the sorghum-Striga hermonthica host-parasite association: Growth and photosynthesis. Plant Cell Environ. 16:237–247.

Davidson, N.J. & Pate, J.S. (1992) Water relations of the mistletoe Amyema fitzgeraldii and its host Acacia acuminata. J. Exp. Bot. 43:1459–1555.

Dawson, J.H., Musselman, L.J., Wolswinkel, P., & Dörr, I. (1994) Biology and control of Cuscuta. In: Reviews of weed science, Vol. 6, S.O. Duke (ed). Imperial Printing Company, Champaign, pp. 265–317.

Ehleringer, J.R., Schulze, E.D., Ziegler, H., Lange, O.L., Farquhar, G.D., & Cowan, I.R. (1985) Xylem-tapping mistletoes: Water or nutrient parasites? Science 227:1479–1481.

Ehleringer, J.R., Ullmann, I., Lange, O.L., Farquhar, G.D., Cowan, G.D., & Schulze, E.-D. (1986) Mistletoes: A hypothesis concerning morphological and chemical avoidance of herbivory. Oecologia 70:234–237.

Einhellig, F.A. & Souza, I.F. (1992) Phytotoxicity of sorgoleone found in grain sorghum root exudates. J. Chem. Ecol. 18:1–11.

Fate, G., Chang, M., & Lynn, D.G. (1990) Control of germination in Striga asiatica: Chemistry of spatial definition. Plant Physiol. 93:201–207.

Govier, R.N., Brown, J.G.S., & Pate, J.S. (1968) Hemiparasitic nutrition in angiosperms. II. Root haustoria and leaf glands of Odontites verna (Bell.) Dum. and their relevance to the abstraction of solutes from the host. New Phytol. 67:863–972.

Graves, J.D., Press, M.C., Smith, S., & Stewart, G.R. (1992) The carbon canopy economy of the association between cowpea and the parasitic angiosperm Striga gesneroides. Plant Cell Environ. 15:283–288.

Harborne, J.B. (1982) Introduction to ecological biochemistry. Academic Press, London.

Jeschke, W.D. & Hilpert, A. (1997) Sink-stimulated photosynthesis and sink-dependent increase in nitrate uptake: nitrogen and carbon relations of the parasitic association Cuscuta reflexa-Ricinus communis. Plant Cell Environ. 20:47–56.

Jeschke, W.D., Bäumel, P., Räth, N., Czygan, F.-C., & Proksch, P. (1994) Modelling of the flows and partitioning of carbon and nitrogen in the holoparasite Cuscuta reflexa Roxb. and its host Lupinus albus. L. II. Flows between host and parasite and within parasitized host. J. Exp. Bot. 45:801–812.

Jeschke, W.D., Bäumel, P., & Räth, N. (1995) Partitioning of nutrients in the system Cuscuta reflexa-Lupinus albus. Aspects Appl. Biol. 42:71–79.

Klaren, C.H. (1975) Physiological aspects of the hemiparasite Rhinanthus serotinus. PhD Thesis, University of Groningen, the Netherlands.

Klaren, C.H. & Janssen, G. (1978) Physiological changes in the hemiparasite Rhinanthus serotinus before and after attachment. Physiol. Plant. 42:151–155.

Klaren, C.H. & Van de Dijk, S.J. (1976) Water relations of the hemiparasite Rhinanthus serotinus before and after attachment. Physiol. Plant. 38:121–125.

Kuijt, J. (1969) The biology of parasitic flowering plants. University of California Press, Berkeley

Kuo, J., Pate, J.S., & Davidson, N.J. (1989) Ultrastructure of the haustorial interface and apoplastic continuum between host and the root hemiparasite Olax phyllanthi (Labill.) R. Br. (Olacaceae). Protoplasma 150:27–39.

Logan, D.C. & Stewart, G.R. (1991) Role of ethylene in the germination of the hemiparasite Striga hermontica. Plant Physiol. 97:1435–1438.

Lynn, D.G. & Chang, M. (1990) Phenolic signals in cohabitation: Implications for plant development. Annu. Rev. Plant Physiol. Plant Mol. Biol. 41:497–526.

Marshall, J.D. & Ehleringer, J.R. (1990) Are xylem-tapping mistletoes partially heterotrophic? Oecologia 84:244–248.

Pate, J.S. (1995) In: Global change and mediterranean-type ecosystems. Ecological Studies 117, J.M. Moreno, & W.C. Oechel (eds). Springer-Verlag, Berlin, pp. 161–180.

Pate, J.S., True, K.C., & Rasins, E. (1991) Xylem transport and storage of amino acids by S.W. Australian mistletoe and their hosts. J. Exp. Bot. 42:441–451.

Pate, J.S., Woodall, G., Jeschke, W.D., & Stewart, G.R. (1994) Root xylem transport of amino acids in the root hemiparasite shrub Olax phyllanthi (Labill) R.Br. (Olacaceae) and its multiple hosts. Plant Cell Environ. 17:1263–1273.

Popp, M., Mensen, R., Richter, A., Buschmann, H., & Von Willert, D.J. (1995) Solutes and succulence in southern African mistletoes. Trees 9:303–310.

Press, M.C. & Graves, J.D. (eds) (1995) Parasitic plants. Chapman, & Hall, London.

Press, M.C., Tuohy, J.M., & Stewart, G.R. (1987) Gas exchange characteristics of the Sorghum-Striga host-parasite association. Plant Physiol. 84:814–819.

Richter, A. & Popp, M. (1992) The physiological importance of accumulation of cyclitols in Viscum album L. New Phytol. 121:431–438.

Richter, A., Popp, M., Mensen, R., Stewart, G.R., & Von Willert, D.J. (1995) Heterotrophic carbon gain of the parasitic angiosperm *Papinanthus oleifolius*. Aust. J. Plant Physiol. 22:537–544.

Schulze, E.-D. & Ehleringer, J.R. (1984) The effect of nitrogen supply on growth and water-use efficiency of xylem-tapping mistletoes. Planta 162:268–275.

Schulze, E.-D., Lange, O.L., Ziegler, H., & Gebauer, G. (1991) Carbon and nitrogen isotope ratios of mistletoes growing on nitrogen and non-nitrogen fixing hosts and on CAM plants in the Namib desert confirm partial heterotrophy. Oecologia 88:457–462.

Shah, N., Smirnoff, N., & Stewart, G.R. (1987) Photosynthesis and stomatal characteristics of *Striga hermonthica* in relation to its parasitic habit. Physiol. Plant. 69:699–703.

Smith, C.E., Dudley, M.W., & Lynn, D.G. (1990) Vegetative/parasitic transition: Control and plasticity in *Striga* development. Plant Physiol. 93:208–215.

Stewart, G.R. & Press, M.C. (1990) The physiology and biochemistry of parasitic angiosperms. Annu. Rev. Plant Physiol. Plant Mol. Biol. 41:127–151.

Taylor, A., Martin, J., & Seel, W.E. (1996) Physiology of the parasitic association between maize and witchweed (*Striga hermonthica*): Is ABA involved? J. Exp. Bot. 47:1057–1065.

Tennakoon, K.U. & Pate, J.S. (1996a) Effects of parasitism by a mistletoe on the structure and functioning of branches of its host. Plant Cell Environ. 19:517–528.

Tennakoon, K.U. & Pate, J.S. (1996b) Heterotrophic gain of carbon from hosts by the xylem-tapping root hemiparasite *Olax phyllanthi* (Olacaceae). Oecologia 105:369–376.

Tuquet, C., Farineau, N., & Sallé, G. (1990) Biochemical composition and photosynthetic activity of chloroplasts from *Striga hermonthica* and *Striga aspera*, root parasites of field-grown cereals. Physiol. Plant. 78:574–582.

Wolswinkel, P. (1978) Phloem unloading in stem parts by *Cuscuta*: the release of 14 C and K^+ to the free space at 0°C and 25°C. Physiol. Plant. 42:167–172.

Wolswinkel, P., Ammerlaan, A., & Peters, H.F.C. (1984) Phloem unloading of amino acids at the site of *Cuscuta europaea*. Plant Physiol. 75:13–20.

Ziegler, H. (1975) Nature of transported substances. In: Encyclopedia of plant physiology, N.S. Vol. 1, M.H. Zimmermann & J.A. Milburn (eds). Springer-Verlag, Berlin, pp. 59–100.

9E. Interactions among Plants

1. Introduction

In previous chapters we have dealt with numerous physical and chemical environmental factors that affect a plant's performance, and with the effects of microsymbionts and microbial pathogens. For many plants, however, the most important factor shaping their environment is other plants. One of the most active debates in ecology focuses on the unresolved question of the mechanisms by which plants interact with one another. Plant–plant interactions range from positive (**facilitation**) to neutral to negative (**competition**) effects on the performance of neighbors (Bazzaz 1996). Competition occurs most commonly when plants utilize the same pool of growth-limiting resources (**resource competition**). Competition may also occur when one individual produces chemicals that negatively affect their neighbors (**interference competition** or **allelopathy**). Competition between two individuals is often highly asymmetric, with one individual having much greater negative impact than the other.

The question of which species wins in competition depends strongly on the time scale of the study. Short-term experimental studies of competition often depend on rates of resource acquisition and growth, whereas equilibrium persistence of a species in a community is affected by rates of resource acquisition, tolerance of ambient resource availability, efficiency of converting acquired resources into biomass, and retention of acquired resources (Goldberg 1990).

The **competitive ability** of a species depends on environment. There are no "super species" that are competitively superior in all environments; rather, there are trade-offs among traits that are beneficial in some environments, but which cause plants to be poor competitors in other environments. For a plant to compete successfully in a particular environment, it must have specific ecophysiological traits that allow effective growth in that environment (the **environmental filter** discussed in the Introduction chapter). We have provided a wealth of examples of physiological traits necessary for ecological success in dry, cold, hot, saline, flooded, or other harsh environments. Only those species that are adapted to or can acclimate to such environmental conditions can survive and compete successfully in these environments. As the saying goes: When the going gets tough, the tough get going. Other plants typically grow in more favorable conditions where abiotic stresses are moderate. Most species can survive in these conditions, but only a small proportion are effective competitors. We have already discussed many of the traits that enable plants to grow rapidly under these conditions.

Although this brief introduction of "plant strategies" provides a context for the present discussion of ecophysiological traits that are important in competitive interactions, the situation is far more complicated (Box 10). Traits that are important for **competitive success** at an early stage of succession may differ greatly from those that are pertinent in later stages. Similarly, plant characteristics that determine the outcome of competition in short-term

Box 10
Plant Ecology Strategy Schemes

Professor Dr. Mark Westoby

School of Biological Sciences
Macquarie University
NSW 2109
Australia

Plant ecology strategy schemes (PESSs) arrange species in categories or along spectra according to their ecological attributes. Long-standing motivations for devising a PESS have been:

1. To express an understanding of the main opportunities and selective forces that shape the ecologies of plants.
2. To describe the plant component of ecosystems in terms of a limited number of types, such that, if vegetation descriptions are condensed into PESS categories rather than listing each species individually, then the most ecologically important information can be retained and the most instructive comparisons between ecosystems can emerge.

Many schemes have been proposed. Some split up species on the basis of a single attribute thought to be important. An example that is still widely used is Raunkiaer's life-form scheme (1907, English translation 1934), which is based on the location of the buds where regrowth arises after the unfavorable season of the year (Fig. 1; many refinements have been proposed by Raunkiaer and subsequently, review in Barkman 1988, but they have not come to be widely used like the bud-location scheme).

Other schemes have an overtly conceptual basis. The best-known is Grime's (1977) "plant strategy theory" triangle (Fig. 2) (see also Sects. 6.1 and 6.3 of growth and allocation). The C–S axis (Competitors-to-Stress tolerators) reflects

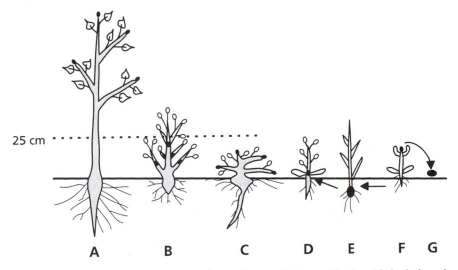

FIGURE 1. **Plant life-forms of Raunkiaer. Perennating organs are shown in black, woody organs in gray, and deciduous organs unshaded. (A) Phanerophyte (tree or tall shrub) with buds more than 25 cm above the ground. (B) Chamaephyte, semishrub (suffrutescent low shrub) with buds less than 25 cm above the ground. (C)** Chamaephyte, semishrub, with buds less than 25 cm above the ground. **(D) Hemicryptophyte, perennial herb with its bud at ground surface. (E) Geophyte, perennial herb with a bulb or other perennating organ below the ground surface. (F) Therophyte, annual plant surviving unfavorable periods only as seed (G).**

continued

Box 10. *Continued*

FIGURE 2. A model that describes the various
equilibria between competition, stress, and
disturbance in vegetation and the location of
primary and secondary strategies. C, competitor; S,
stress-tolerating species; R, ruderal. C–R, S–R, and
C–S–R refers to species with an "intermediate"
strategy. I_c, I_s, and I_d denote the relative importance
of competition, stress and disturbance, respectively
(Grime et al. 1988).

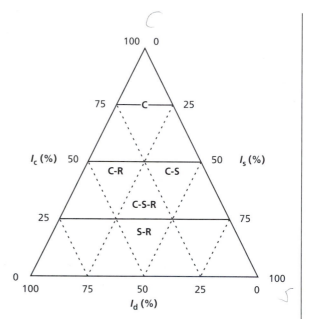

adaptation to favorable versus unfavorable
sites for plant growth, the R-axis (Ruderals) reflects adaptation to disturbance. A given adult
strategy can occur in combination with several
different juvenile strategies.

By presenting itself as a general treatment
of plant strategy concepts, the C–S–R scheme
seems to have precluded universal acceptance
because phenomena that are not accounted for
by the scheme may be seen as reasons to reject it
(e.g., Grubb 1986, 1992). It is actually not reasonable to expect a PESS to express 100% of ecologically relevant variation among species. Rather,
we should have been aiming for a sensible
understanding as to what is and is not expressed
by C–S–R, (e.g., by quantifying how well any
given ecological trait correlates with each
dimension). The C–S–R scheme, however, has
not fostered such quantifications because its dimensions are not defined by any single trait for
use in correlation analysis.

Ecology is presently often likened to chemistry before the periodic table (Begon et al. 1996),
lacking a strong framework to organize the huge
volume of species-specific information. That
this analogy would be used, even though a PESS
would probably be neither periodic, nor tabular,
nor deal mainly with combining ability, testifies
to a perception that strategy schemes are central
for ecology's aspirations to become an effective
applied discipline. Nevertheless, over the last
couple of decades, the research objective of actually agreeing on a consensus PESS seems to have

dropped below the horizon. The attention of
most researchers has been directed at the more
abstract problem of understanding trade-offs
that might underpin ecological strategies; plus,
debates over particular schemes such as C–S–R
have become mired in an expectation that they
should account for all phenomena. The possibility, however, of actually agreeing on a
worldwide consensus PESS—if necessary, at
a lowest-common-denominator level—must
soon come back near the top of plant ecology's
research agenda, driven by two further
motivations currently increasing in importance:

3. Putting ecophysiological function in comparative context. Measurements of growth, allocation, or gas exchange, or other metabolic fluxes
can only be made on a few species in any one
study; studies of more than 20 species are exceptional. Many hundreds of individual studies
have now accumulated, and synthesizing across
species in literature reviews or formal meta-
analyses has become an outstanding problem. Up until now, species have usually been
categorized for meta-analysis into growth form
or life form or annual versus perennial. It seems
reasonable to hope that, if species could be
categorized according to a PESS that captured
a better proportion of ecologically significant
variation between species, then improved meta-
analysis might result. The same problem of

continued

Box 10. *Continued*

generalizing has become urgent for manipulative field experiments on competition, herbivory, and other interactions between species.

4. Vegetation dynamics under global change. Future temperature zones are expected to migrate poleward, atmospheric CO_2 is expected to continue increasing, and land use is expected to continue intensifying. Projecting future vegetation dynamics on a worldwide basis has become an important test for ecology's grasp on substantial applied questions. Global vegetation dynamics models must work with ecological categories of plants; they cannot be parameterized for all 300,000 vascular plant species individually.

To be useful for these purposes, a PESS needs to allow a plant species to be positioned within the PESS by reference to traits that can be measured easily and directly on the plant itself— easily because otherwise the attributes will not come to be reported routinely, and directly because if species are positioned using evidence about the relative distribution or performance they achieve in some particular situation, then cross-relating worldwide becomes very difficult. In other words, we need a PESS as easy to use as the Raunkiaer scheme, but which expresses a larger proportion of ecologically relevant variation among species.

A Proposed L–S–H (Leaf–Seed–Height) Scheme

I have recently (Westoby 1998) proposed an L–S–H scheme that illustrates the potential for worldwide consensus. The scheme would consist of three axes, with the strategy of any one species characterized by a position in the three-dimensional volume:

1. Specific leaf area SLA (of mature leaves, developed in full light, or the fullest light the species naturally grows in)
2. Seed mass
3. Canopy height, meaning typical heights achieved by upper layers of leaves at a late stage during an individual plant's growth

Specific Leaf Area (SLA)

High SLA maximizes future light capture from a given amount of photosynthate available for re-

investment in leaves. (Between-species variation in seedling potential RGR is predominantly associated with the SLA component; Sect. 3.1 of the chapter on growth and allocation). Low SLA maximizes longevity, robustness, and defendedness of light-capture surfaces. The faster payback on investment (Poorter 1994), and higher risk of leaf loss, associated with high SLA, shape many aspects of growth strategy. For example, faster turnover of plant parts permits a more flexible response to the spatial patchiness of light and soil resources (Grime 1994).

Seed Mass

Species having smaller seed mass can produce more seeds with a given reproductive effort. Seed mass is the best easy predictor of seed output per square meter of canopy cover, and therefore of the chance that an occupied site will disperse a propagule to an establishment opportunity. Seed mass is also quite a good indicator of a cotyledon-stage seedling s ability to survive various hazards (Westoby et al. 1996; Sect. 3.1 of life cycles).

Canopy Height

Some disturbances destroy canopy cover and daylight becomes available near the ground. There ensues a race upward for the light. In this race, unlike a standard athletic contest, there is not a single winner determined after a fixed distance. Rather, being among the leaders for a reasonable period at some stage of the race permits a sufficient carbon profit to be accumulated for the species to ensure it runs also in subsequent races (Sect. 5.4 of the chapter on interactions among plants). The entry in subsequent races may occur via vegetative regeneration, via a stored seed bank, or via dispersal to other locations, but the prerequisite for any of these is sufficient carbon accumulation at some stage during vegetative growth. Within a race-series having some typical race duration, one finds successful growth strategies that have been designed by natural selection to be among the leaders early in a race, and other successful strategies that join the leaders at various later stages. Species that achieve most of their lifetime photosynthesis with leaves deployed at 10–50 cm have

continued

Box 10. *Continued*

different stem tissue properties from those designed for 1 to 5 m, and those in turn are different from species that achieve 30 to 40 m. The canopy height that species have been designed by natural selection to achieve is the simplest measure of this spectrum of strategies.

SLA, seed mass, and canopy height are each known to express ecologically important differences between species. They are independent to a substantial degree, both in the sense that they express different aspects of ecological variation between species, and in the sense that each is known to vary widely between species at any given level of the other two.

Summary

The idea of a PESS in many ways encapsulates the whole research agenda of plant ecology (Keddy 1989, Myerscough 1990). Over the last couple of decades the objective of actually reaching a worldwide consensus PESS seems to have been deferred to some future time, but new emerging needs for improved meta-analysis and for global vegetation modeling are making the matter urgent. The proposed L–S–H scheme illustrates that we do not need agreement that three particular PESS dimensions hold the three top ranks for importance, among all conceivable attributes. We need agree only that they might be useful enough dimensions, and easy enough to estimate, that experimentalists should be willing to report them routinely for their study species, with the selfless aim of assisting subsequent synthesis by others.

References and Further Reading

Barkman J.J. (1988) New systems of plant growth forms and phenological plant types. In: Plant form and vegetation structure. Adaptation, plasticity and relation to herbivory, M.J.A. Werger, P.J.M. Van der Aart, H.J. During, & J.T.A. Verhoeven (eds). SPB Academic Publishing, The Hague, pp. 9–44.

Begon, M., Harper, J.L., & Townsend, C.R. (1996) Ecology. 3rd edition. Blackwell Science, Oxford.

Grime, J.P. (1977) Evidence for the existence of three primary strategies in plants and its relevance to ecological and evolutionary theory. Am. Nat. 111:1169–1194.

Grime, J.P. (1994) The role of plasticity in exploiting environmental heterogeneity. In: Exploitation of environmental heterogeneity in plants, M.M. Caldwell & R.W. Pearcy (eds). Academic Press, San Diego, pp. 1–18.

Grime, J.P., Hodgson, J.G., & Hunt, R. (1988) Comparative plant ecology. Unwin-Hyman, London.

Grubb, P.J. (1985) Plant populations and vegetation in relation to habitat, disturbance and competition: problems of generalization. In: The population structure of vegetation, J. White (ed). Junk, Dordrecht, pp. 595–621.

Grubb, P.J. (1992) A positive distrust in simplicity—lessons from plant defences and from competition among plants and among animals. J. Ecol. 80:585–610.

Keddy, P.A. (1989) Competition. Chapman & Hall, London.

Myerscough, P.J. (1990) Comparative plant ecology and the quest for understanding of Australian plants. Proc. Linnean Soc. New South Wales 112:189–199.

Poorter, H. (1994) Construction costs and payback time of biomass: A whole plant perspective. In: A whole-plant perspective on carbon-nitrogen interactions, J. Roy & E. Garnier (eds). SPB Publishing, The Hague, pp. 111–127.

Raunkiaer, C. (1934) The life forms of plants and statistical plant geography. Clarendon Press, Oxford.

Westoby, M. (1998) Towards a consensus plant ecology strategy scheme. Plant Soil, in Press.

Westoby, M., Leishman, M.R., & Lord, J.M. (1996) Comparative ecology of seed size and seed dispersal. Phil. Trans. R. Soc. B 351:1309–1318.

experiments may differ from those that give a species a competitive edge in the long run (Sect. 4). In this text on physiological ecology, however, we will emphasize the physiological mechanisms rather than the community consequences of competition.

An ecophysiologist attempts to explain competitive interactions in terms of the performance of individual plants that make up a community. The challenge then is to scale up from the knowledge that is available at the cell, organ, and whole-plant level to the processes that occur in natural and crop communities.

An important aspect of the functioning of a plant among surrounding competitors may well be to *avoid* any potentially negative effects. That is, rather than producing leaves that are acclimated to shade, or roots that can access sparingly available nutrients, a plant might grow away from its neighbors and make leaves that are acclimated to a high level of irradiance and roots that are adjusted to cope with a favorable nutrient supply. This requires mechanisms, however, that allow a plant to detect the proximity of its neighbors. Such mechanisms do indeed exist (Sect. 3).

2. Theories of Competitive Mechanisms

Several theoretical frameworks have been developed to predict the outcome of plant competition, each of which has implications for the mechanisms by which competition occurs. Grime (1977) suggested that species with high **relative growth rates** are effective competitors because rapid growth enables them to dominate available space and to acquire the most resources (Sect. 6.1 of the chapter on growth and allocation). If correct, then traits that promote rapid resource acquisition and growth should be favored. On the other hand, Tilman (1988) suggested that the species that can draw a resource down to the lowest level (**R***) is the best competitor for that resource because this enables a species to tap that resource at levels below those required by other species. These perspectives are not incompatible (Grace 1990). We expect that, in short-term growth experiments, especially in high-resource environments, traits that contribute to rapid growth contribute to competitive success. At equilibrium, however, when species effects on resource supply should be greatest, the potential of a species to extract scarce resources may be more important than maximum rates of resource acquisition.

If **resource competition** occurs by **depletion of a shared limiting resource**, then there are at least two ways in which a species might be an effective competitor: drawing down resources to a low level (low R*) and/or tolerating low levels of resources (Goldberg 1990). The physiological bases of these two facets of competition are quite different, as will be discussed later. Because of physiological trade-offs, however, traits that promote **resource draw-down** and **tolerance** of low resource supply may be correlated (Sect. 7).

Two major **physiological trade-offs** have been discussed as the basis of broad patterns of competitive ability in different environments. First, there is a trade-off between rapid growth to occupy space

and maximize resource acquisition versus resource conservation through reductions in tissue turnover (Grime 1977) (Sect. 4). Second, there is a trade-off between allocation to roots to acquire water and nutrients versus allocation to shoots to capture light and CO_2 (Tilman 1988) (Sect. 7). Because of these trade-offs, no species can be a superior competitor in all environments, but will specialize instead to grow and compete effectively in a certain restricted set of environments.

Competition in natural environments is the rule rather than the exception (Goldberg & Barton 1992). Resource competition occurs when growth of two species is colimited by the same resource. Grime (1977) has argued that competition is strongest in environments with high resource availability and that, in low-resource environments, tolerance of low-resource supply is more important. Competition, however, is a widespread phenomenon. The effects of competition, as measured experimentally, are equally observable in high- and low-resource environments (Goldberg & Barton 1992, Gurevitch et al. 1992, McGraw & Chapin 1989). Competition is least likely to occur in recently disturbed sites where low plant biomass and/or high resource supply minimize resource limitation. In other cases, coexisting species may be limited by different factors, as when species have radically different phenology, height, or rooting depth. In order for plants to minimize competition, they must adjust growth to tap resources that are not utilized by neighbors.

3. How Do Plants Perceive the Presence of Neighbors?

Plants may perceive the proximity of neighbors as we described in discussing plant growth in shady conditions. First, a reduction in the level of photosynthetically active radiation may reduce the concentration of soluble sugars, which can be sensed by plant cells (cf. Sect. 6.3 of the chapter on photosynthesis). Second, special pigments, **cryptochrome** and **phytochrome**, perceive the level of radiation and the red/far-red ratio of the radiation (cf. Sect. 5.1.1 of the chapter on growth and allocation).

Plants growing in close-spaced rows or high population densities receive lower **red/far-red ratios** than do those growing in wide rows or sparse populations (Kasperbauer 1987). The phytochrome system allows plants to perceive neighboring plants well before these reduce the level of photosynthetically active radiation (Fig. 1, left). An important response of a plant to the proximity of neighbors is an enhanced rate of stem elongation in shade-avoiding plants (Sect. 5.1.1 of the chapter on growth and allocation) (Fig. 1, right).

For *Populus* (poplar) linear relationships exist between stem growth rate, plant spacing, and P_{fr}/P_t calculated from radiation that is propagated horizontally within the canopy. It is quite likely that the dynamics of developing or regenerating canopies is based on phytochrome-mediated perception of the proximity of neighboring plants (Gilbert et al. 1995, Ritchie 1997). Through the phytochrome system, plants, sense cues that indicate current or future shading. How should the plant respond to that information? Shade-avoiding species typically respond with enhanced stem elongation, whereas no such response is found for species naturally occurring under a dense canopy (Sect. 5.1.1 of the chapter on growth and allocation).

Plants are also capable of "smelling" the presence of neighbors above ground, if these produce **chemical signals**, such as jasmonate and possibly other volatiles (see Sect. 2 of the chapter on ecological biochemistry and Sect. 3 of effects of microbial pathogens). Contrary to common expectation, plants have highly sensitive **chemoperception systems** that play a central role in communication with surrounding organisms (Boller 1995). Physically touching surrounding plants is an additional way in which neighbors can be perceived (cf. Sect. 5.7 of the chapter on growth and allocation). Mutants of *Nicotiana tabacum* (tobacco) that lack a receptor for ethylene do not respond to touch by surrounding plants. They no longer respond to "crowding" and enter the private space of their neighbors (Fig. 2).

Plants can also perceive the presence of surrounding plants above ground because of their neighbors' effect on **microclimate**, which is caused by differential heat exchange. This can have a tremendous effect on the outcome of competition (e.g., in frost-prone areas where attempts are made to establish tree seedlings). The tree seedlings may grow well in forest clearings for the first few years, but once a grassy ground cover establishes, the growth of the young trees becomes retarded and more susceptible to frosts. Although some of these effects might well be due to resource competition for nutrients and water, this cannot account for all the observations, including greater frost sensitivity. When seedlings of *Eucalyptus pauciflora* (snow gum) are surrounded by grass, the minimum air temperature that these seedlings experience decreases by as much as 2°C, and they also experience more frosts. Such small effects are large enough to cause a

greater extent of photoinhibition, reduced growth, and, ultimately, a shorter growing season for seedlings surrounded by grass compared to those in bare patches. Thus, the microclimate above grass adversely affects spring growth of juvenile trees and may account for much of the competitive inhibition of tree seedling growth by grass during spring (Fig. 3).

There is convincing evidence that plants can also sense the presence of neighbors below ground. For example, below-ground competition of *Lolium perenne* (perennial ryegrass) with *Plantago lanceolata* (plantain) markedly reduces root mass and root length of *L. perenne*, without any effect on shoot growth. Contrary to the effects of a limiting nutrient supply, competition with *P. lanceolata* does not affect the specific root length. This suggests the perception of the presence of *P. lanceolata* by the grass roots via an allelochemical (Fitter 1976). Roots of two native Californian shrub species, *Haplopappus ericoides* and *H. venetus,* similarly reduce overlap with the roots of an invasive introduced perennial succulent, *Carpobrotus edulis* (iceplant), by redistributing root growth further down in the soil profile (Fig. 4). Removal of *Carpobrotus edulis* from around the native shrubs also results in higher predawn xylem water potentials, which suggests that the invasive succulent uses some water that would have been available for the native shrubs (D'Antonio & Mahall 1991). The change in rooting pattern could partly reflect differential root proliferation in zones of high availability of nutrients or water (cf. Sect. 3.4 of the chapter on plant water relations). The effect also occurs, however, when plants are well provided with water and nutrients, which indicates a specific response to avoid the roots of neighbors (Mahall & Callaway 1991).

A chemical interaction (e.g., the accumulation of **allelochemicals** around roots) is a likely explanation for many of the patterns observed in the field (cf. Sect. 2 of the chapter on ecological biochemistry). When the roots of *Ambrosia dumosa*, whose growth is normally inhibited by the presence of the roots of *Larrea divaricata* (creosote bush), are treated with activated charcoal which adsorbs the allelochemicals, the inhibition is reduced. This agrees with inhibition by a slowly diffusing allelochemical that is released by the roots of *Larrea divaricata* and may account for the dispersed distribution of *L. divaricata* in the Mojave Desert in North America. The intraspecific inhibition of root growth of *Ambrosia dumosa*, however, is not affected by activated charcoal; it obviously depends on **physical contact**. The nature of deterrence by direct contact with the roots of *Ambrosia* is not clear

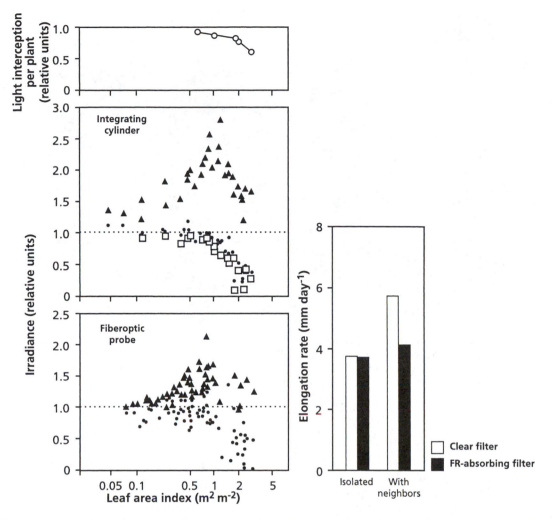

FIGURE 1. (Left) Effects of increasing the leaf area index [LAI, m² (leaf area) per m² (soil surface)] in even-height canopies of dicotyledonous seedlings on (top) light interception and (middle and bottom) the light climate of the stem. Seedling stands of *Sinapis alba* (mustard) and *Datura ferox* of differing densities and plant sizes were used to obtain a range for the leaf area index. The data in the middle figure were obtained with an integrating cylinder, which collects sidelight received by the stem surface. The data in the bottom figure were obtained with a fiberoptic probe, which collects light scattered within the stem tissue. The values are given relative to the measurements obtained for isolated plants (horizontal line). Filled triangles: far-red light; filled circles, red light; open squares, blue light. (Right) Elongation response of the first internode of *Datura ferox* seedlings to the proximity of neighboring plants. The seedlings were placed at the center of an even-height canopy of a leaf area index of approximately 0.9. During the experiment, which ran for 3 days, the seedlings were surrounded by cuvettes containing distilled water (clear filter) or a CuSO₄ solution (far-red-absorbing filter) that maintained the red/far-red radiation near 1.0 (Ballaré et al. 1995; reproduced with the author's permission from HortScience 30:1172–1182).

(Mahall & Callaway 1992), but it might involve **thigmomorphogenetic processes** (Sect. 5.7 of the chapter on growth and allocation). Plants clearly may respond in a different manner to surrounding plants of the same species as they do to plants of a different species (Huber-Sannwald et al. 1996).

It should be kept in mind that a response to the presence of the roots of neighboring plants is not universal (Krannitz & Caldwell 1995).

Climbing plants, which depend on their neighboring plants for support, somehow perceive the presence of mechanical support. The elongation

FIGURE 2. Effects of "crowding" on wildtype and transformed *Nicotiana tabacum* (tobacco) plants. The plants at the right were transformed with a mutant gene from *Arabidopsis thaliana*, which encodes a defective ethylene receptor that results in ethylene insensitivity. The plants at the left are isogenic control plants (courtesy M. Knoester, Utrecht University, the Netherlands).

of tendrils may be suppressed as soon as these contact a supporting structure (Sect. 5.7 of the chapter on growth and allocation). Provided with support other than a neighboring plant, climbing plants grow taller than unsupported individuals. Unsupported plants allocate more resources to their shoot branches, possibly increasing the chance of reaching a supportive structure, and allocate less to their roots. This indicates that it is the support itself, rather than any aspect of the neighboring plant's physiology, which affects the allocation pattern of climbing plants (Den Dubbelden & Oosterbeek 1995, Putz 1984).

There are clearly many ways in which plants perceive their neighbors, both above and below ground. Plants may respond to this in such a way as to avoid competition, or in a manner that makes them superior competitors. That is, plants that are sufficiently plastic for certain traits may well be able to avoid their neighbors and grow in such a way as to tap resources not utilized by neighbors (Sect. 6). When plants are forced to compete for the same pool of limiting resources, what ecophysiological traits determine competitive success?

4. Relationship of Plant Traits to Competitive Ability

4.1 Growth Rate and Tissue Turnover

Evidence from field studies, laboratory experiments, and ecological theory has converged on the conclusion that species from high-resource environments exhibit high **relative growth rate** (RGR), whereas species from low-resource environments will compete most effectively by minimizing tissue loss (greater **tissue longevity**) more than by maximizing resource gain (Sects. 3 and 6 of the chapter on growth and allocation). The ecological advantage of a high potential RGR seems straightforward: fast growth results in the rapid occupation of a large space, which leads to the preemption of limiting resources (Grime 1977). A high RGR may also facilitate rapid completion of the life cycle of a plant, which is essential for ruderals, whose habitat does not persist for a long time. In growth analyses and in short-term competition experiments carried out at a limiting nutrient supply, potentially fast-

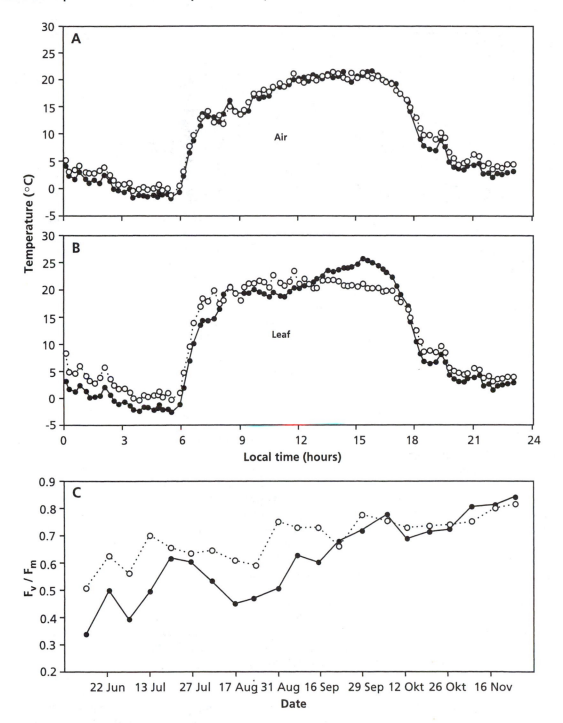

FIGURE 3. Diurnal variation in (A) air temperature and (B) the temperature of the leaves of *Eucalyptus pauciflora* (snow gum) above an open patch (open symbols) and above grass (filled symbols), measured from midnight to midnight on a day in September (early spring). Temperatures were measured 10 cm above ground level for one leaf of a seedling; seedlings were about 2 m apart. (C) Seasonal changes in average weekly midday values for the fluorescence characteristic F_v/F_m, which is an indicator of the quantum yield of photosynthesis, for seedlings of *Eucalyptus pauciflora* grown in an open habitat (open symbols), or above grass (filled symbols) (Ball et al. 1997). Copyright Blackwell Science Ltd.

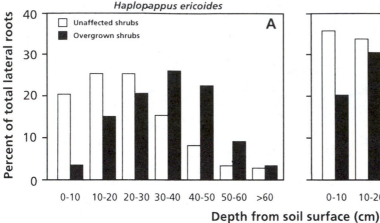
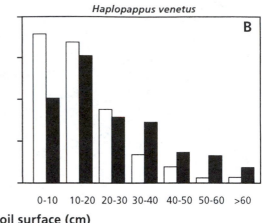

FIGURE 4. Percentage of total number of lateral roots of two shrubs, *Haplopappus ericoides* and *H. venetus* in each 10 cm depth increment below soil surface. Open bars: no competing invasive plants of *Carpobrotus edulis* present; dark bars: competing plants of *V. edulis* present (D'Antonio & Mahall 1991). Copyright Botanical Society of America, Inc.

growing species grow faster and produce more biomass than do slow-growing ones (Lambers & Poorter 1992). Even when growing naturally in a nutrient-poor meadow, in competition with surrounding plants, the species with the highest RGR_{max} grows fastest and produces most biomass, at least in relatively short experiments that last no more than one growing season (Fig. 5). The greater competitive ability in these short-term experiments is associated with a higher leaf area ratio and a higher specific root length, due to their thinner roots (Eissenstat 1992) and lower root mass density (Ryser & Lambers 1995).

Why do plants with a small root diameter and low tissue mass density (i.e., a high specific root length) and with thin leaves and a low tissue mass density (i.e., a high specific leaf area) not become dominant on nutrient-poor sites? For widely different species, such as evergreen and deciduous ones, it has repeatedly been shown that the low **tissue mass density** of fast-growing species is associated with a more rapid turnover of their leaves and a shorter **mean residence time of nutrients** (Sect. 4 of the chapter on mineral nutrition). In a comparison of ecologically contrasting grass species, slower-growing species from nutrient-poor habitats also tend to have a higher tissue mass density and slower turnover rates than do faster-growing ones from more productive sites (Ryser 1996). Turnover of plant parts inevitably causes loss of about half of the leaf nutrients from the plant and reduces the

mean residence time of the nutrients (Reich 1993; Sect. 4 of the chapter on mineral nutrition). Although rapid growth may therefore lead to a competitive advantage in the short term, even when the nutrient supply is severely limiting, there is a penalty associated with this trait in the long run (Berendse & Aerts 1987, Tilman 1988). That is, the losses associated with tissue turnover become so large that they cannot be compensated for by uptake of nutrients from the nutrient-poor environment. As a result, the fast-growing species are outcompeted by the slower-growing ones, once the time scale of the experiment is long enough that differences in tissue loss and mean residence time influence the outcome of competition (Aerts & Van der Peijl 1993).

Why, then, should a low tissue mass density be associated with faster turnover and shorter residence times? Part of the answer is straightforward: A high tissue mass density reflects a large investment in cell walls, sclerenchyma, fibers, and so on, which reduce the palatability and digestibility of the tissue and allow the tissue to withstand abiotic stresses and deter herbivores. As expressed by Eeyore (Milne 1928): "Why do all plants which an animal likes, have the wrong sort of swallow or too many spikes" (Sect. 3.3 of the chapter on growth and allocation).

Although some tissue loss results from herbivory and tissue death under severe conditions, tissue turnover also results from the highly programmed process of **senescence** rather than as a process of collapse and deterioration. The rate of tissue turnover is quite separate from tissue mass density, although it does correlate with it for reasons that will become clear in this section. This programming of cell death is obviously prolonged for leaves with a greater longevity, even though we understand very

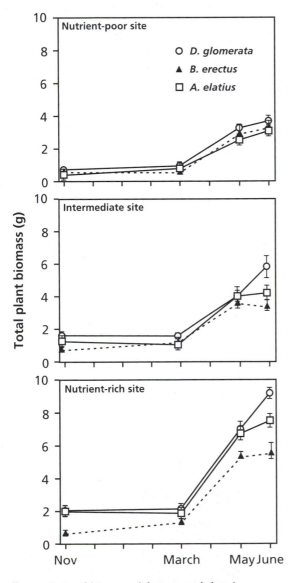

FIGURE 5. Total biomass of three tussock-forming grasses, growing in three meadows that differ in nutrient availability. The grasses differ in their RGR$_{max}$, with *Bromus erectus* (filled triangles) having the lowest RGR$_{max}$, *Arrhenaterum elatius* (open squares) an intermediate RGR$_{max}$ and *Dactylis glomerate* (open circles) the highest (Schläpfer & Ryser 1996). Copyright Oikos.

little of the intricacies of the differences. If the programming, however, is such that the leaves are going to last long, then it is essential to construct the leaves in such a way that biotic and abiotic factors do not prevent longevity. In other words, natural selection for slow turnover and a large investment in defense should go together, which explains the close correlation between the two, without there being a causal link.

There is a third reason for shorter nutrient residence times in faster-growing species at a low nutrient supply (Sect. 7 of the chapter on growth and allocation): Species differ in the manner in which they respond to a limitation by nutrients in the environment. The typical response of a fast-growing species upon sensing nutrient shortage is to promote leaf senescence and so withdraw nutrients from older leaves and use these for its newly developing tissues. A slow-growing species that occurs naturally on nutrient-poor sites will slow down the production of new tissues, with less dramatic effects on leaf senescence and allocation pattern. In other words, the environmentally induced senescence is much stronger in faster-growing species than it is in slower-growing ones. We again understand too little of a plant's physiology to account fully for our ecological observations, but the result is clear: The environmentally induced senescence of the rapidly growing species causes them to lose more nutrients.

4.2. Allocation Pattern, Growth Form, and Tissue Mass Density

In nutrient-rich conditions, *Lychnis flos-cuculi* genotypes with an inherently high **leaf mass ratio** (LMR) achieve higher yields in competition with *Anthoxanthum odoratum* and *Taraxacum hollandicum* (dandelion) than do genotypes with a lower LMR. At a low nutrient supply, this allocation pattern confers no advantage; moreover, genotypes with an inherently high **specific leaf area** (SLA) tend to produce smaller rosettes (Biere 1996). This information on the ecological significance of SLA is consistent with results on African C$_4$ species, which have been introduced into Venezuela. The introduced species with a high SLA outcompeted a native C$_4$ species, which has a low SLA, in relatively fertile places, but not in more infertile habitats (Baruch et al. 1985). On subantarctic islands the introduced grass *Agrostis stolonifera*, with a high SLA, is similarly able to survive in the wind-sheltered places, but it is not found outside these shelters, whereas *Agrostis magellania*, which is characterized by a lower SLA due to more sclerenchyma, occurs in the wind-swept parts of these islands (Pammenter et al. 1986). *Stephanomeria malheurensis*, which is a species with a relatively low SLA that occurs in the same environment as its progenitor *S. exigua* ssp. *coronaria* with a higher SLA, is restricted to sites where it may encounter greater stress. The number of individuals of *S. exigua* ssp. *coronaria* by far exceeds that of *S. malheurensis*, even though their RGR is very similar (Gottlieb 1978). A high SLA again appears to be associated with rapid growth and a low SLA with persistence. This suggests that a high leaf area ratio, due to a high SLA and/or a high LMR, which is associated with a high RGR, is advantageous in productive environments. On the other hand, a low SLA, which is associated with a low RGR, confers a selective advantage in relatively unfavorable environments.

Just as SLA is an important above-ground trait for a plant's competitive ability, the **specific root length** (SRL) is an important below-ground characteristic, determining a plant's ability to compete for resources such as nutrients and water. This can be illustrated using two tussock grasses (*Agropyron*), competing with *Artemisia tridentata* (sagebrush), used as an indicator species (Eissenstat & Caldwell 1988). *Agropyron desertorum* is an introduced species, with a greater competitive ability than the native *A. spicatum*. When *Artemisia* plants are planted among near-monospecific stands of one of the two tussock grasses, they show lower survival, less growth and reproduction, and a more negative water potential during part of the season when surrounded by *A. desertorum* than they do when they compete with *A. spicatum*. *A. desertorum* extracts water more rapidly from the soil profile, but it is remarkably similar in architecture, shoot phenology, root mass distribution in the soil profile, growth rate in various environments, and the efficiency of water and nitrogen use (Eissenstat & Caldwell 1987). Its roots are thinner, however, so that the length per unit mass (SRL) is about twice that of the less-competitive *Agropyron spicatum*. This higher SRL, in combination with more root growth in winter and early spring, allows the more competitive tussock grass to extract water more rapidly from the profile. These traits likely contributed to the observation that *Artemisia tridentata*, growing side by side with the two *Agropyron* species, acquired 86% of all its absorbed labeled phosphate from the interspace shared with *A. spicatum*, and only 14% from the interspace with *A. desertorum* (Fig. 6). Clipping of the tussock grasses enhanced phosphate uptake by sagebrush substantially, confirming that the grasses competed for resources

from the soil before clipping (Caldwell et al. 1987). Because phosphate is highly immobile in the soil, roots of the competing plants or their associated mycorrhizal fungal hyphae must have been very close to each other.

To be a successful competitor above ground as well as below ground, plants would need a high SLA as well as a high SRL, both of which can be realized through a low tissue mass density. Competitive species naturally occurring in productive meadows do indeed have a low leaf mass density as well as a low root mass density (Ryser & Lambers 1996).

4.3 Plasticity

In previous chapters, numerous examples have been given on the acclimation of photosynthesis, respiration, and biomass allocation to environmental factors such as irradiance and nitrogen supply. A great ability to acclimate reflects a genotype's **phenotypic plasticity** for a specific trait; however, a relatively small plasticity for one trait may result from a large plasticity in other traits. For example, the low morphological plasticity (stem length) of an alpine *Stellaria longipes* (Sect. 5.7 of the chapter on growth and allocation) is a consequence of a high physiological plasticity (ethylene production). Both traits are directly related to the same environmental cue (wind stress), and the expressed phenotype has a direct bearing on the plant's fitness (Emery et al. 1994). In addition, a large morphological plasticity in biomass allocation between roots and leaves in response to nutrient supply or irradiance allows a low plasticity of the plant's growth rate, so that this varies relatively little between different environments.

It has been suggested that a high plasticity allows a genotype to maintain dominance in spatially or temporally variable environments by enabling them to explore continuously new patches that have not been depleted, thus sustaining resource capture and maintaining fitness (Grime et al. 1986). By contrast, in habitats of predictably low resource supply, plant production would be restricted to a continuously low level and a strategy of conservation of captured resources, associated with slow growth, would be favored. Such a contention is hard to verify, in view of the fact that greater plasticity for one trait is made possible by smaller plasticity for another.

There are certainly convincing examples of greater plasticity associated with competitive ability in a particular environment. Late-

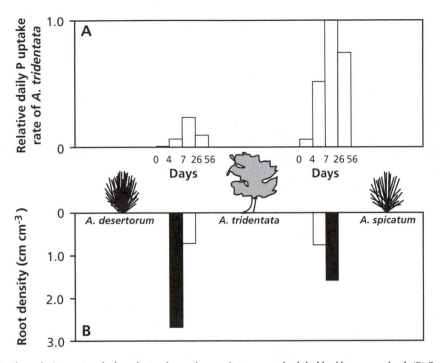

FIGURE 6. (A) The relative rate of phosphate absorption. That is, the average daily uptake of phosphate isotopes, ^{32}P and ^{33}P, by root tips of *Artemisia tridentata* (sagebrush) from soil interspaces shared with one of two tussock grasses, the native *Agropyron spicatum*, or the introduced *Agropyron desertorum* at various times after labeling. The separate labels were injected at either side of *A. tridentata*, in the interspace shared with one of the two *Agropyron* species. This made it possible to assess from which interspace the label had been acquired. (B) Rooting density of the *Agropyron* species (filled bars) and of *A. tridentata* (open bars). Rooting densities were not significantly different in the two interspaces or between the *Agropyron* species, but they were significantly less for *A. tridentata* than they were for the *Agropyron* species (Reprinted with permission from Caldwell et al. 1985). Copyright 1985 American Association for the Advancement of Science.

successional species tend to have a greater potential for adjustment of their photosynthetic characteristics to shade than do early-successional species (Küppers 1984). A classic case is the response of stem elongation to shade light (Sect. 5.1 of the chapter on growth and allocation and Sect. 2 of this chapter). Shade light also suppresses branching in dicotyledonous species and enhances tillering in grasses like *Lolium perenne* and *L. multiflorum*. This plastic response is likely to be important in coping with neighbors (Deregibus et al. 1983). To confirm the importance of the phytochrome system for the perception of neighboring plants and of the following response, Ballaré et al. (1994) used transgenic plants of *Nicotiana tabacum* (tobacco), overexpressing a phytochrome gene. These transgenics showed a dramatically smaller response to the red/far-red ratio of the radiation and to neighboring plants. In a stand of such

transgenics, the small plants of the population were rapidly suppressed by their neighbors. These results indicate that a high degree of plasticity in morphological parameters plays an important role in the competition with surrounding plants.

In addition, for the plastic response to wind, as discussed for alpine and prairie ecotypes of *Stellaria longipes* (Sect. 5.7 of the chapter on growth and allocation), there is supportive evidence. Ecotypic variation for phenotypic plasticity of stem elongation in this species correlates with wind speed along an elevational gradient. The greater the stress by wind, the smaller plasticity for stem elongation (Emery et al. 1994).

With respect to variation in nutrient supply, however, the present information is far from conclusive. Fast-growing species from high-resource environments often show less or a similar change in allocation parameters like root mass ratio and stem

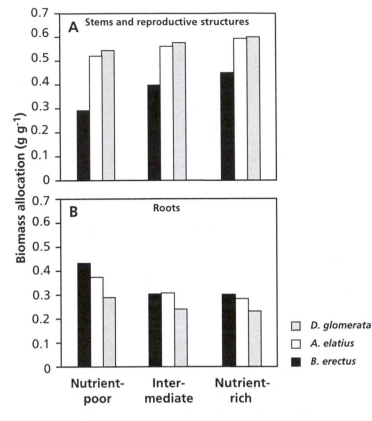

FIGURE 7. Biomass allocation to (A) stems and reproductive structures and (B) roots of three tussock-forming grasses, growing in three meadows that differ in nutrient availability: nutrient-poor, intermediate, and nutrient-rich. The grasses differ in their RGR_{max}, with *Bromus erectus* (filled bars) having the lowest RGR_{max}, *Arrhenaterum elatius* (open bars) an intermediate RGR_{max}, and *Dactylis glomerata* (shaded bars) the highest (Schläpfer & Ryser 1996). Copyright Oikos.

mass ratio as slow-growing ones from nutrient-poor environments (Poorter et al. 1995). This point is illustrated for three tussock-forming grass species, co-occurring in meadows differing in soil fertility (Fig. 7). Interpretation of the data, however, is complicated by the fact that the less plastic species, *Bromus erectus*, has a lower stem mass ratio and lower stature than the other two grasses, *Arrhenaterum elatis* and *Dactylis glomerata*. Hence, at a higher nutrient supply competition for light may play a role. Does this therefore falsify the hypothesis that fast-growing species are more plastic? A survey that includes a vast number of species finds no such correlation (Reynolds & D'Antonio 1996). In one of the examples given here, where plants were compared in a natural meadow, there is an alternative explanation: The competitive and taller fast-growing grasses have a greater capacity to capture light when the nutrient supply is relatively high, as well as a greater ability to absorb nutrients, also from relatively infertile soils where their growth rate is reduced, than do the inherently slow-growing grasses. Hence, when growing at a similar rate, the inherently slow-growing grasses sense a greater shortage of nutrients in their tissue than the inherently faster-growing competitors. That is, they will show a stronger tendency to in-

crease their root mass ratio (Sect. 5.4 of the chapter on growth and allocation). Rather than revealing greater plasticity, this response may show the consequences of a restricted nutrient uptake capacity. This leaves us with the problem of not quite knowing how to test the interesting hypothesis of greater plasticity of fast-growing species in response to nutrient supply.

In summary, it would seem that fast-growing species from high-resource environments are more plastic for some traits, such as photosynthetic characteristics and the rate of stem elongation in response to shade, surrounding plants, and wind. When it comes to below-ground plant traits and morphological plasticity in response to the supply of nutrients, this conclusion is hard to substantiate.

5. Traits Associated with Competition for Specific Resources

5.1 Nutrients

We have now shown the physiological basis for the trade-off between rapid growth and tolerance of low nutrient supply (Sect. 4). What evidence is there that species growing on infertile soils draw down resources below levels used by potential competitors (i.e., low R*) and what might be the processes responsible for this **resource draw-down**? The most explicit test of the R* hypothesis was a field experiment in which several perennial prairie grasses that naturally occur on sites of different soil fertilities were planted in monoculture and in competition on several soils of differing fertility (Wedin & Tilman 1990, Tilman & Wedin 1991). Within 3 years, monocultures of the more slowly growing species reduced the concentration of extractable soil nitrate and ammonium to lower levels than did monocultures of high-RGR species from more fertile sites (Fig. 8). In addition, soil nitrate concentrations were just as low in competition treatments between fast- and slow-growing species as they were in monocultures of the slow-growing species. This coincided with elimination of the more rapidly growing species. The traits most consistently associated with competitive success in these experiments were a high allocation to root biomass and low RGR. High allocation to roots was the plant trait that correlated most strongly with the nitrogen draw-down. The low RGR reduced loss rates and enhanced tolerance of low supply rates. What other nutritional traits might be involved in competition for nutrients?

The **uptake kinetics** of species from infertile soils are unlikely to result in low soil solution concentra-tions. These species typically have a lower I_{max} of nutrient uptake and do not differ consistently in K_m from species that occur on fertile soils (Sect. 2.2.3.1 of the chapter on mineral nutrition). The influence of uptake kinetics on soil solution concentration should be greatest for mobile nutrients (e.g., nitrate) and less pronounced for cations (e.g., ammonium) and for phosphate (Sect. 2.1.2 of the chapter on mineral nutrition). Species from specific nutritional situations often have a capacity to tap sources of nutrients that are unavailable to other species, such as insoluble phosphate or organic nitrogen (Sects. 2.2.4 and 2.2.5 of the chapter on mineral nutrition), but this would not explain the draw-down of extractable inorganic nutrients in the soil. Most species that occur in environments where competition for nutrients is important have well-developed mycorrhizae, so it is unlikely that differences among species in extent of mycorrhizal colonization accounts for differential capacity to reduce soil nutrient concentrations.

The most likely cause of nutrient draw-down by species in infertile soils is **microbial immobilization** of nutrients due to the low litter quality of species adapted to infertile soils (Wedin & Tilman 1990). Litter from these species has low concentrations of nitrogen and phosphorus, leading to low net mineralization rates (Fig. 8). In addition, a large proportion of the litter is produced by roots, which typically have lower tissue nutrient concentrations than do leaves and which are dispersed throughout

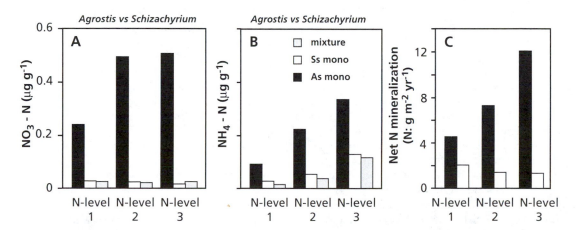

FIGURE 8. Extractable soil nitrate and ammonium and net nitrogen mineralization in experimental monocultures of an early-successional rapidly growing prairie grass (*Agrostis scabra*) and a late-successional slowly growing prairie grass (*Schizachyrium scoparium*), and of the two species growing together in mixture. Plants from a Minne-sota prairie were grown for 3 years in soils that contain three levels of nitrogen, after which soil samples were extracted with 0.01 M KCl for measurement of nitrate and ammonium. Net nitrogen mineralization was measured monthly in the field (Tilman & Wedin 1991, Wedin & Tilman 1990). Copyright Ecological Society of America.

the soil, so that the zone of immobilization coincides with the zone of uptake.

Nutrient-impoverished habitats, such as the heathlands of Western Australia and South Africa, are among the most species-rich habitats. How do so many species coexist where strong competition for nutrients must be critical for survival? There are some specialized traits that enable certain species to tap insoluble phosphorus that is unavailable to other species (Sects. 2.2.4 and 2.2.5 in the chapter on mineral nutrition). Although species differ in preference for different forms of nitrogen, most species have the physiological capability to tap all forms of soluble nitrogen and to adjust their capacities for uptake and assimilation, depending on supply (Eviner & Chapin 1997). Ectomycorrhizae could break down protein nitrogen that would otherwise not be directly available to plants (Read 1991). In most cases, however, competitive coexistence of multiple species in a community is not a simple function of capacity to tap a unique resource or the capacity to draw down a single resource; rather, it involves a wide range of traits and subtle differences in resistance to different environmental circumstances.

5.2 Water

The mechanism by which **drought-resistant plants** draw down soil moisture is well established. The lower the **water potential** that a species can tolerate, the lower the level to which it can reduce soil moisture. When soil water potential falls below the minimum water potential tolerated by potential competitors, they can no longer withdraw water from the soil. The traits that enable a species to maintain activity at a low water potential include osmotic adjustment and a stomatal conductance that is relatively insensitive to root or leaf water potential (Sects. 4.1 and 5.4.1 of the chapter on plant water relations).

Transpiration is the major avenue of water loss to the atmosphere and therefore of soil drying in dense vegetation. The species that cause greatest **moisture draw-down**, however, are not necessarily those with greatest resistance of low water availability. In general, the species with greatest **drought resistance** have a suite of morphological and biochemical traits that enable them to conserve water (e.g., CAM and C_4 photosynthesis, low stomatal conductance, low hydraulic conductance of the stem, and so on). When water is available, most plants maximize stomatal conductance and therefore water loss. In a mixed-species community, the species responsible for the greatest quantity of water loss are not those that are most resistant of water stress. The drought-resistant species are probably most important in the final stages of moisture draw-down, after less-resistant species become dormant. The abundance of different life forms and physiological strategies in deserts indicates that there are many ways of competing effectively in dry environments, only some of which involve extreme resistance of low soil water potential. Other modes of competing effectively in deserts include phenological **avoidance** of drought and rapid growth when water is available.

Roots commonly pass through dry soil layers to deep horizons that contain more moisture. In the dry soil layers the soil matric potential may be more negative than the hydrostatic pressure in the xylem of the roots. Water may then move from the roots to the dry soil, and roots can form a bridge for water transport between soil layers (Sect. 5.2 of the chapter on plant water relations). Stolon-connected plants in separate moist and dry-soil compartments similarly may transport considerable quantities of water from one compartment of the soil to the other (Van Bavel & Baker 1985).

A low conductance between roots and soil or of the soil might preclude substantial efflux of water from roots. A nocturnal down-regulation of water-channels proteins (Sect. 5.2 of the chapter on plant water relations) might reduce water loss to dry soil. Water efflux from roots into soil might be viewed as an undesirable process; however, the water released at night is available for reabsorption during the day. In addition, the moist soil may promote the nutrient acquisition of roots and prolong the activity of symbiotic microorganisms such as mycorrhizal fungi in the upper soil layers. The moist soil may also prevent chemical signals that would otherwise originate from roots in contact with dry soil (cf. Sect. 5.4.1 of the chapter on plant water relations). Some of the hydraulically lifted water will probably be available for shallow-rooted competing plants. As much as 20 to 50% of the water used by a shallow-rooted tussock grass (*Agropyron desertorum*) comes from water that is hydraulically lifted by neighboring sage brush (*Artemisia tridentata*) in the Great Basin desert of western North America (Richards & Caldwell 1987). Sugar maple (*Acer saccharum*) similarly provides 46 to 61% of the water used by strawberries (*Fragaria virginiana*) that grow beneath the tree by hydraulic lift (Dawson 1993). Individuals that are large enough to be quantitatively important in **hydraulic lift** will have predictable access to water and will be taller than the shallow-rooted species; therefore, they may not be severely impacted by this competition.

5.3 Light

Strong competition for light seldom coincides with strong competition for below-ground resources for two reasons. First, high availability of below-ground resources is an essential prerequisite for the development of a leaf canopy dense enough to cause intense light competition, which is strongest under conditions where water and nutrients are not strongly limiting to plant growth. Second, trade-offs between shoot and root competition constrain the amount of biomass that can be simultaneously allocated to acquisition of above- and below-ground resources (Tilman 1988). Those species that are effective competitors for light are trees with a high above-ground allocation.

As with water, the species that most strongly reduce light availability are not necessarily the species that are most tolerant of low light. Species that are tall and have a high leaf area index have greatest impact on light availability, whereas understory plants and late-successional species are generally the most shade-tolerant. Because light is such a strongly directional resource, competition for light is generally quite asymmetric, with the taller species having greatest impact on the shorter species, with often little detectable effect of understory species on the overstory, at least with respect to light competition.

5.4 Carbon Dioxide

Carbon dioxide is relatively well mixed in the atmosphere; therefore, plant uptake creates less localized depletion of CO_2 than of nutrients, water, or light. Nonetheless, photosynthesis is often CO_2-limited, especially in C_3 plants. Plants with contrasting photosynthetic pathways may therefore differ in their competitive ability in relation to atmospheric CO_2 concentration. For example, one might expect the growth of C_4 plants, whose rate of photosynthesis is virtually saturated at 35 Pa CO_2, to respond less to the global rise in atmospheric CO_2 concentration than that of C_3 plants. To test this hypothesis, Johnson et al. (1993) compared the growth of **C_3 and C_4 plants**, while growing in

competition, at CO_2 concentrations ranging from preindustrial levels to the present 35 Pa. As expected, photosynthesis and growth of the C_3 species was enhanced more by high levels of CO_2 than was that of C_4 species. Whereas the C_4 species outyielded the C_3 plants at low CO_2 concentrations, the C_3 plants were superior competitors at elevated [CO_2] (Fig. 9). How, then, can we asess whether a change in competitive ability as suggested by the data in Figure 9 has indeed occurred? To address this question, soil organic matter of known age was analyzed for ^{13}C to estimate changes in the relative abundance of C_3 and C_4 species between the late Pleistocene and the early Holocene in northern Mexico (Cole & Monger 1994). Results showed that there was an increase in C_3 species about 9000 years ago, a time when Antarctic ice cores showed a rapid rise in atmospheric [CO_2]. Plant macrofossils from packrat middens showed that this vegetation change coincided with an increase in aridity, which should have favored C_4 species. The vegetation change therefore was most likely caused by increased atmospheric [CO_2] rather than by climatic change.

Further evidence that C_3 plants profit more from a rise in atmospheric [CO_2] than C_4 species comes from work on an invasive woody C_3 legume, *Prosopis glandulosa* (honey mesquite). This invasive species has increased in abundance in North American C_4-dominated grasslands over the past 150 years. When grown in monoculture, its below-ground biomass, rate of N_2 fixation, and water-use

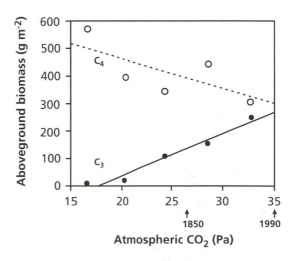

FIGURE 9. Above-ground biomass of C_3 and C_4 species that developed from the seed bank of a Texas savanna soil over a range of CO_2 concentrations from 15 to 35 Pa over a period of 13 weeks (Johnson et al. 1993). Copyright Kluwer Academic Publishing.

efficiency are increased at present-day levels of atmospheric CO_2, in comparison with historically lower levels. In competition with a C_4 grass, *Schizachyrium scoparium* (little bluestem), however, there is no effect on biomass. Rising levels of CO_2 may well have contributed to its success, but the shrub's strategy to avoid competition with neighboring grasses is probably more important (Polley et al. 1994).

Will C_3 species continue to conquer the world at the expense of C_4 species in years to come, while the concentration of CO_2 continues to rise? In experiments using around 34 and 62 Pa CO_2, the competitive ability of *Festuca elatior* or *Triticum aestivum* (wheat) (C_3) was enhanced compared with that of *Sorghum halepense* or *Echinochloa frumentacea* (C_4), respectively (Carter & Peterson 1983, Wong & Osmond 1991). Drake and co-workers studied the effects of elevated $[CO_2]$ on a natural salt-marsh vegetation, consisting of both C_3 (predominantly *Scirpus olneyi*) and C_4 (mainly *Spartina patens*) sedges. After 4 years of exposure to elevated $[CO_2]$, the biomass of *S. olneyi* was greatly enhanced, both on sites where this species occurred as a pure stand and also where it grew in mixtures with *S. patens*. There was very little effect of elevated $[CO_2]$ on the biomass of *S. patens* growing in a monospecific community, whereas it was reduced on sites where it grew in competition with the C_3 sedge (Arp et al. 1993).

The results so far suggest that C_4 plants decreased in competitive ability since the beginning of the industrial revolution. They may well continue to lose ground with a further rise in atmospheric CO_2 concentration. Elevated CO_2 concentrations, however, interact with temperature and affect plant growth in a manner that may be quite different from a plant's response to elevated $[CO_2]$ alone. The climate change caused by elevated $[CO_2]$ may well have an opposite effect on competition between C_3 and C_4 species. Increased temperatures and drier climates might favor C_4 grasses and lead to an expansion of the area occupied by C_4 species in Australia (Henderson et al. 1995).

Elevated atmospheric CO_2 concentrations can alter availability of other environmental resources that can shift competitive balance in unpredictable ways. In a dry North American prairie, elevated $[CO_2]$ causes an increase in soil moisture as a result of the reduction in stomatal conductance and transpiration. The improved soil moisture favors tall C_4 grasses over a subdominant C_3 grass, which is just the opposite to the result expected from direct photosynthetic response to CO_2 (Mo et al. 1992, Owensby et al. 1993). A recent review points out

that 13 of the 14 published studies on the competitive interactions of C_3 and C_4 species have been conducted in relatively fertile soils (Reynolds 1996), where we would expect photosynthetic performance to have the strongest connection to growth and competitive ability. We know almost nothing about how CO_2 will affect competitive interactions on the remaining 90% of the earth's surface.

The results on the outcome of competition between C_3 and C_4 species as dependent on the CO_2 concentration in the atmosphere suggests photosynthesis is a major factor in determining competitive interactions, but is that also the case if we restrict our comparison to C_3 plants only? There is a wealth of information on the photosynthetic traits of "invasive" species as well as on early-succession woody species and the species that ultimately replace these. Succession is far more complicated than can be accounted for by competitive interactions alone. Competition in the succession following a fire or upon canopy destruction by a storm is a race without a single winner, unlike in a standard athletic contest. Rather, any species that is among the leaders at some stage during the race is a winner in that being among the leaders for a reasonable period permits a sufficient carbon and nutrient gain to be accumulated for the species to ensure it runs in subsequent races. The entry in subsequent races may occur via vegetative regeneration, via a stored seed bank, or via dispersal to other locations, but the prerequisite for any of these is sufficient carbon and nutrient accumulation at some stage during vegetative growth. In succession, therefore, competition does play a role and at later stages of succession the early-successional species are very poor competitors. Two exotic vines, *Pueraria lobata* (kudzu) and *Lonicera japonica* (Japanese honeysuckle), are major weed species in the southeastern United States. In comparison with a number of native vines, *Rhus radicans* (poison ivy), *Parthenocissus quinquefolia* (Virginia creeper), *Vitis vulpina* (wild grape), and *Clematis virginiana* (virgin's bower), they have very similar rates of photosynthesis. Thus, the highly prolific growth of the two exotic weedy vines cannot be explained by higher rates of photosynthesis (Carter et al. 1989).

6. Positive Interactions among Plants

Not all plant–plant interactions are competitive. Plants often ameliorate the environment of neighbors and increase their growth and survivorship (**facilitation**), particularly at the seedling stage and where the physical environment or water and nutrients strongly constrain growth (Callaway 1995).

In hot dry environments, seedlings often establish preferentially in the shade of other **nurse plants**. At the seedling stage, barrel cacti (*Ferocactus acanthodes*) suffer high mortality in deserts because of their small thermal mass. Seedlings in the shade of other plants are 11°C cooler than they are in full sun and only survive in shade (Nobel 1984, Turner et al. 1966). Facilitation due to shading also occurs in oak (*Quercus*) savannas by reducing drought and overheating and in salt marshes by reducing soil evaporation and therefore salt accumulation (Callaway 1995). Hydraulic lift by deep-rooted plants often increases water potential and growth of adjacent plants (Sect. 5.2).

A second general category of facilitation involves enhanced nutrient availability. The most dramatic examples of this are establishment of N_2-fixing species in early successional and other low-nitrogen habitats (Calloway 1995, Chapin et al. 1994, Vitousek et al. 1987). Decomposition of high-nitrogen litter of N_2-fixing plants increases nitrogen availability in these environments. In other cases, organic matter enhances the nutrient and water status of understory plants (Calloway 1995). Other facilitative effects of plants include oxygenation of soils, stabilization of soils, physical protection from herbivores, and attraction of pollinators (Calloway 1995).

In the real world, plant–plant interactions involve complex mixtures of competitive and facilitative effects, which often occur simultaneously. For example, at Glacier Bay, Alaska, alder (*Alnus sinuata*) is an early colonizer that has multiple effects on spruce (*Picea sitkensis*), which is the ultimate successional dominant. Alder increases spruce growth by adding nitrogen and organic matter, but negatively affects spruce growth as results of shading and root competition. Alder increases seedling mortality as a result of seedling burial by litter and by providing habitat for seed predators (Chapin et al. 1994). Over the long term, the net effect of alder is to reduce stand density and increase the growth of individual spruce trees. Similar combinations of competitive and facilitative effects have been observed in many studies, with the net effect of one plant on another often changing with time, depending on variation in weather and successional stage (Aguiar et al. 1992, Callaway 1995).

7. Plant–Microbial Symbiosis

The symbiotic relationships between plants and their microbial symbionts can strongly influence the outcome of competition. Many woody species that appear in early phases of succession (e.g., after a fire) are N_2-fixing legumes. When the level of nitrogen in the soil increases, their rates of N_2 fixation decline (Sect. 3.9 of the chapter on symbiotic associations). At later stages during succession, such pioneers may succumb to phytophagous arthropods [e.g., the pioneer *Acacia baileyana* (Cootamundra wattle) in Australia]. The competitive success of *Acacia saligna* (= *A. cyanophylla*), which was introduced into South Africa from Australia to stabilize sand dunes, is partly ascribed to its symbiotic association with rhizobia (Stock 1995).

If competing plants are mycorrhizal, then we also need to consider the ability of their external mycelium to capture nutrients. If they share a common external mycelium, then competition exists between the plants to acquire nutrients from that external mycelium. Can mycorrhizal infection alter the balance between different species? When seedlings of the grass *Festuca ovina* are growing in nutrient-poor sand, either with or without VAM, and in competition with seedlings of other species, they grow less well in the presence of VAM than they do in its absence. Seedlings of many of the competing species, however, grow substantially better, with the exception of nonmycorrhizal species (Grime et al. 1987). Comparing the RGR of the grass *Lolium perenne* with the dicot *Plantago lanceolata*, grown separately or in competition, and with or without VAM, shows similar values for RGR when the plants are growing separately, irrespective of their mycorrhizal status (Fig. 10). When growing in competition, however, the mycorrhizal *P. lanceolata* plants have the higher mean RGR, whereas the opposite is found when the plants are nonmycorrhizal. This suggests that the coexistence of *P. lanceolata* in grasslands may well depend on mycorrhiza (Newman et al. 1992).

Competitive interactions may become complicated when species differ in their mycorrhizal dependency (Koide & Li 1991). For example, of two tallgrass prairie grasses, *Andropogon gerardii* (big bluestem) is 98% dependent on the symbiosis, versus only 0.02% in *Koeleria pyranidata* (junegrass). When competing in pairs, *A. gerardii* dominates in the presence of mycorrhizal fungi, whereas *K. pyranidata* does in the absence of the fungus (Hetrick et al. 1989). In addition, when competing with another tallgrass species with lower

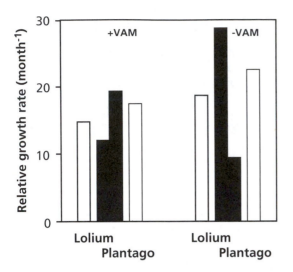

FIGURE 10. Relative growth rates of the grass *Lolium perenne* and the dicot *Plantago lanceolata* grown in a glasshouse in heat-sterilized, nutrient-poor grassland soil that was originally free of mycorrhiza. The plants were grown separately (open bars) or together (filled bars), either without VAM or inoculated at the time that the plants were competing, as judged from the size of the plants (Newman et al. 1992).

mycorrhizal dependency, *Alymys canadensis* (Canada wild rye), the mycorrhizal fungi influence the competitive effects and responses in a similar manner (Hartnett et al. 1993). In their natural habitat *A. gerardii* grows in the warm season, whereas *K. pyranidata* and *E. canadensis* grow most rapidly when soil temperatures are cooler, limiting mycorrhizal nutrient uptake. This phenological separation of growing seasons probably contributes to their coexistence in the tallgrass prairie community .

Some herbaceous pioneers are nonmycorrhizal (Sect. 2.2. of the chapter on symbiotic associations). These plants may grow well in the early phase of succession because of their special ability to release phosphate from sparingly available sources (Sect. 2.2.5 of the chapter on mineral nutrition) or because the availability of phosphate is high. At later stages, mycorrhizal species may arrive and replace nonmycorrhizal species. When growing in competition with the nonmycorrhizal annual crucifer *Brassica nigra*, growth and nutrient uptake of the mycorrhizal grass *Panicum virgatum* are reduced, when plants are of equal size. On the other hand, growth of the nonmycorrhizal *B. nigra* is the same as that in monoculture. The presence of

collembola (Arthropoda) that graze mycorrhizal fungi enhances the competitive advantage of the nonmycorrhizal crucifer. The grazing of VAM hyphae by collembola increases the availability of nitrogen for the nonmycorrhizal *B. nigra*. When seedlings of the nonmycorrhizal *B. nigra* have to compete with the mycorrhizal plants of *Panicum virgatum* that germinated 3 weeks earlier, the situation is reversed: *B. nigra* is negatively affected by competition, whereas the larger and older grass plants are not (Boerner et al. 1991). This may account in part for the gradual replacement of nonmycorrhizal annuals by mycorrhizal perennials. Allelochemicals released by the mycorrhizal fungus may be important in this replacement (Sect. 2.2 of the chapter on symbiotic associations). Thus, germination and seedling growth of nonmycorrhizal species is inhibited by the presence of mycorrhizal hyphae in the rhizosphere (Fig. 11). When phosphate fertilization suppresses the mycorrhizal microsymbiont, the deleterious effects on root growth and functioning of nonmycorrhizal species may become less. This might lead us to the erroneous conclusion that the growth of the plants whose biomass is increased most by phosphate fertilization was limited more by phosphate than was that of the mycorrhizal plants. If we go to the root of the problem, however, then intricate allelochemical interactions that involve mycorrhizal fungi may well account for our field observations (Francis & Reid 1994).

Mycorrhizal fungi can harm nonmycorrhizal plants, but the reverse may also occur. When *Glycine max* (soybean) is grown in the vicinity of the nonmycorrhizal species *Urtica dioica* (stinging nettle), infection of the soybean roots by the mycorrhizal fungus *Glomus mosseae* is inhibited (Fig. 12A). A fungitoxic lectin (Sect. 2.2 of the chapter on symbiotic associations) distinctly inhibits the growth of fungal hyphae (Fig. 12B), which suggests that it might be (partly) responsible for the effect of the presence of nonmycorrhizal species on the performance of mycorrhizal plants. Because the effects of the lectin are strongest 1 hour after the first application and diminish thereafter, despite repeated application, the mycorrhizal fungus may have a mechanism to combat the lectin. There is no further evidence, however, to support this hypothesis. Other lectins [e.g., from mycorrhizal hosts like *Triticum aestivum* (wheat), *Lycopersicon esculentum* (tomato), or *Solanum tuberosum* (potato)] that have a high affinity for chitin have no antifungal properties (Schlumbaum et al. 1986). It remains to be firmly established if the lectin from roots and rhi-

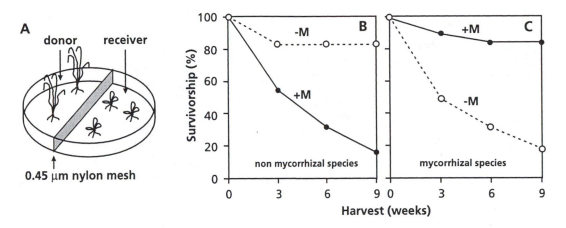

FIGURE 11. (A) Experimental design to assess the effect of the presence of mycorrhizal hyphae on the survival of seedlings of mycorrhizal and nonmycorrhizal species. (B) Effects of mycorrhizal fungi on seedling survival of the nonmycorrhizal *Arenaria serpyllifolia*. (C) Effects of mycorrhizal fungi on seedling survival of the mycorrhizal *Centaurium erythraea* (Francis & Read 1994). Copyright Kluwer Academic Publishing.

zomes of stinging nettle is the major factor that accounts for the effect of this nonmycorrhizal plants on mycorrhizal neighbors.

Herbivory has equally strong effects on competitive interactions, with the effect depending on the selectivity of herbivores. Plants that are selectively grazed, due to low defensive investment or other reasons, always have a reduced competitive ability compared with ungrazed neighbors. In the presence of nonselective grazing (the "lawnmower effect"), however, species that lack well-developed defensive mechanisms are typically more tolerant of grazing (Bryant & Kuropat 1980, Rosenthal & Kotanen 1994).

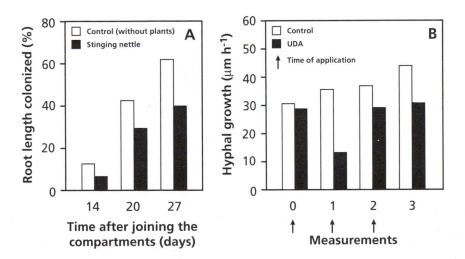

FIGURE 12. (A) Spread of the mycorrhizal fungus *Glomus mosseae* across the rhizosphere of *Urtica dioica* (stinging nettle) or control soil, without stinging nettle. Uncolonized *Glycine max* (soybean) plants were used as acceptor plants. They were separated from well-colonized soybean plants (donor plants) by a test container of soil planted with stinging nettle or a container of soil without plants. (B) Effect of agglutinin (lectin) from *Urtica dioica* on the hyphal growth of *Glomus mosseae*. The growth of hyphae of germinated spores was measured after application of small droplets of purified agglutinin. Application was repeated at 1 h intervals (arrows) (Vierheilig et al. 1996). Copyright Kluwer Academic Publishing.

8. Succession

Successional changes in species composition following disturbance are the net result of different rates of **colonization**, **growth**, and **mortality** of early- and late-successional species (Bazzaz 1979, Chapin 1993, Egler 1954). Competition and facilitation both play strong roles in successional change, and the resulting change in species composition through succession is associated with predictable changes in ecophysiology. The physiology of initial colonizers differs strikingly between **primary succession**, when plants colonize an area for the first time, and **secondary succession**, when plants recolonize previously vegetated areas following disturbances such as fire or agriculture. Soils in primary succession typically have low nitrogen and organic matter content.

Primary-successional soils initially lack a buried seed pool, requiring colonizers to disperse to the site, whereas secondary-successional sites are colonized from both the buried seed pool and from dispersal to the site. Propagules of early colonizers of primary succession have seeds that are as small as, or smaller than, those of species that colonize secondary succession, which, in turn, are smaller than seeds of late-successional species (Fig. 13), perhaps because colonizers of many primary-successional environments have further to travel than do secondary-successional colonizers. The larger **seed size** of late-successional species (see also Sect. 3.1 of the chapter on life cycles) provides reserves to support growth in fully vegetated sites, where competition is likely to be more intense.

When grown under favorable laboratory conditions, early-successional species grow more rapidly than do late-successional species. In addition, colonizers of primary-successional habitats have lower **RGR** than do colonizers of more fertile secondary-successional disturbed sites (Fig. 14), which suggests that among colonizing species low soil fertility has selected for species with traits that cause low RGR.

Early-successional trees or shrubs invariably have higher rates of **photosynthesis** on an area basis than do the ones that appear later during the succession (Owens 1996, Raaimakers et al. 1995; Table 1). When the light-saturated rates of photosynthesis of the shrubs described in Table 1 are compared with those of the final climax tree species, *Fagus sylvatica* (beech), which are only as low as 3 to $4\,\mu mol\,m^{-2}\,s^{-1}$, it is quite obvious that high rates of photosynthesis cannot account for the replacement of early-successional species by later

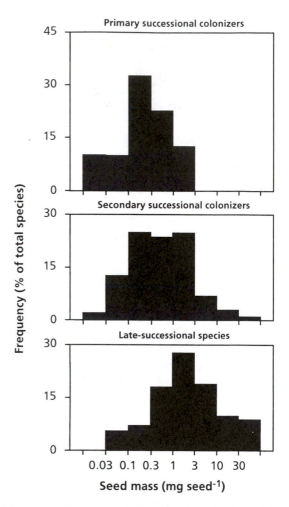

FIGURE 13. Frequency distribution of log (seed mass) for British species that are colonizers of primary successional (skeletal, *n* = 60 species), secondary successional (disturbed, *n* = 88 species), or late-successional (woodland, *n* = 58 species) habitats. Data are calculated from Grime et al. 1981 (Chapin 1993). Copyright Blackwell Science Ltd.

ones. As discussed later in this chapter, however, the late-successional and invasive species have a more positive carbon balance, due to their greater leaf area and better exposure of the leaves. The physiological mechanisms accounting for leaf expansion and leaf exposure are clearly far more important than are the photosynthetic capacity of individual leaves to account for the outcome of competition.

As with photosynthesis, early- and mid-successional species typically have higher potential to absorb nutrients than do late-successional species (Fig. 15). This could reflect the high potential growth rate and, therefore, the high nutrient demands of colonizing species.

Herbivores are often a major cause of plant mortality during succession. Late-successional species, with their long-lived leaves, have higher concentrations of defensive compounds and are therefore less palatable than early-successional species (Fig. 16).

In summary, the changes in ecophysiological traits through succession are identical to those described earlier in species that compete effectively in high- versus low-resource sites, explaining the

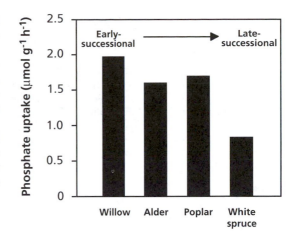

FIGURE 15. Rate of phosphate uptake by excised roots of tree seedlings from an Alaskan primary-successional floodplain sequence grown in a glasshouse (Walker & Chapin 1986). Redrawn from Chapin (1993). Copyright Blackwell Science Ltd.

change in competitive balance that causes species replacement through succession.

9. What Do We Gain from This Chapter?

There is not just one exclusive ecophysiological trait that gives a genotype competitive superiority. The outcome of competition may be due to the occurrence of an event, such as flooding, frost, or drought, with which one genotype is able to cope better and therefore survive, whereas other genotypes may lose out. Superior traits in one environment (e.g., a low tissue density, which is associated with rapid growth when nutrients are plentiful) may be inferior traits in a different environment, when a low tissue density is associated with relatively large losses of nutrients when nutrients are scarce. These trade-offs among suites of physiological traits are critical to understanding patterns of competitive success in different environments.

Competitive advantage may be based on a plant's secondary metabolism (i.e., the exudation of allelochemicals that harm other plants, excretion of compounds that solubilize sparingly available nutrients or precipitate harmful soil components, production of chemicals that sequester heavy metals, or the accumulation of defensive compounds that reduce the effects of herbivore attack and diseases).

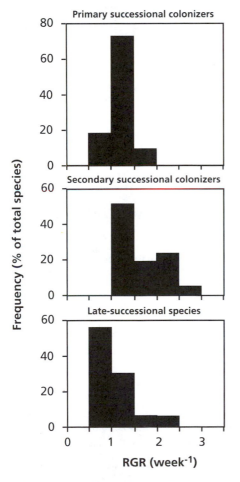

FIGURE 14. Frequency distribution of RGR for British species that are colonizers of primary successional (skeletal), secondary successional (disturbed), or late-successional (woodland) habitats. Calculated from Grime & Hunt (1975) after classifying species according to Grime et al. (1981). Redrawn from Chapin (1993). Copyright Blackwell Science Ltd.

TABLE 1. Photosynthetic characteristics of a number of Central European woody species from a hedgerow.

Photosynthetic trait, units	Species, time of appearance during succession, and competitive ability				
	Rubus corylifolius (blackberry) early pioneer, low competitive ability	*Prunus spinosa* (blackthorn) later pioneer	*Crataegus macrocarpa* (hawthorn) late-successional	*Acer campestre* (field maple) late-successional	*Ribes uva-crispa* (gooseberry) later-successional shrubby undergrowth species
A_{max} $(\mu mol\, m^{-2} s^{-1})$	11–15	9–12	8–12	8–11	6–14
Stomatal conductance at A_{max} $(mmol\, m^{-2} s^{-1})$	150–250	350–450	350–500	150–200	150–350
Photosynthesis per unit leaf N $[\mu mol\, g^{-1}(N)\, s^{-1}]$	8.6–11.6	4.7–6.3	3.6–5.3	4.3–5.9	4.5–10.5
Photosynthesis per unit leaf P $[\mu mol\, g^{-1}(P)\, s^{-1}]$	83–113	56–75	30–45	44–60	62–144

Source: Küppers et al. 1984.

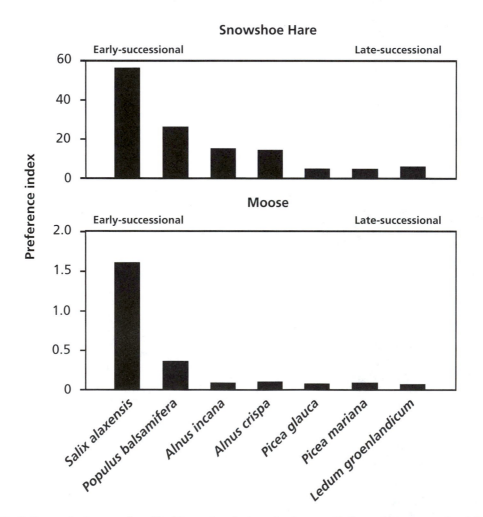

FIGURE 16. Preference by two species of herbivores for plant species from an Alaskan primary-successional floodplain sequence (Bryant & Chapin 1986). Redrawn from Chapin (1993). Copyright Blackwell Science Ltd.

If plants do not produce such defense compounds, then they may be able to grow faster in productive environments. In the longer term, however, such plants may succumb to pests or attack by a pathogenic bacterium, such as *Crataegus* (hawthorn) in Europe and many *Acacia* species in Australia. When released in a foreign environment, where such pests are absent, some species may become invasive species [e.g. *Acacia saligna* (= *A. cyanophylla*) from Australia, which was introduced in South Africa to stabilize sand dunes] (New 1984). Other examples include *Prunus padus* from North America, which was introduced in Western Europe, and *Salix spp.* (weeping willow) from Asia and *Rubus corylifolius* (blackberry) from Europe, both of which now invade river valleys in Australia.

A large phenotypic plasticity for various plant traits (e.g., photosynthetic characteristics, nutrient acquisition, and stem elongation) may also contribute to competitive success. In addition, competitive advantage may be based on a profitable association with another organism, such as a symbiotic N_2-fixing microorganism, a mycorrhizal fungus, or a higher plant, that happens to be a suitable host to parasitize.

References and Further Reading

Aerts, R. & Van der Peijl, M.J. (1993) A simple model to explain the dominance of low-productive perennials in nutrient poor habitats. Oikos 66:144–147.

Aerts, R., Boot, R.G.A., & Van der Aart, P.J.M. (1991) The relation between above- and below-ground biomass allocation patterns and competitive ability. Oecologia 87:551–559.

Aguiar, M.R., Soriano, A., & Sala, O.E. (1992) Competition and facilitation in the recruitment of seedlings in Patagonian steppe. Funct. Ecol. 6:66–70.

Arp, W.J., Drake, B.G., Pockman, W.T., Curtis, P.S., & Whigham, D.F. (1993) Interactions between C_3 and C_4 salt marsh plant species during four years of exposure to elevated atmospheric CO_2. Vegetatio 104/105:133–143.

Ball, M.C., Egerton, J.J.G., Leuning, R., Cunningham, R.B., & Dunne, P. (1997) Microclimate above grass adversely affects growth of seedling snow gum (*Eucalyptus pauciflora*). Plant Cell Environ. 20:155–166.

Ballaré, C.L. (1994) Light gaps: Sensing the light opportunities in highly dynamic canopy environments. In: Exploitation of environmental heterogeneity by plants, M.M. Caldwell & R.W. Pearcy (eds). Academic Press, San Diego, pp. 73–110.

Ballaré, C.L., Scopel, A.L., Jordan, E.T., & Vierstra, R.D. (1994) Signaling among neighboring plants and the development of size inequalities in plant populations. Proc. Natl. Acad. Sci. USA 91:10094–10098.

Ballaré, C.L., Scopel, A.L., & Sanchez, R.A. (1995) Plant photomorphogenesis in canopies, crop growth, and yield. HortScience 30:1172–1182.

Baruch, Z., Ludlow, M.M., & Davis, R. (1985) Photosynthetic responses of native and introduced C_4 grasses from Venezuelan savannas. Oecologia 67:388–393.

Bazzaz, F.A. (1979) The physiological ecology of plant succession. Annu. Rev. Ecol. Syst. 10:351–371.

Bazzaz, F.A. (1990) The response of natural ecosystems to the rising global CO_2 levels. Annu. Rev. Ecol. Syst. 21:167–196.

Bazzaz, F.A. (1996) Plants in changing environments. Cambridge University Press, Cambridge.

Berendse, F. & Aerts, R. (1987) Nitrogen-use efficiency: A biologically meaningful definition? Funct. Ecol. 1:293–296.

Berendse, F. & Elberse, W.T. (1989) Competition and nutrient losses from the plant. Causes and consequences of variation in growth rate and productivity of higher plants, H. Lambers, M.L. Cambridge, H. Konings, & T.L. Pons (eds). SPB Academic Publishing, The Hague, pp. 269–284.

Biere, A. (1996) Intra-specific variation in relative growth rate: Impact on competitive ability and performance of *Lychnis flos-cuculi* in habitats differing in soil fertility. Plant Soil 182:313–327.

Boerner, R.E.J. & Harris, K.K. (1991) Effects of collembola (Arthropoda) and relative germination date on competition between mycorrhizal *Panicum virgatum* (Poaceae) and non-mycorrhizal *Brassica nigra* (Brassicaceae). Plant Soil 136:121–129.

Boller, T. (1995) Chemoperception of microbial signals in plant cells. Annu. Rev. Plant Physiol. Plant Mol. Biol. 46:189–214.

Boot, R.G.A. (1989) The significance of size and morphology of root systems for nutrient acquisition and competition. Causes and Consequences of Variation in Growth Rate and Productivity of Higher Plants, H. Lambers, M.L. Cambridge, H. Konings, & T.L. Pons (eds), pp. 299–311. SPB Academic Publishing, The Hague.

Bryant, J.P & Chapin III, F.S. (1986) Browsing-woody plant interactions during boreal forest plant succession. In: Forest ecosystems in the alaskan taiga. A synthesis of structure and function, K. Van Cleve, F.S. Chapin, III, P.W. Flanagan, L.A. Viereck, & C.T. Dyrness (eds). Springer-Verlag, New York, pp. 213–225.

Bryant, J.P. & Kuropat, P.J. (1980) Selection of winter forage by subarctic browsing vertebrates: The role of plant chemistry. Annu. Rev. Ecol. Syst. 11:261–285.

Caldwell, M.M., Eissenstat, D.M., Richards, J.H., & Allen, M.F. (1985) Competition for phosphorus: Differential uptake from dual-isotope-labeled soil interspaces between shrub and grass. Science 229:384–386.

Caldwell, M.M., Richards, J.H., Manwaring, J.H., & Eissenstat, D.M. (1987) Rapid shifts in phosphate acquisition show direct competition between neigbouring plants. Nature 327:6123–6124.

Callaway, R.M. (1995) Positive interactions among plants. Bot. Rev. 61:306–349.

Carter, D.R. & Peterson, K.M. (1983) Effects of a CO₂-enriched atmosphere on the growth and competition interaction of a C₃ and a C₄ grass. Oecologia 58:188–193.

Carter, G.A., Teramura, A.H., & Forseth, I.N. (1989) Photosynthesis in an open field for exotic versus native vines of the south-eastern United States. Can. J. Bot. 67:443–446.

Chapin III, F.S. (1980) The mineral nutrition of wild plants. Annu. Rev. Ecol. Syst. 11:233–260.

Chapin III, F.S. (1993) Physiological controls over plant establishment in primary succession. In: Primary succession, J. Miles & D.W.H. Walton (eds). Blackwells, Oxford, pp. 161–178.

Chapin III, F.S. , Walker, L.R., Fastie, C.L., & Sharman, L.C. (1994) Mechanisms of primary succession following deglaciation at Glacier Bay, Alaska. Ecol. Monogr. 64:149–175.

Coleman, J.S. & Bazzaz, F.A. (1992) Effects of CO₂ and temperature on growth and resource use of co-occurring C₃ and C₄ annuals. Ecology 73:1244–1259.

D'Antonio, C.M. & Mahall, B.E. (1991) Root profiles and competition between the invasive, exotic perennial, *Carpobrotus edulis* and two native shrub species in California coastal shrub. Am. J. Bot. 78:885–894.

Dawson, T.E. (1993) Water sources of plants as determined from xylem-water isotopic composition: Perspectives on plant competition, distribution, and water relations. In: Stable isotopes and plant carbon-water relations, J.R. Ehleringer, A.E. Hall, & G.D. Farquhar (eds). Academic Press, San Diego, pp. 465–496.

Den Dubbelden, K.C. & Oosterbeek, B. (1995) The availability of external support affects allocation patterns and morphology of herbaceous climbing plants. Funct. Ecol. 9:628–634.

Deregibus, V.A., Sanchez, R.A., & Casal, J.J. (1983) Effects of light quality on tiller production in *Lolium* spp. Plant Physiol. 72:900–902.

Egler, F.E. (1954) Vegetation science concepts. I. Initial floristic composition, a factor in old-field vegetation development. Vegetatio 4:414–417.

Eissenstat, D.M. (1992) Costs and benefits of constructing roots of small diameter. J. Plant Nutr. 15:763–782.

Eissenstat, D.M. & Caldwell, M.M. (1987) Characteristics of successful competitors: An evaluation of potential growth rate in two cold desert tussock grasses. Oecologia 71:167–173.

Eissenstat, D.M. & Caldwell, M.M. (1988) Competitive ability is linked to rates of water extraction. A field study of two aridland tussock grasses. Oecologia 75:1–7.

Emery, R.J.N., Chinnappa, C.C., & Chmielewski, J.G. (1994) Specialization, plant strategies, and phenotypic plasticity in populations of *Stellaria longipes* along an elevational gradient. Int. J. Plant Sci 155:203–219.

Eviner, V.T. & Chapin III, F.S. (1997) Plant-microbial interactions. Nature 385:26–27.

Fitter, A.H. (1976) Effects of nutrient supply and competition from other species on root growth of Lolium perenne in soil. Plant Soil 45:177–189.

Francis, R. & Read, D.J. (1994) The contribution of mycorrhizal fungi to the determination of plant community structure. Plant Soil 159:11–25.

Gilbert, I.R., Seavers, G.P., Jarvis, P.G., & Smith, H. (1995) Photomorphogenesis and canopy dynamics. Phytochrome-mediated proximity perception accounts for the growth dynamics of canopies of *Populus trichocarpa x deltoides* "Beaupré". Plant Cell Environ. 18:475–497.

Goldberg, D.E. (1990) Components of resource competition in plant communities. In: Perspectives on plant competition, J.B. Grace & D. Tilman (eds). Academic Press, San Diego, pp. 27–49.

Goldberg, D.E. & Barton, A.M. (1992) Patterns and consequences of interspecific competition in natural communities: A review of field experiments with plants. Am. Nat. 139:771–801.

Gottlieb, L.D. (1978) Allocation, growth rates and gas exchange in seedlings of *Stephanomeria exigua* ssp. *coronaria* and its recent derivative *S. malheurensis*. Am. J. Bot. 65:970–977.

Grace, J.B. (1990) On the relationship between plant traits and competitive ability. In: Perspectives on plant competition, J.B. Grace & D. Tilman (eds). Academic Press, San Diego, pp. 51–65.

Grime, J.P. (1977) Evidence for the existence of three primary strategies in plants and its relevance to ecological and evolutionary theory. Am. Nat. 111:1169–1194.

Grime, J.P. & Hunt, R. (1975) Relative growth rate: Its range and adaptive significance in a local flora. J. Ecol. 63:393–422.

Grime, J.P., Mason, G., Curtis, A.V., Rodman, J., Band, S.R., Mowforth, M.A.G., Neal, A.M., & Shaw, S. (1981) A comparative study of germination characteristics in a local flora. J. Ecol. 69:1017–1059.

Grime, J.P., Crick, J.C., & Rincon, E. (1986) The ecological significance of plasticity. In: Plasticity in plants, D.H. Jennings (ed). Company of Biologists, Cambridge, pp. 5–29.

Grime, J.P., Mackey, J.M.L., Hillier, S.H., & Read, D.J. (1987) Floristic diversity in a model system using experimental microcosms. Nature 328:420–422.

Grime, J.P., Campbell, B.D., Mackey, J.M.L., & Crick, J.C. (1991) Root plasticity, nitrogen capture and competitive ability. In: Plant root growth. An ecological perspective. Special Publication of the British Ecological Society. D. Atkinson (ed). Blackwell Scientific Publications, London, pp. 381–397.

Gurevitch, J., Morrow, L.L., Wallace, A., & Walsh, J.S. (1992) A meta-analysis of competition in field experiments. Am. Nat. 140:539–572.

Hartnett, D.C., Hetrick, B.A.D., Wilson, G.W.T., & Gibson, D.J. (1993) Mycorrhizal influence on intra- and interspecific neighbor interactions among co-occurring prairie grasses. J. Ecol. 81:787–795.

Henderson, S., Hattersley, P., Von Caemmerer, S., & Osmond C.B. (1995) Are C₄ pathway plants threatened by global climatic change? In: Ecophysiology of photosynthesis, E.-D. Schulze & M.M. Caldwell (eds), Springer-Verlag, Berlin, pp. 529–549.

Hetrick, B.A.D., Wilson, G.W., & Hartnett, D.C. (1989) Relationship between mycorrhizal dependence and competitive ability of two tallgrass prairie species. Can J. Bot. 67:2608–2615.

Houghton, J.T., Jenkins, G.J., & Ephraums, J.J. (1990) Climate Change, The IPCC Scientific Assessment. Cambridge University Press, Cambridge.

Huber-Sannwald, E., Pyke, D.A., & Caldwell, M.M. (1996) Morphological plasticity following species-specific competition in two perennial grasses. Am. J. Bot. 83:919–931.

Johnson, H.B., Polley, H.W., & Mayeux, H.S. (1993) Increasing CO_2 and plant-plant interactions: Effects on natural vegetation. Vegetatio 104/105:157–170.

Kasperbauer, M.J. (1987) Far-red reflection from green leaves and effects on phytochrome-mediated assimilate partitioning under field conditions. Plant Physiol. 85:350–354.

Koide, R.T. & Li, M. (1991) Mycorrhizal fungi and the nutrient ecology of three oldfield annual plant species. Oecologia 85:403–412.

Krannitz, P.G. & Caldwell, M.M. (1995) Root growth responses of three Great Basin perennials to intra- and interspecific contact with other roots. Flora 190:161–167.

Küppers, M. (1984) Carbon relations and competition between woody species in a central European hedgerow. I. Photosynthetic characteristics. Oecologia 64:332–343.

Lambers, H. & Poorter, H. (1992) Inherent variation in growth rate between higher plants: a search for physiological causes and ecological consequences. Adv. Ecol. Res. 23:187–261.

Mahall, B.E. & Callaway, R.M. (1991) Root communication among desert shrubs. Proc. Nat. Acad. Sci. (USA) 88:874–876.

Mahall, B.E. & Callaway, R.M. (1992) Root communication mechanisms and intracommunity distribution of two Mojave Desert shrubs. Ecology 73:2145–2151.

Mahmoud, A. & Grime, J.P. (1976) An analysis of competitive ability in three perennial grasses. New Phytol. 77:431–435.

McGraw, J.B. & Chapin III, F.S. (1989) Competitive ability and adaptation to fertile and infertile soils in two Eriophorum species. Ecology 70:736–749.

Milne, A.A. (1928) The house at Pooh Corner. Dutton, New York.

Mo, H. Kirkham, M.B., He, H., Ballou, L.K., Caldwell, F.W., & Kanemasu, E.T. (1992) Root and shoot weight in a tallgrass prairie under elevated carbon dioxide. Env. Exp. Bot. 32:193–201.

New, T.R. (1984) A Biology of acacias. Oxford University Press, Melbourne.

Newman, E.I., Eason, W.R., Eissenstat, D.M., & Ramos, M.I.F.R. (1992) Interactions between plants: The role of mycorrhizae. Mycorrhiza 1:47–53.

Nobel, P.S. (1984) Extreme temperatures and thermal tolerances for seedlings of desert succulents. Oecologia 62:310–317.

Owens, M.K. (1996) The role of leaf and canopy-level gas exchange in the replacement of Quercus virginiana (Fagaceae) by Juniperus ashei (Cupressaceae) in semiarid savannas. Am. J. Bot. 83:617–623.

Owensby, C.E., Coyne, P.I., Ham, J.M., Auen, L.M., & Knapp, A.K. (1993) Biomass production in a tallgrass prairie ecosystem exposed to ambient and elevated CO_2. Ecol. Appl. 3:644–653.

Pammenter, N.W., Drennan, P.M., & Smith, V.R. (1986) Physiological and anatomical aspects of photosynthesis of two Agrostis species at a sub-antarctic island. New Phytol. 102:143–160.

Polley, H.W., Johnson, H.B., & Mayeuz, H.S. (1994) Increasing CO_2: Comparative responses of the C_4 grass Schizachyrium and grassland invader Prosopis. Ecology 75:976–988.

Poorter, H., Van de Vijver, C.A.D.M., Boot, R.G.A., & Lambers, H. (1995) Growth and carbon economy of a fast-growing and a slow-growing grass species as dependent on nitrate supply. Plant Soil 171:217–227.

Putz, F.E. (1984) The natural history of lianas on Barro Colorado Island, Panama. Ecology 65:1713–1724.

Raaimakers, D., Boot, R.G.A., Dijkstra, P., Pot, S.,, & Pons T.L. (1995) Photosynthetic rates in relation to leaf phosphorous content in pioneer versus climax tropical rainforest species. Oecologia 102:120–125.

Read, D.J. (1991) Mycorrhizas in ecosystems. Experientia 47:376–391.

Reich, P. (1993) Reconciling apparent discrepancies among studies relating life span, structure and function of leaves in contrasting plant life forms and climates: "The blind man and the elephant retold." Funct. Ecol. 7:721–725.

Reynolds, H.L. (1996) Effects of elevated CO_2 on plants grown in competition. In: Carbon dioxide, populations, and communities, C. Körner & F.A. Bazzaz (eds). Academic Press, San Diego, pp. 273–286.

Reynolds, H.L. & D'Antonio, C. (1996) The ecological significance of plasticity in root weight ratio in response to nitrogen. Opinion. Plant Soil 185:75–97.

Richards, J.H. & Caldwell, M.M. (1987) Hydraulic lift: substantial nocturnal water transport between soil layers by Artemisia tridentata roots. Oecologia 73:486–489.

Ritchie, G.A. (1997) Evidence for red: far red signaling and photomorphogenic growth response in Douglas-fir Pseudostuga menziesii) seedlings. Tree Physiol. 17:161–168.

Rosenthal, J.P. & Kotanen, P.M. (1994) Terrestrial plant tolerance to herbivory. Trends Ecol. Evolu. 9: 145–148.

Ryser, P. (1996) The importance of tissue density for growth and life span of leaves and roots: A comparison of five ecologically contrasting grasses. Funct. Ecol. 10:717–723.

Ryser P. & Lambers H. (1995) Root and leaf attributes accounting for the performance of fast- and slow-growing grasses at different nutrient supply. Plant Soil 170:251–265.

Schläpfer, B. & Ryser, P. (1996) Leaf and root turnover of three ecologically contrasting grass species in relation to their performance along a productivity gradient. Oikos 75:398–406.

Schlumbaum, A., Mauch, F., Vögeli, U., & Boller, T. (1986) Plant chitinases are potent inhibitors of fungal growth. Nature 324:365–367.

Stock, W.D., Wienand, K.T., & Baker A.C. (1995) Impacts of invading N_2-fixing *Acacia* species on patterns of nutrient cycling in two Cape Ecosystems: Evidence from soil incubation studies and ^{15}N natural abundance values. Oecologia 101:375–382.

Tilman, D. (1988) Plant Strategies and the Dynamics and Function of Plant Communities. Princeton University Press, Princeton.

Tilman, D. (1990) Mechanisms of plant competition for nutrients: The elements of a predictive theory of competition. In: Perspective on plant competition, J.B. Grace & D. Tilman (eds). Academic Press, San Diego, pp. 117–141.

Tilman, D. & Wedin, D. (1991) Dynamics of nitrogen competition between successional grasses. Ecology 72:1038–1049.

Turner, R.M., Alcorn, S.M., Olin, G., & Booth, J.A. (1966) The influence of shade, soil, and water on saguaro seedling establishment. Bot. Gaz. 127:95–102.

Van Bavel, C.H.M. & Baker, J.M. (1985) Water transfer by plant roots from wet to dry soil. Naturwissenschaften 72:606–607.

Vierheilig, H., Iseli, B., Alt, M., Raikhel, N., Wiemken, A., & Boller, T. (1996) Resistance of *Urtica dioica* to mycorrhizal colonization: A possible involvement of *Urtica dioica* agglutinin. Plant Soil 183:131–136.

Walker, L.R. & Chapin III, F.S. (1986) Physiological controls over seedling growth in primary succession on an Alaskan floodplain. Ecology 67:1508–1523.

Wedin, D.A. & Tilman, D. (1990) Species effects on nitrogen cycling: A test with perennial grasses. Oecologia 84:433–441.

Wong, S.-C. & Osmond, C.B. (1991) Elevated atmospheric partial pressure of CO_2 and plant growth. III. Interactions between *Triticum aestivum* (C_3) and *Echinochloa frumentacea* (C_4) during growth in mixed culture under different CO_2, N nutrition and irradiance treatments, with emphasis on below-ground responses estimated using ^{13}C value of root biomass. Aust. J. Plant Physiol. 18:137–152.

9F. Carnivory

1. Introduction

Since the classic work of Charles and Francis Darwin (1875, 1878) more than a century ago on the carnivorous habit of *Drosera*, considerable information has accumulated on the significance of captured animal prey in the nutrition of carnivorous plants. Carnivory includes the catching and subsequent digestion of the freshly trapped prey. This is a common form of nutrition in the animal kingdom, but it is rather rare in plants, with only a few hundred species from seven families (Table 1). Carnivorous plants are distributed worldwide, but they are generally restricted to sunny, wet, and nutrient-poor environments. This distribution pattern suggests that there are conditions where carnivory has a major benefit for plant survival. The restricted distribution of carnivorous plants, however, also suggests that the **costs** involved in the carnivorous habit exclude carnivores from most habitats. These costs include a reduced level of photosynthesis of the carnivorous tissues (Knight 1992).

2. Structures Associated with the Catching of the Prey and Subsequent Withdrawal of Nutrients from the Prey

Carnivorous plants invariably have highly specialized structures, such as **adhesive hairs** [in *Drosera* (sundew) and *Byblis*], bladderlike **suction traps** [in *Utricularia* (bladderwort)], **snapping traps** [in *Dionaea*, (Venus' fly trap); these entrap the prey after it touches sensitive parts of the trap], or **pitfalls** [e.g., of *Nepenthes* (pitcher plant) and *Sarracenia*] (Table 1). These pitfall traps mostly contain water and are an ecological niche for numerous small animals, of which some (e.g., the larvae of many Diptera) are exclusively associated with this habitat. Although it is strictly speaking not a carnivorous species, *Capsella bursa-pastoris* (shepherd's purse) has a mucous layer that surrounds the germinating seeds, which has the capacity to catch and digest nematodes, protozoa, and bacteria (Barber 1978).

Some carnivorous plants attract their prey by the production of **nectar** at the edge of the trap. This nectar is actively secreted and the carbon costs of this process may amount to 4 to 6% of the plant's total carbon budget (Pate 1986, as cited in Karlsson 1991). In addition, they secrete adhesive substances by special glands [e.g., *Drosera* and *Pinguicula* (butterwort)]. All carnivorous plants are green and capable of C_3 photosynthesis. Hence, carbon is unlikely to be a major element to be withdrawn from their prey, although it is certainly incorporated. Carnivorous species naturally occur on nutrient-poor, wet, and acidic soils, with the exception of *Pinguicula*, which grows on a chalky substrate. Most species have a poorly developed root system and at least some species (*Drosera*) are **nonmycorrhizal** (Sect. 2.2 of the chapter on symbiotic associations). It is generally assumed that carnivory is an adaptation to nutrient-poor soils

TABLE 1. Carnivorous plant families and genera with their geographical distribution and trapping mechanisms.

Family • genus	N	Geographical distribution	Suction trap	Snapping trap	Pitfall pitcher	Adhesive trap	Movement involved
Nepenthaceae							
• *Nepenthes*	70	Madagascar, Seychelles, Sri Lanka, Assam, S. China, Indochina, Malaysia, N. Queensland, New Caledonia			x		
Sarraceniaceae							
• *Sarracenia*	8	Altantic N-America			x		
• *Heliamphora*	5	Guyana Highlands			x		
• *Darlingtonia*	1	California, Oregon			x		
Dioncophyllaceae							
• *Triphyophyllum*	1	W-Africa				x	
Droseraceae							
• *Drosera*	80	Worldwide				x	x
• *Drosophyllum*	1	W-Mediterranium				x	
• *Dionaea*	1	N- and S-Carolina		x			x
• *Aldrovanda*	1	Central Europe, Asia, NE-Australia		x			x
Byblidaceae							
• *Byblis*	2	Australia, New Guinea				x	
• *Roridula*	2	S-Africa				x	
Cephalotaceae							
• *Cephalotus*	1	SW-Australia			x		
Lentibulariaceae							
• *Pinguicula*	46	Northern-Hemisphere				x	x
• *Genlisea*	16	Central America, West India, tropical and S. Africa, Madagascar	x				
• *Utricularia*	180	Worldwide	x				x
• *Polypompholyx*	3	Australia					

Source: Lüttge 1983, Mabberley 1993.
Note: N is the number of specics in each genus.

and that inorganic nutrients are largely derived from the prey. There is a positive effect of supplementary feeding with prey on growth, even when this is done in the plant's natural habitat (Thum 1988). Carnivorous plants, however, respond to a supply with prey both at low and at high soil-nutrient levels, which suggests that the positive effect of the prey on growth may not be due entirely to a supply of extra nutrients (Karlsson et al. 1991).

Nitrogen is a major element withdrawn from the prey (Schulze & Schulze 1990), especially in fairly tall, erect, or climbing *Drosera* species, which may derive approximately 50% of all their nitrogen from insect feeding (Fig. 1). *Drosera* species with a rosette habit derive less nitrogen from insects (12 to 32%, depending on site). Carnivorous plants are likely to derive other elements from their prey as well (Tables 2 and 3).

When insects are scarce (e.g., in the extremely nutrient-poor habitat of the tuberous sundew *Drosera erythrorhiza* in Western Australia) the input of nitrogen from the catch of arthropods by the glandular leaves may be very small. It was established by using a stable-isotope technique (^{15}N; cf. Sects. 2.4 and 3.6 of the chapter on symbiotic associations) that 76% of all the nitrogen in the prey is

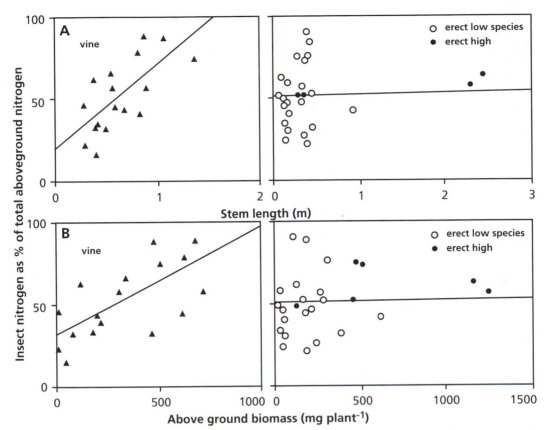

FIGURE 1. The relative proportion of nitrogen derived from insects in climbing and erect *Drosera* species as dependent on stem length and above-ground plant biomass. The study was carried out in southwestern Australia; estimates are based on an analysis of δ^{15}N values (^{15}N/^{14}N natural isotope composition) of plants and insects (Schulze et al. 1991).

transferred to various plants parts of *Drosera erythrorhiza*. The total amount of nitrogen acquired from the prey by this sundew species in its natural habitat, however, is only 11 to 17% of its total nitrogen requirement (Dixon et al. 1980). In this habitat, there are specialized small beetles that do not stick to the glandular hairs; rather, they consume the prey stuck to the leaf hairs. The glandular hairs may function to deter herbivores or be a mere vestige from the past. This example serves to illustrate that very little quantitative work has been done on the actual significance of the carnivorous habit in acquring nutrients from a prey in a natural habitat.

3. Some Case Studies

This section deals with some examples of carnivorous plants that have been the subject of more detailed ecophysiological studies.

3.1 *Dionaea muscipula*

One of de most fascinating traps is that of *Dionaea muscipula* (Venus' fly trap), which is a species from North America. The trap consists of two lobes that are attached to a petiole. There are three "trigger

TABLE 2. The effect of feeding inorganic nutrients (N, S, P) and prey animals (*Drosophila melanogaster*) on the growth of *Drosera binata*.

Supplies in root medium (inorganic nutrients)			Supply of insects	Dry mass increment (%)
N	S	P		
+	+	+	+	100
+	+	+	−	65
−	+	+	+	145
−	+	+	−	50

Source: Chandler & Anderson 1976.

TABLE 3. The effect of feeding *Utricularia gibba* (bladderwort) with *Paramecium* on magnesium and potassium deficiency.

Media	Internodes		Number of bladders formed
	Number	Length	
	(% of control)		
Complete medium (= control)	100	100	
Complete plus feeding	96	104	
Complete minus Mg^{2+}	38	33	85
Complete minus Mg^{2+} plus feeding	53	42	151
Complete medium (= control)	100	100	66
Complete plus feeding	136	139	86
Complete minus K^+	72	52	66
Complete minus K^+ plus feeding	100	83	104

Source: Sorenson & Jackson 1968, as cited in Lüttge 1983.

hairs" on each lobe. Mechanical stimulation of these hairs leads to rapid closure of the trap; this is one of the fastest movements known of plants, and it is sufficiently rapid to catch even the most alert insects (Hodick & Sievers 1988, 1989). One of the six hairs inside the trap must be stimulated twice within 20 seconds; stimulation of two diffrent hairs within the same time frame has the same effect (Fig. 2). Touching of one of the hairs leads to an action potential, which is propagated over the surface of one of the lobes (Fig. 3). At least two action potentials are required to close the trap. Calcium in the cell wall is a prerequisite for any action potential to develop. In the absence of available calcium [e.g., when it is experimentally bound by a strong chelator] no action potential is produced. Inhibitors of the cytochrome path, uncouplers (Sect. 2.3.3 of the chapter on plant respiration), and compounds that block calcium channels also inhibit the trap's excitability. We do not yet know what role, either direct or indirect, calcium plays in the signal transduction that leads to the closure of the trap.

Trap closure involves an increased wall extensibility of the lower epidermal cells (Table 4). Increased extensibility of the lower epidermis, in combination with the **tissue tension** (Sect. 4 of the chapter on plant water relations), leads to trap closure. The tissue tension is due to the relatively elastic walls of the mesophyll cells ("swelling tissue"), compared with that of the epidermal cells. The presence of the relatively rigid upper and lower epidermal cells prevents the cells in the swelling tissue from reaching full turgor while the trap is open. The changes in extensibility that allow closure of the trap are not due to cell-wall acidification, as is the case when auxin induces similar changes in

cell-wall properties (Sect. 2.2.2 of the chapter on growth and allocation).

Closure of the trap occurs in two steps. The first step is a movement triggered by **mechanical stimulation**. In the absence of a **chemical stimulation** after the mechanical stimulation, the trap gradually opens, due to the growth of the upper epidermis. If a chemical stimulus (in the form of chemical compounds from the hemolymph of the trapped insect) does occur, then the trap closes tightly and **digestive enzymes** (proteases, phosphatases, DNAase) and fluid are secreted by special glands (Table 5). Trap opening is a much slower process, requiring extension (growth) of the upper epidermis (Fagerberg & Howe 1996). The opening and closing of the trap can only occur for a limited number of times: until both the upper and the lower epidermis have achieved their maximum length.

3.2 *Utricularia*

Many *Utricularia* (bladderwort) species are aquatic plants, occurring in nutrient-poor shallow water. Small, submerged bladders are produced on the shoots. Small aquatic animals (e.g., *Daphnia* species) touch one of the hairs of a trap door, causing the "door" to snap open inward. Because the **hydrostatic pressure** inside before opening is less than that outside, water flows in, carrying the prey with it (Fig. 4). Then the trap door closes again. The entire process takes 10 to 15 milliseconds. The role of the hairs might be that of a "lever," but action potentials may also play a role. The low hydrostatic pressure in the bladder, compared with that outside the plant, is the result of **active transport** of

chloride from the lumen of the bladder to the cells that surround it (across the membrane A in Fig. 5). Sodium follows passively. Active transport is probably via the "two-armed glands" depicted in Figure 5. Transport of NaCl to the cells that surround the lumen of the bladder causes a gradient in water potential between the lumen and these cells. As a result, water flows from the lumen to these cells, causing an increase in turgor, which in turn promotes the passive transport of Na^+, Cl^-, and water out of the cells in the direction of the medium that surrounds the *Utricularia* plants. The membrane of the cells facing the lumen has a lower permeability for Cl^-, cations, and water than does the membrane

facing the opposite side. The cuticle (membrane C in Fig. 5) appears to be highly permeable for ions and water.

Digestion of the prey presumably requires the secretion of enzymes and possibly a lowering of the pH, as in other carnivorous plants, but details of these processes are as yet unknown for *Utricularia*.

3.3 The Tentacles of *Drosera*

The organs of **adhesive traps** such as those of *Drosera* (sundew) and *Pinguicula* (butterwort) can also move after mechanical and/or chemical trig-

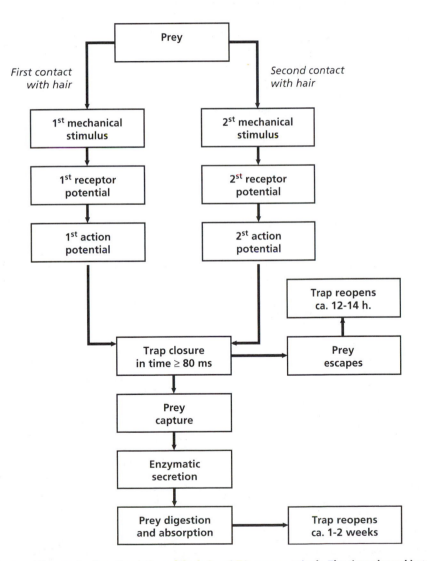

FIGURE 2. **A scheme of the events after stimulation of the hairs of *Dionaea muscipula*. The time elapsed between the first and second stimulus cannot exceed 20 seconds** (Lüttge 1983).

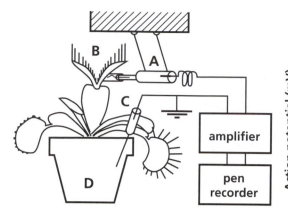

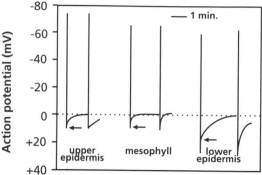

FIGURE 3. (Left) A scheme of the experimental design to determine action potentials from the surface of *Dionaea muscipula*. One electrode (A), suspended by a thread pendulum to maintain electrical contact with the leaf (B) during movement. The reference electrode (C) is inserted into the substratum and earthed. (Right) Extracellular recordings of action potentials during trap closure after touching a trigger hairs (arrows). Trigger hairs (not shown), the upper and lower epidermis, as well as the mesophyll produce action potentials (Hodick & Sievers 1988).

gering. The tentacles on the leaves of *Drosera* function in catching and digestion of the prey, so that the tentacles, or even the entire leaf, may surround the prey as a result of their movement. Part of these (faster) movements are triggered by action potentials. Two action potentials within 1 minute are required to trigger bending of a tentacle. The slower movements require a chemical stimulus.

Digestion of the prey by carnivorous plants requires specific digestive enzymes. Some of these enzymes are produced by the plant, in response to chemical stimuli from the prey, but it is likely that some hydrolytic enzymes are also produced by microorganisms and that the carnivorous plant takes advantage of the presence of such microorganisms (Table 6).

4. The Message to Catch

Carnivory is a rare trait in the plant world and it is predominantly associated with nutrient-poor habitats. There are **benefits** of the carnivorous habit, in that the prey provides an extra source of nutrients ("fertilizing effect"), but their importance has rarely been measured in situ. There are also **costs**. These

TABLE 4. **The relative extensibilities of the upper and lower sides of the trap of** *Dionaea muscipula*, **measured as reversible (elastic) and irreversible (plastic) extension, induced by the application of a constant load for 10 minutes.***

	Upper side	Lower side
Trap closed		
Elastic extensibility	3.5	6.9
Plastic extensibility	1.6	11.4
Trap closed and then paralysed		
Elastic extensibility	n.d.	8.1
Plastic extensibility	n.d.	12.6
Trap open and then paralysed		
Elastic extensibility	3.5	2.7
Plastic extensibility	1.8	1.8

Source: Hodick & Sievers 1989.
*Tissue strips were extended perpendicular to the midrib (n.d. = not determined). In some of the experiments, the trap was paralyzed with $LaCl_3$, which blocks Ca^{2+} channels and prevents excitability in whole leaves.

TABLE 5. **The effect of chemical compounds normally present in the hemolymph of the prey, on the secretion of digestive fluids by** *Dionaea muscipula*.*

N-compound	Protein secretion (% of control)	Volume secretion (% of control)
Whole fly (*Calliphora*)	100	100
Uric acid	63	107
Ammonia	44	121
Glutamine	20	94
Urea	9	27
Phenylalanine	—	16

Source: Robins 1976, as cited in Lüttge 1983.
* Data are expressed as a percentage of the volumes found for whole flies.

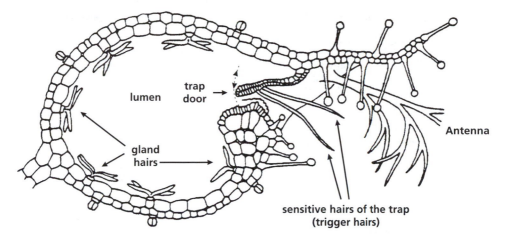

FIGURE 4. A schematic drawing of a longitudinal section of the trap of *Utricularia* (Schmucker & Linnemann 1959, as cited in Lüttge 1983).

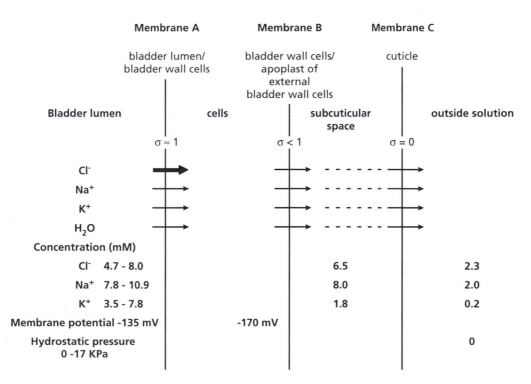

FIGURE 5. A model of solute and water flow in the resetting of the bladder of *Utricularia*. Heavy arrow: active transport; thin arrows: passive transport; dotted lines connecting the arrows: bulk flow of solution through the subcuticular space to the outside. The values of ionic concentrations and pressure in the bladder lumen show the range between triggered and reset bladder. The potential difference in the lumen, apart from a very rapid change at the time of triggering, is constant during resetting (Sydenham & Finlay 1975, as cited in Lüttge 1983).

TABLE 6. The effect of sterilization on insect-enhanced growth of *Drosera* plants.

	Inorganic nutrients in root medium	Insect feeding	Sterilization	Dry mass increase (mg plant^{-1})
D. whittakeri				
field grown	Low	−	−	5.9
growth period	Low	−	+	6.1
7 weeks	Low	+	−	9.1
	Low	+	+	4.9
D. binata				
axenic culture	−	−	Axenic	0.46
growth period	−	+	Axenic	0.43
14 weeks	+	−	Axenic	1.71
	+	+	Axenic	1.65

may be the costs of secreting nectar, slime, and enzymes, but these would seem relatively small. It is more likely that the carnivorous habit incurs costs in terms of a reduced photosynthetic capacity and that this precludes carnivorous species from nutrient-rich sites where competition plays a role.

References and Further Reading

Barber, J.T. (1978) *Capsella bursa-pastoris* seeds. Are they "carnivorous"? Carniv. Plant Newslett. 7:39–42.

Darwin, C. (1875) Insectivorous plants. Murray, London.

Darwin, F. (1878) Experiments on the nutrition and growth of *Drosera rotundifolia*. J. Linn. Soc. Bot. 17:17–23.

Dixon, K.W., Pate, J.S., & Bailey, W.J. (1980) Nitrogen nutrition of the tuberous sundew *Drosera erythrorhiza* Lindl, with special reference to catch of arthropod fauna by glandular leaves. Aust. J. Bot. 28:283–297.

Chandler, G.E. & Andersson, J.W. (1976) Studies on the nutrition and growth of *Drosera* species with reference to the carnivorous habit. New Phytol. 76:129–141.

Fagerberg, W.R. & Howe, D.G. (1996) A quantitative study of tissue dynamics in Venus' fly trap *Dionaea muscipula* (Droseraceae). II. Trap reopening. Am. J. Bot. 83:836–842.

Hodick, D. & Sievers, A. (1988) The action potential of *Dionaea muscipula* Ellis. Planta 174:8–18.

Hodick, D. & Sievers, A. (1989) On the mechanism of trap closure of Venus flytrap (*Dionaea muscipula* Ellis). Planta 179:32–42.

Karlsson, P.S., Nordell, K.O., Carlsson, B.A., & Svensson, B.M. (1991) The effect of soil nutrient status on prey utilization in four carnivorous plants. Oecologia 86:1–7.

Knight, S.E. (1992) The costs of carnivory in the common bladderwort, *Utricularia macrorhiza*. Oecologia 89:348–355.

Lüttge, U. (1983) Ecophysiology of carnivorous plants. In: Encyclopedia of plant physiology, N.S. Vol. 12C, O.L. Lange, P.S. Nobel, C.B. Osmond, & H. Ziegler (eds). Springer-Verlag, Berlin, pp. 489–517.

Mabberley, D.J. (1993) The plant-book. A portable dictionary of the higher plants. Cambridge University Press, New York.

Schulze, E.-D., Gebauer, G., Schulze, W., & Pate, J.S. (1991) The utilization of nitrogen from insect capture by different growth forms of *Drosera* from southwest Australia. Oecologia 87:240–246.

Schulze, W. & Schulze, E.-D. (1990) Insect capture and growth of the insectivorous *Drosera rotundifolia* L. Oecologia 82:427–429.

Thum, M. (1988) The significance of carnivory for the fitness of *Drosera* in its natural habitat. 1. The reactions of *Drosera intermedia* and *D. rotundifolia* to supplementary insect feeding. Oecologia 75:472–480.

10
Role in Ecosystem and Global Processes

10A. Decomposition

1. Introduction

Decomposition of plant litter involves the physical and chemical processes that reduce litter to CO_2, water, and mineral nutrients. It is a key process in the **nutrient cycle** of most terrestrial ecosystems, and the amount of carbon returned to the atmosphere by decomposition of dead organic matter is an important component of the global carbon budget (Vitousek 1982, Vitousek et al. 1994) (Sect. 2.6 of the chapter on ecosystem and global processes). Sooner or later, most plant material is decomposed, although a small proportion of recalcitrant organic matter becomes stabilized for thousands of years as humus. Most root-released material (exudates and other root-derived organic matter) is incorporated in the soil microbial biomass or lost as CO_2 within weeks, at least at a high nutrient supply. When nutrients are limiting for growth, soil microorganisms utilize the root-derived material more slowly because their growth is limited by nutrients rather than by carbon (Merckx et al. 1987). Some of it may end up as peat, or even coal. In that case carbon is temporarily removed from the global carbon cycle. The rate of carbon sequestration in peatlands is mainly determined by low rates of decomposition of dead organic matter rather than by high rates of primary production. Due to the relatively large peat cover on earth, changes in the extent to which peat bogs act as a CO_2-sink will affect the global carbon budget (Gorham 1991).

2. Litter Quality and Decomposition Rate

2.1 Species Effects on Litter Quality: Links with Ecological Strategy

In a comparison of 125 British vascular plant species, which cover a wide range of life forms, leaf habits, and taxa, the rate of leaf litter decomposition can be predicted from a limited number of whole-plant traits, which reflect the plants' physiological adaptation to its environment (Cornelissen 1996) (Fig. 1). These traits include life form, deciduous versus evergreen habit, autumn coloration of the leaf litter, family, and a species' success in disturbed and productive habitats. In this wide comparison of species from the British Isles, as in other comparisons (Aerts 1995, 1997), there is a negative relationship between decomposition rate and **leaf life-span**. For example, leaves of woody climbers and ramblers, which tend to have short-lived leaves with little investment in quantitatively important defense compounds, decompose more readily than do those of subshrubs, which invest more in chemicals that reduce leaf digestibility and palatability, such as lignin and tannins (cf. Sect. 3.2 of the chapter on ecological biochemistry).

Comparing tree species, the **specific leaf area** (SLA) of litter is a good predictor of the rate of decomposition. Such a trend is harder to discern among herbaceous plants, if these include widely

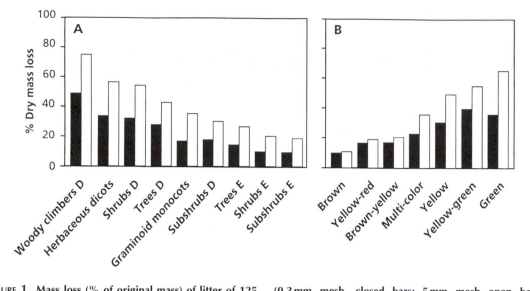

FIGURE 1. Mass loss (% of original mass) of litter of 125 British species as related to (A) growth form and duration of decomposition period (8 weeks, closed bars; 20 weeks, open bars) or (B) litter color (for deciduous woody species only) and mesh size of the bag that contained the litter (0.3 mm mesh, closed bars; 5 mm mesh open bars). Litter was buried in leaf mould near Sheffield, England. D = deciduous; E = evergreen. Means were calculated from mean values of individual species (based on information in Cornelissen 1996).

different functional types such as succulents and sclerophylls (i.e., plant traits other than litter SLA are also important). Long-lived leaves, with relatively large investments in quantitatively important chemical defense, tend to have a lower SLA. This accounts for the positive correlation between rate of litter decomposition and litter SLA. The association between autumn colors and decomposition is also a reflection of the leaf's secondary chemistry (Fig. 1). Brown colors are associated with phenolics, which slow down the rate of decomposition, in a manner similar to their effects on protein digestion (Sect. 3.2 of the chapter on ecological biochemistry).

To explain the biochemical basis of variation in leaf litter decomposition, more information is required about **leaf chemistry**: Rates of litter decomposition are negatively correlated with both the **lignin:nutrient ratio** (N or P) and with the **lignin concentration** (Berendse et al. 1987, 1989, Berg & Staaf 1981, Fox et al. 1990). For example, *Sphagnum* species occur in extremely nutrient-poor and wet environments. Decomposition of *Sphagnum* litter is remarkably slow, because of the anoxic acidic conditions in the bog environment (Sect. 2.2), and because of the chemical composition of the litter (Johnson & Damman 1993). *Sphagnum* species all produce a number of phenolic compounds, with the most important one being sphagnum acid (Fig. 2). Leachates from *Sphagnum* species reduce the

decomposition rate of litter from other plants as well (Verhoeven & Toth 1995). Decomposition rate often correlates inversely with carbon:nitrogen ratio, especially among herbaceous species, which vary in nitrogen and phosphorus concentrations but typically have low concentrations of quantitative defensive compounds in leaves. Thus, both carbon and nutrient chemistry influence the decomposition rate, although their relative importance may vary across species.

Species differences in allocation strongly influence decomposition because of the strikingly different chemistry of leaves, wood, and roots. Stems and roots, with their high lignin and low nitrogen concentration, decompose more slowly than do leaves. Species differences in litter quality due to differences in allocation to wood versus leaves often ex-

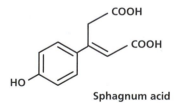

FIGURE 2. The chemical structure of sphagnum acid (Wilschke et al. 1990).

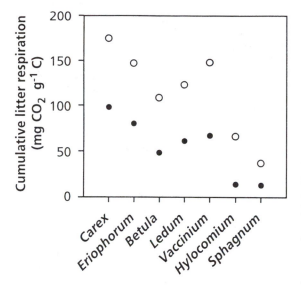

FIGURE 3. Cumulative respiration of litter of seven Alaskan tundra species incubated in the laboratory with tundra soil over 149 days at either 4°C (filled circles) or 8°C (open circles). Each litter bag had the same initial mass, but included leaves, stems, and roots in proportion to their production in the field. Litter respiration is estimated by subtracting the average respiration in the controls (soil only) from the total respiration in those incubations amended with litter (Hobbie 1996). Copyright by the Ecological Society of America.

ceed differences due to the variation in leaf quality (Hobbie 1995). For example, dwarf birch (*Betula nana*), which is a tundra deciduous shrub, has a low overall decomposition rate, when all plant parts are included, despite relatively rapid rates of leaf decomposition (Fig. 3). These differences in allocation and the decomposition rates of litter other than leaves have not been adequately considered in evaluating the overall effect of plant species on carbon and nutrient cycling. Woody stems of slowly growing, late-successional species, with their higher concentrations of quantitative defense compounds, decompose more slowly than do woody stems of rapidly growing species. We unfortunately know relatively little about species differences in root decomposition, despite the large proportion of litter production that occurs below ground.

2.2 Environmental Effects on Decomposition

Environment affects decomposition both because of its effect on the quality of litter produced and its direct effects on microbial activity. The direct effects

of environment on microbial activity are similar to effects on plant production, which results in highest rates of decomposition in warm, moist environments (Sect. 2.5 of the chapter on ecosystem and global processes). In anaerobic soils (e.g., in peatlands) decomposition is more restricted than is plant production, which results in substantial carbon accumulation (Sect. 1). Plant species can strongly influence decomposition through their effect on environment. For example, in the Arctic, mosses are effective thermal insulators, which results in cold soils that retard decomposition even more than might be expected from their low litter quality (Hobbie 1995). Some mosses, such as *Sphagnum*, produce litter that is highly acidic, further reducing decomposition (Clymo & Hayward 1982).

Microbial respiration associated with the decomposition of surface litter is often enhanced at night because dew provides moisture for microbial activity and decreases during the day as the litter dries out (Edwards & Sollins 1973). In dry environments, the moister conditions beneath vegetation may favor decomposition.

Environment also affects tissue chemistry and therefore litter quality. As expected, the higher tissue nitrogen or phosphorus concentrations in plants grown on fertile soils results in high litter nutrient concentrations (Fig. 38 in the chapter on mineral nutrition) and therefore high rates of decomposition. Among woody plants, growth in infertile soils also increases quantities of quantitative defense compounds, further contributing to the slow decomposition of litter produced on these soils (Sect. 4.1 of the chapter on ecological biochemistry). Reciprocal transfers of litter among forests that differ strongly in litter quality and environment often show that litter quality exerts a stronger effect on decomposition than do differences in temperature or moisture (Flanagan & Van Cleve 1983).

3. The Link Between Decomposition Rate and Nutrient Mineralization

3.1 Effects of Litter Quality on Mineralization

A major reason for interest in decomposition is its close link to **nutrient supply**. In most ecosystems the nutrients released during decomposition provide more than 90% of the nitrogen and phosphorus supply to plants (Table 1 in the chapter on mineral nutrition). When litter or soil organic matter contains nutrients in excess of microbial demands, nitrogen and phosphorus are excreted by soil microorganisms (**net mineralization**) during the decomposition process and become available for plant uptake (Fig. 2 in the chapter on mineral nutrition). On the other hand, if the organic matter is low in nutrients, then microorganisms meet their nutrient demand by absorbing nutrients from the soil solution (**immobilization**), which results in competition for nutrients between soil microorganisms and plants. After nutrient resorption (Sect. 4.3.2 of the chapter on mineral nutrition), plant litter often has a higher C:N ratio than microbial biomass. Empirical observations suggest that above a critical C:N ratio of about 20:1, microorganisms absorb nutrients from the soil solution, which causes a net immobilization (Paul & Clark 1989). The C:N ratio of litter declines as microorganisms decompose the organic matter and respire carbon to meet respiratory demands for growth and maintenance. Net mineralization occurs when the C:N ratio falls below the critical 20:1 ratio. The result is that fresh litter often initially increases in nutrient concentration due to microbial immobilization, before net mineralization occurs.

It is not surprising that tissue nutrient concentration strongly affects the mineralization rate. Net immobilization occurs to a greater extent and for a longer time where plants produce litter with low tissue nitrogen and phosphorus concentrations. In those ecosystems where plant growth is N-limited, litter C:N ratios strongly govern decomposition and nitrogen immobilization, with P being mineralized more quickly, whereas in areas of heavy **N deposition**, such as in the Netherlands, C:P ratios exert stronger control over decomposition, and N is mineralized more quickly (Aerts & De Caluwe 1997). The high litter nitrogen and phosphorus concentrations of plants on fertile soils, with their high growth rate and SLA, thus promote nutrient mineralization, whereas there is slower mineralization in ecosystems that are dominated by slow-growing plants with low SLA (Chapin 1991, Hobbie 1992, Van Breemen 1993).

If nutrient concentration affects mineralization so strongly, then will the low tissue nutrient concentrations caused by elevated atmospheric CO_2 concentrations reduce litter nutrient concentrations and, therefore, decomposition rate? In most cases studied to date, differences in leaf chemistry caused by elevated CO_2 diminish during senescence, perhaps due to respiration or resorption of accumulated starch, so that litter quality and, therefore, decomposition and mineralization rates are similar for litter produced under elevated and ambient $[CO_2]$ (Franck et al. 1997). The CO_2 effects that do occur are often species-specific.

Species differences in carbon quality (that is, the types of carbon compounds they contain) magnify differences in mineralization rate due to litter nutrient concentration. The high concentrations of quantitative secondary metabolites in species with long-lived leaves (Sect. 2.1) retard decomposition because of both the toxic effects on microorganisms and the difficulty of breakdown of secondary metabolites. **Phenolic** compounds that are decomposed slowly include lignins and tannins (cf. Sect. 3.2 in the chapter on ecological biochemistry). High tannin concentrations reduce the rate of **mineralization** of the litter; therefore, most of the nitrogen in the soil occurs as complexes of organic nitrogen and tannin, rather than as nitrate, ammonium, or amino acids (Northup et al. 1995). In many species, including *Pinus* (pine), the concentration of tannin and lignin is enhanced when plants are grown under nitrogen limitation as compared with a high supply (Bryant et al. 1983, Gershenzon 1984). As a result, the availability of nitrogen is even further reduced, at least for plants lacking mechanisms to release nitrogen from the tannin-organic nitrogen complexes (Aerts & De Caluwe 1997, Northup et al. 1995). Some **mycorrhizal associations**, especially ectomycorrhizae and ericoid mycorrhizae, can use complex organic nitrogen compounds (Sect. 2.4 of the chapter on symbiotic associations), possibly including the complexes produced under pine stands that grow under nutrient-poor conditions. Tannins and other protein-binding phenolics also inhibit nitrification, the microbial conversion of ammonia, via nitrite, to nitrate (Baldwin et al. 1983).

In contrast to situations where the nitrogen availability is decreased by plants that produce phenolics, invasion of grasses (*Molinia caerulea*, *Deschampsia flexuosa*) into nutrient-poor habitats dominated by ericaceous dwarf shrubs (*Calluna vulgaris*, *Erica tetralix*) may enhance rates of **mineralization**. Such invasions are made possible by nitrogen deposition, due to **acid rain**, or damage by the heather beetle (*Lochmaea suturalis*). They may enhance the rate of nitrogen cycling in the system because organic nitrogen contained in the litter of the grasses is mineralized faster than that in residues of the dwarf shrubs.

3.2 Root Exudation and Rhizosphere Effects

The presence of living roots can greatly enhance litter decomposition and mineralization, either by directly using organic matter in the litter through associations with ectomycorrhizal fungi (Sect. 4.2 of the chapter on symbiotic associations), or by providing a carbon source that either stimulates or retards the growth and activity of soil microorganisms and nematodes (Cheng & Coleman 1990, Griffiths et al. 1992, Zhu & Ehrenfeld 1996). There have been too few studies to be certain of general patterns, but results suggest several mechanisms for both positive and negative effects of roots on mineralization.

Root exudates stimulate mineralization only if microorganisms consume the exuded carbohydrates and, in addition, decompose soil organic matter in the rhizosphere. Gram-negative bacteria with high growth rates but a low capability to degrade complex substrates are generally the major microorganisms that are stimulated by exudation. For example, when wheat (*Triticum aestivum*) or rye (*Secale cereale*) plants are grown in soil with ^{14}C-labeled straw, only 6% of the microbial biomass is labeled with ^{14}C. This microbial biomass, however, is highly active in releasing of $^{14}CO_2$, which indicates a "priming" of decomposition by the exudates (Cheng & Coleman 1990, Merckx et al. 1985, Van Veen et al. 1989). Under conditions of low nutrient availability, this priming effect is often less pronounced, perhaps because bacteria have insufficient nutrients to grow and attack soil organic matter (Van Veen et al. 1989) and because plants may intensely compete with soil microorganisms for nutrients under these conditions (Norton & Firestone 1996). This may explain why agricultural and other mineral soils often show a positive effect of roots on nitrogen mineralization (Bottner et al. 1991, Van Veen et al. 1989), whereas these effects are less pronounced or negative in infertile or highly organic soils (Harris & Riha 1991, Parmelee et al. 1993, Tate et al. 1991). For example, roots of tree seedlings stimulate nitrogen mineralization in fertile mull soils, but they decrease mineralization in infertile highly organic mor soils (Bradley & Fyles 1996). These authors suggested that exudates are effective in priming mineralization of mull soil organic matter because of its relatively labile carbon. By contrast, lignolytic activity may control soil nitrogen turnover in organic mor soils, where bacteria that are stimulated by root exudates lack the enzymatic capacity to degrade lignin. Thus, soil fertility could determine the nutritional consequences of root exudation both through its effect on the C/N balance of bacteria and through effects on the recalcitrance of soil organic matter.

Roots may also promote mineralization as a result of more intense grazing of bacteria by protozoa.

The increased growth of bacteria in response to exudates in the rhizosphere attracts protozoa, which use the bacterial carbon to support their growth and maintenance; the protozoa excrete the mineralized nutrients, which are then available for uptake by the plant (Clarholm 1985). We expect this nutrient release by bacterial grazers to be most pronounced in fertile soils, where bacterial growth rates would be highest. The rapid bacterial growth in response to root exudates can also positively affect the plant by outcompeting microorganisms that have detrimental effects on plants. There are obviously major technical difficulties in studying the complex biotic interactions that may occur in the rhizosphere (Dormaar 1990).

Elevated $[CO_2]$ can influence mineralization through its effects on rhizosphere processes. In a relatively fertile soil elevated $[CO_2]$ stimulated nitrogen mineralization (Zak et al. 1993), whereas the opposite was observed in a less fertile grassland soil (Diaz et al. 1993). In other cases, $[CO_2]$ effects on microbial processes varied, depending on the plant species present, which suggests that plant species composition could influence how soil nitrogen cycling will respond to further increases in $[CO_2]$ (Hungate et al. 1996).

The nature of the rhizosphere community affects the quantity and quality of root exudates, with much higher exudation rates occurring in soils than in solution culture. Different populations of soil bacteria and fungi can have distinct effects on the quantity and quality of root exudates (Leyval & Berthelin 1993, Rygiewicz & Andersen 1994) and, therefore, on patterns of mineralization in the rhizosphere.

4. The End-Product of Decomposition

Decomposition of plant litter is a key process of the nutrient cycles of most terrestrial ecosystems. Rates of decomposition strongly depend on chemical composition, with slower rates being associated with high concentrations of phenolics (tannin, lignin) and low concentrations of nitrogen. Because plants in nutrient-poor habitats tend to accumulate more of these secondary plant compounds and have low nitrogen concentrations, their litter is decomposed rather slowly, thus aggravating the low-nutrient status in these habitats. Mycorrhizal associations may prove to be pivotal to access nitrogen in litter containing high concentrations of phenolics (Fig. 4).

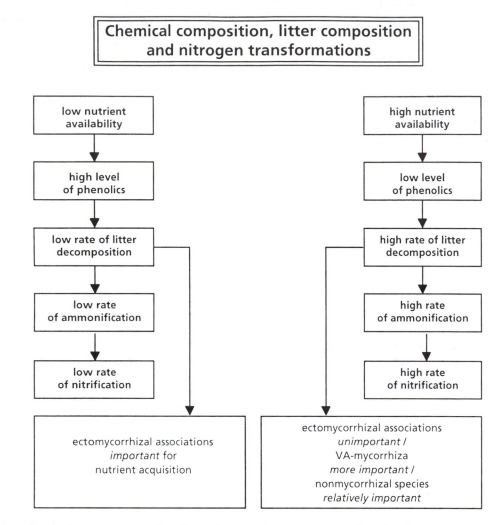

FIGURE 4. Generalized scheme to account for effects of chemical composition of biomass and litter on rates of decomposition and nitrification. Only two extreme situations are shown, with many intermediate habitats or stages of succession occurring between these extremes. Rates of litter decomposition depend on nutrient supply in the habitat. In slow-growing species growing in low-nutrient soils (left), the concentrations of nitrogen and phosphorus tend to be low and phenolics accumulate. These phenolics act as digestibility-reducing defense compounds in the living plant. They also reduce the rate of litter decomposition, thus reducing the rate at which nutrients become available for plant growth. Mycorrhizal associations, but not VAM, may access complexes of organic nitrogen and phenolics, thus making nitrogen available for plant growth. Such nitrogen-acquiring mycorrhizal associations are absent in all herbaceous species and some woody species, but are common in ericaceous and coniferous species. Ectomycorrhizal associations predominate among the woody species. Nonmycorrhizal species (e.g., Proteaceae) may occur, when phosphate occurs mainly in an inorganic insoluble form. In plants growing in high-nutrient soils (right), the concentration of nutrients is high and that of phenolics is low. The litter of these plants also contains low concentrations of phenolics and relatively higher nutrient levels. As a result, it is readily decomposed, releasing ammonium that is either absorbed by plant roots or used by soil microorganisms. Some of these microorganisms (*Nitrosomonas*) use ammonium as an energy source, oxidizing it to nitrite, which is then further oxidized by other soil microorganisms (*Nitrobacter*). The entire process, from ammonium to nitrate, is called nitrification. It occurs more rapidly in the high-nutrient environment of the faster-growing plants. Vesicular-arbuscular mycorrhizae are more common in the higher-nutrient environment of the faster-growing species. Nonmycorrhizal species may dominate during the earliest stages of succession, when soil nutrient levels are relatively high and phosphate is readily available.

References and Further Reading

Aerts, R. (1995) The advantages of being evergreen. Trends Ecol. Evol. 10:402–407.

Aerts, R. (1997) Climate, leaf litter chemistry and leaf litter decomposition in terrestrial ecosystems: A triangular relationship. Oikos 79:439–449.

Aerts, R. & De Caluwe, H. (1997) Nutritional and plant-mediated controls on leaf litter decomposition of *Carex* species. Ecology 78:244–260.

Baldwin, I.T., Olson, R.K., & Reiners, W.A. (1983) Protein-binding phenolics and the inhibition of nitrification in subalpine balsam fir soils. Soil Biol. Biochem. 15:419–423.

Berendse, F., Berg, B., & Bosatta, E. (1987) The effect of lignin and nitrogen on the decomposition of litter in nutrient-poor ecosystems: A theoretical approach. Can. J. Bot. 65:1116–1120.

Berendse, F., Bobbink, R., & Rouwenhorst, G. (1989) A comparative study on nutrient cycling in wet heathland ecosystems. II. Litter decomposition and nutrient mineralization. Oecologia 78:338–348.

Berg, B. & Staaf, H. (1981) Leaching, accumulation and release of nitrogen in decomposing forest litter. Ecol. Bull. 33:163–178.

Bottner, P., Cortez, J., & Sallih, Z. (1991) Effect of living roots on carbon and nitrogen of the soil microbial biomass. In: Plant root growth, D. Atkinson (eds). Blackwell Scientific, London, pp. 201–210.

Bradley, R.L. & Fyles, J.W. (1996) Interactions between tree seedling roots and humus forms in the control of soil C and N cycling. Biol. Fertil. Soils 23:70–79.

Bryant, J.P., Chapin III, F.S., & Klein, D.R. (1983) Carbon/nutrient balance of boreal plants in relation to herbivory. Oikos 40:357–368.

Chapin III, F.S. (1991) Effects of multiple environmental stresses on nutrient availability and use. In: Response of plants to multiple stresses, H.A. Mooney, W.E. Winner, & E.J. Pell (eds). Academic Press, San Diego, pp. 67–88.

Cheng, W. & Coleman, D.C. (1990) Effect of living roots on soil organic matter decomposition. Soil Biol. Biochem. 22:781–787.

Clarholm, M. (1985) Interactions of bacteria, protozoa and plants leading to mineralization of soil nitrogen. Soil Biol. Biochem. 17:181–187.

Clymo, R.S. & Hayward, P.M. (1982) The ecology of *Sphagnum*. In: Bryophyte ecology, A.J.E. Smith (ed). Chapman and Hall, London, pp. 229–289.

Cornelissen, J.H.C. (1996) An experimental comparison of leaf decomposition rates in a wide range of temperate plant species and types. J. Ecol. 84:573–582.

Diaz, S.A., Grime, J.P., Harris, J., & McPherson, E. (1993) Evidence of a feedback mechanism limiting plant response to elevated carbon dioxide. Nature 364:616–617.

Dormaar, J.F. (1990) Effect of active roots on the decomposition of soil organic materials. Biol. Fertil. Soils 10:121–126.

Edwards, N.T. & Sollins, P. (19973) Continuous measurement of carbon dioxide evolution from partitioned forest floor components. Ecology 54:406–412.

Flanagan, P.W. & Van Cleve, K. (1983) Nutrient cycling in relation to decomposition and organic matter quality in taiga ecosystems. Can. J. For. Res. 13:795–817.

Fox, R.H., Myers, R.J.K., & Vallis, I. (1990) The nitrogen mineralization rate of legume residues in soil as influenced by their polyphenol, lignin, and nitrogen contents. Plant Soil 129:251–259.

Franck, V.M., Hungate, B.A., Chapin III, F.S., & Field, C.B. (1997) Decomposition of litter produced under elevated CO_2: Dependence on plant species and nutrient supply. Biogeochemistry 36:223–237.

Gershenzon, J. (1984) Changes in the levels of plant secondary metabolites under water and nutrient stress. In: Phytochemical adaptations to stress, recent advances in phytochemistry, Vol. 18, B.N. Timmermann, C. Steelink, & F.A. Loewus (eds). Plenum Publishing Corporation New York, pp. 273–320.

Gorham, E. (1991) Northern peatlands: Role in the carbon cycle and probable responses to climate warming. Ecol. Appl. 1:182–195.

Griffiths, B.S., Welschen, R., Van Arendonk, J.J.C.M., & Lambers, H. (1992) The effects of nitrogen supply on bacteria and bacterial-feeding fauna in the rhizosphere of different grass species. Oecologia 91:253–259.

Harris, M.M. & Riha, S.J. (1991) Carbon and nitrogen dynamics in forest floor during short-term laboratory incubations. Soil Biol. Biochem. 23:1035–1041.

Hobbie, S.E. (1992) Effects of plant species on nutrient cycling. Trends Ecol. Evolu. 7:336–339.

Hobbie, S.E. (1995) Direct and indirect effects of plant species on biogeochemical processes in arctic ecosystems. In: Arctic and alpine biodiversity: Patterns, causes and ecosystem consequences, F.S. Chapin III & Ch. Körner (eds). Springer-Verlag, Berlin, pp. 213–224.

Hobbie, S.E. (1996) Temperature and plant species control over litter decomposition in Alaskan tundra. Ecol. Monogr. 66:503–522.

Hungate, B.A., Canadell, J.C., & Chapin III, F.S. (1996) Plant species mediate changes in microbial N in response to elevated CO_2. Ecology 77:2505–2515.

Johnson, L.C. & Damman, A.W.H. (1993) Decay and its regulation in *Sphagnum* peatlands. Adv. Bryol. 5:249–296.

Leyval, C. & Berthelin, J. (1993) Rhizodeposition and net release of soluble organic compounds by pine and beech seedlings inoculated with rhizobacteria and ectomycorrhizal fungi. Biol. Fertil. Soils 15:259–267.

Merckx, R., Den Hartog, A., & Van Veen, J.A. (1985) Turnover of root-derived material and related microbial biomass formation in soils of different texture. Soil Biol. Biochem. 17:565–569.

Merckx, R., Dijkstra, A., Den Hartog, A., & Van Veen, J.A. (1987) Production of root-derived material and associated microbial growth in soil at different nutrient levels. Biol. Fertil. Soils 5:126–132.

Northup, R.R., Yu, Z., Dahlgren, R.A., & Vogt, K.A. (1995) Polyphenol control of nitrogen release from pine litter. Nature 377:227–229.

Norton, J.M. & Firestone, M.K. (1996) N dynamics in the rhizosphere of *Pinus ponderosa* seedlings. Soil Biol. Biochem. 28:351–362.

Parmelee, R.W., Ehrenfeld, J.G., & Tate, R.L., III (1993) Effects of pine roots on microorganisms, fauna, and nitrogen availability in two soil horizons of a coniferous forest spodosol. Biol. Fertil. Soils 15:113–119.

Paul, E.A. & Clark, F.E. (1989) Soil microbiology and biochemistry. Academic Press, San Diego.

Rygiewicz, P.T. & Andersen, C.P. (1994) Mycorrhizae alter quality and quantity of carbon allocated below ground. Nature 369:58–60.

Van Breemen, N. (1993) Soils as biotic constructs favouring net primary productivity. Geoderma 57:183–211.

Van Veen, J.A., Merckx, R., & Van de Geijn, S.C. (1989) Plant- and soil related controls of the flow of carbon from roots through the soil microbial biomass. Plant Soil 115:179–188.

Van Vuuren, Aerts, R., Berendse, F., & De Visser, W. (1992) Nitrogen mineralization in heathland ecosystems dominated by different plant species. Biogeochemistry 16:151–166.

Verhoeven, J.T.A. & Toth, E. (1995) Decomposition of *Carex* and *Sphagnum* litter in fens: effect of litter quality and inhibition by living tissue homogenates. Soil Biol. Biochem. 27:271–275.

Vitousek, P.M. (1982) Nutrient cycling and nutrient use efficiency. Am. Nat. 119:553–572.

Vitousek, P.M., Turner, D.R., Parton, W.J., & Sanford, R.L. (1994) Litter decomposition on the Mauna Loa environmental matrix, Hawaii: Patterns, mechanisms, and models. Ecology 75:418–429.

Wilschke, J., Hoppe, E., & Rudolph, H.-J. (1990) Biosynthesis of sphagnum acid. In: Bryophytes: Their chemistry and chemical Taxonomy, H.D. Zinsmeister & R. Mues (eds). Oxford Science Publications, Oxford, pp. 253–263.

Zak, D.R., Pregitzer, K.S., Curtis, P.S., Teeri, J.A., Fogel, R., & Randlett, D.A. (1993) Elevated atmospheric CO_2 and feedback between carbon and nitrogen cycles. Plant Soil 151:105–117.

Zhu, W. & Ehrenfeld, J.G. (1996) The effects of mycorrhizal roots on litter decomposition, soil biota, and nutrients in a spodosolic soil. Plant Soil 179:109–118.

10B. Ecosystem and Global Processes: Ecophysiological Controls

1. Introduction

In previous chapters we emphasized the integration among processes from molecular to whole-plant levels, and we considered the physiological consequences of interactions between plants and other organisms. In this chapter we will move up in scale to consider relationships between **plant ecophysiological processes** and those occurring at **ecosystem to global scales**. Plant species differ substantially in their responses to environment and to other organisms. It is not surprising that these physiological differences among plants contribute strongly to functional differences among ecosystems.

2. Ecosystem Biomass and Productivity

2.1 Scaling from Plants to Ecosystems

The supply rates of light, water, and nutrients that govern ecosystem processes are functions of ground area and soil volume. A critical initial step in relating the processes in individual plants to those in ecosystems, therefore, is to determine **how plant size and density relate to stand biomass**. In sparse stands of plants there is no necessary relationship between size and density, so that plants increase in mass without changes in density (Fig. 1). As plants begin to compete, however, mortality

reduces plant density in a predictable fashion. Communities in approximate equilibrium with their environment show an inverse relationship between ln(biomass) and ln(density), with a slope of about $-3/2$. This **self-thinning line** was initially derived empirically (Yoda et al. 1963), but it has been observed in a wide array of studies, both experimental and in the field, in ecosystems ranging from meadows to forests (Silvertown 1982, Weller 1987, White 1980). The slope and intercept of the self-thinning line vary among species and experimental conditions more than was originally thought (Weller 1987); nonetheless, it provides an empirical basis to extrapolate from individuals to stands of vegetation. Given that

$$\ln(b) = -3/2\ln(d) \qquad (1)$$

it follows that

$$b = (d)^{-3/2} \quad \text{or} \quad d = (b)^{-2/3} \qquad (2)$$

$$B = b \cdot d = d^{-1/2} = b^{1/3} \qquad (3)$$

where b is individual biomass (g plant^{-1}), d is density (plants m^{-2}), and B is stand biomass (g m^{-2}). These relationships indicate that increasing stand biomass is typically associated with increased plant size and reduced density.

2.2 Physiological Basis of Productivity

Net primary productivity (NPP) is the net biomass gain by vegetation per unit time. The main plant

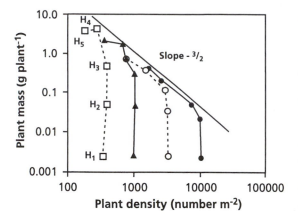

FIGURE 1. Self-thinning in four populations of *Lolium perenne* that were planted in glasshouse beds at four densities. H1 to H5 are replicates harvested at five successive intervals. After germination, plant biomass increases without change in density due to increased size of individual plants. As plants begin to compete, smaller individuals die, causing a decrease in density and a slower rate of increase in average plant biomass. From this point onward, the biomass–density relationship follows a self-thinning line in which ln biomass and ln density have a slope of −3/2 (Kays & Harper 1974). Copyright Blackwell Science Ltd.

traits that govern NPP $(g m^{-2} yr^{-1})$ are **biomass** $(g m^{-2})$ and **RGR** $(g g^{-1} yr^{-1})$:

$$NPP = Biomass \cdot RGR \qquad (4)$$

In woody communities biomass exerts the predominant influence over NPP, with forests being more productive than shrublands or grasslands, despite their typically low RGR (Table 1). By contrast, in herbaceous communities, where maximum biomass is constrained by the lack of woody support structures, RGR is generally more important than biomass per individual in determining NPP.

At the **global scale, climate** is the major determinant of biomass and productivity (Fig. 2) because of constraints on both the growth of individual plants and the types of species that can compete effectively. Tropical rain forests have the highest biomass and productivity, where warm moist conditions favor plant growth and development of a large plant size; lowest values are in desert and tundra, where low precipitation or temperature, respectively, constrain growth (Table 2). In the tropics, where temperature is not a constraint, rainforests have greater productivity than do dry deciduous forests, which are more productive than are savannas (i.e., productivity declines with reduced water availability). Where moisture is less limiting to growth, biomass and productivity are similarly governed by temperature, decreasing from tropical to temperate to boreal forests, and finally to tundra. As discussed earlier, the most productive ecosystems are tree-dominated and have a low RGR, despite favorable conditions for growth, because trees have an inherently low growth rate—they are productive because of their large size, not their high RGR.

At **local to regional scales, climate** continues to be important, with strong differences in productivity associated with altitudinal gradients in temperature and precipitation and with temperature differences between north- and south-facing slopes. At regional scales, however, variation in soil moisture and nutrients, due to topographic variations in drainage and erosional transport of soils and to differences in **parent material** (the rocks that give rise to soils), become increasingly strong controls over productivity. For example, marshes are among the most productive habitats in most climate zones,

TABLE 1. Above-ground biomass, production, and nitrogen flux in major temperate ecosystem types; maximum height, and relative growth rate of species typical of these ecosystem types.*

Parameter	Grassland	Shrubland	Deciduous forest	Evergreen forest
Above-ground biomass[a] $(kg m^{-2})$	0.3 ± 0.02	3.7 ± 0.05	15 ± 2	31 ± 8
Above-ground NPP[a] $(kg m^{-2} yr^{-1})$	0.3 ± 0.02	0.4 ± 0.07	1.0 ± 0.08	0.8 ± 0.08
N flux[a] $(g m^{-2} yr^{-1})$	2.6 ± 0.2	3.9 ± 1.6	7.5 ± 0.5	4.7 ± 0.5
Canopy height[b] (m)	1	4	22	22
Field RGR[c] (yr^{-1})	1.0	0.1	0.07	0.03
Laboratory[b] RGR (wk^{-1})	1.3	0.8	0.7	0.4

Source: Chapin 1993.
*Note: Data are means ± SE.
[a]Bokhari & Singh 1975, Cole & Rapp 1981, Gray & Schlesinger 1981, Sala et al. 1988.
[b]Grime & Hunt 1975, Tilman 1988.
[c]Above-ground production/above-ground biomass.

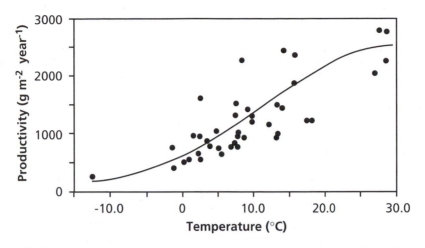

FIGURE 2. Relationship between NPP determined by harvest and mean annual temperature for 52 locations throughout the world (Lieth 1975).

because of high moisture and nutrient availability. As discussed in the chapter on growth and allocation, low-moisture and low-nutrient environments are typically dominated by slowly growing species with low specific leaf area, low rates of photosynthesis per unit leaf mass, and low leaf area ratios. These plant traits, sometimes combined with low plant density, result in low biomass and productivity.

At the local scale, there can still be important differences in biomass and productivity because of differences in species traits, even with the same climate and parent material. Species introductions often result in strikingly different species dominating adjacent sites. For example, in California, *Eucalyptus* forests have been planted on sites that would otherwise be grasslands. The *Eucalyptus*

TABLE 2. Primary production and biomass estimates for the world.

Ecosystem type	Area (10^6km^2)	Mean biomass (kg C m^{-2})	Total biomass (10^9 ton C)	Mean NPP $(\text{g C m}^{-2}\text{yr}^{-1})$	Total NPP $(\text{Gt C yr}^{-1})^a$	RGR (yr^{-1})
Tropical rainforest	17.0	20	340	900	15.3	0.045
Tropical seasonal forest	7.5	16	120	675	5.1	0.042
Temperate evergreen forest	5.0	16	80	585	2.9	0.037
Temperate deciduous forest	7.0	13.5	95	540	3.8	0.040
Boreal forest	12.0	9.0	108	360	4.3	0.040
Woodland and shrubland	8.0	2.7	22	270	2.2	0.100
Savanna	15.0	1.8	27	315	4.7	0.175
Temperate grassland	9.0	0.7	6.3	225	2.0	0.321
Tundra and alpine meadow	8.0	0.3	2.4	65	0.5	0.217
Desert scrub	18.0	0.3	5.4	32	0.6	0.107
Rock, ice, and sand	24.0	0.01	0.2	1.5	0.04	—
Cultivated land	14.0	0.5	7.0	290	4.1	0.580
Swamp and marsh	2.0	6.8	13.6	1125	2.2	0.165
Lake and stream	2.5	0.01	0.02	225	0.6	22.5
Total continental	149	5.5	827	324	48.3	0.058
Total marine	361	0.005	1.8	69	24.9	14.1
Total global	510	1.63	829	144	73.2	0.088

Source: Schlesinger 1991.
[a] Gigatons (Gt) are 10^{15} g.

forest has a biomass and productivity much greater than that of the grassland, despite the same climate and parent material. *Eucalyptus* has deeper roots that tap water unavailable to the grasses, thus supporting the larger biomass and productivity (Robles & Chapin 1995). Once the grassland or forest is established, it is difficult for species of contrasting life form to colonize. As a result, there can be **alternative stable community types** with strikingly different biomass and productivity in the same environment. In deserts deep-rooted **phreatophytes** can similarly tap the water table and support a larger biomass and productivity than do shallow-rooted species. Thus, although climate and resource supply govern large-scale patterns of biomass and productivity, the actual productivity on a site depends strongly on historical factors that govern the species present on a site (Sect. 3 of introduction chapter).

2.3 Disturbance and Succession

Stand age modifies environmental controls over biomass and productivity. Following **disturbance**, the most common initial colonizers are herbaceous weedy species that have high reproductive allocation, effective dispersal and are commonly well represented in the buried seed pool (Sect. 3.1 of the chapter on life cycles). There is initially an exponential increase in plant biomass because of the exponential nature of plant growth (Sect. 2.1 of the chapter on growth and allocation). RGR declines as plants get larger and begin to compete with one another. In addition, as succession proceeds, there is often a replacement of rapidly growing herbaceous species by woody species that grow more slowly, are taller and shade out the initial colonizers. This causes a further decline in relative growth rate (Table 1), despite the increase in biomass and productivity through time. In some ecosystems, productivity declines in late succession due to declines in soil nutrient availability and, in some forests, to declines in leaf area and photosynthetic capacity associated with reduced hydraulic conductance of old trees (Sect. 5.2.2 of the chapter on plant respiration and Sect. 5.1 of the chapter on plant water relations). Thus, changes in productivity through succession are governed initially by rates of colonization and RGR, followed by a gradual transition to a woody community that has lower RGR, but whose larger plant size results in further increases in productivity.

Disturbance regime determines the relative proportion of early- and late-successional stands in a region. For example, fire is a natural agent of disturbance that is common at intermediate moisture regimes. In deserts there is insufficient fuel to carry a fire and few convective storms to provide the lightening that ignites fires. By contrast, in temperate and tropical ecosystems with high precipitation or in arctic ecosystems with low evapotranspiration, soils and vegetation are too wet to carry a fire. In grasslands fire occurs so frequently that woody plants rarely establish, so the region is dominated by herbaceous vegetation with high relative growth rate and modest productivity. These vegetation characteristics are favorable to mammalian grazers, which act as an additional disturbance to prevent colonization by woody plants. Most grasslands have sufficient water and nutrients to support growth of woody plants. It is primarily the disturbance regime that maintains the high-RGR, low-wood nature of grasslands.

Plant traits strongly influence the disturbance regime of ecosystems. In grasslands, grasses produce an abundant fine-structured fuel that burns readily when dry, because of the high SLA, high leaf production rate, and low leaf longevity. Abundant below-ground reserve storage and meristem pools allow grasses to recover after grazing or fire. Thus, it is **adaptations to tolerate grazing** that result in traits that promote fire in grasslands. Introduction of grasses into forest or shrubland ecosystems can increase fire frequency and cause a replacement of forest by savanna (D'Antonio & Vitousek 1992). Once the grasses create this disturbance regime with high **fire frequency**, tree and shrub seedlings can no longer establish. Boreal conifers also create a fire regime that favors their own persistence. They are more flammable than deciduous trees because of their large leaf and twig surface area, low moisture content, and high resin content—an antiherbivore/pathogen defense (Van Cleve et al. 1991). Thus, there is an increase in fire probability as succession proceeds. Because of high fire frequency in these forest types many conifers such as lodgepole pine (*Pinus contorta*) produce **serotenous cones** (i.e., cones that open only after exposure to the heat of a fire). The invasion of the northern hardwood forests by hemlock (*Tsuga*) in the early Holocene caused an increase in fire frequency (Davis et al. 1992), with associated changes in both plant and animal communities (Slobodchikoff & Doyen 1977), clearly showing the role of plant traits in determining fire regime.

2.4 Photosynthesis and Absorbed Radiation

One scaling approach is to extrapolate directly from leaf carbon exchange to the ecosystem level based on the relationship between photosynthesis and absorbed radiation. This approach was pioneered in agriculture (Monteith 1977) and has been extended to estimate patterns of carbon exchange in natural ecosystems (Field 1991). The fraction of incident photosynthetically active radiation that is absorbed by plants (**APAR**) is either converted to new biomass (NPP) or is respired. APAR depends on total leaf area, its vertical distribution and its photosynthetic capacity. As discussed in the chapter on photosynthesis, both light and nitrogen decline in a predictable fashion through the canopy (Sect. 3.1 of the chapter on photosynthesis and Box 7 attached to the chapter scaling-up), with nitrogen preferentially allocated to the tops of canopies to maximize light capture (Hirose & Werger 1987, Terashima & Hikosaka 1995). Thus, as an initial simplification, the plant canopy can be treated as a **big leaf** whose photosynthetic capacity depends on total canopy nitrogen (Farquhar 1989, Field 1991) (Sect. 2 of the chapter on scaling-up). In unstressed crops, dry matter accumulation is roughly proportional to integrated radiation interception over the growing season with a conversion efficiency of about $1.4\,g\,MJ^{-1}$ (Monteith 1977). Natural ecosystems vary 10- to 100-fold in productivity (Table 1). Most of this variation is due to variation in APAR rather than in conversion efficiency, which varies about two fold among studies. There are no striking ecological patterns in reported values of conversion efficiency, with much of the variation among studies likely due to differences in methodology, rather than inherent differences among ecosystems (Field 1991).

Most of the variation in APAR is due to variation in LAI (>50-fold variation among ecosystems), although leaf nitrogen concentration can vary fivefold among ecosystems. Thus, carbon gain and NPP are reduced in unfavorable environments due to the small amount of leaf biomass that can be supported and leaf nitrogen concentration that can be attained, as discussed in Section 5 of the chapter on growth and allocation.

The relatively consistent conversion of APAR into plant production among ecosystems provides a tool for estimating global patterns of NPP. APAR can be estimated from satellite-borne sensors, using the normalized difference vegetation index (**NDVI**):

$$NDVI = (NIR - VIS)/(NIR + VIS) \qquad (5)$$

where NIR ($W\,m^{-2}$) is reflectance in the near infrared, and VIS ($W\,m^{-2}$) is reflectance in the visible. NDVI uses the unique absorption spectrum of chlorophyll, which differs from that of clouds, water, and bare soil, to estimate absorbed radiation. Stands with high rates of photosynthesis have a high NDVI because they have low values of reflected VIS and high values of reflected NIR. NDVI is an excellent predictor of APAR and daily net photosynthesis in short-term plot-level studies (Fig. 3). It also provides good estimates of NPP using satellites (Fig. 4). The consistency of this relationship supports the argument that there may be a relatively constant efficiency of converting absorbed radiation into plant biomass. One reason for

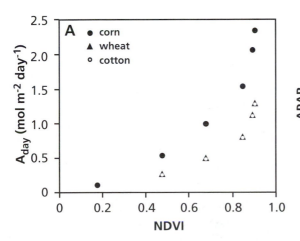
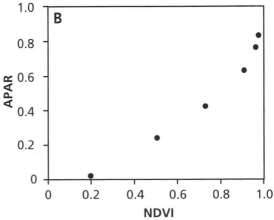

FIGURE 3. Relationship of normalized difference vegetation index (NDVI) to (A) daily net rate of CO_2 assimilation (A_{day}) and to (B) the fraction of absorbed photosynthetically active radiation (APAR). These relationships were simulated based on data collected from wheat, maize, and cotton (Field 1991, as redrawn from Choudhury 1987). Copyright Remote Sensing of the Environment.

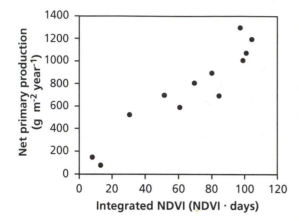

FIGURE 4. Relationship between mean net primary production (NPP) for several biomes and the seasonally integrated normalized difference vegetation index (NDVI) measured from satellites. Each point represents a different biome (Field 1991, as redrawn from Goward et al. 1985). Copyright Kluwer Academic Publishers.

the modest variation in conversion efficiency between APAR and NPP may the the similarity of growth respiration across plant tissues and species (Sect. 5.2 of the chapter on plant respiration). From a pragmatic perspective, the strong relationship between NDVI and NPP is important because it allows us to estimate NPP directly from satellites (Fig. 4). In this way, we can estimate **regional and global patterns of NPP** in ways that avoid the errors and biases that are associated with the extrapolation of harvest data to the global scale.

Any factor that alters the leaf area of an ecosystem or the availability of water or nitrogen changes the capacity of that ecosystem for carbon gain by moving vegetation along the generalized APAR-NPP relationship. Climate has obvious effects on LAI and leaf nitrogen. The physiological differences among plant species that we have discussed throughout the book also have pronounced effects on the leaf area and leaf nitrogen that can be supported in any environment, as mediated by competitive interactions, herbivores, and pathogens. In general, the sorting of species among habitats by competition over the long term probably maximizes APAR and NPP, whereas pathogens and herbivores tend to reduce APAR and NPP. Disturbance regime also influences regional APAR and NPP, as does human land conversion of natural ecosystems to pastures and agriculture.

Satellite-based measurements of NDVI provide evidence for several large-scale changes in NPP. In the tropics, there have been decreases in NDVI associated with forest clearing and conversion to agri-

culture. Sub-Saharan Africa also shows reductions in NDVI associated with over-grazing. At high latitudes, however, NDVI appears to have increased (Myneni et al. 1997). This high-latitude increase in NDVI is particularly intriguing because it is remote from areas of large-scale anthropogenic land-use change and could reflect broad biospheric responses to changes in climate. There has been an increase in early spring temperature in continental regions of northern Canada, Alaska, and Siberia in the past 30 years (Chapman & Walsh 1993) that coincides broadly with the NDVI increases. This warming may have increased NPP through increased length of growing season, direct temperature effects on growth, or increased disturbance (causing a shift in stand age toward more productive stands). There are also many potential artifacts associated with the lack of calibration of satellite sensors among years; this complicates the interpretation. The striking trends in changes in NDVI nevertheless strongly suggest that global NPP can be substantially altered over broad regions of the globe.

2.5 Net Carbon Balance of Ecosystems

Net ecosystem production (NEP, $gCm^{-2}yr^{-1}$) of carbon by an ecosystem depends on the balance between **net primary production** (NPP, $gCm^{-2}yr^{-1}$) and **heterotrophic respiration** (R_h, $gCm^{-2}yr^{-1}$), or between **gross photosynthesis** (P_g, $gCm^{-2}yr^{-1}$) and **total ecosystem respiration** (R_e, $gCm^{-2}yr^{-1}$), which is the sum of R_h and plant respiration (R_p, $gCm^{-2}yr^{-1}$).

$$NEP = NPP - R_h = P_g - R_e \qquad (6)$$

NEP is important because it is a short-term measure of the increment in carbon stored by an ecosystem. We have discussed the plant physiological and environmental constraints on NPP (Sects. 2.2 and 2.4). Decomposers account for most of the heterotrophic respiration. Their respiration is controlled by moisture and temperature and by the quantity, quality and location (above or below ground) of organic matter produced by plants (Sect. 3 of the chapter on decomposition). In general, conditions that favor high NPP also favor high R_h. For example, both NPP and decomposition are higher in the tropics than in the arctic and higher in rainforests than in deserts, due to similar environmental sensitivities of NPP and R_h. Species that are highly productive (high RGR and/or large biomass) similarly produce more litter or higher-quality litter than do species of low potential productivity. Habitats dominated by productive species, therefore, are characterized by high decomposition rates (Hobbie 1992). There is a necessary functional linkage between NPP and R_h. NPP provides the organic material that fuels R_h, and R_h releases the minerals that support NPP (Harte & Kinzig 1993). For all these reasons, NPP and R_h tend to be closely matched in ecosystems at steady state (Odum 1969, Wofsy et al. 1993). At steady state, therefore, by definition, NEP and changes in carbon storage are small and show no correlation with NPP or R_h. In fact, peat bogs, which are among the least productive ecosystems, are ecosystems with the greatest long-term carbon storage.

NEP, which is the net carbon exchange of ecosystems, is a small difference between two large fluxes, **gross photosynthesis** (P_g) and **ecosystem respiration** (R_e) (Fig. 5). Although NEP is on average close to zero in ecosystems at steady state, it shows large enough seasonal variation to cause seasonal fluctuations in atmospheric [CO_2] at the global scale in the northern hemisphere (Fig. 56 in the chapter on photosynthesis), with decreases in atmospheric [CO_2] during summer and increases in winter. The most clearcut causes of variation among ecosystems in NEP are successional cycles of disturbance and recovery. Most disturbances initially cause a negative NEP. Fire releases carbon directly by combustion and causes even larger carbon losses by promoting conditions for R_h (Kasischke et al. 1995). For example, removal of vegetation typically reduces transpiration, which causes an increase in soil moisture and increases soil temperature due to greater radiation absorption (lower albedo and greater penetration of solar radiation to the soil surface). The warmer, moister soils enhance R_h and the reduction in plant biomass reduces NPP, which results in negative NEP for years ofter a fire (Kasischke et al. 1995). Agricultural tillage breaks up soil aggregates and increases access of soil microbes to soil organic matter, which results in a similar increase in R_h and negative NEP after conversion of natural ecosystems to agriculture. Prairie soils often lose half their soil carbon within a few decades after conversion to agriculture (Davidson & Ackerman 1993). During succession after a disturbance there is typically an increase in plant biomass and soil organic matter because NPP increases more rapidly during succession than does R_h.

NEP can also vary substantially among years, due to different environmental responses of photosynthesis and respiration. For example, northern ecosystems are a net carbon source in warm years and a carbon sink in cool years (Oechel et al. 1993, Zimov et al. 1996) because heterotrophic respiration responds to temperature more strongly than does photosynthesis in cold climates.

2.6 The Global Carbon Cycle

Recent large-scale changes in the global environment (e.g., regional warming, nitrogen deposition, and elevated CO_2 concentrations) can alter NEP, if they have differential effects on photosynthesis and respiration. For example, photosynthesis responds more strongly to atmospheric CO_2 concentration than does heterotrophic respiration; therefore, the terrestrial biosphere might increase net CO_2 uptake in response to the increases in atmospheric [CO_2] caused by fossil fuel combustion and biomass burning associated with land-use change. This [CO_2] effect on plant carbon balance could be further magnified if plant respiration is inhibited by [CO_2] (Sect. 4.7 of the chapter on plant respiration). In most terrestrial ecosystems, however, NPP is nutrient-limited, which strongly constrains the capacity of vegetation to respond to elevated [CO_2]. The clearest evidence for increases in NPP in response to elevated [CO_2], therefore, is in regions of nitrogen deposition, where there are widespread increases in tree growth (Kauppi et al. 1992). NPP, however, is only half the story: NPP must change more strongly than R_h, if there is to be an increase in NEP.

Only half of the annual anthropogenic input of CO_2 remains in the atmosphere, with the rest being removed by the oceans or the terrestrial biosphere (Fig. 55 in the chapter on photosynthesis). The location of this **missing sink** of atmospheric CO_2 is difficult to identify by direct measurement because its global magnitude ($4.1\,GtC\,yr^{-1}$) is only about 5% of global NPP, which is much smaller than measurement errors and typical interannual variability. Isotopic discrimination by photosynthesis (Sect. 4.3 of the chapter on photosynthesis and Box 2) has provided an important key to identifying the magnitude and location of the missing sink. Atmospheric transport models can be run in "inverse mode" (i.e., opposite to the direction of cause to effect) to estimate the global distribution of CO_2 sources and sinks that are required to match the observed geographic and seasonal patterns of concentrations of CO_2 and $^{13}CO_2$ in the atmosphere (Ciais et al. 1995, Denning et al. 1995, Tans et al. 1990) (Fig. 56 in the chapter on photosynthesis). Carbon uptake by the terrestrial biosphere can be distinguished from the CO_2 that dissolves in the ocean because of the strong isotopic discrimination during photosynthesis. Atmospheric stoichiometry between CO_2 and O_2 similarly separate biological

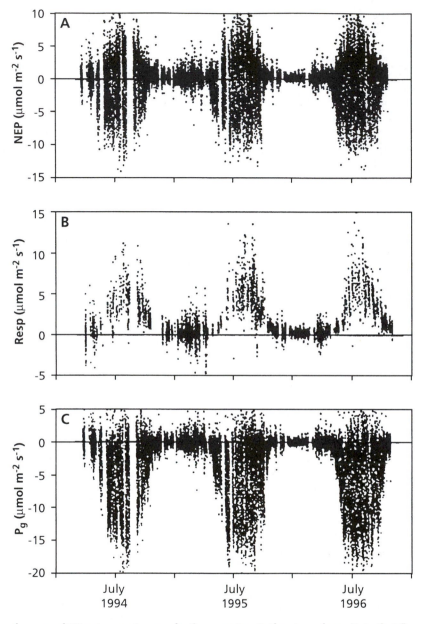

FIGURE 5. Annual course of (A) net ecosystem production (NEP), (B) ecosystem respiration (Resp), and (C) gross CO_2 assimilation (Pg) in an old-growth black spruce forest in northern Canada. Positive values are fluxes from the eco- system to the atmosphere. Note that fluxes vary consider- ably from day to day, with largest fluxes of both photosyn- thesis and respiration in summer (Goulden et al. 1997). Copyright Journal of Geophysical Research.

from physical causes of changing atmospheric [CO_2]. Although there are still many uncertainties, these models suggest that terrestrial ecosystems account for about half (2.1 Gt yr^{-1}) of the missing sink and that these terrestrial sinks are concentrated at mid- to high-northern latitudes (Schimel 1995). Current estimates are that half of the terrestrial

component of the missing sink is a result of the photosynthetic response to rising atmospheric [CO_2], 25% to increased plant growth due to nitro- gen deposition, and 25% to regrowth of mid- latitude forests after forest clearing in the nineteenth and early twentieth centuries (Schimel 1995).

3. Nutrient Cycling

3.1 Vegetation Controls over Nutrient Uptake and Loss

The controls over nutrient uptake and loss by stands of vegetation are basically the same as those described for individual plants (Sect. 2.2 of the chapter on mineral nutrition). Plants control **nutrient uptake** directly by root biomass and the kinetics of ion uptake and indirectly by influencing nutrient supply rate. **Root biomass** is the major factor governing stand-level nutrient uptake because a large root biomass is the major mechanism by which plants minimize diffusional limitations of nutrient delivery to the root surface (Sect. 2.2.1 of the chapter on mineral nutrition). The absolute magnitude of root biomass is probably greatest in high-resource environments, where there is a large total plant biomass (e.g., forests; Table 1). Root biomass, however, varies less across ecosystems (Table 5 of the chapter on plant water relations) than does total biomass because proportional allocation to roots increases in low-resource environments (Sect. 5.4.4 of the chapter on growth and allocation). I_{max} of ion uptake is generally greatest in plants that grow rapidly and would therefore contribute to the high nutrient uptake in high-resource environments. In low-nutrient environments, vegetation maximizes nutrient acquisition through high root biomass (an acclimation response rather than adaptation), symbiotic associations (mycorrhizae and symbiotic nitrogen fixation), and by solubilizing insolu-ble inorganic phosphates or organic nitrogen or phosphorus (Sect. 2.2 of the chapter on mineral nutrition and Sects. 2.3 to 2.5 and 3.7 of the chapter on symbiotic associations). Overall, there is a strong correlation between NPP and nutrient uptake by vegetation (Fig. 6).

Annual **nutrient loss** from vegetation is greatest in high-nutrient environments where NPP and biomass are high, and where there is a low mean residence time of nutrients in plants (rapid leaf and perhaps root turnover) and high nutrient concentrations in litter (Aerts 1995). Thus, for both plant nutrient uptake and loss, the differences observed among ecosystems are the same as would be predicted by the patterns of acclimation and adaptation of individual plants, but are more pronounced because of the larger size of plants in favorable environments.

3.2 Vegetation Controls over Mineralization

The effects of climate and resource availability on nutrient supply are similar to those described for decomposition (Sect. 2.5 of this chapter; Sect. 2.1.1 of the chapter on mineral nutrition), with high rates of nutrient supply under favorable environmental conditions. Within these environmental constraints, however, **plant traits** strongly influence nutrient supply through their effects on root exudation, microenvironment, and litter quality. Litter quality differs among ecosystems and strongly influences mineralization rates (Sect. 3.1 of the chapter on decomposition). Root exudation provides a labile carbon source of sugars, organic acids, and amino acids that can either enhance (Zak et al. 1993) or reduce (Diaz et al. 1993) mineralization. We expect that carbon-rich exudates will promote nitrogen immobilization in infertile soils, whereas they might stimulate breakdown of additional carbon substrates in more fertile soils, as discussed in Section 3.2 of the chapter on decomposition. Over longer time scales, successional development of vegetation modifies soil temperature (shading), soil moisture (transpiration), and the quantity and quality of organic matter inputs (litter and root exudates), as we have already discussed (Sect. 2.2 of the chapter on decomposition.

Over long time scales (decades to centuries), patterns of **nutrient input and loss** exert additional influences over nutrient supply. There is only fragmentary understanding of these long-term controls, although we know that abundance of dinitrogen-fixing plants strongly influences nitrogen inputs

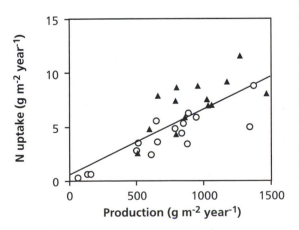

FIGURE 6. Relationship between net primary production (NPP) and nitrogen uptake of temperate and boreal coniferous (circles) and deciduous (triangles) forests (Chapin 1993). Copyright Academic Press.

(Vitousek & Howarth 1991). For example, introduction of the nitrogen-fixing tree *Myrica faya* into the Hawaiian Islands greatly increased nitrogen inputs, N supply and annual rates of N cycling (Vitousek et al. 1987). Replacement of perennial grasses by annual grasses, with their shorter period of physiological activity, may account for autumn nitrogen losses from California grasslands. Anthropogenic inputs of nitrogen from industrial fixation and planting of legume crops now exceeds inputs by natural fixation at the global scale (Vitousek 1994), which suggests that there may be substantial changes in the regulation of inputs and outputs of nitrogen in natural ecosystems.

4. Ecosystem Energy Exchange and the Hydrologic Cycle

4.1 Vegetation Effects on Energy Exchange

4.1.1 Albedo

Energy exchange at the ecosystem scale is influenced not only by the properties of individual leaves and stems (e.g., albedo and the partitioning of dissipated energy between sensible and latent heat (Sect. 2.1 of the chapter on the plant's energy balance) but also by any contrasts between plant properties and those of the underlying surface. The atmosphere is nearly transparent to the shortwave radiation that is emitted by the sun, so **air temperature** at local to global scales is primarily determined by the amount of energy absorbed and dissipated by the earth's surface. The influence of vegetation on surface reflectance (**albedo**), therefore, can have substantial effects on climate. For example, snow and sand have higher albedos than does vegetation; therefore, they reduce absorption of radiation at the surface. In tundra, any increase in plant height relative to snow depth or increased density of tall shrubs or trees will mask the snow and reduce the albedo (i.e., increase absorbed energy and the energy dissipated to the atmosphere), thus raising the temperature of the overlying air (Bonan et al. 1992). Model simulations suggest that if the boreal forest were converted to snow-covered tundra, then this would reduce annual average air temperature in the boreal zone by 6°C and that this temperature effect would be large enough to extend into the tropics (Bonan et al. 1992, 1995). Similarly, when temperature warmed at the last thermal maximum, 6000 yr ago, the **treeline** moved northward, which reduced the regional albedo and increased energy absorption. Approximately half of the climatic warming that occurred at that time is estimated to be due to the northward movement of treeline, with the remaining climate warming due to increased solar input (Fig. 7). The warmer regional climate, in turn, favors tree reproduction and establishment at treeline (Payette & Filion 1985), which provides a positive feedback to regional warming.

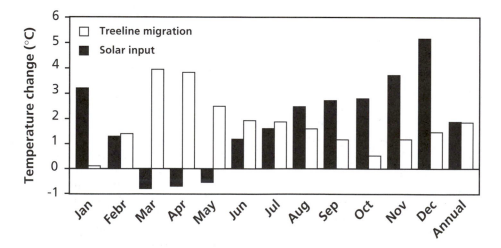

FIGURE 7. The change in arctic air temperature at the last thermal maximum caused directly by changes in solar inputs and caused by the change in albedo associated with northward movement of treeline, as simulated by a general circulation model. Reprinted from Chapin et al. (1997), as redrawn from Foley et al. (1994). Copyright Blackwell Science Ltd.

Thus, changes in vegetation height, relative to snow depth, could exert a large effect on regional climate.

Vegetation effects on albedo also influence regional climate in arid areas. **Overgrazing** in the Middle East reduced plant density, exposed more light-colored soil, and thus reduced absorbed radiation. This reduced heating and convective uplift of the overlying air, which resulted in less advection of moisture from the Mediterranean and reduced precipitation (Charney et al. 1977). This increase in drought acted as a positive feedback to further reduce plant production and biomass, resulting in a permanent drying of regional climate.

Differences in albedo among **vegetated surfaces** are more subtle than are those between vegetation and snow or soil. Vegetation albedo depends primarily on phenology. Leaf appearance in deciduous ecosystems increases albedo if the soil surface is dark and reduces albedo over light-colored surfaces. Evergreen communities show minimal seasonal change in albedo. Even the small differences in albedo among plant species could be climatically important. For example, grasslands typically have higher albedo than do forests because of their more rapid leaf turnover and retention of dead reflective leaves in the canopy.

4.1.2 Energy Partitioning

Differences among plant species in **energy partitioning** between **latent and sensible heat** can have large-scale consequences. Leaf area index (LAI) is the strongest determinant of **evapotranspiration** because it determines (1) the amount of precipitation that is intercepted by the canopy and quickly evaporates after a rain and (2) the size of the transpiring surface (Fig. 8). LAI of undisturbed communities is determined primarily by plant size and secondarily by phenology because evergreen plants have a transpiring surface present for more of the year (Chapin 1993). Although evergreen plants have a larger proportion of biomass in leaves than do deciduous species of the same size, this attribute is counterbalanced by the lower stomatal conductance of evergreens; therefore, evergreen and deciduous trees of a given size may have similar rates of water loss under the same environmental conditions (Sect. 5.2 of the chapter on interactions among plants). Plant biomass indirectly influences evapotranspiration because of its correlation with the quantity of litter on the soil surface, which strongly influences the partitioning of water between surface **run-off** and **infiltration** into the soil. Surface **run-off** is negligible in forests and other communi-

ties with a well-developed litter layer (Running & Coughlan 1988).

In dry environments, **stomatal conductance** and **rooting depth** exert additional influence over evapotranspiration. Drought-tolerant species keep their stomata open at times of lower water availability; therefore, they support greater evapotranspiration during dry periods than do species typical of more mesic environments (Schulze & Hall 1982). Tall plants such as trees generally transpire more water than do herbs because of their more extensive root systems and greater leaf area. As a result, forest harvest reduces evapotranspiration and increases run-off (Bormann & Likens 1979), especially during seasons of rapid plant growth. In summary, plant size, which is a function of resource availability in the environment, is the major determinant of canopy water loss, although the response of stomatal conductance to plant water status becomes important under dry conditions.

Differences among species in evapotranspiration can have climatic consequences. Simulations suggest that conversion of the Amazon basin from forest to pasture would cause a permanent warming and drying of South America because the shallower roots of grasses would reduce evapotranspiration and cause greater energy dissipation as sensible heat (Shukla et al. 1990). These drier conditions would favor persistence of grasses. In Mexico, the reduction in transpiration that resulted from **overgrazing** increased sensible heat flux, causing regional warming (Balling 1988). Summer air masses that move from the Arctic Ocean into arctic

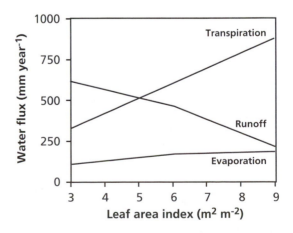

FIGURE 8. The effect of leaf area index (LAI) on simulated evapotranspiration, transpiration, and runoff in a Florida forest [Chapin (1993), as calculated from Running & Coughlan (1988)]. Copyright Academic Press.

Canada carry only enough moisture to account for 25% of the precipitation that occurs on land (Walsh et al. 1994). Thus, the remaining 75% of precipitation must originate from evapotranspiration over land. In other words, recycling of moisture between the land surface and the atmosphere accounts for most of the precipitation in this part of the Arctic (Chapin et al. 1997).

Environmental conditions could influence **vegetation feedbacks** to precipitation. For example, global warming caused by a doubling of atmospheric [CO_2] is predicted to increase precipitation by 8%. The reduction, however, in stomatal conductance caused by this rise in CO_2 concentration (Sect. 10.1 of the chapter on photosynthesis) should reduce the magnitude of the expected precipitation increase to only 5% (Henderson-Sellers et al. 1995). On the other hand, increased plant growth and stomatal conductance caused by nitrogen deposition might increase evapotranspiration and therefore precipitation. Thus, the interaction among environmental factors that influence plant growth and physiology modulate many of the terrestrial feedbacks to climate.

In most ecosystems, there is a close correlation of evapotranspiration with gross photosynthesis because a high leaf area and high stomatal conductance promote both processes. In low-resource communities, however, canopies are sparse, and the soil or surface mosses contribute substantially to evapotranspiration (Chapin et al. 1997). Below an LAI of 4, evapotranspiration becomes increasingly uncoupled from photosynthesis due to proportional increase in surface evaporation (Schulze et al. 1994).

4.1.3 Surface Roughness

The **roughness of the canopy** surface determines the degree of **coupling** between plants and the atmosphere, as well as the extent to which stomatal conductance influences the partitioning between latent and sensible heat (Sect. 2 of the chapter on scaling-up). Roughness is determined primarily by topography and patch structure of vegetation rather than by traits of individual plants. On broad flat landscapes, however, such as prairies, plant size has a major effect on surface roughness, with small-statured grasses having lower roughness and therefore a thicker boundary layer than would occur over forests. The lower roughness of short-statured vegetation reduces the influence of stomatal regulation by individual leaves on overall conductance of the canopy to water loss, especially under moist conditions.

4.2 Vegetation Effects on the Hydrologic Cycle

If vegetation affects evapotranspiration, it must also affect stream **run-off**, which is the difference between precipitation and evapotranspiration. The most dramatic vegetation effects are the increased run-off observed following forest harvest (Bormann & Likens 1979); however, the same plant traits that influence evapotranspiration (Sect. 4.1.3) also influence soil moisture and run-off. Thus, high rates of evapotranspiration dry the soil and reduce the amount of water that enters streams. Grasslands generally promote greater run-off than forests, in the same climate zone. In northern regions, species characteristic of steppe vegetation have higher rates of evapotranspiration than do mosses and other vegetation characteristic of tundra (Table 3). Either of these vegetation types can persist under the climate typical of tundra, with the higher transpiration rate of steppe plants maintaining the low soil moisture that favors these species and the lower transpiration rate of tundra species causing higher soil moisture that favors tundra species (Zimov et al. 1995). Zimov et al. hypothesize that extirpation of megaherbivores by humans at the end of the Pleistocene shifted the competitive balance from steppe species that tolerate grazing to tundra species. The resulting reduction in evapotranspiration would have increased soil

TABLE 3. **Average evapotranspiration rate of tundra and steppe plants from weighing lysimeters under field conditions in northeast Siberia during July.***

	Evapotranspiration rate ($mm\,day^{-1}$)	
Surface type	Field capacity	Natural precipitation
Tundra plants		
Lichen	1.6	0.9
Moss	2.8	1.0
Steppe plants		
Agropyron	6.7	2.5
Eriophorum	5.3	3.0
Equisetum	4.0	1.6
Artemesia	6.1	2.3
Probability of tundra—steppe difference	0.03	0.02

Source: Zimov et al. 1995.
*Lysimeters were either maintained at field capacity by twice-daily watering or were given access only to natural precipitation.

moisture, contributing to the shift from steppe to tundra that occurred at the end of the Pleistocene.

5. Scaling from Physiology to the Globe

Physiological differences among species have important predictable consequences for ecosystem and global processes. Environments with favorable climate and high resource availability support growth forms that are highly productive due to either large size or high RGR, depending on time since disturbance. By contrast, unfavorable environments support slowly growing plants, whose well-developed chemical defenses minimize rates of herbivory and decomposition. Rapidly growing plants have high rates of photosynthesis, transpiration (on a mass basis, less consistently true on a leaf area basis), tissue turnover, herbivory, and decomposition. Plant size is one of the major determinants of exchanges of carbon, nutrients, energy, and water. Vegetation differences in size and growth rate feed back to reinforce natural environmental differences, largely because large plants reduce soil moisture, and rapidly growing plants produce litter that enhances nutrient availability.

At regional scales large size and high stomatal conductance promote evapotranspiration and therefore precipitation, whereas small size or sparse vegetative cover dissipates more energy as sensible heat, which leads to higher air temperatures. At high latitudes, large size reduces albedo by covering the snow with a dark surface, thereby promoting regional warming during winter and spring. The increasing recognition of the importance of plant traits in influencing ecosystem processes and climate provide a central role for physiological ecology in studies of ecosystem and global processes.

References and Further Reading

Aerts, R. (1995) Nutrient resorption from senescing leaves of perennials: Are there general patterns? J. Ecol. 84:597–608.

Balling, R.C. (1988) The climatic impact of a Sonoran vegetation discontinuity. Clim. Change 13:99–109.

Bokhari, U.G. & Singh, J.S. (1975) Standing state and cycling of nitrogen in soil-vegetation components of prairie ecosystems. Ann. Bot. 39:273–285.

Bonan, G.B., Pollard, D., & Thompson, S.L. (1992) Effects of boreal forest vegetation on global climate. Nature 359:716–718.

Bonan, G.B., Chapin III, F.S., & Thompson, S.L. (1995) Boreal forest and tundra ecosystems as components of the climate system. Clim. Change 29:145–167

Bormann, F.H. & Likens, G.E. (1979) Pattern and process in a forested ecosystem. Springer-Verlag, New York.

Chapin III, F.S. (1993) Functional role of growth forms in ecosystem and global processes. In: Scaling physiological processes: Leaf to globe, J.R. Ehleringer & C.B. Field (eds). Academic Press, San Diego, pp. 287–312.

Chapin III, F.S., McFadden, J.P., & Hobbie, S.E. (1997) The role of arctic vegetation in ecosystem and global processes. In: Ecology of arctic environments, S.J. Woodin & M. Marquiss (eds). Blackwell Scientific, Oxford, pp. 121–135.

Chapman, W.L. & Walsh, J.E. (1993) Recent variations of sea ice and air temperature in high latitudes. Bull. Am. Meteor. Soc. 74:33–47.

Charney, J.G., Quirk, W.J., Chow, S.-H., & Kornfield, J. (1977) A comparative study of effects of albedo change on drought in semiarid regions. J. Atmos. Sci. 34:1366–1385.

Choudhury, B.J. (1987) Relationships between vegetation indices, radiation absorption, and net photosynthesis evaluated by a sensitivity analysis. Rem. Sens. Env. 22:209–233.

Ciais, P., Tans, P.P., Trolier, M., White, J.W.C., & Francey, R.J. (1995) A large northern hemisphere terrestrial CO_2 sink indicated by the $^{13}C/^{12}C$ ratio of atmospheric CO_2. Nature 269:1098–1102.

Cole, D.W. & Rapp, M. (1981) Elemental cycling in forest ecosystems. In: Dynamic properties of forest ecosystems, D.E. Reichle (ed). Cambridge University Press, Cambridge, pp. 341–409.

D'Antonio, C.M. & Vitousek, P.M. (1992) Biological invasions by exotic grasses, the grass-fire cycle, and global change. Annu. Rev. Ecol. Syst. 23:63–87.

Davidson, E.A. & Ackerman, I.L. (1993) Changes in soil carbon inventories following cultivation of previously untilled soils. Biogeochemistry 20:161–164.

Davis, M.B., Sugita, S., Calcote, R.R., & Frelich, L. (1992) Invasion of forests by hemlock coincided with change in disturbance regime. Bull. Ecol. Soc. Am. 73:155.

Denning, A.S., Fung, I.Y., & Randall, D. (1995) Latitudinal gradient of atmospheric CO_2 due to seasonal exchange with land biota. Nature 376:240–243.

Diaz, S.A., Grime, J.P., Harris, J., & McPherson. E. (1993) Evidence of a feedback mechanism limiting plant response to elevated carbon dioxide. Nature 364:616–617.

Farquhar, G.D. (1989) Models of integrated photosynthesis of cells and leaves. Phil. Trans. R. Soc. Lond. Series B 323:357–367.

Field. C.B. (1991) Ecological scaling of carbon gain to stress and resource availability. In: Integrated responses of plants to stress, H.A. Mooney, W.E. Winner, & E.J. Pell (eds). Academic Press, San Diego, pp. 35–65.

Foley, J.A., Kutzbach, J.E., Coe, M.T., & Levis, S. (1994) Feedbacks between climate and boreal forests during the Holocene epoch. Nature 371:52–54.

Goulden, M.L., Daube, B.C., Fan, S.-M., Sutton, D.J., Bazzaz, A., Munger, J.W., & Wofsy, S.C. (1997) Physiological responses of a black spruce forest to weather. J. Geophys. Res. 1020:28987–28996.

Goward, S. N., Tucker, C.J., & Dye, D.G. (1985) North American vegetation patterns observed with the NOAA-7 advanced very high resolution radiomater. Vegetatio 64:3–14.

Gray, J.T. & Schlesinger, W.H. (1981) Nutrient cycling in Mediterranean type ecosystems. In: Resource use by chaparral and matorral, P.C. Miller (ed). Springer-Verlag, New York, pp. 259–285.

Grime, J.P. & Hunt, R. (1975) Relative growth rate: Its range and adaptive significance in a local flora. J. Ecol. 63:393–422.

Harte, J. & Kinzig, A.P. (1993) Mutualism and competition between plants and decomposers: Implications for nutrient allocation in ecosystems. Am. Nat. 141:829–846.

Henderson-Sellers, A., McGuffie, K., & Gross, C. (1995) Sensitivity of global climate model simulations to increased stomatal resistance and CO_2 increase. J. Climat. 8:1738–1756.

Hirose, T. & Werger, M.J.A. (1987) Maximizing daily canopy photosynthesis with respect to the leaf nitrogen allocation pattern in the canopy. Oecologia 72:520–526.

Hobbie, S.E. (1992) Effects of plant species on nutrient cycling. Trends Ecol. Evolu. 7:336–339.

Kasischke, E.S., Christensen, N.L., & Stocks, B.J. (1995) Fire, global warming, and the carbon balance of boreal forests. Ecol. Appl. 5:437–451.

Kauppi, P.E., Mielikäinen, K., & Kuusela, K. (1992) Biomass and carbon budget of European forests, 1971 to 1990. Science 256:70–74.

Lieth, H. (1975) Modeling the primary productivity of the world. In: Primary productivity of the biosphere, H. Lieth & R.H. Whittaker (eds). Springer-Verlag, Berlin, pp. 237–263.

Monteith, J.L. (1977) Climate and the efficiency of crop production in Britain. Phil. Trans. R. Soc. Lond. B 281:277–294.

Myneni, R.B., Keeling, C.D., Tucker, C.J., Asrar, G., & Nemani, R.R. (1997) Increased plant growth in the northern high latitudes from 1981–1991. Nature 386:698–702.

Odum, E.P. (1969) The strategy of ecosystem development. Science 164:262–270.

Oechel, W.C., Hastings, S.J., Vourlitis, G., Jenkins, M., Riechers, G., & Grulke, N. (1993) Recent change of Arctic tundra ecosystems from a net carbon dioxide sink to a source. Nature 361:520–523.

Payette, S. & Filion, L. (1985) White spruce expansion at the tree line and recent climatic change. Can. J. For. Res. 15:241–251.

Robles, M. & Chapin III, F.S. (1995) Comparison of the influence of two exotic species on ecosystem processes in the Berkeley Hills. Madroño 42:349–357.

Running, S.W. & Coughlan, J.C. (1988) A general model of forest ecosystem processes for regional applications. I.

Hydrologic balance, canopy gas exchange and primary production processes. Ecol. Modelling 42:125–154.

Sala, O.E., Parton, W.J., Joyce, L.A., & Lauenroth, W.K. (1988) Primary production of the cental grassland region of the United States. Ecology 69:40–45.

Schimel, D.S. (1995) Terrestrial ecosystems and the carbon cycle. Global Change Biol. 1:77–91.

Schlesinger, W.H. (1991) Biogeochemistry: An analysis of global change. Academic Press, San Diego.

Schulze, E.-D. & Hall, A.E. (1982) Stomatal responses, water loss and CO_2 assimilation rates of plants in contrasting environments. In: Encyclopedia of plant physiology, Vol. 12B, O.L. Lange, P.S. Nobel, C.B. Osmond, & H. Ziegler (eds). Springer-Verlag, Berlin, pp. 181–230.

Schulze, E.-D., Kelliher, F.M., Körner, C., Lloyd, J., & Leuning, R. (1994) Relationship among maximum stomatal conductance, ecosystem surface conductance, carbon assimilation rate, and plant nitrogen nutrition: A global ecology scaling exercise. Annu. Rev. Ecol. Syst. 25:629–660.

Shukla, J., Nobre, C., & Sellers, P. (1990) Amazon deforestation and climate change. Science 247:1322–1325.

Silvertown, J.W. (1982) Introduction to plant population ecology. Longman, London.

Slobodchikoff, F.S. & Doyen, J.T. (1977) Effects of *Ammophila arenaria* on sand dune arthropod communities. Ecology 58:1171–1175.

Tans, P.P., Fung, I.Y., & Takahashi, T. (1990) Observational constraints on the global CO_2 budget. Science 247:1431–1438.

Terashima, I. & Hikosaka, K. (1995) Comparative ecophysiology of leaf and canopy photosynthesis. Plant Cell Environ. 18:1111–1128.

Tilman, D. (1988) Plant strategies and the dynamics and function of plant communities. Princeton University Press, Princeton.

Van Cleve, K., Chapin III, F.S., Dryness, C.T., & Viereck, L.A. (1991) Element cycling in taiga forest: State-factor control. BioScience 41:78–88.

Vitousek, P.M. (1994) Beyond global warming: Ecology and global change. Ecology 75:1861–1876.

Vitousek, P.M. & Howarth, R.W. (1991) Nitrogen limitation on land and in the sea: How can it occur? Biogeochemistry 13:87–115.

Vitousek, P.M., Walker, L.R., Whiteacre, L.D., Mueller-Dombois, D., & Matson, P.A. (1987) Biological invasion by *Myrica faya* alters ecosystem development in Hawaii. Science 238:802–804.

Walsh, J.E., Zhou, X., Portis, D., & Serreze, M. (1994) Atmospheric contribution to hydrologic variations in the arctic. Atmosphere-Ocean 32:733–755.

Weller, D.E. (1987) A reevaluation of the -3/2 power rule of plant self-thinning. Ecol. Monogr. 57:23–43.

White, J. (1980) Demographic factors in populations of plants. In: Demography and evolution in plant populations, O.T. Solbrig (eds). Blackwell Scientific, Oxford, pp. 21–48.

Wofsy, S.C., Goulden, M.L., Munger, J.W., Fan, S.-M., Bakwin, P.S., Daube, B.C., Bassow, S.L., & Bazzaz, F.A.

(1993) Net exchange of CO_2 in a mid-latitude forest. Science 260:1314–1317.

Yoda, K., Kira, T., Ogawa, H., & Hozumi, K. (1963) Self-thinning in overcrowded pure stands under cultivated and natural conditions. J. Biol. Osaka City Univ. 14:107–129.

Zak, D.R., Pregitzer, K.S., Curtis, P.S., Teeri, J.A., Fogel, R., & Randlett, D.A. (1993) Elevated atmospheric CO_2 and feedback between carbon and nitrogen cycles. Plant Soil 151:105–117.

Zimov, S.A., Chuprynin, V.I., Oreshko, A.P., Chapin III, F.S., Reynolds, J.F., & Chapin, M.C. (1995) Steppe-tundra transition: An herbivore-driven biome shift at the end of the Pleistocene. Am. Nat. 146:765–794.

Zimov S.A., Davidov S.P., Voropaev Y.V., Prosiannikov S.F., Semiletov I.P., Chapin M.C., & Chapin F.S. III (1996) Siberian CO_2 efflux in winter as a CO_2 source and cause of seasonality in atmospheric CO_2. Clim. Change 33:111–120.

Glossary

Abaxial   upper (side of a leaf)

Abiotic   not directly caused or induced by organisms

Absorbance   fraction of radiation incident on a surface that is absorbed

Abscisic acid, ABA   one of the six classes of phytohormones; although it derives its name from its involvement in leaf abscission, it has many other effects, including signaling adverse conditions in the root environment to the leaves, where, for example, it reduces stomatal conductance

Acclimation   morphological and physiological adjustment by individual plants to compensate for the decline in performance following exposure to unfavorable levels of **one** environmental factor

Acclimatization   morphological and physiological adjustment by individual plants to compensate for the decline in performance following exposure to unfavorable levels of **multiple** environmental factors, as occurs in natural environments

Accumulation   build-up of storage products resulting from an excess of supply over demand; also termed *interim deposition*

Acidifuge   avoiding acid soils; with a preference for a substrate that does not have a low pH

Active (or reactive) oxygen species   hydrogen peroxide (H_2O_2), soperoxide radicals (O_2^-), and hydroxyl radicals (OH.)

Active transport   transport of molecules across a membrane against an electrochemical gradient through expenditure of metabolic energy

Acyanogenic   not releasing cyanide

Adaptation   genetically determined trait that enhances the performance of an individual in a specific environment

Adaxial   lower (side of a leaf)

Adsorption   binding of ions or molecules to a surface (e.g., of a soil particle or a root)

Advection   net horizontal transfer of gases

Aerenchyma   gas-transport tissue within plants

Agglutinin   synonym for lectin

Albedo   fraction of the incident short-wave radiation reflected from the earth's surface

Alkaloid   secondary plant compound characterized by its alkaline reaction and a heterocyclic ring; examples include nicotine, caffeine, and colchicine

Allelochemical   plant secondary metabolite that negatively affects other organisms

Allelopathy   suppression of growth of one plant species by another due to the release of toxic substances

Allocation   proportional distribution of products or newly acquired nutrients among different organs or functions in a plant

Alternative oxidase   mitochondrial enzyme catalyzing the transfer of electrons from ubiquinol (the reduced form of ubiquinone) to O_2

Alternative pathway   nonphosphorylating electron-transport pathway in the inner membrane of plant mitochondria, transporting electrons from ubiquinol (the reduced form of ubiquinone) to O_2, catalyzed by the alternative oxidase

Amphistomatous with stomata at both the adaxial (upper) and abaxial (lower) sides of the leaf

Amylase starch-hydrolyzing enzyme

Anion negatively charged ion

Anisotropic not equal in all directions; for example, the longitudinal walls of anisotropic cells have different chemical and biophysical properties from those of the radial walls

Anoxia conditions in complete absence of oxygen

Annual plant that lives for 1 year or less

Antiport cotransport of one compound in one direction coupled to transport of another compound (mostly H^+) in the opposite direction

Apoplast space in a plant's tissue outside the symplast; the apoplast includes the walls of each cell and the dead tissues of the xylem, but excludes the cytoplasm

Apoplasmic phloem loading occurs in plants in which sucrose moves from the cytoplasm of the mesophyll cells of the leaves to the apoplast, after which it is taken up from the apoplast by transfer cells in the phloem; from the transfer cells it moves via plasmodesmata to the sieve tubes

Aquaporin water-channel protein

Assimilation incorporation of an inorganic resource (e.g., CO_2 or NH_4^+) into organic compounds; often synonymous with net photosynthesis

ATPase enzyme catalyzing the hydrolysis of ATP, producing ADP and P_i; the energy from this hydrolysis is used to pump protons across a membrane (e.g., plasma membrane, tonoplast), thus generating an electrochemical gradient

ATPase/ATP synthase enzyme complex in the inner membrane of mitochondria and the thylakoid membrane of chloroplasts catalyzing the formation of ATP, driven by the pmf

Autotoxicity deleterious effect of a chemical compound released by plants of the same species

Auxin one of the six classes of phytohormones; the name *auxin* literally means enhancing, and is derived from the growth-promoting action of these compounds; indoleacetic acid is the first auxin and first phytohormone discovered (by F.A.F.C Went, in Utrecht, the Netherlands)

Avoidance plant strategy of resisting adverse conditions by preventing deleterious effects of

these conditions; examples include germination of desert ephemerals only after a sufficient amount of precipitation, salt-excluding and salt-excreting mechanisms and compartmentation of heavy metals in compartments where they do not affect plant metabolism

Bacteroid state of rhizobia after they have penetrated the root and the symbiosis has been established

Biennial species whose individuals typically live 2 years, vegetative growth in the first year and seed production in the second year

Biomass plant dry mass

Biomass density dry mass of plant tissue per unit of fresh mass or volume (in the first case, the presence of intercellular air spaces is not taken into account)

Biotic caused or induced by organisms

Biotic filter biotic interactions which eliminate species that can survive the physical environment of a site

Blue-light receptor flavin-containing pigment absorbing in the blue region; this receptor senses the level of radiation and thus affects morphogenesis

Boundary layer thin layer around the leaf or root in which the conditions differ from those in the atmosphere or soil, respectively

Boundary layer conductance/resistance conductance/resistance for diffusion of CO_2, water vapor or heat between the leaf surface and the atmosphere

Bowen ratio the ratio between sensible heat loss and heat loss due to transpiration

Bulk density mass of dry soil per unit volume

Bulk soil soil beyond the immediate influence of plant roots

Bundle sheath cells cells surrounding the vascular bundle of a leaf

C_3 **photosynthesis** photosynthetic pathway in which CO_2 is initially fixed by Rubisco and converted to a three-carbon intermediate

C_4 **photosynthesis** photosynthetic pathway in which CO_2 is initially fixed by PEP carboxylase during the day, producing a four-carbon acid

Calcicole plant with a preference for calcareous or high-pH soils

Calcifuge plant that typically occupies acidic soils but is absent from calcareous or high-pH soils

Callose β-(1–3)-polymer of glucose, synthesized in functioning sieve elements in response to damage, sealing of the sieve tubes; callose is also produced in other cells upon microbial attack, thus providing a physical barrier

Calmodulin ubiquitous Ca^{2+}-binding protein whose binding to other proteins depends on the intracellular Ca^{2+} concentration; component of signal-transduction pathways

Calvin cycle (Calvin-Benson cycle) pathway of photosynthetic carbon fixation beginning with carboxylation by Rubisco

Canopy conductance/resistance conductance/resistance for diffusion of CO_2, water vapor or heat between the plant canopy and the atmosphere

Carbamylation reaction between CO_2 and an amino-group; in many species Rubisco is activated by carbamylation, catalyzed by Rubisco activase

Carbonic anhydrase enzyme catalyzing the equilibrium between bicarbonate and CO_2

Carboxylate organic acid minus its protons

Carboxylation binding of a CO_2 molecule to a CO_2-acceptor molecule

Carboxylation efficiency initial slope of the CO_2-response curve of photosynthesis

Carotenoid accessory photosynthetic pigment; carotenoids of the xanthophyll cycle play a role in dissipation of excess energy

Carrier protein involved in ion transport across a membrane

Casparian band strip waxy suberin impregnation on the radial and transverse wall of endodermis and exodermis cells that renders the wall impermeable to water

Cation positively charged ion

Cavitation breakage of a water column due to entry of air

Cellulose structural polymer of glucose; major component of plant cell walls giving tensile strength

Cell wall structural matrix surrounding plant cells; part of the apoplast

Cell-wall elasticity reversible extension of cell walls

Cell-wall extensibility irreversible extension of cell walls, due to chemical changes

Chaperones group of stress proteins that are encoded by a multigene family in the nucleus; chaperones bind to and stabilize an otherwise unstable conformation and, thus, mediate the correct assembly of other proteins

Chelate combine reversibly, usually with high affinity, with a metal ion (e.g., iron, copper, or calcium)

Chelator cation-binding organic molecule, such as citric acid, malic acid, and phytometallophores

Chemiosmotic model theory accounting for the synthesis of ATP driven by a proton-motive force

Chilling injury/tolerance injury/tolerance caused by cooling of tissues to temperatures above 0°C

Chitin polymer of N-acetylglucosamine; component of the exoskeleton of arthropods and the cell wall of fungi, but **not** that of plants

Chitinase chitin-hydrolyzing enzyme, which breaks down fungal cell walls

Chlorenchyma tissue containing chloroplasts

Chlorophyll green pigment involved in light capture by photosynthesis

Chloroplast organelle in which photosynthesis occurs

Chromophore light-absorbing constituent of a macromolecule that is responsible for light absorption

Citric acid cycle Krebs cycle

Climax species species that are a major component of the vegetation at later stages of succession

Clonal growth asexual production of physiologically complete plants

Cluster roots bottle-brushlike structures in roots, releasing organic molecules and solubilizing slightly available compounds in the soil

CO_2-compensation point CO_2 concentration at which the rate of CO_2 assimilation in photosynthesis is balanced by the rate of CO_2 production in respiration

Coevolution evolution of two (or more) species of which at least one depends on the other as a result of selection by mutual interactions

Cofactor inorganic ion or coenzyme required for an enzyme's activity

Cohesion theory accounts for the ascent of sap in the xylem due to the cohesive forces between ascending water molecules and the adhesive forces between water and capillaries in the wall of xylem conduits

Companion cell cell type in the phloem, adjacent to sieve element, involved in phloem loading

Compartmentation isolation of compounds in specific cells, organelles or parts of the cell, such as storage of secondary metabolites in vacuoles

Compatible interaction response of a susceptible host to a virulent pathogen

Compatible solute solute that has no deleterious effect on metabolism at high concentrations

Compensation point temperature, [CO_2] or light level at which net carbon exchange by a leaf is zero (i.e., photosynthesis equals respiration)

Competition interactions among organisms (of the same or different species) which utilize common resources that are in short supply (resource competition), or which harm one another in the process of seeking a resource, even if the resource is not in short supply (interference competition)

Competitive ability probability of winning in competition with another species in a particular environment

Conductance flux per unit driving force (e.g., concentration gradient); inverse of resistance

Constitutive produced in constant amount (as opposed to regulated) (e.g., genes can be expressed constitutively)

Constitutive defense background level of plant defense in the absence of induction by herbivory

Construction cost the carbon or nutrients required to produce new tissue, including the respiration associated with the biosynthetic pathways

Contractile roots mature roots that decrease in length, while increasing in diameter, thus pulling the plant deeper in the soil as in geophytes

Convection heat transfer by turbulent movement of a fluid (e.g., air or water)

Coupling factor ATP-synthetase in thylakoid membrane of chloroplasts and inner membrane of mitochondria

Crassulacean acid metabolism photosynthetic pathway in which stomates open, and carbon is fixed at night into a four-carbon acid. During the day stomates close, C_4 acids are decarboxylated, and CO_2 is fixed by C_3 photosynthesis

Crista fold of the inner mitochondrial membrane

Critical daylength length of the night triggering flowering

Cryptochrome blue-light-absorbing plant pigment

Cuticle waxy coating of external plant surfaces

Cuticular conductance/resistance conductance/resistance for diffusion of CO_2 or water vapor movement through the cuticle

Cutin waxy substances that coat external plant surfaces; polymer consisting of many long-chain hydroxy fatty acids that are attached to each other by ester linkages, forming a rigid three-dimensional network

Cyanogenic releasing cyanide

Cytochrome colored, heme-containing protein that transfers electrons in the respiratory and photosynthetic electron transport chain

Cytochrome oxidase mitochondrial enzyme catalyzing the final step in the transfer of electrons from organic molecules to O_2

Cytochrome pathway phosphorylating electron-transport pathway in the inner membrane of plant mitochondria, transporting electrons from NAD(P)H or $FADH_2$ to O_2, with cytochrome oxidase being the terminal oxidase

Cytokinin one of the six classes of phytohormones; involved in the delay of leaf senescence, cell division, cell extension and dormancy of buds.

Cytoplasm contents of a cell that are contained within its plasma membrane, but outside the vacuole and the nucleus

Cytosol cellular matrix in which cytoplasmic organelles are suspended

Dark reaction carbon fixation during photosynthesis; does not directly require light but uses the products of the light reaction

Deciduous shedding of leaves in response to specific environmental cues, such as occur during or preceding unfavorable seasons

Decomposition breakdown of organic matter

Defense compound secondary metabolite conferring some degree of protection from pathogens or herbivores

Delayed greening pattern of leaf initiation typical of shade-tolerant rain-forest species; leaves are initially white, red, blue, or light-green during the stage of leaf expansion, reflecting their low chlorophyll concentration

Demand requirement; the term is used in the context of the control of the rate of a process (e.g., nutrient uptake) by the amount needed

Demand function dependence of net CO_2 assimilation rate on p_i, irrespective of the supply of CO_2

Denitrification microbial conversion of nitrate to gaseous nitrogen (N_2 and N_2O); nitrate is used as an electron acceptor

Desiccation resistance resistance to extreme water stress, with recovery of normal rates of photosynthesis and respiration shortly following rehydration

Desorption the reverse of adsorption

Diaheliotropism solar tracking in which the leaf remains perpendicular to incident radiation

Differentiation cellular specialization

Diffuse porous wood in which wide and narrow xylem vessels are randomly distributed throughout each growth ring

Diffusion net movement of a substance along a concentration gradient due to random kinetic activity of molecules

Diffusion shell zone of nutrient depletion around individual roots caused by active nutrient uptake at the root surface and diffusion to the root from the surrounding soil

Disulfide bond   covalent linkage between two sulfhydryl groups on cysteines

Dormancy   state of seeds or buds that fail to grow when exposed to a favorable environment

Down-regulation   decrease of the normal rate of a process, sometimes involving suppression of genes encoding enzymes involved in that process

Ecophysiology   study of the physiological mechanisms by which plants cope with their environment

Ecosystem   ecological system that consists of all the organisms in an area and the physical environment with which they interact

Ecosystem respiration   sum of plant and heterotrophic respiration

Ecotone   boundary between two plant communities

Ecotype   genetically differentiated population that is restricted to a specific habitat

Ectomycorrhiza   mycorrhizal association in some trees in which a large part of the fungal tissue is found outside the root

Efficiency   rate of a process per unit plant resource

Elastic modulus   force needed to achieve a certain change in cell volume

Embolism   see cavitation

Emissivity   coefficient that describes the thermal radiation emitted by a body at a particular temperature relative to the radiation emitted by a black body

Endocytosis   uptake of material into a cell by an invagination of the plasma membrane and its internalization in a membrane-bound vesicle

Endodermis   innermost layer of root cortical cells; these cells are suberized and surrounded by a Casparian strip

Endomycorrhiza   mycorrhizal association in many herbaceous species and some trees in which a large part of the fungal tissue is found inside the root

Ephemeral   short-lived

Epidermis   outermost cell layer of an organ

Epinasty   downward bending of a plant organ

Epiphyte   plant living on another plant, without a symbiotic or parasitic association

Ethylene   ethene (C_2H_4); one of the six classes of phytohormones; among others, ethylene is involved in aerenchyma formation

Evapotranspiration   water loss from an ecosystem by transpiration and surface evaporation

Evergreen   strategy of plants to have green leaves throughout the year, as opposed to deciduous plants

Exclusion   prevention of net entry of a molecule; it may be due to low permeability for a molecule or to extrusion of the molecule

Excretion   active secretion of compounds (e.g., salt from leaves)

Exodermis   outer cortical cell layer in roots, immediately below the epidermis; these cells are suberized and surrounded by a Casparian strip

Expansin   cell-wall enzyme involved in cell expansion

Extensin   rigid cell-wall glycoprotein, rich in hydroxyproline, that represents 5 to 10% of the dry weight of most primary cell walls; significant component of the secondary walls of sclerenchyma cells

Extinction coefficient   coefficient describing the exponential decrease in irradiance through a leaf or canopy

Extrusion   ion transport from root cells to the external medium, dependent on respiratory metabolism

Exudate   compounds released by plants (mostly by roots); also xylem fluid that appears when the stem is severed from the roots; also phloem fluid that appears when a cut is made in the stem of a plant that does not have a mechanism to block the phloem with callose

Exudation   release of exudates, or the appearance of fluid from cut roots, due to root pressure

Facilitation   positive effect of one plant on another

Facultative CAM plants   plants that photosynthesize by CAM during dry periods and by C_3 or C_4 photosynthesis at other times

Feedback   response in which the product of one of the final steps in a chain of events affects one of the first steps in this chain; fluctuations in rate or concentration are minimized with negative feedbacks or amplified with positive feedbacks

Feedforward   response in which the rate of a process is affected before any deleterious effect of that process has occurred; for example, the decline in stomatal conductance before the water potential in leaf cells has been affected

Fermentation   anaerobic conversion of glucose to organic acids or alcohol

Field capacity   water content that a soil can hold against the force of gravity

Flavan, flavine, flavone   subclass of flavonoids

Flavonoid   one of the largest classes of plant phenolics, in which two aromatic rings are connected by a carbon link to a third phenyl ring; these compounds play a role in the symbiosis between rhizobia and legumes, as phytoalexins, as antioxidants, and in the colors of flowers

Fluence response response to a dosage of light

Fluorescence light produced when excited electrons return to the ground state

Frost hardening metabolic changes that occur in autumn that make a plant frost hardy

Frost hardiness/tolerance physiological condition that allows exposure to subzero temperatures without cellular damage

Geotropism growth response of plant organs with respect to gravity

Germination process during which a seed absorbs water, followed by emergence of the radicle through the seed coat

Gibberellin one of the six classes of phytohormones; the first gibberellin was found in the fungus *Gibberella fujikora*, from which these phytohormones derive their name

Giga- prefix denoting 10^9

Glass virtually immobile state of molecules in a fluid; the best example of a glass is "glass" as we know it from everyday life (which is **not** a solid, but a fluid, as apparent from the gradually changing properties of glass when it gets old); glass formation, rather than the formation of ice crystals, is essential to prevent damage incurred by the formation of ice crystals

Glaucousness shiny appearance (of leaves), due to the presence of specific wax compounds

Glucoside (or glycoside) compound in which a side chain is attached to glucose by an acetal bond

Glucosinolate secondary sulfur-containing metabolite in Brassicaceae (cabbage family), which gives these plants a distinct sharp smell and taste

Glycolipid membrane lipid molecule with a short carbohydrate chain attached to a hydrophobic tail

Glycolysis ubiquitous metabolic pathway in the cytosol in which sugars are degraded to pyruvate and/or malate with production of ATP and NADH (when pyruvate is the endproduct)

Glycophyte species restricted to nonsaline soils

Glycoprotein any protein with one or more covalently linked oligosaccharide chains

Glycoside (or glucoside) compound in which a side chain is attached to a sugar by an acetal bond

Grana stacked membrane discs (thylakoids) in chloroplasts that contain chlorophyll

Gross photosynthesis total amount of carbon fixed by photosynthesis per unit ground area

Growth increment in plant mass, volume, length, or area

Growth respiration the amount of respiration required per unit growth; it is **not** a rate

Guard cells specialized epidermal cells that surround the stomata and regulate the size of their opening

Guttation water exuded by leaves due to root pressure

Halophyte species that typically grows on saline soils

Hartig net hyphal network of ectomycorrhizal fungi that have penetrated intercellularly into the cortex of a higher plant

Haustorium organ that functions in attachment, penetration, and transfer of water and solutes from a host to a parasitic plant

Heat-shock protein protein produced upon heat stress

Heavy metal metal with a mass density exceeding $5 \, g \, ml^{-1}$

Heliotropism solar tracking; movement of a leaf or flower that follows the angle of incident radiation

Heme cyclic organic molecule that contains an iron atom in the center which binds O_2 in leghemoglobin and carries an electron in cytochromes

Hemicellulose heterogeneous mixture of neutral and acidic polysaccharides, which consists predominantly of galacturonic acid and some rhamnose; these cell-wall polymers coat the surface of cellulose microfibrils and run parallel to them

Heterodimer protein complex composed of two different polypeptide chains

Heterotrophic respiration respiration by non-autotrophic organisms, mainly microorganisms and animals

Hexokinase enzyme catalyzing the phosphorylation of hexose sugars while hydrolyzing ATP; a specific hexokinase is involved in sugar sensing

Historical filter historical factors that prevent a species from arriving at a site

Homeostasis tendency to maintain constant internal conditions in the face of a varying external environment

Homodimer protein complex composed of two identical polypeptide chains

Hormone organic compound produced in one part of a plant and transported to another, where it acts in low concentrations to control processes

Humic substances high molecular-weight polymers with abundant phenolic rings and variable side chains found in humus

Humus amorphous soil organic matter

Hydraulic lift upward movement of water from deep moist soils to dry surface soils through roots along a water potential gradient

Hydrenchyma water-storing tissue; during dehydration of a plant, water is predominantly lost from the cells in the hydrenchyma, while other cells lose relatively less water

Hydrolysis cleavage of a covalent bond with accompanying addition of water, -H being added to one product and -OH to the other

Hydrophyte plant that typically grows in water

Hydrotropism morphogenetic response (of roots) to a moisture gradient

Hygrophyte species typically occurring on permanently moist sites

Hypostomatous with stomates at the abaxial (lower) side of the leaf only

Hypoxia low oxygen level

Immobilization nutrient absorption by soil microorganisms from the soil solution

Incompatible interaction response of a resistant host to an avirulent pathogen

Induced defense increased levels of plant secondary metabolites in response to herbivory or pathogen attack

Infiltration movement of water into the soil

Infrared radiation radiation with wavelengths greater than 740 nm; long-wave radiation

Interception acquisition of nutrients by roots as a result of growing through soil; the nutrients contained in the soil volume displaced by the growing root

Intercrop plant used in combination with a crop plant

Interference competition competition mediated by production of allelochemicals by a plant

Intermediary cell phloem cell in plants with a symplasmic pathway of phloem loading; sucrose moves from the mesophyll into these cells, where it is processed to form oligosaccharides that move to the sieve tube

Internal conductance/resistance conductance/resistance for CO_2 diffusion between the intercellular spaces and the site of Rubisco

Ion channel or **ion-selective channel** pore in a membrane made by a protein, through which ions enter single file along an electrochemical potential gradient; channels are specific and either open or closed, depending on membrane potential or the presence of regulatory molecules

Isohydric maintaining a constant water status

Isoprene small unsaturated hydrocarbon, containing five carbon atoms; volatile compound, synthesized from mevalonic acid and precursor of other isoprenoids

Isotope discrimination differentiation by some enzymes or processes between different isotopes (e.g., of hydrogen, carbon, nitrogen, oxygen)

Isotope effect end-result of various processes that have different rate constants for different isotopes

Isotope fractionation occurs when different isotopes contribute to a reaction with different rate constants

Isotropic similar in all directions

Jarowization vernalization (from the Russian word for *spring*)

Jasmonic acid novel secondary plant compound [3-oxo-2-(2'-*cis*-pentenyl)-cyclopropane-1-acetic acid], named after its scent from jasmine; stress-signaling molecule in plants as well as **between** plants

Juvenile phase life stage of a plant between the seedling and reproductive phases

k_{cat} catalytic constant of an enzyme: rate of the catalyzed reaction expressed in moles per mole enzyme (rather than per unit protein, as in V_{max})

K_i concentration of an inhibitor which reduces the activity of an enzyme to half the rate of that in the absence of that inhibitor

K_m substrate concentration at which a reaction proceeds at half the maximum rate

K strategy suite of traits that enables a plant to persist in a community at equilibrium

Kranz anatomy specialized leaf anatomy of C_4 species with photosynthetic bundle sheath cells surrounding vascular bundles

Krebs cycle tricarboxylic acid cycle; metabolic pathway in the matrix of the mitochondrion oxidizing acetyl groups derived from imported substrates to CO_2 and H_2O

Latent heat heat absorbed or released by evaporation or condensation, respectively, of water

Law of the minimum concept that plant growth is always limited at any point in time by one single resource

Leaf area index total leaf area per unit area of ground

Leaf area ratio (LAR) ratio between total leaf area and total plant biomass

Leaf conductance/resistance conductance/resistance for diffusion of CO_2 or water vapor of the leaf (it includes the stomatal and the boundary layer conductance/resistance)

Leaf-mass density leaf dry mass per unit of fresh mass or volume (in the first case, the presence of intercellular air spaces is not taken into account)

Leaf mass per unit leaf area (LMA) leaf mass expressed per unit leaf area

Leaf mass ratio (LMR) ratio of leaf mass and total plant biomass

Leaf turnover replacement of senescing leaves by new ones, without leading to an increase in leaf area

Lectin protein with noncatalytic sugar-binding domains; lectins are involved in defense and cellular interactions

Leghemoglobin hemoglobinlike protein in nodules, that associates with O_2 by means of a bound heme group

Light-compensation point irradiance level at which the rate of CO_2 assimilation in photosynthesis is balanced by the rate of CO_2 production in respiration

Light-harvesting complex complex of molecules of chlorophyll, accessory pigments, and proteins in the thylakoid membrane which absorbs quanta and transfers the excitation energy to a reaction center

Light reaction transfer of energy from light to ATP and NADP(H) during photosynthesis

Light saturation range of light intensities above which the rate of CO_2 assimilation is insensitive to light intensity

Lignan phenolic compound with antifungal, antifeeding, and antitumor activity; minor component in most plants and tissues, but quantitatively more important in the wood of hardwood species such as redwood

Lignin large amorphous polyphenolic compound that confers woodiness to stems

Litter dead plant material that is sufficiently intact to be recognizable

Litter quality chemical properties of litter that determine its susceptibility to decomposition, largely determined by concentrations of secondary metabolites and nutrients

Lockhart equation equation that describes cell expansion in terms of turgor pressure and cell-wall properties

Long-day plant plant whose flowering is induced by exposure to short nights

Long-wave radiation radiation with wavelengths greater than 740 nm; infrared radiation

Lumen cavity, such as the space surrounded by the thylakoid membrane or the trap of *Utricularia* surrounded by cells

Luxury consumption uptake of nutrients beyond the rate that enhances plant growth rate

Macronutrient inorganic nutrients that a plant requires in relatively large quantities: K, Ca, Mg, N, S, P, (Fe)

Macrosymbiont the larger partner (i.e., higher plant) in a symbiosis with a microorganism

Maintenance respiration respiration required to maintain the status quo of plant tissues

Mass flow bulk transport of solutes due to the movement of water

Matric potential component of the water potential that is due to the interaction of water with capillaries in large molecules (e.g., clay particles in soil)

Matrix compartment inside mitochondria or chloroplasts, not including the membrane system

Mean residence time time a nutrient remains in the plant, before being lost due to leaf shedding, consumption by a herbivore, and so on

Mega- prefix denoting 10^6

Membrane channel transmembrane protein complex that allows inorganic ions, small molecules or water to move passively across the lipid bilayer of a membrane

Mesophyll photosynthetic cells in a leaf; the palisade and spongy parenchyma cells

Mesophyte plant that typically grows in environments without severe environmental stresses

Metallophyte species that typically grows in areas with high concentrations of certain heavy metals in the soil

Metallothionein low-molecular mass metal-binding protein

Micro- prefix denoting 10^{-6}

Microclimate climate modified by an object (e.g., the environment immediately adjacent to a leaf or within a forest)

Microfibril structural component in cell walls, consisting of bundles of around 50 cellulose molecules, that provides the tensile strength of the wall

Micronutrient inorganic nutrients that a plant requires in relatively small quantities

Microsymbiont the smaller partner (i.e., microorganism) in a symbiosis with a higher plant

Mimicry resemblance of an organism to another organism or object in the environment, evolved to deceive predators or prey

Mineralization breakdown of organic matter, which releases mineral nutrients

Mistletoe xylem-tapping stem parasite

Mitochondrion organelle in which part of the respiratory process (TCA cycle, respiratory electron transport) occurs

Mycorrhiza structure arising from a sym-biotic association between a mycorrhizal fungus and the root of a higher plant (from the Greek words for fungus and root, respectively)

Mycorrhizal dependency the ratio of dry mass of mycorrhizal plants to that of nonmycorrhizal plants

Nano- prefix denoting 10^{-9}

Net assimilation rate (NAR) rate of plant biomass increment per unit leaf area

Net ecosystem production annual increment in organic matter in an ecosystem per unit ground area; equals gross photosynthesis minus ecosystem respiration or net primary production minus heterotrophic respiration

Net primary production quantity of new plant material produced annually per unit ground area

Nitrification microbial process that transforms ammonia, via nitrite, into nitrate

Nitrogen assimilation incorporation of inorganic nitrogen (nitrate, ammonium) into organic compounds

Nitrogen fixation reduction of dinitrogen gas to ammonium

Nod factor product of *nod* genes

***Nod* gene** rhizobial gene involved in the process of nodulation

Nodulation formation of nodules in symbiotic N_2-fixing plants

Nodulins class of plant proteins that are synthesized upon infection by rhizobia

Normalized difference vegetation index (NDVI) greenness index used to estimate aboveground net primary production from satellites, based on reflectance in the visible and near-infrared

Nuclear magnetic resonance (NMR) spectroscopy technique used to make a spectrum of molecules with a permanent magnetic moment, due to nuclear spin; the spectra are made in a strong magnetic field that lines up the nuclear spin in all the molecules; because the site of the peak in a spectrum depends on the pH around the molecule, among others, NMR spectroscopy is used to measure the pH in different cellular compartments invivo

Nutrient productivity rate of plant biomass increment per unit nutrient in the plant

Nutrient-use efficiency growth per unit of plant nutrient, which equals nutrient productivity times mean residence time; ecosystem nutrient-use efficiency is the ratio of litterfall mass to litterfall nutrient content (i.e., the amount of litter produced per unit of nutrient lost in senescence)

Opportunity costs diminished growth resulting from diversion of resources from alternative functions that might have yielded greater growth

Osmoregulation maintenance of the concentration of osmotic solutes at a stable value

Osmotic potential component of the water potential that is due to the presence of osmotic solutes

Overflow hypothesis theory indicating that respiration proceeds via the nonphosphorylating alternative path only in the presence of high availability of respiratory substrate; it considers the alternative path as a "coarse control" of carbohydrate metabolism

Oxidative pentose phosphate pathway metabolic pathway that oxidizes glucose and generates NADPH for biosynthesis

Oxidative phosphorylation formation of ATP (from ADP and P_i) coupled to a respiratory electron-transport chain and driven by a proton-motive force

Oxygenation reaction catalyzing the binding of O_2, without changing the redox state of O

Palisade mesophyll uppermost layer(s) of vertically oriented photosynthetic cells in a leaf

Paraheliotropism leaf movement that positions the leaf more or less parallel to the incident radiation throughout the day

Parent material rock and other substrates that generate soils through weathering

Pectin cell-wall polymer rich in galacturonic acid

Perennial species whose individuals typically live more than 2 years

Peribacteroid membrane plant-derived membrane that surrounds one or more bacteroids in root nodules

Pericarp matured ovulary wall in a seed

Pericycle circle of outermost stelar cells, adjacent to the endodermis

Permanent wilting point equilibrium soil moisture at which a plant can no longer absorb water from the soil; generally taken to be -1.5 MPa for many herbaceous species

Phenol compound that contains a hydroxyl group on an aromatic ring

Phenolics aromatic hydrocarbons, many of which have antimicrobial properties

Phenology time course of periodic events in an organism that are correlated with climate (e.g., budbreak or flowering)

Phenotypic plasticity range of expression of a trait in an individual organism in response to environmental variation

Phloem long-distance transport system in plants for mass flow of carbohydrates and other solutes

Phosphatase enzyme hydrolyzing organic phosphate-containing molecules

Phospholipid major category of membrane lipids, generally composed of two fatty acids linked through glycerol phosphate to one of a variety of polar groups

Phosphorylation process involving the covalent binding of a phosphate molecule; many enzymes change their catalytic properties when phosphorylated

Photodamage/photodestruction decline in photosynthesis upon exposure to high irradiance, due to damage/destruction of components of the photosynthetic apparatus

Photoinhibition decline in photosynthesis upon exposure to high irradiance, due to adjustment in the photosynthetic apparatus

Photon discrete unit of light that describes its particlelike properties; light also has wavelike properties

Photoperiod length of the daylight period each day

Photoperiodic responding to the length of the night

Photorespiration light respiration; production of CO_2 due to the oxygenation reaction catalyzed by Rubisco

Photosynthesis process in which light energy is used to reduce CO_2 to organic compounds

Photosynthetic nitrogen-use efficiency rate of photosynthesis expressed per unit (organic) nitrogen in the photosynthesizing tissue

Photosynthetic quotient ratio between CO_2 uptake and O_2 release in photosynthesis

Photosynthetic water-use efficiency ratio between photosynthetic carbon gain and transpirational water loss

Photosynthetically active radiation (PAR) part of the electromagnetic spectrum (400 to 700 nm) that drives photosynthesis

Photosystem group of pigments and associated proteins that capture light energy

Phototropism growth response of plant organs with respect to light

Phreatophyte plant species that accesses deep layers of water

Phyllosphere immediate surroundings of the leaf

Phylogenetic constraint genetic constitution of a population or species that prevents evolution of particular traits

Physiological filter physiological limitations due to intolerance of the physical environment, which prevent survivorship of plant species that arrive at a site

Phytate inositol hexaphosphate; organic P-storage compound in seeds and endodermis of some plant species and major fraction of organic P in soils

Phytoalexin plant defense compound against microorganism, whose synthesis is triggered by components of microbial origin

Phytoanticipin constitutively produced plant defense compound against microorganisms

Phytochelatin sulfur-rich peptide which binds (heavy) metals

Phytochrome plant pigment absorbing red or far-red radiation (depending on its configuration); this pigment is involved in measuring light quality and daylength and is the first step in a signal-transduction pathway leading to morphogenetic events

Phytohormone plant compound produced in one part of the plant and having its effect in another part at minute concentrations (nanomolar and picomolar range)

Phytometallophore metal-chelating organic molecule in grasses

Phytoremediation use of green plants to remove, contain, or render harmless environmental contaminants

Phytosiderophore iron-chelating organic molecule in grasses

Pico- prefix denoting 10^{-12}

Pioneer species that is a major component of the vegetation at early stage of succession

Pit-membrane pore pore in the walls of xylem vessels or tracheids

Plasmalemma plasma membrane; external membrane surrounding the cytoplasm

Plasmodesma(ta)   minute membrane-lined channels that traverse the plant cell wall to provide a cytoplasmic pathway for diffusion of small molecules between neigboring cells; it has recently been shown that these connections also transport very large molecules

Plasmolysis   separation of the cytoplasm from the cell wall due to water loss

Pneumatophore   specialized root that protrudes vertically from waterlogged soil and that is used for ventilation of the root

Poikilohydric   plants or plant parts (seeds, pollen) that can dry out without losing their capacity to function upon rehydration

Post-illumination CO_2 fixation CO_2 fixation that occurs briefly after a light pulse

ppb   part per billion; $1 \, nmol \cdot mol^{-1}$; $1 \, ng \cdot g^{-1}$; $nl \cdot l^{-1}$;

PPFD   photosynthetic photon flux density (referring only to photosynthetically active radiation, 400 to 700 nm)

ppm   part per million; $1 \, \mu mol \cdot mol^{-1}$; $1 \, \mu g \cdot g^{-1}$; $\mu l \cdot l^{-1}$;

Pressure chamber   piece of equipment used to determine the water potential in the xylem of plant stems

Pressure potential   pressure component of the water potential; it is positive in nonplasmolyzed living plant cells and negative in the xylem of transpiring plants

Pressure probe   microcapillary that is injected in a living cell to measure cell turgor

Protease/proteinase   protein-hydrolyzing enzyme

Protein turnover   breakdown and synthesis of proteins, without a net change in protein concentration

Proteoid root   cluster root; the name stems from the family of the Proteaceae

Proton cotransport   transport mechanism that allows movement of a compound against the electrochemical potential gradient for that molecule, using the proton-motive force

Proton-motive force   driving force across cell membranes due to a membrane potential and/or proton gradient

Protoplasmic streaming   circulation of the cytoplasm, mediated by the cytoskeleton

Protoplast   cellular material contained inside the cell wall

Pulvinus   "joint" in a petiole that allows the movement of a leaf, due to transport of ions between cells in the pulvinus, followed by changes in turgor

Q_{10}   change in rate of a reaction in response to a 10°C change in temperature

Qualitative defense compound   secondary plant metabolite that gives some protection against attack at low concentration, due to its high toxicity

Qualitative long-day plant   plant that will not flower unless the length of the night gets below a critical value

Qualitative short-day plant   plant that will not flower unless the length of the night gets above a critical value

Quantitative defense compound   secondary plant metabolite that gives some protection against attack when present in large amounts

Quantitative long-day plant   plant whose flower induction is promoted by exposure to short nights

Quantitative short-day plant   plant whose flower induction is promoted by exposure to long nights

Quantum yield   moles of CO_2 fixed, O_2 produced, or electrons generated in the photosynthetic process per mole of quanta absorbed

Receptor   protein with a high affinity and specificity for a signaling molecule (e.g., a phytohormone) that is the start of a signal-transduction chain

Reductive pentose phosphate pathway   metabolic pathway that utilizes NADPH produced in the light reaction of photosynthesis and produces triose phosphate

Reflectance   fraction of radiant energy incident on a surface that is reflected

Relative humidity   water vapor concentration of air relative to the maximum water vapor concentration at that temperature

Relative water content   water content of a plant tissue relative to the water content at full hydration

Reserve formation   build-up of storage products that results from diversion of plant resources to storage from alternative allocations, such as growth

Resistance   plant capacity to minimize the impact of stress factors in the environment, either by the presence of tolerance mechanisms or by avoidance of the stress

Resorption   translocation of nutrients and soluble organic compounds from senescing tissues prior to abscission

Resource competition   use of the same pool of growth-limiting resources by two or more plants

Respiratory quotient   ratio between CO_2 release and O_2 consumption in respiration

Resurrection plant plant that withstands complete dehydration and resumes functioning upon rehydration

Rhizobia collective term for symbiotic bacteria of the genera *Rhizobium*, *Bradyrhizobium*, *Azorhizobium*, *Mesorhizobium* and *Sinorhizobium*

Rhizosphere zone of soil influenced by the presence of the root

Ring porous wood in which xylem vessels produced early in the growing season are longer and wider than those produced in late wood, resulting in distinct annual growth rings

Rock phosphate phosphate compound with very low solubility

Root density total root length per unit soil volume

Root-mass density see biomass density

Root mass ratio ratio between root biomass and total plant biomass

Root pressure positive water potential in the xylem due to ion transport into the xylem of roots

Root shoot ratio ratio between root biomass and shoot biomass

Root turnover replacement of (old) roots by new ones, without changing the total amount of roots

Roughness unevenness of a surface that creates turbulence and enhances convective exchange between the surface and the atmosphere

Rubisco ribulose-1,5-bisphosphate carboxylase/oxygenase; Calvin-cycle enzyme catalyzing the attachment of CO_2 to a CO_2-acceptor molecule

Rubisco activase enzyme catalyzing the carbamylation of Rubisco

Ruderal species species that flourish on disturbed sites and complete their life cycle relatively rapidly

Run-off gravitational water loss from an ecosystem; the difference between precipitation and evapotranspiration

Saline soils soils with high salt concentration

Salt gland group of cells involved in salt excretion

Saponin secondary plant compound with soaplike properties

Sapwood most recent wood, with open xylem conduits that still function in water transport

Sclerenchyma tissue that consists of two types of cells: sclereids and fibers, which both have thick secondary walls and are frequently dead at maturity

Scleromorph containing a relatively large amount of sclerenchyma

Sclerophyllous condition of leaves that are thick, tough, and have a thick cuticle

Secondary metabolites compounds produced by plants that are not essential for normal growth and development

Seedling phase life stage of recently germinated plants that still have cotyledons

Self-thinning reduction in plant density due to competition

Senescence programmed series of metabolic events that involve metabolic breakdown of cellular constituents and transport of the breakdown products out of the senescing organ

Serotinous state of cones on a tree that remain closed with release of seeds delayed or occurring gradually

Serpentine soil soils that naturally contain high levels of various heavy metals and magnesium, but low concentrations of calcium, nitrogen, and phosphate

Short-day plant plants whose flowering is induced by exposure to long nights

Signal-transduction pathway chain of events by which an extracellular chemical (e.g., a phytohormone) or physical (e.g., radiation) signal is sensed and relayed into a response

Sink part of the plant that shows a net import of a compound (e.g., a root is a sink for carbohydrates and a leaf is a sink for nutrients)

Soil texture relative proportions of sand, silt, and clay in a soil

Solar tracking movement of leaf or flower that positions this organ at a more or less constant angle relative to the incident radiation throughout the entire day

Source part of the plant that shows a net export of a compound (e.g., a leaf is a source for carbohydrates and a root is a source for nutrients)

Specific leaf area (SLA) leaf area per unit leaf mass

Specific leaf mass leaf mass per unit leaf area

Specific root length (SRL) root length per unit root mass

Spongy mesophyll loosely packed photosynthetic cells in the lower portion of a leaf

Starch polymer of glucose; storage compound in plastids

Stomata structures in the leaf epidermis formed by specialized epidermal cells; mostly the term refers to the pores as well as to the stomatal apparatus

Stomatal conductance/resistance conductance/resistance for diffusion of CO_2 or water vapor through the stomata

Stomatal pore openings in the leaf epidermis formed by stomata

Storage buildup of a metabolically inactive pool of compounds that can subsequently serve to support growth or other physiological functions; see reserve storage and accumulation

Strategy complex suite of traits allowing adaptation to a particular environment

Stratification breaking of seed dormancy by exposure of moist seeds to low temperatures

Stress environmental factor that reduces plant performance

Stress protein protein that is produced only or in greater quantities upon stress

Stress response the immediate detrimental effect of a stress on a plant process; it generally occurs over a time scale of seconds to days, during which the net effect on the process is a decline in performance

Stroma matrix within the chloroplast containing Calvin-cycle enzymes and in which the grana are suspended

Strophiole preformed weak site in the seed coat

Suberin polymer containing long-chain acids, hydroxy acids, alcohols, dicarboxylic acid, and phenols; the exact structure is not fully understood; cell-wall component in many locations, especially in the Casparian strip

Subsidiary cell cell type in many stomata, located distally and laterally to a guard cell

Succession replacement of populations in a habitat through a regular procession toward a stable state

Succulence thick fleshy state of herbaceous tissues due to high water content; for leaves, it is quantified as the volume of water in the leaf at a relative water content of 100% divided by the leaf area

Succulent plant with a large volume of water in the leaf or other plant part (at a relative water content of 100%) per unit area

Sugar sensing mechanism in cells by which the concentration of sugars is "measured" after which it affects physiological events (e.g., via gene transcription); sugar sensing may involve a specific hexokinase

Summer annual species whose seeds germinate after winter, while the life cycle is completed before the start of the next winter

Sunfleck short period of high irradiance that interrupts a general background of low diffuse radiation

Supercooling refers to the noncrystalline state of water at low temperatures

Supply function equation describing CO_2 diffusion from the atmosphere into the leaf, supplying substrate for photosynthesis

Symbiosis intimate association between two organisms of different species (in this text the term is used when both symbionts derive a long-term selective advantage)

Symbiosome membrane-surrounded space containing one or more rhizobia in an infected cell of a root nodule in a legume

Symplasmic phloem loading occurs in plants in which photosynthates move from the cytoplasm of the mesophyll cells of the leaves, via plasmodesmata, to intermediary cells; after chemical transformation into oligosaccharides, these move, again via plasmodesmata, to the sieve tubes

Symplast space comprising all the cells of a plant's tissues connected by plasmodesmata and surrounded by a plasma membrane.

Symport cotransport of one compound in one direction coupled to transport of another compound (mostly H^+) in the same (uniport) or opposite (antiport) direction

Systemic resistance resistance that is induced by a microorganism that has colonized a plant at a location that differs from the plant part that has become reistant; the microorganisms that induce the resistance may be either pathogenic or have a growth-promiting effect

Tannin class of protein-precipitating polymeric phenolic secondary plant compound

Terpenoid class of secondary plant compounds containing C and H, produced from the precursor mevalonic acid

Testa seed coat

Thermogenic respiration respiration that increases the temperature of an organ, such a the flowers of *Arum* lilies

Thigmotropism growth response of plant organs to physical contact (touch, wind, vibrations, rain, turbulent water flow)

Thylakoid membrane-bound sack within the chloroplast containing photosynthetic pigments and electron-transport chain

Tissue-mass density dry mass per unit volume of a tissue

Tissue tension result of differences in water relations between different cells in a tissue; the existence of tension may become "visible" when an organ is cut; for example, longitudinally split roots of *Vicia faba* bend inward, due to compression of the external tissue by the internal tissues; tissue tension may be the result of differences in turgor or, more likely, in wall elasticity between different cells; tissue tension plays an important role in the closing mechanism of the carnivorous Venus fly trap (*Dionaea*)

Tolerance   endurance of unfavorable environmental conditions

Tracheid   cell type in the xylem

Trade-off   a balancing of factors, all of which cannot be attained at the same time (e.g., investments in protective strucures vs. investments in photosynthetic machinery)

Transfer cell   cell involved in transport that has a proliferation of the plasma membrane (surface enlargement); transfer cells are found in the phloem of plants using the apoplasmic phloem-loading pathway, in the epidermis of aquatic plants using bicarbonate, and so on

Translocation   transport of solutes through the phloem

Transmittance   fraction of the global solar radiant energy incident on a body that passes through the body

Transpiration   water loss from leaves due to evaporation within the leaf

Tricarboxylic acid cycle   Krebs cycle

Trichome   epidermal hair on a leaf or stem

Trypsin   protein-hydrolyzing enzyme (in animals)

Turgor   positive hydrostatic pressure in live cells

Uncoupler   chemical compound which enhances the membrane conductance for protons and so uncouples electron transport from phosphorylation

Unit leaf rate (ULR)   net assimilation rate (NAR)

Up-regulation   increase of the normal rate of a process, sometimes involving increased transcription of genes coding for enzymes involved in that process

V_{max} substrate-saturated rate (expressed per unit protein, rather than per mole, as in k_{cat})

Vacuole   membrane-bound sac within a cell; used for storage

Vapor pressure deficit   difference in actual vapor pressure and the vapor pressure in air of the same temperature and pressure that is saturated with water vapor

Vapor pressure difference   difference in vapor pressure between the intercellular spaces and the atmosphere

Vegetative reproduction   clonal growth

Vegetative storage protein   proteins accumulating in vegetative plant parts (leaves and

hypocotyls) at a high supply of nitrogen (e.g., in *Glycine max*)

Vernalization   induction of flowering by exposure to low temperatures (from the Latin word ver = spring)

Vescicular-arbuscular mycorrhiza (VAM)   kind of endomycorrhiza

Vessel   a conducting element of the xylem

Viscoelastic creep   mixture of viscous and elastic processes during cell wall expansion; also unsavory character met in dark alleys

Wall loosening   refers to the process during which covalent or noncovalent bonds between cellulose microfibrils and other macromolecules are broken, so that the cell under turgor can expand

Water channel   pore in membrane made by a protein, through which water enters single file

Water potential   chemical potential of water divided by the molar volume of water, relative to that of pure water at standard temperature and pressure

Water status   loose term referring to aspects of the plant's water potential, stomatal conductance, and so on

Water stress   stress due to the lack of water

Water-use efficiency   ratio between the gain of (above-ground) biomass in growth or CO_2 in photosynthesis and transpirational water loss

Wilting point   water potential at which turgor pressure is zero

Winter annual   species whose seeds germinate before winter, while the life cycle is completed before the start of the next summer

Xanthophyll cycle   chemical transformations of a number of carotenoid molecules in the chloroplast that avoid serious damage by high levels of radiation

Xerophyte   plant that typically grows in dry environments

Yield coefficient   a proportionality constant in the Lockhart equation that refers to the plasticity of cell walls

Yield threshold   minimum turgor pressure for cell expansion

Zeatin   a kind of cytokinin (occurring in *Zea mays* and other plants)

Index